PHYSICAL GEOGRAPHY

EDITION
11

PHYSICAL GEOGRAPHY

JAMES F. PETERSEN
Texas State University

DOROTHY SACK
Ohio University

ROBERT E. GABLER
Western Illinois University, Emeritus

CENGAGE
Learning™

Australia • Brazil • Mexico • Singapore • United Kingdom • United States

Physical Geography, **Eleventh Edition**
James F. Petersen, Dorothy Sack,
Robert E. Gabler

Product Director: Dawn Giovanniello

Product Manager: Morgan Carney

Content Developers: Nedah Rose, Jeffrey Hahn

Associate Content Developer: Kellie N. Petruzzelli

Product Assistant: Victor Luu

Marketing Manager: Ana Albinson

Content Project Manager: Hal Humphrey

Senior Designer: Michael Cook

Manufacturing Planner: Becky Cross

Production Service: Rebecca L. Lazure, SPi Global

Photo Researcher: Veerabhagu Nagarajan, Lumina Datamatics

Text Researcher: Kachana Vijayarangan, Lumina Datamatics

Copy Editor: Lynda Griffiths

Illustrators: Accurate Art, Precision Graphics, Rolin Graphics, Lumina Datamatics, SPi Global

Text Designer: Liz Harasymczuk

Cover Designer: Michael Cook

Cover and Frontispiece Image: Kirkjufellsfoss (Church Mountain Falls) and Kirkjufell mountain, Iceland. Robert Postma/Design Pics, Axiom Photographic Agency/Getty.

Compositor: SPi Global

For product information and technology assistance, contact us at **Cengage Learning Customer & Sales Support, 1-800-354-9706.**

For permission to use material from this text or product, submit all requests online at **www.cengage.com/permissions.** Further permissions questions can be e-mailed to **permissionrequest@cengage.com.**

Library of Congress Control Number: 2015960923

Student Edition:

ISBN: 978-1-305-65264-4

Cengage Learning
20 Channel Center Street
Boston, MA 02210
USA

Cengage Learning is a leading provider of customized learning solutions with employees residing in nearly 40 different countries and sales in more than 125 countries around the world. Find your local representative at **www.cengage.com**.

Cengage Learning products are represented in Canada by Nelson Education, Ltd.

To learn more about Cengage Learning Solutions, visit **www.cengage.com**.

Purchase any of our products at your local college store or at our preferred online store **www.cengagebrain.com**.

Printed in Canada
Print Number: 01 Print Year: 2016

Preface

Physical geography concerns understanding Earth as an integrated system, how it functions, and how it and its environments vary over space and time. Earth is a complex environmental system driven by interactions among factors that include climate, weather and the atmosphere, organisms and their communities, water, landforms, and soils. Understanding these factors and how they change and interact is crucial for making informed decisions about the use and preservation of Earth's environments and resources. Our interactions with the environment are also important, as they can either benefit or endanger our own living conditions as well those of future generations. The more we know about the Earth system and how it operates, the more effective we can be in working toward preservation, stewardship, and sustainability on a local, regional, and global basis.

Recognition of geography's importance to society as a major field of inquiry, and as a professional career focus, is growing along with environmental awareness. Geographic knowledge, skills, and techniques are increasingly valued in the workplace. Physical geographers use the latest technologies to observe, study, map, and measure features and processes and their interactions as parts of the Earth system. They work on understanding and modeling environmental responses and interactions. Physical geographers analyze digital images from satellites and aircraft, and employ digital mapmaking techniques (cartography), geographic information science (GIS), the global positioning system (GPS), and other tools for analysis and problem solving.

Physical Geography 11ᵗʰ edition's focus on relevance is supported by explanations of geographic tools and methods. Practical applications that use geographic data, maps, satellite images, and landscape photographs are provided in this book. These activities, along with the text material, encourage spatial thinking and give students opportunities to apply effectively the geographic knowledge they are acquiring.

At the college level, physical geography is an ideal science course for students who want to make informed decisions that consider environmental limits and possibilities as well as people's wants and needs. Today's students will be the decision makers of tomorrow, dealing with issues that affect what people need and want, while considering environmental limits and possibilities. It is for these students that *Physical Geography* has been written.

Features

Comprehensive View of the Earth System
Physical Geography introduces all major aspects of the Earth system, identifying physical phenomena and natural processes and stressing their characteristics, relationships, interactions, and distributions.

The text covers a wide range of topics, including the atmosphere, the solid Earth, oceans and other water bodies, and the living environments of our planet.

Engaging Graphics
Visual aids greatly enhance the study of geography; therefore this text includes an array of illustrations and photographs that make the concepts come alive. There are more than 200 new or significantly improved images, figures, and graphics. Clear, easy to understand, and colorful diagrams illuminate important concepts. Stunning photographs that show excellent examples bring the physical world to the student. Locator maps accompany selected photographs to provide a spatial context and help students identify the place or feature's geographic position on Earth. Questions are included with most figures to invite interpretive thinking about Earth's features, environments, and concepts.

Clear Explanations
The text uses an easily understandable, narrative style to explain the origins, development, significance, and distribution of processes, physical features, and events that occur within, on, or above Earth's surface. The writing style is targeted toward rapid comprehension and making the study of physical geography meaningful and enjoyable.

Introduction to the Geographer's Tools
Digital technologies have revolutionized our abilities to study Earth's natural processes and environments. A full chapter is devoted to maps, digital imagery, data, and technologies used by geographers. Illustrations include maps and images with interpretations provided for the environmental attributes shown in the scenes. There are also introductory discussions of many techniques that geographers use for displaying and analyzing environmental features and processes, including remote sensing, geographic information systems, cartography, and global positioning systems.

Focus on Student Interaction
The text continually encourages students to think, conceptualize, hypothesize, and interact with the subject matter of physical geography. Activities at the end of each chapter can be completed either individually or by a group, and are designed to engage students and promote active learning. Review questions reinforce concepts and prepare students for exams, and practical application assignments require active solutions, such as sketching a diagram, performing data calculation and analysis, or exploring geographic features using Google Earth®. Questions following most figure captions prompt students to think beyond, or to use, the map, graph, diagram, or image, and to give further consideration to the topic. Latitude/longitude coordinates for selected

photographs and climatic location examples invite students to further explore physical geography using satellite imagery provided by Google Earth or other Internet system for viewing Earth locations. Detailed learning objectives at the beginning of the chapters provide a means for assessing comprehension of the material.

Three Unique Perspectives Physical geography is a subject that seeks to develop an understanding and appreciation of our Earth and its environmental diversity. In approaching this goal, this textbook employs feature boxes that illustrate three major perspectives of physical geography. Through a **spatial science perspective**, physical geography focuses on understanding and explaining the locations, distribution, and spatial interactions of natural phenomena. Physical geography can also be approached from a **physical science perspective**, which applies the knowledge and methods of the natural and physical sciences, for example, by using the scientific method and systems analytical techniques. Through an **environmental science perspective**, physical geographers consider impacts, influences, and interactions among human and natural components of the environment. Basically, this means understanding how Earth's environments and environmental processes influence human life and how humans affect environments on scales from local to global.

Map Interpretation Series Developing map interpretation skills is a priority in a physical geography course, and beneficial in many career fields. To meet the needs of students who do not have access to a laboratory setting, this text includes map activities with, full-color maps generally printed at their original scale, satellite images, and interpretation questions. These maps give students an opportunity to develop valuable map-reading skills. In courses that have a lab section, the map interpretation features offer an excellent supplement to lab activities, where each student has access to the same map, image and activity. These interpretive activities provide strong links to lectures, the textbook content, and practical lab applications.

New Features

Understanding Map Content Thematic maps have the ability to present a great deal of geographic data in graphic form. The goal of the **Understanding Map Content** feature is to help students understand the information, geographic-spatial representations, and data presented on a thematic map. Students are encouraged to answer questions based on a map's content. This opportunity for practice will increase students' comprehension of how much useful information is contained in a thematic map through the visual presentation of geographic data. These activities illustrate the usefulness of being able to read and really understand the information that maps present, not only while studying geography but also in their daily lives.

Thinking Geographically Most chapters dealing with Earth surface processes and landforms include map activities in the **Map Interpretation** series. These full-size topographic map excerpts are presented at the end of chapters that discuss elevation

(Chapter 2), lakes (Chapter 16), rivers (Chapter 17), deserts (Chapter 18), glaciers (Chapter 19), and coasts (Chapter 20). There is also a weather map and satellite image interpretation activity at the end of Chapter 7. Many chapters also include new images to interpret in the **Thinking Geographically** series. Students are asked to interpret an image that features a scene related to the chapter content. These images offer additional practice in looking closely and visually evaluating a landscape by recognizing the geographic features they see, their significance, how they may be related, and to think about how landscapes and landforms develop.

Four Major Objectives

Ever since the first edition of this book, the authors have sought to accomplish four major objectives:

To Meet the Academic Needs of the Student In content and style, *Physical Geography*, 11th edition, was written specifically to meet the needs of students, the end-users of this textbook. Students can use the knowledge and understanding obtained through the text and its activities to help them make informed decisions involving the environment at the local, regional, and global scales. The book also considers the needs of beginning students—those with little or no background in the study of physical geography or other Earth sciences. Examples from throughout the world illustrate important concepts and help students bridge the gap between theory and practical application.

To Integrate the Illustrations with the Written Text The photographs, maps, satellite images, scientific visualizations, block diagrams, graphs, and line drawings have been carefully chosen to illustrate important concepts in physical geography. Figures are called out in the conceptual discussions so that students can easily make the connection between an illustration and its related text. Some examples of topics that are clearly explained through the integration of visuals and text include: map and image interpretation (Chapter 2), the seasons (Chapter 3), Earth's energy budget (Chapter 4), surface wind systems (Chapter 5), storms (Chapter 7), soils (Chapter 12), plate tectonics (Chapter 13), rivers (Chapter 17), glaciers (Chapter 19), and coastal processes (Chapter 20).

To Communicate the Nature of Geography The nature of geography and three major perspectives of physical geography (spatial science, physical science, and environmental science) are discussed in Chapter 1. In subsequent chapters, important geographic topics are discussed that involve all three of these perspectives. For example, location is a dominant topic in Chapter 2 and remains an important theme throughout the text. Spatial distributions are stressed, providing context to the elements of climate in Chapters 4 through 6. The changing Earth system is a central focus in Chapter 8. Characteristics of climate regions and their associated environments constitute Chapters 9 and 10. Spatial interactions are demonstrated in explanations of weather systems (Chapter 7), soils (Chapter 12), and volcanic and tectonic activity (Chapters 13 and 14). Feature boxes present interesting and important examples of each geographic perspective.

To Fulfill the Major Requirements of Introductory Physical Science College Courses The Earth as a system and the physical processes that are responsible for the location, distribution, and spatial relationships of physical phenomena beneath, at, and above Earth's surface are examined in detail. Scientific method, hypotheses, theories, and explanation are stressed. End-of-chapter questions that involve understanding, analyzing, and interpreting graphs of environmental data (or graphing data for analysis), quantitative transformation or calculation of environmental variables, and/or hands-on map analysis all support science learning. Models and systems are frequently cited in the discussion of important concepts, and scientific classification is presented in several chapters—some of these topics include air masses, tornadoes, and hurricanes (Chapter 7), climates (Chapters 8, 9, and 10), biogeography (Chapter 11), soils (Chapter 12), rivers (Chapter 17), and coasts (Chapter 20).

Physical geography plays a central role in understanding environmental aspects and issues, human–environment interactions, and approaches to environmental problem solving. The beginning students in this course include the professional geographers of tomorrow. Spreading the message about the importance, relevance, and career potential of geography in today's world is essential to the strength of geography at educational levels from precollegiate through university. *Physical Geography* seeks to reinforce that message.

Eleventh Edition Revision

This new edition has been revised so that the latest and most important information is presented to those who are studying physical geography. Not only is our planet ever-changing, but so are the many ways that we study, observe, measure, and analyze Earth's characteristics, environments, and processes. New scientific findings and new ways of communicating those findings are continually being developed. The eleventh edition introduces or expands discussions of technology, such as LIDAR imaging and satellite weather maps. Students have greater access to geographic tools than ever before; therefore, latitude/longitude coordinates are provided for places shown in photographs and maps as well as climate stations that so students may further explore an area or phenomenon on their own, using an interactive mapping tool such as Google Earth.

As authors, we continually seek to include physical geography topics that will spark student interest. In this edition, climate change is given a greater emphasis, and sustainable living is discussed, along with global movements to better manage environmental impacts on Earth and its climate. Attention has been given to recent environmental concerns, findings, and occurrences of natural hazards—we explain the events, the conditions and processes that led to those events, and how they are related to physical geography.

This edition has been updated to include discussions of the most recent natural disasters, including Superstorm Sandy, flooding in the Atacama Desert (the world's driest place) and in other areas, heavy lake effect snowfalls in the eastern United States, wildfires throughout California and the West, recent volcanic eruptions, strong storms, and tornado outbreaks. Recent disasters are discussed in terms of human impacts and efforts toward minimizing, or avoiding the effects of such tragedies in the future. These events and others are addressed as examples of Earth processes and human–environment interactions.

New and Revised Text Revising *Physical Geography* for an eleventh edition involved thoughtful consideration of the input from many reviewers with varied opinions. Many topics have been rewritten for greater clarity, discussions have been expanded where more explanation was required, and new feature boxes have been developed. Great concern has been given to recent occurrences of unusual weather conditions and the impacts of climate change. Earth systems approaches are reinforced with new content, illustrations, and examples.

Most sections of the book contain new material, line art, and photographs. For example, the following are some examples of the discussions and features that are new or improved in this edition: the importance of sustainable development (Chapter 1); geographic information systems, 3-D digital landscape models, and LIDAR (Chapter 2); star circles and Earth rotation (Chapter 3); photosynthesis, the greenhouse effect, and urban heat islands and green roofs (Chapter 4); global isobar map, volcanoes and air travel, and recent El Niño/La Niña conditions (Chapter 5); Hurricane/Superstorm Sandy, 3-D block diagrams of fronts and midlatitude cyclonic storms, and hazard warning systems on cell phones (Chapter 7); recent atmospheric CO_2 levels, climate change, Saharan dust and the Amazon rainforest (Chapter 9); lake effect snowfalls (Chapter 10); plant succession (Chapter 11); continental growth by accretion (Chapter 13); 2015 earthquake in Nepal (Chapter 14); deadly huge landslide in Oso, Washington, and lethal earthquake-induced avalanches on Mt. Everest (Chapter 15); cars in a sudden sinkhole collapse at the National Corvette Museum in Kentucky (Chapter 16); and removal of large dams on rivers and catastrophic flash floods in central Texas (Chapter 17).

Enhanced Illustration Program More than 200 new and updated figures are included in this edition. Textbook illustrations have been revised, updated, or improved, and many have been replaced by excellent new examples. Topics that are new or expanded in this edition required new, updated or improved figures, including numerous photographs, satellite images, maps, and all climographs. Selected photographs include a locator map to provide a spatial reference for the feature or place pictured. Additionally, two new illustrated features invite student inquiry: **Understanding Map Content** and **Thinking Geographically**. An example of one feature or the other appears in most chapters.

Locate and Explore Activities Exercises at the end of many chapters, titled Locate and Explore require you to use Google Earth or a similar technology. Google Earth allows you to interactively locate, display, and investigate geographic imagery and data from anywhere in the world. To perform these exercises, you should have the latest version of Google Earth installed on your computer. Some of the exercises require you to use some data layers that are included with Google Earth, as well as some additional data layers that you must download. For detailed instructions about using Google Earth, and to download the necessary data, visit the book's Geography MindTap at www.cengagebrain.com.

Acknowledgments

This edition of *Physical Geography* would not have been possible without the encouragement and assistance of editors, friends, and colleagues from throughout the country. Great appreciation is extended to Martha, Emily, and Hannah Petersen, and Greg Nadon, for their patience, support, and understanding. The teaching and learning acumen of Dr. Robert E. Gabler, the original author of this text, guided us as we worked on this edition, and will continue to do so in future editions.

Special thanks go to the splendid freelancers and staff members of Cengage Learning. These include Morgan Carney, Earth Sciences Product Manager; Jeffrey Hahn, Nedah Rose, and Aileen Berg, Development Editors; Rebecca L. Lazure of SPi Global, Production Manager; Kellie N. Petruzzelli, Associate Content Developer; Hal Humphrey, Content Project Manager; Michael C. Cook, Senior Designer; Liz Harasymczuk, Interior Designer; Denise Davidson, Cover Designer; Veerabhagu Nagarajan of Lumina Datamatics, Photo Researcher; illustrators SPi Global, Graphic World Inc., Lumina Datamatics, Accurate Art, Precision Graphics, and Rolin Graphics; Victor Luu, Product Assistant; and Dr. Chris Houser, creator of our wonderful *Locate and Explore* activities.

Colleagues who reviewed the plans and manuscript for this and previous editions include Peter Blanken, University of Colorado; Brock Brown, Texas State University; Greg Carbone, University of South Carolina; Richard A. Crooker, Kutztown State University; Joanna Curran, Texas State University; J. Michael Daniels, University of Wyoming; Ben Dattilo, University of Nevada, Las Vegas; Leland R. Dexter, Northern Arizona University; James Doerner, University of Northern Colorado; Percy "Doc" Dougherty, Kutztown State University; Daniel Dugas, New Mexico State University; Richard Earl, Texas State University; Tom Feldman, Joliet Junior College; Mark Fonstad, University of Oregon; Roberto Garza, San Antonio College; Greg Gaston, University of North Alabama; Beth L. Hall, Towson University; Perry Hardin, Brigham Young University; David Helgren, San Jose State University; Chris Houser, University of West Florida; Kenneth Hundreiser, Roosevelt University; Karen Johnson, North Hennepin Community College; Fritz C. Kessler, Frostburg State University; Elizabeth Lawrence, Miles Community College; Jeffrey Lee, Texas Tech University; Michael E. Lewis, University of North Carolina, Greensboro; Elena Lioubimtseva, Grand Valley State University; Kangshu Lu, Towson University; John Lyman, Bakersfield College; Charles Martin, Kansas State University; Christopher F. Meindl, University of South Florida; Debra Morimoto, Merced College; Andrew Oliphant, San Francisco State University; James R. Powers, Pasadena City College; Joyce Quinn, California State University, Fresno; George A. Schnell, State University of New York, New Paltz; Peter Siska, Austin Peay State University; Richard W. Smith, Harford Community College; Ray Sumner, Long Beach City College; Michael Talbot, Pima Community College; Colin Thorn, University of Illinois at Urbana-Champaign; David L. Weide, University of Nevada, Las Vegas; Thomas Wikle, Oklahoma State University, Stillwater; Glynis Jean Wray, Ocean County College; Amy Wyman, University of Nevada, Las Vegas; and Craig-ZumBrunnen, University of Washington, Redmond.

The photographs throughout the book are courtesy of: Rainer Duttmann, University of Kiel; Richard Earl, Texas State University; Dan Satterfield, WOBC, Salisbury, MD; Erin Himmel/National Park Service; Delphine Farmer, Colorado State University; Lynn Betts/NRCS; Melissa Gabrielson, Chuck Young, and Fred Broerman, U.S. Fish and Wildlife Service; Bob DeGross, Everglades National Park; J. Good, National Park Service; Michael McCollum/McCollum Associates; Jason Neely, Polar Field Services; Christoph W. Borst and Gary L. Kinsland, University of Louisiana at Lafayette; VORTEX II/Sean Waugh, NO-AA/NSSL; Michael Studinger, NASA; John Shea, FEMA; USGS Alaska Volcano Observatory, D. Josefczyk; National Scenic Byways/Digital Library; Sasan Saatchi, NASA/JPL-Caltech; Wind Cave National Park; Emily Petersen; Parv Sethi; Martha Moran, White River National Forest; Mark Muir, Fishlake National Forest; National Park Service, Cape Cod National Seashore; Mark Reid, USGS; Dawn Endico; Gary P. Fleming, Virginia Natural Heritage Program; Tessy Shirakawa, Mesa Verde National Park; Bill Case, Chris Wilkerson, and Michael Vanden Berg, Utah Geological Survey; Center for Cave and Karst Studies, Western Kentucky University; Hari Eswaran, USDA Natural Resources Conservation Service; Richard Hackney, Western Kentucky University; David Hansen, University of Minnesota; Susan Jones, Nashville, Tennessee; David Hansen, University of Minnesota; L. Elliot Jones, U.S. Geological Survey; Greg Nadon, Ohio University; Anthony G. Taranto Jr., Palisades Interstate Park—New Jersey Section; Justin Wilkinson, Earth Sciences, and NASA Johnson Space Center. Spectacular images taken by two government photographers, Bob Wick of the Bureau of Land Management, and Tim Rains of the National Park Service, appear in several places. Michael Kuhwald, University of Kiel, Germany, and Bob Stafford, Texas State University produced digital landscape models for us. We would also like to thank Matt Melancon and Matthew Murphy, Texas State University, and Shawn Trueman, Central Lakes College, Ohio.

The detailed comments, suggestions, and imagery from the preceding individuals have been instrumental in bringing about the many changes and improvements incorporated in this latest edition of the book. Countless others, both known and unknown, deserve heartfelt thanks for their interest and support over the years.

Despite the painstaking efforts of all reviewers, there will always be questions of content, approach, and opinion associated with the text. The authors wish to make it clear that they accept full responsibility for all that is included in the eleventh edition of *Physical Geography*.

James F. Petersen
Dorothy Sack
Robert E. Gabler

Foreword to the Student

Why Study Geography?

In this global age, the study of geography is absolutely essential to an educated citizenry of a nation whose influence extends throughout the world. Geography deals with location, and a good sense of where places or features are, especially in relation to other places or features in the world. Such knowledge is an invaluable asset whether you are traveling, conducting international business, or browsing the Internet.

Geography examines the characteristics of places and areas on Earth, their roles as part of the Earth system, how they interact with other locations, and the changes and processes involved in these interactions. Geography also gives strong consideration to the relationships between humans and their environments. Today, everyone shares the responsibility of learning more about our physical environment so that we can preserve and protect it for future generations.

Geography provides essential information about the distribution of features on Earth's surface and the interconnections of places. The distribution pattern of volcanoes, for example, provides an excellent indication of where Earth's great crustal plates come in contact with one another; and the violent thunderstorms that plague Illinois on a given day may be directly associated with the low-pressure system spawned in Texas two days before. Geography, through a study of regions, provides a focus and a level of generalization that allows people to examine and understand the immensely varied environmental characteristics of Earth.

As you will note when reading Chapter 1, there are many approaches to the study of geography. Some courses are regional in nature; they may include an examination of one or all of the world's political, cultural, economic, or physical regions. Some courses are topical or systematic in nature, dealing with human geography, physical geography, or one of the major subfields of the two.

A great advantage of taking a general course in physical geography is the permanence of the knowledge learned. Although change is constant and often sudden in human aspects of geography, alterations of the physical environment on a global scale are generally slow unless they are influenced by human intervention. Theories and explanations may differ, but the broad patterns of atmospheric and oceanic circulation and of world climates, landforms, soils, natural vegetation, and physical landscapes will be the same tomorrow as they are today.

Keys to Successful Study

Good study habits are essential if you are to master science courses such as physical geography, where the topics, explanations, and terminology are often complex and unfamiliar. To help you succeed in the course in which you are currently enrolled, we offer the following suggestions.

Reading Assignments

- Read the assignments before the material is covered in class by the instructor.
- Compare what you have read with the instructor's presentation in class. Pay particular attention if the instructor introduces new examples or course content not included in the reading assignment.
- Ask questions in class and seek to understand any material that was not clear from your first reading of the assignment.
- Reread the assignment as soon after class as possible, concentrating on areas that were emphasized in class. Highlight only those items or phrases that you now consider to be important, and skim those sections already mastered.
- Record in your class notes important terms, your own comments, and summarized information from each reading assignment.

Understanding Vocabulary

- Mastery of basic vocabulary often becomes a critical issue in the success or failure of students in a beginning science course.
- Focus on terms that appear in boldface type in your reading assignments. Do not overlook additional terms that the instructor may introduce in class.
- Develop your own definition of each term (representing a feature, process, or characteristic) or phrase and associate it with related terms in physical geography.
- Identify any physical processes associated with the term. Often, knowing the process will help to define the term.
- Whenever possible, associate terms with location.
- Consider the significance to humans of terms you are defining. Recognizing the significance of terms and phrases can make them relevant and easier to recall.

Learning Earth Locations

- A good knowledge of place names and of the relative locations of physical and cultural phenomena on Earth is fundamental to the study of geography.
- Take personal responsibility for learning locations on Earth. Your instructor may identify important physical features and place names, but you must learn their locations for yourself.
- Thoroughly understand latitude, longitude, and the Earth grid. They are fundamental to finding and describing locations on maps and globes. Practice locating features by their latitude and longitude until you are entirely comfortable using the system.

- Develop a general knowledge of the world political map. The most common way of expressing the location of physical features is by identifying the political unit (state, country, or region) in which it can be found.
- Make liberal use of outline maps. They are a key to learning the names of states and countries and the locations of specific physical features. Personally placing features correctly on an outline map is often the best way to learn location.
- Cultivate the habit of using an atlas. The atlas does for the individual who encounters place names or the features they represent what the dictionary does for the individual who encounters a new vocabulary word.

Utilizing Textbook Illustrations

The secret to making good use of maps, diagrams, and photographs lies in understanding why an illustration has been included in the textbook or incorporated as part of your instructor's presentation.
- Concentrate on the instructor's discussion. Taking notes about photographs, maps, graphs, and other illustrations will allow you to follow the same line of thought at a later date.
- Study all the textbook illustrations. Be sure to note which of them received considerable attention in the lecture. Do not quit examining an illustration until it makes sense to you, until you can read the map or graph, or until you can recognize what a diagram or photograph has been selected to explain.
- Hand-copy important diagrams and graphs. Few of us are graphic artists, but you might be surprised at how much better you understand a graph or line drawing after you reproduce it yourself.
- Read the captions of photographs and illustrations thoroughly and thoughtfully. If the information is included, be certain to note where a photograph was taken and in what way it is representative. What does it tell you about the feature, process, region, or site being illustrated?
- Attempt to place the concept that is being illustrated in new situations. Seek other opportunities to test your skills at interpreting similar maps, graphs, and photographs and think of other examples that support the text.
- Remember that all illustrations are reference tools, particularly tables, graphs, and diagrams. Refer to them often.

Taking Class Notes

A good set of class notes is based on selectivity. You cannot, and should not, try to write down every word uttered by your classroom instructor.
- Learn to paraphrase. With the exception of specific quotations or definitions, put the instructor's ideas, explanations, and comments into your own words. You will understand them better when you read them over at a later time.
- Be succinct. Never use a sentence when a short phrase will do, and never use a phrase when a word will do. Start your recall process with your note taking by forcing yourself to rebuild an image, an explanation, or a concept from a few words.

- Outline where possible. Preparing an outline helps you to discern the logical organization of information. As you take notes, organize them under main headings and subheadings.
- Take the instructor at his or her word. If the instructor takes the time to make a list, then you should do so, too. If he or she writes something on the board, it should be in your notes. If the instructor's voice indicates special concern, take special notes.
- Come to class and take your own notes. Notes trigger the memory, but only if they are your notes.

Doing Well on Tests

Follow these important study techniques to make the most of your time and effort preparing for tests.
- Practice distillation. Do not try to reread but skim the assignments carefully, taking notes in your own words that record as economically as possible the important definitions, descriptions, and explanations. Do the same with any supplementary readings, handouts, and laboratory exercises. It takes practice to use this technique, but it is much easier to remember a few key phrases that lead to ever-increasing amounts of organized information than it is to memorize all your notes. And the act of distillation in itself is a splendid memory device.
- Combine and reorganize. Merge all your notes into a coherent study outline.
- Become familiar with the type of questions that will be asked. Knowing whether the questions will be objective, short-answer, essay, or related to diagrams and other illustrations can help in your preparation. Some instructors make old tests available so that you can examine them or discover their evaluation styles if you inquire. If not, then turn to former students; there are usually some around the department or residence halls who have already experienced the instructor's tests.
- Anticipate the questions that will likely be on the test. The really successful students almost seem to be able to predict the test items before they appear. Take your educated guesses and turn them into real questions.
- Try cooperative study. This can best be described as role-playing and consists very simply of serving temporarily as the instructor. So go ahead and teach. If you can demonstrate a technique, illustrate an idea, or explain a process or theory to other students so that they can understand it, there is little doubt that you can answer test questions over the same material.
- Avoid the "all-nighter." Use the early evening hours the night before the test for a final unhurried review of your study outline. Then get a good night's sleep.

The Importance of Maps

Like graphs, tables, and diagrams, maps are excellent reference tools. Familiarize yourself with the maps in your textbook to better judge when it is appropriate to seek information from these important sources.

Maps are especially useful for comparison purposes and to illustrate relationships or the possible associations of two features, areas, or processes. But the map reader must beware. The associations of phenomena by their locations and distributions are not necessarily cause-and-effect relationships. In some instances the similarities in a distribution result from another factor that has not been considered or mapped. For example, a map of worldwide volcano distribution is almost exactly congruent with one of incidence of earthquakes, yet volcanoes are not the cause of major earthquakes. A third factor, the location of tectonic plate boundaries, explains the first two phenomena.

Finally, remember that a map is the most important tool of the professional geographer. It is also useful to all natural and social scientists, engineers, politicians, military planners, road builders, farmers, and countless others, but maps are essential expressions of the geographer's primary concern with location, distribution, and spatial interaction.

About Your Textbook

This textbook has been written for you, the student. It has been written so that the text can be read and understood easily. Explanations are as clear, concise, and uncomplicated as possible. Illustrations have been designed to complement the text and to help you visualize the processes, places, and phenomena being discussed. In addition, the authors do not believe it is sufficient to offer you a textbook that simply provides information to pass a course. We urge you to think critically about what you read in the textbook and hear in class.

As you learn about the physical aspects of Earth environments, ask yourself what they mean to you and to people throughout the world. Make an honest attempt to consider how what you are learning relates to the problems and issues of today and tomorrow. Practice using your geographic skills and knowledge in new situations so that you will continue use them in the years ahead. Your textbook includes several special features that will support learning, encouraging you to go beyond memorization and to reason geographically.

Chapter Activities At the end of each chapter, Consider and Respond questions and Practical Applications require you to go beyond a routine chapter review. The questions and problems are designed so that you can apply your knowledge of physical geography and, on occasion, personally respond to critical issues in society today. Locate and Explore activities (found at the end of many chapters) teach you how to use the Google Earth application as an exploratory learning tool. Check with your instructor for answers to the problems.

Caption Questions Most illustrations and photographs in your textbook, have a caption that links the image with the chapter text it supports. Read the captions carefully because they explain the illustrations and may also contain new information. Wherever appropriate, questions at the ends of captions have been designed to help you seize the opportunity to consider your own personal reaction to the subject under consideration.

Map Interpretation Series A major goal of your textbook is to help you become an adept map reader, and the Map Interpretation Series in your text has been designed to help you reach that goal.

Environmental Systems Viewing Earth as a system comprising many subsystems is a fundamental concept for researchers, instructors, and students in physical geography. The concept is introduced in Chapter 1 and reappears frequently throughout your textbook. The interrelationships and dependencies among the variables and components of Earth systems are important. Many of the illustrations included in the text will help you visualize how systems work. There are also diagrams designed to help you understand how human activities can affect the delicate balance that exists within many Earth systems.

As authors of your textbook, we wish you well in your studies. It is our fond hope that you will become better informed about our home—Earth and its varied environments—and that you will enjoy the study of physical geography.

Brief Contents

Contents

5 Atmospheric Pressure, Winds, and Circulation Patterns   111

6 Humidity, Condensation, and Precipitation   141

7 Air Masses and Weather Systems   171

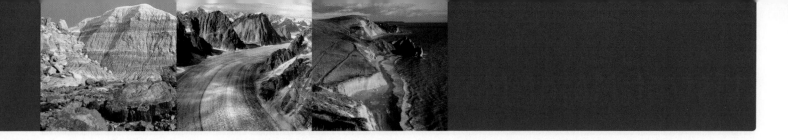

List of Major Maps

The World

The Ocean

The Contiguous United States

North America

Author Biographies

J. Petersen

James F. Petersen James F. Petersen is Professor of Geography at Texas State University, in San Marcos, Texas. A broadly trained physical geographer with strong interests in geomorphology and Earth Science education, he enjoys writing about topics related to physical geography for the public. Involved in environmental interpretation, he has written guidebooks and a number of field and trail guides for parks. Professor Petersen is a past President of the National Council for Geographic Education (NCGE), a recipient of a national teaching award, and received the NCGE's highest honor for distinguished service to geography education. He has written or served as a senior consultant for nationally published educational materials at levels from middle school through university. He has published articles that deal with education in geomorphology, climate history and climate change, the environmental history of central Texas, the role of field methods in geography, earthquake hazards, and geographic education in general. Believing in the value of learning in the field, every year since Professor Petersen began teaching he has taken students on extended field excursions to learn about Earth environments, locations, processes, and features through firsthand experience.

J. Petersen

Dorothy Sack Dorothy Sack, Professor of Geography at Ohio University in Athens, Ohio, is a physical geographer who specializes in geomorphology. Her primary research interests emphasize arid region landforms, particularly the geomorphic evidence of paleolakes, which contributes to our understanding of Earth's paleoclimate. Other research themes include the impact of off-road vehicles on the landscape and the history of geomorphology. Professor Sack has published research in a variety of professional journals, academic volumes, and Utah Geological Survey publications. Her work has been funded by the National Geographic Society, National Science Foundation, Association of American Geographers (AAG), American Chemical Society, and other groups. She is active in professional organizations, having served as chair of both the Geomorphology and the History of Geography Specialty Groups of the AAG, and in other offices for the AAG, Geological Society of America, and History of Earth Sciences Society. She also serves on the editorial boards of *Geomorphology, Physical Geography,* and *Earth Sciences History.* Professor Sack enjoys teaching as well as research, and has received the Outstanding Teacher Award from Ohio University's College of Arts and Sciences.

J. Petersen

Robert E. Gabler During his nearly five decades of professional experience, Professor Gabler has taught geography at Hunter College, City of New York, Columbia University, and Western Illinois University, in addition to 5 years in public elementary and secondary schools. At times in his career at Western Illinois he served as Chairperson of the Geography and Geology Department, Chairperson of the Geography Department, and University Director of International Programs. Professor Gabler received three University Presidential Citations for Teaching Excellence and University Service, served two terms as Chairperson of the Faculty Senate, edited the *Bulletin of the Illinois Geographical Society,* and authored numerous articles in state and national periodicals. He is a past President of the Illinois Geographical Society, former Director of Coordinators and past President of the National Council for Geographic Education (NCGE), and the recipient of the NCGE George J. Miller Distinguished Service Award.

PHYSICAL GEOGRAPHY

PHYSICAL GEOGRAPHY: EARTH ENVIRONMENTS AND SYSTEMS

■ OBJECTIVES

WHEN YOU COMPLETE THIS CHAPTER YOU SHOULD BE ABLE TO:

- ■ 1.1 Explain why physical geography examines both the natural world and human interaction with the natural world.
- ■ 1.2 Discuss important ways in which geographic information and techniques are useful in different careers.
- ■ 1.3 Describe the three major perspectives of physical geography: spatial science, physical science, and environmental science.
- ■ 1.4 Think of Earth as a system of interacting parts that respond to both natural processes and human actions.
- ■ 1.5 Illustrate with examples how some interactions between people and their environment are advantageous, whereas others are detrimental or hazardous.
- ■ 1.6 Summarize how knowledge of physical geography contributes to a better understanding of the environment.
- ■ 1.7 Recognize that every physical environment offers an array of advantages and challenges to human life and living conditions.
- ■ 1.8 Explain how physical geography is relevant to your everyday life.

PLANET EARTH IS HOME TO a large and complex set of living organisms, including humans. Earth's surface is our natural habitat, providing us with air, water, nutrients, and shelter. Earth also receives enough sunlight to maintain livable temperatures and to power food production by plants through photosynthesis. Surrounded by the emptiness of space, life on Earth is dependent on the planet's self-contained natural resources that are extensive, but not limitless.

Only recently in the history of human life have we been able to view Earth in its entirety from space, giving us a fresh perspective on characteristics of the ocean, atmosphere, land masses, and natural vegetation. In addition to being able to see the physical Earth as a whole, technological advances in communication, transportation, and information sciences over the last several decades have led to an increasingly global social perspective. These developments, combined with population growth, make the world seem smaller, and have helped heighten awareness of the finite nature of Earth, its environments, and its resources.

Earth, our home, our natural habitat.
NASA/NOAA/GSFC/Suomi NPP/VIIRS/Norman Kuring

In physical geography we seek to understand as much as possible about the natural Earth—its characteristics, materials, and processes—so that humans (and Earth's other organisms) can live on this planet long into the future in a healthy and sustainable way and with minimal risk from its inherent hazards. **Physical geography** is this study of the natural aspects of Earth as our habitat and home.

Earth is beautiful and intriguing—a dynamic life-giving planetary oasis of great environmental diversity. It is important for people to gain an understanding of the planet that sustains us—to learn about the components and processes that change and regulate Earth's environmental conditions. In all fields of study, asking questions is an important step toward acquiring knowledge, explanations, and understanding. In physical geography, many of our questions are directed toward how Earth's matter, energy, and processes interact to create the environmental diversity that exists on our planet.

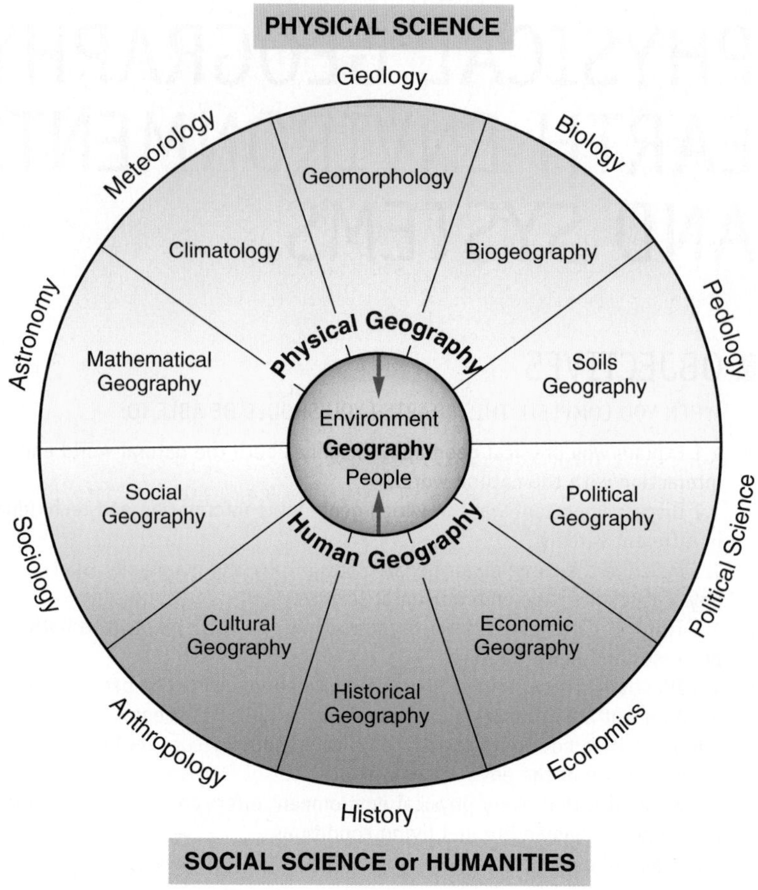

The Study of Geography

Physical geography is a major part of the field of **geography**, which is the study of all aspects of Earth in its role as the home of people. The word *geography* comes from the Greek language. *Geo* refers to Earth, and *graphy* means picture or writing, thus the word itself designates a broad field of study. Geography as a whole includes the examination, description, and explanation of cultural as well as natural physical variables on Earth. Geography emphasizes how physical and cultural attributes vary from place to place, how places and features change over time, and the processes and interactions responsible for these variations. Geography is commonly considered the **spatial discipline** (the study of locational space) because it includes analyzing and explaining the locations, distributions, patterns, variations, and similarities or differences among phenomena on Earth's surface. Geography is

> an integrative discipline that brings together the physical and human dimensions of the world in the study of people, places, and environments. Its subject matter is the Earth's surface and the processes that shape it, the relationships between people and environments, and the connections between people and places.

—*Geography for Life*, Geography Education Standards Project

Geographers are concerned with how physical and human processes affect, have affected, or will affect our planet and its natural

● **FIGURE 1.1** Geography has many subdivisions that are related to other disciplines with which they share some of their interests. Geographers, however, apply their own distinct perspectives and approaches to these areas of study.

What advantage might a geographer have when working with other physical scientists seeking a solution to a problem?

and human environments. They study processes that influenced Earth's physical and cultural landscapes in the past, how processes affect landscapes today, how a landscape may change in the future, and the significance of process and landscape changes over space and time. Geography is distinctive among academic disciplines in its definition and central purpose and can involve studying any topic related to the analysis of natural or human processes on or near Earth's surface. Because geography embraces the study of virtually any Earth phenomenon, the subject has several subdivisions. Typically, geographers specialize in one or more of these subfields (● Fig. 1.1). Geography also encompasses multiple approaches; some geographers are natural scientists, some are social scientists, and others use a humanities approach.

The broadest divisions of the field of geography are physical geography, which is the focus of our study, and human geography. Whereas physical geographers use a natural science approach to analyze the nonhuman elements of Earth's environments, **human geography** concerns the nature, processes, and variations in space and time of human-generated phenomena, including culture; thus, human geographers use approaches from social science or the humanities. Human geographers are, for example, concerned with such topics as population distributions, migration patterns, cultural patterns, the spread of ideas, cities and urbanization, industrial and commercial location, natural resource use, and transportation networks.

Geographers gather, organize, and analyze many kinds of data and information in seeking to explain locations, areas, patterns, distributions, and relationships over the surface of Earth. Many individual geographers, whether human or physical, solve problems or answer questions at the local or regional scale, rather than consider the entire planet at once. As a result, geographers are interested in defining meaningful **regions**, which are areas identified by distinctive characteristics that distinguish them from surrounding areas. The distinctive characteristics used to define a region can be physical, human, or a combination of factors, and a given location can belong to various types and sizes of regions depending on the criteria used and purposes for identifying the regions. *Regional geography* emphasizes the characteristics of a region or of multiple regions.

Physical Geography

Physical geography focuses on understanding the natural processes and features of Earth. Physical geography, however, cannot exclude the human element because people affect, and are affected by, natural processes and features. Geographers are excellent observers of the world around them and generally take a *holistic approach* to problem solving, meaning that they are open to all factors that might be involved in the solution, including human factors. Being concerned with nearly all aspects of Earth, physical geographers are trained to view a natural environment and consider how it functions in its entirety (● Fig. 1.2). Yet, as in other fields of study, most individual physical geographers develop focused expertise in one or two subfields. For example, physical geography includes the study of weather and climate and some physical geographers are meteorologists or climatologists. *Meteorologists* consider the processes that affect daily weather, and they forecast weather conditions. *Climatologists* are interested in the averages and extremes of long-term weather data, regional climates, large-scale atmospheric circulation processes,

Martha Moran, Aspen Ranger District, White River National Forest

● FIGURE 1.2 Physical geographers study all processes, features, and characteristics of the natural environment, including those related to weather, climate, rock structures, landforms, soils, vegetation, animals, water, and human impacts, several of which are represented in this view from the White River National Forest in Colorado.

What elements of physical geography can you recognize in this scene?

climate-related hazards, understanding climate change, and how climate and climate change impact people and the environment.

Geomorphology, another major subfield of physical geography, is the study of landforms and how the interactions of Earth's processes and surface materials contribute to landform development and modification. *Geomorphologists* work to understand variations in landforms, the processes that produce them, and the hazards that they pose for people. *Biogeographers* study the geographic ranges, distribution patterns, and assemblages of plant and animal species, seeking to discover the natural and human-induced environmental factors that influence them. *Biogeography* includes analyzing plant and animal distributions of the present and past and how they may change in the future. Other physical geographers, known as *soil scientists*, analyze and map soil types, determine the suitability of different soils for various uses, and work to conserve soil as a natural resource.

Water plays a critical role in many natural processes on Earth. Meteorology, climatology, geomorphology, biogeography, and soil science each involve water in some way. Physical geographers participate widely in the study of water, water bodies, and water resources, and physical geographers may also serve as *hydrologists*, *oceanographers*, or *glaciologists*. Many geographers contribute to the effective management of water resources to ensure that lakes, watersheds, springs, and groundwater sources meet human and environmental needs in terms of quality and quantity.

Geographic Tools and Technology

Rapid advances over the last couple of decades in the *global positioning system (GPS)*, communication, and information science technologies have greatly enhanced our abilities to learn about Earth's physical geography. We locate points on Earth's surface more quickly, easily, and accurately; measure attributes of those ground points more precisely; and gather, manipulate, and share larger data sets than ever before. We obtain much data about Earth's surface remotely, such as from sophisticated instruments and sensors carried on board satellites and aircraft. Some problems, however, also require in-person fieldwork, often employing specialized instruments to aid in data collection, and fieldwork will probably always play an important role in physical geography (● Fig. 1.3).

Maps are essential components of geography. Maps function as sources of geographic data, tools to aid in the analysis and interpretation of geographic data, and means for displaying results of geographic studies. Geographers use satellite-based GPS technology to determine the precise location of points on Earth's surface, and digital technologies for mapmaking (**cartography**) and for conducting many aspects of map analysis.

Continuous satellite imaging of Earth has been ongoing for more than 40 years, which gives us an important perspective on environmental change. By recording from space various energy signals from Earth and producing images, we are able to measure, monitor, and map Earth processes and their effects, many of which are invisible to the naked eye. As methods of data processing and visual representation techniques improve, the geographer's ability to visualize environmental data and their change over time is

● **FIGURE 1.3** In addition to using the latest technological innovations for locating, observing, measuring, imaging, and mapping Earth and its environments, fieldwork remains an important element of information gathering in much of physical geography.

continually being enhanced with greater image resolution, increasingly sophisticated three-dimensional graphics, and animations that are more vivid and striking than ever (● Fig. 1.4).

Technology greatly aids in the collection, storage, analysis, and visualization of geographic data, as well as in the effective communication of results. But solving geographic problems also requires knowledgeable people who understand the content, scope, and nature of geographic information, methods, and techniques. Many geographers are gainfully employed in positions that apply technology to help understand our planet better, and their numbers are certain to increase in the future.

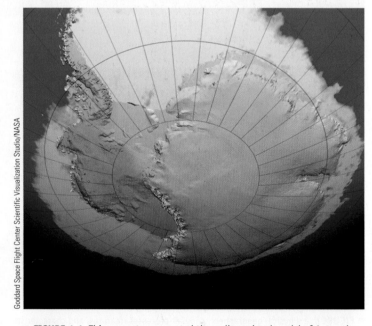

● **FIGURE 1.4** This computer-generated three-dimensional model of Antarctica was made by combining a 50-year history of temperature records from locations on the continent with recent satellite images of the ice surface. The red area has undergone the largest temperature increase in response to global warming.

Major Perspectives in Physical Geography

Physical geographers use the **scientific method** to guide their learning about the processes and features on Earth. The scientific method entails developing valid explanations about the issue being studied by objectively testing hypotheses and analyzing all pertinent evidence and facts (● Fig. 1.5). Using the scientific method, new ideas or proposed answers to questions are accepted as valid only if they withstand rigorous objective testing.

This textbook highlights three major perspectives, all using the scientific method, fundamental to physical geography: *spatial science*, *physical science*, and *environmental science*. This chapter introduces all three, whereas subsequent chapters vary in which perspective they emphasize. As you progress through the book, take note of how each perspective relates to the unique nature of geography as a discipline.

Spatial Science Perspective

Physical geography uses the scientific method to study variations over space, thus it is a *spatial science*. Specific interests vary widely among physical geographers, but they share the common goals of understanding and explaining spatial variations on Earth's surface. The following five spatial topics—location, characteristics of places, spatial distribution and pattern, spatial interaction, and change over space and time—illustrate factors that geographers typically consider and problems that they address. Because learning the types of questions that geographers ask is a first step toward understanding the field of study, example questions are included for each topic.

Location Geographic studies often begin with locational information. Describing a location usually uses one of two methods: **absolute location**, which is expressed by a coordinate system, and **relative location**, which identifies where a feature lies in relation to something else, usually a fairly well-known site. For example, Pikes Peak, in the Rocky Mountains of Colorado, with an elevation of 4302 meters (14,115 ft), has a location of latitude 38°51′N (north) and longitude 105°03′W (west). This kind of global address is an absolute location. However, another way to report its location would be as 36 kilometers (22 mi) west of Colorado Springs (● Fig. 1.6). This is an example of relative location because the position of the peak is given in relation to the city of Colorado Springs. Typical spatial questions involving location include the following: *Where are volcanoes (or other type of Earth feature) found, and where are they not found? Why are volcanoes located where they are? What methods can we use to locate volcanoes on Earth? How can we describe their location? What is the most likely or least likely location for a volcano?*

Characteristics of Places Physical geographers are interested in the environmental features and processes that make a place unique, as well as in the shared or similar characteristics

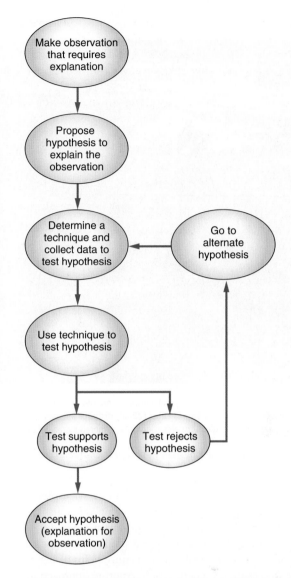

● **FIGURE 1.5** The scientific method, widely applicable in physical geography, follows these steps. For example:

1. **Make an observation that requires an explanation.** On a trip to the mountains, you feel colder at higher elevations than you do at lower elevations. Is that just a result of local conditions and weather changes on the day you were there, or is it a universal relationship?

2. **State the observation in the form of a testable hypothesis.** For example: Within a single mountain range, higher elevations have colder temperatures than lower elevations.

3. **Determine a technique or strategy for testing the hypothesis and for collecting the data needed to conduct the test.** You could test the hypothesis by selecting a mountain range to study and collecting temperature data for many days from weather stations at different elevations. Use temperature measurements made at the same time of day at all weather stations on any given day because the evidence must be collected under similar circumstances to minimize bias.

4. **Apply the technique or strategy to test the validity of the hypothesis.** Here you discover if the collected data support the hypothesis. The technique or strategy will indicate either accepting or rejecting the hypothesis. If the hypothesis is rejected, you could devise and test a different hypothesis to explain the initial observation. If the hypothesis is accepted, you have a reasonable explanation for the differences you felt in temperature.

Pikes Peak   Colorado Springs

Goddard Space Flight Center/Earth Observatory/NASA

● **FIGURE 1.6** A three-dimensional digital model shows the relative location of Pikes Peak to Colorado Springs, Colorado. Because this is a perspective view, the 36 kilometer (22 mi) distance appears to be shorter than its actual ground distance. A satellite image was merged with elevation data gathered by radar from the space shuttle to create this scene.

What can you discern about the physical geographic characteristics of this place from the image?

between places. For example, we might determine what physical geographic features give a particular mountain range its distinctive appearance or describe how one mountain range differs from another. Assessing the differences and similarities between two examples of the same general type of feature helps us better understand the processes influencing each. Another approach to learning about the characteristics of places is through analyzing the environmental advantages and challenges that exist in a locality. Example questions about the characteristics of places are: *How does an Australian desert compare to the Sonoran Desert of the southwestern United States? How do the grasslands of the North American Great Plains compare to the grasslands of Argentina? What environmental conditions make one county more agriculturally productive than another? How does the climate of Maine differ from the climate of Alaska, and why? What weather can we expect when traveling to a different country?*

Spatial Distribution and Pattern Two fundamental characteristics describe how features or events are arranged over space. **Spatial distribution** refers to the extent of the area or areas where the feature of interest exists. Tropical rainforests, for example, cover particular expanses of Earth's surface, thus they are spatially distributed. Likewise, the parts of the United States where rain fell yesterday, or regions that have high potential for damaging earthquakes, have spatial distribution. **Spatial pattern** refers to how multiple individuals of the same type of feature or event

are arranged over Earth's surface: Are they regularly or randomly spaced, clustered together, or far apart from each other? Population can be dense or sparse (● Fig. 1.7). The spatial pattern of earthquakes may be aligned on a map because earthquake faults display linear patterns. Some questions relevant to spatial distribution and spatial pattern are: *Where are certain features abundant, and where are they rare? How are particular factors or elements of physical geography arranged in space, and what spatial patterns exist, if any? What processes are responsible for these distributions or patterns? If a spatial pattern exists, what does it signify?*

Spatial Interaction Few processes on Earth operate in isolation because areas on our planet are interconnected, that is, they are linked to conditions elsewhere on Earth. A **spatial interaction** exists if a condition, an occurrence, or a process in one place has an impact on other places. Excessive rainfall in Minnesota and Wisconsin, for example, might lead to flooding along the Mississippi River in Missouri. Reducing the size of the tropical rainforest may have a widespread impact on world climates. The exact nature of many spatial interactions is often difficult to establish with complete confidence. It is much easier to determine that a condition at one location is *associated* with a condition at another without knowing if one event actually *causes* the other. For instance, it is well established that atypical weather in some parts of the world accompanies the presence of abnormally warm ocean waters off South America's west coast, a

Image by Craig Matthew (NASA/GSFC) and Robert Simmon (NASA/GSFC)

● **FIGURE 1.7** A nighttime satellite image provides good illustrations of distribution and pattern, shown here for part of North America. *Spatial distribution* is where features are located (or perhaps, absent). *Spatial pattern* refers to their arrangement. Geographers seek to explain observed spatial relationships.

Can you locate two distributions and two patterns in this scene and propose possible explanations for each?

condition called El Niño. Climatologists, however, are still working to clarify the nature of that link, and it is possible that circumstances leading to an El Niño separately induce the atypical weather conditions. Spatial interactions exist at all geographic scales: global, regional, and local. Physical geographers consider problems such as: *What effect will stricter pollution controls in North America have on the future size of the ozone hole over Antarctica? How do two spatially interacting variables affect each other? What important interconnections link the ocean to the atmosphere and the atmosphere to the land surface?*

Change over Space and Time Earth's features and landscapes are continuously changing over space and time and at a variety of spatial and temporal scales. Storm conditions intensify or weaken over time and travel from one region to another. Weather varies from day to day, over the seasons, and from year to year. Landslides, volcanic eruptions, and floods of different types and sizes modify the landscape at different rates. Coastlines are altered by storm waves, tsunamis, human actions, or changes in sea level. Desert areas expand and contract over the decades. Vegetation and wildlife communities return to areas once devastated by wildfires (● Fig. 1.8).

U.S. Fish and Wildlife Service

● **FIGURE 1.8** Earth's features and landscapes change continually, sometimes at spatial and temporal scales that make the change difficult for people to notice. Changes to vegetation caused by a regional fire are rapid and catastrophic (top and middle photos), whereas reestablishment of plant and animal communities (bottom photo) proceed much more slowly.

Natural Regions

The term *region* has a precise meaning and special significance to geographers. Simply stated, a region is an area that is defined by a certain shared characteristic (or a set of characteristics) existing within its boundaries. The concept of a region is a tool for thinking about and analyzing logical divisions of areas based on their geographic characteristics. Geographers not only study and explain regions, including their locations and characteristics, but they also strive to delimit them—to outline their boundaries on a map. An unlimited number of regions can be derived for each of the four major Earth subsystems.

Regions help us understand the arrangement and nature of areas on our planet. Regions can also be divided into subregions. For example, the region of North America can be subdivided into many subregions based on natural characteristics. These include the Atlantic Coastal Plain (similarity of landforms, geology, and locality), the Prairies (ecological type), the Sonoran Desert (climate type, ecological type, and locality), the Pacific Northwest (general locality), and Tornado Alley (region of high potential for these storms).

Three important points should be kept in mind about natural regions. Each of these points has endless applications and adds considerably to the questions that the process of defining regions based on spatial characteristics seeks to answer.

- **Natural regions can change in size and shape over time in response to environmental changes.** An example is desertification, the expansion of desert regions that has occurred in recent years. Using images from space, we can see and monitor changes in the areas covered by deserts and other natural regions.
- **Boundaries separating different natural or environmental regions tend to be indistinct or transitional rather than sharp.** For example, on a climate map, lines separating desert from nondesert regions do not imply that extremely arid conditions instantly appear when the line is crossed. When people travel to a desert, the environment is likely to get progressively more arid as they approach their destination.
- **Regions are spatial models devised by humans for geographic analysis, study, and understanding.** Natural regions are conceptual models that help us comprehend and organize spatial relationships and geographic distributions. Learning geography is an invitation to think spatially, and regions provide an essential, extremely useful conceptual framework in that process.

Understanding regions through an awareness of how areas can be divided into geographically logical units, and why it is useful to do so, is essential in geography. Regions help us understand, reason about, and make sense of the spatial aspects of our world.

USDA Forest Service

Natural Regions: The Great Basin of the Western United States, outlined here in white, is a landform region that is clearly defined based on an important physical geographic characteristic. No rivers flow to the ocean from this arid and semiarid region of mountains and topographic basins. The rivers and streams that exist flow into enclosed basins where the water evaporates away from temporary lakes, or they flow into lakes, such as Great Salt Lake, that have no outlet to the sea. Topographic features called *drainage divides* (mountain ridges) form the outer edges of the Great Basin, defining and enclosing this natural region.

1913

W. C. Alden (USGS), Courtesy of Glacier National Park Archives

2005

Blase Reardon (USGS), Courtesy of Glacier National Park Archives

● FIGURE 1.9 Photographs taken 92 years apart in Montana's Glacier National Park show that Shepard Glacier, like other glaciers in the park, has dramatically receded during that time. This retreat is in response to climate warming and droughts.

What other kinds of environmental change might require long-term observations and recording of evidence?

Global, regional, and local climates have changed naturally throughout Earth's history, with attendant shifts in the distribution of plant and animal life. Today, most of Earth's glaciers are shrinking in response to global warming, and alterations in Earth's climates and environments are complicated by the impact of human activities (● Fig. 1.9). Continuous change at different spatial and temporal scales of multiple variables, many of which are complexly interrelated, makes their direction and impact a challenge to determine. As a result, geographers ask questions such as: *How are Earth features changing in ways that can be documented in a spatial sense? What processes contribute to those changes? What are the rates of change? Do changes occur in cycles? Can humans witness a particular change as it is taking place, or is long-term study required to recognize the change? Do all places on Earth experience the same levels of change or is there spatial variation?*

Physical Science Perspective

In addition to being a spatial science, physical geographers consider the natural aspects of Earth from a physical science perspective. We've already seen that, like other scientists, physical geographers use the scientific method in their investigations of the characteristics and processes acting on Earth's surface. Like other scientists, geographers observe phenomena, collect and analyze data, answer questions, and find solutions to problems related to natural processes acting on Earth. Geographers also draw from and contribute to the larger body of physical science knowledge, and results of their research are of interest to many other physical scientists. Physical geographers who specialize in climatology or meteorology exchange many ideas and much information with atmospheric physicists. Geomorphologists work at the interface between physical geography and geology, and they communicate readily across the disciplinary boundary. Soil geographers apply their knowledge of chemistry to the study of soil properties and soil formation. Biogeographers are concerned about geographic aspects of plant and animal communities, thus they share fundamental interests with many biologists. Despite these areas of shared knowledge, physical geography differs from those other fields because of its emphasis on problems of a spatial nature and because geographers tend to take a holistic approach to their studies.

Physical geography considers all spatial scales, from local to global, and all component parts of the natural environment: the atmosphere, earth materials, plants and animals, and water. By examining the whole set of factors, features, and processes that work together at Earth's surface, physical geographers tend to ask different questions than physicists, geologists, chemists, or biologists and provide different insights into the planet's dynamic nature. Because of its holistic approach to the study of Earth's surface characteristics and processes, physical geography is especially well suited to using the scientific notion of systems in studying Earth.

Earth as a System A **system** is any organized entity that consists of interrelated and interacting components. Physical systems have matter and energy as well as an organizational structure consisting of pathways and linkages between components. Our planetary environment, the **Earth system**, includes interactions among a vast combination of factors. The individual components of a system interact with each other as parts of a functioning unit. System attributes that can change value are called *variables*. The status or magnitude of one variable commonly influences the nature of other variables in the system, and a change in one variable typically leads to changes in others. For example, in the environmental system of a mountainous region, the distribution of higher and lower elevations influences temperature and rainfall patterns, which in turn affect the density, type, and variety of vegetation. As the moisture, temperature, plants, and also the type of underlying rock vary over the region, so too will the nature of the soil that forms. The systems concept provides scientists in any field with a useful framework for analyzing how something works; it provides physical geographers with an excellent means for studying the interacting variables and processes that affect Earth's surface characteristics and environments.

Scientists use the notion of a system to focus their attention on just the variables, processes, and relationships of interest. A system can consist of something as large as Earth or the solar system or something as small as an organism, cell, or molecule, depending on our purpose. It can also be helpful to consider a system as composed of multiple **subsystems**, which are functioning units within a system that demonstrate strong internal connections. For example, the human body is a system that is composed of many subsystems, such as the respiratory, circulatory, and digestive systems. Examining the Earth system as consisting of a set of interdependent subsystems is an important and useful approach to studying physical geography from the physical science perspective.

Earth's Four Major Subsystems The Earth system consists of four principal subsystems (● Fig. 1.10). The **atmosphere** is the gaseous blanket of air that envelops, shields, and insulates Earth. Variations in atmospheric components and processes create the changing conditions that we know as weather and climate. The **lithosphere** consists of the solid Earth—landforms, minerals, rocks, and soils. Next is the **biosphere**, which is composed of people, animals, plants, and all other living things. Last is the

hydrosphere, which includes the waters of Earth, such as the ocean, lakes, rivers, glaciers, and water in the soil, atmosphere, and organisms. These major subsystems and the interactions among them nurture the conditions necessary for life on Earth, but the impact and intensity of those interactions are not equal everywhere. This inequality leads to our planet's environmental diversity and produces the wide variety of geographic patterns on Earth.

Earth's four major subsystems do not function in isolation of each other. Instead, they interact, overlap, and are complexly interconnected. Water of the hydrosphere flows through organisms, including people, and provides a habitat for aquatic plants and animals within the biosphere. The hydrosphere interacts with the lithosphere in many ways, such as streams, waves, and currents shaping landforms. Water also has important connections with the atmosphere through evaporation, condensation, and the effects of ocean temperatures on climate. Many other examples exist of connections and interactions among the four major Earth subsystems. Soil can be examined from the perspective of the lithosphere, biosphere, hydrosphere, or atmosphere because soils typically contain minerals, organisms, water, and gases. The water in clouds is simultaneously a component of both the hydrosphere and the atmosphere, and water stored in plants and animals is part of both the hydrosphere and the biosphere. The fact that we cannot draw sharp boundaries between these subsystems underscores the interrelatedness among various components of the Earth system.

Earth System Dynamics The Earth system is *dynamic*, responding to continuous change. Some of these changes occur frequently or proceed at an appropriate speed for people to easily observe them. We can directly observe seasonal changes, ocean tides, earthquakes, floods, volcanic eruptions, and at times even the creation of new volcanic islands (● Fig. 1.11). We generally have less information on processes that are relatively rare, occur at rates that vary widely over time, or that take many years compared to a human lifetime to accumulate enough modification so that people can recognize their effects. Long-term changes on our planet are often difficult to understand or forecast with certainty. The evidence must be studied carefully and scientifically to determine what is occurring and what the potential consequences might be. Changes of this type include shifts in climate, expansion and contraction of deserts, varying composition of the oceans or atmosphere, and changes in global sea level.

Some changes within the Earth system are natural, others are human induced, and some result from a combination of these factors. Today, much of the concern about environmental change, such as global warming, centers on the increasing impact that human activities are exerting on Earth's natural systems. To understand

● FIGURE 1.10 Earth's four major subsystems. Studying Earth as a system is central to understanding changes in our planet's environments and adjusting to or dealing with these changes. Earth consists of many interconnected subsystems.

How do these four subsystems overlap? For example, how does the atmosphere overlap with the hydrosphere or the biosphere?

● **FIGURE 1.11** This new volcanic island formed in the Red Sea starting on December 23, 2011, when volcanic eruptions from the seafloor began to reach the surface. The island at this time was about 500 meters (0.3 mi) long and growing.

After the volcanic island cools, what other environmental changes could slowly begin to take place?

Earth and the effects that humans have on it, whether at the global or local scale, we must continue to expand our knowledge about the natural characteristics, components, and processes of the Earth system. Understanding how our planet works is critical to the existence of humans and the other Earth organisms. We will return to the important topic of systems later in this chapter.

Environmental Science Perspective

In the broadest sense, the **environment** is our surroundings, consisting of all physical, social, and cultural aspects of the world that affect our growth, our health, and the way we live. The physical environment is the primarily natural part of the environment, including weather, climate, landforms, rocks, soil, water, plants, and animals, as well as their characteristics, processes, and interconnections. Because these are major topics of study in physical geography, it should come as no surprise that geography is often described as the original environmental science. In addition, just as physical geographers share common ground with other disciplines through the physical science perspective, they also share interests, knowledge, results, and understanding with other disciplines through the environmental science perspective. Physical geography's holistic approach enhances the ability of geographers to study the environment because important factors and processes are considered not only individually but also as integral parts of a functioning environmental system. Poised at the interface between Earth and human existence, geography has much to offer for understanding how the environment affects

human lives and how our ways of living affect the environment. Geographic data collection, scientific analyses, and environmental monitoring play crucial roles in identifying connections between environmental change and human activities and in determining what we can do to minimize or eliminate environmental problems.

Ecosystems Like environmental biologists, many physical geographers are fascinated by the living part of the natural Earth system. The study of relationships between organisms and their environment is the branch of biology known as **ecology**. The word **ecosystem** (a contraction of ecological system) refers to a community of organisms and the relationships of those organisms to one another and to their environment (● Fig. 1.12).

The ecosystem concept can be applied on almost any scale from local to regional or global, in virtually any geographic location. Your backyard, a farm pond, a grass-covered field, a marsh, a forest, a lagoon, or a desert sand dune can be viewed as an ecosystem. Ecosystems exist wherever there is an exchange of materials among living organisms and functional relationships between organisms and their natural surroundings. Certain

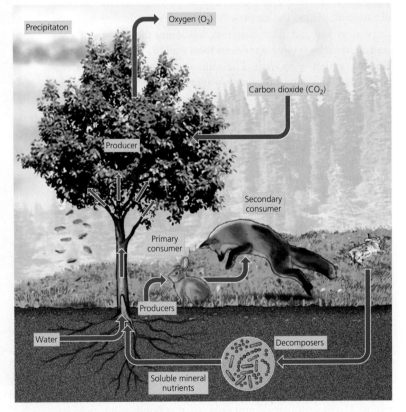

● **FIGURE 1.12** Ecosystems are an important aspect of natural environments, which are affected by the interaction of many processes and components.

How do ecosystems illustrate the interactions in the environment?

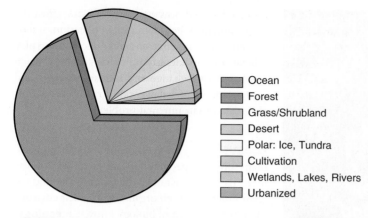

● **FIGURE 1.13** Percentages of the different land and water categories on Earth. Habitable land is a limited resource on our planet.

Ocean
Forest
Grass/Shrubland
Desert
Polar: Ice, Tundra
Cultivation
Wetlands, Lakes, Rivers
Urbanized

What options do we have for future settlement of Earth's lands?

ecosystems, such as a lake or a desert oasis, may have relatively clear-cut boundaries, but the limits of many others are not so precisely defined. Typically, the spatial boundary between adjacent ecosystems is transitional, being crossed gradually over distance.

Ecosystems are dynamic in that their various parts are always changing. Plants grow, rain falls, animals eat, and soils develop. All of these actions and processes affect ecosystems. Because each component of an ecosystem interacts with other components of that system, action or change in one element often leads to action or change in others, which transforms the ecosystem. For example, a change in the ecosystem's weather from sunshine to rain can benefit the soil, plants, and animals. Very heavy rains, however, could carry away soil and plant nutrients, hindering the growth of vegetation, thereby leaving a reduced food supply for the animals that depend on that vegetation. The supply of large amounts of moisture to the remaining soil might preferentially benefit a few types of plants. As those plants grow and thrive, the increased shade that they create could restrict the growth of other types of plants that would otherwise be competing for the remaining soil nutrients.

The capability of different environments to adequately support a human population or absorb human impacts varies widely, and some land areas on Earth do not provide a suitable ecosystem for people (● Fig. 1.13). The geographic distribution of human population densities around the world, varying from uninhabited to dense settlements, reflects this environmental disparity.

Human–Environment Interactions Physical geography includes considering environmental relationships that involve humans and human activities. Human–environment interactions are two-way relationships; the environment influences human activities and human activities affect the environment. We will first consider ways in which the environment affects human activities.

In our spatial distribution over Earth, we live and work in some locations that are subject to potentially hazardous natural conditions or events. News reports depict disasters as people are exposed to violent natural processes, such as floods and tornadoes. It is not uncommon to witness the consequences of a major earthquake somewhere in the world. In recent years, major earthquakes have occurred in Nepal, Japan, China, New Zealand, Chile, Haiti,

Indonesia, and Italy, to name just a few examples. Earthquakes in Japan in 2011 and Indonesia in 2004 generated devastating tsunamis (● Fig. 1.14).

The term **natural hazard** refers to any natural process, typically of unusual intensity, that puts environments and human life or property at risk of damage or destruction. In addition to earthquakes, tsunamis, floods, and tornadoes, natural hazards include events such as volcanic eruptions, hurricanes and other severe storms, coastal erosion, wildfires, and landslides. Sometimes these events are comparatively minor and cause little damage, but occasionally extremely large and powerful examples strike that, unfortunately, are responsible for the loss of human life. In 2015, the earthquake in Nepal, and avalanches and landslides that it triggered, killed or injured thousands, destroyed a large number of homes and other buildings, and left some rural communities cut off from assistance for days due to blocked or damaged roads. In 2012, winds, waves, and a surge of ocean water from Hurricane Sandy

U.S. Navy Photo/Alexender Tidd

(a)

FEMA/Patsy Lynch

(b)

● **FIGURE 1.14** Environmental hazards: (a) The devastated port town of Wakuya, Japan, after being battered by a powerful tsunami generated by a massive earthquake in 2011. (b) What was left of a home on the New Jersey shore after Hurricane Sandy's storm surge swept the coast in 2012.

caused widespread destruction to coastal areas in the Caribbean and in the northeastern United States, leaving many communities without electricity or running water. If an especially powerful event has not occurred for a long time people might underestimate the potential hazard. It is important to be aware of the types of natural hazard that may affect the area where you live and to know how to avoid or respond to their possible effects. Because they happen at Earth's surface and affect people and the environment, many geographers are interested in the causes, characteristics, intensity, and distribution in space and time of natural hazards, as well as in finding ways for communities to be better prepared for their occurrence. Geographers locate, classify, and map hazard zones and recommend ways to prepare for safeguarding life and property during the occurrence of a natural hazard.

The environment affects people, but people also impact the environment. Human occupation of an environment results in consumption of fresh water, food, fuel, and other natural resources, alteration of plant and animal communities, nutrient depletion and erosion of soils, and reduction of air and water quality. Through activities such as urbanization, road construction, and mining, human occupation even changes the nature and configuration of the land surface itself. Human activities will always affect the environment in some way, but if we understand the factors and processes involved, we can work to minimize detrimental impacts.

Many geographers specialize in identifying and reducing **environmental degradation**, which is environmental damage caused by human activities. Human action, for example, can lead to **pollution**, an undesirable or unhealthy contamination in the environment. Critical resources, such as air, water, and soil, can be polluted to the point where they become unusable, toxic, or even lethal to some life forms. Air pollution is a serious problem for urban areas around the world (● Fig. 1.15). Pollution of waterways can kill off important native fish species, allowing less desirable species to increase in number. Acid rain, caused by airborne pollutants from industries and power plants, has damaged forests and killed fish in freshwater lakes. Leaking pipelines and degrading toxic waste containers can deliver contaminants to the soil. Many pollutants are often transported by winds and waterways hundreds or even thousands of kilometers from their source. Lead from automobile exhaust has been found in the ice of Antarctica, as has the insecticide DDT. Pollution is a global problem that does not stop at political, or even continental, boundaries.

The ability of humans to alter the landscape has been increasing over time. Consider the example of the Everglades in Florida. The interconnected Kissimmee River-Lake Okeechobee-Everglades ecosystem was one of the most productive wetland regions in the world 125 years ago, but marshlands and slow-moving water stood in the way of urban and agricultural development. Intricate systems of ditches and canals were built, and eventually half of the original 1.6 million hectares (4 million acres) of the Everglades disappeared. The Kissimmee River was channeled into an arrow-straight ditch, and wetlands were drained (● Fig. 1.16). Levees have prevented water in Lake Okeechobee from flowing to the Everglades, and highway construction disrupted the natural drainage patterns.

EPA, South Florida Water Management Division

● **FIGURE 1.16** (a) The Kissimmee River in Florida originally flowed in sweeping bends on its floodplain for 160 kilometers (100 mi) from Lake Kissimmee to Lake Okeechobee. (b) In the 1960s and early 1970s, the Kissimmee River was artificially straightened, disrupting the previously existing ecosystem at the expense of plants, animals, and water supplies for human as well as the plant and animal communities. After much work on habitat restoration that continues today, the Kissimmee is reestablishing its wetland environments and the form of its natural channel.

What factors should be considered prior to any attempts to return rivers and wetland habitats to their original condition?

Feng Li/Getty Images

● **FIGURE 1.15** Air pollution is a major problem in many of the world's largest cities. This episode of dangerous air quality in Beijing, China, occurred in 2013.

Human–Environment Interactions

Earth's environmental characteristics support all life on our planet. Yet the effects of natural processes on humans, as well as human impacts on the environment, have become topics of increasing concern. Certain environmental processes can be hazardous to human life and property, and certain human activities threaten to cause major, and possibly irrevocable, damage to Earth environments.

Environmental Hazards

The environment becomes a hazard to humans and other life–forms when, occasionally and often unpredictably, a natural process operates in an unusually intense or violent fashion. Molten rock and gases move upward toward the surface and suddenly trigger massive eruptions that can blow apart volcanic mountains. Rain showers can become torrential rains that occur for days or weeks and cause flooding. Some tropical storms gain strength and reach coastlines with great intensity, such as Hurricane Sandy in 2012. The extremely powerful earthquake in Nepal in 2015 or the earthquake and accompanying tsunami in Japan in 2011 provide other examples of the potential for occasional occurrences of natural processes to far exceed our expectable norm.

In September 2008, after Hurricane Ike became a powerful storm in the Atlantic Ocean, it passed over several islands in the Caribbean Sea, causing great damage, and continued into the Gulf of Mexico. Moving northwest, Ike made landfall near Galveston, Texas, a coastal city that had been rebuilt after being almost completely destroyed by a hurricane in 1900. Ike

Jocelyn Augustino, FEMA News Service

Natural Hazards: Hurricane Ike caused great damage in 2008 and devastated this coastal area near Galveston, Texas. This house is the only one left standing in a beach community on the Texas coast of the Gulf of Mexico after Hurricane Ike made landfall.

Can you cite some examples of natural processes that can affect the area where you live?

brought violent winds, high waves, and a massive 4.5- to 6.5-meters (15- to 22-ft) high surge of seawater that swept low-lying coastal areas for several kilometers inland.

A natural process that operates in an extraordinary fashion is a noteworthy environmental event, but it is not considered a natural hazard unless people or their properties are affected. Many natural hazards exist because people live where potentially catastrophic environmental events can occur. Nearly every populated area of the world is associated with a natural hazard or perhaps several hazards. Forested regions are subject to fire; earthquakes, landslides, and volcanic activities plague mountain regions; violent storms threaten interior plains; and many coastal regions experience periodic hurricanes or severe winter storms.

Environmental Degradation

Just as the environment can pose a danger to humans, human activities can constitute a serious threat to the environment. Issues such as greenhouse gas emissions and global warming, acid precipitation, deforestation, endangered species, deterioration of the ozone layer, and desertification have risen to the top of agendas at international conferences and meetings of world leaders, and are common topics of news and information on multimedia. But as population pressures mount, human activities are exacting an increasing toll on the air, water, soils, forests, and wildlife. Environmental deterioration is a problem of worldwide concern, and effective solutions must involve international cooperation as well as individual efforts.

Examining environmental issues from the physical geographic perspective requires that characteristics of the environment and the humans involved in those issues be given strong consideration. As will become apparent in this study of geography, physical environments are constantly changing, and all too often human activities result in negative environmental consequences. In addition, across the globe, humans live in constant threat from various environmental hazards, such as earthquakes, fires, floods, and storms—environmental hazards that vary in their geographic distribution. The natural processes involved are a part of the physical environment, but causes and solutions are imbedded in the human–environmental interactions that include the economic, political, and social characteristics of the people involved. Geography's holistic approach contributes to understanding these issues because it takes both natural and human factors into consideration in dealing with environmental concerns.

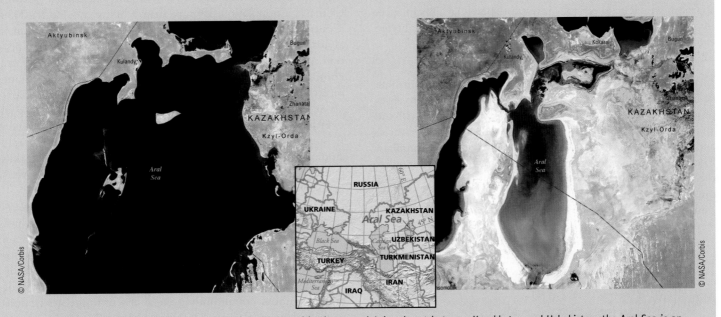

Environmental Degradation: The shrinking Aral Sea. Located in the central Asian desert between Kazakhstan and Uzbekistan, the Aral Sea is an inland lake that does not have an outlet stream. Water that flows in is eventually lost by evaporation to the air. Before the 1960s, rivers flowing into the lake supplied enough water to maintain what was the fourth largest body of inland water in the world. Since that time, agricultural diversion of river water upstream from the lake has caused the Aral Sea to shrink dramatically. The result has been the disappearance of many species that relied on the lake for survival, frequent dust storms, and an economic disaster for the local economy. With less area, the lake has a reduced ability to moderate temperatures in the surrounding landscape; winters have become colder and summers hotter.

As a result of humans altering the Kissimmee River-Lake Okeechobee-Everglades ecosystem, fires in the Everglades have been more frequent and more destructive (● Fig. 1.17), and biotic communities have been eliminated by lowered water levels. When the area receives too much rain, portions of the Everglades have been deliberately flooded to prevent drainage canals from overflowing, causing some animals to drown and eliminating sites for birds to rest and reproduce. Southern Florida's wading bird population has decreased by 95% over the past century. Without the natural purifying effects of wetland systems, water quality in southern Florida has deteriorated; with lower water levels, saltwater encroachment is a serious problem in coastal areas.

The extent and severity of the environmental problems in southern Florida have not gone unnoticed. Substantial efforts on multiple fronts are underway by government agencies and scientists to restore the ecosystems of the area. The goals are to allow the Kissimmee River to flow naturally across its former floodplain, to return agricultural land to wetlands, and to restore water flow through the Everglades. The problems of southern Florida provide useful lessons. Alleviating environmental problems that result from human actions may be possible, but doing so often takes considerable time, effort, and financial resources. A wiser approach would be to carefully investigate potential consequences of proposed human alterations to the natural environment before carrying out the alterations.

Environmental Sustainability
The most unique attribute of Earth is its ability to support life. On Earth, natural processes produce an adequate supply of oxygen; the sun interacts with the atmosphere, ocean, and land to maintain tolerable temperatures; and photosynthesis and other processes provide food supplies. Except for the input of energy from the sun, Earth itself supplies the necessary materials and conditions that allow life to exist on our planet. If a critical part of this system changes significantly, organisms may no longer be able to survive on Earth.

● **FIGURE 1.17** An ongoing problem resulting from human action in the Kissimmee River-Lake Okeechobee-Everglades area of southern Florida is the invasion of weedy plants that causes a serious fire hazard during the dry season. Controlled burns by the U.S. Fish and Wildlife Department are necessary to avoid catastrophic fires and to help restore the natural vegetation.

The substances needed to support life, or that are used by humans to improve our living conditions, are **natural resources**. Despite the wealth of natural resources available on Earth, they can be abused, wasted, or exhausted. A serious concern is that humans are depleting, or have the potential to deplete, nonrenewable natural resources, which are those that cannot be replaced, as well as those natural resources that are renewable but regenerate far slower than the rate of consumption. Overconsumed resources include forests, agricultural soils, water resources, and food sources from the ocean. This problem of using more of a resource in a year than its annual renewal, growth, or replacement is known as **environmental overshoot**. Throughout most of human history, natural resource depletion has occurred primarily at the local or regional scale; new supplies of a depleted resource could be found in another locality or region. Now, however, with the human population larger than ever before and with unprecedented rates of resource consumption, we must realize that Earth is a finite space and its natural resources of finite extent.

The capacity of Earth to support the growing numbers of humans may have an ultimate limit. The continually increasing world population has passed 7 billion, and the United Nations estimates a population of more than 9 billion people by 2050 if current growth rates continue. Earth's human population is annually using 50% more natural resources than the Earth system is able to renew and resupply, meaning that people are consuming significant resource reserves that are not being replaced. We would need 1.5 Earths to sustain our current levels of consumption indefinitely. It is obvious that this trend of overshoot, which continues to increase, cannot go on forever. Resource consumption, moreover, is not equally distributed among Earth's human inhabitants. More than half of the world's people already suffer substandard living conditions and insufficient food supplies.

In recent years, there have been increasing efforts to encourage **environmental sustainability**, or *sustainable living*, which means consuming resources at a level that our planet can sustain indefinitely, while still developing the resources that we need for adequate living conditions. The United Nations reports that this will require practicing "development that meets the needs of the present without compromising the ability of future generations to meet their own needs." It will not be possible for humans to continue to consume, abuse, or destroy natural resources at current rates, and it is irresponsible to exhaust the natural resources that will be needed by future generations. We need to understand the impact of our individual and collective actions on the complex environmental systems of our planet.

As the world population grows, productive lands continue to diminish. Prime agricultural lands are sometimes built on or paved over for other uses while the crops are relegated to areas more poorly suited to agriculture. To sustain acceptable living standards for generations to come, it is essential to realize that environments do not change their nature to accommodate humans; thus we must use our lands, and all of our resources, wisely.

Geography has much to offer in understanding the factors involved in sustainable living. We have the responsibility of helping to maintain our present and future habitat—the Earth system.

Using Models and Systems

We cannot separately study every individual molecule on Earth's surface, nor should we try to do so. Instead, scientists generalize, looking for commonalities in numerous individuals, to explain whole categories of phenomena. As physical geographers work to describe and explain the innumerable and complex features of Earth and its environments, they support these efforts, as other scientists do, by developing generalized representations of the real world. These useful simplifications of complex reality are **models**. Models permit prediction, and every model is designed with a specific purpose in mind. A map or a globe, for example, is a model; each is a simplified representation of part or all of Earth that provide us with useful information. Another example of a model is a specialized computer program capable of handling large amounts of data and performing extensive mathematical calculations to simulate Earth surface responses to different characteristics, conditions, and processes.

Scientists use many kinds of models (● Fig. 1.18). **Pictorial and graphic models** include pictures, maps, graphs, diagrams, and drawings. A world globe or other three-dimensional replica of part of Earth, such as the terrain of a mountain range or national park, is a **physical model**. **Mathematical and statistical models** are used to predict an outcome or the probability of an outcome. These models might suggest the influence of climate change on a region's daily weather or calculate the probability of a flood or earthquake. **Computer-generated models** often create **visualizations**, which are three-dimensional images or animations showing the nature of a landscape or how it might respond to changes in processes. Words, language, and the definitions of terms or ideas can also serve as models.

Also important are **conceptual models**, the mind imagery that we use for understanding our surroundings and experiences. Focus for a moment on the image that the word mountain (or waterfall, cloud, tornado, beach, forest, desert) generates in your mind. Most likely what you visualize (conceptualize) in your mind is sketchy rather than detailed, but sufficient to convey a mental idea of the feature. This image is a conceptual model. A particularly important conceptual model is the **mental map**, which people use to think about places, travel routes, and the distribution of features over space. The notion of a system, consisting of components, variables, linkages, processes, and boundaries, serves as a type of conceptual model to help scientists focus their attention on and analyze a selected process or portion of reality. How could we even begin to understand our world without conceptual models and, in terms of spatial understanding, without mental maps?

Systems Analysis

If you try to think about all aspects of Earth in its entirety, or to understand everything involved in the functioning of just a part of the Earth system, you will probably conclude that there are just too many factors to envision. Our planet is much too complex for a single model to explain all of its environmental components and how they affect one another. To begin to comprehend Earth as a whole or to understand its environmental components, physical geographers have adopted from physics and other sciences the notion of systems, which was introduced earlier in this chapter, and **systems analysis**, a powerful method of examining

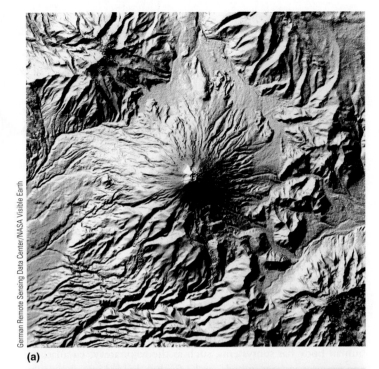

(a)

(b)

● **FIGURE 1.18** Models help us understand Earth and its subsystems by focusing our attention on major features or processes. (a) A computer-generated elevation model of Cotopaxi, a large volcano in the Andes Mountains of Ecuador. The terrain and the drainage pattern of stream channels are well illustrated in this model. (b) This table-sized physical model of part of the Illinois River uses a channel and flowing water to study the behavior of the river under various conditions.

systems. In systems analysis, understanding how something works is approached using the following strategy:

1. Clearly define the system that you wish to study and delineate the system boundaries.
2. Identify any matter and energy that enters or leaves the system.
3. Inventory component parts of the system and processes acting within the system.
4. Examine how the system components and processes interact with one another.
5. Determine how the component interactions affect the operation of the system.

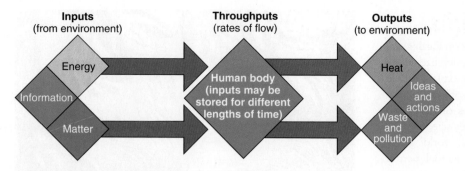

Inputs (from environment) **Throughputs** (rates of flow) **Outputs** (to environment)

Energy

Information

Matter

Human body (inputs may be stored for different lengths of time)

Heat

Ideas and actions

Waste and pollution

● **FIGURE 1.19** The human body is an example of a system, with inputs, storage, and outputs of energy and matter. Interactions between energy and matter drive the system.

What characteristics of the human body as a system are similar to those of Earth as a system?

The human body is an excellent example of a physical system. It has boundaries and interacting component parts. Energy and matter enter the system as **inputs**, are stored in the system, and leave the system as **outputs** (● Fig. 1.19). Like most systems, the human body has subsystems, such as the respiratory, circulatory, and nervous systems, which can be further divided into their own subsystems—for example, the lungs, heart, and brain, respectively.

The systems approach is a beneficial strategy for studying processes and conditions of Earth. The atmosphere, lithosphere, biosphere, and hydrosphere each function as a major subsystem of the Earth system, and are further divided into smaller subsystems to focus attention on understanding a particular part of the whole. Examples of subsystems of the atmosphere, lithosphere, biosphere, and hydrosphere examined by physical geographers include climate systems, storm systems, stream systems, ecosystems, the water cycle, and the systematic heating of the atmosphere and ocean. A great advantage of systems analysis in physical geography is that it can be applied to environments at virtually any spatial scale from global to microscopic.

Open and Closed Systems

Energy and matter move within Earth and its subsystems by means of various processes. As shown in ● Figure 1.20 for the water cycle subsystem, sunlight (*energy*) warms (*process*) a body of water (*matter*) and the water evaporates (*process*) into the atmosphere. Later, the water condenses (*process*) back into rain (*matter*) and the rain falls (*process*) on the land and runs off (*process*) downslope back to the sea. In a systems model, geographers can trace inputs of energy and matter into the system, their storage in the system, their output from the system, and the interactions between components within the system.

Open systems allow inputs and outputs across their boundaries, whereas **closed systems** do not. Some systems are open to energy but closed to inputs and outputs of matter. Most of Earth's subsystems are open to both energy and matter, allowing both to enter and leave the system (● Fig 1.21a). A stream system like the one shown in Figure 1.20 is an excellent illustration of an open subsystem: Matter

● **FIGURE 1.20** The water cycle provides examples of interactions between energy and matter, their storage in the system, and the processes involved. Being aware of energy and matter and the interactive processes that link them is important in understanding how environmental systems operate.

Can you think of another environmental system and break it down into its components of energy, matter, storage in the system, and processes?

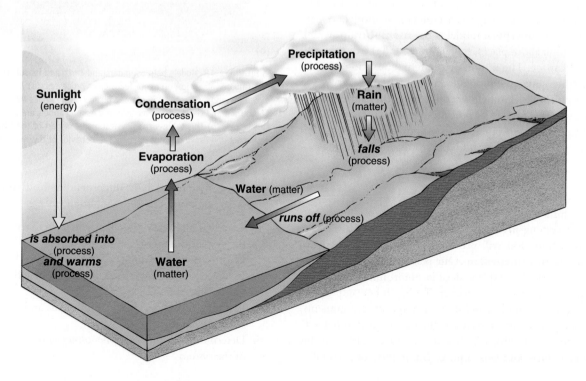

Precipitation (process)

Sunlight (energy)

Condensation (process)

Rain (matter)

Evaporation (process)

falls (process)

Water (matter)

runs off (process)

is absorbed into (process) and warms (process)

Water (matter)

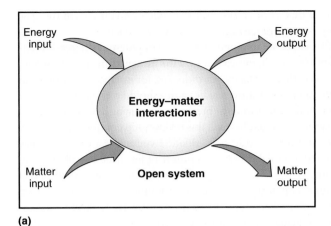

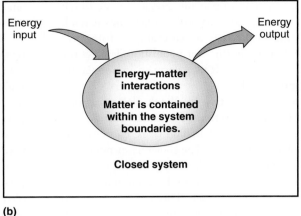

(a)   **(b)**

● **FIGURE 1.21** The two basic types of systems. (a) Most of Earth's subsystems are open in terms of both energy and matter. Open systems undergo inputs and outputs across the system boundaries. (b) Earth is an open system for energy but is virtually closed in terms of matter because so little matter enters and leaves the system. Solar energy (input) enters the Earth system, and that energy is dissipated (output) to space mainly as heat. Because natural resources do not enter the Earth system and waste does not naturally leave, humans face limits to available natural resources and must be mindful of the accumulation of waste and pollutants.

Think of an example of an open system and outline some of the matter and energy inputs and outputs involved in that system.

in the form of soil particles, rock fragments, precipitation, and water running off the land enter the stream, as does solar energy. The same materials, and some energy, such as heat, leave the stream primarily where it empties into the ocean or other body of water.

Planet Earth is an open system to energy. Solar energy enters the Earth system and heat energy leaves the system. Technically, Earth is also an open system to matter because meteorites reach Earth's surface, some atmospheric gas molecules escape to space, and a few moon rocks have been brought back by astronauts. The amount of matter entering and leaving the Earth system, however, is so small that we usually consider it closed to the input or output of matter (Fig. 1.21b).

When we describe Earth as a system or as a set of interrelated systems, we are using conceptual models to help organize our thinking about what we are observing. Throughout this book, we will use the systems concept, and many other kinds of models, to simplify and illustrate complex features of Earth's natural physical environment.

Equilibrium in Earth Systems

We often hear about the balance of nature. *Balance of nature* implies that natural systems tend to have built-in mechanisms to counterbalance, or accommodate, change without dramatically affecting the system. If the inputs entering the system are balanced by outputs, the system is said to have reached a state of **equilibrium**. Over a very short time, the balanced state might appear to be unchanging, or *static*. Observations over a longer interval will reveal that most systems are continually adjusting slightly one way then another as they react to variations in inputs. This change within a range of tolerance is called **dynamic equilibrium**—that is,

conditions are not static, but instead oscillate somewhat around a typical or average state. Over very long periods of time the typical or average condition may itself be slowly changing. A reservoir contained by a dam is a good example. In the short term, such as days, despite inflows and outflows, the water level may not change. During a year's time, the water will oscillate in a dynamic equilibrium around an average level as inflows and outflows vary. Over decades, the average water level may rise as accumulating sediment builds up the bottom of the reservoir (● Fig. 1.22).

An adjustment in one part of a system in response to a change that occurred in another part is called **feedback**. Two kinds of

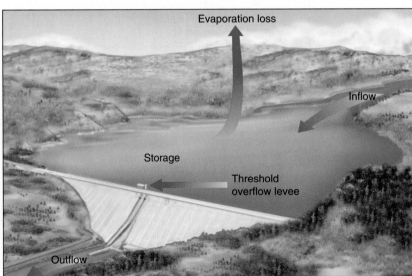

● **FIGURE 1.22** A reservoir provides a useful illustration of equilibrium in systems. The amount of water coming in may increase or decrease over time, but it must be equaled by the water going out, or the level of the lake will rise or fall. If the input–output balance is not maintained, the reservoir will get larger or smaller as the system adjusts to storing more or less water. A state of equilibrium (balance) will exist between inputs, outputs, and storage in the system.

feedback relationships operate in systems. **Negative feedback** works to offset or counteract the original change, thus it tends to maintain equilibrium in the system. An example of negative feedback is when an increase in the deer population depletes the animals' food supplies, which leads to a decrease in population back toward the original number.

Systems can also exhibit **positive feedback**, that is, respond in a way that reinforces the direction of an initial change. For example, several times in the past 2.6 million years, Earth experienced episodes of cooler temperatures. The cooler temperatures led to the growth of extensive glacial ice sheets. The highly reflective surfaces of these massive ice sheets increased the amount of solar energy that Earth reflected back to space. This, in turn, enhanced the cooling trend leading to further growth of glaciers through positive feedback. Ultimately, however, evaporation of water from the ocean decreased as the climate became colder and because of the expanded cover of sea ice. Decreased evaporation reduced the supply of moisture to storms that supplied snow to the glaciers. A reduction in snowfall caused the glaciers to shrink, which led to less reflection of solar energy, contributing to a warmer climate, thus beginning another cycle. The cold-climate circumstances that caused the reduction of snowfall is an example of a **threshold**, a

condition that causes a system to change dramatically, in this case ending, or even reversing, the positive feedback.

Thresholds are important regulators of system processes. Reaching threshold conditions can cause a fundamental change in a system and the way that it behaves. Earthquakes, for instance, will not occur until the buildup of stress reaches a threshold level that exceeds the strength of the rocks to resist breaking. Fertilizing a plant will help it grow faster and larger, but will this relationship continue indefinitely if more and more fertilizer is added? Eventually, a threshold may be reached at which the large amount of fertilizer actually poisons the plant. With environmental systems, an important question that geographers often try to answer is how much change can a system tolerate without becoming drastically or irreversibly altered, particularly if the change has negative consequences.

Systems analysis allows the investigation of how natural or human-induced changes in one variable or process affect others. In fact, systems analysis, along with the scientific method, can be a powerful tool for helping to identify environmental consequences of human actions. For example, when people began using chlorofluorocarbons (CFCs) many years ago, it was not known that releasing those compounds would destroy protective ozone in the upper atmosphere. ● Figure 1.23 uses that example to illustrate

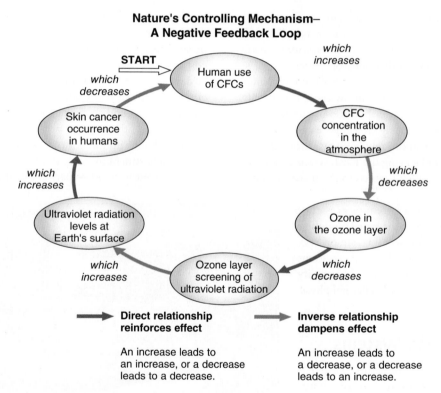

Nature's Controlling Mechanism– A Negative Feedback Loop

Direct relationship reinforces effect
An increase leads to an increase, or a decrease leads to a decrease.

Inverse relationship dampens effect
An increase leads to a decrease, or a decrease leads to an increase.

● **FIGURE 1.23** A feedback loop illustrates how negative feedback tends to maintain system equilibrium. This example shows relationships between the ozone layer, which screens out harmful (cancer-related) ultraviolet (UV) energy from the sun, chlorofluorocarbons (CFCs), and potential impacts on life on Earth. CFCs have been used in air conditioning/refrigeration systems and, if leaked to the atmosphere, they cause ozone depletion. *Positive feedback* (direct relationship) means that either an increase or a decrease in the first variable will lead to the same effect on the next. *Negative feedback* (inverse relationship) means that a change in one variable will cause an opposite change in the next. After one pass through the negative-feedback loop, all subsequent changes in the next cycle will be reversed. A second pass through the feedback loop (reversing each increase or decrease interaction) illustrates how this works. The last link between skin cancer and human use of CFCs would likely result in people acting to reduce the problem.

What might be the potential (extreme) result of ozone depletion if humans fail to take sufficient corrective action?

a **feedback loop**, which is a set of feedback operations that can be repeated as a cycle. In natural systems, the eventual result of a feedback loop is generally negative feedback because the sequence of operations tends to reverse the direction of change in the initial element.

Figure 1.23 also illustrates six of the most important factors related to damage to the ozone layer by CFCs. Each of these factors is linked by a feedback interaction to the next variable in the loop. To illustrate a feedback loop, consider a simplified hypothetical scenario of what might happen if human-caused damage to the atmosphere's ozone layer continues unimpeded. In fact, substantial efforts have been undertaken in recent decades to reduce the use of known ozone-destroying substances, often by substituting them with compounds considered less harmful that serve the same purpose. The example, however, will show you how to think about Earth processes and interactions as a functioning system. First, consider some facts:

1. The ozone layer in the upper atmosphere protects us by blocking much of the sun's harmful ultraviolet (UV) radiation, which causes skin cancer and cell mutations.
2. Chlorofluorocarbons, and some related chemicals that have been widely used in air conditioners, can migrate to the upper atmosphere and cause chemical reactions that destroy ozone.

Follow Figure 1.23, starting with the human use of CFCs at the top of the diagram, and trace the feedback links. Then, consider these questions: What is likely to happen to the amount of CFCs used if skin cancer continues to increase? Will humans act to correct the problem, or not? What would be the potential outcome in each case? What are some other examples of feedback operations in natural systems?

Negative feedback loops are generally beneficial to the environment because they regulate a system through a tendency toward balance. Positive feedback loops normally do not operate for extended periods in nature because environmental limiting factors (thresholds) act to return the system to a state of equilibrium.

It is essential to remember that systems are models, so they are not the same as reality. They are products of the human mind and are only one way of looking at the real world. Examining various Earth subsystems helps us understand the natural processes involved in the development of the atmosphere, lithosphere, biosphere, and hydrosphere. Models also help us simulate past events or predict future changes. But we must be careful not to confuse simplified models with the actual complexities of the real world.

Physical Geography and You

The physical environment affects our everyday lives. The principles and perspectives of physical geography help us be environmentally aware, assess environmental situations, analyze the factors involved, and make informed decisions about courses of action. What are the environmental advantages and disadvantages of a particular homesite? What season would be the best time to plant a tree? What environmental impacts might be expected from a proposed development? What potential natural hazards—flooding, landslides, earthquakes, hurricanes, and tornadoes—should you be aware of where you live? What can you do to minimize potential damage to your family, friends, or household from a natural hazard?

You might wonder what people with a background in physical geography do in the workplace and what kinds of jobs they hold. By applying their knowledge, skills, and techniques to real-world problems, physical geographers make major contributions to human well-being and to environmental stewardship. According to the U.S. Department of Labor, people in any career field that deals with maps, location, spatial data, or the environment would benefit from an educational background in geography (● Fig. 1.24).

Geography is a way of looking at the world and of observing its features. It involves asking questions about the nature of those features and appreciating their beauty and complexity. It encourages you to seek explanations; gather information; and use geographic skills, tools, and knowledge to solve problems. Just as you see a painting differently after an art course, after this course you should see sunsets, waves, storms, deserts, valleys, rivers, forests, prairies, and mountains with a geographically educated eye. You will see greater variety in the landscape because you will have learned to observe Earth differently, with greater awareness and a deeper understanding.

● **FIGURE 1.24** A geographer uses his laptop to access, collect, record, and analyze the data, map information, and imagery in the field.

How is a geography background and an understanding of mapping and spatial concepts valuable in the workplace?

CHAPTER 1 ACTIVITIES

TERMS FOR REVIEW

physical geography
geography
spatial discipline
human geography
region
cartography
scientific method
absolute location
relative location
spatial distribution
spatial pattern
spatial interaction
system
Earth system
subsystem
atmosphere

lithosphere
biosphere
hydrosphere
environment
ecology
ecosystem
natural hazard
environmental degradation
pollution
natural resource
environmental overshoot
environmental sustainability
model
pictorial/graphic model
physical model
mathematical/statistical model

computer-generated model
visualization
conceptual model
mental map
systems analysis
input
output
open system
closed system
equilibrium
dynamic equilibrium
feedback
negative feedback
positive feedback
threshold
feedback loop

QUESTIONS FOR REVIEW

1. What are some of the subfields of physical geography, and what do geographers study in those areas of specialization?
2. Why is geography known as the spatial discipline? What are some topics in physical geography that illustrate the role of geography as a spatial science? How do physical geography's three major perspectives make it unique among the sciences?
3. What does a holistic approach mean in terms of thinking about an environmental problem?
4. What are the four major divisions of the Earth system, and how do the divisions interact with one another?

5. How does the study of systems relate to the role of geography as a physical science?
6. How does the examination of human–environment relationships in ecosystems serve to illustrate the role of geography as an environmental science?
7. How do open and closed systems differ? Under what circumstances might you consider Earth as a closed system?
8. What is a threshold in a system? What are some examples that relate to physical geography?
9. How does negative feedback maintain a tendency toward balance in a system?

CONSIDER AND RESPOND

1. Give examples from your local area that demonstrate each of the five topics considered with respect to spatial science.
2. List some possible sources of pollution in your city or town. Could these kinds of pollution affect your life? What are some potential solutions to these problems?
3. Give one example of an environment in your local area that has been altered by human activity. In your opinion, was the change good or bad? What values are you using in making such a judgment?

4. There are advantages and disadvantages to the use of models and the study of systems by scientists. List and compare the advantages and disadvantages from the point of view of a physical geographer.
5. How can an understanding of physical geography be of value to you now and in the future? What steps should you take if you wish to seek employment as a physical geographer? What advantages might a physical geographer have when applying for a job?

▮ PRACTICAL APPLICATIONS

1. From memory, draw on paper the mental map that you envision when you think about the geography of your local natural environment (or neighborhood). Try to maintain some reasonable geographic (spatial) relationships for the sizes of areas and distances between features.

2. Draw a simple, circular feedback loop to illustrate the interactions between components and processes involved in some system that you are familiar with. Label the positive (direct) and negative (inverse) feedback relationships. What threshold conditions exist and how do they enact change?

 MindTap—Make the most of your study time by accessing everything you need to succeed in one place. Read your textbook, take notes, review flashcards, watch videos, complete activities, take practice quizzes, and more online with MindTap. Log in at **www.cengagebrain.com**.

REPRESENTATIONS OF EARTH

■ OBJECTIVES

WHEN YOU COMPLETE THIS CHAPTER YOU SHOULD BE ABLE TO:

- ■ 2.1 Explain the ways that Earth and its regions, places, and locations can be represented on a variety of visual media—maps, aerial photographs, and other imagery.
- ■ 2.2 Assess the nature and useful applications of maps and map-like presentations of the planet, or parts of Earth, citing some examples.
- ■ 2.3 Find and describe the locations of places using coordinate systems, use topographic maps to find elevations, and understand the three types of map scales.
- ■ 2.4 Demonstrate knowledge of tecÚiques that support geographic investigations, including mapping, spatial analysis, global positioning systems (GPS), geographic information systems (GIS), and remote sensing.
- ■ 2.5 Evaluate the advantages and limitations of different kinds of representations of Earth and its areas.
- ■ 2.6 Understand how the proper tecÚiques, images, and maps can be used to best advantage in solving geographic problems.

DESCRIBING INFORMATION ABOUT A LOCATION, where it is located, how to get there, and what that place will be like is a fundamental part of human communication, essential in our everyday lives. One way to do this is with a verbal explanation, but it is often difficult to explain where some place is, and how to get there, with words alone. Many times, too, the person receiving that information may have difficulty remembering all of it by the time it is needed. Have you ever tried to draw a simple diagram in order to clarify that kind of information for someone? A graphic record with diagrams and words is a very effective way to communicate these kinds of spatial information.

The earliest concept of a map probably began by drawing diagrams scratched into the soil, but these sketches left no permanent record because of exposure to the elements. What is thought to be the oldest known map is a rock tablet with scratches on the surface that represent simple landscape features. Discovered in a cave in Spain, it is thought to be 14,000 years old, and the scene must have been quite important—enough to make it permanent, etched in stone. Other ancient maps were drawn on clay tablets, metal plates, papyrus, linen, or silk or were constructed of sticks (● Fig. 2.1). Maps are powerful tools for finding locations, selecting a route to a destination, navigating to that place, learning what to expect along the way, and

● **FIGURE 2.1** This clay tablet preserves an ancient map from Mesopotamia, located in today's Iraq. Estimated to be 3500 years old, it is thought to be the world's oldest map of a city.

understanding a few things about your destination. This kind of information has been fundamental since ancient times, and is an essential part of geography today.

Many of the basic mapmaking principles that have been known for centuries are still applicable today. Yet, the technologies applied to creating maps, acquiring map-like images of our planet as well as many extraterrestrial bodies, and analyzing the information contained in these graphic representations is rapidly improving and changing. Digital and satellite technologies are widely used for locating, describing, storing, and analyzing spatial data and information. Maps are becoming increasingly more common, on television, on the Internet, in vehicles, and on our mobile devices such as cell phones and tablets. We might say that today, maps are virtually everywhere. Today, computer systems support the development of complex maps and three-dimensional (3-D) displays of geographic features that would have been nearly impossible or extremely time-consuming to produce two or three decades ago. Geographers use these technologies to help them understand environmental concerns and to facilitate spatial problem solving. They also work to improve methods for mapping,

analyzing, and displaying spatial information. Because maps are so widely used as a form of visual communication, it is important to be able to read and interpret them correctly. An informed knowledge of spatial representation and the ability to effectively communicate locational information is useful in our daily lives and is essential in the study of physical geography.

Maps and Location on Earth

Cartography is the science and profession of mapmaking. Geographers who specialize in cartography design maps and globes to ensure that mapped information and data are accurate and effectively presented. The purpose of a map is to communicate spatial and locational information. Maps, through graphic representations and symbols, "a language of location," are very efficient visual media that can relay a vast amount of information in a short amount of time. Maps are an essential resource for navigation, community planning, surveying, history, meteorology, geology, political science, and many other career fields. On television, on the Internet, and in weather reports, maps contribute to our understanding of current events. Think of all the places where you encounter maps. In travel, recreation, education, the media, entertainment, and business, maps communicate important information.

Computer tecŸology has revolutionized cartography and digital imagery. Maps that once had to be hand drawn (● Fig. 2.2) are now produced digitally and can be displayed or printed in a short time. Computer-assisted mapping allows easy revision, which was a time-consuming process when maps were hand drawn. Information that was once gathered little by little from ground observations and field surveys can now be collected instantly by satellites that flash recorded data back to Earth at the speed of light. Environmental data and information are still collected in the field, but the process is greatly facilitated by digital tecŸology. Many high-tech locational and mapping tecŸologies are now in widespread use by the public on cell phones, personal computers, and satellite-based systems that display locations and directions for use in hiking, traveling, and virtually any means of transportation.

> Cartographers can now gather spatial data and make maps faster than ever before—within hours—and the accuracy of these maps is excellent. Moreover, digital mapping enables mapmakers to experiment with a map's basic characteristics (for example, scale or projections), to combine and manipulate map data, to transmit entire maps electronically, and to produce unique maps on demand.
>
> —*Exploring Maps,* United States Geological Survey (USGS)

Maps are an essential resource in navigation, political science, community planning, meteorology, surveying, history, geology, and many other career fields. Maps communicate important information for travel, recreation, education, the media, business, and entertainment.

Maps are ever-present in newspapers, on news and weather broadcasts, on the Internet, and on cell phones and vehicle navigation systems. How many maps do you see in a typical day? How many would that equal in a year? How do these maps affect your daily life?

PHYSIOGRAPHIC DIAGRAM
Erwin Raisz, 1954

● **FIGURE 2.2** When maps had to be hand drawn, artistic talent was required in addition to knowledge of cartographic principles. Erwin Raisz, a famous and talented cartographer, drew this map of U.S. landforms in 1954, when there were only 48 states.

Are maps like this still valuable for learning about landscapes, or are they obsolete?

Earth's Shape and Size

Scientists were first able to image our planet's shape from space in the 1960s, but Greek philosophers theorized that our planet was a sphere as early as 540 B.C. About 240 B.C. Eratosthenes, a philosopher with an interest in geography, not only understood that the Earth was round but he also estimated the planet's circumference fairly accurately. (How he accomplished this will be illustrated in the next chapter.)

Describing global locations and creating maps require knowledge of our planet's form and its features. Earth can generally be considered as a sphere, with an equatorial circumference of 39,840 kilometers (24,900 mi), but the centrifugal force caused by Earth's daily *rotation* bulges the equatorial region outward and slightly flattens the polar regions, forming an overall shape that approximates an **oblate spheroid**. Recent precise satellite measurement has revealed that the Southern Hemisphere is also very slightly larger than the Northern Hemisphere. On a planetary scale these deviations of Earth's shape from a true sphere are relatively minor. Earth's diameter at the *equator* is 12,758 kilometers (7927 mi), whereas from pole to pole it is 12,714 kilometers (7900 mi). On a 30.5-centimeter (12-in.) globe, this difference of 44 kilometers

(27 mi) is about as thick as the wire in a paper clip. The variations from a spherical shape are less than one third of 1% and not noticeable in views of Earth from space (● Fig. 2.3). Nevertheless, people working in very precise navigation, surveying, aeronautics, and cartography must consider Earth's deviations from a perfect sphere.

Landforms also cause deviations from true sphericity. Mount Everest in the Himalayas is the highest point on Earth at 8850 meters (29,035 ft) above sea level. The lowest point is the Challenger Deep in the Mariana Trench of the Pacific Ocean, southwest of Guam, at 11,033 meters (36,200 ft) below sea level. The difference between these two elevations, 19,883 meters, or just more than 12 miles (19.2 km), would also be insignificant when reduced in scale on a 12-inch (30.5 cm) globe.

When ancient peoples began to sail the oceans, they recognized a need for ways to find directions and describe locational positions. Long before the first compass was developed, people understood that the positions of the sun and the stars—rising, setting, or circling in the sky—could provide locational information. Observing relationships between the sun and stars to find a position on Earth is a basic skill in **navigation**, the science of location and wayfinding. Navigation is basically the process of getting from where you are to where you want to go. High-tech navigational and mapping

NASA/JSC

● **FIGURE 2.3** The lunar astronauts were the first humans to see a spherical Earth from space, and referred to their view of the planet as the "blue marble." This scene shows Earth from the moon, as seen in an "Earthrise," above the horizon.

tecŸologies are now in widespread public use on personal computers and hand–held systems that display maps, locations, and directions for use in hiking, traveling, and virtually any means of transportation. Almost everyone's cell phone uses these tecŸologies, with the capability to display locations, routes, and maps.

Globes and Great Circles

World globes have essentially the same geometric form as our planet, so they represent geographic features and spatial relationships virtually without distortion. A world globe correctly displays the relative shapes, sizes, and comparative areas of Earth's features, landforms, water bodies, and distances between locations. Globes also preserve true compass directions. If we want to view the entire world, a globe provides the most accurate representation. Being familiar with the characteristics of a globe also helps us understand maps and how they are constructed.

An imaginary circle drawn in any direction on Earth's surface and whose plane passes through the center of Earth is a **great circle** (● Fig. 2.4a). It is called "great" because this is the largest circle that can be drawn around Earth that connects any two points on the surface. Every great circle divides Earth into equal halves called **hemispheres**. An important example of a great circle is the *circle of illumination*, which divides Earth into light and dark halves—a day hemisphere and a night hemisphere. Any circle on Earth's surface that does not divide the planet into equal halves is called a **small circle** (Fig. 2.4b).

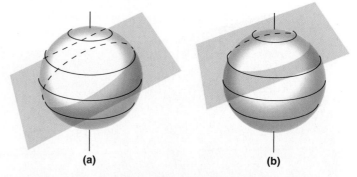

(a)   (b)

● **FIGURE 2.4** (a) Any imaginary geometric plane, which cuts through Earth and divides it into two equal halves, forms a great circle on Earth's surface. This plane can be oriented in any direction as long as it defines two (equal) hemispheres. (b) The plane shown here slices the globe into unequal parts, so the line of intersection with Earth's surface is a small circle.

The shortest route between two places can be located by finding the great circle that connects them. Great circles are useful to navigation, because the line that follows any great circle on a map marks the shortest travel route on land, sea, or air, between any two locations on Earth's surface. Connect any two cities, such as Beijing and New York, San Francisco and Tokyo, New Orleans and Paris, or Kansas City and Moscow, by stretching a large rubber band around a globe so that it touches both cities and divides the globe in half. The rubber band then marks the shortest distance between these two locations. Navigators chart *great circle routes* for aircraft and ships because traveling the shortest distance saves time and fuel. The farther away two points are on Earth, the greater the travel distance savings will be by following the great circle route that connects them.

Latitude and Longitude

Imagine that you are traveling by car and you want to visit the Pro Football Hall of Fame in Canton, Ohio. Using the Ohio road map, you look up Canton in the map index and find that it is located at "G-6." The letter *G* and the number *6* meet in a box marked on the map. In box G-6, you locate Canton (● Fig. 2.5). What you have used is a **coordinate system** of intersecting lines, a system of *grid cells* on the map. Without a locational coordinate system, it would be difficult to describe a location. A problem that a sphere presents, however, is deciding where the starting points should be for a grid system. Without reference points, either natural or arbitrary, a sphere is a geometric form that looks the same from any direction and has no natural beginning or end points. Earth's coordinate system of *latitude and longitude* is based on a set of reference lines that are naturally defined based on its planetary *rotation* and another set of lines that was arbitrarily defined by international agreement.

Measuring Latitude The North Pole and the South Pole provide two natural reference points because they mark the opposite positions of Earth's *rotational* axis, around which it turns in 24 hours. The **equator**, halfway between the poles, forms a great circle that divides the planet into the Northern and Southern Hemispheres. The equator is the reference line for measuring **latitude** in degrees north or degrees south—the equator is 0° latitude. North or south

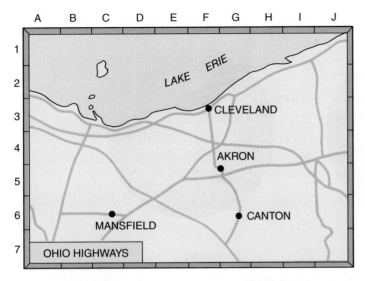

● **FIGURE 2.5** Using a simple rectangular coordinate system to locate a position. This map uses an alphanumeric location system, similar to that used on many local and campus maps.

What are the rectangular coordinates of Mansfield? What is at location F-3?

© Bernie Bernard, TDI-Brooks International, Inc.

● **FIGURE 2.6** Finding latitude by celestial navigation. A traditional way to determine latitude is to measure the angle between the horizon and a celestial body with a sextant. Today, the satellite-based global positioning system (GPS) supports most air and sea navigation (and land travel).

With high-tech location systems like the GPS available, why would it still be important to know how to use a sextant?

of the equator, the angles and their arcs increase until they reach the North or South Pole at the maximum latitudes of 90° north or 90° south. A **sextant** can be used to determine latitude by celestial navigation (● Fig. 2.6). This instrument measures the angle between the *horizon*, the visual boundary line between the sky and Earth, and a celestial body, such as the noonday sun or the North Star (Polaris).

To locate the latitude of Los Angeles, imagine two lines radiating outward from the center of Earth. One goes straight to Los Angeles and the other goes to the equator at a point directly south of the city. These two lines form a 34° angle that is the latitudinal distance (in degrees) that Los Angeles lies north of the equator, so the latitude of Los Angeles is about 34°N (● Fig. 2.7a). Because Earth's circumference is approximately 40,000 kilometers (25,000 mi) and there are 360 degrees in a circle, we can divide 40,000 kilometers by 360° to find that 1° of latitude equals about 111 kilometers (69 mi).

A single degree of latitude covers a relatively large distance, so degrees are further divided into minutes (′) and seconds (″) of arc. There are 60 minutes of arc in a degree. Actually, Los Angeles is located at 34°03′N (34 degrees, 3 minutes north latitude). We can get even more precise: 1 minute is equal to 60 seconds of arc. We could locate a different position at latitude 23°34′12″S, which we would read as 23 degrees, 34 minutes, 12 seconds south latitude. A minute of latitude equals 1.85 kilometers (1.15 mi), and a second is about 31 meters (102 ft). The latitude of a location, however, is only half of its global address. Los Angeles is located approximately 34° north of the equator, but an infinite number of points exist on the same line of latitude.

Measuring Longitude To accurately describe the location of Los Angeles, we must also determine where it is situated along the line of 34°N latitude. However, to describe an east or west position, we must have a starting line, just as the equator provides

our reference for latitude. To find a location east or west, we use longitude lines, which run from pole to pole, each one forming half of a great circle. The global position of the 0° east–west line for longitude is arbitrary, and this is the grid reference line that was established by international agreement. The longitude line passing through Greenwich, England (near London), was accepted as the **prime meridian**, or 0° longitude, in 1884. **Longitude** is the angular distance east or west of the prime meridian.

Longitude is also measured in degrees, minutes, and seconds. Imagine a line drawn from the center of Earth to the point where the north–south running line of longitude that passes through Los Angeles crosses the equator. A second imaginary line will go from the center of Earth to the point where the prime meridian crosses the equator (this location is 0°E or W and 0°N or S). Figure 2.7b shows that these two lines drawn from Earth's center define an angle, the arc of which is the angular distance that Los Angeles lies west of the prime meridian (118°W longitude). Figure 2.7c provides the location of Los Angeles by latitude and longitude.

Longitude increases as we go farther east or west from 0° at the prime meridian. Traveling eastward from the prime meridian, we will eventually be halfway around the world from Greenwich, in the middle of the Pacific Ocean at 180°. Longitude is measured in degrees up to a maximum of 180° east or west of the prime meridian. Along the prime meridian (0°E–W) or the 180° meridian, the E–W designation does not matter, and along the equator (0°N–S), the N–S designation does not matter and is not needed for indicating location.

Decimal Degrees Instead of using minutes and seconds of arc, **decimal degrees** of longitude and latitude are also used to describe the location of a point. Decimal degrees, expressed to many decimal

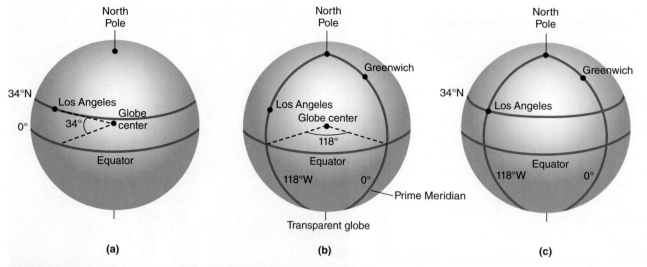

(a)   **(b)**   **(c)**

● **FIGURE 2.7** Finding a location by latitude and longitude. (a) The geometric basis for the latitude of Los Angeles, California. Latitude is the angular distance in degrees north or south of the equator. (b) The geometric basis for the longitude of Los Angeles. Longitude is the angular distance in degrees east or west of the prime meridian, which passes through Greenwich, England. (c) The location of Los Angeles is 34°N, 118°W.

What is the latitude of the North Pole, and how does it relate to longitude?

places, are very precise in pinpointing a location. Computer mapping systems handle decimals much more readily than degrees, minutes, and seconds. Some computer systems require coordinate locations in decimal format, and many use decimal degrees without the N, S, E, W letters to indicate cardinal directions. Instead, north latitudes are designated as positive numbers, and south latitudes are designated as negative numbers (with a minus sign). For longitudes, east is positive and west is negative. For example, the Statue of Liberty in New York is located at 40.6894, −74.0447. Latitude is always listed first and the degree symbol (°) is generally not necessary with decimal degrees. Some of the latitude/longitude coordinates found in this book are in decimal degrees with letters to indicate cardinal direction. The practical exercises found at the end of chapters in this book use either degrees, minutes and seconds, or decimal coordinates to familiarize you with their locational notation system.

The Geographic Grid

Every point on Earth's surface can be located by its latitude north or south of the equator in degrees and its longitude east or west of the prime meridian in degrees. This locational reference system is called the **geographic grid** (● Fig. 2.8), the set of imaginary lines that run east and west around the globe to mark latitudes and the lines that run north and south from pole to pole to indicate longitudes.

Parallels and Meridians

Lines of latitude run east–west, completely circle the globe, are evenly spaced, and are parallel to the equator and each other. This is why they are known as **parallels**. The equator is the only parallel that is a great circle; all other lines of latitude are small circles. One degree of latitude equals about 111 kilometers (69 mi) anywhere on Earth.

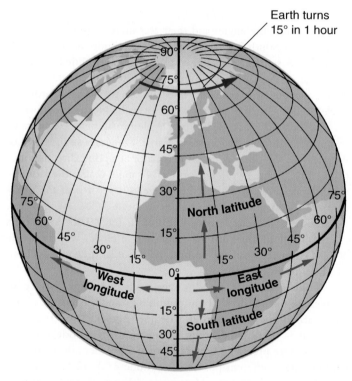

● **FIGURE 2.8** A globelike representation of Earth that shows the geographic grid with the parallels of latitude and meridians of longitude at 15° intervals.

In what ways are the latitude lines (parallels) and longitude lines (meridians) different?

Lines of longitude, called **meridians**, run north and south, converge at the poles, and measure longitudinal distances east or west of the prime meridian. Each meridian of longitude, when joined with its mate on the opposite side of Earth, forms a great circle. Meridians at any given latitude are evenly spaced, although meridians get closer together as they move toward the poles and away from the equator. At the equator, meridians separated by 1° of longitude are about 111 kilometers (69 mi) apart, but at 60°N or 60°S latitude, they are only half that distance apart, about 56 kilometers (35 mi).

Longitude and Time

The world's **time zones** were established based on the relationships among longitude, Earth's *rotation*, and time. Until the late nineteenth century, each town or area used what was known as *local time*. **Solar noon** was determined by the precise moment in a day when a vertical stake cast its shortest shadow. This meant that the sun had reached its highest angle in the sky for that day at that location—noon—and local clocks were set to that time. Because of Earth's *rotation*, noon occurs earlier in a town toward the east, and towns to the west experience noon later.

By the late 1800s, advances in travel and communication made the use of local time by each community impractical. In 1884, the International Meridian Conference in Washington, DC, set standardized time zones and established the longitude passing through Greenwich, England, as the prime meridian (0° longitude). Earth was divided into 24 time zones, one for every hour of the day because Earth turns 15° of longitude in an hour (360° ÷ 24 hours). Ideally, each time zone spans 15° of longitude. The prime meridian is the *central meridian* of its time zone, and the time when solar noon occurs at the prime meridian was established as noon for all places in its time zone, between 7.5°E and 7.5°W of that meridian. The same pattern was followed around the world. Every line of longitude evenly divisible by 15° is the central meridian for a time zone extending 7.5° of longitude on either side. However, time zone boundaries, as shown in ● Figure 2.9, do not follow meridians exactly. In the United States, time zone boundaries commonly follow state lines or international borders. It would be inconvenient and confusing to have a time zone boundary dividing a city or town into two time zones, so jogs in the lines were established to avoid most of these problems.

The time of day at the prime meridian, known as *Greenwich Mean Time* (GMT, also called Universal Time, UTC, or Zulu Time), is used as a worldwide reference. Times to the east or west of Greenwich can be easily determined by comparing them with GMT. A location 90°E of the prime meridian would be 6 hours later

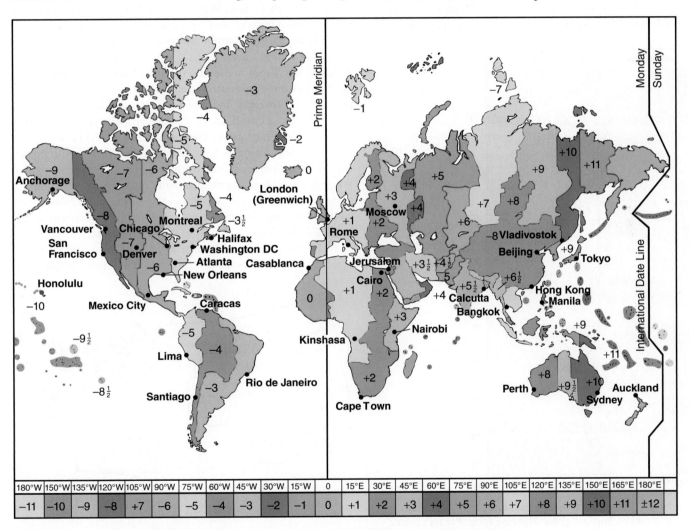

● **FIGURE 2.9** World time zones are based on the fact that Earth rotates 15° of longitude each hour. Thus, time zones are approximately 15° wide. Political boundaries usually prevent the time zones from following a meridian perfectly.

How many hours of difference are there between the time zone where you live and Greenwich, England? Is it earlier in England or later?

($90° \div 15°$ per hour), and in the Pacific Time Zone of the United States and Canada, whose central meridian is $120°$W, the time would be 8 hours earlier than GMT. For navigation, longitude can be determined with a *chronometer*, an extremely accurate clock. Two chronometers are used, one set on Greenwich time, and the other on local time. The number of hours between them, earlier or later, determines longitude (1 hour = $15°$ of longitude). Until the advent of electronic navigation by surface and satellite-based systems, the sextant and chronometer were a navigator's basic tools for determining location.

The International Date Line

On the opposite side of Earth from the prime meridian is the **International Date Line** that generally follows the 180th meridian, except for jogs to separate Alaska and Siberia and to skirt some Pacific island groups (● Fig. 2.10). At the International Date Line if we are traveling east, we turn our calendar back a full day or forward a full day if we are traveling west. Thus, if we are going east from Tokyo to San Francisco and it is 4:30 p.m. Monday just as we cross the International Date Line, it will be 4:30 p.m. Sunday on the other side. If we are traveling west from Alaska to Siberia and it is 10:00 a.m. Wednesday when we reach the International

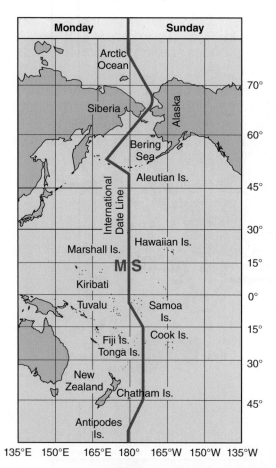

● **FIGURE 2.10** The International Date Line (IDL). West of the line is a day later than east of the line. Maps and globes often have either "Monday | Sunday" or "M | S" shown on opposite sides of the line to indicate the direction of the day change. This is the IDL as it is officially accepted by the United States.

Why does the International Date Line deviate from the 180° meridian in some places?

Date Line, it will be 10:00 a.m. Thursday once we cross it. It may seem strange but the time of day does not change, but the day of the week does change. As a way of remembering this relationship, many world maps and globes have Monday and Sunday (M | S) labeled in that order on the opposite sides of the International Date Line. To find the correct day, you just substitute the current day for Monday or Sunday, and use the same relationship.

The International Date Line was not established officially until the 1880s, but the need for such a line on Earth to adjust the day was inadvertently discovered by Magellan's crew who, from 1519 to 1521, were the first to circumnavigate the world. Sailing westward from Spain when they returned from their voyage, the crew noticed that 1 day had apparently been missed in the ship's log. What actually happened is that in going around the world in a westward direction, the crew had experienced one less sunset and one less sunrise than had occurred in Spain during their absence.

The U.S. Public Lands Survey System

The longitude and latitude system was designed to locate precise *points* where those lines intersect. A different system is used in much of the United States to define and locate land *areas*. This is the **U.S. Public Lands Survey System**, or the *Township and Range System*, developed for parceling public lands west of Pennsylvania. Lands in the eastern United States had already been surveyed into irregular parcels at the time this system was established. The Township and Range System divides land areas into parcels based on north–south lines called **principal meridians** and east–west lines called **base lines** (● Fig. 2.11). Base lines were surveyed along parallels of latitude. The north–south meridians, although perpendicular to the base lines, had to be adjusted (jogged) along their length to accommodate Earth's curvature. If these adjustments were not made, the north–south lines would tend to converge and land parcels defined by this system would be smaller in northern regions of the United States.

The Township and Range System forms a grid of nearly square parcels called *townships* laid out in horizontal *tiers* north and

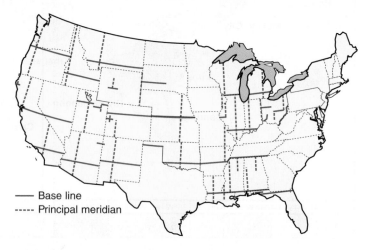

—— Base line
----- Principal meridian

● **FIGURE 2.11** Principal meridians and base lines of the U.S. Public Lands Survey System (Township and Range System).

Why wasn't the Township and Range System applied throughout the eastern United States?

south of the base lines and in vertical *columns* ranging east and west of the principal meridians. A **township** is a square plot 6 miles on a side (36 sq mi, or 93 sq km). As illustrated in ● Figure 2.12, townships are first labeled by their position north or south of a base line; thus, a township in the third tier south of a base line will be labeled Township 3 South, which is abbreviated T3S. However, a township is also labeled according to its *range*—its location east or west of the principal meridian for the survey area. Thus, if Township 3 South is in the second range east of the principal meridian, its full location can be given as T3S/R2E (Range 2 East).

The Public Lands Survey System divides townships into 36 **sections** of 1 square mile, or 640 acres (2.6 sq km, or 259 ha). Sections are designated by numbers from 1 to 36, beginning in the northeastern most section with section 1, snaking back and forth across the township, and ending in the southeast corner with section 36. Sections are divided into four *quarter sections*, named by their location within the section—northeast, northwest, southeast,

and southwest, each with 160 acres (65 ha). Quarter sections are also subdivided into four *quarter-quarter sections*, sometimes known as *forties*, each with an area of 40 acres (16.25 ha). These quarter-quarter sections, or 40-acre plots, are also named after their position in the quarter: the northeast, northwest, southeast, and southwest forties. Thus, we can describe the location of the 40-acre tract that is shaded in Figure 2.12 as the SW ¼ of the NE ¼ of Sec. 14, T3S/R2E, which we can find if we locate the principal meridian and the base line. The order is consistent from smaller division to larger, and township location is always listed before range (T3S/R2E).

The Township and Range system has exerted an enormous influence on landscapes in many areas of the United States and gives most of the Midwest and West a checkerboard appearance from the air or from space (● Fig. 2.13). Road maps in states that use this survey system strongly reflect its grid, and many roads follow the regular and angular boundaries between square parcels of land.

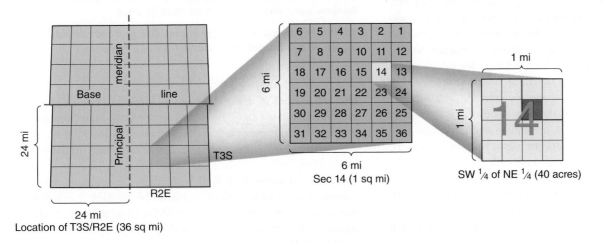

● **FIGURE 2.12** Describing the locations for *areas* of land according to the U.S. Public Lands Survey System.

How would you describe the extreme southeastern 40 acres of section 20 in the middle diagram?

● **FIGURE 2.13** Rectangular field patterns resulting from the Public Lands Survey System strongly influence the landscape of the Midwest and the western United States.

How do you know this photo was not taken in the U.S. Midwest?

Global Navigational Satellite Systems

Global Navigational Satellite Systems (GNSS) is the general term for satellite-based tecŸologies that are used for determining locations on Earth. Several countries and the European Union are either developing their own GNSS tecŸologies or already have them operational. The first one, which was developed by the United States is the **global positioning system** (GPS), originally created for military applications, is now widely used in the United States. Today, GPS and other GNSS are being adapted to many public uses, from surveying and mapping to aircraft and marine navigation. The global positioning system is also used to locate where you are with your cell phone, or when using a vehicle navigation system. The GPS uses radio signals, basically radar, transmitted by a network of satellites orbiting 17,700 kilometers (11,000 mi) above Earth (● Fig. 2.14).

By accessing signals from several satellites, a GPS receiver calculates the distances from those satellites to its location on Earth. Satellite navigation systems are based on the principle of *triangulation*, which means that by finding the distance to your position measured from four or more satellites in different orbital positions, you can determine your location. The distances are calculated by measuring the time it takes for a signal, broadcast at the speed of light from a satellite, to arrive at the receiver. A GPS receiver performs these calculations and displays a locational readout in latitude, longitude, and elevation, or displays that information on a digital map. Small GPS receivers are useful to travelers, hikers, and backpackers who need to keep track of their location, find a destination, and navigate to that site along the way (● Fig. 2.15). In mountain areas, hikers can use a GPS to monitor their elevation to help them understand the relationship between changes in elevations and environments. Global positioning systems are integrated

● **FIGURE 2.15** A global positioning system receiver can provide many kinds of readouts based on its latitudinal and longitudinal position. Small handheld units provide an accuracy that is acceptable for many uses, and can also display locations, topography, roads, and travel routes on a map.

What uses can you think of for a small GPS unit that displays its longitude and latitude as it moves from place to place?

into cell phones so that a caller can be located in case of an emergency. Map-based navigation systems are becoming popular and widely used in motor vehicles, as well as on boats and aircraft.

Satellite navigation systems are very important in surveying, mapmaking and environmental monitoring based on locational data gathered in the field. With sophisticated equipment and techniques, it is possible to find and describe locational coordinates within small fractions of a meter. Many geography students are studying GPS applications and applying their capabilities to real-world problems (● Fig. 2.16).

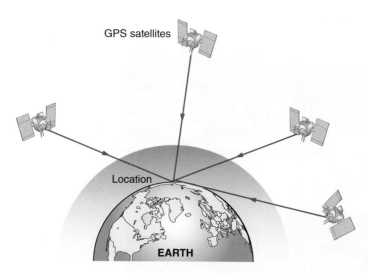

● **FIGURE 2.14** The global positioning system (GPS) uses signals from a network of satellites to determine a position on Earth. A GPS receiver on the ground calculates the distances from several satellites (a minimum of four) to find its location by longitude, latitude, and its elevation. With the distances from these satellites, a position can be located within meters, but with more satellite signals and with more sophisticated equipment, the position can be located more precisely—within centimeters.

● **FIGURE 2.16** An intern in Wyoming helps the U.S. Bureau of Land Management with field surveying by collecting accurate locations with a global positioning system receiver. Many geography students serve in summer intern positions with federal government agencies.

Maps and Map Projections

Maps are extremely versatile—they can be reproduced easily, can depict the entire Earth or show a small area in great detail, are easy to handle and transport, and can be displayed on a computer monitor, cell phone, or tablet. Yet, it is not possible for one map to fit all uses. The many different varieties of maps have qualities that can be either advantageous or problematic, depending on the application. Knowing some basic concepts concerning the characteristics of maps and cartography greatly enhances a person's ability to effectively use a map and to select the right one for a particular task.

Advantages of Maps

If a picture is worth a thousand words, then a map is worth a million. Because they are graphic representations and use symbolic language, often along with words, maps show spatial relationships and portray geographic information with great efficiency. As visual representations, maps can supply an enormous amount of information that would take many pages to describe in words, most likely with less success.

Imagine trying to explain to someone all of the information that a map of your city, county, state, or campus provides: sizes, areas, distances, directions, street patterns, railroads, bus routes, hospitals, schools, libraries, museums, highway routes, business districts, residential areas, population centers, and so forth. Maps can display true courses for navigation and accurate shapes of Earth features. They can be used to measure areas or distances, and they can show the best route from one place to another. The potential applications of maps are practically infinite, even "out of this world." Our space programs have produced detailed maps of the moon, Mars, and other extraterrestrial features (● Fig. 2.17). Cartographers can develop maps to illustrate almost any relationship in the environment. For many reasons, whether it is presented on paper, on a computer screen, or in the mind, the map is the geographer's most important tool. Further, having an understanding of maps and some of the decisions that go into their production is advantageous in almost any career today, certainly for those involved with human or environmental issues and concerns.

Limitations of Maps

A globe has essentially the same shape as our planet, and this is the reason why we can directly compare the size, shape, and area of the Earth's features and we can measure distances, shortest routes, and true cardinal directions. Because maps are flat and not spherical, we cannot accurately compare or measure all of these properties on a single map. It is impossible to present a spherical planet on a flat (two-dimensional) surface and accurately maintain all of its geometric properties. The problem of representing a sphere on a flat map has been described as trying to flatten out an eggshell.

Earth's curvature causes apparent and pronounced distortion, especially on maps that show large regions of the world, but on maps that depict only a small area, the distortion should be negligible. If we use a state park map while hiking, the distortion is too small to affect us. To be knowledgeable and skilled map users, we must know which properties a certain kind of map depicts accurately, which features it distorts, and for what purpose a map is best suited. By being aware of these map characteristics, we can make accurate comparisons and better understand the information that a map conveys.

Map Projections

Transferring a spherical grid onto a flat surface produces a **map projection**. Although maps are not actually made this way, certain projections can be demonstrated by putting a light inside a transparent globe so that the grid lines are projected onto a flat surface (plane), a cone, a cylinder, or some other geometric forms that are flat or can be cut and flattened out (● Fig. 2.18). Today, map projections are developed mathematically, using computers to fit the geographic grid to a surface, and there are hundreds of different kinds. Map projections will always distort certain characteristics displayed on the map, such as the shape of mapped features, accurate area relationships, true compass directions, uniform distance relationships, or some combination of these factors.

Planar Projections
Projecting the grid lines onto a plane, or flat surface, produces a map called a **planar projection** (Fig. 2.18a). These maps are most often used to show the polar regions, with the pole centrally located on the circular map, which displays one hemisphere. Maps of the Arctic Ocean with its surrounding polar region, as well as maps of Antarctica most commonly use this kind of projection.

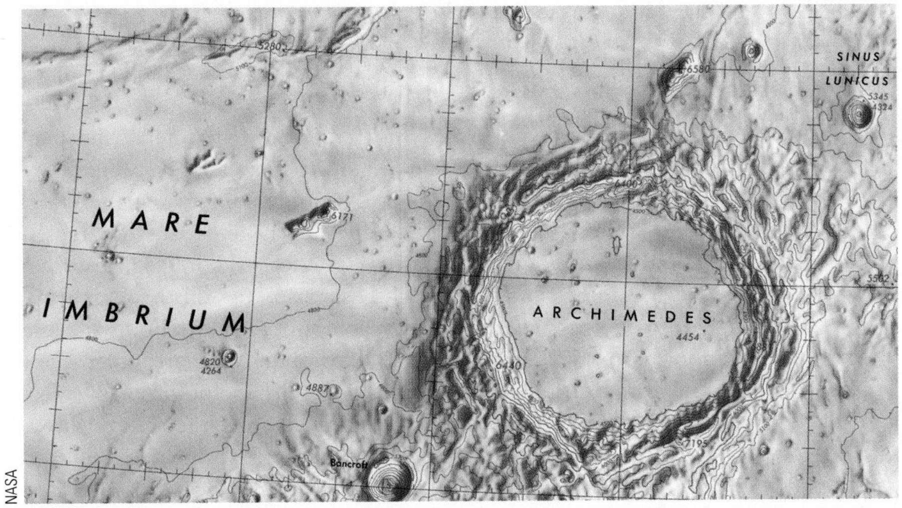

● **FIGURE 2.17** Lunar geography. A detailed map shows a major moon crater that is 120 kilometers (75 mi) in diameter. Even the side of the moon that never faces Earth has been mapped in considerable detail.

How were we able to map the moon in such detail?

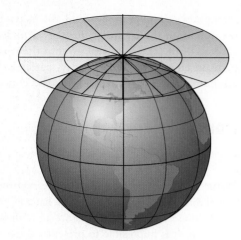

(a) Planar projection

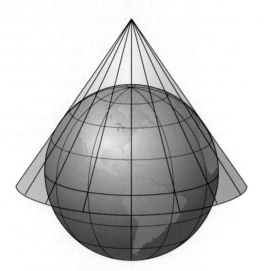

(b) Conical projection

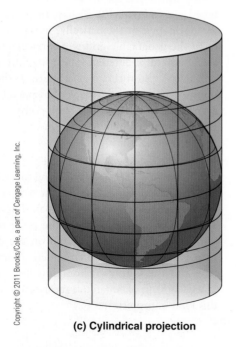

(c) Cylindrical projection

● **FIGURE 2.18** The concepts behind the development of (a) planar, (b) conic, and (c) cylindrical projections. Although projections are not actually produced this way, they can be demonstrated by projecting light from a transparent globe.

Why are different map projections used to map different areas and regions?

Conic Projections Maps of regions in the middle latitudes, such as the contiguous United States, are typically based on a **conic projection**, which portrays these latitudes with minimal distortion. In a simple conic projection, a cone is fitted over the globe with its pointed top centered over a pole (Fig. 2.18b). Parallels of latitude on a conic projection appear as curved arcs that become smaller toward the pole, and meridians appear as straight lines radiating toward the pole.

Cylindrical Projections A well-known example of a **cylindrical projection** (Fig. 2.18c) is the **Mercator projection**, once widely used in schools for wall maps and textbook maps because it was a very common world map and its rectangular outline fits well on a page or poster. The Mercator world map is a mathematically adjusted cylindrical projection on which meridians appear as parallel lines instead of converging at the poles. Obviously, there is enormous east–west distortion of areas in the high latitudes because the distances between meridians are stretched to the same width that they are at the equator (● Fig. 2.19). The spacing of latitudinal parallels on a Mercator projection is also not equal, as they are on Earth. This projection does not display all areas accurately, and size distortion increases toward the poles.

Gerhardus Mercator devised this map in 1569 to provide a property that no other projection of the whole would possesses. A straight line drawn anywhere on a Mercator projection is a line of true compass direction, called a *rhumb line*, which was very important in navigation (see Fig. 2.19). On Mercator's map, navigators could draw a straight line between their location and the place where they wanted to go, and then follow a constant compass direction to get to their destination.

Properties of Map Projections

The geographic grid has four important geometric properties: (1) parallels of latitude are always parallel, (2) parallels are evenly spaced, (3) meridians of longitude converge at the poles, and (4) longitude and latitude lines always cross at right angles. There are many ways to transfer a spherical grid onto a flat surface, but it is impossible to have all these properties on the same map. Closely examining a map's grid system to determine how these four properties are affected will help you discover areas of greatest and least distortion. Cartographers carefully and purposefully select the projection that is best for what the map is intended to show. The decision involves considering which map projection will preserve the characteristics that are most important, which would be compromised, and what degree of compromise is acceptable. Cartographers must decide which properties to preserve at the expense of others.

Today, map projections are developed mathematically, using computers to fit the geographic grid to a surface. Distortions in

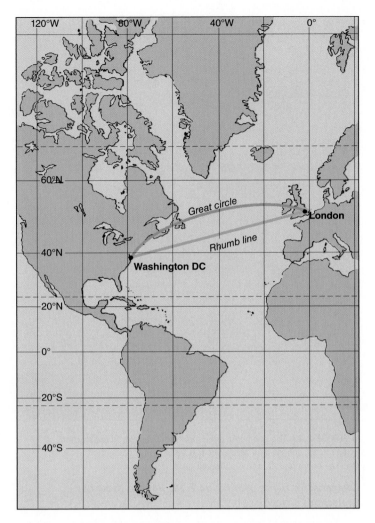

Compare the sizes of Greenland and South America on this map to their sizes on a globe. Is the distortion great or small?

in size. Because it was once commonly used as a world map in schools and textbooks, the Mercator projection's distortions led generations of students to believe incorrectly that Greenland is as large as South America. On Mercator's projection, Greenland is shown as being about equal in size to South America, but South America is more than eight times larger.

Area Cartographers can create a world map that maintains correct area relationships—that is, areas on the map have the same size proportions to each other as they have in reality. Thus, if we cover any two parts of the map with, let's say, a quarter, no matter where the quarter is placed it will cover equivalent areas on Earth. Maps drawn with this property, called **equal-area maps**, should be used if size comparisons are being made between two or more areas. The property of equal area is also essential when examining spatial distributions. So long as the map displays equal area and a symbol represents the same quantity throughout the map, we can get a good idea of the distribution of any feature—for example, people, earthquakes, hurricanes, rainfall, or volcanoes. However, equal-area maps distort the shapes of map features (● Fig. 2.20) because it is impossible to show both equal areas and correct shapes on the same map.

Distance As you've learned, no flat map can maintain a constant distance scale over the world's entire surface. The scale on a map that depicts a large area cannot be applied equally everywhere on that map. On maps of small areas, however, distance distortions will be minor, and the accuracy will usually be sufficient for most purposes. Maps can be made with the property of **equidistance** in specific instances. That is, on a world map, the equator may have equidistance (a constant scale) along its length, and all meridians

● **FIGURE 2.20** This world map uses an equal-area projection, which preserves area relationships but distorts the shapes of landmasses.

Which kind of world map would you prefer, one that preserves area or one that preserves shape? Why?

the geographic grid that are required to make a projection can affect the geometry of several characteristics of the areas and features that a map portrays.

Shape Flat maps cannot depict large world regions without distorting either their shape or their comparative sizes. However, there are map projections that will depict the true shapes of continents, regions, mountain ranges, lakes, islands, and bays. Maps that maintain the correct shapes of areas are **conformal maps**. To preserve the shapes of Earth's features, meridians and parallels on a conformal map always cross at right angles, just as they do on the globe.

The Mercator projection does present correct shapes, so it is a conformal map, but areas away from the equator are exaggerated

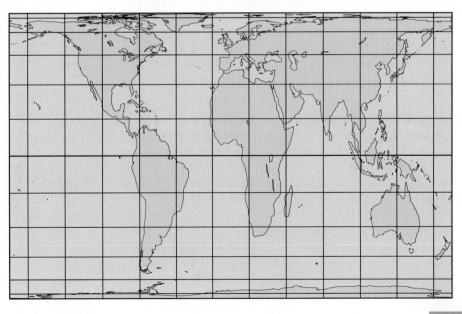

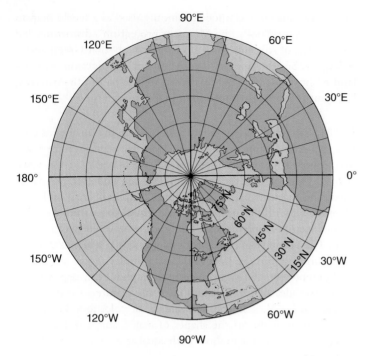

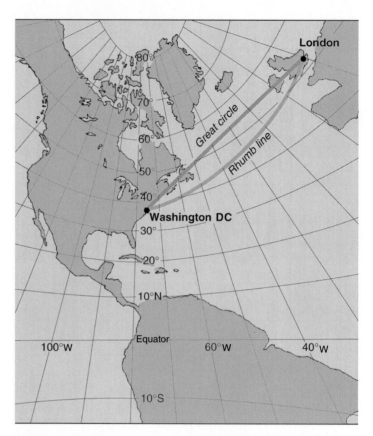

● **FIGURE 2.21** An azimuthal map centered on the North Pole. Although a polar view is the conventional orientation of these maps, it could be centered anywhere on Earth. Azimuthal maps show true directions between all points from the center, but can only show one hemisphere.

● **FIGURE 2.22** The gnomonic projection produces extreme distortion of distances, shapes, and areas, yet it is valuable for navigation because it is the only projection that shows all great circles as straight lines.

Compare this figure with Figure 2.19. How do these two projections differ?

may have equidistance, but not the parallels. On another map, all straight lines drawn from the center may have equidistance, but the scale will not be constant unless lines are drawn from the center.

Direction Because longitude and latitude directions run in straight lines, not all flat maps can show true directions as straight lines. Thus, lines of latitude or longitude that curve on maps are not drawn as true compass directions. **Azimuthal projections** produce maps that show true directions as straight lines. These projections are drawn with a central focus, and all straight lines drawn from that center are true compass directions (● Fig. 2.21). **Gnomonic projections** are planar projections that present the grid with great distortion. Despite their distortion, gnomonic projections (● Fig. 2.22) are the only maps that display all arcs of great circles as straight lines. Navigators can draw a straight line between their location and this line will be the shortest route between the two places.

Compromise Projections What if you want a world map that shows the continents and oceans fairly well in terms of shape and relative sizes? Cartographers can create a map that shows both area and shape fairly well but it won't be really correct for either property. These world maps are **compromise projections** that are neither conformal nor equal area, but an effort is made to balance distortion to produce an "accurate looking" global map (● Fig. 2.23a). An *interrupted projection* can also be used to reduce the distortion of landmasses (Fig. 2.23b) by moving much of the distortion to the oceanic regions. If your interest was centered on the world ocean, however, the projection could be interrupted in the continental areas to minimize distortion of the ocean basins.

Map Basics

Maps not only contain spatial information and data; they also display essential information about the map itself. This information and certain graphic features (often in the margins) are intended to facilitate using and understanding the map. Among these items are the map title, date, legend, scale, and direction. A map should have a title that tells what area is depicted and what subject the map concerns. For example, a hiking and camping map for Yellowstone National Park should have a title such as "Yellowstone National Park: Trails and Camp Sites." Most maps should also indicate when they were published and the date to which its information applies. For instance, a population map of the United States should tell when the census was taken to let the map user know if the map information is current or outdated or whether the map is intended to show historical data.

Legend A map should have a **legend**—a key to symbols used on the map. For example, if one dot represents 1000 people or the symbol of a pine tree represents a roadside park, the legend should explain this information. If color shading is used on the map to represent elevations, different climatic regions, or other factors, then a key to the color coding or shading should be provided. Map symbols can be designed to represent virtually any feature (see Appendix B).

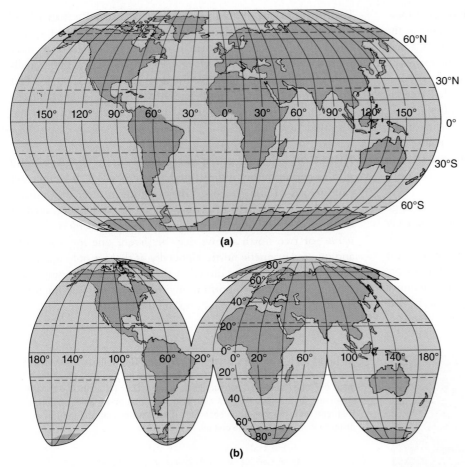

(a)

(b)

● **FIGURE 2.23** Compromise projections. (a) The Robinson projection is considered a compromise projection because it departs from equal area to show both area and shape reasonably well, although neither property is truly accurate. (b) Distortion of certain areas can be also reduced with an interrupted projection—in this case the continents appear fairly accurately, at the expense of the oceans.

edge of a piece of paper and mark the distance between any two points on the map. Then use the graphic scale to find the equivalent distance on Earth's surface. Graphic scales have two major advantages:

1. It is easy to determine distances on the map because the graphic scale can be used like a ruler to make measurements.

2. Graphic scales are applicable even if the map is reduced or enlarged because the graphic scale (on the map) will also change proportionally in size. This is particularly useful because maps can be reproduced or presented easily in a reduced or enlarged scale using computers or photocopiers. The map and the graphic scale, however, must be enlarged or reduced together (the same amount) for the graphic scale to be applicable. Given the ease of reproduction and changes in the size of maps that exists today, it is best to check stated or representative fraction scales against the graphic scale, before making any important measurements from a map that is not in its original size.

A **representative fraction (RF) scale** is a ratio between a unit of distance on the map to the distance that unit represents in reality (expressed in the same units). Because a ratio is also a fraction, the units of measure, being the same in the numerator and denominator, cancel each other out. An RF scale is therefore free of

Scale Obviously, maps depict features smaller than they actually are. If the map is used for measuring sizes or distances, or if the size of the area represented might be unclear to the map user, it is essential to know the map scale (● Fig. 2.24). A **map scale** is an expression of the relationship between a distance on the ground and the same distance as it appears on the map. Knowing the map scale is essential for measuring distances and for determining areas. Map scales can be conveyed in three basic ways: verbal, graphic (or bar), and representative fraction.

A **verbal scale** is a statement on the map that indicates, for example, "1 centimeter to 100 kilometers" (1 cm represents 100 km) or "1 inch to 1 mile" (1 in. on the map represents 1 mi on the ground). Stating a verbal scale tends to be how most of us would refer to a map scale in conversation. A verbal scale, however, will no longer be correct if the original map is reduced or enlarged. When stating a verbal scale, it is acceptable to use different map units (centimeters, inches) to represent another measure of true length it represents (kilometers, miles).

A **graphic scale**, or **bar scale**, is useful for making distance measurements on a map. Graphic scales are graduated lines (or bars) marked with map distances that are proportional to distances on the Earth. To use a graphic scale, take a straight

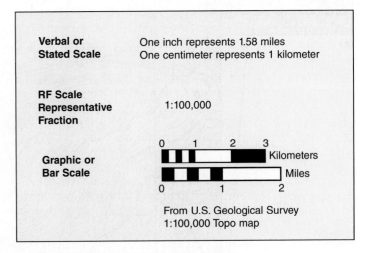

Verbal or Stated Scale	One inch represents 1.58 miles One centimeter represents 1 kilometer
RF Scale Representative Fraction	1:100,000
Graphic or Bar Scale	(graphic bar scales in Kilometers and Miles) From U.S. Geological Survey 1:100,000 Topo map

● **FIGURE 2.24** Map scales. A *verbal scale* states the relationship between a map measurement and the distance it represents on the ground. Verbal scales generally mix units (centimeters per kilometer or inches per mile). A *representative fraction (RF)* scale is a ratio between a distance on a map (1 unit) and its actual length on the ground (here, 100,000 units). An RF scale must be expressed in the same units of measurement both on the map and on the ground. A *graphic scale* is used for making measurements on the map manually, in terms of distances on the ground.

units of measurement and can be used with any unit of linear measurement—meters, centimeters, feet, and inches—so long as the same unit is used on both sides of the ratio. As an example, a map may have an RF scale of 1:63,360, which can also be expressed 1/63,360. This RF scale can mean that 1 inch on the map represents 63,360 inches on the ground. It also means that 1 centimeter on the map represents 63,360 centimeters on the ground. Knowing that 1 inch on the map represents 63,360 inches on the ground may be difficult to conceptualize unless we realize that 63,360 inches is equal to 1 mile. Thus, the representative fraction 1:63,360 means the map has the same scale as a map with a verbal scale of 1 inch to 1 mile.

Maps are often described as being of small, medium, or large scale (● Fig. 2.25). *Small-scale* maps show large areas in a relatively small size, include little detail, and have large denominators in their representative fractions. *Large-scale* maps show small areas of Earth's surface in greater detail and have smaller denominators in their representative fractions. To avoid confusion, remember that 1/2 is a *larger* fraction than 1/100 and *small scale* means that

Earth features are shown very small. A large-scale map would show the same features larger. Maps with representative fractions larger than 1:25,000 are large scale. Medium-scale maps have representative fractions between 1:25,000 and 1:250,000. Small-scale maps have representative fractions less than 1:250,000. This classification follows the guidelines of the U.S. Geological Survey (USGS), publisher of many maps for the federal government and for public use.

Direction The orientation and geometry of the geographic grid give us an indication of direction because parallels of latitude are east–west lines and meridians of longitude run directly north–south. Many maps have an arrow pointing to north as displayed on the map. A north arrow may indicate either *true north* or *magnetic north*—or two north arrows may be given, one for true north and one for magnetic north. Generally, if there is only one north arrow, it represents the direction to true north. On some maps where the direction to north is obvious, and north is at the top of the map, a north arrow might not be used.

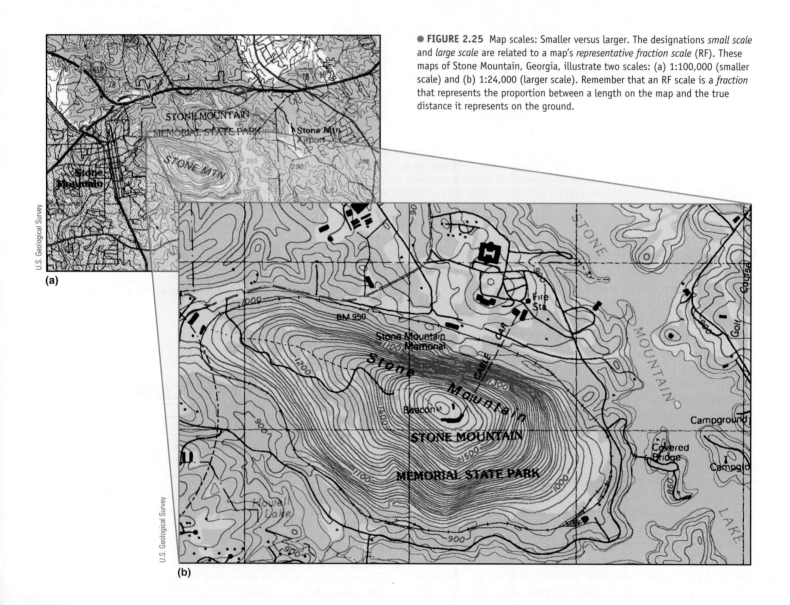

● **FIGURE 2.25** Map scales: Smaller versus larger. The designations *small scale* and *large scale* are related to a map's *representative fraction scale* (RF). These maps of Stone Mountain, Georgia, illustrate two scales: (a) 1:100,000 (smaller scale) and (b) 1:24,000 (larger scale). Remember that an RF scale is a *fraction* that represents the proportion between a length on the map and the true distance it represents on the ground.

U.S. Geological Survey

(a)

U.S. Geological Survey

(b)

Earth has a magnetic field that makes the planet act like a giant bar magnet, with a magnetic north pole and a magnetic south pole, each with opposite charges. Although the magnetic poles shift position slightly over time, they are located in the Arctic and Antarctic regions and do not coincide with the geographic poles. Aligning itself with Earth's magnetic field, the north-seeking end of a compass needle will point toward the magnetic north pole.

If we know the **magnetic declination**, the angular difference between magnetic north and true geographic north (for our location), we can compensate for this difference (● Fig. 2.26a). Thus, if our compass points north and we know that the magnetic declination for our location is 20°E, we can adjust our course knowing that our compass is pointing 20°E of true north. To do this, we should turn 20°W from the direction indicated by our compass to face true north. Magnetic declination varies from place to place and changes through time. For this reason, magnetic declination maps are revised periodically, so if you are navigating by compass, it is important to have a recent map. A map of magnetic declination is called an *isogonic map* (Fig. 2.26b), and *isogonic lines* connect locations that have equal declination.

Compass directions can be given by either the azimuth system or the bearing system. In the **azimuth** system, direction is given in degrees of a full circle (360°) clockwise from north. That is, if we imagine the 360° of a circle with north at 0° (and at 360°) and read the degrees clockwise, we can describe a direction by its number of degrees away from north. For instance, straight east would have an azimuth of 90°, and due south would

be 180°. The **bearing** system divides compass directions into four quadrants of 90° (N, E, S, W), each numbered by directions in degrees away from either north or south. Using this system, an azimuth of 20° would be north, 20° east (20° east of due north), and an azimuth of 210° would be south, 30° west (30° west of due south). Both azimuths and bearings are used for mapping, surveying, and navigation for both military and civilian purposes.

Thematic Maps

Thematic maps are designed to focus attention on the spatial distribution and pattern of one feature (or a few related ones). Examples include maps of climate, vegetation, soils, earthquake epicenters, and tornadoes. There are two major types of spatial data: *discrete* and *continuous*.

Discrete and Continuous Data
Discrete data refers to phenomena that are each located at a particular place, but do not exist everywhere—for example, hot springs, tropical rainforests, rivers, lightning strikes, tornado paths, or earthquake faults. Discrete data are represented on maps by point, area, or line symbols to show their locations and spatial distributions (● Figure 2.27a–c). *Regions*, discrete areas that exhibit a common characteristic or set of characteristics within their boundaries, are typically represented by various colors, shading, or other symbols to differentiate one region from another. Physical geographic regions include areas of similar soil, climate, vegetation, landform type, or many other characteristics.

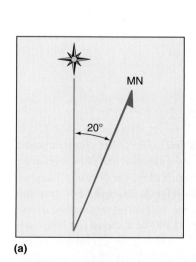

(a)

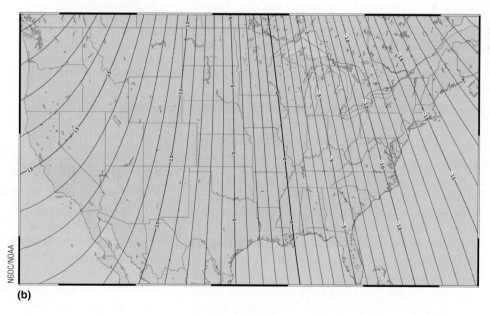

(b)

● **FIGURE 2.26** (a) Map symbol showing direction to true north on the map, symbolized by a star representing Polaris (the North Star), and magnetic north, symbolized by an arrow. The example indicates 20°E magnetic declination. (b) Isogonic map of the conterminous United States, showing the magnetic declination that must be added or subtracted from a compass reading to determine true directions.

What is the magnetic declination of your hometown to the nearest degree?

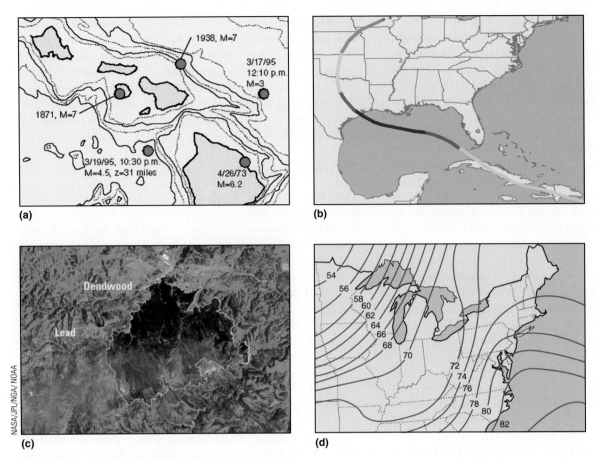

(a)

(b)

NASA/JPL/NGA/NOAA

(c)

(d)

● **FIGURE 2.27** Discrete and continuous spatial data (variables). *Discrete variables* represent features existing at certain locations but do not exist everywhere. The locations, distributions, and patterns of discrete features are of great interest in understanding spatial relationships. Discrete variables can be (a) *points* representing locations of large earthquakes in Hawaii (or places where lightning has struck or locations of water-pollution sources); (b) *lines* as in the path taken by the hurricane that struck Galveston, Texas, in 1900 (or river channels, tornado paths, or earthquake fault lines); or (c) *areas* like the land burned by a wildfire (or clear-cuts in a forest, or the area where an earthquake was felt). *A continuous variable* means that every location has a certain measurable characteristic—for example, everywhere on Earth has an elevation, even if it is zero (at sea level) or below (a negative value). Changes in a continuous variable over an area can be represented by isolines, shading, or colors or with a 3-D appearance. The map (d) shows the continuous distribution of temperature variation in part of eastern North America.

What are some other environmental examples of discrete and continuous variables?

Continuous data means that a numerical value exists everywhere within the area of interest for a certain characteristic; for example, every location has a measurable elevation (or temperature, or air pressure, or population density). The distribution of continuous data is often shown using isolines—lines on a map that connect points with the same numerical value (Fig. 2.27d). **Isolines** that we will be discussing later include *isotherms*, which connect points of equal temperature; *isobars*, which connect points of equal barometric pressure; *isobaths* (also called *bathymetric contours*), which connect points with equal water depth; and *isohyets*, which connect points receiving equal amounts of precipitation. Another way of showing spatial variation in continuous data is by displaying it as a continuous surface, with varying heights, and often colors corresponding to the data values and levels of the characteristic shown.

Topographic Maps

Topographic contour lines are isolines that connect points on a map that are at the same elevation above mean sea level (or below sea level, such as in Death Valley, California). They are lines that follow constant elevation levels on a map. For example, if we walk around a hill along the 1200-foot *contour line* shown on a map, we would always be 1200 feet above sea level, and maintain a constant elevation by walking on a level line. Contour lines are an excellent means for showing elevation changes, numerical elevations, and the form of the land surface. The arrangement, spacing, and shape of contours give a map reader a mental image of what the topography (the "lay of the land") is like (● Fig. 2.28).

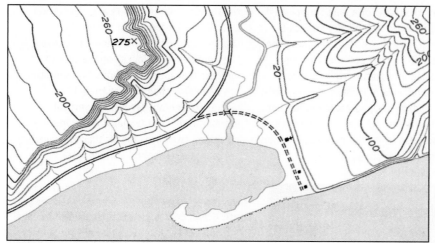

● **FIGURE 2.28** (Top) A perspective view of a river valley and surrounding hills, shown on a shaded-relief diagram. Note that a river flows into a bay partly enclosed by a spit. The hill on the right has a rounded surface form, but the one on the left forms a cliff along the edge of an inclined but flat upland. (Bottom) The same features represented on a contour map.

If you had only a topographic map, could you visualize the terrain shown in the shaded-relief diagram?

● Figure 2.29 illustrates how contour lines portray a land surface. The bottom portion of the diagram is a simple *contour map* of an asymmetrical hill. Note that the elevation difference between adjacent contour lines on this map is 20 feet. The constant difference in elevation between adjacent contour lines is called the **contour interval**. Whenever you examine a topographic map to understand the landscape, it is essential to determine the map's contour interval. Contour intervals are not the same on all maps; rather, they vary widely, depending on the terrain.

If we hiked from point A to point B, what kind of terrain would we cover? We start from sea level at point A and immediately begin to climb. We cross the 20-foot contour line, then the 40-foot, then the 60-foot, and, near the top of the hill, the 80-foot contour level. After walking over a relatively broad summit that is above 80 feet but not as high as 100 feet (or we would cross another contour line), we once again cross the 80-foot contour line, which means we must be starting down. During our descent, we cross each lower level in turn until we arrive back at sea level (point B).

In the top portion of Figure 2.29, a **profile** (side view) helps us visualize the topography we covered in our walk. We can see why the trip up the mountain was more difficult than the trip down. Closely spaced contour lines near point A represent a steeper slope than the more widely spaced contour lines near point B. Actually, we have discovered something that is true of all isoline maps: The closer together the lines are on the map, the steeper the **gradient** (the greater the rate of vertical

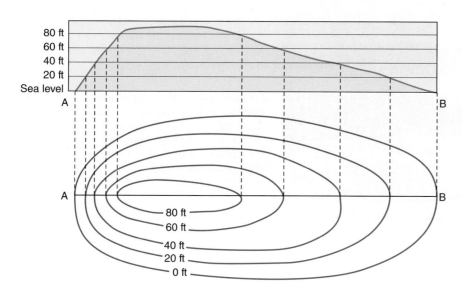

● **FIGURE 2.29** A topographic profile and contour map. Topographic contours connect points of equal elevation relative to mean sea level. The upper part of the figure shows the topographic profile (side view) of an island. Horizontal lines mark 20-foot intervals of elevation above sea level. The lower part of the figure shows how these contour lines look in map view.

Study this figure and the map in Figure 2.25. What is the relationship between the spacing of contour lines and the steepness of slope?

Using Vertical Exaggeration to Portray Topography

Most maps present a landscape as if viewed from directly overhead, looking straight down. This perspective is referred to as a *planimetric*, or *plan*, *view* (like architectural house plans). On a flat map, illustrating terrain and showing elevation differences require some kind of symbol to display elevations. Topographic maps use contour lines, which can also be enhanced by relief shading.

For many purposes, either a profile view (side view), or an oblique (perspective) view of the terrain helps us visualize the landscape. Information about the subsurface can also be included. Block diagrams and 3-D models of Earth's surface, show the topography from a perspective view, similar to looking out an airplane window or from a high vantage point.

A topographic profile (see Fig. 2.29) illustrates the shape of a land surface as if viewed directly from the side. Profiles are basically graphs of elevation changes over distance along a transect line. Elevation and distance information collected from a topographic map or from other data can be used to draw a topographic profile that depicts the terrain. If the geology of the subsurface also is represented, such profiles are called *geologic cross sections*.

Block diagrams, profiles, and cross sections are often drawn with *vertical exaggeration*, which enhances changes in elevation. This makes slopes appear steeper, mountains taller, valleys deeper, and the terrain more rugged. Vertical exaggeration is used so that important yet subtle changes in the terrain are more noticeable. Most profiles and block diagrams should indicate how much the vertical presentation has been stretched so that there is no misunderstanding. "Two times vertical exaggeration" means that the features are presented two times higher than they really are, but the horizontal scale is correct.

USGS/digital elevation model by Steve Schilling; geo-referenced by Frank Trusdell

This digital elevation model (DEM) of Anatahan Island (16°22" N, 145°40' E) and the surrounding Pacific Ocean floor is presented in 3-D and colorized according to elevation and seafloor depth relative to sea level. This image of Anatahan has three times vertical exaggeration. The vertical scale has been stretched three times compared to the horizontal scale. The vertical scale bar represents a distance of 3800 meters, so taking the vertical exaggeration into account, what horizontal distance would the same scale bar length represent in meters?

Compare the vertically exaggerated elevation model to this natural scale (not vertically exaggerated) version. This is how the island and the seafloor actually look in terms of slope steepness and relief.

change per unit of horizontal distance). When studying a contour map, we should understand that the slope between contours almost always changes gradually, and it is unlikely that the land drops off downslope in steps as the contour lines might suggest.

Topographic maps use symbols to show many other features in addition to elevations (see Appendix B)—for instance, water bodies such as streams, lakes, rivers, and oceans, or cultural features such as towns, cities, bridges, and railroads. The USGS produces topographic maps of the United States at several different scales. Some of these maps—1:24,000, 1:62,500, and 1:250,000—use English units for their contour intervals. Many recent maps are produced at scales of 1:25,000 and 1:100,000 and use metric units. The National Ocean Service produces contour maps of undersea topography, called bathymetric charts, in the United States.

Modern Mapmaking

Today, nearly all maps are developed and produced using computer tecŸologies. Computer systems are faster, more efficient, and less expensive than the hand-drawn cartographic tecŸiques they have replaced. This is why we now see many more maps and more different kinds of maps than we did in the past. Computer systems make it easy to make changes in projection, contour interval, symbols, colors, map scale, and directions of view (rotating the orientation). Being able to make on-the-spot modifications is especially important for showing changing phenomena such as weather systems, flooding rivers, air pollution alerts, volcanic eruptions, and forest fires. Digital maps can be disseminated instantly and shared via television and the Internet, which is a great advantage when spatial information is rapidly changing or mapped information must be communicated as soon as possible.

Geographic Information Systems

A **geographic information system (GIS)** is a versatile computer-based system used to enter, store, access, manipulate, display, and analyze data and information. The different kinds of data are aligned and located according to their specific positions on a coordinate system, in a step called *georeferencing*. A GIS assists the user in the spatial analysis and display of geographic information derived from combining **digital map layers**, each composed of a specific *thematic map* (● Fig. 2.30). Spatial data in these map layers that represent elevations, depths, river systems, vegetation types, temperatures, road networks and any mappable factor or characteristic can be mapped, analyzed, and displayed in any combination of overlays in a GIS. A GIS is used to standardize the map projection and scale of the different map layers, thus allowing spatially matched overlays of information from several data layers to be combined into a meaningful composite map. Geographic information systems are especially useful to geographers as they work to address problems that require large amounts of spatial data from a variety of sources.

What a GIS Does Imagine that you are in a giant map library with hundreds of paper maps, all of the same area, but each map shows a different characteristic that exists at that location. One map shows roads, another highways, another trails, another rivers (or soils, or vegetation, or slopes, or rainfall, and so on, almost to infinity). The maps were originally produced at different sizes, scales, and projections (including some maps that do not preserve shape or area). These cartographic factors make it difficult to directly compare and *visually* overlay the spatial information among these different maps. You also have digital elevation models and satellite images that you would like to transform into the same scale and projection so that you can directly compare them with the maps. Furthermore, because few characteristics of the environment involve only one factor and they seldom exist in spatial isolation, you want to be able to combine a selection of these geographic factors on a single composite map. You have a spatial-geographic problem, and to solve

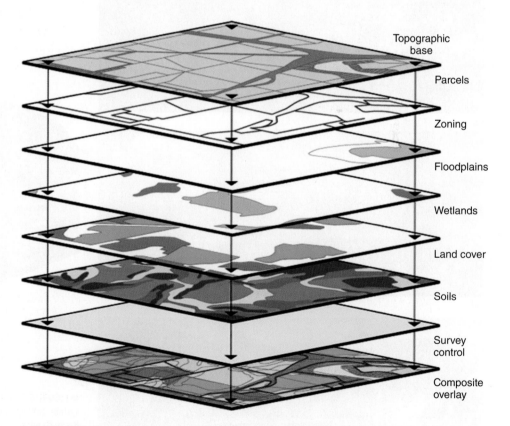

● **FIGURE 2.30** Geographic information systems (GIS) store spatial information and data as individual map layers. GIS technology is widely used in geographic and environmental studies when multiple variables need to be assessed and compared spatially to solve a problem. No longer limited to specialized fields, there is an abundance of mapping applications that can be found on the Internet, powered by GIS.

What are some applications for geographic information systems that would benefit your home area?

that problem, you need a way to make multiple representations of a part of Earth directly comparable. What you need is a GIS and the knowledge required to use this system.

Data and Attribute Entry

The first step in using a geographic information system to solve a geographic problem is to acquire the relevant spatial data. Data acquisition from paper maps requires scanning them into a computer system to digitize the mapped information. Digitized maps and images, along with other data sources are entered into a GIS, storing them in a database as digital files of individual map layers (see Fig. 2.30). Each category of spatial data and information is fitted to a common system of mapped grid coordinates.

Registration, Display, and Analysis

A GIS can display any layer or any combination of layers, geometrically rectified (fitted) to any map projection and at any scale that you specify. The maps, images, and data sets can now be directly compared at the same size, shape, and location based on the standardized grid coordinates, map projections, and map scales. A GIS can create digital *overlays of* any combination of thematic map layers that are needed. If you want to see the locations of *homes* on a *river floodplain*, a GIS can be used to provide a map by combining and displaying the *home locations* and *floodplain* map layers simultaneously. If you want to see *earthquake faults* and *artificial landfill areas* in relation to locations of *fire stations* and *police stations*, that composite map will require four layers, but this is no problem for a GIS to display.

Digital Elevation Models

Computer-generated three-dimensional (3-D) views of elevation data, called **digital elevation models (DEMs)**, are particularly effective at displaying topography in a way that simulates a 3-D view. Digital elevation data can generate many types of terrain displays and maps, including color-scaled contour maps, where areas between contours are assigned a certain color; conventional topographic maps, and shaded-relief maps. An example in ● Figure 2.31 shows an aerial photograph and a shaded DEM as map layers in a GIS and how they can be

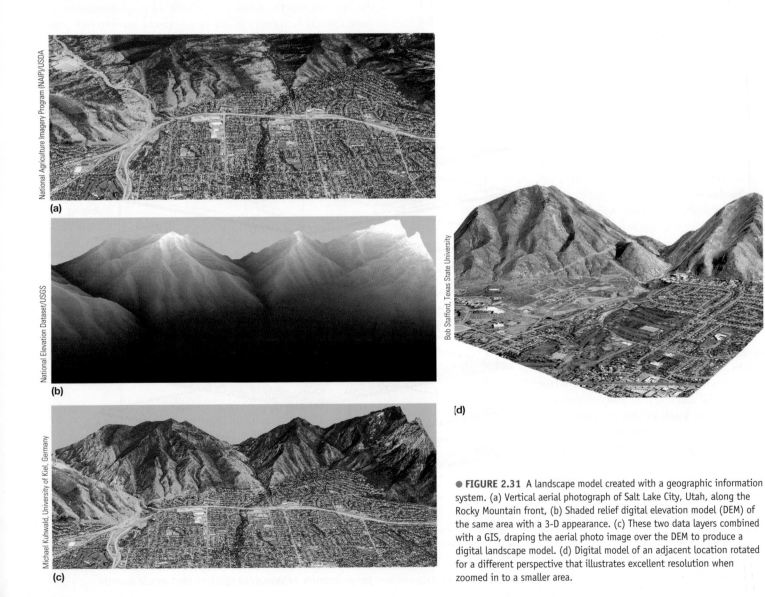

● **FIGURE 2.31** A landscape model created with a geographic information system. (a) Vertical aerial photograph of Salt Lake City, Utah, along the Rocky Mountain front, (b) Shaded relief digital elevation model (DEM) of the same area with a 3-D appearance. (c) These two data layers combined with a GIS, draping the aerial photo image over the DEM to produce a digital landscape model. (d) Digital model of an adjacent location rotated for a different perspective that illustrates excellent resolution when zoomed in to a smaller area.

combined into a perspective view producing a landscape model. In this case, it is part of the Rocky Mountain front at Salt Lake City, Utah, based on real image and elevation data. This process is called **draping** (like draping cloth over some object), but the scale and the perspective are accurately registered among the map layers.

Digital elevation models may be designed with **vertical exaggeration** by stretching the vertical scale of the display to enhance the relief of an area (see the boxed feature on vertical exaggeration in this chapter). Actually, any geographic factor represented by continuous spatial data can be displayed either as a two-dimensional contour map or as a three-dimensional surface to enhance the visibility of the spatial variation that it conveys (● Fig. 2.32).

Visualization Models Also referred to as *visualizations*, **visualization models** are computer-generated image models designed to illustrate and explain complex processes and features. Most visualizations are presented as 3-D images and/or as animations. Visualizations can combine and present several components of the Earth system in stunning 3-D views, based on actual environmental data and satellite images or air photos (● Fig. 2.33). Visualization models help us conceptualize and understand many environmental processes and features. Today, the products and tecŸiques of cartography are vastly different from their beginnings and they continue to be improved, but the goal of making a representation of Earth remains the same—to effectively communicate geographic and spatial knowledge in a visual format.

Geographic Information System in the Workplace

A simple example can illustrate the utility of a geographic information system. Suppose you are a geographer working for the Natural Resources Conservation Service. Your current problem is to control erosion along the banks of a reservoir. You know that erosion is a function of many environmental variables, including soil types, slope steepness, vegetation characteristics, and others. Using a GIS, you would enter map data for each of these variables as a separate data layer. You could analyze these variables individually, or you could overlay information from individual layers (soils, slope, vegetation, and so on) to identify the locations most susceptible to erosion. Your resources and personnel could then be directed toward controlling erosion in those target areas. In physical geography and the Earth sciences, GIS is being used to analyze potential coastal flooding from sea level rise, areas in need of habitat restoration, flood hazard potential, and earthquake distributions, just to list a few examples. The spatial analysis capabilities of a GIS are nearly unlimited and are applicable in almost any career field.

Many geographers are employed in careers that apply GIS along with other digital mapping and imaging tecŸologies. The capacity of a GIS to integrate and analyze a wide variety of geographic information, from census data to landform characteristics, makes it useful to both human and physical geographers. With nearly unlimited applications in geography and other disciplines, GIS will continue to be an important tool for understanding our environment and making important decisions based on spatial information.

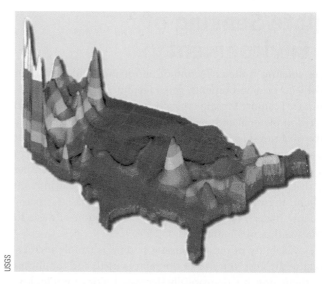

USGS

● **FIGURE 2.32** Earthquake hazard in the conterminous United States: a continuous variable displayed as a continuous surface in 3-D perspective. Here it is easy to develop a mental map of how potential earthquake danger varies across this part of the United States.

Are you surprised by any of the locations that are shown to have substantial earthquake hazard? Can you think of other applications for geographic information systems?

R.B. Husar, Washington University; the land layer from the SeaWiFS Project; fire maps from the European Space Agency; the sea surface temperature from the Naval Oceanographic Office's Visualization Laboratory; and cloud layer from SSEC, U. of Wisconsin

● **FIGURE 2.33** A visualization model of Earth systems. This computer-generated visualization is based on actual satellite measurements of Earth features and characteristics. Differences in ocean temperature are colorized and cloud heights are greatly exaggerated. The landforms and vegetated areas are also shown.

Remote Sensing of the Environment

Remote sensing is the collection of information and data about distant objects or environments, generally, but not always to produce images. Remote sensing typically involves the gathering and interpretation of aerial and space imagery, images that have many map-like qualities. Using remote sensing systems, we can detect objects and scenes that are not visible to humans and display them in a form that we can visually interpret. There are remote sensing systems that use ultraviolet (UV) light, visible light, near infrared (NIR) light, thermal infrared (TIR) energy (heat), lasers, microwave energy (radar), or sound, or some combination of these to gather data and produce images.

The availability of remote sensing systems is rapidly increasing, because the cost, the size of the sensors used, and the means to carry them aloft has dramatically decreased. For example, low cost drones (unmanned aerial vehicles—UAVs), that can collect images and video from above are widely available to the public. More sophisticated remote sensing systems mounted on drones can collect thermal images and map the landscape using lasers. The civilian, commercial, professional, and research applications of remotely-sensed images, video, and data collected from UAVs continue to expand.

Digital and Photographic Images

Today, with the recent widespread use of digital cameras, the use of the term *photograph* is changing in common usage, but in technical terms, **aerial photographs** are made by using cameras to record a picture on *film*. Digital cameras or image scanners produce a **digital image**—a scene that is converted into a grid of numerical data to make a picture. Most images returned from space are digital because digital data can be easily broadcast back to Earth. Digital imagery also offers the advantages of computer-assisted data processing, image enhancement, and image sharing, and can also provide thematic layers to a GIS. Digital images consist of **pixels**, short for "picture element," the smallest area resolved in a digital picture. A digital image is similar to a mosaic, made up of grid cells (pixels) with varying colors or tones that form a picture (● Fig. 2.34). Each cell has a locational address within the grid and a value that represents the brightness, intensity, or color of the area that the pixel represents. The digital values in a grid of cells or several grid layers are translated into an image by computer technology.

A key factor in digital images is **spatial resolution**, expressed as the size of the area imaged, that each pixel represents. For example, a satellite image with a resolution of 15 by 15 or 30 by 30 meters provides discernable detail about a city or small region. Satellites that image an entire hemisphere at once, or large continental areas, typically use resolutions that are much more coarse (e.g., 1 km by 1 km) to produce a more generalized scene. In general, when smaller areas are being imaged, finer spatial resolution is applied, although this varies depending on the type of remote sensing systems. One reason is that the greater the number of pixels,

USGS

● **FIGURE 2.34** A digital image. Pixels produce an image much like a mosaic, and each pixel has a value of a color or tone to make a digital image. Here, an enlargement allows us to see the pixels that produced the image of a compass (above). Without magnification, and viewed at the original size, the pixels are too small to discern.

the larger the digital file, and another is that coarser resolution tends to generalize characteristics of the scene, and fine resolution may not be necessary.

Digital cameras express resolution in *megapixels*, or how many million pixels make an image. The more megapixels a camera or digital scanner can image, the better the resolution and, generally, the images will be sharper. But if the original image is enlarged or reduced, the spatial resolution will not change. When pixel sizes are very small on the finished image, the mosaic effect will be either barely noticeable or not visible without magnification.

Remote Sensing Systems

There are two fundamental ways of collecting remotely sensed images and data. Remote sensing systems that make use of available energy where the image is taken are called *passive systems*. *Active systems*, in comparison, emit their own energy source, and record its reflected return from a surface. Taking a picture with a camera is an example of a passive approach, recording light reflections or emissions that exist when the image is taken. However, a camera using a flash is an active sensor, as the resulting image is of light from the flash, reflected back to the camera.

Passive Systems Aerial photography provided us with "bird's-eye" views via kites and balloons even before airplanes were invented, and aircraft led to a tremendous increase in the availability of aerial photography. Aerial photos and digital images may be *oblique* (● Fig. 2.35a), taken at an angle other than perpendicular to the surface, or *vertical* (Fig. 2.35b), looking straight down. Image interpreters use aerial photographs and imagery to examine and describe relationships among objects on the surface.

A device called a *stereoscope* uses overlapped pairs of images (such as air photos) taken from different positions to provide viewing of features in three dimensions.

Near-infrared (NIR) energy—light energy at wavelengths that are too long for our eyes to see—cuts though atmospheric haze better than visual light does. Natural-color photographs taken from very high altitudes or from space can have low contrast or appear hazy. Natural-color pictures and color near-infrared pictures of the same scene each display different visual information about a landscape or environment. Comparing these two different kinds of images is typically better than using a single format for understanding an environment (● Fig. 2.36a and b). Near-infrared systems tend to provide very clear images when taken from high altitude or space. Color-infrared images are sometimes referred to being in "false color" because on NIR, healthy grasses, trees, and most plants will show up as bright red, rather than green (Fig. 2.36b). Near-infrared images have many applications for environmental study, particularly for water resources, vegetation, and crops. An incorrect, but widely held, notion of NIR tecŸiques is that they image heat or temperature variations.

(a)   **(b)**

● **FIGURE 2.35** (a) Oblique photos provide a "natural view," like looking out of an airplane window. This oblique aerial photograph in natural color shows farmland, countryside, and forest. (b) Vertical photos provide a map-like view that is more useful for mapping and making measurements (as in this view of Tampa Bay, Florida).

What are the benefits of an oblique view compared to a vertical view?

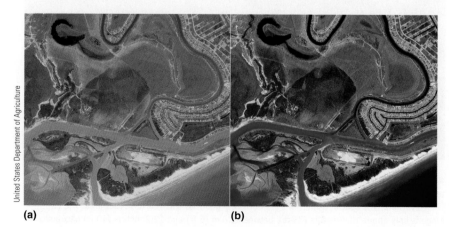

(a)   **(b)**

● **FIGURE 2.36** (a) A natural color aerial photograph. (b) The same scene in false color near-infrared. Red tones indicate vegetation; dark blue—clear, deep water; and light blue—shallow or muddy water. This is a coastal wetlands area in Texas on the Gulf of Mexico.

If you were asked to make a map of vegetation or water features, which image would you prefer to use and why?

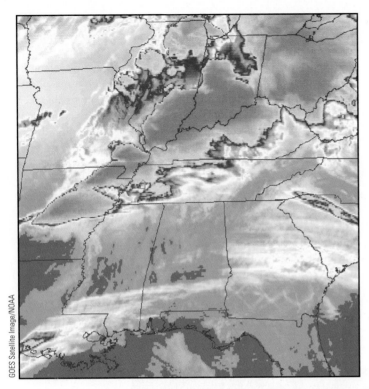

GOES Satellite Image/NOAA

● **FIGURE 2.37** Thermal infrared weather images show patterns of heat and cold. This image shows part of the southeastern United States beamed back from a U.S. weather satellite. Reds, oranges, and yellows show where the storm is most intense, and blues less intense.

Why are the storm patterns on thermal weather images important to us?

Near-infrared energy is *light*. Its images are produced by measuring reflections off of surfaces, and not by imaging radiated heat.

Thermal infrared (TIR) images show patterns of heat and temperature instead of light, and can therefore be taken either day or night by thermal sensors. Spatially recorded heat patterns are digitally converted into a visual image. Hot objects show up in light tones, and cool objects will be dark, but thermal images are often enhanced to emphasize heat differences by colorizing the image. Some TIR applications include finding volcanic hot spots and geothermal sites, locating forest fires through dense smoke, finding leaks in building insulation or in pipelines, and detecting thermal pollution in lakes and rivers.

Weather satellites also use thermal infrared imaging, and we often see thermal images of cloud patterns and storms on television broadcasts. We have all seen these TIR images on television when the meteorologist says, "Let's see what the weather satellite shows." Clouds are depicted in black on the original thermal image because they are colder than their background, the surface of Earth below. Because we don't like to see black clouds, the image tones are reversed, like a photo negative, so that the clouds appear white. These images may also be colorized to show cloud heights, because clouds are progressively colder at higher altitudes (● Fig. 2.37).

Active Remote Sensing Systems **Radar** (*RA*dio *Detec*tion *A*nd *R*anging) transmits microwaves (radio waves) to produce an image by reading the energy signals that are reflected back. Radar systems can operate day or night and can see through clouds. Radar systems are also used to monitor and track thunderstorms, hurricanes, and tornadoes. **Weather radar** systems produce map-like images of precipitation. These radar systems penetrate most clouds (day or night) but the transmitted signal reflects back from raindrops and other precipitation, producing an image on the radar screen.

Imaging radar systems are used to scan the surface (topography, rock, water, ice, sand dunes, and so forth) and convert radar reflections into a map-like images. Data from radar mapping systems can be used to generate digital elevation models because imaging radar mainly measures changes in elevation. Excellent for mapping terrain and water features, radar is used to map remote, inhospitable, inaccessible, cloud-covered regions and heavily forested areas, because it does not register the trees, but rather the terrain.

Lidar (*LI*ght *Detec*tion *A*nd *R*anging) is a remote-sensing technology that is similar to radar but uses discrete pulses of laser light (from visible to near infrared). Linked to GPS locational data, lidar is used to derive the elevation (or distance to) and geographic location of each lidar pixel. Lidar has many advantages over other active systems because of its high resolution and capability of collecting very accurate elevation data. It is being used to measure and map features on the ground or under shallow water, and can also image tree tops and heights, or penetrate to the ground below without imaging the vegetation cover (● Fig. 2.38). In addition to natural landscape features, it is also applicable to the built environment.

Lidar systems are usually mounted on aircraft, and more recently, drones, but they also support ground-based applications. These systems can produce surveys and accurate images and measurements of features such as bridges, street grids, and buildings, with a much higher spatial resolution compared to radar. Because of its high spatial resolution, lidar images are being taken as a base for determining environmental changes that accumulate in small increments over time. Lidar is also being used to compare beaches

Christopher Borst, Center for Advanced Computer Studies, University of Louisiana at Lafayette

● **FIGURE 2.38** Lidar image. This digital elevation model was made using lidar and shows the terrain, built-up areas, and elevation differences along the Vermillion River in Lafayette, Louisiana. The elevational and spatial resolutions are both very good, as this is a low relief area. The elevation range is only about 10.7 meters (35 ft), between 1.5 m (5 ft) and 12.2 meters (40 ft) above sea level. In Google Earth fly to 30°13′ N 92°02′ W.

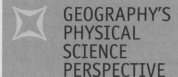
Polar versus Geostationary Satellite Orbits

Satellite systems that return images from orbit are designed to produce many different kinds of Earth imagery. Some of these differences are related to the type of orbit the satellite system follows while scanning the surface. Two distinctively different orbits are commonly used: the *polar orbit* and the *geostationary orbit* (sometimes called a *geosynchronous orbit*), each with a different purpose.

The polar orbit was developed first, and as its name implies, the satellite orbits Earth from pole to pole. This orbit has some distinct advantages. It is typically a low orbit for a satellite, usually varying in altitude from 700 kilometers (435 mi) to 900 kilometers (560 mi). At this height, but also depending on the equipment used, a polar-orbiting satellite can produce clear, close-up images of Earth. Polar orbiters travel at an average of around 27,400 kilometers/hour (17,000 mph), traveling completely around Earth in about

90 minutes. While the satellite orbits from pole to pole, Earth rotates on its axis below, so each orbit views a different path along the surface. Thus, polar orbits will at times cover the dark side of the planet. To adjust for this, a slightly modified polar orbit was developed, called a *sun synchronous orbit*. If the polar orbit is tilted a few degrees off of the poles, it can remain over the sunlit part of the globe at all times. Most modern polar-orbiting satellites are sun synchronous (a near-polar orbit).

The geostationary orbit, developed later, offered some innovations in satellite image gathering. A geostationary orbit must have three characteristics: (1) it must travel in the same direction as Earth's rotation; (2) it must orbit exactly over the equator; and (3) the orbit must be perfectly circular. The altitude of the orbit must be very close to 35,786 kilometers (22,236 mi), and the orbital velocity is 11,120 kilometers/hour (6900 mph). Under these conditions, the

satellite's orbit is synchronized with Earth's rotation, and the satellite is always located over the same spot above Earth.

A geostationary orbit offers some advantages. First, at its great distance, a geostationary satellite can view an entire Earth hemisphere in one image (the half it is always facing—a companion satellite images the other hemisphere). Another advantage is that geostationary satellites can send back a continuous stream of images for monitoring changes in our atmosphere and oceans. An animation made from successive geostationary images is what we see on TV weather broadcasts when we see motion in the atmosphere. Geostationary satellite images give us broad regional presentations of an entire hemisphere at once. Satellites in a near-polar orbit take image after image in a swath and rely on Earth's rotation to cover much of the planet over a time span of about a week and a half.

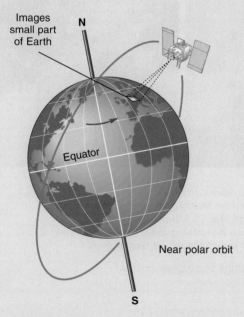

Near polar orbit

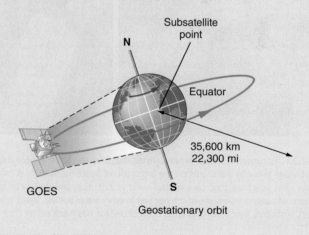

Geostationary orbit

Polar orbits circle Earth approximately from pole to pole and use the movement of Earth as it turns on its axis to image small areas (perhaps 100 × 100 km) to gain good detail of the surface. This orbital technique yields nearly full Earth coverage in a mosaic of images, and the satellite travels over the same region every few days, always at the same local time. (*Note:* Not to scale.)

Geostationary orbits are used with satellites orbiting above the equator at a speed that is synchronized with Earth rotation so that the satellite can image the same location continuously. Many weather satellites use this orbit at a height that will permit imaging an entire hemisphere of Earth. (*Note:* Not to scale.)

before and after storms to determine erosion, in vegetation and forestry studies, for mapping floodplains, and in urban planning. Lidar can even monitor certain types of air pollution.

Sonar (*SO*und *NA*vigation and *R*anging) is an active system that uses reflections from emitted sound waves to probe the ocean depths. Much of our understanding of sea floor topography, and mapping of the sea floor, has been a result of sonar applications.

Multispectral Remote Sensing

Multispectral remote sensing means using and comparing more than one kind of image of the same place—for example, radar and TIR images, or NIR and normal color photos. Common on imaging satellites, *multispectral scanners* sense many kinds of energy simultaneously and relay them to receiving stations as separate digital images. Each part of the energy spectrum yields different information about aspects of the environment (● Fig. 2.39).

The separate images, like thematic map layers in a GIS, can be combined, depending on which ones are needed for analysis.

Computer tecŸologies for mapping, modeling, and imaging our planet's features continue to provide us with data and information that contribute to our understanding of the Earth system and to help us address environmental concerns. Through continuous monitoring, global, regional, and even local changes can be detected and mapped. Digital mapping, GPS, GIS, and remote sensing have revolutionized the field of geography, but the fundamental principles concerning maps and cartography remain basically unchanged. Whether they are on paper, displayed on a computer monitor, hand drawn in the field, or stored as a mental image, maps and various kinds of representations of Earth continue to be essential tools for geographers and other scientists. Many geographers are gainfully employed in positions that apply spatial tecŸologies to understanding our planet and its environments, and their numbers are certain to increase in the future.

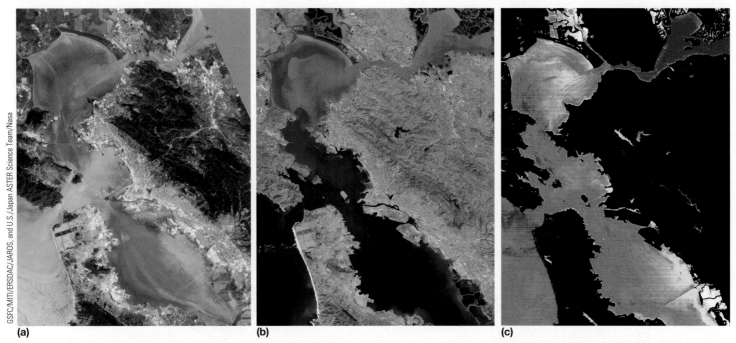

(a) (b) (c)

● **FIGURE 2.39** Multispectral remote sensing. These images of the San Francisco Bay Area each reveal different information. (a) A natural color digital image taken by astronauts on the International Space Station. (b) A color infrared image shows healthy growing vegetation as red, urban areas as a blue-gray, deep or clear water as dark blue, and shallow or muddy water as light blue. (c) A thermal infrared image shows patterns of water temperature in the bay and in other water bodies. Rivers bring colder water into the northeast arm of the bay, which warms in shallow areas. The land looks black because it was much colder than the water bodies when this image was taken perhaps.

What other aspects of the environment or the landscape can you recognize by comparing these three images?

■ TERMS FOR REVIEW

cartography
oblate spheroid
navigation
great circle
hemisphere
small circle
coordinate system
equator
latitude
sextant
prime meridian
longitude
decimal degrees
geographic grid
parallel
meridian
time zone
solar noon
International Date Line
U.S Public Lands Survey System
principal meridian
base line
township
section
Global Navigational Satellite Systems (GNSS)

global positioning system (GPS)
map projection
planar projection
conic projection
cylindrical projection
Mercator projection
conformal map
equal-area map
equidistance
azimuthal projection
gnomonic projection
compromise projection
legend
map scale
verbal scale
graphic (bar) scale
representative fraction (RF) scale
magnetic declination
azimuth
bearing
thematic map
discrete data
continuous data
isoline
topographic contour line

contour interval
profile
gradient
geographic information system (GIS)
digital map layers
digital elevation model (DEM)
draping
vertical exaggeration
visualization models
remote sensing
aerial photograph
digital image
pixel
spatial resolution
near-infrared (NIR)
thermal infrared (TIR)
radar
weather radar
imaging radar
lidar
sonar
multispectral remote sensing

■ QUESTIONS FOR REVIEW

1. Why is a great circle useful for navigation?
2. What are the latitude and longitude coordinates of the place (town, city) where you live?
3. Approximately how precise in meters could you be if you tried to locate a building in your city to the nearest second of latitude and longitude? What about using the global positioning system (GPS)?
4. What time zone are you in? What is the time difference between Greenwich Mean Time and your time zone?
5. If you fly across the Pacific Ocean from the United States to Japan, how will the International Date Line affect the day change?
6. How has the use of the Public Lands Survey System affected the landscape of the United States? Has your local area been affected by its use? If so, how?
7. What is the difference between a conformal map and an equal-area map?

8. What is the difference between a representative fraction and a verbal map scale?
9. What does a GIS do to make overlays that can be directly compared, if they were originally derived from maps of different scales and different map projections?
10. What does the concept of *thematic map layers* mean in a geographic information system?
11. What specific advantages do computers offer to the mapmaking process?
12. What does a weather radar image show to help us understand weather patterns?
13. What are some useful environmental applications for color infrared images, thermal infrared images, digital elevation models, and lidar images?

CONSIDER AND RESPOND

1. Select a place within the United States that you would most like to visit for a vacation. You have a highway map, a USGS topographic map, and a satellite image of the area. What kinds of information could you get from one of these sources that are not displayed on the other two? What spatial information do they share (visible on all three)?

2. If you were an applied geographer and wanted to use a geographic information system to build an information database about the environment of a park (pick a state or national park near you), what are the five most important layers of mapped information that you would want to have? What combinations of two or more layers would be particularly important to your purpose?

PRACTICAL APPLICATIONS

1. If it is 2:00 a.m. Tuesday in New York (EST), what time and day is it in California (PST)? What time is it in London (GMT)? What is the date and time in Sydney, Australia (151° East)?

2. A topographic profile has a linear scale of 1:2400, and a vertical scale of 1 inch equals 100 feet. How many feet does 1 inch equal on the linear scale? If there is vertical exaggeration, what amount is the vertical stretched?

3. If 10 centimeters (3.94 in.) on a map equal 1 kilometer (3281 ft) on the ground, what is the map's RF scale (rounded to the nearest thousand)? Here is the formula for this kind of scale conversion.

$$\frac{\text{Map distance}}{\text{Earth distance}} = \frac{1}{\text{RFD}}$$

where RFD is the representative fraction denominator.

Be sure to use the same units of measurement for the map and Earth distances.

LOCATE AND EXPLORE

1. The coordinate system used on a globe is latitude and longitude, representing angular distance (in degrees) north and south of the equator and angular distance east and west of the prime meridian that passes through Greenwich, England. Using the Search window in Google Earth, fly to the heart of the following cities and identify the latitude and longitude. Measure the latitude and longitude using decimal degrees, and round off to two decimal places (for example, 41.89 N as opposed to 41°88′54.32″ N). Make sure that you correctly note whether the latitude is North (N or +) or South (S or −) of the equator and whether the longitude is East (E or +) or West (W or −) of the prime meridian.

    Tip: Change the latitude/longitude setting in the Tools > Options dialog box.

London, England	Cape Town, South Africa
Paris, France	Moscow, Russia
New York City	Beijing, China
San Francisco, California	Sydney, Australia
Buenos Aires, Argentina	Your hometown

    Now reverse the latitude and longitude of your hometown (N to S, and W to E) and note where you are in the world.

2. To find your location on the surface of the Earth, you can use a global positioning system (GPS) device, which gives location in latitude and longitude. Enter the following coordinates into Google Earth to identify the location:

41.89 N, 12.492 E	27.175 N, 78.042 E
33.857 S, 151.215 E	27.99 N, 86.92 E
29.975 N, 31.135 E	40.822 N, 14.425 E
90.0 N, 0 E	48.858 N, 2.295 E
90.0 S, 90.0 W	

    Tip: Use the zoom, tilt, rotate, and elevation exaggeration functions of Google Earth to help view and interpret the landform object shown in the browser.

3. When looking at a topographic profile or using the terrain feature in Google Earth, you can control the elevation exaggeration, which is calculated as the ratio of the units on the horizontal (*x*) axis to the units on the vertical (*z*) axis. In Google Earth you can adjust the elevation exaggeration between 0.5 and 3, thereby making subtle objects more noticeable. (Go to Tools > Options to set the elevation exaggeration.) In Google Earth turn on the Elevation Exaggeration Layer and then Fly to Mount Everest, the Nebraska Sand Hills, and the Goosenecks of the San Juan River. Adjust the exaggeration

from 0.5 to 3 and notice the change in the terrain. In which of these landscapes is a higher level of vertical exaggeration most useful in interpreting the natural terrain?

Tip: Use the zoom, tilt, and rotate functions of Google Earth to help view and interpret the landform object shown in the browser.

4. Download the Stone Mountain Topographic Layer from this book's online resources found at www.cengagebrain.com.

Using this layer as a reference, draw a topographic profile (similar to Fig. 2.29 in your text) from Point A to Point B along the line shown. Use the contour lines and Google Earth's elevation exaggeration and tilt features to decipher the landform.

Tip: Adjust the transparency of the topographic layer to see the image below. Turn off unnecessary layers for better visibility.

 MindTap—Make the most of your study time by accessing everything you need to succeed in one place. Read your textbook, take notes, review flashcards, watch videos, complete activities, take practice quizzes, and more online with MindTap. Log in at **www.cengagebrain.com**.

Map Interpretation

TOPOGRAPHIC MAPS

The Map

South Menan Butte on the Snake River Plain in southern Idaho.
Topographic maps show elevation changes with contour lines that connect points of equal elevation. The character of the landscape and the landforms can be interpreted from these maps if you understand how to interpret contour lines. Elevations, hiking routes, slope steepness, and the direction of stream flow can be determined by reading a topographic map. Standard symbols are used on USGS topographic maps to represent mapped features and information (a guide is in Appendix B).

The elevation difference between adjacent contour lines, called the *contour interval*, depends on the map scale and how much elevation change exists in the mapped area (the local relief). Elevations represented by topographic contours are typically divisible by 10. Mountainous areas are shown with greater intervals to keep contours from visually merging. Low relief areas use a smaller contour interval to show subtle relief features. It is important to note the map scale and the contour interval when first examining a topographic map.

Keep in mind several rules when interpreting contours:

- Closely spaced contours indicate a steep slope, and widely spaced contours indicate a gentle slope.
- Evenly spaced contours indicate a uniform slope.
- Closed contour lines represent a hill or a depression (if there are tick marks on the downslope side).
- Contour lines never cross but may converge along a vertical cliff.
- Contour lines bend upstream when they cross a valley.

Interpreting the Map

1. The topographic map scale is now 1:12,000. What would 1 inch on the map represent in feet on the Earth's surface? What was the representative fraction scale of the original map scale before enlargement? What is the contour interval on this map?
2. What are the highest and lowest elevations on the map? Where are they located? What is the crater depth from the highest point on the rim to the lowest point in the crater?
3. Slope ratio is calculated by dividing elevation change by horizontal distance. For example, a 1000-meter-high elevation rise over a horizontal distance of 3000 meters would have a slope ratio of 1:3. What is the slope ratio for the slope of South Menan Butte from the 5289 elevation point down to the Snake River (below the "R" in river)?
4. Why would it be useful to have both a map and an aerial photograph when studying landforms? What is the chief advantage of each? If you were using a compass with this map, what is the magnetic declination as shown here?

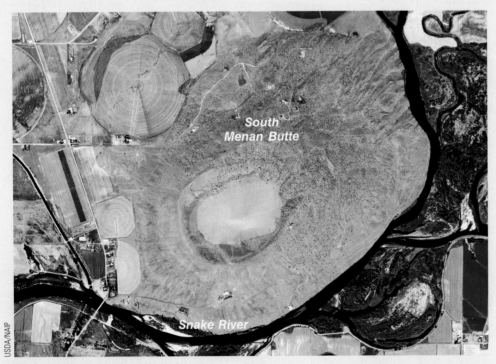

A natural color aerial photograph of South Menan Butte.

⊕ In Google Earth fly to: 43.76°N, 111.97°W.

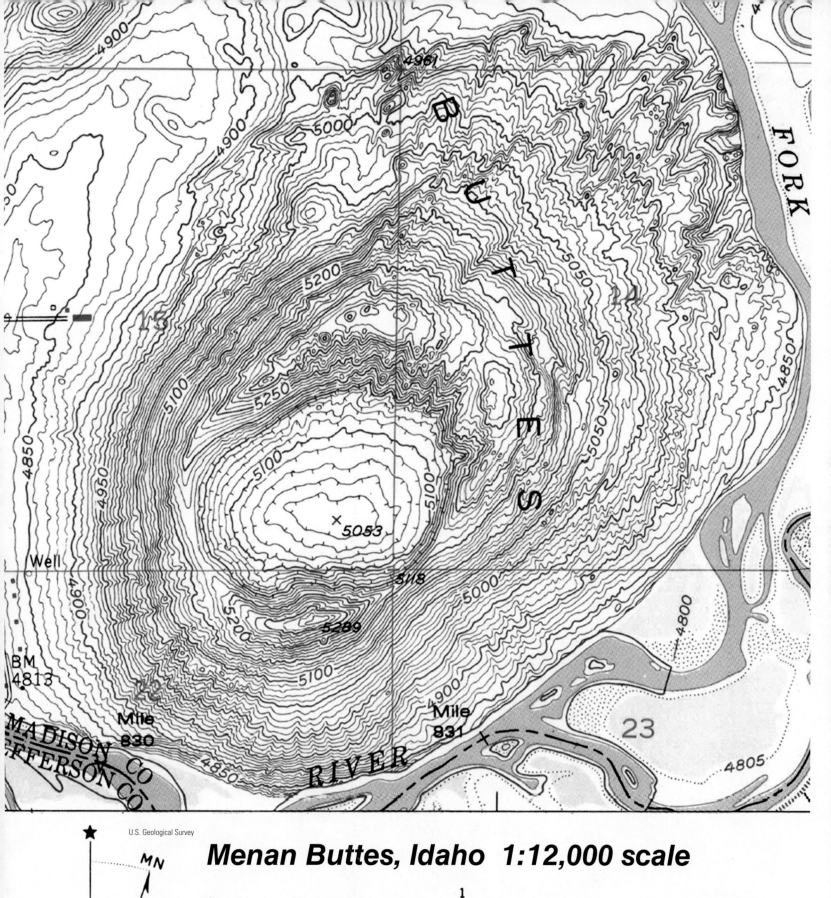

FORK

BUTTES

×5053

Well

5118

5289

BM
4813

Mile
830

Mile
831

MADISON CO
JEFFERSON CO

RIVER

15

14

22

23

4900

4900

5000

4961

5000

5200

5250

5100

5100

5100

5200

5200

5100

4950

4900

4850

4850

4800

4805

4850

5050

5050

5050

4900

5000

4800

4850

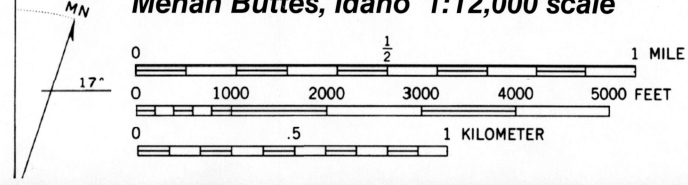

U.S. Geological Survey

MN

17^

Menan Buttes, Idaho 1:12,000 scale

| 0 | | | | ½ | | | | 1 MILE |

| 0 | 1000 | 2000 | 3000 | 4000 | 5000 FEET |

| 0 | | .5 | | 1 KILOMETER |

SOLAR ENERGY AND EARTH–SUN RELATIONSHIPS

■ OBJECTIVES

WHEN YOU COMPLETE THIS CHAPTER YOU SHOULD BE ABLE TO:

- ■ 3.1 Explain why the sun is the ultimate energy source that drives the various components of the Earth system.
- ■ 3.2 Discuss the types of energy emitted by the sun as they are represented in the electromagnetic spectrum and their use or impact on Earth.
- ■ 3.3 Describe how Earth–sun relationships affect the receipt and distribution of solar energy on our planet.
- ■ 3.4 List each of the four astronomical seasons and their occurrence in terms of Earth–sun relationships.
- ■ 3.5 Diagram Earth's positions in its annual orbit around the sun during the solstice and equinox days, including the orientation of the axis.
- ■ 3.6 Explain how and why the amount of daylight hours varies by latitude and by hemisphere.
- ■ 3.7 Outline the significant relationships of the Tropics of Cancer and Capricorn and the Arctic and Antarctic Circles to solar radiation and the seasons.

THE SUN'S RADIUS is 110 times that of Earth, its mass is 330,000 times greater, and it reigns as the center of our solar system. The gravitational pull of this star, a fierce and stormy ball of gas, holds Earth in orbit, and its energy emissions power the Earth systems on which our lives depend. As the ultimate source of almost all the energy in our world, the sun holds the key to many of our questions about the Earth, the sky, and our weather and climate.

Everyone has wondered about environmental changes that occur throughout the year and from place to place on our planet's surface. Perhaps when you were young you wondered why summer was warmer than it was in winter and why the summertime hours of daylight were long but in other seasons they were shorter. These questions and many like them are probably as old as the earliest human thoughts, and the answers to them provide us with a better understanding of our world's physical geography and the daily and seasonal changes that directly influence our lives.

Gaining this understanding requires examining certain factors that are beyond the confines of our planet, particularly to consider the sun's characteristics and how Earth functions as a part of the solar system. Geographers examine relationships between the sun and Earth to explain phenomena such as the alternating periods

◀The sun at midnight in June over a fjord in Skånland, Troms, Norway, located just north of the Arctic Circle. Daylight lasts 24 hours here during the longest day of the year. Stocktrek Images, Inc./Alamy

⊕ In Google Earth fly to the fjord at 68.87°N, 60.90°E

of light and dark that we know as day and night. Other relationships between Earth and the sun help to explain seasonal weather changes and long-term variations in climate. An examination of how Earth interacts with its ultimate energy source is vital to understanding the environments that support life as we know it.

The Solar System and Beyond

When you look at the sky on a clear night, all of the stars that you can see are a part of the Milky Way Galaxy. A **galaxy** (● Fig. 3.1) is an almost incomprehensible cluster of stars, dust, and gases, surrounded by space, and is like an enormous island in the universe. Our sun is one of the billions of stars in the Milky Way Galaxy. In addition, the vast part of the universe that we are able to observe contains billions of other galaxies.

Distances within the universe are so great that it is necessary to use a large unit of measure termed a **light-year** (the distance that light travels in 1 year). A light-year is equal to 9.5 trillion kilometers (6 trillion mi). Light travels at the amazing speed of 298,000 kilometers per second (186,000 miles per second). Thus, in a single second, light could travel seven times around the Earth's circumference. Although that may seem like a great distance, the closest star to Earth, other than the sun, is Proxima Centauri, which is 4.2 light-years away, and the closest galaxy to ours is the Canis Major Dwarf Galaxy, at 25,000 light-years away.

The Solar System

The sun is the center and the unifying body of our solar system. A **solar system** can be defined as all heavenly bodies surrounding a particular star because of the star's dominant mass and gravity. **Gravity** is the attractive force that one body has for another. The

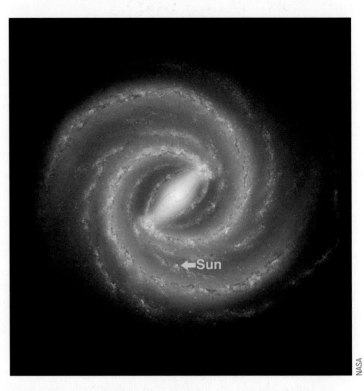

NASA

● **FIGURE 3.1** The Milky Way Galaxy. Based on scientific information, an artist at the National Aeronautical and Space Administration (NASA) produced this image of the Milky Way. The sun's location shows the vastness of our galaxy, one of billions in the universe.

greater the **mass** (the amount of matter a body has), the greater the gravitational pull it will exert on other bodies. The major celestial bodies in our solar system are the eight planets (● Fig. 3.2). A **planet**, as defined by the International Astronomical Union in 2006, is a celestial body in orbit around the sun, with sufficient gravitational attraction to overcome rigid body forces, has a nearly spherical shape, and has cleared the neighborhood around its orbit. Under this relatively new definition, Pluto, formerly designated as

The International Astronomical Union/Martin Kornmesser

Mercury Venus Earth Mars Jupiter Saturn Uranus Neptune

● **FIGURE 3.2** The solar system, with the planets in order, outward from the sun. The comparative sizes of the planets are shown; however, the planetary orbital distances are greatly compressed.

What is Pluto categorized as today?

TABLE 3.1
Comparison of the Planets

	Mercury	Venus	Earth	Mars	Jupiter	Saturn	Uranus	Neptune
Distance from Sun (in millions)	57.9 km (36.0 mi)	108.2 km (67.0 mi)	149.6 km (93.0 mi)	227.9 km (141.7 mi)	778.3 km (483.5 mi)	1427.0 km (888.8 mi)	2870.7 km (1783.7 mi)	4497.0 km (2797.70 mi)
Diameter	4878 km (3031 mi)	12,104 km (7521 mi)	12,756 km (7926 mi)	6794 km (4222 mi)	142,984 km (88,846 mi)	120,536 km (74,900 mi)	51,118 km (31,763 mi)	49,532 km (30,779 mi)
Day (Earth = 1)	59 days	243 days	1 day	1.03	9 hrs, 55 mins	10 hrs, 39 mins	17 hrs, 14 mins	16 hrs, 7 mins
Year (1 solar orbit)	88 days	224 days	365.25 days	687 days	11.86 years	29 years	84 years	164.8 years
Gravity (Earth's = 1.0)	0.38	0.9	1	0.38	2.64	1.16	1.11	1.21
Average Temperature (°C)	167°C (333°F)	464°C (867°F)	15°C (59°F)	−65°C (−85°F)	−110°C (−202°F)	−140°C (−220°F)	−195°C (−319°F)	−200°C (−328°F)
Tilt of the Rotational Axis	0.01°	177.4°	23.5°	25.2°	3.1°	26.7°	98.8°	28.3

our ninth planet, has been reclassified as a *dwarf planet*. It is generally agreed that Pluto is a large body captured by the sun's gravitational pull from the Kuiper Belt, a disk-shaped region of space containing small icy bodies that lie beyond the orbit of Neptune.

The Planets

The four planets closest to the sun (Mercury, Venus, Earth, and Mars) are called the **terrestrial planets** because they are composed of rock and metallic minerals (Earthlike materials). They are also relatively small compared with the other four planets and are warmed by their proximity to the sun. They all have solid surfaces that record evidence of geologic forces in the form of craters, mountains, valleys, and volcanoes. The four planets that are more distant from the sun (Jupiter, Saturn, Uranus, and Neptune) are much larger. Composed primarily of lighter ices, liquids, and gases, these planets are termed either the **giant planets** or the **gas planets**. Although they have solid cores at their centers, they are more like huge balls of gas and liquid with no solid surface on which to walk.

The eight major planets, however, share several common phenomena. From a distant point, far out in space above the sun's "north pole," they would all appear to move around the sun in a counterclockwise direction. Their orbits follow an elliptical, but almost circular, path. Each of the planets rotates, or spins, on an axis. With the exception of Venus and Uranus, all rotate in the same counterclockwise direction. The planets' orbits all follow closely to the same plane (the plane of the ecliptic) passing through the sun's equator. All planets have an atmospheric layer of gases with the exception of Mercury, which is not dense or heavy enough for its gravity to hold appreciable amounts of gases (Table 3.1).

Our solar system also includes approximately 146 moons (like Earth's moon, these are bodies orbiting a planet); numerous **asteroids**, which are smaller rocky bodies with a diameter of less than 800 kilometers (500 mi); and comets and meteors. **Comets** are made up of a head of solid fragments held together by ice and a tail composed of gases, which can be millions of miles long (● Fig. 3.3). **Meteors** are small, stone-like or metallic bodies that melt and burn when encountering oxygen and friction with Earth's atmosphere and may appear as a streak of light, or "shooting star." A meteor that survives the fall through the atmosphere and strikes Earth's surface is called a **meteorite** (● Fig. 3.4).

European Southern Observatory

● **FIGURE 3.3** The comet Ison in 2013 viewed from the La Silla, Chile, observatory in the Atacama Desert. Dry desert air offers excellent conditions for viewing objects in space. This special composite image enhances the view of the comet as it streaks across space.

Courtesy of Jason Utas, UCLA/Wikimedia Commons

● **FIGURE 3.4** This iron meteorite, found in Morocco, shows the effect of partial melting from friction as it passed through Earth's atmosphere. Many meteorites burn up completely before reaching Earth's surface.

The Earth–Sun System

Earth receives only about one two-billionth (1/2,000,000,000) of the radiation given off by the sun, but even this tiny fraction of the sun's energy is enough to drive the biological and physical characteristics of Earth's surface. Other bodies in the solar system also receive some of the sun's radiant energy, but most of this energy travels out through space unimpeded. Solar energy is the most important factor in determining environmental conditions on Earth. With the exception of geothermal heat sources (such as volcanic activity, geysers, and hot springs) and heat emitted by the decay of radioactive minerals, the sun remains the energy source for Earth's atmospheric, oceanic, and biological systems.

The intimate, life-producing relationship between Earth and the sun results from the amount and distribution of radiant energy received from the sun. Our planet's size, its distance from the sun, the nature of its atmosphere and surface, its annual orbit around the sun, and its rotation on its axis affect the amount of solar radiation that Earth receives. Although some processes that affect our physical environment result from Earth forces that are not related to the sun, these processes would have little relevance were it not for life-sustaining solar energy.

Earth orbits around the sun at an average distance of 150 million kilometers (93 million mi) at an average velocity of 107,300 kilometers per hour (67,062 mph). The magnitude of the sun's size and its distance from Earth may be difficult to comprehend. About 130 million Earths could fit inside the sun, and an airplane flying at 500 miles per hour would take 21 years to reach the sun. Distances are vast—if the sun were the size of a softball, Earth would be the size of a poppy seed 10.7 meters (35 ft) away.

As far as we know with certainty, within our solar system, only Earth receives the amount of solar energy to be able to support life—to create something that can grow, develop, reproduce, and eventually die. Yet there remains a possibility of life, or at least the basic organic building blocks, on Mars and perhaps on one or two of Saturn's moons. What fascinates scientists, geographers, and philosophers alike, however, is the possibility that among the millions of planets like Earth in the universe, forms of life might have developed that could even be more advanced than humans.

The Sun and Solar Energy

A slightly less than average-sized star, the sun, like all stars, is a self-luminous sphere of gases that emits radiant energy. Solar radiation is the source of almost all the light and heat for the various celestial bodies in our planetary system. The energy emitted by the sun comes from **fusion (thermonuclear) reactions** taking place at its core. There, under extremely high pressure and temperatures exceeding 15 million°C (27 million °F), two hydrogen atoms fuse together to form one helium atom in a process similar to that of a hydrogen bomb explosion (● Fig. 3.5). Nuclear fusion releases tremendous amounts of energy that radiate out from the core to just below the sun's surface where countless convection currents act like water boiling in a pot (hotter gases rising and cooler gas sinking). These convective currents give the sun a "grainy appearance" that can be seen in special remotely sensed imagery.

The *photosphere* (sphere of light) is what we see as the sun's surface and it is the densest layer. It has an estimated temperature of between 5500°C and 6100°C (10,000°F and 11,000°F). The *chromosphere* (sphere of color) is a thin layer of gases above the photosphere and appears reddish in color. Last, the *corona* (or *crown*) is the outermost layer of the sun's atmosphere. Its constantly changing shape is caused by charged particles trapped by the

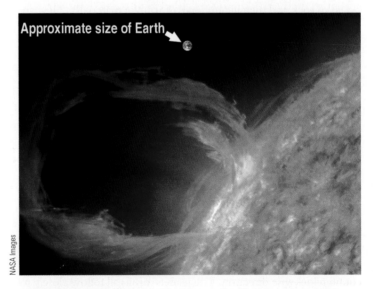

● **FIGURE 3.5** The sun is powered by thermonuclear fusion reactions similar to those of exploding hydrogen bombs. This gigantic solar prominence expelled charged particles from the sun and into space. Note the comparative size of Earth.

What elements drive a nuclear fusion reaction?

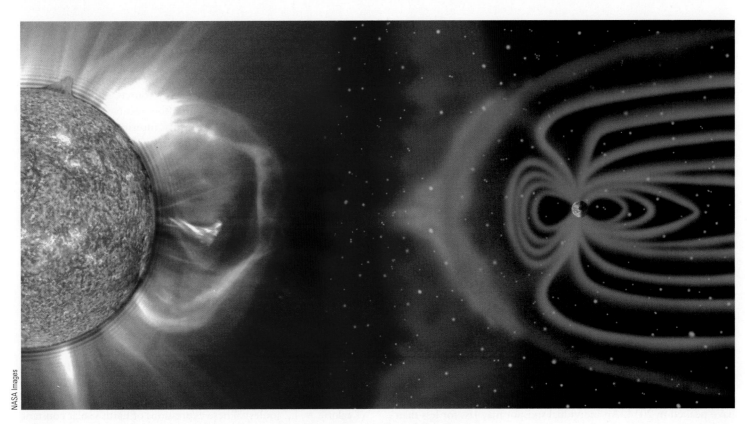

● **FIGURE 3.6** Earth's magnetic field protects our planet's surface from the harmful effects of the solar wind.

What impacts can the solar wind still have on our lives despite the magnetic field?

sun's magnetic field. Charged particles (mainly protons and electrons) from the corona can flow along lines of the sun's magnetic field millions of miles into space in what is called the **solar wind** (● Fig. 3.6). Unlike solar radiation, which moves at the speed of light, a solar wind travels about 400 kilometers (250 mi) per second and takes more than 4 days to reach our planet. When these solar winds reach Earth, they are prevented from harming the surface by Earth's magnetic field and are confined to the upper atmosphere. During these times, however, they can disrupt cell phone, radio, and television communications and may disable orbiting satellites.

The Earth's magnetic field tends to direct the solar wind into the outer atmosphere and to the areas surrounding our planet's magnetic poles (see Fig. 3.6). When this happens, the solar wind energizes ions in the outer atmosphere, resulting in an amazing light show known as an **aurora** (● Fig. 3.7). The **aurora borealis**, often called the *northern lights* (● Fig. 3.8), and the **aurora australis**, the *southern lights*, happen simultaneously in the northern and southern latitudes, respectively.

Sunspots, the best-known solar feature, influence the intensity of solar winds. Visible on the photosphere, **sunspots** appear as dark regions that are about 1500°C to 2000°C (2700°F to 3600°F) cooler than the temperatures of the surrounding areas of the sun. During the 1600s, Galileo began recording sunspots, and for many years they have been used as an indicator of solar activity. Sunspots seem to occur on an 11-year cycle from one maximum (where 100

● **FIGURE 3.7** Solar winds form ring-shaped auroras over and around Earth's magnetic poles in both hemispheres. This is the Northern Hemisphere's aurora borealis.

What is the Southern Hemisphere's aurora called?

Joshua Strang/U.S. Air Force

● **FIGURE 3.8** The aurora borealis, here in Alaska, interrupts the winter sky's darkness with a beautiful and amazing light show. Energy interactions between the solar wind and ions in the atmosphere produce the auroras.

Solar Energy and Atmospheric Dynamics

As previously noted, our sun is the Earth system's major energy source, either directly or indirectly. Earth does receive certain amounts of heat from its interior (volcanoes, hot springs, and geysers). When compared with the amount of energy received from the sun, however, these other sources are insignificant.

The sun radiates energy into space at an almost steady rate. The rate of a planet's receipt of solar energy is known as the **solar constant** and has been measured with great precision outside Earth's atmosphere by satellites. The amount of energy received at the outer edge of Earth's atmosphere is about 1370 watts per square meter (in SI units, this is 1370 joules per second per square meter). These are both expressions of power. The solar constant has also been defined as being slightly less than two calories per square centimeter per minute. A **calorie** is the quantity of energy required to raise the temperature of 1 gallon of water 1°C (1 calorie = 4.2 joules). These expressions of Earth's solar constant in different units, and their usage varies depends on the application or the location of a scientific study. The atmosphere affects the amount of solar radiation received at the Earth's surface because clouds absorb some energy, some is reflected back to space, and some is refracted away. If Earth did not have an atmosphere, the solar energy striking the surface at a certain location at a particular time would be a constant value, determined by the latitudinal position.

Of course, the value of the solar constant for other planets varies with their distance from the sun as the same amount of energy radiates outward into larger areas. When Earth is closest to the sun in its orbit, its solar constant is slightly higher than the yearly average, and when it is farthest away, the solar constant is slightly lower than average. However, this difference does not have a very significant effect on Earth's temperatures. The solar constant also varies slightly with changes in the sun's activity; during intense sunspot or solar storm activity, for example, the solar constant will be slightly higher than usual (about 0.2%).

The sun emits **electromagnetic energy** that travels at the speed of light in a range of varying wavelengths, called the **electromagnetic spectrum** (● Fig. 3.10). It takes about 8.3 minutes for this energy to reach Earth. Approximately 9% of solar energy consists of radiation with wavelengths that are shorter than visible light, which includes gamma rays, X-rays, and ultraviolet radiation. These short wavelengths cannot be seen but can affect tissues of the human body. It is well known that absorbing too much X-ray energy can be dangerous. Exposure to excessive ultraviolet energy results in sunburned skin and is a primary cause of skin cancer. About 41% of the solar energy spectrum comes in the form of visible light rays—the

or more may be visible) to the next. The latest sunspot cycle may have peaked in 2014. An individual sunspot may last for less than a day or as long as 6 months, and they can be larger than Earth (● Fig. 3.9). The largest sunspot in 24 years occurred in 2014, and was more than 10 times Earth's size. Just how sunspots might affect our atmosphere is still a matter of controversy. Proving direct connections between sunspot numbers and weather or climate is difficult, but such relationships have been proposed and supported by evidence. Times of minimum sunspots lasting tens of years have been known to be times of cooler temperatures.

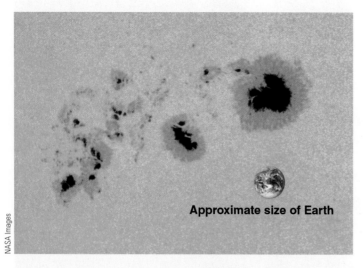

Approximate size of Earth

NASA Images

● **FIGURE 3.9** Sunspots as they appear on the solar surface. The relative size of Earth to these sunspots is shown for comparison.

About how many Earth diameters can fit east to west across this sunspot?

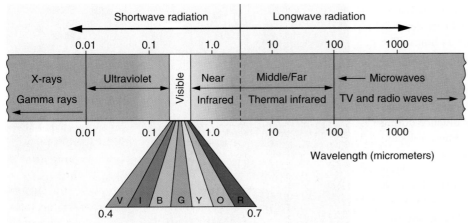

● FIGURE 3.10 Solar energy travels toward Earth in a wide spectrum of wavelengths, here measured in micrometers (μm) (1 μm is one millionth of a meter). Human vision sees wavelengths between 0.4 and 0.7 micrometers (visible light). Solar radiation is in short wavelengths (mainly light), whereas terrestrial (Earth) radiation is in long wavelengths (thermal infrared, sensed as heat).

What colors represent the shortest, and the longest, wavelengths that humans can see?

visible spectrum—where each color is distinguishable by its specific wavelength band. Wavelengths longer than visible light are invisible to the human eye. About 49% of the sun's radiant energy is emitted in wavelengths that are longer than the range of human vision, from 0.4 to 0.7 micrometers. Although these wavelengths are invisible, they can sometimes be sensed by human skin. Near-infrared energy, emitted in wavelengths of light that are longer than the range of human vision, is harmless to living organisms. Longer wavelength energy in the infrared part of the spectrum, called *thermal infrared*, can be felt as heat. The remaining 1% of solar radiation falls into the very long wavelength regions, which includes microwaves and those used as television signals and radio waves. Collectively, gamma rays, X-rays, ultraviolet rays, visible light, and near-infrared energy are known as **shortwave radiation**. Types of energy including the thermal infrared range and longer wavelengths (microwaves and radio waves) are referred to as **longwave radiation** (see Fig. 3.10).

Through advances in technology, we use certain bands of electromagnetic wave energy for a great variety of useful applications. In the field of communications, we use radio waves for television and radio broadcasts and cell phone connections. Microwaves are also used to transmit information. X-rays and gamma rays are used in diagnostic health care and in transportation security checks. The digital and film cameras people generally use record images from reflected light energy. In the remote sensing field, radar uses microwaves to map elevations and terrain as well as detect weather patterns, aircraft, vehicles, and ships. Thermal infrared scanners make images of heat emissions, basically recording differences in temperature. Many of these applications of electromagnetic energy were discussed in more detail in Chapter 2.

Movements of Earth

Our planet continually undergoes three movements: galactic movement, rotation, and revolution. **Galactic movement** is the motion of the Earth, the sun, and the entire solar system in an orbit around the center of the Milky Way Galaxy. This movement does not affect the changing environments of Earth and is generally the concern of astronomers rather than geographers. The other two Earth movements, **rotation** on its axis and its **revolution** around the sun, are important in understanding physical geography. The consequences of these movements include the phenomena of day and night, variations in day length (sunrise to sunset) over the year, and the changing seasons.

Rotation *Rotation* refers to the turning of Earth on its axis, an imaginary line extending from the North Pole to the South Pole. Earth rotates on its axis at a uniform rate, making one complete turn with respect to the sun in 24 hours. Earth rotates in an eastward direction, as illustrated in ● Figure 3.11. This direction of rotation is why the sun "rises" in the east and then appears to move westward across the sky, but it is actually Earth, rather than the sun, that is moving, rotating eastward toward the morning sun.

Earth rotation causes the apparent east-west movement of the sun and moon as well as the stars across the sky. If we look down on a globe from above the North Pole, the rotational direction is counterclockwise. This eastward direction of rotation moves the zones of daylight and darkness across Earth's surface and also affects the circulatory movements of the atmosphere and oceans.

The rotational velocity on Earth's surface varies with the distance of a place from the equator. Every location on Earth completes a full rotation (360°) in 24 hours. Thus, the angular velocity

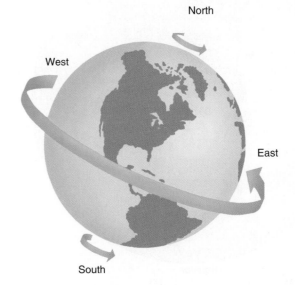

● FIGURE 3.11 Earth turns on a tilted axis as it orbits around the sun. The rotation direction is from the west and toward the east, making sun appear to rise in the east and set in the west.

for all locations on Earth is the same—360° per 24 hours; therefore, Earth rotates 15° per hour. However, the linear velocity depends on the distance (not the angle) covered during that 24 hours. The linear velocity at the poles is zero. You can see this by turning a globe with a sticker affixed to the North Pole. There, the sticker rotates 360° but covers no distance and therefore has no linear velocity. If you place the sticker anywhere between the North and South Poles, however, it will cover a measurable distance during one Earth rotation. The equator has the greatest linear velocity, where the distance traveled by a point in 24 hours is longest on Earth. At Kampala, Uganda, near the equator, the velocity is about 460 meters (1500 ft) per second, or approximately 1660 kilometers (1038 mi) per hour (● Fig. 3.12). In comparison, at St. Petersburg, Russia (60°N latitude), where the distance traveled during one complete rotation of Earth is about half that at the equator, Earth rotates at approximately 830 kilometers (519 mi) per hour.

We are unaware of the speed of rotation because everything around us is rotating along with us. Other reasons include (1) the angular velocity is constant for each latitude, (2) the atmosphere rotates with Earth, and (3) there are no nearby objects moving at a different rate with respect to Earth's motions for comparison for us to be aware of Earth's rotation. Without these references, we cannot perceive the rotational speed at our latitudinal position.

Our alternating days and nights can be demonstrated by shining a light on a globe while rotating the globe slowly toward the east. You can see that one-half of the sphere is always illuminated while the other half is dark. But as Earth rotates, new longitudes are continually moving into the illuminated half of the planet while others are moving into the darkened hemisphere. Earth

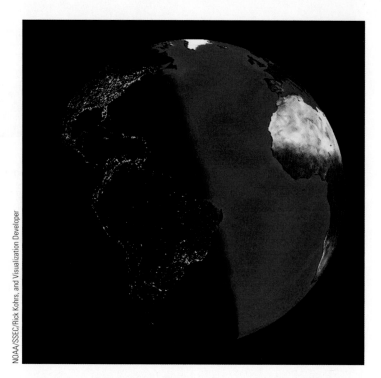

● **FIGURE 3.13** The circle of illumination, which separates day from night, is clearly seen on this image.

Which way is the circle of illumination moving across Earth's surface?

rotation and the way that solar energy strikes our spherical planet mean that half of Earth receives daylight and the energy of solar radiation while the other half is in darkness. As noted in Chapter 2, the great circle that separates day and night, moving westward, is known as the **circle of illumination** (● Fig. 3.13).

Revolution As Earth rotates on its axis, it also revolves around the sun in a slightly elliptical orbit with an average distance from the sun of about 150 million kilometers (93 million mi). On about January 3, Earth is closest to the sun, at a time called **perihelion** (from Greek: *peri*, close to; *helios*, sun); its distance from the sun at that time is approximately 147.5 million kilometers (91.5 million mi) (● Fig. 3.14). Around July 4, Earth is about 152.5 million kilometers (94.5 million mi) from the sun, at its farthest point away from the sun, and is said to be at **aphelion** (from Greek: *ap*, away; *helios*, sun). Because of this great distance from Earth to the sun, the annual difference of 5 million kilometers is relatively insignificant (about 3.5%), and this amount of variation in distance has only a minimal effect on Earth's receipt of solar energy. Therefore, the changing distance between Earth and the sun has virtually no relationship to the seasons. In fact, it is important to remember that the northern and southern hemisphere seasons are reversed—when it is winter in the North America, it is summer in Australia.

The period of time that Earth takes to make one revolution around the sun determines the length of a year. Earth makes 365 rotations on its axis during the time it takes to complete one revolution around the sun; therefore, a year has 365 days (actually 365.25 days). Because of the difficulty of dealing with a fraction of a

● **FIGURE 3.12** Earth's rotational speed varies with the latitudinal distance of a location from the equator.

How much faster does a point on the equator move than a point at 60°N latitude?

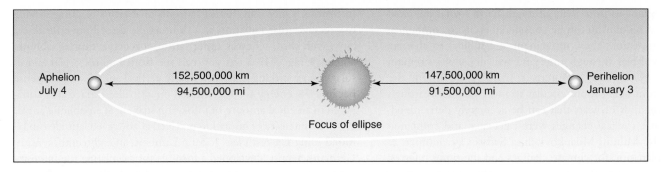

Aphelion
July 4

152,500,000 km
94,500,000 mi

Focus of ellipse

147,500,000 km
91,500,000 mi

Perihelion
January 3

● **FIGURE 3.14** An oblique view of Earth's elliptical orbit around the sun. Earth is closest to the sun at perihelion and farthest away at aphelion. Note that in July, during the Northern Hemisphere summer, Earth is farther from the sun than at any other time of the year.

During what northern hemisphere season is Earth closest to the sun?

day, it was decided that a year would have 365 days, and every fourth year, called a *leap year*, an extra day would be added as February 29.

Plane of the Ecliptic, Inclination, and Parallelism

In its orbit around the sun, Earth moves along a constant, imaginary plane, called the **plane of the ecliptic**. The axis of rotation is not, however, oriented at a 90° angle to this imaginary surface that Earth tracks along during its annual revolution. The Earth's axis is tilted $23\frac{1}{2}°$ away from being perpendicular to the plane of the ecliptic (● Fig. 3.15). The tilt of Earth's axis away from being perpendicular to the orbital plane is called the **angle of inclination**, which is $23\frac{1}{2}°$. In addition to the angle of inclination, the Earth's tilt maintains another characteristic called **parallelism of the axis**. As Earth revolves around the sun, the planetary axis remains parallel to all of its former positions—that is, at every position in Earth's orbit, the axis points toward the same spot in the sky. For the North Pole, that spot is close to the **North Star**, called **Polaris**. Earth's axis always points in the same direction in respect to stars outside our solar system, but its orientation in respect to the sun does change during its annual orbit. Photographs of the northern sky taken over

a few hours at night show the stars as circular trails with the North Star in the center (● Fig. 3.16) This configuration occurs because the northern end of Earth's axis points directly toward Polaris. A camera, fixed in terms of Earth location, moves with the surface of our rotating planet. The stars have not moved relative to Earth, but rotation moves Earth positions relative to the stars.

● **FIGURE 3.16** Proof of Earth rotation. The light from stars forms circles around the North Star in this long-exposure photograph. Polaris is at the center and does not form a circle because it is directly above the North Pole. These photographic star trails result from a fixed camera location moving with Earth's rotation, not the stars' movement in space.

$23\frac{1}{2}°$

$66\frac{1}{2}°$

Equator

Sun

Plane of ecliptic

$23\frac{1}{2}°$

Plane of equator

$66\frac{1}{2}°$

Axis

● **FIGURE 3.15** The plane of the ecliptic is an imaginary surface formed by Earth's orbit around the sun. The $23\frac{1}{2}°$ inclination of Earth's rotational axis causes the plane of the equator to cut across the plane of the ecliptic.

O-iieb-in/Shutterstock.com

THE EARTH-SUN SYSTEM 69

THE EARTH-SUN SYSTEM 69

It is important to note that, although Earth's rotation, axial inclination, and revolution can be considered as being constant in our current discussion, these movements are subject to change. Earth's axis wobbles through time and will not always remain at an angle of exactly 23½°. Moreover, Earth's orbit around the sun changes from being more circular to more elliptical through a cycle of thousands of years that can be accurately determined. These and other cyclical changes were calculated and compared in the 1940s by Milutin Milankovitch, a Serbian astronomer, as a possible explanation for climate changes and the times referred to as *ice ages*. Since then, the Milankovitch cycles have often been used by climatologists who seek to explain climate changes. Variations in climate over time will be discussed in more detail along with other theories of climatic change in Chapter 8.

Insolation, Sun Angle, and Duration

Understanding Earth–sun relationships leads to a discussion of how and why the sun's rays vary in intensity from place to place throughout the year and into an examination of seasonal changes on Earth. Solar radiation received by the Earth system, called **insolation** (short for "incoming solar radiation"), is the energy source for the Earth system. The seasonal variations in temperature that we experience primarily result from the impacts of these fluctuations in insolation.

What causes these variations in insolation and brings about seasonal changes? It is true that Earth's atmosphere affects the amount of insolation received. Heavy cloud cover, for instance, will decrease the amount of solar radiation that reaches Earth's surface, and a clear sky allows more insolation through to the surface. However, cloud cover is irregular and unpredictable from day to day and over the seasons.

The real answer behind variations in insolation lies with two major phenomena that vary regularly for a given location as Earth rotates on its axis and revolves around the sun: the duration of daylight and the angle of the sun's rays. The amount of daylight hours affects the length of time that insolation is received, and the angle that the sun's rays strike the surface affects the intensity of solar radiation. Together, the intensity and duration of solar radiation are two major factors that influence the warming and cooling of places on Earth during the days and over the seasons.

A location on Earth will receive more insolation if (1) the sun's rays strike at a steep angle, (2) the sun shines longer, or (3) both. The intensity of solar radiation received at any one time also varies from place to place because Earth presents a spherical surface to

insolation. Only one line of latitude on Earth's rotating surface can receive solar radiation at a right angle, directly overhead, on any given date, whereas other latitudes receive varying oblique angles (● Fig. 3.17a). As you can see from Figures 3.17b and c, solar energy striking Earth at a nearly vertical angle is more intense because the energy is concentrated on a smaller surface area compared to an equal amount of insolation striking at an oblique angle.

The intensity of insolation received at any given latitude can be found using *Lambert's law*. Johann Lambert, an eighteenth-century German scientist, developed a formula for calculating the intensity of insolation using the sun's zenith angle (that is, the sun angle deviating from 90° directly overhead). Using Lambert's law, one can

● **FIGURE 3.17** The angle at which the sun's rays strike Earth's surface determines the amount of solar energy received per unit of surface area and changes with the seasons. Diagram (a) represents June, when solar radiation strikes the surface perpendicularly at 23½° North, creating summer conditions in the Northern Hemisphere. On the same day in the Southern Hemisphere, the sun's rays are more oblique and spread over larger areas, thus receiving less energy per unit of area, making this the winter hemisphere. (b) In summer, the sun appears high in the sky and its rays hit Earth more directly, spreading out less. (c) In winter, the sun appears lower in the sky with oblique rays that spread out over a wider area, so they are less effective at heating the ground.

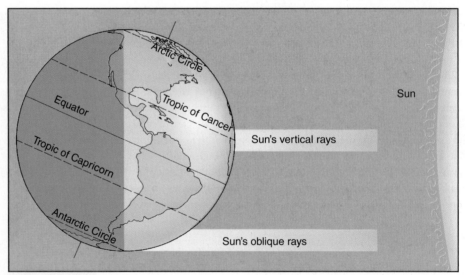

(a)

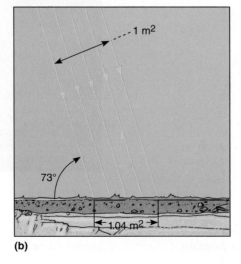

(b)

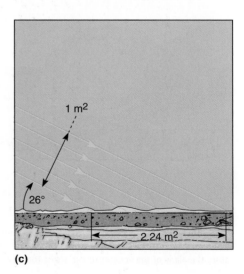

(c)

Passive Solar Energy— An Ancient Concept

When we consider using solar energy to heat buildings or to produce electrical power, we look toward developing technology to find the way. In fact, photovoltaic (PV) cells that convert sunlight into electricity and solar thermal towers that convert water into steam to drive electrical turbines are excellent ways to harness this inexhaustible energy source. However, long before these were invented, people wished to be comfortable in their homes and used a passive form of solar energy. This can easily be done, assuming you know the seasonal sun angles that affect your home.

The concept is very simple; flood your home with solar energy in the wintertime. This makes better use of indoor sunlight and adds more heat during the cold season. Then restrict the amount of insolation entering the home in summer; this keeps the interior cooler during the hottest months, while still allowing daylight to illuminate the interior. Environmentally conscious home designers can do this by adjusting the number and placement of windows in the home and controlling the length and angle of the eaves (or roof overhang).

This concept was understood in ancient times. The Cliff Palace in Mesa Verde National Park, Colorado, is a wonderful example of an 800-year-old Native American cliff dwelling. Here, the cave roof and overhang performed the same service as a passive solar home design. More direct sunlight enters the structures during the winter, and indirect sunlight enters during the summer.

How did these people know about sun angles? For millennia, ancient people observed the sun and its changes in angle over the seasons. Early cultures in China, in addition to Egyptians, Greeks, and Romans, designed their architecture to take advantage of solar energy. In the Americas, the ancient Maya, Incas, Aztecs, and North American tribes used their knowledge of sun angles as guides to erect buildings, temples, and pyramids to their chief god—the sun. Designing dwellings and other structures in accordance with the sun and solar energy is not a new concept.

Using this knowledge, we can use simple strategies to help us save money on future heating and air conditioning costs. Keep your home or apartment shaded in the summer and sunlit in the winter. Window curtains, shades, and blinds can do a lot to save on energy use.

The Cliff Palace at Mesa Verde National Park in Colorado shows that early Native Americans understood the use of passive solar energy in locating their cliff dwellings under natural overhangs.

⊕ In Google Earth, fly to 37.166°N, 108.472°N.

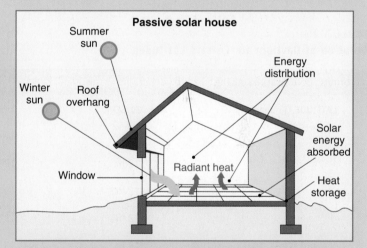

Passive solar house designs take into account the seasonal changes in sun angles. The diagram shows the maximum and minimum noon sun angles experienced at a hypothetical location over the seasons. The low-angle winter sun passes through the windows, causing the room to be warm. The higher-angle rays of the summer sun are blocked by an overhang.

identify, based on latitude, where greater or lesser solar radiation is received on Earth's surface. ● Figure 3.18 shows the intensity of total solar energy received at various latitudes, when the most direct radiation (from 90° angle rays) is striking directly on the equator.

The amount of solar radiation that reaches Earth's surface can also be diminished to some extent by atmospheric gases. Because oblique rays must pass through a greater distance of atmosphere compared to vertical rays, more insolation will be lost in the process. In 1854, German scientist and mathematician August Beer established a relationship to calculate the amount of solar energy lost as it interacts with atmospheric gases. *Beer's law*, as it is called, is strongly affected by the atmospheric thickness through which the energy must pass. Of course, clouds and particulates, such as dust in the air, also diminish the receipt of insolation.

Because insolation is not received at night, the duration of solar energy is related to the number of daylight hours at a location, and duration varies over the seasons (Table 3.2). The change in daylight hours over the seasons is also a function of latitude.

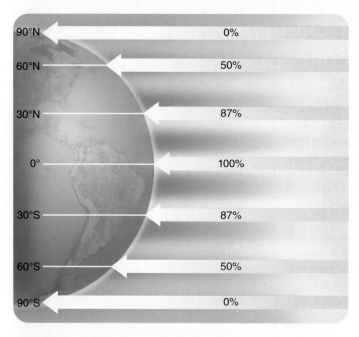

FIGURE 3.18 The percentage of incoming solar radiation (insolation) striking at various latitudes during an equinox date according to Lambert's law.

How much less solar energy does 60° latitude receive compared to the equator's receipt at this time?

We experience days that are longer in the summer and shorter in the winter. Of course, every day is 24 hour long, but what we are referring to is the length of time between sunrise and sunset. Higher latitudes experience greater seasonal differences in the length of time between sunrise and sunset. Obviously, if more hours of daylight are experienced, the amount of solar radiation received at a location should be greater. In the next section, we will discuss why periods of daylight vary in length over the seasons and from place to place.

The Seasons

Many people erroneously believe that the seasons are caused by changes in the distance between Earth and the sun during the year. As noted earlier, in terms of percentage, this change in distance is very small. Further, for people in the Northern Hemisphere, Earth is actually closest to the sun in January and farthest away in July. This is exactly opposite of that hemisphere's seasonal temperatures. The seasons, however, are caused by the $23\frac{1}{2}°$ tilt of Earth's axis away from being perpendicular to the plane of the ecliptic, and positions of the axis remain parallel and pointing to Polaris as Earth orbits the sun.

About June 21, Earth is in an orbital position where the north polar axis is inclined toward the sun at an angle of $23\frac{1}{2}°$. On this date, at noon, at $23\frac{1}{2}°$N latitude, the sun's rays will be directly overhead,

TABLE 3.2
Duration of Daylight for Certain Latitudes

Length of Day (Northern Hemisphere) (read down)			
LATITUDE (in degrees)	MAR. 20/SEPT. 22	JUNE 21	DEC. 21
0°	12 hr	12 hr	12 hr
10°	12 hr	12 hr 35 min	11 hr 25 min
20°	12 hr	13 hr 12 min	10 hr 48 min
23.5°	12 hr	13 hr 35 min	10 hr 41 min
30°	12 hr	13 hr 56 min	10 hr 4 min
40°	12 hr	14 hr 52 min	9 hr 8 min
50°	12 hr	16 hr 18 min	7 hr 42 min
60°	12 hr	18 hr 27 min	5 hr 33 min
66.5°	12 hr	24 hr	0 hr
70°	12 hr	24 hr	0 hr
80°	12 hr	24 hr	0 hr
90°	12 hr	24 hr	0 hr
LATITUDE	MAR. 20/SEPT. 22	DEC. 21	JUNE 21
Length of Day (Southern Hemisphere) (read up)			

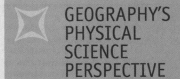
Using the Sun's Rays to Measure the Spherical Earth—2200 Years Ago

About 240 B.C. in Egypt, Eratosthenes, a Greek philosopher and geographer, who already knew that Earth was spherical, used geometry and solar observations to make a remarkably accurate estimate of the planet's circumference. As a librarian in Alexandria, Egypt, he read about a water well in Syene (today Aswan), located to the south about 800 kilometers (500 mi) on the Nile River. On June 21 (summer solstice), this account stated that the sun's rays shined to the bottom of the well, illuminating the water. The well was vertical, so this meant that the sun was directly overhead on that day. Syene was located very near the Tropic of Cancer, the latitude of the subsolar point on that date.

Eratosthenes made many observations of the sun's angle over the year, so he knew that the sun's rays were never vertical in Alexandria, and at noon on that June day a vertical column near the library formed a shadow. Measuring the angle between the column and a line from the column top to the shadow's edge, he found that the sun angle was 7.2° away from vertical.

Assuming that the sun's rays strike Earth's spherical surface in a parallel fashion, Eratosthenes knew that Alexandria was located 7.2° north of Syene. Dividing the number of degrees in a circle (360°) by 7.2°, he calculated that the two cities were separated by 1/50 of Earth's circumference. The distance between Syene and Alexandria was 5000 stades (a stade was the distance around the running track at a stadium of that time). Therefore, 5000 stades times 50 meant that Earth must be 250,000 stades in circumference.

Unfortunately, stade lengths used in various regions were different, so it is not certain what distance Eratosthenes used. A commonly cited stade length is about 0.157 kilometers (515 ft). By using this measure, the estimate of Earth's circumference would be 39,250 kilometers (24,388 mi). This distance is very close to the actual circumference of the great circle that would connect Alexandria and Syene.

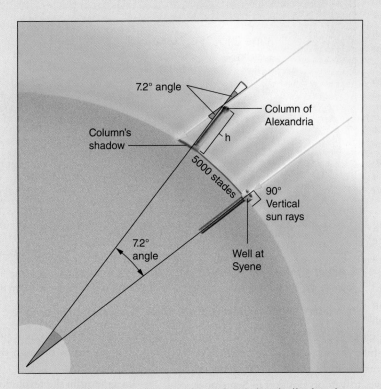

By observing the noon sun angle cast by a column where he lived, and knowing that no shadow was cast that same day in Syene to the south, Eratosthenes used geometry to estimate the Earth's circumference. On a spherical Earth, a 7.2° difference in angle also meant that Syene was 7.2° south in latitude from Alexandria or 1/50 of Earth's circumference.

In Google Earth, fly to 31.213°N, 29.944°E (Alexandria) and 24.088°N, 32.899°E (Syene, near Aswan today) and strike the surface at 90°. In the Northern Hemisphere, this day during Earth's orbit is called the summer **solstice**. In ● Figure 3.19, position A, note that the Northern Hemisphere as well as the Southern Hemisphere receive unequal amounts of light from the sun, because as Earth rotates under these conditions, a larger portion of the Northern Hemisphere remains in daylight. Conversely, a larger area of the Southern Hemisphere remains in darkness. The time of year and the latitude of a location each affect the length of daylight, as shown in Table 3.2. Thus, Repulse Bay, Canada, north of the Arctic Circle, experiences a full 24 hours of daylight at the June solstice. On the same day, New York City will experience a longer period of daylight than darkness. However, Buenos Aires, Argentina, will have a longer period of darkness than daylight on June 21. This day is called the *winter solstice* in the Southern Hemisphere. Therefore, June 21 is the longest day of the year in the Northern Hemisphere (with the highest yearly sun angle), and in the Southern Hemisphere it is the shortest day, with the lowest noon sun angle of the year.

Imagine Earth moving from its position at the June solstice toward a position a quarter of a year later, in September. As Earth moves toward that new position, we can imagine the changes that will

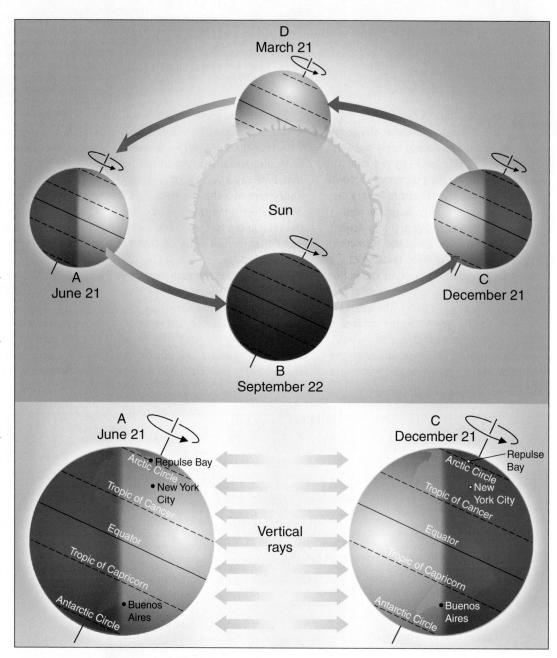

● **FIGURE 3.19** The geometric relationships between Earth and the sun during the June and December solstices. Note the opposite day lengths during the summer and winter solstices between the Northern Hemisphere and the Southern Hemisphere.

In relation to the sun, how is the Northern Hemisphere tilted on June 21?

be taking place in our three cities. In Repulse Bay, the time of darkness will increase through July, August, and September. In New York, sunset will be arriving earlier. In Buenos Aires, the situation will be reversed; as Earth moves toward its position in September, the daylight hours in the Southern Hemisphere will begin to get longer and the nights will become shorter.

On or about September 22, Earth will reach a position known as an **equinox** (from Latin: *aequus*, equal; *nox*, night). On this date (the **autumnal equinox** in the Northern Hemisphere), day and night will be of equal length at all locations on Earth. Thus, on the equinox, daylight hours are identical for both hemispheres and at all latitudes. As you can see in ● Figure 3.20, position B, Earth's axis points neither toward nor away from

the sun (imagine the axis is pointed at the reader); the circle of illumination passes through the poles, and it cuts Earth in half along its axis.

Imagine again the revolution and rotation of Earth while moving from about September 22 toward a new position another quarter of a year later in December. We can see that in Repulse Bay, Canada the nights will be getting longer until, on the winter solstice, which occurs on or about December 21, this northern town will experience 24 hours of darkness (Fig. 3.19, position C). The only natural light in Repulse Bay will be a faint glow at noon refracted from the sun below the horizon. Also, in New York, the days will get shorter and the sun will set earlier. Again, we can see that in Buenos Aires the situation is reversed. Around December 21, that city will experience its summer solstice, and the daylight hours will be much like they were in New York City in June.

Moving from late December through another quarter of a year to late March, Repulse Bay will increase its hours of daylight, as will New York, while in Buenos Aires the nights will be getting longer. Then, on or about March 21, Earth will again be in an equinox position (the **vernal equinox** in the Northern Hemisphere) similar to the one in September (Fig. 3.20, position D). Again, days and nights will be equal all over Earth, with 12 hours of daylight and 12 hours of darkness.

Finally, moving through another quarter of the year toward the June solstice, where we began, Repulse Bay and New York City are both experiencing more hours of daylight than darkness. The sun is setting earlier in Buenos Aires until on or about June 21 (its winter solstice), when Repulse Bay and New York City will have their longest day of the year and Buenos Aires will have its shortest. Furthermore, we can see that on or around June 21 a point on the Antarctic Circle in the Southern Hemisphere will experience a winter solstice similar to that which Repulse Bay had around December 21 (see Fig. 3.19, position A). During this June winter solstice day there will be no daylight in 24 hours at the Antarctic Circle, except what appears at noon as a glow of twilight on the horizon seen in the image at the beginning of this chapter.

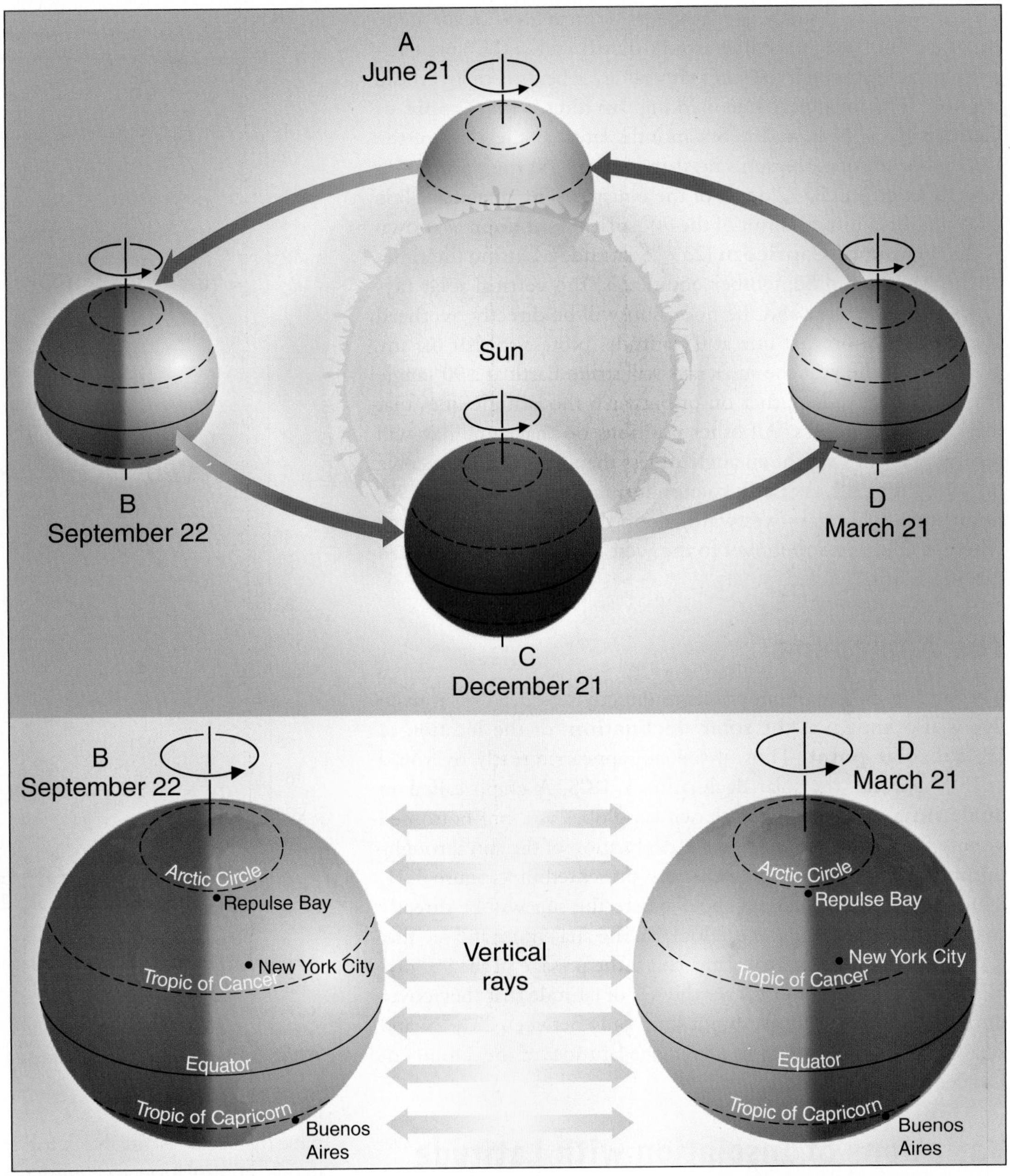

● **FIGURE 3.20** The geometric relationships between Earth and the sun during the March and September equinoxes. Periods of daylight and darkness are 12 hours everywhere because the circle of illumination passes through both poles.

If Earth were not inclined on its axis, would there still be latitudinal temperature variations? Would there be seasons?

Latitude Lines Delimiting Solar Energy

The diagrams of Earth in its various positions as it revolves around the sun illustrate the importance of the angle of inclination. At noon on June 21, solar radiation is at 90° directly on 23½°N latitude. On this date, the sun's rays can reach 23½° beyond the North Pole. The **Arctic Circle**, an imaginary line drawn around Earth 23½° south of the North Pole (66½° north of the equator),

marks this limit, and the entire region north of the Arctic Circle can be bathed in sunlight. You can see from the diagram that all points on or north of the Arctic Circle will experience no darkness on the June solstice and that all points south of the Arctic Circle will experience some darkness on that day. The **Antarctic Circle** in the Southern Hemisphere (23½° north of the South Pole, or 66½° south of the equator) marks a similar limit.

The diagrams also show that the sun's **direct (vertical) rays**, those that strike Earth's surface at right angles, also shift latitudinal position in relation to the poles and the equator over the seasons. At the time of the June solstice, the sun's rays are vertical—that is, directly overhead—at noon at 23½° *north* of the equator. This imaginary line around Earth marks the northernmost latitude at which the sun's rays can ever be directly overhead at noon. The imaginary line marking this limit is the **Tropic of Cancer** (23½°N latitude). Six months later, at the time of the December solstice, the sun's rays are vertical and the noon sun is directly overhead 23½° *south* of the equator. The imaginary line marking this southern limit of the 90° solar rays at noon is known as the **Tropic of Capricorn** (23½°S latitude). During the times of the March and September equinoxes, the vertical solar rays strike at the equator and the noon sun will be directly overhead at all locations on that line at 0° latitude. Note also that on any given day of the year the sun's rays will strike Earth at a 90° angle at only one latitude, either on or between the latitude lines that mark the two tropics. All other positions on that same day will receive the sun's rays at an angle of less than 90° (or may receive no sunlight at all). As Earth rotates during a solstice day, the 90° noon sun angle tracks westward, along the tropical latitude line where it falls, as noon moves to the west because of the eastward turning Earth.

The Analemma

The latitude where the noon sun is directly overhead on a given day is also known as the **solar declination**, or the location of the **subsolar point**. Thus, if the sun appears directly overhead at 18°S latitude, the solar declination is 18°S. A graph called an **analemma**, which is often drawn on globes as a big-bottomed "figure 8," can be used to find the declination of the sun throughout the year. A modified analemma is presented in ● Figure 3.21. Thus, if you would like to know where the sun will be directly overhead on April 25, you can look on the analemma and see that it will be at 13°N. The analemma charts the passage of the 90° rays of the sun north and south over the 47° of latitude that they cover during a year. The subsolar point shifts only between 23½°N and 23½°N during the course of a year, the latitudes of the Tropics of Cancer and Capricorn.

Variations of Insolation with Latitude

Neglecting for the moment the atmosphere's influence on variations in insolation during a 24-hour period, the receipt of solar energy begins after sunrise and increases as Earth rotates toward the time of solar noon. The time of the day when the sun reaches its **zenith**, the highest angle in the sky for that day, is called *solar noon*. Solar noon is when a location will receive its greatest insolation as the sun reaches its highest position, and angle, above the horizon for that day. Insolation then decreases as the sun angle lowers in the afternoon toward sunset and the darkness after sunset. Remember that this does not apply when a polar location is receiving 24 hours of light or 24 hours of darkness. Because solar radiation is not received during hours of darkness, nighttime tends to be cooler than it is during daylight hours.

The intensity of insolation received at any location on Earth varies with latitude, however, three distinct geographic patterns exist based on the seasonal distribution of solar energy. These

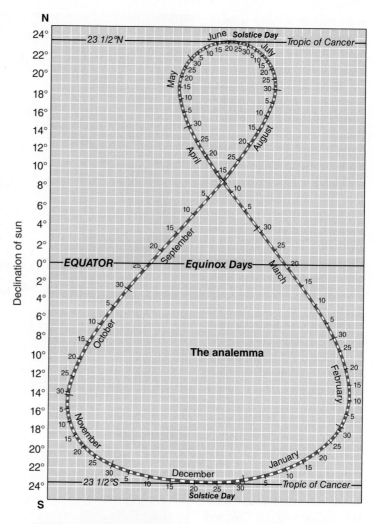

● **FIGURE 3.21** An analemma is a graph used to find the solar declination (latitudinal position) of the vertical (90°) sun's rays at noon (the subsolar point) for every day of the year.

What is the declination of the sun on your birthday?

patterns serve as the basis for recognizing six latitudinal zones, bands, of insolation and temperature that circle Earth (● Fig. 3.22).

Looking first at the Northern Hemisphere, the Tropic of Cancer and the Arctic Circle are the dividing lines for three of these distinctive zones. The area between the equator and the Tropic of Cancer is called the north **tropical zone**. Here, insolation is always high but is greatest at the time of year when the sun is directly overhead at noon. This occurs twice a year, and these dates vary according to latitude (see Fig. 3.21). The north **midlatitude zone** is the wide band between the Tropic of Cancer and the Arctic Circle. In this belt, insolation is greatest on the June solstice when the sun reaches its highest noon angle and the period of daylight is longest. Insolation is least at the December solstice when the sun is lowest in the sky and the period of daylight is the shortest. The north **polar zone**, or Arctic zone, extends from the Arctic Circle to the pole. In this region, as in the north midlatitude zone, insolation is greatest at the June solstice, but it ceases during the period when the sun is blocked entirely because of the tilt of Earth on its

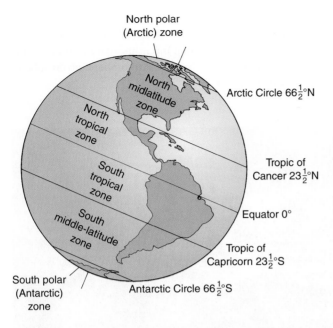

North polar (Arctic) zone

North midlatitude zone

North tropical zone

South tropical zone

South middle-latitude zone

South polar (Antarctic) zone

Arctic Circle $66\frac{1}{2}$°N

Tropic of Cancer $23\frac{1}{2}$°N

Tropic of Capricorn $23\frac{1}{2}$°S

Equator 0°

Antarctic Circle $66\frac{1}{2}$°S

● **FIGURE 3.22** The equator, the Tropic of Cancer, the Tropic of Capricorn, the Arctic Circle, and the Antarctic Circle define six latitudinal zones that have distinctive characteristics of insolation receipt.

Which zone(s) would have the least annual variation in insolation? Why?

axis. This period lasts for 6 months at the North Pole but is as short as 1 day directly on the Arctic Circle. The equinoxes are the days when sunrise and sunset occur the North and South Poles.

Similarly, there is a south tropical zone, a south midlatitude zone, and a south polar zone, or Antarctic zone, all separated by the Tropic of Capricorn and the Antarctic Circle in the Southern Hemisphere. These areas receive their greatest amounts of insolation at opposite times of the year from the northern zones.

Despite various patterns in insolation amounts received in these zones, we can make some generalizations. For example, annual insolation tends to decrease from lower latitudes to higher latitudes (Lambert's law together with Beer's law). The closer a place is to the poles, the greater will be its seasonal variation caused by fluctuations in insolation. Furthermore, total annual insolation at the top of the atmosphere over a particular latitude remains nearly constant from year to year (the solar constant). Factors such as elevation and humidity also affect the amount of solar radiation received at a location, even on cloudless days. Although latitude remains a very important factor, global patterns of annual insolation receipt do not follow latitudinal zones exactly. Worldwide yearly variation in insolation is shown in ● Figure 3.23. Here you can see the sunniest and least sunny places in the world in terms of average annual receipt of solar energy at ground level, and you can see that factors other than latitude also affect insolation. This map was derived by the United

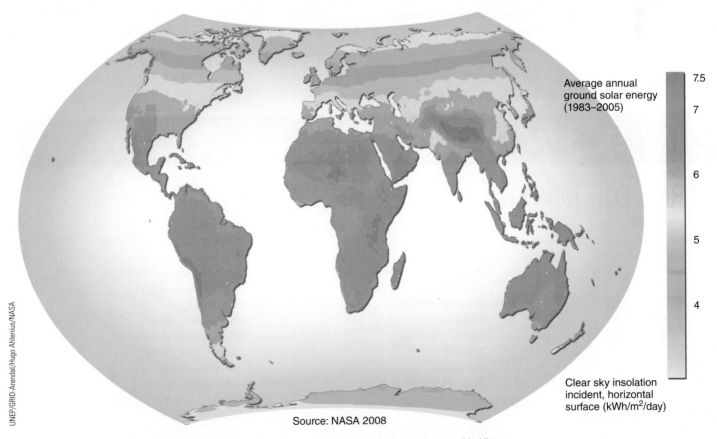

Average annual ground solar energy (1983–2005)

7.5

7

6

5

4

Clear sky insolation incident, horizontal surface (kWh/m²/day)

Source: NASA 2008

UNEP/GRID-Arendal/Hugo Ahlenius/NASA

● **FIGURE 3.23** The world's geographic distribution of average annual solar energy receipt on the ground in kilowatt hours per square meter per day during clear days.

Which world areas seem to have the most complex patterns, and which have the most regular patterns? Why?

Nations, using NASA satellite data, in an effort to understand the geographic distribution of the potential for harnessing solar power as an energy resource around the world.

The amount of insolation received by the Earth system is an important factor in understanding atmospheric and oceanic dynamics and the geographic distributions of climate, soils, and vegetation. Climatic elements—such as temperature, precipitation, and wind—are controlled in part by the amount of insolation received at various locations on Earth. People depend on certain levels of solar radiation for physical comfort, and plant life is especially sensitive to the availability and seasonality of sunlight. You may have noticed that plants can wilt if they receive too much sun or they might become brown if they do not receive adequate sunlight. Over a longer period, certain plants have an annual cycle of budding, flowering, leafing, and losing their leaves. The increase and decrease in solar radiation that accompanies the changing seasons apparently determine this cycle. Animals also respond to seasonal changes. Some animals hibernate during cold seasons; many North American birds fly south toward warmer weather as winter approaches; and many animals breed at this time so that their offspring will be born in the spring, when warm weather is approaching.

Ancient civilizations in Egypt, China, Mexico, and in many other regions around the world, realized the importance of the sun's energy to life. Many societies worshiped the sun as chief among their gods (● Fig. 3.24). This should not be too surprising—they were well aware that the sun was vital to their survival. Today, most people do not worship the sun, but it is extremely important to understand and appreciate its role as the ultimate energy source for the vast majority of Earth's natural subsystems. Solar radiation is the only external input necessary to support our otherwise self-contained Earth System: Atmosphere, Hydrosphere, lithosphere, and Biosphere.

● **FIGURE 3.24** El Castillo, a Mayan pyramid at Chichén Itzá, Mexico, was oriented to the annual changes in sun angle because the Maya worshiped the sun. Each side has 91 steps—the number of days between the solstice and equinox days. On the vernal equinox, the afternoon sun casts a shadow that makes it appear like a giant snake is crawling down the pyramid's north side.

Why would ancient civilizations worship the sun?

CHAPTER 3 ACTIVITIES

■ TERMS FOR REVIEW

galaxy
light-year
solar system
gravity
mass
planet
terrestrial planet
giant planet (gas planet)
asteroid
comet
meteor
meteorite
fusion (thermonuclear) reaction
solar wind
aurora
aurora borealis
aurora australis
sunspot

solar constant
calorie
electromagnetic energy
electromagnetic spectrum
visible spectrum
shortwave radiation
longwave radiation
galactic movement
rotation
revolution
circle of illumination
perihelion
aphelion
plane of the ecliptic
angle of inclination
parallelism of the axis
North Star (Polaris)

insolation
solstice (summer and winter)
equinox
autumnal equinox
vernal equinox
Arctic Circle
Antarctic Circle
direct (vertical) rays
Tropic of Cancer
Tropic of Capricorn
solar declination
subsolar point
analemma
zenith
tropical zone
midlatitude zone
polar zone

QUESTIONS FOR REVIEW

1. How is a solar system defined? What bodies constitute our solar system?
2. How is the energy emitted from the sun produced?
3. Name the terrestrial planets. What do they have in common? Name the giant planets. What do they have in common?
4. Which planets are capable of maintaining a gaseous atmosphere?
5. The electromagnetic spectrum displays various types of energy by wavelengths. Where is the general division between long-wave and shortwave energy? In what ways do humans use electromagnetic energy?
6. Is the amount of solar energy reaching Earth's outer atmosphere constant? What might make it change?
7. Describe briefly how Earth's rotation and revolution affect insolation and influence life on Earth.
8. If the sun is closest to Earth on January 3, why isn't the winter in the Northern Hemisphere warmer than winter in the Southern Hemisphere?
9. What are the two major factors that cause regular variations in insolation throughout the year? How do they combine to cause the seasons?
10. Examine the world map of average annual insolation in Figure 3.23. What are some reasons why patterns of insolation receipt do not always correspond with belts of latitude?

CONSIDER AND RESPOND

1. Given what you know about the sun's relation to life on Earth, explain why the solstices and equinoxes have been so important to cultures all over the world.
2. Use the discussion of solar angle, including Figure 3.17, to explain why we can look directly at the sun at sunrise and sunset but not at the noon hour.
3. Describe in your own words the relationship between insolation and latitude.
4. Use the analemma presented in Figure 3.21 to determine the latitude where the noon sun will be directly overhead on February 12, July 30, November 2, and December 30. At what latitude is the sun directly overhead on your birthday?

PRACTICAL APPLICATIONS

1. Imagine you are at the equator on March 22. The noon sun would be directly overhead. However, for every degree of latitude that you travel to the north or south, the noon solar angle would decrease by the same amount. For example, if you travel to 40°N latitude, the solar angle would be 50°. Explain this relationship.

   Develop a formula or set of instructions to generalize this relationship.

   What would be the solar angle at 40°N on June 21? On December 21?

   Lambert's law can be expressed as: $I = I_0 \cos g$

   Where I = −Intensity of solar radiation received at the surface.

   I_0 = −Intensity of solar radiation received from a 90° angle.

   $\cos$ = Cosine of g

   g = the sun's zenith angle

   The zenith angle is measured from 90° straight above, down to the sun's position in the sky.

2. Using Lambert's law, calculate the intensity of the insolation on September 22 at the following locations:
   a. 0° latitude
   b. 27°S latitude
   c. 40°N latitude
   d. 65°S latitude
   e. 83°N latitude

 MindTap—Make the most of your study time by accessing everything you need to succeed in one place. Read your textbook, take notes, review flashcards, watch videos, complete activities, take practice quizzes, and more online with MindTap. Log in at **www.cengagebrain.com**.

THE ATMOSPHERE, TEMPERATURE, AND EARTH'S ENERGY BUDGET

■ OBJECTIVES

WHEN YOU COMPLETE THIS CHAPTER YOU SHOULD BE ABLE TO:

■ 4.1 List several reasons why our planet's atmosphere is essential to life on Earth.

■ 4.2 Outline the major gases found in the atmosphere.

■ 4.3 Conceptualize the ways that solar energy interacts with Earth's atmosphere and surface.

■ 4.4 Explain how the solar energy that reaches Earth's surface is transferred to the atmosphere.

■ 4.5 Discuss the processes and the important roles that water plays in heat transfer.

■ 4.6 Outline the characteristics of the temperature-based layers of the atmosphere.

■ 4.7 Describe the controls on horizontal distribution of Earth's surface temperatures.

■ 4.8 Explain the major inputs, outputs, and processes in Earth's energy budget and why it is in a state of energy equilibrium.

THE ATMOSPHERE SURROUNDS OUR PLANET like an ocean of air. Held in place by our Earth's gravitational force, its unique composition of gases maintains life on our planet. As a dynamic system, its interactions with the biosphere, hydrosphere, and lithosphere are essential to the life-giving environmental conditions existing on Earth. Atmospheric processes drive characteristics of weather and climate through changes in winds, variations in air pressure, and the movement of moisture from the oceans to the land. The atmosphere profoundly affects the solar energy that heats Earth's surface and it blocks harmful radiation from space. Absorbing and storing energy, along with redistribution and transfer of heat, are other important atmospheric interactions with the environment. Our atmosphere's composition, structure, dynamic air circulation, and winds play critical roles in the operation of the Earth system and in producing environmental conditions that support life on our planet.

The atmosphere provides oxygen and water that are vital for the survival of animals and humans as well as the water and carbon dioxide that plants require. Compared to our planet's size, the atmosphere is a rather thin film of air (● Fig. 4.1), yet it serves as an effective insulator, maintaining the viable temperatures we experience on Earth. Most living things cannot survive extreme temperatures,

Watching a brilliant sunset can be a reminder of the atmosphere's importance to life on Earth, as well as the beauty of the sky.
NOAA

● **FIGURE 4.1** The atmosphere is a relatively thin envelope of gases that surrounds our planet as seen in this image. The thickness of the envelope can be compared to the thickness of an apple's skin as compared to the fruit.

What does this comparison say about the potentially fragile nature of the atmosphere?

nor can they live long if exposed to large doses of harmful radiation. The atmosphere, the envelope of air surrounding our planet, supplies most of the oxygen and carbon dioxide for life and helps to maintain viable environmental conditions on Earth.

Life on Earth exists within rather narrow ranges of temperature. Without the atmosphere, our planet would experience temperature extremes of as much as 260°C (500°F) between day and night. Serving as a filter or shield, the atmosphere blocks much of the sun's gamma rays, X-rays, and ultraviolet (UV) radiation, and protects us from many meteor impacts.

Other aspects of the atmosphere's importance can be illustrated by comparing conditions on Earth with conditions on the moon, which is virtually without an atmosphere. Without a space suit that supplies air, a person on the moon would immediately die due to lack of oxygen. An astronaut on the moon would also notice an "unearthly" silence. On Earth we are able to hear noises and speech because sound waves move by vibrating gas molecules in the air. Without an atmosphere or air molecules to transmit sound waves, a lunar visitor would not hear any sounds or be able to speak with an audible voice; only radio communications are possible. Also, because there is no atmosphere, it is not possible to fly airplanes or helicopters, and using a parachute would result in a fatal plummet to the surface. The moon's lack of

atmosphere also means that it receives no protection from being bombarded by meteors, which is why its surface has so many craters. On Earth, however, most meteors burn up before reaching the surface because of friction with the atmosphere and oxygen, which encourages burning. Our lunar astronauts recorded temperatures of up to 204°C (400°F) on the hot, sunlit side of the moon, and on the dark side, they recorded temperatures approaching −121°C (−250°F). These temperature extremes would kill an unprotected human. Without a protective atmosphere, ultraviolet rays from the sun would also burn a person on the moon. In comparison, we are protected to a large degree from UV radiation because the ozone layer in the upper atmosphere absorbs a major portion of this harmful solar radiation. Gas molecules and particles in the air interact with sunlight in a process called **scattering**. (See the box titled "White Clouds, Blue Skies, Red Sunsets, and Rainbows" in this chapter.) Scattering is the diffusion of various wavelengths of light, which gives us blue skies and the beautiful red and orange sunrises and sunsets. Without scattering, which diffuses light in many directions, the sky would appear black, as it does from the moon, and as it does in space (● Fig. 4.2).

We can see that, compared with our stark, silent, and lifeless moon, Earth harbors a hospitable environment for life almost solely because of the nature of our atmosphere. For example, many plants reproduce by pollen and spores that

● **FIGURE 4.2** The moon's lack of an atmosphere would be deadly to an unprotected astronaut.

Why are shadows and the sky so dark on the moon?

are carried by winds. Birds can fly only because of the air. Atmospheric processes maintain the cycling of water from the oceans to the land and back. The atmosphere is also an important medium for moving thermal energy from one place to another.

The atmosphere's interactions with other Earth systems also support a tendency toward equilibrium conditions. Weather changes result from atmospheric processes that tend to equalize air temperature and pressure differences by transferring energy and moisture and through interactions with oceanic circulation.

The Nature of Our Atmosphere

The atmosphere has a profound effect on the amount of solar energy received to heat Earth and power its environmental systems. Knowing a bit about the atmosphere's composition, structure, and its dynamic processes helps us understand the interactive relationships between solar radiation, the air, and Earth's surface.

The atmosphere's thickness can be defined in a variety of ways, and because gas molecules continue to become thinner with altitude, it does not end in a sharp definable boundary. The atmosphere

is often defined as extending to approximately 480 kilometers (300 mi) above Earth's surface. Certain interactions with energy occur within this altitude, such as the auroras, and Earth's magnetic field continues beyond.

Gas molecule density decreases rapidly with altitude; some 99% of the air is concentrated in about the lowest 30 kilometers (18.6 mi). It may be surprising to think of the atmosphere as a rather thin film of air surrounding Earth. An altitude of 100 kilometers (62 mi), called the **Karman line**, is also not a sharp boundary, but is an internationally accepted definition for the beginning of space—the realm of astronauts. This definition would make the atmosphere reach to a height of only 1.6% of Earth's radius; 99.999% the atmosphere's mass lies below this level.

Because air has mass, the atmosphere exerts pressure on Earth's surface. At sea level, this pressure is about 1034 grams per square centimeter (14.7 lb/sq in.), but as the altitude *increases*, atmospheric pressure progressively decreases. In Chapter 5, we will examine relationships between atmospheric pressure and altitude in more detail.

Atmospheric Composition

The atmosphere is composed of a variety of gases (Table 4.1) and because of mixing by winds, most of these gases remain in the same proportions regardless of atmospheric density. About 78% of the atmosphere's volume is nitrogen and nearly 21% consists of oxygen. Argon comprises most of the remaining 1%. The percentage of carbon dioxide in the atmosphere has risen over time to the highest level in known history, but it is approximately 0.04% by volume. There also are traces of other gases: ozone, hydrogen, neon, xenon, helium, methane, nitrous oxide, krypton, and others.

Nitrogen, Oxygen, Argon, and Carbon Dioxide Of the four most abundant atmospheric gases, nitrogen (N_2) makes up the largest proportion of air, and nitrogen is very important for plant growth. In addition, several other atmospheric gases are vital to the development and maintenance of life. One of the most critical of these atmospheric gases is, of course, oxygen (O_2), which people and other animals use to breathe and to oxidize (burn) the food they consume. Oxidation, which is the chemical combination of oxygen with other substances to create new products, also occurs in many other situations. Rapid oxidation takes place when we burn fossil fuels, wood, or refuse, which releases large amounts of heat. The decay of certain minerals or organic debris and the development of rust are examples of slow oxidation. These processes depend on the presence of oxygen in the atmosphere. The third most abundant gas in our atmosphere is argon (Ar). It is not a chemically active gas and therefore neither helps nor hinders life on Earth. Argon is an inert, heavy, and stable product from the radioactive decay of minerals.

TABLE 4.1
Composition of the Atmosphere Near the Earth's Surface

Permanent Gases			Variable Gases			
GAS	SYMBOL	PERCENT (BY VOLUME) DRY AIR	GAS (AND PARTICLES)	SYMBOL	PERCENT (BY VOLUME)	PARTS PER MILLION (PPM)*
Nitrogen	N_2	78.08	Water vapor	H_2O	0 to 4	
Oxygen	O_2	20.95	Carbon dioxide	CO_2	0.400	400*
Argon	Ar	0.93	Methane	CH_4	0.00017	1.7
Neon	Ne	0.0018	Nitrous oxide	N_2O	0.00003	0.3
Helium	He	0.0005	Ozone	O_3	0.000004	0.04†
Hydrogen	H	0.00006	Particles (dust, soot, etc.)		0.000001	0.01–0.15
Xenon	Xe	0.000009	Chlorofluorocarbons (CFCs)		0.00000002	0.0002

*For CO_2, 400 parts per million means that out of every million air molecules, 400 are CO_2 molecules.

†Stratospheric values at altitudes between 11 and 50 kilometers are about 5 to 12 ppm.

Carbon dioxide is the fourth most abundant atmospheric gas. The involvement of carbon dioxide in the system known as the *carbon cycle* has been studied for generations. Plants, through **photosynthesis**, use sunlight (mainly ultraviolet radiation) as the driving force to combine carbon dioxide and water to produce carbohydrates (sugars and starches), in which energy, derived originally from the sun, is stored and used by vegetation (● Fig. 4.3). Oxygen is given off as a by-product. Animals then use the oxygen to oxidize the carbohydrates, releasing the stored energy. A by-product of this process in animals is the release of carbon dioxide, which completes the cycle when it is in turn used by plants in photosynthesis.

Water Vapor and Aerosols Water in the gaseous state is called **water vapor** and is always mixed in some proportion with the air of the lower atmosphere. Water vapor is the most variable of the atmospheric gases and can range from 0.02% by volume in cold, dry climates to more than 4% in the humid tropics. Water vapor in the air will be discussed in some detail in Chapter 6 under the broad topic of humidity, but it is important to note that variations in the percentage of water vapor over time and place are important to understanding weather and climate.

Water vapor absorbs and stores heat in the atmosphere, preventing its rapid escape from Earth. Thus, like carbon dioxide, water vapor plays a large role in the insulating properties of the atmosphere. Water occurs naturally in all three states in the atmosphere—as a solid, a liquid, and a gas. Liquid water occurs as rain and as fine droplets in clouds, mist, and fog. Solid water (ice) is found in the atmosphere as ice crystals, snow, sleet, and hail.

Aerosols are tiny solids or liquids suspended in the atmosphere—including dust and pollutants, and also including tiny liquid droplets and/or ice crystals composed of chemicals other than water. For example, sulfur dioxide crystals (SO_2) are atmospheric aerosols. **Particulates** (a variety of aerosols) is a term

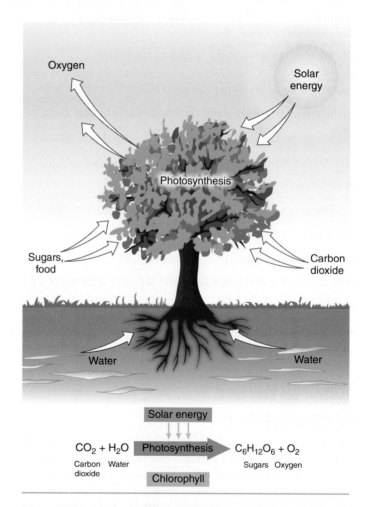

● FIGURE 4.3 The equation for photosynthesis shows how plants use solar energy to manufacture food from atmospheric carbon dioxide and water, liberating oxygen in the process. The stored food is consumed by animals that also need the oxygen released by photosynthesis for respiration.

$$CO_2 + H_2O \xrightarrow{\text{Photosynthesis}} C_6H_{12}O_6 + O_2$$

Solar energy

Carbon dioxide Water

Chlorophyll

Sugars Oxygen

sometimes used for solid particles in the air. Aerosols, including particulates, which vary greatly in composition, can emanate from both natural and human-generated sources. Aerosols from natural sources have always existed in our atmosphere (● Fig. 4.4). Dust, smoke, pollen, spores, volcanic emissions, bacteria, and salts from ocean spray are aerosols that play important roles in the absorption or blocking of energy and in the formation of rain and snow. They have a wide range of impact on our weather and climate, and can be transported thousands of miles by winds. Atmospheric **pollutants** are certain aerosols such as smog, which can be a health hazard, depending on their concentration in the air.

Atmospheric Environmental Issues

Two gases in our atmosphere are particularly relevant to environmental issues. One is carbon dioxide, which is related to increases in global temperatures. The other is ozone, which forms a layer in the upper atmosphere that protects Earth from excessive UV radiation but is endangered by other gases associated with industrialization. In recent years, scientists have directly linked carbon dioxide with long-term changes in Earth's atmospheric temperatures and climates. The major concern is with the relationship between atmospheric carbon dioxide levels and rising global temperatures.

● **FIGURE 4.4** Volcanic eruptions, like this dramatic one from the Calbuco volcano in Chile in April 2015, add a variety of gases, particulates, and water vapor into our atmosphere.

What other processes add particles to the atmosphere?

The Greenhouse Effect We are all familiar with what happens to the inside of a parked car left in the sun with the windows closed. Shortwave radiation (mainly visible light) from the sun penetrates the glass windows with ease. When insolation strikes the interior of the car, it is absorbed, heating exposed surfaces. Energy, emitted from these surfaces as longwave thermal radiation (heat), cannot escape through the glass as freely (● Fig. 4.5). The result is that the vehicle's interior can become hot in a short amount of time. In a university experiment, the temperature in a closed car rose almost 20°F in only 10 minutes, by almost 30°F degrees in 20 minutes, and almost 45°F in an hour. Many drivers recognize this as a blast of hot air as the car door opens and seats that are very hot. Temperatures can become hot enough in vehicles with closed windows (or even if they are only slightly open) that they can pose a deadly threat to small children and pets left behind.

Termed the **greenhouse effect**, this process is the primary reason for the moderate temperatures observed on Earth. A greenhouse (glass structure that houses plants) will behave in a somewhat similar manner to a closed vehicle parked in the sun. Insolation (shortwave radiation) goes through the transparent glass roof and walls of the greenhouse and helps the plants inside to thrive, even in a cold outdoor environment. After being absorbed by the materials in the greenhouse, the shortwave energy is re-radiated as longwave thermal energy, which cannot escape rapidly, thus warming the interior of the greenhouse.

Like the glass of a greenhouse, atmospheric carbon dioxide, water vapor, and other greenhouse gases are largely transparent to incoming solar radiation, but they can impede the escape of thermal infrared radiation by absorbing it and then radiating it back to Earth. For example, carbon dioxide emits about half of its absorbed thermal energy back to Earth's surface. Although the results are similar, the processes involving the glass of a car or greenhouse and the atmosphere are significantly different. The heat of a closed car, or a greenhouse, increases because the inside air is trapped and cannot circulate to the outside environment. Our atmosphere is free to circulate heated air to higher altitudes and also to move heat laterally over the surface.

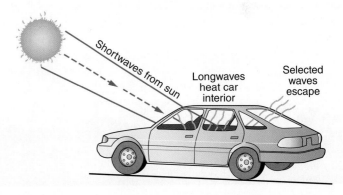

● **FIGURE 4.5** Sunlight (shortwave energy), streaming through car windows is absorbed into the interior and is re-radiated as longwave thermal infrared energy, which heats the car's interior. The closed windows do not allow the thermal radiation to escape and the trapped heat can make the interior become very hot. This effect is somewhat like the atmospheric greenhouse effect.

How might you prevent your car's interior from becoming very hot on a summer day?

The greenhouse effect is a natural process, essential to environmental conditions on Earth because without greenhouse gases in our atmosphere, our planet would be too cold to sustain human life (or perhaps any life). The greenhouse process helps maintain the warmth of our planet and is a factor in Earth's *heat energy budget* (discussed later in this chapter). However, serious environmental issues have arisen in recent years, as growing concentrations of greenhouse gases have been related to measurable increases in global temperatures.

Since the Industrial Revolution began in the late 1800s, burning of fossil (carbon) fuels has added increasing amounts of carbon dioxide to the atmosphere. During this time Earth has undergone widespread deforestation, as forests and other vegetation are removed for agriculture and urban development. These human activities have combined to increase carbon dioxide in the atmosphere. Vegetation uses large amounts of carbon dioxide in photosynthesis (see Fig. 4.3), and the extensive removal of vegetation permits more CO_2 to remain in the atmosphere. The rise in atmospheric carbon dioxide over time is a result of human contributions to this increase. Because carbon dioxide absorbs thermal infrared radiation emitted from Earth's surface and restricts its loss to space, increasing amounts of atmospheric CO_2 increase the greenhouse effect and produce a global rise in temperatures (● Fig. 4.6). Human influences on increasing carbon dioxide in the atmosphere have resulted in what has been referred to as the **enhanced greenhouse effect**.

The environmental impacts of global warming are many and varied. It is feared that continued temperature increases will significantly raise sea levels. Higher average temperatures will increase the melting of glaciers in Greenland and Antarctica and in mountain glaciers worldwide. Meltwaters will add to rising ocean levels, as less of Earth's water would be stored on land in glaciers, and the water in the oceans will increase. Rising sea levels are of great concern to nearly all coastal regions but especially to areas that lie at or below sea level, like the Netherlands and such cities as New Orleans, Houston, and New York. Increasing temperatures can cause vegetation regions to shift as plants begin to grow in new locations that previously lacked suitable growing conditions or they die out in other places that have become too warm or dry for growth. With temperature shifts, scientists also expect changes in bird and animal migration patterns and the movement of insect-borne diseases to new areas. Also, major global temperature changes will cause a parallel shift in regional and local weather systems. Temperature and rainfall patterns may change worldwide, but the impact will be to differing degrees and with varying consequences among the world's regions. Although certain places may benefit and other locations may experience detrimental change, the process of adapting to the impacts of significant climate change would be difficult worldwide. Currently, researchers are closely monitoring the trends in global and regional temperatures and are studying the impacts of global warming related to the *enhanced greenhouse effect*. The issue of climate change will be discussed in more detail in Chapter 8.

The Ozone Layer Another vital gas in Earth's atmosphere is **ozone**. The ozone molecule (O_3) is related to the oxygen molecule (O_2), except it is made up of three oxygen atoms, whereas oxygen gas consists of only two. Ozone is formed in the upper atmosphere where the sun's ultraviolet radiation splits an oxygen molecule into two oxygen atoms (O). Then, the free oxygen

● **FIGURE 4.6** The *greenhouse effect* (a-b): (a) Earth's surface absorbs solar radiation that passes through the atmosphere and is not reflected back to space. (b) Earth re-radiates this absorbed energy as thermal infrared radiation, warming the surface and the atmosphere, although much of that thermal energy is lost to space. Atmospheric heating results from heat absorption by water vapor and greenhouse gases, and some of that thermal energy is re-radiated from the atmosphere back to Earth. The *enhanced greenhouse effect* (c): (c) With a growing atmospheric proportion of heat-absorbing greenhouse gases, less thermal energy escapes to space. More heat is trapped at the surface, and more thermal energy is cycled back to Earth from the atmosphere. This process causes the global temperature to increase.

atoms join two oxygen gas molecules to form two molecules of ozone gas consisting of three oxygen atoms each:

$$(1) \ 2O_2 + 2O^- \rightarrow 2O_3 \ \text{(using UV radiation)}$$

In the lower atmosphere, ozone is formed by electrical discharges (such as high-tension power lines and lightning strokes), and engine exhausts interacting with incoming shortwave solar radiation. It is a toxic pollutant and a major component of urban smog, which can cause sore and watery eyes, soreness in the throat and sinuses, and difficulty breathing. Near the surface of Earth, ozone is a menace that can hurt life forms. Here, the ozone gas is often referred to as "bad ozone." However, in the upper atmosphere, ozone is essential to both terrestrial and marine life because it absorbs large amounts of the sun's ultraviolet radiation. The same ozone gas in the upper atmosphere is sometimes referred to as "good ozone."

In the upper atmosphere, UV radiation is consumed as it breaks the chemical bonds of ozone (O_3) to form an oxygen gas molecule (O_2) and an oxygen atom (O):

$$(2) \ 2O_3 \rightarrow 2O_2 + 2O \ \text{(using UV radiation)}$$

After this process, more UV radiation is consumed to recombine the oxygen gas and the oxygen atom back into ozone, as in Formula 1. This process is repeated over and over, involving large amounts of ultraviolet energy that would otherwise reach Earth's surface. The chemistry of Formulas 1 and 2, repeating back and forth, creates a very efficient UV filter.

Without the ozone layer of the upper atmosphere, excessive UV radiation reaching Earth would severely burn human skin, increase the incidence of skin cancer and optical cataracts, destroy certain microscopic forms of marine life, and damage plants. Ultraviolet radiation is also responsible for suntans and painful sunburns, depending on an individual's skin tolerance and the length of exposure to the sun.

For many years there has been concern that chlorofluorocarbons (CFCs) and nitrogen oxides (NO_x), added to the atmosphere by human activities, may permanently damage Earth's fragile **ozone layer**. Chlorofluorocarbons have been used extensively in refrigeration and air conditioning. As refrigeration became more common in around the world, CFCs entered the upper atmosphere, and removed ozone through chemical processes threatening the ozone layer, our natural UV filter. The reaction that destroys ozone works in the following way: chlorine atoms (Cl), found in CFCs, split off and enter the stratosphere. There, they bond to oxygen atoms (O) to form chlorine monoxide and oxygen gas.

$$(3) \ Cl + O_3 \rightarrow ClO + O_2$$

Oxygen atoms that have now bonded with the chlorine cannot be used to replenish the original amount of ozone (O_3), as in Formula 1. This and other chemical reactions attack the ozone layer and threaten our natural UV filter. Nitrogen oxide compounds (NO_x), emitted with exhaust from vehicles and jet engines, also have the ability to enter the stratosphere and destroy our ozone shield.

The UV radiation that the ozone layer allows to reach Earth does serve useful purposes. For instance, it has a vital function in the process of photosynthesis (see Fig. 4.3); it is important in the production of certain vitamins (especially vitamin D); it can help treat certain types of skin disorders; and it helps the growth of some beneficial viruses and bacteria. However, increasing amounts of UV radiation reaching the surface can become a serious problem for Earth's environments, and thus the ozone layer must be protected from the pollutants that threaten its existence.

The Ozone "Hole" Actually, no real hole exists in our protective ozone layer. The ozone "hole" refers to the roughly circular zones in the polar stratosphere where the ozone concentration is lower than expected. From the surface to the outer reaches of the atmosphere, there is always some ozone present (keep in mind that ozone is only a trace gas). If all the ozone in our atmosphere were forced down to sea level, the atmospheric pressure there would compress the ozone into a worldwide layer ranging from 3 to 4 millimeters, about the thickness of two pennies stacked together (<1/8 of an inch) thick. An ozone hole is really an area where the amount of ozone is considerably less than it should be. ● Figure 4.7 displays the ozone hole (the area of maximum ozone loss) centered over the South Pole and Antarctica.

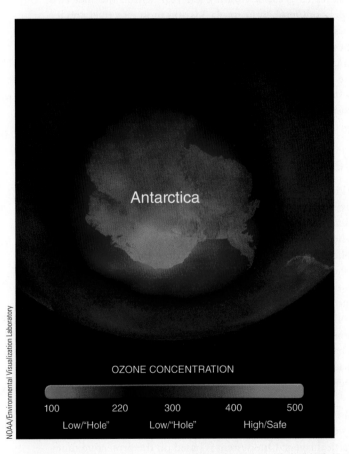

NOAA/Environmental Visualization Laboratory

Antarctica

OZONE CONCENTRATION

| 100 | 220 | 300 | 400 | 500 |
| Low/"Hole" | | Low/"Hole" | | High/Safe |

● **FIGURE 4.7** Satellite sensors monitor changes in the ozone hole *(shaded in red)* over Antarctica. This image shows the spatial extent of the ozone hole in September 2010. A value of 300 DU is the approximate worldwide average for the ozone layer.

What are the potential effects of ozone depletion on the world's human population?

Atmospheric ozone levels are measured in Dobson units (DU), named for a scientist who devised an ozone-measuring spectrometer in the late 1920s. A range of 300 to 400 DU indicates a sufficient amount of ozone to prevent damage to Earth's life-forms. To understand these units better, consider that 100 DU equals 1 millimeter of thickness (1/32 in) that ozone would have at sea level. Ozone measured inside the "hole" has dropped below 90 DU in recent years, and the Southern Hemisphere area of the ozone-deficient atmosphere (a more accurate description than a "hole") has exceeded that of the North American continent.

Ozone-destroying pollutants enter a circulation pattern in the stratosphere that transports them over the tropics, over the middle latitudes, and toward the poles. Near the South Pole, ozone and its destroyers (CFCs and NO_x) become trapped in an atmospheric system called the *Southern Hemisphere polar vortex*. This is a closed flow of extremely cold air that circles around the South Pole during the dark winter months (June, July, and August). During the Southern Hemisphere Polar spring (September, October, and November), incoming solar radiation starts to dissolve this circulation but also energizes the ozone-destroying reactions shown in Formula 3. The Southern Hemisphere ozone hole usually reaches its greatest extent each year within the first two weeks of October. Formula 3 works most effectively on ice crystal surfaces, and ice clouds in the stratosphere provide the perfect laboratory. Affected by the circulation in the stratosphere, the Southern Hemisphere polar vortex, and the polar stratospheric clouds, the area of maximum ozone destruction appears as a swirling elliptical or circular region similar to those seen in satellite images from ozone mapping spectrometers. (See Fig. 4.7.)

Vertical Layers of the Atmosphere

Although people live and function primarily in the lowest level of the atmosphere, there are times, such as when we fly in an aircraft, climb a mountain, or travel to highland areas, when we leave our normal elevation. The thinner atmosphere may affect us if we are not accustomed to the atmospheric conditions at higher elevations. Visitors to Inca ruins in the high Andes or climbers in the Himalayas may experience altitude sickness, and even skiers in the Sierra Nevada or in the Rocky Mountains near mile-high Denver may need time to adjust. The air at these levels is much *thinner* than what most of us are used to experiencing. Thin air means that there is more empty space between air molecules so there is less oxygen and other gases in every breath we take.

Much like the varying approaches to defining the altitude of the outer limit of the atmosphere, several different criteria can be used to divide the atmosphere into distinctive vertical layers. One system is based on the protective function that the layers provide. An example of a layer in this system is the *ozonosphere*, another name for the ozone layer. A second system, often used by chemists and physicists, divides the atmosphere into layers based on chemical composition. A third system, most often used by meteorologists and climatologists, identifies four layers divided according to differences in temperature and rates of temperature change.

Atmospheric Layers by Temperature Characteristics

The system based on temperature characteristics divides the atmosphere into four main layers (● Fig. 4.8). The lowest layer is the **troposphere** (from Greek: *tropo*, turn—the turning or mixing zone). The troposphere extends about 6 to 20 kilometers (4 to 12 mi) above Earth. Its thickness, which varies with latitude, is least at the poles and greatest at the equator. It is within the troposphere that virtually all weather takes place.

Two distinct characteristics differentiate the troposphere from the other atmospheric layers. One is that water vapor and particulates are concentrated in this layer, so these constituents are rarely found above the troposphere. The other characteristic is that temperature normally decreases with increased altitude. The average rate of temperature decrease in the troposphere with altitude is called the **normal lapse rate** (or the *environmental lapse rate*); it amounts to −6.5°C per 1000 meters (−3.6°F/1000 ft).

The altitudinal zone where temperature ceases to drop with increased altitude is called the **tropopause**. This is the boundary separating the troposphere from the **stratosphere**—the next layer of the atmosphere. The temperature of the lower stratosphere

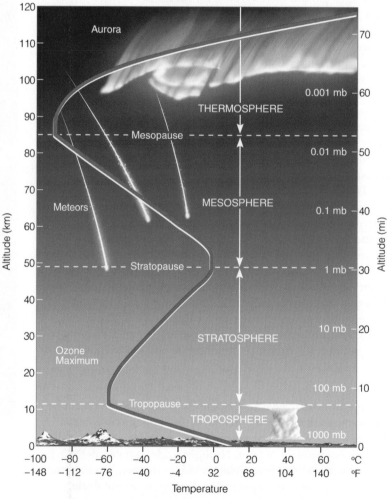

● **FIGURE 4.8** Vertical temperature changes in Earth's atmosphere are the basis for its subdivision into the troposphere, stratosphere, mesosphere, and thermosphere.

At what altitude is our atmosphere the coldest?

remains fairly constant (about −57°C, or −70°F) to an altitude of about 32 kilometers (20 mi). Some water exists in the stratosphere, but it appears as stratospheric ice clouds. These thin veils of ice clouds have no effect on weather as we experience it. It is in the stratosphere that we find the ozone layer and its conditions that protect the surface from excessive UV radiation. As ozone absorbs UV radiation, the absorbed energy results in releases of heat, increasing temperatures in the upper stratosphere. Temperatures at the stratopause, the upper limit of the stratosphere, which is at an altitude of about 50 kilometers (30 mi), are about the same as temperatures on the surface, but almost none of that heat can be transferred because the air is so thin.

Above the stratopause is the **mesosphere**, where temperatures again tend to drop with increased altitude; the mesopause (the last boundary) separates the mesosphere from the **thermosphere**, where temperatures increase until they approach 1100°C (2000°F) at noon. Again, the air is so thin at this altitude that there is practically a vacuum and little heat can be transferred.

Atmospheric Layers by Functional Characteristics

Astronomers, geographers, and communications experts sometimes use another method for separating the atmosphere above the troposphere into layers based on their protective functions. This system divides the atmosphere into two distinct layers, the lowest being the **ozonosphere**, approximately the zone between 15 and 50 kilometers (10 and 30 mi) above the surface. The ozonosphere is another name for the ozone layer mentioned previously. Here, ozone effectively filters the UV energy from the sun and gives off heat (as thermal energy) instead.

From about 60 to 400 kilometers (40 to 250 mi) above the surface is the **ionosphere**. This name denotes the ionization of molecules and atoms in this layer, mostly as a result of UV energy, X-rays, and gamma radiation. *Ionization* refers to the process whereby atoms are changed to ions through the removal or addition of electrons, giving them an electrical charge. The ionosphere helps to shield Earth from harmful shortwave radiation. This electrically charged layer also aids in transmitting communication and broadcast signals to distant regions on Earth. It is in the ionosphere that the auroras (defined in Chapter 3) occur. The ionosphere gradually gives way to interplanetary space (● Fig. 4.9).

It is interesting to note that, when watching television reports from the International Space Station, the orbiting astronauts are still in the atmosphere by these criteria. The background appears black, making it look like interplanetary space, but in reality these missions still take place within the realm of the outer atmosphere. One should also keep in mind that these different layering systems can focus on the same regions. For example, note that the thermosphere and the ionosphere occupy the same altitudes above Earth—that is, from 80 kilometers (50 mi) and outward. The names are different because of the criteria used in the differing systems.

Energy Transfer Processes

The Earth system, as we have learned, is self-contained. The only external input that it requires is energy from the sun in order to function as it does. Within the components of the Earth

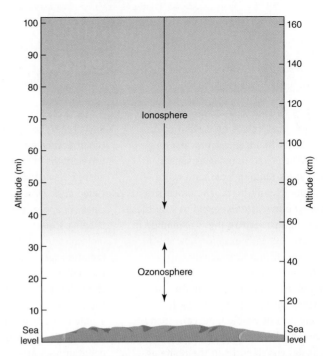

● **FIGURE 4.9** Vertical changes based on functions of the gases subdivide the atmosphere into two layers: the ozonosphere and the ionosphere.

How do these layers protect life on Earth?

system, energy is stored and transferred through many interactions that are important to understanding our planet's weather and climate.

Several important processes are responsible for the transfer of energy, from one form to another and exchanging from one place to another. These are radiation, conduction, convection, advection, and latent heat exchange (● Fig. 4.10).

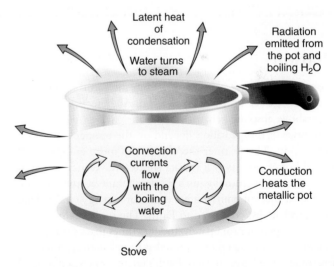

● **FIGURE 4.10** Heat transfer mechanisms. Conduction by contact with a heat source heats the pot and the pot heats the water. Convection occurs as hotter water flows upward, and cooler water sinks, forming a convective current in boiling water. Radiation, emitted as heat (thermal energy), radiates outward from the hot water and the pot. Also, latent heat of condensation is released as water vapor turns back into a liquid as tiny droplets in steam.

How is advection involved in this small system?

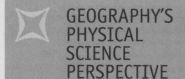
White Clouds, Blue Skies, Red Sunsets, and Rainbows

When we look at the sky, we are used to seeing certain colors: brilliant blue on a clear day, vibrant reds and oranges at sunset, and clouds ranging from pure white to ominous gray-black. The process of *atmospheric scattering* explains why these colors appear. Scattering is explained in the text, but let's examine how it affects the colors of our atmosphere.

As sunlight is scattered in our atmosphere, certain wavelengths are scattered more than others. Each wavelength of visible light corresponds to one of the spectrum of colors. For example, wavelengths in the range of 0.45 to 0.50 micrometers (1 micrometer is 1/1,000,000 of a meter) correspond to the color blue, and 0.66 to 0.70 micrometers correspond to the color red. These scattered wavelengths, when viewed by the human eye, are transmitted to the brain and translated into the colors we see. The particular wavelengths that are being scattered in the sky at our location are the colors that we see in the atmosphere.

In 1871, radiation scientist Lord Rayleigh explained that gas molecules in the atmosphere tend to scatter the shorter wavelengths of visible light and also UV energy. Named after him, this process is called *Rayleigh scattering*, which makes a clear sky appear blue because blue light is being scattered preferentially by gas molecules in the air. In 1908, another radiation specialist, Gustav Mie, explained that oblique (or sharp) sun angles coming through the atmosphere tend to cause the scattering of longer wavelengths of visible light. Also, larger particles in the atmosphere—such as dust, smoke, and pollutants—scatter the longer visible wavelengths, unless the accumulation of these particles is so thick that they block visibility. *Mie scattering* explains why sunsets bring about red- and orange-colored skies. Longer wavelengths in the visible spectrum also penetrate the atmosphere better than shortwave energy does, even in fog and haze. This is one reason why stoplights and brake lights are red—they are easier to see under various weather conditions.

Another form of scattering, called *nonselective scattering*, is caused by tiny water droplets and ice crystals in clouds that tend to scatter all wavelengths of visible light with equal intensity. When all the wavelengths are scattered equally, white light emerges, which shows most clouds as white. When we see dark-gray to black storm clouds, we are actually seeing white clouds plus shadows cast by the height and thickness of these massive clouds.

Rainbows are caused by a different process: the refraction (bending) of light by water droplets suspended in the atmosphere. The droplets act like tiny glass prisms, which take the incoming sunlight, bend the wavelengths differentially, and project the light as the colors of the spectrum. The factors necessary for seeing a rainbow include the sun angle and direction, the altitude of the water droplets, their density, and the observation angle. When you see a rainbow from a moving vehicle, it is only a matter of time until your observation angle changes and the rainbow disappears.

Radiation

Electromagnetic energy is transferred from the sun through space to Earth through the process of **radiation**. All objects with a temperature above absolute zero emit electromagnetic radiation. The characteristics of that radiation depend on the temperature of the radiating body. The warmer the object, the more energy it will emit, and hotter objects will radiate shorter wavelengths at peak emission output. Because the sun's absolute temperature is 20 times that of Earth, the sun emits a great deal more energy, and radiates shorter wavelengths, compared to Earth. The sun's energy output per square meter is approximately 160,000 times that of our planet. The majority of solar energy is emitted at wavelengths shorter than 0.7 micrometers, whereas most of Earth's energy is radiated at much longer wavelengths of around 9 to 10 micrometers (see Fig. 4.10).

Earth is a heat engine powered by solar energy. It is the *shortwave solar radiation (light)* reaching the surface that is absorbed and re-radiated to heat the Earth. Because the Earth is cooler than the sun, it radiates energy in the form of longwave radiation *(thermal infrared), which we sense as heat*. This *thermal infrared energy*, **terrestrial radiation**, is emitted from Earth's surface (including the oceans) and warms the atmosphere, accounting for the heat of the day. Basically, the source of atmospheric heat is terrestrial radiation from the surface, so the atmosphere mainly heats from the bottom upward (see Fig. 4.6).

Conduction

The transfer of energy between two objects in contact with each other is called **conduction**. Heat flows from warmer objects or substances to the cooler ones they are in contact with, which tends to equalize temperatures. If you touch something hot, heat flows from the hot object to your hand; if you touch something cold, the heat flow is from you to the cooler object. Conduction actually occurs as thermal energy (heat) is passed from one molecule to another in a chainlike fashion.

Atmospheric conduction occurs at the interface (contact zone) between the atmosphere and Earth's surface. Conduction, however, is actually a relatively minor heat transfer process in terms of atmospheric warming because it affects only the layers of air closest to the surface. This is because air is a poor conductor of heat. In fact, air is a good insulator—the opposite of a good conductor. This

A beautiful sunset over the Pacific Ocean illustrates the role of scattering in producing the colorful skies. The blue sky aloft results because the air is thin at that altitude and gas molecules are scattering blue light. The sunset is red–orange because of longer wavelengths scattering in the thicker and denser atmosphere at that level. Parts of the clouds are dark and in shadow, and the other parts are white because of nonselective scattering.

A rainbow colors the sky in Denali National Park, Alaska. Because of refraction as sunlight passes through water droplets in the atmosphere, light separates into the colors of the spectrum to project a rainbow. The colors are in wavelength order, with shorter wavelengths on the inside of the arc.

property explains why an air layer is sometimes put between two panes of glass to help insulate a window. Insulation that contains a great deal of air-filled cavities is also used in sleeping bags and ski parkas. In fact, if air conducted heat well, our kitchens would become unbearably hot whenever we turn on the stove or oven.

Convection

As parcels of air near the surface are heated, they expand in volume and become less dense than the surrounding air so they rise. This vertical transfer of heat through the atmosphere is called **convection**; it is much the same process that circulates boiling water in a pot. The water at the bottom, nearest to the heat source, becomes lighter and less dense as it warms. As the hot water rises, cooler, denser surface water replaces it by flowing downward. The water that flows down to the bottom is then warmed, so it flows upward to be replaced by cooler water moving downward. This vertical, circular movement of a fluid is called a *convection current*. These currents, set into motion by the heating of a fluid (liquid or gas), make up a convectional system. Convection currents account for much of the vertical transfer of heat within the atmosphere and

the oceans and they have a great impact on the development of clouds and precipitation.

Advection

The transfer of heat in a horizontal direction is called **advection**. On Earth, there are two major advection processes, and both are key factors in weather and climate. These are the winds and ocean currents, which are extremely important because they move heat energy from areas where it is warmer, to cooler places, with a moderating effect on both areas. Winds and ocean currents help to distribute heat horizontally from the equatorial and tropical regions to the higher latitude and polar regions. Advection of heat has a great influence on weather conditions and on the world's climates.

Latent Heat Exchange

Water is the only substance that exists in all three states of matter—as a solid, a liquid, and a gas—within the normal temperature range of Earth's surface and the troposphere. The energy transfer that occurs with the changes in water from one state to another is called **latent**

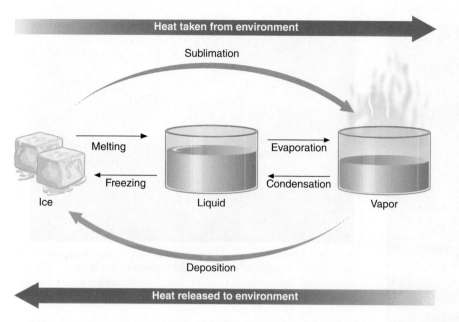

● **FIGURE 4.11** Energy exchanges between the three physical states of water. This diagram shows where energy comes from and where it goes during these exchanges. When water freezes into ice, thermal energy comes out of water as it freezes and heat enters the environment. To evaporate water into vapor, thermal energy goes into the water as it evaporates, removing some heat from the surrounding environment.

Why are some of these energy exchanges referred to as "latent heat"?

heat exchange. This exchange involves heat changing from *sensible heat* (which we can feel) to latent heat, which is a stored form of energy that we cannot feel, and also the reverse process—from latent to sensible heat. These exchanges from one state to another, which occur under natural conditions, are illustrated in ● Figure 4.11.

Molecules of a gas move faster than those of a liquid. During the process of **condensation**, when water vapor changes to liquid water, its molecules slow down and release some of their energy (590 cal/g, or 2470 joules/g) as sensible heat. Molecules in a solid move much more slowly than those of a liquid, and when water undergoes **freezing**, changing into ice, less energy is released (80 cal/g, or 333 joules/g). When these processes are reversed, sensible heat becomes stored as latent heat. **Melting** takes up 80 calories per gram (333 joules/g), and **evaporation** takes up 590 calories per gram (2470 joules/g); the latter is called **evaporative cooling**.

When water evaporates, energy is stored in water vapor as latent heat. We are cooled by perspiration evaporating from our skin, thereby lowering our temperatures. When water condenses from a gas to a liquid, the stored latent energy is released as sensible heat. In climate regions that experience enough evaporation during warm, dry seasons, this effect can be used to cool buildings through *evaporative coolers* (often called *swamp coolers*). A fan draws air through wet filters, evaporative cooling occurs, and the cooler air is blown indoors by the fan.

All of these major energy transfer processes play important roles in moving heat within the Earth system. Light energy from the sun is absorbed by the Earth and re-radiated as thermal infrared energy, which we sense as heat. Conduction transfers heat by contact from warmer to colder substances. *Convection* moves heat in fluids (air, water) upward and moves cooler fluids downward. *Advection* moves heat to cooler areas and moves cool (air or water) to warmer areas.

Evaporation helps cool the atmosphere, and the release of latent heat of condensation helps warm the atmosphere. The massive amounts of heat released from condensation in the atmosphere provide a source of energy that drives many kinds of storms. These energy transfers are the major processes that drive Earth's dynamic weather systems, and they have a tremendous impact on climates.

Earth's Energy Budget

The amount of solar energy that reaches the outer atmosphere is affected by a variety of processes that cause a little more than half of that insolation to be lost. This amount of solar energy loss is a global annual average. The insolation amount actually received at a particular location, however, depends on the processes involved as well as latitude, time of day, and time of year (these factors are related to the angle at which the sun's rays strike). The transparency of the atmosphere, related to the amount of cloud cover, moisture, carbon dioxide, and solid particles in the air, is another very important factor.

Several processes affect the sun's energy as it interacts with the atmosphere. A determination (and inventory) of the incoming solar energy, its storage, and the outgoing energy processes is called the **Earth's energy budget**. The following figures represent approximate averages for the entire Earth for an average year (● Fig. 4.12):

27% reaches Earth's surface as direct radiation.

20% reaches the surface as diffuse radiation after being scattered.

26% of the energy is reflected directly back to space by clouds and the surface *(land and oceans)*.

8% is *scattered* by minute atmospheric particles and returned to space as diffuse radiation.

19% is *absorbed* by the ozone layer and water vapor in the clouds.

In other words, on a worldwide annual average, about 47% of the incoming solar radiation eventually reaches the surface, 19% is retained in the atmosphere, and 34% is returned to space. Because Earth's energy budget is in equilibrium, the 47% received at the surface is ultimately returned to the atmosphere and out to space by processes that we will now examine.

Heating the Atmosphere

The 19% of direct solar radiation retained by the atmosphere is stored in clouds and the ozone layer and eventually is dissipated back to space. Other sources must be found to explain how energy heats the atmosphere. The explanation lies in the 47% of incoming solar energy that reaches Earth's surface (land and water) and in the transfer of energy from Earth to the atmosphere, which is

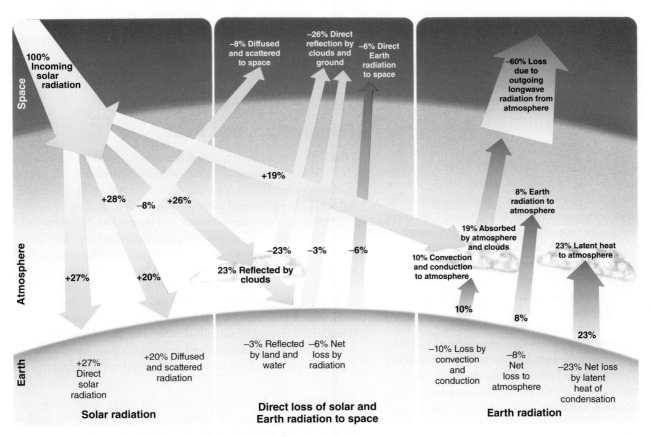

● FIGURE 4.12 Earth's energy budget indicates that a long-term global balance, or equilibrium, exists. Only about 47% of the incoming solar radiation reaches and is absorbed by Earth's surface. Eventually, solar radiation received is re-radiated away by the Earth system (terrestrial radiation). However, the energy budget is dynamic. There is growing concern that increasing greenhouse gases from human activities are causing the atmosphere to absorb and re-radiate more thermal radiation, thus raising global temperatures.

accomplished through the energy transfer processes of conduction, convection, and latent heat exchange. These processes, however, are driven by the ultimate source of atmospheric heat, which is *terrestrial radiation*, the thermal infrared energy converted from the solar radiation received and re-radiated by Earth.

Earth's Surface Energy Budget Now that we are familiar with the various means of heat transfer, we can examine what happens to the 47% of solar energy that reaches, and is absorbed into, Earth's surface (see Fig. 4.12). Approximately 14% of this energy is re-radiated by the Earth in the form of longwave *terrestrial radiation*. This 14% includes a net loss of 6% (of the total originally received by the atmosphere) directly to outer space; the other 8% is absorbed into the atmosphere. In addition, there is a net transfer back to the atmosphere (by conduction and convection) of 10% of the 47% that reached Earth. The remaining 23% is returned to the atmosphere by the release of latent heat of condensation. Thus, the 47% of the sun's original insolation that reached Earth's surface is returned to other components of the energy system and ultimately is lost back to space. There has been no long-term gain or loss.

The Energy Budget in the Atmosphere About 60% of the solar energy intercepted by the Earth system is temporarily retained by the atmosphere and stored. This includes 19%

of direct solar radiation absorbed by the clouds and the ozone layer, 8% emitted by *thermal radiation* from the Earth's surface, 10% transferred from the surface by *conduction* and *convection*, and 23% released by the *latent heat of condensation*. Some of this energy is recycled back to the surface, but eventually it is lost to outer space as more solar energy is received. Hence, just as was the case at Earth's surface, the energy budget in the atmosphere is in balance over long periods—a dynamic but generally stable system.

Energy Balance

The Earth's energy budget helps us understand the *open energy system* involved in heating the atmosphere. The *input* to the system is the incoming shortwave solar radiation (light) that reaches Earth's surface; this is balanced by the *output* of longwave terrestrial radiation (thermal energy) back to the atmosphere and lost to space. As these functions adjust to remain in balance, there is an overall **energy balance** in the energy budget of Earth, but it is in a state of *dynamic equilibrium*. Therefore, on Earth, over the long term, the energy budget is in balance. What this means is that, because of this tendency toward equilibrium, Earth has gotten neither continually warmer nor continually colder year after year through the entire history of the planet.

Of course, the percentages mentioned earlier are generalized and refer to *net* inputs and outputs of energy averaged over a long period of time. Earth also stores energy, re-radiates part of it to heat the atmosphere, and then some of this energy circulates back to Earth in a chain of cycles before it is finally lost into space. Today, there is much evidence that an enhanced greenhouse effect is causing an imbalance in the energy budget, with the Earth system storing more heat, resulting in the associated impacts of increasing global temperatures.

Variations in the Energy Budget The figures for the energy budget are averages for the whole Earth system, over the entire year. For any particular location, the energy budget is most likely not balanced. Some places receive a surplus of incoming solar energy compared to outgoing energy loss to space, and others receive less solar energy than the amount of energy they return to space. That may sound strange, but although energy from the sun is the ultimate source, regions with an insolation deficit receive heat indirectly through the energy transfer processes. Whether a location receives a surplus or a deficit in absorbed energy depends on latitude and seasonal fluctuations (● Fig. 4.13).

In the tropical and subtropical zones, where insolation is high throughout the year, more solar energy is received at Earth's surface and in the atmosphere than can be radiated back to space. In contrast, the high latitude regions receive little insolation during the winter, although Earth is still emitting thermal energy, causing a large deficit for the year in those places. Locations in the midlatitude zones have lower deficits or surpluses. The latitudes of 38° North and South mark the general division between equatorward zones that receive a surplus of solar energy and the poleward zones that run an annual deficit. If it were not for heat transfers within the atmosphere, the land, and the oceans, particularly through advection, the tropical zones would continually get hotter and the polar zones would get colder over time.

During the year at any location, the energy budget varies over the seasons, with a tendency toward a heat surplus in the summer or high-sun season and a tendency toward a deficit 6 months later. Seasonal differences are small near the equator, but they are generally large in the midlatitude and polar regions.

Air Temperature
Temperature and Heat

Heat and temperature are highly related, but they are not the same. **Heat** is energy being transferred from one substance or medium to another as a result of their temperature differences. All substances consist of molecules that are constantly in motion (vibrating and colliding) and therefore possess kinetic energy—the energy of motion. This energy is experienced as heat. **Temperature**, in comparison, is a measure of the average kinetic energy of individual molecules in a substance. When a substance is heated, its atoms and molecules vibrate faster, increasing its temperature. It is important to know that the amount of heat energy also depends on the mass of a substance, whereas the temperature refers to the energy of individual molecules. A burning match has a high temperature but minimal heat; the oceans have moderate temperatures but a high content of heat—thermal energy.

Temperature Scales Three different scales are generally used to measure temperature. The one that Americans are most familiar with is the **Fahrenheit scale**, devised in 1714 by Daniel Fahrenheit, a German scientist. On the Fahrenheit scale, the boiling temperature of water at sea level is 212°F and the temperature freezing temperature is 32°F. This temperature scale is used in the English system of measurements.

The **Celsius scale** (also called the **centigrade scale**) was devised in 1742 by Anders Celsius, a Swedish astronomer. Celsius temperatures are used as a part of the International System of Units (the metric system). The freezing temperature of water at sea level on this scale was arbitrarily set at 0°C, and the boiling temperature of water was designated as 100°C.

The United States is one of the few countries that still makes widespread use of the Fahrenheit scale, and even in the United States, a majority of the scientific community uses the Celsius scale. For these reasons, this book presents comparable Celsius and Fahrenheit figures given side by side for temperatures. Similarly, whenever important figures for distance, area, weight, or speed are given, the text states the metric system followed by the English system. Appendix A has the formulas for comparisons and conversions between the two systems.

● Figure 4.14 can help you compare the Fahrenheit and Celsius scales as you encounter temperatures that you wish to convert. In addition, the following formulas can be used for conversion from Fahrenheit to Celsius or vice versa:

$$°C = (°F − 32) ÷ 1.8$$
$$°F = (°C × 1.8) + 32$$

A third temperature scale, used primarily by scientists, is the **Kelvin scale**. Lord Kelvin, a British scientist, believed that negative temperatures were not proper and should not be used. In his

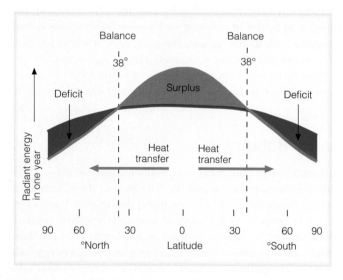

● **FIGURE 4.13** The energy budgets for specific locations are not necessarily balanced. Low latitudes receive more solar energy than they lose by re-radiation back to space (energy surplus). High latitudes receive less energy from the sun than they radiate back to space (energy deficit).

How is surplus energy from the low latitudes transferred to higher latitudes?

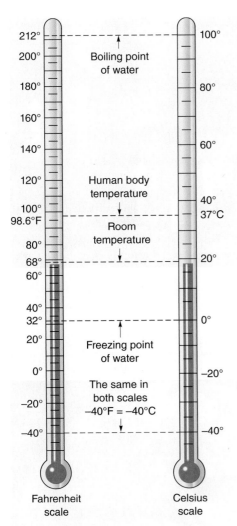

When it is 70°F, what is the temperature in Celsius degrees?

mind, no temperature should ever go below zero. This scale is based on the fact that temperature is related to molecular movement. As the temperature is reduced, molecular motion slows. There is a temperature at which molecular motion stops and no further cooling is possible. This temperature, approximately −273°C (−460°F), is termed **absolute zero**. The Kelvin scale uses absolute zero as its starting point, and does not use the degree symbol (°).

Thus, 0K equals −273°C. Conversion of Celsius to Kelvin is expressed by the following formula:

$$K = °C + 273$$

Short-Term Temperature Variations

Local changes in air temperature can result from a wide range of causes. These are related to the receipt and dissipation of energy from the sun, to various properties of Earth's surface and to processes in the atmosphere.

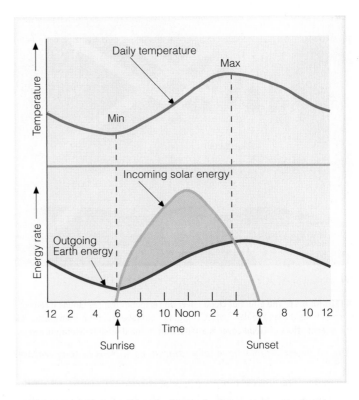

Why does temperature continue to rise even after the input of solar energy declines?

The Daily Effects of Insolation As we noted earlier, the amount of insolation at any particular location varies both throughout the year (annually) and throughout the day (diurnally). Annual fluctuations are mainly associated with the changing declination of the sun with the seasons. **Diurnal** changes are related to Earth's daily rotation. Diurnal typically refers to some cycle of day-to-night change. Insolation receipt begins at sunrise, reaches a maximum at noon (local solar time), and returns to zero at sunset. The exceptions to this occur in the polar regions during times of the year when the sun never rises above the horizon or during the times when it never sets.

Although insolation is greatest at noon, when the sun angle is highest, you probably know that temperatures usually do not reach their maximum until about 2:00 to 4:00 p.m. (● Fig. 4.15); this is called a **temperature lag**. This lag occurs because the insolation that is received between sunrise and the afternoon hours exceeds the energy being lost through terrestrial radiation. After sunrise, as Earth and atmosphere continue to gain energy, temperatures normally show a gradual increase. Later in the afternoon, outgoing Earth radiation begins to exceed insolation, and temperatures begin to decrease. The **daily temperature lag** of a few hours between the time of the highest sun angle (solar noon) and the warmest time of day results from the time it takes to heat Earth's surface to its daily maximum and for this energy to be radiated to the atmosphere.

Insolation receipt ends after sunset, and much of the energy that has been stored in Earth's surface during the day is lost during

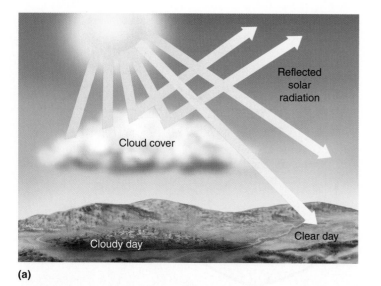

(a)

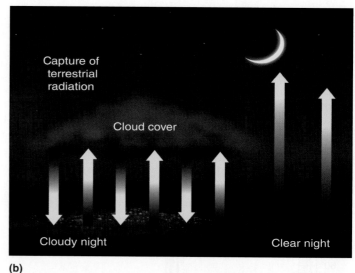

(b)

● **FIGURE 4.16** The effect of cloud cover on temperatures. (a) Clouds intercept solar radiation and lower air temperatures during the day. (b) By trapping longwave radiation from Earth, clouds increase air temperatures at night. The overall effect of cloud cover is a reduction in the diurnal temperature range.

Why do desert regions have large diurnal variations in temperature?

the night. The lowest temperatures occur around dawn, when the maximum amount of terrestrial energy has been emitted and before the replenishment from the sun occurs after dawn. Thus, disregarding (for now) other factors that can cause temperature changes, we can see a predictable hourly temperature change over the day called the **daily march of temperature**. There is a decline from midafternoon until dusk and a rapid increase in the hours from dawn the next day until the maximum temperature is reached in the afternoon.

Cloud Cover The extent of cloud cover is another factor that affects the temperatures of the surface and the atmosphere (● Fig. 4.16). A heavy cloud cover reduces the amount of insolation a place receives, so daytime temperatures are generally higher when the sky is clear and lower on cloudy days as cloud tops reflect solar radiation back to space. Cloud cover also affects the natural greenhouse effect, because clouds, which are composed mainly of water droplets, absorb thermal infrared radiation from Earth, and store heat. Clouds keep nighttime temperatures near the surface warmer than they would otherwise be. The general effect of cloud cover is to moderate temperatures by lowering the maximum and raising the minimum temperatures. In other words, cloud cover makes for cooler days and warmer nights. Weather satellites have shown that, at any time, clouds cover about 50% of our planet (● Fig. 4.17).

Differential Heating of Land and Water Bodies of water heat and cool slowly in comparison to the rates of heating and cooling of land surfaces. The air directly above the surface is heated or cooled in part by the surface characteristics beneath the air. Therefore, temperatures over water bodies or on land areas that are subjected to oceanic winds (**maritime** locations) tend to be more moderate than those of inland places at the same latitude. In contrast, the greater the **continentality** of a location (the distance removed from a large body of water), the greater the variations in daily and seasonal temperature tend to be. Areas deep within

Courtesy of NOAA/NASA/GSFC, GOES-13

● **FIGURE 4.17** This view of Earth shows that much of the planet is covered by clouds.

Why do the clouds appear so white and the oceans appear so dark?

a continent or distant from the ocean or winds from the ocean, particularly in regions outside the tropics, tend to experience larger temperature differences from day to night and from summer to winter, compared to what coastal locations experience.

Reflection The ability of a surface to reflect solar energy (expressed as a percentage) is called its **albedo**; a high albedo surface reflects a high percentage of the energy that it receives.

Albedo and absorption of energy are related because energy that is not reflected from a surface is absorbed. The higher the albedo, the lower the amount of energy absorbed and vice versa. The solar energy that the Earth system reflects back to space reduces the amount of insolation that will be absorbed for heating the atmosphere, oceans, and land surfaces.

As you may know from experience, snow and ice are good reflectors; they have an albedo of 90% to 95%. This is one reason why glaciers on high mountains do not melt away in the summer or why there may still be snow on the ground on a sunny day in the spring—the solar radiation is reflected and not absorbed. Forests, however, have albedos of about 10% to 12%, which is good for the trees because they need solar energy for photosynthesis. The albedo of cloud cover varies, from 40% to 80%, according to the cloud thickness. The high albedo of cloud tops is why much solar radiation is reflected directly back to space from the tops of clouds.

The albedo of a water body varies greatly, depending on the water depth and the angle of the sun's rays. If the sun's rays are at a high angle, smooth water will reflect very little and absorb a great deal of energy. In fact, if the sun is vertical over a calm lake or ocean, the albedo will be only about 2%. However, low sun angles during sunrise and sunset, in winter, or at high latitudes can cause an albedo of more than 90% from the same water body (● Fig. 4.18). Likewise, in winter, when solar angles are lower, a snow surface can reflect up to 95% of the energy striking it, and skiers must be aware of the danger of severe sunburns and possible snow blindness from intense reflected solar radiation.

● **FIGURE 4.18** At low sun angles, water reflects most of the solar radiation that strikes it.

Why is it so difficult to assign a single albedo value to a water surface?

Horizontal Air Movement Advection is the major mechanism of horizontal heat transfer over Earth's surface. Winds, whether they are on a large or small scale, can have significant short-term effects on temperatures. Winds blowing from an ocean to land will generally bring cooler temperatures in summer and warmer temperatures in winter. Large masses of air moving from polar regions into the middle latitudes can cause sharp drops in temperature, whereas tropical air moving poleward will usually bring warmer temperatures.

Vertical Temperature Distributions

Normal Lapse Rates You have learned that Earth's atmosphere is primarily heated from the ground up as a result of terrestrial thermal radiation, conduction, and convection. Temperatures in the troposphere are usually highest at low altitudes and decrease as the altitude increases. As noted earlier in the chapter, this *average* decrease in air temperature with altitude is approximately 6.5°C per 1000 meters (3.6°F/1000 ft) and is referred to as the *normal lapse rate*. In this case, "normal" refers to the expected, worldwide average conditions in the troposphere.

The actual lapse rate at a particular place can vary. Smaller lapse rates (which means less change with altitude) can exist if cold, dense air drains into a valley from a higher elevation or if winds bring air in from a cooler region at the same elevation. In each case, the surface is cooled so that its temperature is closer to that at higher altitudes directly above it. However, if the surface is heated strongly on a hot summer afternoon, the air near the surface will be disproportionately warm and the lapse rate will be steep (great change with altitude).

Temperature Inversions Under certain circumstances, the normal decrease of temperature with increased altitude might be reversed and temperature may actually *increase* with altitude for several hundred meters. This atmospheric condition is called a **temperature inversion**, where a layer of warmer air aloft interrupts the altitudinal decrease in temperature. The altitudinal boundary between the cooler air below and the warmer air aloft can be abrupt and appear as a horizontal line across the sky, marking the **inversion layer**. Inversions tend to stabilize the air, causing less turbulence, which discourages the development of precipitation and storms. There are two main kinds of inversions, which form in different ways and at different altitudes: upper-level and surface (or ground).

Upper-Level Inversions Air moving downward from the upper atmosphere can cause an **upper-level inversion** (● Fig. 4.19). This kind of atmospheric condition involves an inversion layer typically about 1000 or 2000 meters above the ground. Descending air is compressed as it sinks, causing increases in temperature and atmospheric stability. Inversions caused by descending air are most common at about 30° to 35° north and south latitudes.

An upper-level inversion, common to some coastal areas of California, results when cool marine air drifting in from the Pacific Ocean moves under stable, warmer, and lighter air aloft. This inversion layer tends to be stable—the cold underlying air is heavier and cannot rise through the warmer air above. The cold

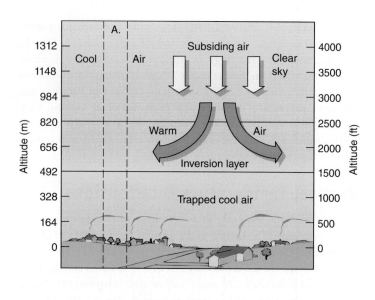

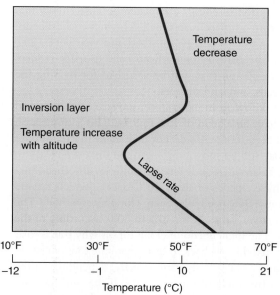

● **FIGURE 4.19** (Left) Upper-level inversions are caused by air subsiding from aloft, warming as it sinks to form a stable inversion layer over cooler air below (note column of air A). (Right) Lapse rate associated with the column of air (A) during a temperature inversion.

What problems might occur with a strong temperature inversion, particularly if the inversion lasts for days?

air resists rising or mixing with the warmer air aloft, and pollutants created at the surface, such as smoke, dust, and automobile exhaust, fail to disperse. During these conditions, pollutants and smoke accumulate in the stable air below the inversion layer. Upper-level inversions are particularly acute in the Los Angeles area, which is located in a topographic basin surrounded by mountainous areas. Cooler air drifts into the basin from the ocean and then cannot escape either horizontally, because of the landform barriers, or vertically, because of the inversion. The Los Angeles area is affected by smoggy conditions frequently during times of upper-level inversions, and because of its coastal and topographic setting is often cited as a classic example of this problem. Many other urbanized areas, however, also experience smog problems related to the accumulation of pollutants below an inversion layer.

Surface Inversions Some of the most noticeable temperature inversions are those that occur close to the surface, when the ground is cold because of heat loss from terrestrial radiation and then cools the lowest layer of air. These are called **surface inversions**, or **radiation inversions** (● Fig. 4.20). In this situation, the coldest air is also close to the surface, but the inversion layer is much lower than those of an upper-level inversion, and the temperature rises with altitude. Ground inversions most often occur in the middle and high latitudes on clear, cool (or cold) nights with calm conditions. Snow cover or the advection of cool, dry air into an area may strengthen an inversion. Clear skies under these conditions produce rapid cooling of Earth's surface at night because through terrestrial radiation much of the thermal energy (heat) accumulated from daytime insolation is lost. The air layer that is closest to Earth is cooled by conduction, because of the drop in temperature of air near the cold surface. As the

sun rises in the morning and heats the ground, ground inversions tend to break up, and fogs tends to "burn off," but they may form again the next night if it is cool and the air is clear and calm.

Because they occur when it is cold, ground inversions often cause the formation of dew, fog, and frost. Areas located in a topographic basin can have these inversions last for days or weeks, partly because of cold air flowing downslope from the surrounding mountains, which supplies more cold air near the ground level. Cold air on the valley floors and other low-lying areas sometimes produces fog, or, if temperatures are cold enough, a killing frost

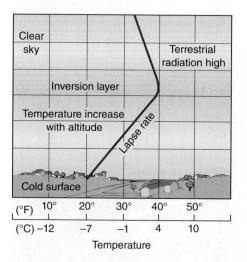

● **FIGURE 4.20** Graph showing a surface temperature inversion caused by the rapid cooling of air above a cold land surface. Cooling occurs because of loss of thermal radiation to space through a clear sky.

How are the temperatures trending in the lowest layer, in the inversion layer, and above the inversion layer?

● **FIGURE 4.21** A wintertime radiation inversion in the Salt Lake City, Utah, area. The inversion layer is clearly visible, with warmer air above and cold air below. Radiation inversions are caused by the rapid cooling of air under clear skies and above a cold land surface. Fog has formed at the inversion layer.

What problems might occur with a strong temperature inversion, particularly if the inversion lasts for days?

⊕ In Google Earth fly to: 40.75°N, 111.85°W (Salt Lake City).

(● Fig. 4.21). A variety of methods are used to prevent frosts from destroying crops during a ground inversion. Fruit trees are often planted on warmer hillsides at elevations that tend to be above the inversion layers instead of down in the valleys. Farmers may also put blankets of straw, cloth, or other insulating materials over their plants. This prevents the escape of heat to space, insulates the plants from the cold air, and keeps the plants warmer. Large fans are often used in an effort to mix colder surface air with the warmer air aloft, thus disturbing the inversion (● Fig. 4.22). Huge orchard heaters that warm the air can also be used to disturb the temperature layers. Smudge pots, an older method of preventing frost, used oil fires to put heat into the air and emit particulates that encouraged condensation, which released some latent heat. However, smudge pot use has declined because of their air pollution potential.

Controls of Earth's Surface Temperatures

There are several major **controls of temperature**: (1) latitude, (2) land and water distribution, (3) ocean currents, (4) elevation, (5) landform barriers, and (6) human activities.

Latitude The most important control of temperature variations is latitude. Because of changes in sun angle and length of daylight hours, latitudinal patterns are related to the seasonal and annual receipt of solar energy. In general, annual insolation tends to decrease from the lower latitudes to the higher latitudes. Table 4.2 shows the average annual temperatures for several Northern Hemisphere locations. We can see that, responding to insolation (with one exception), a poleward decrease in temperature exists for these locations. Areas near the equator are the exception. Because of heavy cloud cover in equatorial regions, annual temperatures there tend to be lower than in bordering regions to the north and south, where skies tend to be clearer.

● **FIGURE 4.22** Powerful wind machines can break up a ground inversion to avoid crop damage from frost. These fan towers blow warm air downward from above an inversion layer, which moderates the temperature below to protect these vineyards from frost.

TABLE 4.2
Average Annual Temperatures in the Northern Hemisphere

Location	Latitude	(°C)	(°F)
Libreville, Gabon	0°23′N	26.5	80
Ciudad Bolivar, Venezuela	8°19′N	27.5	82
Mumbai, India	18°58′N	26.5	80
Xiamen, China	24°26′N	22.0	72
Raleigh, North Carolina	35°50′N	18.0	66
Bordeaux, France	44°50′N	12.5	55
Goose Bay, Labrador, Canada	53°19′N	−1.0	31
Markova, Russia	64°45′N	−9.0	15
Point Barrow, Alaska	71°18′N	−12.0	10
Mould Bay, NWT, Canada	76°17′N	−17.5	0

Land and Water Distribution

Oceans, seas, and large lakes serve as store-houses of water that store tremendous amounts of heat. The widespread extent and distribution of the oceans make them a major influence on atmospheric and weather conditions. Different substances heat and cool at varying rates, and land heats and cools much faster than water does. There are three reasons for this phenomenon. First, the **specific heat** of water is greater than that of land. Specific heat refers to the amount of heat necessary to raise the temperature of 1 gram of any substance 1°C. Water has a specific heat of 1 calorie per gram per degree Celsius (4.19 joules/g °C) and must absorb more thermal energy than land (which has specific heat values of about 0.2 calories/g °C, or 0.84 joules/g °C) to be warmed by the same number of degrees in temperature.

Second, water is transparent. Solar energy passes through the surface and down into the water below, yet the energy is concentrated on the surface of opaque materials such as soil and rock. Thus, a given amount of heat will spread through a greater volume of water in comparison to land's surface heating. Third, liquid water circulates and mixes, which transfers heat among the various depths within its mass. As summer changes to winter, land cools more rapidly than bodies of water, and as winter becomes summer, the land heats more rapidly. Air gets much of its heat from the surface below, so the differential heating of land and water also causes temperature differences in the atmosphere above these two surfaces.

Seattle, Washington, has a mean temperature in July of 18°C (64°F), whereas the mean temperature during the same month in Minneapolis, Minnesota, is 21°C (70°F). Because these two cities are at similar latitudes (● Fig. 4.23), their annual pattern and receipt of solar energy should also be similar. Therefore, their different July temperatures must be related to a factor other than latitude. Much of this temperature difference exists because Seattle is near the Pacific coast, and Minneapolis is in the heart of a large continent, far from moderating oceanic influences. Seattle stays cooler than Minneapolis in the summer because the ocean warms up slowly, keeping the air relatively cool. Minneapolis, in comparison, is in the center of a large landmass that warms very quickly and to a greater degree. The opposite is true because Seattle is warmer then Minneapolis in winter, because temperatures there are moderated by its nearness to ocean water, but Minneapolis is distant from the moderating influence of a large water body. The mean temperature in January is 4.5°C (40°F) in Seattle and −15.5°C (4°F) in Minneapolis.

Not only do water and land heat and cool at different rates, but so do various land surface materials. Soil, forest, grass, and rock surfaces all heat and cool differentially and can affect the temperatures of overlying air.

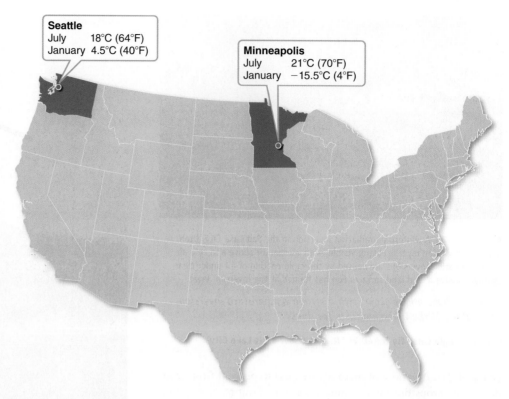

Seattle
July   18°C (64°F)
January 4.5°C (40°F)

Minneapolis
July   21°C (70°F)
January −15.5°C (4°F)

● **FIGURE 4.23** Location map of Seattle, Washington, and Minneapolis, Minnesota, showing their similar latitudinal position but much different seasonal temperature characteristics.

Ocean Currents Surface ocean currents are large movements of water flowing much like rivers in the ocean. They flow from places of warm temperatures to places of cooler temperatures and vice versa. These movements are another example of processes that tend to keep Earth systems in equilibrium—in this instance, a balance of temperature and density.

Earth rotation affects the movements of winds, which in turn affect the flow of ocean currents. In general, ocean currents move in a clockwise direction in the Northern Hemisphere and in a counter-clockwise direction in the Southern Hemisphere (● Fig. 4.24). The ocean temperature greatly affects the air temperature above it. An ocean current that moves warm tropical water toward the poles (a **warm current**) or cold polar water toward the equator (a **cold current**) can significantly modify the air temperatures of regions near to where these currents flow. Ocean currents that pass close to land and are accompanied by onshore winds can have a significant impact on coastal climates.

The Gulf Stream, with its extension, the North Atlantic Drift, is an example of an ocean current that moves warm water poleward. This relatively warm water keeps the coasts of Great Britain, Iceland, and Norway ice-free in winter and moderates the climates of nearby areas (● Fig. 4.25). We can see the Gulf Stream's impact if we compare winter conditions in the British Isles with those of Labrador in northeastern Canada. Although these locations are at the same latitude, the average temperature in Glasgow, Scotland, in January is 4°C (39°F), whereas during the same month it is −21.5°C (27°F) in Nain, Labrador.

The California Current off the west coast of North America moderates the climate of that coastal region as it brings cold water

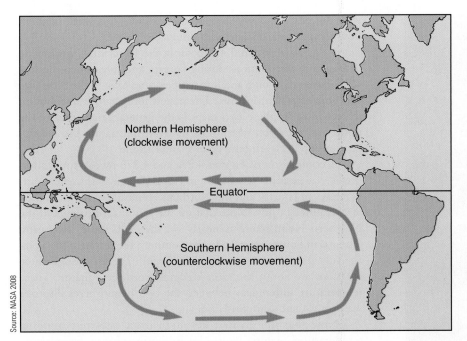

● FIGURE 4.24 A simplified map of surface currents in the Pacific Ocean shows their basic circular pattern. Major currents move clockwise in the Northern Hemisphere and counterclockwise in the Southern Hemisphere. A similar pattern exists in the Atlantic Ocean.

What direction would a hurricane forming off of west Africa typically take as it approaches the United States?

south. As this current forces ocean water southwest and away from the California coast, cold bottom water is drawn to the surface, causing further chilling of the air above. San Francisco's cool summers (July average: 14°C, or 58°F) show the effect of this current and the associated upwelling of cold waters.

Landform Barriers Large mountain ranges can block air movements from one place to another and thus can affect temperatures. During the winter, the Himalayas block the cold, Northern Asiatic air from flowing south, which gives most of the Indian subcontinent a year-round tropical climate. The geographic orientation of mountains is an important factor. In North America, for example, southern slopes face the sun and tend to be warmer than shady north-facing slopes. Snow on Northern Hemisphere mountains that has accumulated on south-facing slopes, experiences more melting and is typically restricted to higher elevations than the snow cover on north-facing slopes. North-facing slopes retain more snow for a longer season and the snow cover typically extends to lower elevations in comparison to south-facing slopes. In the Southern Hemisphere, the same is true but for south-facing slopes.

Human Activities Various human activities can also have an impact on the temperature. Congregations of people in cities and other built environments can result in what is referred to as an *urban heat island* (see Geography's Spatial Science Perspective box titled "The Urban Heat Island"). What this means is that urbanized areas tend to be warmer in both summer and in winter in comparison to the surrounding countryside. In cities, many thousands of automobiles add heat to the air, whereas rural areas experience much fewer vehicles. Air conditioning units and industries also expel warmth into the urban atmosphere and environment. Even

Elevation As you know, temperatures in the troposphere decrease as the altitude or elevation increases. Anyone who has hiked upward of 500 or 1000 meters (1640 to 3280 ft), or higher in midsummer has experienced a decline in temperature with increasing elevation (● Fig. 4.26). Even if it is hot on the valley floor, you may need warm clothes once you climb to a high elevation. In Southern California you can find snow for skiing if you go to an elevation of 2400 to 3000 meters (8000 to 10,000 ft) in winter. The summits of Mauna Loa and Mauna Kea, the high volcanoes on the island of Hawaii, receive snow in the winter, and you can snowboard and ski there. Mount Kenya, 5199 meters (17,058 ft) high and located at the equator, is cold enough to have glaciers (now shrinking) on its summit. The city of Quito, Ecuador, only 1° south of the equator, has an average annual temperature of only 13°C (55°F) because it is located about 2900 meters (9500 ft) in elevation.

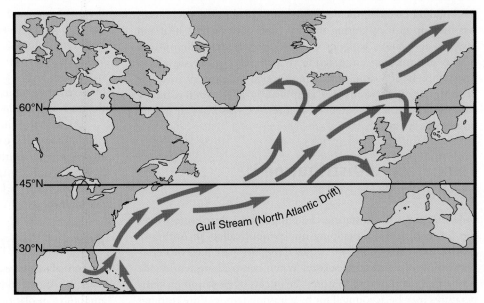

● FIGURE 4.25 The Gulf Stream (called the *North Atlantic Drift* farther north and eastward) is a warm current that moderates the climate of northern Europe.

Use this figure and the currents shown in Figure 4.24 to determine the route that sailing ships would likely follow from the United States to England and back.

At the left margin: Bryan Reckard/U.S. Navy photo

● **FIGURE 4.26** Snow-capped mountains show visual evidence of the temperature decrease with elevation. This is the summit of Mount Fuji in Japan (3776 m/12,388 ft) in autumn. Note the fall colors on the slope and the elevation where trees cannot grow.

At what rate per 1000 meters do temperatures decrease with height in the troposphere?

In Google Earth fly to: 35.356°N, 138.734°W (Mt. Fuji).

the building material in cities (concrete, asphalt, glass, and metal) can heat up quickly during the day (with low specific heat and low albedo values), as opposed to the grass, trees, and croplands of the surrounding countryside, which react more slowly to insolation. In addition, destroying forests, draining swamps, or creating large reservoirs can significantly affect local climatic patterns. The massive extent of deforestation also has an impact on global climate.

Temperature Distribution at the Surface

One method for illustrating the temperature distribution over the Earth is to use a mapping line symbol called an isotherm. **Isotherms** (from Greek: *isos*, equal *therm*; heat) are lines that connect points of equal temperature. When constructing worldwide isothermal maps of temperature, the effects of elevation on temperatures may be accounted for by adjusting temperature readings to what they would be at sea level. This adjustment means adding 6.5°C for every 1000 meters of elevation (the normal lapse rate). The amount of temperature change over distance on an isothermal map is the **temperature gradient**. Closely spaced isotherms indicate a steep temperature gradient (high temperature change

with distance), and widely spaced lines indicate a weak gradient (low temperature change with distance).

● Figures 4.27a and 4.27b show the horizontal, worldwide temperature distributions during January and July, when the seasonal extremes of high and low temperatures exist in the Northern and Southern hemispheres. The easiest temperature distribution to recognize on these two maps is the general orientation of the isotherms; they run nearly east to west, illustrating the general relationship between temperature and latitude.

A comparison of Figures 4.27a and 4.27b reveals some additional relationships. The Southern Hemisphere has its highest temperatures in January and the Northern Hemisphere has its highest temperatures in July. Note that on the July map, Portugal in the Northern Hemisphere is nearly on the 20°C (68°F) isotherm, whereas in South Australia in the Southern Hemisphere the average July temperature is around 10°C (50°F), even though these locations are about the same distance from the equator. These temperature differences between the two hemispheres illustrate seasonal variations in insolation.

Note that the greatest deviation from east-west trending temperatures occurs where the isotherms change from crossing landmasses to crossing oceans. As these isotherms leave the land, they usually bend rather sharply poleward in the winter and toward the equator in the summer. This isotherm pattern is a result of the differential heating and cooling of land and water. Continents tend to be warmer than the oceans in the summer and colder than the oceans in the winter. Another interesting geographic pattern can be seen on the January and July maps. Note that the isotherms poleward of 40° latitude have a more regular east-west orientation in the Southern Hemisphere compared to the Northern Hemisphere. This is because the Southern Hemisphere (sometimes referred to as the *water hemisphere*) has very little land south of 40°S latitude, other than Antarctica, so there are few land and water contrasts. Note also that the temperature gradients are much steeper in winter than in summer in both hemispheres. This is because the tropical zones have high and relatively even year-round temperatures, but the polar zones have large seasonal differences. The difference in temperature between the tropical and polar zones is much greater in winter than it is in summer.

As a final point, observe the sharp swing of the isotherms off the coasts of eastern North America, southwestern South America, and southwestern Africa in January, as well as off of Southern California in July. In these locations, the bending of the isotherms is related to the presence of either warm or cool ocean currents.

Annual Temperature Changes

Isothermal maps are commonly plotted for January and July because there is a lag of about 30 to 40 days after the solstices, when the amount of insolation is at a minimum or maximum (depending on the hemisphere) to the time of minimum or maximum temperatures. This **annual temperature lag** behind insolation is similar to the daily temperature lag. Temperatures continue to rise for a month or more after the summer solstice because insolation continues to exceed radiation loss. Temperatures continue to fall after

The Urban Heat Island

In 1820, Sir Luke Howard, an amateur meteorologist, observed that temperatures in the city of London, England, were warmer than those of the surrounding countryside. This is an early reference to an *urban heat island*, a measurable dome of warm air produced by a large urban area. Research on urban heat islands in the United States began in the mid-1950s, but serious urban climate monitoring started in the early 1970s. Since then, *urban heat island* has been the standard name for this phenomenon.

Several factors contribute to the temperature increase in cities, which are generally 1 to 6°C (2 to 10°F) warmer than the surrounding rural areas. Densely urbanized areas have a high concentration of sources that release heat into the environment, warming the city.

1. **Heating and cooling:** During the summer air conditioners pump heat into the environment, and in the wintertime some of the heat from buildings is lost to the outside. In a rural area, these factors also exist but in a much lower density.

2. **Automobiles:** In urban areas, great numbers of automobile and truck engines generate heat and expel hot exhaust. In the country, fewer vehicles—cars, trucks, and farm machinery—are operating at any given time.

3. **Industry:** Factories that operate large machinery give off heat and they tend to be located in or near urban areas. Fewer industries exist in the countryside.

4. **Materials and buildings:** The built environment of cities is dominated by concrete, asphalt, glass, and metal, which can heat up quickly during the day (with low specific heat and low albedos), as opposed to the grass, trees, and croplands of rural areas. Although urban building materials may cool down rather quickly at night, buildings, roads, and highways also continue to radiate heat until well after sunset.

5. **Pollution levels:** Particulates, aerosols, and greenhouse gas levels are higher in cities than in the surrounding countryside. A dome of pollution and haze can form over large cities. Certain pollutants reflect away some of the incoming solar radiation, shading the city somewhat by day, but this pollution may also absorb the heat trying to escape and re-radiate it back to the surface, especially at night.

6. **Surface water:** In the countryside, evaporation from lakes, ponds, streams, and water loss from vegetation can help cool the surroundings by the latent heat of evaporation (590 cal/g, or 2470 joules/g). Urban areas are designed to drain surface water quickly and efficiently with gutters, drains, and sewers.

It has been estimated that by the year 2025, about 80% of the world's population will reside in urban areas. As world population grows and urbanization increases, these six factors will likely intensify, and the urban heat island will directly affect more people than the general climatic conditions of the region.

What is the solution? Planting more trees, shrubs, and grass along with adding park space to cities can help moderate urban heat island conditions. Roofs of buildings, surfaces of paved areas, and roads can be made from materials that are more reflective of solar radiation, reducing absorption and re-radiation of heat. The practice of constructing "green roofs," gardens planted on the tops of buildings, is also increasing. This is another way to help reduce the impact of an urban heat island.

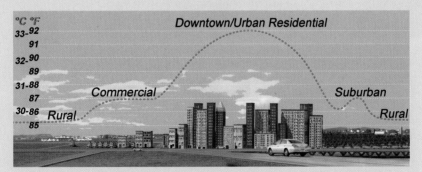

A temperature profile of an urban "heat island" shows the increase in temperature with increasing development and urbanization.

One method for reducing radiant heat from buildings is the installation of a green roof, basically growing plants on the flat tops of structures. The vegetation and soil insulate the buildings, making them cooler. Evaporation of dew, rainwater, moisture held in the soil, and moisture loss by plants adds a cooling effect. Temporary storage of precipitation received on the roof also reduces water runoff, which can be a problem in highly built-up and paved areas. Another benefit is that most people in large cities find the addition of green roofs aesthetically pleasing. The thermal infrared image (right) shows that the warmest areas, in red or yellow tones, are bare, compared to the vegetated or shaded cooler areas.

The urban heat island in Salt Lake City, Utah. The natural color view on the upper image shows areas with many buildings close together, and areas that include residential neighborhoods with lawns and trees, and vegetated areas like parks.

In comparison, the daytime thermal infrared (heat) image below shows the thermal radiation from buildings, roads, and pavement emitting heat in the densely urbanized areas of the city. The hottest areas are in red and the coolest areas are in blue. Red tones dominate in areas with high densities of buildings. Note the cool areas of the mountains on the right, the park areas, and along the river.

Comparing relative temperature values that the colors represent on the thermal image to the same areas on the natural color image provides a good illustration of the urban heat island effect.

January °C

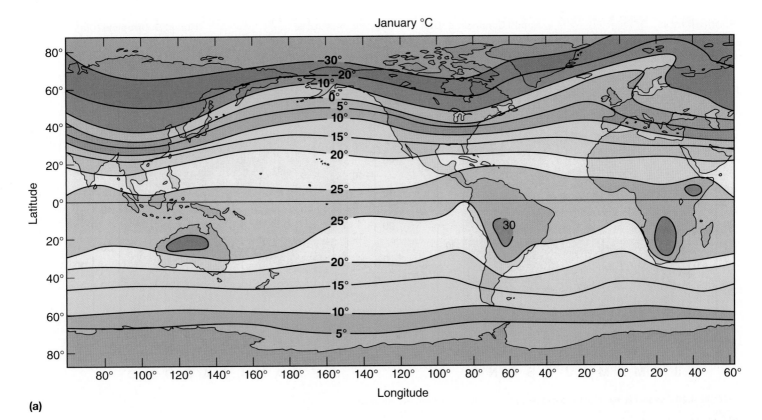

(a)

July °C

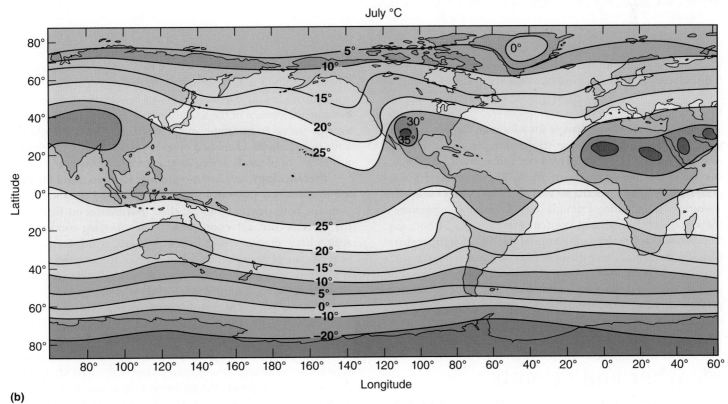

(b)

● **FIGURE 4.27** (a) Average sea-level temperatures in January (°C). (b) Average sea-level temperatures in July (°C).

Observe the temperature gradients between the equator and northern Canada in January and July. Which is greater and why?

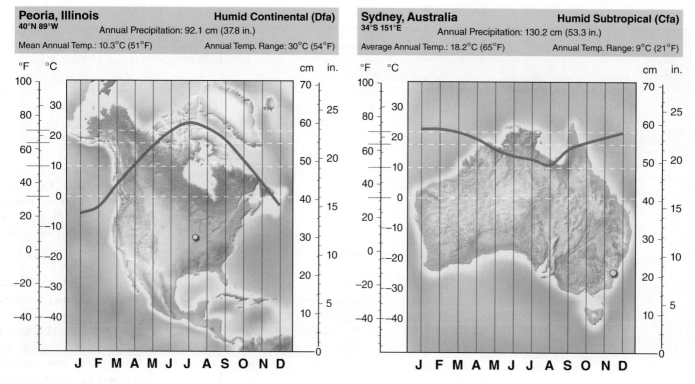

Peoria, Illinois   **Humid Continental (Dfa)**
40°N 89°W   Annual Precipitation: 92.1 cm (37.8 in.)
Mean Annual Temp.: 10.3°C (51°F)   Annual Temp. Range: 30°C (54°F)

Sydney, Australia   **Humid Subtropical (Cfa)**
34°S 151°E   Annual Precipitation: 130.2 cm (53.3 in.)
Average Annual Temp.: 18.2°C (65°F)   Annual Temp. Range: 9°C (21°F)

● **FIGURE 4.28** The annual march of temperatures at Peoria, Illinois and Sydney, Australia.

Why do these two locations have opposite seasonal temperature curves?

the winter solstice until the increase in insolation finally matches Earth's radiation. In short, these lags exist because it takes time for Earth to heat or cool and transfer those temperature changes to the atmosphere.

Annual temperature changes for a location can be plotted in a graph (● Fig. 4.28). The mean temperature for each month in a place such as Peoria, Illinois, is recorded and a curved line is drawn to connect the 12 monthly temperatures. The mean monthly temperature is the average of the mean daily temperatures recorded at a weather station during a month. The mean daily temperature is the average of the high and low temperatures for a 24-hour day. Temperature graphs depict the **annual march of temperature**, basically the temperature changes over the year—the temperature decrease from midsummer to midwinter and the temperature increase from midwinter to midsummer. The shape of the temperature curve shows how much and at what rates the temperature changes over the year.

Weather and Climate

We use the words *weather* and *climate* in general conversation, but it is important to distinguish between these terms (● Fig. 4.29a and 4.29b). **Weather** refers to the atmospheric conditions at a given time and for a specific area. The analysis of many weather observations for a place over a period of at least 30 years provides us with a description of its climate. **Climate** describes an area's average weather, but it also includes deviations from the normal or average

that are likely to occur and the atmospheric processes involved. Climatic study is also concerned with the potential for extreme weather or climatic events, which can be significant. We could describe the climate of the southeastern United States in terms of average annual temperatures and precipitation, but we would also have to include the likelihood of severe events such as hurricanes, tornadoes, or blizzards during certain periods of the year.

Meteorology involves monitoring, analyzing, and forecasting the weather. **Climatology** is the study and classification of climates, both past and present; their distribution on Earth; and the processes that influence climate. The changing conditions of atmospheric elements such as temperature, wind, rainfall, and other forms of precipitation affect soils and vegetation, modify landforms, cause flooding, and in a multitude of other ways influence the function of Earth's physical environments.

Five fundamental **atmospheric elements** serve as the indicators of weather and climate: (1) solar energy (insolation), (2) temperature, (3) atmospheric pressure, (4) wind, and (5) precipitation. We must examine these elements to understand and categorize weather and climate. A weather forecast will generally include the current temperature, the probable daily temperature range, a description of the cloud cover, the chance of precipitation, wind speed and direction, and air pressure.

The amount of insolation a place receives is the most important weather element. The other four elements depend in part on the intensity and duration of solar energy.

The temperature of the atmosphere on or near Earth's surface is largely a function of insolation, but it is also influenced by other

(a)   (b)

● **FIGURE 4.29** (a) Weather, such as a snowy day, is a short-term meteorological event. (b) These extensive glaciers in Alaska's Kenai Fjords National Park, however, have formed as a result of a cold climate over the long term.

Can you make the same weather versus climate comparisons using rainfall?

factors, such as elevation and the distribution of land and water. Unless precipitation is occurring, the air temperature may be the first weather element that we describe when someone asks us what it is like outside.

However, if it is raining, foggy, or snowing, we will probably mention that condition first. We tend to be less aware of the amount of water vapor or moisture in the air unless it is very dry or very humid. However, moisture in the air is another vital weather element that plays an important role in the likelihood of precipitation.

At some locations, or during certain times of the year, the weather can change almost daily. At other locations, there may be weeks of uninterrupted sunshine, blue skies, and moderate temperatures followed by weeks of persistent rain. A few places experience only minor differences in the weather throughout the year. The language of the original people of Hawaii is said to have no word for weather because conditions there varied so little.

Complexity of Earth's Energy Systems

This chapter has explained Earth's energy systems and dynamic interactions. These systems and processes vary as a result of complex interrelationships between the Earth and its atmosphere, especially the energy gained, stored, and lost by the Earth system. The geographic distributions of energy and temperature vary horizontally across Earth's surface and vertically through the atmosphere. They also vary over both daily and seasonal time frames.

Variations in Earth's energy systems impose both diurnal and annual rhythms on virtually all environments and on our agriculture, recreational pursuits, clothing styles, architecture, and energy bills. Human activities are constantly influenced by temperature changes, which reflect the input-output patterns of Earth's energy systems.

CHAPTER 4 ACTIVITIES

■ TERMS FOR REVIEW

scattering
Karman Line
photosynthesis
water vapor
aerosols
particulates
pollutants
greenhouse effect
enhanced greenhouse effect
ozone
ozone layer
troposphere
normal lapse rate (environmental lapse rate)
tropopause
stratosphere
mesosphere
thermosphere
ozonosphere
ionosphere
radiation

terrestrial radiation
conduction
convection
advection
latent heat exchange
condensation
freezing
melting
evaporation
evaporative cooling
Earth's energy budget
energy balance
heat
temperature
Fahrenheit scale
Celsius (centigrade) scale
Kelvin scale
absolute zero
diurnal
temperature lag
daily temperature lag

daily march of temperature
maritime
continentality
albedo
temperature inversion
inversion layer
upper-level inversion
surface inversion (radiation inversion)
controls of temperature
specific heat
warm current
cold current
isotherm
temperature gradient
annual temperature lag
annual march of temperature
weather
climate
meteorology
climatology
atmospheric elements

■ QUESTIONS FOR REVIEW

1. Why is it useful to think of the atmosphere as a thin film of air? As an ocean of air?
2. Name the gases of the atmosphere and give the percentage of the total that each supplies.
3. What function does ozone play in supporting life on Earth? Where and how is ozone formed?
4. How is the atmosphere subdivided? In what levels do you live? Have you been in any of the other levels?
5. How does Earth's atmosphere affect incoming solar radiation (insolation)? What processes prevent insolation from reaching Earth's surface? What percentages are involved in a generalized situation? What percentages reach the surface and by what processes?
6. Discuss the role of water in energy exchange. What characteristics of water make it so important?
7. How is the atmosphere heated from Earth's surface? What processes and percentages are involved in a generalized situation?
8. What is meant by Earth's energy budget? List and define the important energy exchanges that keep it in balance.
9. What is the temperature in Fahrenheit degrees today in your area? In Celsius degrees?

10. At what time of day does insolation reach its maximum and its minimum? Compare this to the typical times when the daily high and low temperatures occur.
11. How would albedo affect your selection of outdoor clothes on a hot, sunny day? On a cold, clear winter day?
12. What is a temperature inversion? Explain the two major types of temperature inversions.
13. Why do vineyards and orchards use wind machines and heaters? What methods are used to prevent frost damage to plants in your area, including landscaping plants?
14. Would you expect places like Seattle to have a milder or a harsher winter than Grand Forks, North Dakota? Why?
15. Describe the behavior of the isotherms in Figures 4.27a and 4.27b. What factors cause the greatest deviation from an east-west trend? What factors cause the greatest temperature shifts between the January and July maps (for the same location)?
16. What is the difference between meteorology and climatology?
17. What are the basic *atmospheric elements* of weather and climate?
18. What factors cause variations in the elements of weather and climate?

CONSIDER AND RESPOND

1. Convert the following temperatures to Fahrenheit: 20°C, 30°C, and 15°C.
2. Convert the following temperatures to Celsius: 60°F, 15°F, and 90°F.
3. Convert the temperatures in items 1 and 2 to Kelvin.
4. Refer to Figure 4.12. List the major means by which the Earth gains and loses the energy that heats the Earth's atmosphere, oceans and surface.

5. What are the major controls of weather and climate that operate in your area?
6. The normal lapse rate is 6.5°C/1000 meters. If the surface temperature is 25°C, what is the air temperature at 10,000 meters above Earth's surface? Convert your answer to degrees Fahrenheit.
7. Refer to Figures 4.27a and 4.27b. What location on Earth's surface exhibits the greatest annual range of temperature? Why?

 MindTap—Make the most of your study time by accessing everything you need to succeed in one place. Read your textbook, take notes, review flashcards, watch videos, complete activities, take practice quizzes, and more online with MindTap. Log in at **www.cengagebrain.com**.

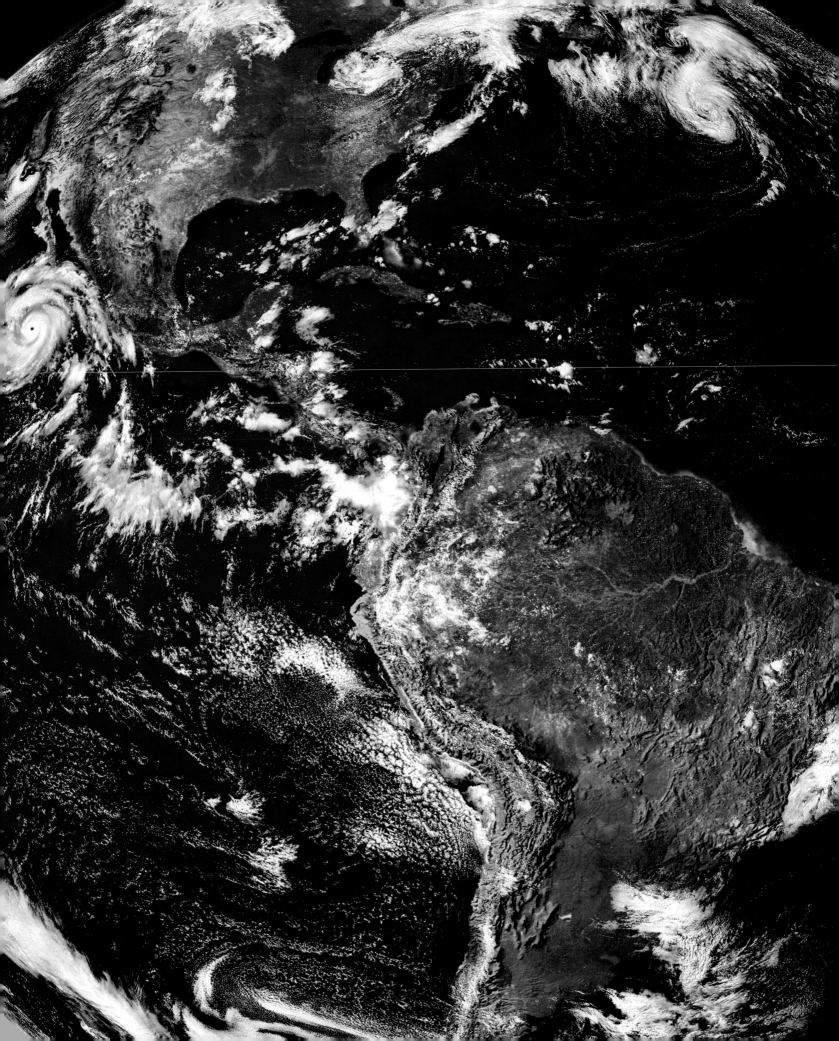

ATMOSPHERIC PRESSURE, WINDS, AND CIRCULATION PATTERNS

5

■ OBJECTIVES

WHEN YOU COMPLETE THIS CHAPTER YOU SHOULD BE ABLE TO:

- 5.1 Explain why atmospheric pressure declines with altitude and generally varies with latitude because of air temperature differences and vertical air movement.
- 5.2 Associate precipitation, clouds, and windy conditions with the rising air of low pressure systems and clear, calm conditions with the subsiding air of high pressure systems.
- 5.3 Describe why latitudinal changes in sun angle and duration of daylight hours with the seasons cause temperature variations that influence atmospheric pressure.
- 5.4 Provide examples of how differences in atmospheric pressure affect wind velocity and direction.
- 5.5 Articulate why the Coriolis effect apparently causes winds and ocean currents to bend to the right of the direction of motion in the Northern Hemisphere and to the left in the Southern Hemisphere.
- 5.6 Outline the major latitudinal pressure systems and wind belts and their influence on the circulation of global winds and ocean currents.
- 5.7 Discuss examples of how interchanges between the atmosphere and oceans influence weather systems.

AN INDIVIDUAL GAS MOLECULE IN the atmosphere weighs almost nothing. So it may be surprising to learn that as huge numbers of air molecules collide with anything, they exert an average pressure of 1034 grams per square centimeter (14.7 lb/sq in.) at sea level. The reason we are not crushed by this atmospheric pressure is that the air and water inside us—in our blood, tissues, and cells—exert an equal outward pressure that balances the atmospheric pressure.

Variations in atmospheric pressure exert a major influence on our weather and climate. Differences in *atmospheric pressure* circulate the air and create our *winds*, which cycle water from the oceans to the landmasses. Winds are also an important factor in driving the world's ocean currents. Winds transport dust, soil particles, sand, pollutants, and the emissions from volcanoes, sometimes for thousands of miles. Dispersing seeds and pollen in the biosphere is another wind process that is important to the Earth's system.

◄ The areas that are cloudy or relatively cloud-free in this image result from weather conditions related to geographic patterns of wind circulation and differences in atmospheric pressure.   NOAA

Atmospheric Pressure

Evangelista Torricelli, a student of Galileo, performed an experiment in 1643 that was the basis for inventing the **mercury barometer**, an instrument that measures **barometric pressure** (atmospheric pressure). Torricelli filled a long glass tube, which was closed on one end, with mercury and inverted it in an open pan of mercury. The mercury inside the tube fell until it was at a height of about 76 centimeters (29.92 in.) above the mercury in the pan, leaving a vacuum bubble at the closed, upper end of the tube. The atmospheric pressure exerted on the mercury in the open pan was equal to the pressure from the mercury trying to drain from the tube. When the atmospheric pressure increased, it pushed the mercury to a higher level in the tube, and as the air pressure decreased, the mercury level in the column dropped proportionately.

In the strictest sense, a mercury barometer does not actually measure the pressure exerted by the atmosphere, but instead measures the response to atmospheric pressure. That is, when the atmosphere exerts a specific pressure, the mercury will respond by rising to a specific height (● Fig. 5.1a). An **aneroid barometer** (Fig. 5.1b), a different device for measuring air pressure, uses a hollow metal cell that is sealed after removing some of the air inside. The cell, typically accordion shaped, expands and contracts with pressure changes, moving a needle pointer that indicates the atmospheric pressure on a calibrated scale. Although we usually hear atmospheric pressure reported in inches of mercury, meteorologists typically work with actual pressure units, most often the millibar (mb). **Standard sea level pressure** is equal to 1013.2 millibars and will support a level of 76 centimeters (29.92 in.) of mercury in a barometer. These values are important because they provide the cutoff measurements used to define areas (often called *cells*) of low atmospheric pressure and areas of high pressure.

Our study of the atmospheric elements that produce weather and climate has considered variations in the receipt of solar energy and their influence on temperature patterns from the equator to the poles. In this chapter we learn that temperature differences are one of the major causes of the development of higher and lower pressure areas that are often linked to latitude. In addition, we examine patterns of another kind—movement—the circulations of the atmosphere and oceans, which move energy and matter through Earth's subsystems.

Geographers are particularly interested in circulation patterns because they illustrate spatial interactions, one of geography's major themes and introduced in Chapter 1. Patterns of movement from one place to another reveal shared relationships, prompting efforts to understand the nature and effects of the interactions

● **FIGURE 5.1** (a) A mercury barometer. Standard sea level pressure of 1013.2 millibars causes the mercury to rise 76 centimeters (29.92 in.) in the tube. (b) An aneroid barometer works by an accordion-shaped cell that expands or contracts in response to air pressure changes, which moves an indicator needle up or down.

What are some advantages of an aneroid barometer?

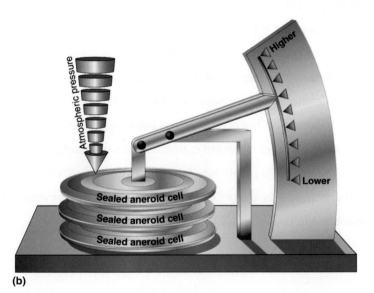

(a)

(b)

involved. It is also important to understand the causes and results of spatial interactions taking place. As we examine the circulation patterns featured later in this chapter, you should make an effort to trace each pattern back to the fundamental influence of solar energy.

Air Pressure, Altitude, and Elevation

Air pressure decreases with increasing *elevation* on Earth's surface and with greater *altitudes* above it because as we go higher the air molecules become more widely spaced and diffused. The increased space between gas molecules in the atmosphere results in lower air density and lower air pressure (● Fig. 5.2). In fact, at the top of Mount Everest (elevation 8848 m, or 29,028 ft), the air pressure is only about one third of the pressure at sea level (337 mb). Note that **altitude** usually refers to a height in the air (above sea level), and **elevation** refers to height on the surface above (or below) sea level.

People are usually not sensitive to small, gradual changes in air pressure. However, when we climb to high elevations or fly to altitudes significantly above sea level, we become aware of the effects of air pressure on our bodies. While flying at altitudes around 10,000 to 12,000 meters (33,000 to 35,000 ft), commercial airliners are pressurized to maintain air pressure at safe levels for the passengers and crew. At cruising altitude, aircraft cabins are typically pressurized to an equivalent pressure that would be experienced at an elevation of about 2100 to 2400 meters (7000 to 8000 ft). Still, pressurization may vary, so our ears may pop as they adjust to rapid pressure changes when ascending or descending.

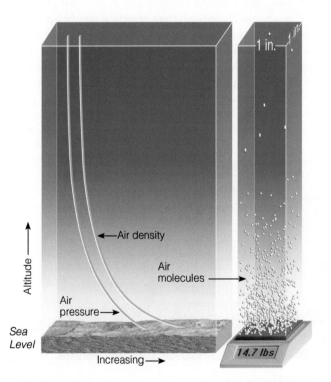

● **FIGURE 5.2** Both air pressure and air density decrease rapidly with increasing altitude.

Hiking or skiing in locations that are several thousand feet in elevation will affect us if we are used to sea level pressure. Reduced air pressure means that there are fewer air molecules in a given volume of air, so less oxygen will be contained in each breath. Thus, we get out of breath much more easily at high elevations until our bodies adjust to the drop in oxygen level because of the reduced air pressure.

Altitude and elevation are not the only factors that cause air pressure to change. On Earth's surface, pressure variations are also related to the intensity of heating from insolation, as well as from changes in local humidity and in global or regional air circulations. When the air pressure changes at a given location, it typically indicates that weather changes are occurring or coming. Different kinds of weather systems can be classified by their tendency toward changes in atmospheric pressure, as well as their structure.

Horizontal Pressure Variations

There are two main causes why air pressure changes horizontally (from one place to another). One is *thermal* (determined by air temperature) and the other is *dynamic* (related to air motion in the atmosphere). In reality, although one of these processes may dominate, the two operate together.

Thermally induced changes in air pressure are relatively easy to understand. Both air movement and air density are related to temperature differences caused by the unequal distribution of insolation, differential heating of land and water, and the varying albedos that exist on Earth's land and water surfaces. A scientific law states that the pressure and density of a gas varies inversely with its temperature. Thus, as daytime heating warms the air in contact with Earth's surface, the air expands in volume and decreases in density. As the heated air decreases in density, it will rise and consequently lower the atmospheric pressure at the surface. Thermally induced rising of warm air is a major cause of thunderstorms and contributes to the typically low pressures that dominate the equatorial regions. When air becomes cold, the volume decreases and the density increases, causing the air to sink and the atmospheric pressure to rise. As a result, the polar regions regularly experience high pressure. The low atmospheric pressure in the equatorial zone and the high pressures at the polar regions are considered thermally induced because the air temperature plays a dominant role in creating these pressure conditions.

Knowing this, we might expect that a gradual increase in air pressure would accompany the general decline in temperature from the equator to the poles. However, sea level barometer readings indicate that atmospheric pressure does not increase in a regular pattern from the poles to the equator. Instead, dynamic air movements result in regions of high pressure in the subtropics and areas of low pressure in the subpolar regions.

The dynamic processes that affect air pressure are related to broad regional patterns of atmospheric circulation. For example, at the equator when rising air encounters the tropopause, the air splits into two currents flowing in opposite, poleward directions (north and south). When these high-level air currents reach the subtropical regions, they encounter similar air currents flowing equatorward from the midlatitiudes. As these opposing upper air currents merge, the air aloft will "pile up" and descend, producing high pressure in the subtropical regions. Downward sinking air forms a high pressure cell where air will flow outward from the high toward areas of lower pressure, where the air rises, reinforcing the low atmospheric pressure in those areas.

At the surface in the subpolar regions, air flowing out of the polar high pressure zones encounters air flowing out of the subtropical high pressure regions. The collision of these converging winds causes the air to rise, creating a dynamically induced zone of low pressure that is common in the subpolar regions. High pressure in the subtropical regions and low pressure in the subpolar regions are dominantly the result of dynamic air movement. These examples describe global scale horizontal pressure variations, which we will further investigate later in this chapter.

Cells of High and Low Pressure

An area where the pressure is lower than standard sea level pressure is often simply called a **low**, but a **cyclone** is also a general term for a low pressure area. A high pressure area is called a **high**, or an **anticyclone**. Low and high pressure areas are often referred to as *cells* and are represented by the capital letters *L* and *H* on weather maps that we see on TV, on the Internet, and in the newspaper.

A cyclone (a low) is an area where air is rising. As air moves upward, it relieves pressure on that surface and barometer readings will fall. The situation in a high, or anticyclone, is just the opposite. Under conditions of high pressure, air descends and barometer readings rise, indicating an increased atmospheric pressure on the surface. Lows and highs are illustrated in ● Fig. 5.3.

Convergent and Divergent Circulation

Winds blow toward the center of a cyclone—that is, they converge on a low pressure cell. The center of a low pressure system serves as the focus for **convergent wind circulation**. In contrast, the winds blow outward and away from the center of an anticyclone, diverging from the high pressure cell. The center of a high pressure cell, where the pressure is highest, serves as the source for **divergent wind circulation**. Given existing pressure differences, we would expect converging and diverging winds to move in straight paths, as shown in Figure 5.3. Yet, wind motions are much more complex because they are influenced by other factors, which will be explained in the section on wind.

Mapping Pressure Distributions

Maps are an excellent medium that geographers and meteorologists use to visually track and analyze the areas of high and low pressure that affect our weather. But in addition to these horizontal cells of air pressure, we know that elevation also has a strong influence on atmospheric pressure.

When atmospheric pressure is mapped or presented in weather broadcasts, the air pressure reported for every surface location is adjusted to reflect what it would be if that place was at sea level. Adjusting air pressure to its sea level equivalent is important because the variations resulting from elevation are far greater than those caused by changes in the weather or in the air temperature. Without this adjustment, elevation differences would mask the regional differences, which are more important to understanding the weather. For example, if meteorologists did not adjust barometer readings to the sea level equivalent, Denver, Colorado (the "Mile High City") would always be reporting low pressure conditions, yet the barometric pressure there varies up and down, just like it does at every other place.

Isobars (from Greek: *isos*, equal; *baros*, weight) are lines displayed on maps that connect location points of equal pressure. When isobars are closely spaced, they portray a significant difference in pressure over a short distance—in other words, a strong **pressure gradient**. Isobars that are widely spaced indicate a weak pressure gradient. In this sense, isobars and their spacing on a map are much like topographic contours that display land surface areas that are steep or relatively flat (● Fig. 5.4). High and low pressure cells are outlined on atmospheric pressure maps by roughly concentric isobars that form closed cells around centers of high or

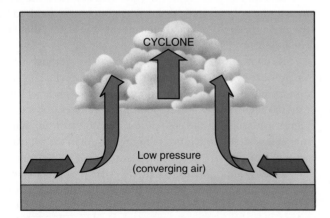
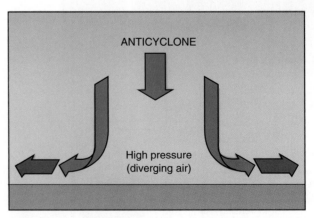

● **FIGURE 5.3** Winds converge and ascend in cyclones (low pressure centers) and descend and diverge from anticyclones (high pressure centers).

What happens to the temperature of air if it rises, and what happens if it descends?

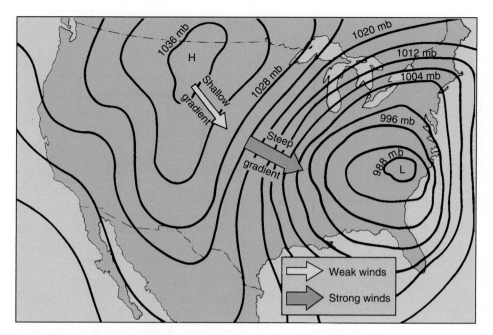

● **FIGURE 5.4** The relationship of winds to the pressure gradient: The steeper the pressure gradient, the stronger the wind.

What is the highest pressure shown on this map, and what is the lowest?

low pressure. In map view, cells of high and low pressure vary in shape from circular to elongated. If pressure cells are somewhat elongated, they may be referred to as either a *trough (low)* or a *ridge (high)*.

Wind

Wind is air that moves in response to atmospheric pressure differences. Winds play a major role in correcting the temperature imbalances that occur over Earth's surface. On average, locations below 38° latitude receive more radiant energy from the sun than they lose back to space, and locations poleward of 38° radiate away more thermal energy than they have gained through absorbed insolation (see Fig. 4.13). This makes the advection of energy by winds (and ocean currents) extremely important in affecting temperatures worldwide. The global wind system also strongly influences the surface ocean currents, which transport great quantities of thermal energy from areas that receive a surplus to regions where a deficit exists. Over time, without advection of heat by winds and ocean currents, the equatorial regions would continually get hotter and the polar regions would continually get colder.

In addition to the advectional (horizontal) transport of heat energy, winds also transport water vapor from the air over bodies of water where water has evaporated to land surfaces where it condenses to produce precipitation. Without moisture-laden winds, virtually all land areas would be arid and barren. The wind also influences evaporation rates. Furthermore, as technology increases the methods that are used to supply our energy

needs, generating electricity by wind is becoming more important, along with other natural sources such as water power and solar energy. These are clean, abundant, and renewable energy sources.

Pressure Gradients and Wind

Winds are the atmosphere's means of attempting to balance uneven pressure distributions, and they vary in velocity, duration, and direction. The velocity and strength of a wind depends on the intensity of the pressure gradient that produces the wind. As noted previously, a pressure gradient is the rate of change in atmospheric pressure between two points, typically indicated on weather maps by the spacing of isobars. When the pressure gradient is steep, with a large pressure change over a short distance, the winds will be fast and strong (see Fig. 5.4). Winds tend to flow down a pressure gradient from high pressure toward lower pressure, just as water flows downslope from a high point to a low one. A useful little rhyme, "Winds always blow from high to low," is a reminder of the direction of winds. Steep pressure gradients produce faster and stronger winds, yet the wind does not generally flow in a straight line directly from high to low.

The Coriolis Effect and Wind

Two factors related to Earth's rotation greatly influence wind direction. First, the fixed-grid system of latitude and longitude is constantly rotating at 15° per hour along with everything else on Earth. Thus, our frame of reference for tracking the path of any free-moving object—whether it is an aircraft, a missile, or the wind—is constantly changing its position. Second, the rotational speed of Earth's surface increases as we move equatorward and decreases as we move toward the poles (see Fig. 3.11). Referring to our previous example, someone in St. Petersburg, Russia (60°N latitude), where the distance around a parallel of latitude is about half of that at the equator, moves at about 840 kilometers per hour (525 mph) as the Earth rotates, whereas someone in Kampala, Uganda, near the equator, moves at about 1680 kilometers per hour (1050 mph).

Because of Earth's rotation, anything moving horizontally appears to be deflected to the right of its direction of travel in the Northern Hemisphere and to the left in the Southern Hemisphere. This apparent deflection is the **Coriolis effect**. The amount of deflection, or apparent curvature of the travel path, is a function of the object's speed and its latitudinal position. As latitude increases, so does the impact of the Coriolis effect (● Fig. 5.5). The Coriolis effect decreases at lower latitudes, and it has no impact along the equator. Also, as the distance of travel increases, so does the apparent deflection of the travel path resulting from the Coriolis effect.

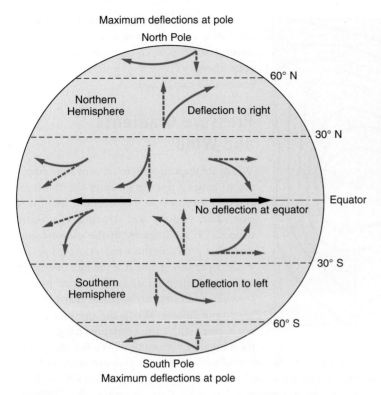

Maximum deflections at pole
North Pole

Northern
Hemisphere

Deflection to right

60° N

30° N

No deflection at equator

Equator

30° S

Southern
Hemisphere

Deflection to left

60° S

South Pole
Maximum deflections at pole

● **FIGURE 5.5** Schematic illustration of the apparent deflection (Coriolis effect) of a moving object caused by Earth's rotation when the object (or the wind) moves north, south, east, or west in either hemispheres.

Are the winds between 30°N and the equator apparently deflected to the right? How can you tell?

The flow of winds and ocean currents both experience Coriolis effect deflection. Winds in the Northern Hemisphere moving from high to low pressure are apparently deflected to the right of their expected path, and to the left in the Southern Hemisphere. In addition, when considering surface winds, we must also take into account the effect of **wind friction** with the Earth's surface features. Friction, the pressure gradient, and the Coriolis effect interact by varying degrees to influence the winds and their direction.

Friction and Wind At altitudes of about 1000 m (3281 ft) or more above the surface, frictional drag has little impact on the winds. At this level, with virtually no frictional drag, the wind will initially flow down the pressure gradient but then turn 90° in response to the Coriolis effect. When the Coriolis effect is equally counterbalanced by the pressure gradient force, the resulting wind, termed a **geostrophic wind**, will flow parallel to the isobars, instead of across the isobars (● Fig. 5.6).

At or near the surface, where winds encounter obstacles such as trees, buildings, topography, and slower moving layers of air, frictional drag becomes an additional factor because it reduces the wind speed. A lower wind speed reduces the Coriolis effect, but the pressure gradient is not affected. If the pressure gradient and the Coriolis effect are not in balance, the wind will not flow between the isobars like its upper-level counterpart. Instead, surface winds typically flow obliquely (at about a 30° angle) across the isobars and into areas of lower pressure.

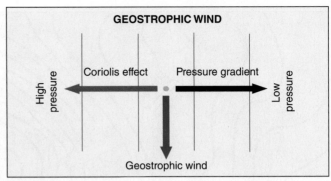

GEOSTROPHIC WIND

High pressure

Coriolis effect

Pressure gradient

Low pressure

Geostrophic wind

● **FIGURE 5.6** This Northern Hemisphere example illustrates that in a geostrophic wind, the Coriolis effect causes it to veer to the right until the pressure gradient and the Coriolis effect reach an equilibrium and the wind flows parallel to and in between the isobars.

Wind Terminology Winds are named after the direction or location that they come from. Thus, a wind that comes out of the northeast is called a *northeast wind*. A wind from the south, even though blowing toward the north, is called a *south*, or *southerly*, *wind*. It is helpful to use the phrase "out of" when describing a wind direction to help people understand the correct direction and avoid confusion about the origin of the wind.

The side of any object that faces the direction from which a wind is coming is called the **windward** side. A windward slope, then, is the side of a mountain or hill that the wind blows against (● Fig. 5.7). **Leeward** means the sheltered, downwind side that faces in the direction the wind is blowing toward. So, when the wind is coming from the west, the leeward slope of a mountain would be the eastern slope. Winds can blow from any direction, but at some locations or during certain seasons at a particular location, winds may tend to blow more from one direction than any other. Winds that commonly or seasonally flow from the same direction are referred to as **prevailing winds**.

Cyclones, Anticyclones, and Wind Directions

Imagine a high pressure cell (anticyclone) in the Northern Hemisphere in which the air is moving outward from the center in all directions down pressure gradients. As it moves, the air will be deflected to the right, no matter which direction it was originally going. Therefore, the wind moving out of an anticyclone in the Northern Hemisphere will move away from the center of high pressure in a clockwise spiral (● Fig. 5.8).

In response to the pressure gradient, air from all directions tends to flow toward the center of a low pressure area (cyclone). Despite the fact that winds are apparently deflected to the right in the Northern Hemisphere, strong pressure gradients in a low pressure cell cause the winds to flow into the center of the low in a counterclockwise spiral. These spirals are reversed in direction in the Southern Hemisphere, where winds and currents are apparently deflected to the left. Thus, in the Southern Hemisphere, winds moving away from an anticyclone do so in a counterclockwise spiral, and winds moving into a cyclone move in a clockwise spiral.

● **FIGURE 5.7** *Windward* means facing into the wind, and *leeward* means facing away from the wind.

How might vegetation differ on the windward and leeward sides of an island?

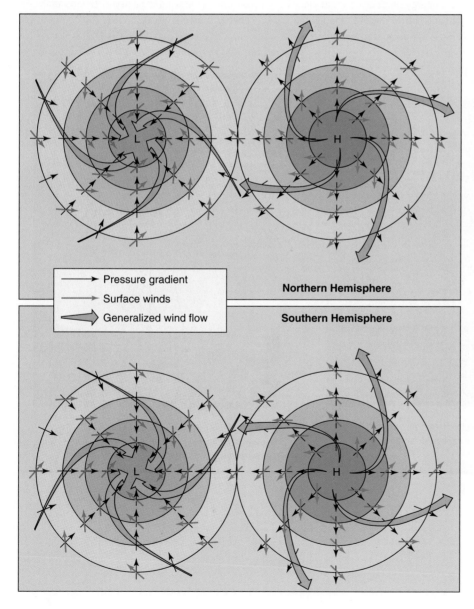

Pressure gradient
Surface winds
Generalized wind flow

Northern Hemisphere

Southern Hemisphere

● **FIGURE 5.8** The surface wind movement associated with low pressure centers and high pressure centers in the Northern and Southern hemispheres. Northern Hemisphere winds circulate clockwise out of a high but counterclockwise into a low. These rotational directions are the opposite for highs and lows in the Southern Hemisphere.

Which way do winds from a Southern Hemisphere high pressure circulate, and which way do they circulate into a low pressure cell?

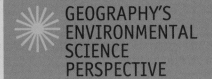
Harnessing the Wind's Energy

For centuries, windmills provided the power to pump water and grind grain throughout the world. But the widespread availability of electricity changed the role of most windmills to that of nostalgic tourist attractions. Today, air pollution problems associated with fossil fuels encourage the use of clean energy sources such as wind power to generate electricity. Although the cost of wind-produced electrical energy depends on favorable sites for locating wind turbines, wind power is cost competitive with power produced from fossil fuels.

There are two particularly important criteria for locating wind turbines: (1) the site must experience strong winds and (2) it must be located where the electricity can be linked to an electrical grid system. In the United States, individual wind turbines can be found producing electricity (such as those located on farms and ranches), but most wind power is generated from wind farms. These are long rows or groups of many turbines. Each turbine can extract up to 60% of the wind's energy at minimum wind speeds of 20 kilometers (12 mi) per hour, although higher wind speeds are desirable. Because the power generated is proportional to the cube of the wind speed, a doubling of wind velocity increases energy production eight times.

Although North America lags behind Europe in wind-generated energy, the continent has great potential. Excellent sites for wind farms exist in the open plains of North America's interior and along its coasts from the Maritime Provinces of Canada to Texas and from California to the Pacific Northwest. The sites are available, the technology has been developed, the costs are competitive, and the resolve to shift from fossil fuels is growing. Between 2010 and 2014, electricity generated by wind power in the United States increased by about 40%, but this supplied only a small percentage of the nation's current energy needs. The potential for expanding the generating capacity of this natural energy source is great.

J. Petersen

Wind turbines like this one in Germany generate electricity as the winds spin the propellers. The field of yellow flowers is planted with rapeseed, an agricultural crop that is harvested to make biofuels.

Julie Stone/US Army Corps of Engineers

Windmills like this have been used for years to pump well water on ranches and farms in the semiarid and arid regions of North America.

Global Pressure and Wind Systems
A Model of Global Pressure

Using what we have learned about atmospheric pressure on Earth's surface, we can construct a simplified and generalized model of the pressure belts of the world (● Fig. 5.9). Later, we will see how conditions depart from our model and why these differences occur by examining global pressure in more detail.

Centered approximately over the equator in our model is a belt of low pressure, or a **trough**. Because this is the region on Earth of greatest annual heating, we can conclude that the low pressure of this area, the **equatorial low (equatorial trough)**, is determined primarily by thermal factors, which cause the air to rise.

North and south of the equatorial low, centered on about 30°N and 30°S, cells of relatively high pressure dominate. These are the **subtropical highs**, which result from dynamic air motion related to the sinking of convectional cells initiated at the equatorial low.

Poleward of the subtropical highs in both hemispheres are large belts of low pressure that extend through the upper mid-latitudes. Pressure decreases through these **subpolar lows** until about 65° latitude. Dynamic factors dominate the formation of the subpolar lows, as opposing winds collide and cause air to rise.

The Arctic and Antarctic regions are dominated by high pressure systems called the **polar highs**. Extremely cold temperatures and the consequent sinking of dense cold air create the higher pressures found in the polar regions.

This generalized model of pressure belts is very useful for understanding global wind circulations and dominant pressure patterns. Yet, both temperatures and atmospheric pressures change from month to month, day to day, or hour to hour at any location. These local changes are not reflected in this model of global pressure belts, but the model does give a general idea of the latitudinal patterns of surface atmospheric pressure that influence Earth's weather and climate regions.

As our idealized model indicates, processes in the atmosphere tend to form latitudinal belts of high and low pressure, but this model does not consider the alternating influences of ocean basins and continents that these belts traverse. In latitudes where

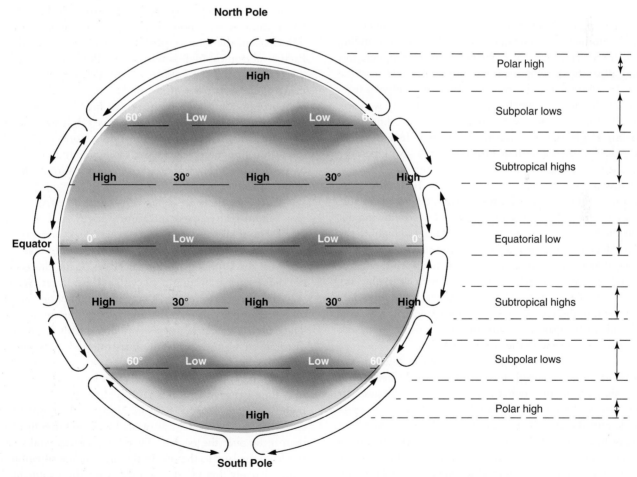

● **FIGURE 5.9** Idealized world pressure belts. Note the arrows on the perimeter of the globe that illustrate the vertical flows of air associated with the surface pressure belts.

Why do some of these pressure belts occur in pairs, one on each side of the equator?

continental landmasses are separated by ocean basins, pressure belts tend to break up to form cellular pressure systems. These cells of high and low pressure develop because the belts are affected by the differential heating of land and water. Landmasses also affect air movement and the development of pressure systems through surface friction and air flowing up and down mountains.

Seasonal Pressure Differences

In general, the atmospheric pressure belts shift northward in July and southward in January, following the changing position of the sun's direct rays as they migrate between the Tropics of Cancer and Capricorn. Thus, thermally induced seasonal variations affect the pressure patterns, as seen in ● Fig. 5.10. Seasonal differences tend to be minimal at low latitudes, where little temperature variation occurs. At high latitudes, seasonal temperature differences are larger because of greater annual variations in daylight length and angle of the sun's rays. Landmasses also alter the pattern of seasonal pressure variations for certain latitudes. This is a particularly important factor in the Northern Hemisphere, where land accounts for 40% of surface area, as opposed to less than 20% in the Southern Hemisphere.

January During the Northern Hemisphere winter, middle and high latitude continents become much colder than the surrounding oceans. Figure 5.10a shows that in the Northern Hemisphere this variation leads to the development of high pressure cells over the land areas. In contrast, also during wintertime, the subpolar lows develop over the oceans, which are are comparatively warmer. Over eastern Asia, a strong, very cold, anticyclone, known as the **Siberian High**, develops during the winter. Its equivalent in North America, the **Canadian High**, is not as strong because North America is smaller than the Eurasian continent.

Two low pressure centers also develop: the **Icelandic Low** in the North Atlantic and the **Aleutian Low** in the North Pacific. These low pressure cells result from the clash of winds flowing out of the polar high to the north and from the subtropical highs to the south. The Aleutian and Icelandic lows are associated with cloudy, unstable weather and are a major source of winter storms, whereas the Canadian and Siberian highs are associated with clear, blue-sky days, calm, starry nights, and cold, stable weather. Therefore, during the winter months, cloudy and sometimes dangerously stormy weather tends to be associated with the two oceanic lows, and clear, but cold, weather with the continental highs.

We can also see that in January the polar high in the Northern Hemisphere is well developed, primarily because of thermal cooling during the coldest time of the year. The subpolar lows have developed into the Aleutian and Icelandic low pressure cells described earlier. The subtropical highs of the Northern Hemisphere have moved slightly south of their average annual position as the sun's rays migrate toward the Tropic of Capricorn. The equatorial low also shifts south of its average annual position at the equator.

In the Southern Hemisphere during January (where it is summer), the subtropical high pressure belt breaks into three cells centered over the oceans because the warmer continents produce lower pressures compared to those over the oceans. Because there is virtually no land between 45°S and 70°S latitude, the subpolar low circles Earth over ocean waters as a belt of low pressure rather than being divided into cells. There is little seasonal change in this belt of low pressure other than in the Southern Hemisphere summer (January) when it shifts a few degrees north of its July (winter) position.

July The high pressure over the North Pole weakens during the summer, primarily because of the lengthy (24-hour daylight) heating in that region (Fig. 5.10b). The Aleutian and Icelandic Lows also weaken and shift poleward from their winter position. The large landmasses of North America and Eurasia, which developed high pressure cells during the cold winter months, develop extensive low pressure cells slightly to the south during the summer.

In Asia, a low pressure system develops, but it is divided into two separate cells by the Himalayas. The low pressure cell over northwest India is strengthened because it combines with the equatorial low, which has moved north of the position it had 6 months earlier. The subtropical highs in the Northern Hemisphere are stronger over the oceans than over the landmasses, and in the summer they have migrated northward from their winter position. In this more northerly location they have a strong influence on the climate of the nearby landmasses. The North Pacific subtropical high is termed the **Pacific High** (also called the *Hawaiian High*), a pressure system that greatly affects the climates of the west coast of North America. North Americans call the corresponding high pressure cell in the North Atlantic the **Bermuda High**, but it is the **Azores High** to Europeans and West Africans. The equatorial low moves north in July, following the seasonal migration of the sun's rays, and the subtropical highs of the Southern Hemisphere move equatorward of their January (summer) locations.

Essentially there are seven belts of pressure (two polar highs, two subpolar lows, two subtropical highs, and one equatorial low). These pressure belts, however, break into pressure cells in some places primarily because of alternating continents and oceans at the latitudes that they occupy. Again, this is because of the differences in heating between land and water. We have also seen that these belts and cells vary in size, intensity, and location with the seasons and with the migration of the sun's vertical rays. These global-scale pressure systems form a fairly regular latitudinal distribution, so they are sometimes referred to as *semipermanent pressure systems*.

As seen in Fig. 5.10 there are great land–sea contrasts in pressure (often related to temperature) throughout the year in the higher latitudes, especially in the Northern Hemisphere. During cold winters, land areas are associated with pressures that are higher than those over the oceans. Therefore, strong, cold winds blow out from the land to the sea. In the summer the situation changes, with relatively low pressure existing over the continents because of higher temperatures. Wind directions change and the pattern reverses so that in summer the winds affected by these conditions tend to flow from the sea toward the land.

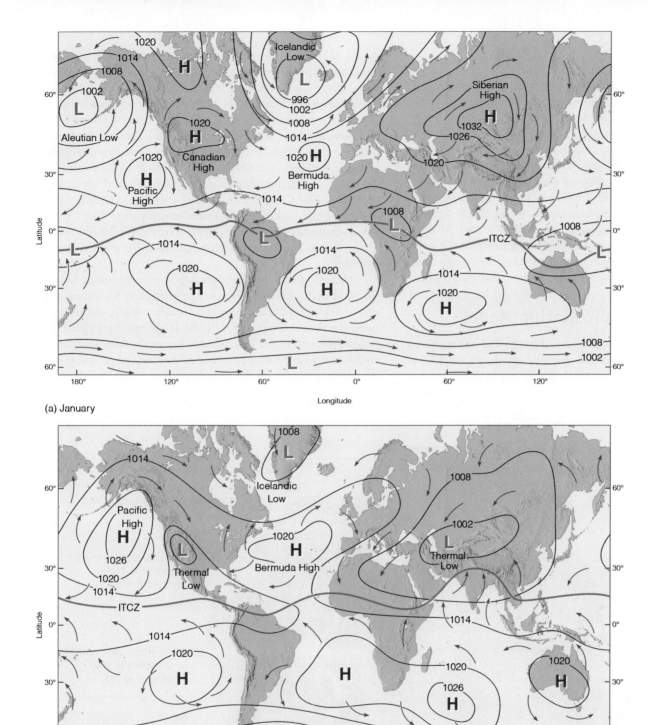

(a) January

(b) July

● **FIGURE 5.10** (a) Average sea level pressure (in millibars) in January. (b) Average sea level pressure (in millibars) in July.

UNDERSTANDING MAP CONTENT *As you examine these two maps that at first glance appear similar, you will discover that the secret to understanding the maps lies in comparing the two. In what latitudes are the most prominent high pressure systems located in January? In July? What explains the seasonal differences in the locations of high pressure systems? What explains the pressure differences between January and July over major landmasses? Which hemisphere in both maps exhibits the most regular east–west trend in the isobars? Why?*

What is the difference between the January and July average sea level pressures at your location? Why do they vary?

Global Wind Systems

Earth's planetary wind system is a response to the global pressure patterns and also plays a role in the maintenance of those regions of high and low pressures. Winds, the major means of transport for energy and moisture through the atmosphere, may also be examined on a global scale. The influences of land and water differences, variations in elevation, and seasonal changes will temporarily be disregarded for now. These factors will be addressed later. This allows the opportunity to construct a generalized model of the atmosphere's global circulation. This model can also help explain specific climate features, such as the rain and snow of the mountains of western North America and the arid regions directly to the east of those mountains. A basic wind model will provide insights for understanding the movement of ocean currents, which are driven, in part, by the global wind systems.

An Atmospheric Circulation Model

Winds transport heat and moisture through the atmosphere, locally, regionally, and globally. Because winds are caused by pressure differences, a global wind system can be based on the model of atmospheric pressures. Convergence and divergence of winds are both very important to understanding global wind patterns. Knowing that surface winds originate from and flow out of high pressure areas and blow toward low pressure areas, the model of pressure belts can be used to develop a model of the global wind system (● Fig. 5.11). Combining the worldwide pressure belts with winds flowing between them produces a simplified global atmospheric circulation model, which takes into account the effects of differential heating, Earth rotation, and atmospheric dynamics. The Coriolis effect is also considered because winds will not blow along straight north–south paths. The Coriolis effect's deflection to the right in the Northern Hemisphere and to the left in the Southern Hemisphere increases with higher latitudes.

This idealized model of atmospheric circulation includes six wind belts affected by the seven previously identified pressure zones (the Siberian High, the Canadian High, the Icelandic Low, the Aleutian Low, the Pacific High, the Bermuda High, and the Azores High). Two wind belts, one in each hemisphere, are located where winds blow out of the polar highs and toward the lower pressure of the subpolar lows. These winds in the

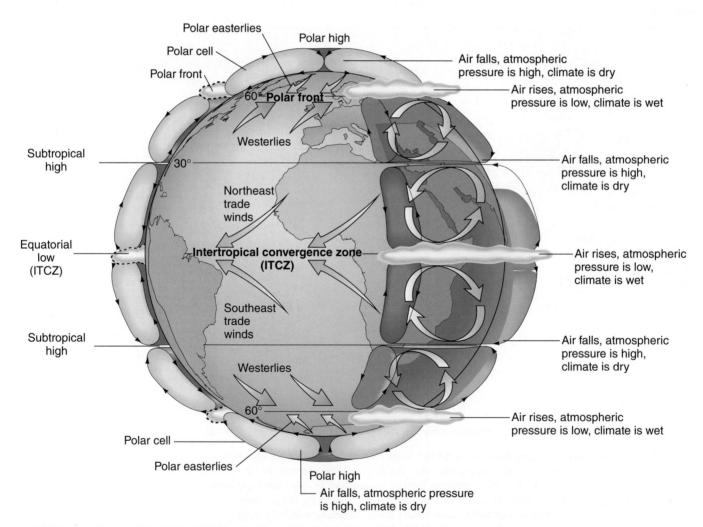

● **FIGURE 5.11** The general circulation of Earth's atmosphere, with wind directions and pressure belts.

Are the cloudy areas associated with winds converging or diverging?

high latitudes are strongly affected by apparent Coriolis deflection to the right in the Northern Hemisphere and to the left in the Southern Hemisphere, where they become the **polar easterlies**.

The divergent winds out of the subtropical highs flow both equatorward and poleward and provide the source for the remaining four wind belts. The subtropical highs are very important to weather outside of the polar regions, because four (two-thirds) of the world's six major wind belts flow out of the subtropical highs and into the subpolar lows and equatorial lows. Because of the strong apparent bending of winds flowing from the higher latitude sides of the subtropical highs, the general wind movement is from the west. These winds of the upper midlatitudes are the **westerlies**. The winds blowing from the subtropical highs toward the equator are the **trade winds**. The Coriolis effect bending in lower latitudes is minimal, so these winds are the **northeast trades** in the Northern Hemisphere and the **southeast trades** south of the equator. It is important to remember that winds are named after the direction from which they come, and that Figure 5.11 illustrates global and regional circulation, and necessarily the local area winds.

The model provides a basic concept of global atmospheric circulation, although many other factors, local and regional, also influence the winds. The pressure systems, and consequently the winds, move in response to heating differences that change with the seasonal position of the sun's rays. The winds are also influenced by continent and ocean differences between the Northern and Southern Hemispheres that affect high and low pressure zones.

Winds in Latitudinal Zones

Trade Winds
A good place to begin a more detailed examination of winds and associated weather patterns is in the vicinity of the subtropical highs. The trade winds, which blow out of the subtropical highs toward the equatorial low in both the Northern and Southern Hemispheres, are generally located between latitudes 5° and 25°. As a result of the Coriolis effect, the northern trade winds move away from the subtropical high in a clockwise direction, blowing out of the northeast. In the Southern Hemisphere, the trades diverge out of the subtropical high toward the equatorial low from the southeast in a counterclockwise direction. Because the trade winds tend to blow out of the east, they are also known as the **tropical easterlies**.

The trade winds tend to be consistent in their direction, constant, and steady. The location of the trade winds, however, does shift somewhat during the seasons, moving seasonally a few degrees of latitude north and south, with the sun's rays. Near their source in the subtropical highs, the weather of the trade winds is typically clear and dry, but after crossing large expanses of warm ocean water, the trade winds have a high potential for stormy weather. Early Spanish sailing ships depended on the northeast trade winds to drive their galleons from Europe to Central and South America and plotted a course that would use the westerlies to the north to return to the Old World. The trade winds are one of the reasons why the Hawaiian Islands are a popular tourist destination; the steady winds help keep temperatures pleasant, even though Hawaii is located south of the Tropic of Cancer.

The Intertropical Convergence Zone
The equatorial low, where the trade winds converge, coincides with the latitudinal belt that receives the world's heaviest precipitation and most persistent cloud cover. This area is called the **Intertropical Convergence Zone** (ITCZ or ITC) because it is where the trade winds from both hemispheres converge on the equatorial region. In the ITCZ, humid air heated over a warm surface expands and rises, which maintains low pressure in the area and a high potential for rainfall.

The strongest belt of convergence is located close to the equator, roughly between 5°N and 5°S, experiences rising air, heavy rainfall, and calm winds that have no prevailing direction. This zone is known as the **doldrums**. Without adequate winds for sailing, ships could remain stranded in the doldrums for days. It is interesting to note that the word *doldrums* in English means a listless condition or a depressed state of mind. The sailors were in the doldrums in more ways than one.

Subtropical Highs
The subtropical high pressure regions, generally located between latitudes 25° and 35°N and S, are the source of winds that blow generally poleward as the westerlies and equatorward as the trade winds. These subtropical belts of variable or calm winds have been called the "**horse latitudes**." This name comes from sailors having to resort to either eating their horses or throwing them overboard to conserve drinking water or to lighten their sailing ships when they became stranded in these latitudes. The centers of the subtropical highs, like the doldrums, are also regions where there are no strong prevailing winds. The doldrums are characterized by convergence, rising air, and heavy rainfall. In contrast, the subtropical highs are areas of air descending from higher altitudes. Weather conditions there are typically clear, sunny, dry, and rainless, especially over the eastern sides of the ocean basins (along west coasts) where the high pressure cells are strongest.

Westerlies
The winds that flow poleward out of the subtropical highs in the Northern Hemisphere are strongly deflected to the right and blow from the southwest. The Southern Hemisphere westerlies are strongly deflected by the Coriolis effect to the left and blow out of the northwest. This is why these winds are correctly labeled the *westerlies*. They tend to be less consistent in direction in comparison to the trade winds, but they are usually strong winds and may be associated with stormy weather. The westerlies occur between about 35° and 65°N and S latitudes. In the Southern Hemisphere, where there is less land than in the Northern Hemisphere to affect wind development, the westerlies attain their greatest consistency and strength. In the Northern Hemisphere, much of western Europe, Canada, and most of the United States (except Florida, Hawaii, and northern Alaska) are influenced by weather brought by the westerlies.

Here is a smaller version of the image at the beginning of this chapter. How well do the latitudinal patterns of cloudy and clear areas compare to the simplified models of pressure and winds shown in this chapter? Find the Intertropical Convergence Zone (Equatorial Low), and, for *each hemisphere*, locate the Subtropical Highs and the Polar Fronts (Subpolar Lows). Other than latitude, what atmospheric conditions can help you find those "belts" in this actual image of Earth? Explain any similarities and differences that you recognize.

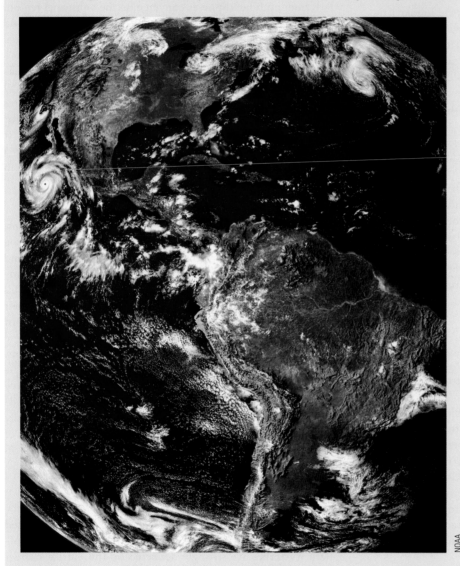

NOAA

Polar Winds Accurate surface observations of pressure and wind are sparse in the polar regions. Much of our information is derived from satellite data and imagery. Atmospheric pressure tends to be consistently high throughout the year at the poles. The polar highs feed prevailing winds (the polar easterlies) that circle the polar regions and blow easterly toward the subpolar low pressure systems. This circulation around the poles expands outward in winter and contracts in summer in response to seasonal temperatures.

Polar Front Despite our limited knowledge of wind systems in the polar regions, we do know that these winds can be highly variable, blowing at times with great speed and intensity. When the cold air flowing out of the polar regions meets the warmer air of the westerlies, these air masses meet like two warring armies. One does not absorb the other, but the denser, heavier cold air pushes warm air upward, forcing it to rise rapidly. The line along which these two great wind systems battle is appropriately known as the **polar front**, basically the zone of the *subpolar low*. The weather that results from the convergence and collision of cold polar air and warmer air from the subtropics can be very stormy. In fact, most of the storms that move through the midlatitiudes are following the path of the prevailing westerlies along the polar front.

Latitudinal Migration with the Seasons

Just as insolation, temperature, and pressure systems migrate north and south, Earth's wind systems also migrate with the seasons. In general, the pressure belts, and their associated winds, shift northward in July and southward in January between the Tropics of Cancer and Capricorn, following the seasonal migration of the sun's high-angle rays. During the Northern Hemisphere summer, maximum insolation is received north of the equator. This condition causes the pressure belts to move north also, and the wind belts of both hemispheres shift accordingly. Six months later, when maximum heating takes place south of the equator, the wind systems shift south in response to the migrating pressure systems. Thus, seasonal variations in wind and pressure conditions are important factors of the atmosphere's circulation patterns and this shifting modifies our idealized model.

The boundary zones between two wind or pressure systems are the regions that are most strongly influenced by these seasonal migrations. During the winter months, these boundary zones are generally subject to the impact of one system. As summer

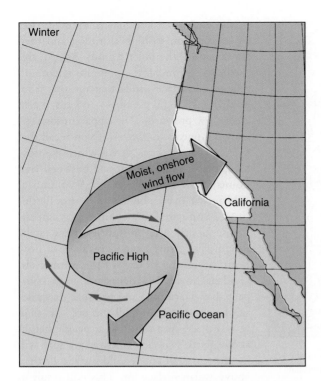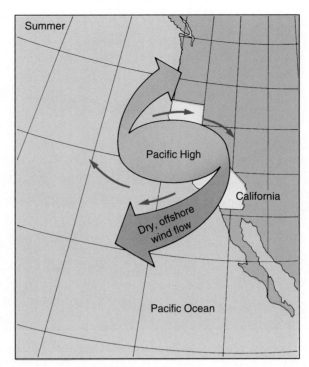

● **FIGURE 5.12** Winter and summer positions of the Pacific High off California. In the winter, this anticyclone lies to the south and feeds the westerlies that bring storms and rain from the North Pacific to California. The Pacific High dominates during the summer, blocking storms and producing warm, sunny, and dry conditions.

In what ways would the seasonal migration of the Pacific High affect agriculture in California?

approaches, the system that dominated in the winter will migrate poleward and the equatorward system will move in to influence the region. Two of these boundary zones in each hemisphere experience these distinctive summer-to-winter climatic fluctuations. The first lies between latitudes 5° and 15°, where the wet equatorial low of the high-sun season (summer) alternates with the dry subtropical high of the low-sun season (winter). The second occurs between latitudes 30° and 40°, where the subtropical high dominates in summer but is replaced by wetter winter weather delivered by the westerlies and the polar front.

California is an example of a region located within a transition zone between two wind and pressure systems (● Fig. 5.12). During the winter, this region is under the influence of the westerlies blowing out of the Pacific High. These winds, turbulent and full of moisture from the ocean, bring winter rains and storms to "sunny" California. As summer approaches, however, the subtropical high and its associated westerlies move north. As California comes under the influence of the calm and steady subtropical high pressure system, it experiences the climate for which it is famous: day after day of warmth, with clear, blue skies. This alternation between moist winters and dry summers is typical of the western sides of all landmasses between about 30° and 40° latitudes both north and south of the equator.

Longitudinal Variations in Pressure and Wind

In addition to latitudinal variations in pressure and winds, significant longitudinal differences also occur, especially in the regions of the subtropical highs. As was previously noted, the subtropical high pressure cells, which are generally centered over the oceans, are much stronger on their eastern sides than on their western sides. Thus, over the eastern portions of the oceans (west coasts of the continents) in the subtropics, subsidence and divergence are especially noticeable. Upperlevel temperature inversions, associated with the subtropical highs, develop air that is clear and calm. Air flowing equatorward from this eastern side of the subtropical high produces steady trade winds with clear, dry weather.

Over the western sides of the ocean basins (east coasts of continents), conditions are markedly different. While passing over the ocean, the air gradually warms and picks up moisture along with the turbulent and stormy weather conditions that can develop. As indicated in ● Fig. 5.13, winds in the western portions of the subtropical anticyclones, influenced by the Coriolis effect, bend poleward and toward landmasses. The trade winds in these areas are especially weak or nonexistent much of the year.

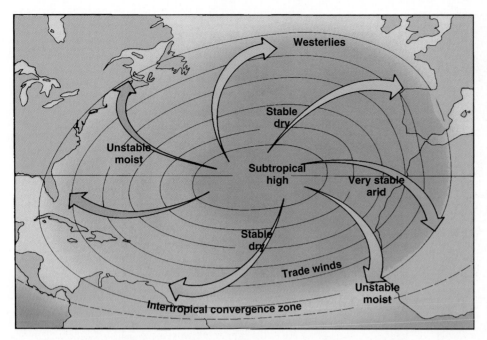

● **FIGURE 5.13** Circulation pattern in a Northern Hemisphere subtropical anticyclone. Subsidence of air is strongest in the eastern part of the anticyclone, producing calm air and arid conditions over adjacent land areas. The southern margin of the anticyclone feeds the northeast trade winds.

What wind system flows from the northern margin of this high pressure area?

Upper Air Winds and Jet Streams

We have examined the atmosphere's surface wind patterns, but upper level winds are also important—particularly the winds at altitudes above 5000 meters (16,500 ft) and at higher levels in the upper troposphere. The formation, movement, and decay of cyclones and anticyclones in the midlatitiudes depend to a great extent on air flowing at high altitudes in the atmosphere.

Upper air wind circulation is less complex compared with the circulation of surface winds. In the upper troposphere, an average westerly flow, the upper air westerlies, is maintained poleward of about 15° to 20° latitude in both hemispheres. Because of the reduced frictional drag high above the surface, the upper air westerlies blow much faster than their surface counterparts. This characteristic of upper air winds became apparent during World War II when high-altitude aircraft flying eastward covered distances faster than when they flew toward the west. Pilots had encountered the upper air westerlies or perhaps even the jet streams—very strong air currents embedded within the upper air westerlies. The **jet streams** are high-altitude examples of geostrophic winds, flowing parallel between isobars in response to a balance between the Coriolis effect and the pressure gradient.

The upper air westerlies form as a response to temperature difference between warm tropical air and cold polar air. Air in the equatorial latitudes is warmed, rises by convection to high altitudes, and then flows toward the polar regions. At first, this seems to contradict the previous statement, relative to surface winds, that air flows from cold areas (high pressure) toward warm areas (low pressure). This apparent discrepancy disappears, however, if you recall that the pressure gradient, down which the flow takes place, must be assessed between two points at the same elevation. A column of cold air will exert a higher pressure at Earth's surface compared to a column of warm air. Consequently, the pressure gradient established at the surface will result in a flow from the cooler air toward the warmer air. Cold air, however, is denser and more compact than warm air, and pressure decreases with height more rapidly in cold air than in warm air. As a result, at a specific height above Earth's surface, a lower pressure will be encountered above cold surface air than above warm surface air. This will result in an air flowing from the warmer surface air toward the colder surface air at that height. ● Fig. 5.14 illustrates this concept.

As the upper air winds flow from the equator toward the poles (down the pressure gradient), they are turned eastward because of the Coriolis effect. The net result is a broad circumpolar flow of westerly winds throughout most of the upper atmosphere (● Fig. 5.15). Because the upper air westerlies form in response to the temperature gradient between tropical and polar areas, it is not surprising that they are strongest in winter (the low-sun season) when the thermal contrast is greatest. During summer (the high-sun season) when the temperature contrast over the hemisphere is much reduced, the upper air westerlies move more slowly. The temperature gradient between tropical and polar air, especially in winter, is concentrated where the warm tropical air meets cold polar air. This boundary is the polar front, which has a strong pressure gradient.

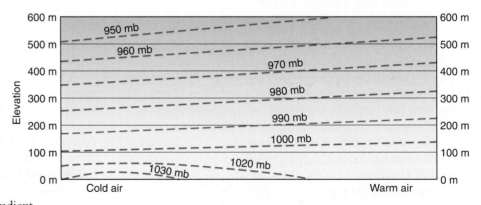

● **FIGURE 5.14** Variations of pressure with height. Note that the horizontal pressure gradient is from cold to warm air at the surface and in the opposite direction at higher elevations (such as 400 m).

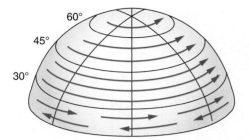

● **FIGURE 5.15** The upper air westerlies form a broad circumpolar flow.

What would you call these kinds of winds based on their relationships to the pressure gradient?

The best-known jet stream is the **polar jet stream**, which flows in the tropopause above the polar front, the area of the subpolar low. Ranging from 40 to 160 kilometers (25 to 100 mi) in width and up to 2 or 3 kilometers (1 to 2 mi) in depth, the polar jet stream is a faster, internal current of air within the upper air westerlies. Whereas the polar jet stream flows over the mid-latitidues, another westerly **subtropical jet stream** flows above the subtropical highs in the lower midlatitudes (● Fig. 5.16). Both jet streams are best developed in the winter, when temperatures exhibit their steepest gradient, and weaken in the summer. During the summer, the subtropical jet stream frequently disappears completely and the polar jet stream migrates poleward. ● Fig. 5.17 shows the position of these jet streams as they relate to the vertical and surface circulation of the atmosphere.

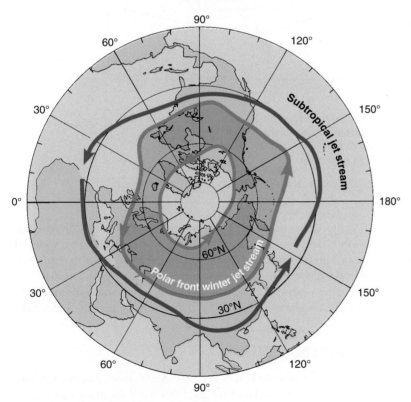

● **FIGURE 5.16** Approximate location of the subtropical jet stream and area of activity of the polar front jet stream (*shaded*) in the Northern Hemisphere winter.

Which jet stream is most likely to affect your home state?

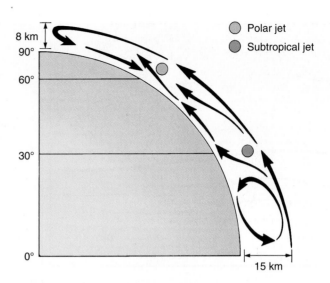

● **FIGURE 5.17** A schematic cross section of the average circulation in the atmosphere at the surface and aloft.

In general, the upper air westerlies and the associated polar jet stream flow in a fairly smooth pattern (● Fig. 5.18a). At times, however, the upper air westerlies develop a wave form, called *long waves* or **Rossby waves**, named after the Swedish meteorologist who discovered their existence (Fig. 5.18b). Rossby waves result from cold polar air pushing into the lower latitudes, forming troughs of low pressure, while warm tropical air moves into higher latitudes, forming ridges of high pressure. It is when the upper air circulation is in this configuration that midlatitiude surface weather is most influenced. We will examine this influence in more detail in Chapter 7.

Eventually, the upper air Rossby waves can become so elongated that "tongues" of air are cut off, forming warm and cold cells in the upper air (Fig. 5.18c and d). This process helps maintain a net poleward exchange of heat from equatorial and tropical areas. The cells eventually dissipate, and the normal pattern returns (Fig. 5.18a). A complete cycle takes approximately 4 to 8 weeks. Although it is not completely clear why the upper atmosphere goes into these oscillating patterns, researchers are currently gaining additional insights. One possible cause is variations in temperatures of the ocean surface. If the oceans in the northern Pacific or near the equator become unusually warm or cold (for example, during El Niño or La Niña, discussed later in this chapter), this apparently triggers oscillations, which continue until the ocean surface temperature returns to normal conditions. Other causes are also being investigated.

In addition to their influence on weather, jet streams are important for other reasons. They can carry pollutants, such as radioactive or volcanic dust (called *ash*), over great distances and at relatively rapid rates. It was the polar jet stream that carried volcanic ash from the 1980 eruption of Mount St. Helens eastward across the United States and southern Canada. The westerly flow of the jet stream also carried volcanic ash from an eruption in Iceland in 2010 to Europe, disrupting air travel (see *Geography's Spatial*

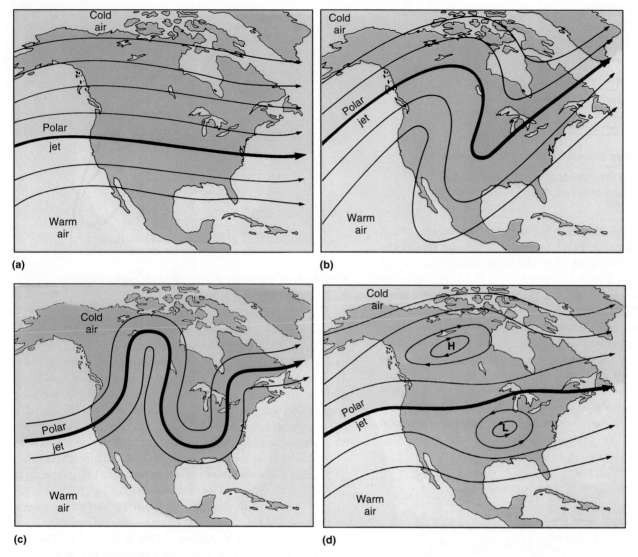

● **FIGURE 5.18** Development and dissipation of Rossby waves in the upper air westerlies. (a) A fairly smooth flow prevails. (b) Rossby waves form, with a ridge of warm air extending into Canada and a trough of cold air extending down to Texas. (c) The trough and ridge may begin to elongate and turn back on themselves. (d) The trough and ridge are cut off into cells that will eventually dissipate. The flow will then return to a pattern similar to (a).

How are Rossby waves associated with the changeable weather of the central and eastern United States?

Science Perspective: Volcanic Eruptions, Upper Air Winds, and Aviation Routes). Nuclear fallout from the Chernobyl incident in the former Soviet Union could be monitored in succeeding days as it crossed the Pacific, and later the United States, in the jet streams. Because of a westerly tailwind from the jet stream, flying times on commercial airplanes from North America to Europe may be significantly shorter than those in the reverse direction.

Regional and Local Wind Systems

The global wind system illustrates general circulation patterns that reflect latitudinal imbalances in temperature. On a regional or local scale, additional wind systems also develop in response to

temperature differences. Many regions experience differences in wind directions and conditions over the seasons. Monsoon winds are an example that are subcontinental in size and develop in response to seasonal variations in temperature and pressure. At the smallest scale are the local winds, which develop in response to the diurnal (daily) variations in heating and other local effects on pressure and winds.

Monsoon Winds

The term *monsoon* comes from the Arabic word *mausim*, meaning "season." Arab sailors have used this word for many centuries to describe seasonal reversals in wind direction across the Arabian Sea between Arabia and India. As a meteorological term, **monsoon** refers to the directional reversal of winds from one season to another. Usually, a monsoon occurs when a wind blowing warm,

Volcanic Eruptions, Upper Air Winds, and Aviation Routes

Dust-sized lava fragments erupted by volcanoes, called *volcanic ash*, can be blasted high into the atmosphere and carried by the upper air winds for hundreds or thousands of miles. Volcanic ash particles are harder than steel and very abrasive. When they come in contact with aircraft, these particles can cause problems, and if sucked into a jet engine, the resulting damage may cause engines to lose power. Clouds of volcanic ash are also generally invisible to aircraft pilots flying in their vicinity.

In 1982, a passenger jet flying from Malaysia to Western Australia at about 11,000 m (36,000 ft) flew through a plume of volcanic ash from an Indonesian volcano. Damage from the abrasive ash caused all four engines of the aircraft to fail. The plane rapidly descended to an altitude of 3810 m (12,500 ft), when the pilots were able to restart the engines and safely land at a nearby airport despite the jet engine damage. A terrible accident was averted, and attention was focused on how to avoid aircraft encounters with the damaging effects of dense atmospheric clouds of volcanic ash.

A geographic-spatial aspect was related to this problem. Many worldwide air travel

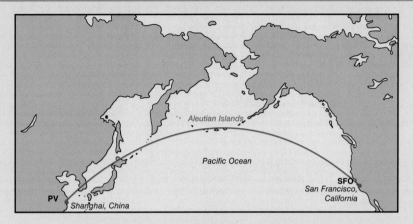

Many great circle route-flight paths between the U.S. West Coast and East Asia traverse the vicinity of active volcanoes.

routes, particularly around the Pacific Ocean, pass near active volcanoes with the potential to erupt great volumes of ash. Seeking the shortest distances between places on Earth's surface, aircraft follow great circle routes. The Aleutian Islands have many active volcanoes, as well as the east coast of Asia and the Japanese islands. The upper air westerly winds, steered in Rossby waves, often carry ash from erupting volcanoes into and across heavily used airline routes. Note the air route between

San Francisco and Shanghai, China, seen on the map here.

Today many systems are in place from the ground, from satellites, and on commercial aircraft to avoid ash plumes by detection, mapping, and tracking volcanic ash in air routes. If flights cannot avoid the ash plumes with a change in route, they are generally canceled until the ash problem clears, typically within a few days, as eruptions change and wind directions shift.

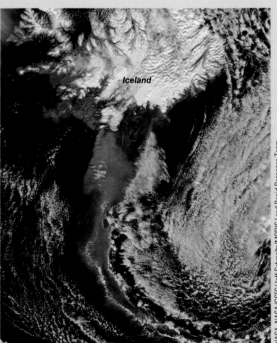

NASA, NASA/GSFC/Jeff Schmaltz/MODIS Land Rapid Response Team

The volcanic ash cloud from this volcano in Iceland was well detected, as it crossed busy airline routes, delaying flights in Europe and between Europe and North America for several days in April 2010.

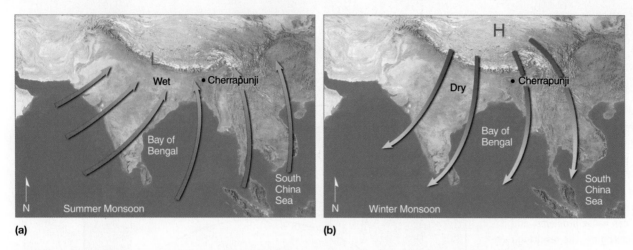

● **FIGURE 5.19** Seasonal wind reversals create the Asiatic monsoon system. (a) The "wet monsoon," with its onshore flow of humid tropical air in summer, is characterized by heavy precipitation. (b) The winter offshore flow of dry continental air creates the "dry monsoon" and drought conditions in southern Asia. Despite having a dry season, Cherrapungi, India, is one of the world's rainiest places.

How do the seasonal changes of wind direction in Asia compare to those of the southern United States?

humid air from the ocean toward the land in summer shifts to a dry, cooler wind blowing toward the sea from the land in winter. A monsoon, or monsoonal conditions, typically involves an opposite seasonal change in wind direction.

The best-known monsoon is characteristic of southern Asia, although it also occurs on other continents. During the summer the Asian continent becomes very warm and develops a large low pressure cell. Warm, humid, tropical air from the ocean where air pressure is higher, is attracted into this low (● Fig. 5.19a). This is reinforced by a poleward shift of moist tropical air to a position over southern Asia. Full of water vapor, any turbulence, convection, or landform barrier that makes this air rise will bring about the heavy precipitation and wet summer weather associated with the monsoon. Extremely heavy rainfall and flooding can occur during summertime in the Himalayan foothills, in other parts of India, and in areas of Southeast Asia.

In winter the large Asian landmass cools to temperatures that are much colder than the surrounding oceans. The continent develops a strong high pressure cell, which causes an outflow of air. These cold, dry winds blow southward from the continent toward the tropical low that exists over the warmer oceans (● Fig. 5.19b). During this time, the winter monsoon is a dry season dominated by low humidity air coming in from a dry continental area.

In the lower latitudes, a monsoonal wind shift can be related to the migration of the high-angle rays of the sun. For example, winds in the equatorial zone migrate during the summer months northward toward the southern coast of Asia, bringing with them warm, moist, turbulent air. The winds of the Southern Hemisphere also migrate north with the sun, with some crossing the equator. These winds bring warm, moist air from the ocean to the southern and southeastern coasts of India. In the winter months, the equatorial and tropical winds migrate south, leaving southern Asia under the influence of the dry, calm winds of the tropical Northern Hemisphere. Asia and northern Australia are true monsoon areas, with a 180° wind shift with changes from summer

to winter. Other regions, such as the southern and southwestern United States and West Africa, have "monsoonal tendencies" but are not monsoons in the true meaning of the term.

The phenomenon of monsoon winds and their characteristic seasonal shifting cannot be fully explained, however, by the differential heating of land and water or the seasonal shifting of tropical and subtropical wind belts. Some aspects of the monsoon system— for example, its "burst" or sudden transition between dry and wet in southern Asia—must have other causes. Meteorologists looking for a more complete explanation of the monsoon are examining the role played by the jet stream and other wind movements of the upper atmosphere.

Local Winds

Despite affecting a much smaller area, local winds are very important. They are often a response to terrain, or land–water differences in heating and cooling, much like larger wind systems.

Land Breeze–Sea Breeze The **land breeze–sea breeze** is a diurnal (daily) cycle of local winds that occurs in response to pressure differences caused by the differential heating of land and water (● Fig. 5.20). During the day, the land and the air above it heats quickly, warming to a higher temperature than that of the nearby ocean. The warm daytime air above the land expands and rises, creating a local area of low pressure, and the rising air is replaced by denser, cooler air from over the body of water. Thus, a sea breeze of cool, moist air blows in over the land during the day.

Alleviation of the heat by a cooling sea breeze explains one reason why seashores are so popular in summer. In some situations, however, sea breezes are responsible for afternoon cloud cover and light rain, which could spoil an otherwise sunny day at the shore. A sea breeze can cause a 5°C to 9°C (9°F to 16°F) reduction in temperature along the coast and a lesser influence on land perhaps as far from the ocean as 15 to 50 kilometers (9 to 30 mi). Large

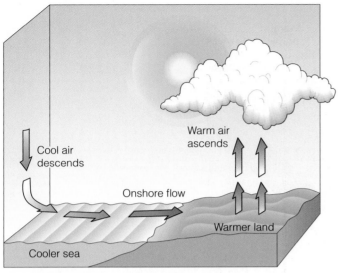

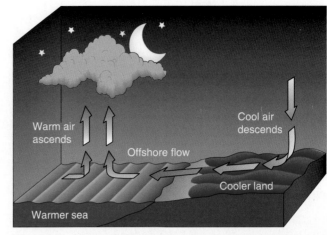

(a) In the afternoon, the land is warmer than the ocean surface, and warm air rising from the land is replaced by an onshore sea breeze.

(b) At night, as the land cools, the air over the ocean is now warmer than the air over the land. The ocean air rises. Air flows offshore to replace it, generating an offshore flow (a land breeze).

● **FIGURE 5.20** Land and sea breezes. This day-to-night reversal of winds is a consequence of the different rates of heating and cooling of land and water areas. The land becomes warmer than the sea during the day and colder than the sea at night; the air flows from the cooler to the warmer area.

What is the impact of the sea breeze on daytime coastal temperatures?

lakes and their shorelines can also experience this kind of local wind to cool cities such as Chicago, Milwaukee, and Cleveland during hot summer days.

At night, the land and the air above it cools more quickly and to a lower temperature than the nearby water body and the air above it. Consequently, the pressure builds higher over the land and air flows out to the lower pressure over the water, creating a land breeze. For thousands of years, sailboats have left their coasts at dawn, when there is still a land breeze, and have returned with the sea breeze of the late afternoon. The land–sea breeze cycle is a classic example of a diurnal wind shift.

Mountain Breeze–Valley Breeze Under the calming influence of a high pressure system, there is a daily **mountain breeze–valley breeze** cycle (● Fig. 5.21), which is somewhat similar in mechanism to the land breeze–sea breeze. During the day, when the sun heats the valleys and mountain slopes, the high exposed slopes heat faster than the lower and shadier valley. Air on the slope expands and rises, drawing air from the valley and up the mountain sides. This warm daytime breeze is the *valley breeze*, named, like most winds, for its place of origin. Clouds, which can often be seen obscuring mountain peaks, can actually be the visible evidence of condensation in the warm air rising from the

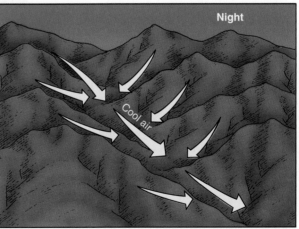

● **FIGURE 5.21** Mountain and valley breezes. This daily reversal of winds results from the heating of mountain slopes during the daytime and their cooling at night. Warm air is drawn up the slopes during the day, and cold air drains down the slopes at night.

How might a green, shady valley floor and a bare, rocky mountain slope contribute to these changes?

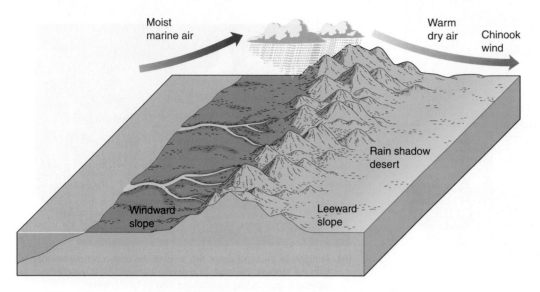

● **FIGURE 5.22** Chinook (or Foehn) winds result when air ascends the windward side of a mountain barrier, becoming cooler as it expands and losing some of its moisture through condensation and precipitation. As the air descends the leeward side of the range, its relative humidity decreases as the air compresses and warms. This produces the relatively warm, dry conditions characteristic of Chinook winds.

The term Chinook means "snow eater." How did this name came about?

valleys. At night, the valley and mountain slopes cool considerably because of radiation loss of heat energy to space through the clear air, yet they receive no solar radiation. The air cools, sinks, and flows down into the valley as a cool *mountain breeze*.

Chinooks and Other Warming Winds

One type of local wind is known by several names in different parts of the world—for example, **Chinook** in the Rocky Mountain area and **Foehn** (pronounced "fern") in the Alps. Chinook-type winds occur when air passes over a mountain range. After crossing the mountains, as these winds flow down the leeward slope, the air compresses and heats at a greater rate compared to the degree to which it cooled when it ascended the windward slope (● Fig. 5.22). Thus, the air flows into the valley below as warm, dry winds. The rapid temperature rise brought about by such winds has been known to damage crops, increase forest fire hazard, and set off avalanches.

An especially hot and dry wind is the **Santa Ana wind** of Southern California. It forms when high pressure develops over the desert regions of Southern California and Nevada. The clockwise circulation out of the high pressure cell drives warm dry air from the desert over the mountains of eastern California, and the air becomes warmer and more arid as it moves down the western slopes. These warm, dry, Santa Ana winds blow down from nearby high-desert regions, becoming warmer and drier as they descend into the coastal lowlands. Santa Ana winds are most common in the fall and winter, with wind speeds that can be 50 to 90 kilometers per hour (30 to 50 mph). Stronger wind gusts can reach 160 kilometers per hour (100 mph). The hot, dry Santa Ana winds are

notorious for fanning wildfires, which plague the southwestern United States, especially in California (● Fig. 5.23).

Drainage Winds

Also known as **katabatic winds**, **drainage winds** are local to mountainous regions and occur under cold, calm, and clear conditions. Cold, dense air can accumulate in a high valley, plateau, or snowfield within a mountainous area.

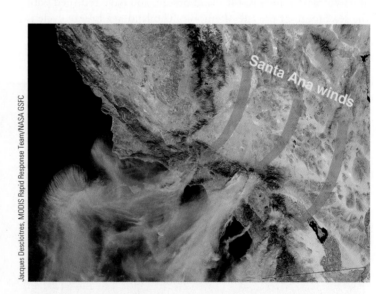

Jacques Descloitres, MODIS Rapid Response Team/NASA GSFC

● **FIGURE 5.23** The geographic setting and wind directions for the Santa Ana winds, generated by a high pressure cell to the east of Southern California. The Santa Ana winds fan wildfires during dry seasons, and blow plumes of smoke across Los Angeles and miles out to sea.

Because cold air is very dense, it tends to flow down slopes and valleys and pour out onto the land below. The katabatic winds that result from cold air accumulating over ice sheets in Greenland and Antarctica can be extremely cold and strong. Drainage winds in other regions are known by many local names; for example, on the Adriatic coast, they are called the *bora*; in France, the *mistral*; and in Alaska, the *Taku*.

There is no question that winds, both local and global, are effective elements of atmospheric conditions and dynamics. We know that a hot, breezy day is not nearly as unpleasant as a hot day without any wind. This difference exists because winds increase evaporation, resulting in the removal of heat from our bodies, the land, the air, animals, and plants. In contrast, wind on a cold day increases our discomfort, mainly by conducting heat away from our bodies and advecting heat away from our surroundings.

Ocean–Atmosphere Interactions

The majority of Earth's surface area is an interface between two fluids: air in the atmosphere and water in the oceans. Gases and water are both characterized as fluids because they can flow, so the dynamics of gases and water follow the same laws of physics. The major difference is in their densities. Water molecules are much more closely packed together, so water's density is 800 times higher than the density of air at sea level. Yet, atmospheric air motion can cause or strongly affect the movement of water in the oceans, and the oceans influence the atmosphere in many ways. Some of these interactions affect only small or local areas, some are regional, and others can have a global impact. In recent decades, oceanographers, geographers, and atmospheric scientists have combined their expertise to research some of these ocean–atmosphere relationships.

One of the best-known ocean–atmosphere interactions is the wind motion that creates waves and affects surface ocean currents. Because of the high density of water compared to air, faster movements in the atmosphere are reflected as much slower movements in the oceans. Ocean–atmosphere interactions are complex and exist at many scales with respect to both time and geographic area, and it will take many years of study before they are fully understood. The remainder of this chapter will discuss some of these very important relationships.

Ocean Currents

Like the global wind system, ocean currents play a significant role in helping to moderate latitudinal temperature imbalances. **Surface ocean currents** are fairly steady flows of seawater that move in a prevailing direction, behaving somewhat like rivers in the ocean. The surface water temperatures of ocean currents flowing along the coasts have a great influence on the climate of coastal locations.

Earth's surface wind system is a primary factor in the flow of these ocean currents. Other factors include the Coriolis effect and

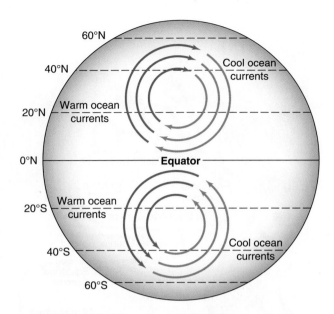

● **FIGURE 5.24** The major ocean currents flow in broad gyres in opposite directions in the Northern and Southern Hemispheres. In the midlatitiudes, cold currents flow equatorward along west coasts and warm currents flow poleward along east coasts.

What is the main control on the direction of these gyres?

the size, shape, and depth of the basin of a sea or ocean. Ocean currents are also influenced and driven by differences in density that result from temperature, evaporation, salinity, and the action of waves and tides.

The major ocean currents move in broad circulatory patterns, called **gyres**, that flow around the subtropical highs. Because of the Coriolis effect and the direction of flow around a cell of high pressure, the gyres follow a clockwise direction in the Northern Hemisphere and a counterclockwise direction in the Southern Hemisphere (● Fig. 5.24). As a general rule, surface currents do not cross the equator where the impact of Coriolis effect is minimal.

Waters near the equator in both hemispheres are driven westward by the tropical easterlies or the trade winds. The resulting ocean current is called the *Equatorial Current*. At the western margin of the ocean (east coasts), its warm, tropical waters are deflected poleward along the coastline. As these warm waters move into higher latitudes, they move through waters cooler than the water flowing in the current; therefore, they are identified as *warm currents* (● Fig. 5.25).

In the Northern Hemisphere, warm currents, such as the Gulf Stream and the Kuroshio (Japan) Current, are deflected strongly to the right (or east) because of the increasing apparent impact of the Coriolis effect that occurs with higher latitudes. At about 40°N, the westerlies begin to drive these warm waters eastward across the ocean, forming the North Atlantic Drift and the North Pacific Drift. Eventually, these currents encounter landmasses at the eastern margin of the ocean (west coasts), and most of the waters are deflected toward the equator. After flowing across the ocean basin in higher latitudes, the currents lose much

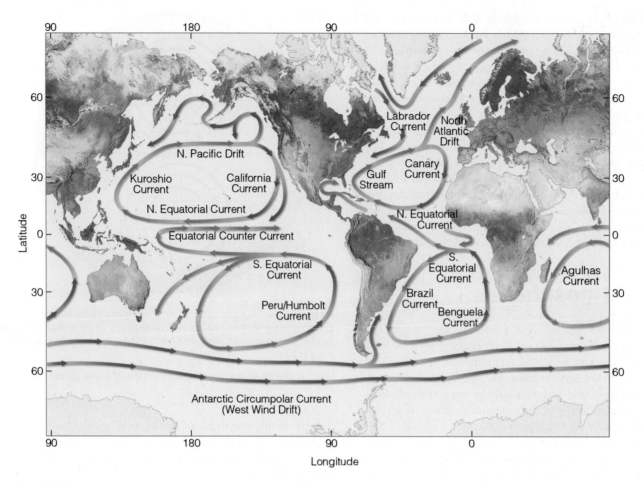

● **FIGURE 5.25** Map of the major world ocean currents showing the locations of warm and cold currents.

UNDERSTANDING MAP CONTENT What evidence on this map documents that Earth's wind and pressure systems are largely responsible for the movement and direction of flow of Earth's ocean currents? What global pressure systems are most involved? In which general directions in regard to the equator and the poles are the major warm currents flowing? The major cool currents? How does the map demonstrate the influence of the Coriolis effect on ocean currents?

How does this map of ocean currents help explain the mild winters in London, England?

of their warmth, become cooler than the adjacent waters and flow equatorward, as cold currents into the subtropical latitudes. Nearing the equator, these currents complete the circulation pattern when they rejoin the westward-flowing Equatorial Current.

On the eastern side of the North Atlantic Ocean, the North Atlantic Drift flows into the seas north of the British Isles and around Scandinavia, keeping those areas warmer than their latitudes would suggest. Some Norwegian ports north of the Arctic Circle remain ice free because of the moderating effect of this relatively warmer water. Cold polar waters in the Labrador Current and in the Oyashio (Kurile) Current flow southward into the ocean basins of the Atlantic and Pacific oceans and along the margins of the continents.

Oceanic circulation in the Southern Hemisphere is comparable to that in the Northern Hemisphere except that the gyres flow in a counterclockwise direction. Because there is little land poleward of 40°S in the Southern Hemisphere, the Antarctic Circumpolar Current (also called the *West Wind Drift*) circles Antarctica as a cool

(cold) current across the Southern Ocean almost without interruption. It is cooled by the influence of its high latitude location and cold air flowing from the polar high on the Antarctic ice sheet.

In general, warm currents flow poleward, carrying warm tropical waters into the cooler waters of higher latitudes, as in the case of the Gulf Stream or the Brazil Current. Cold currents flow equatorward along west coasts of continents, and two examples are the California Current and the Humboldt Current. Warm currents tend to bring humidity and warmth to the east coasts of continents along which they flow, and cool currents tend to have a drying and cooling effect on the west coasts.

The contact between the atmosphere and ocean currents is one reason why subtropical highs have a strong side and a weak side. Subtropical highs on the west coast of continents are in contact with cold ocean currents, which cool the air and stabilize and strengthen the eastern side of a subtropical high. On the east coasts of continents, contacts with warm ocean currents cause the western sides of subtropical highs to be less stable and weaker.

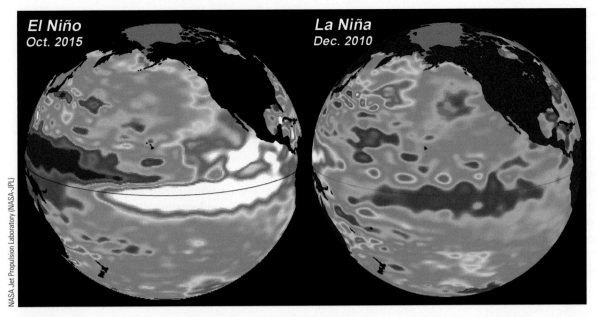

El Niño
Oct. 2015

La Niña
Dec. 2010

NASA Jet Propulsion Laboratory (NASA-JPL)

● **FIGURE 5.26** These thermal infrared satellite images show El Niño (*left*) and La Niña (*right*) episodes in the tropical Pacific. The red and white shades display the warmer sea surface temperatures, and the blue or purple colors mark areas of cooler temperatures.

What are some impacts of an El Niño?

Upwelling, the vertical movement of cold bottom waters upward to the surface, generally reinforces cold currents flowing along west coasts in subtropical latitudes. The Coriolis effect tends to shift equatorward-flowing winds in these latitudes away from the coast, and the wind's frictional drag on the ocean surface displaces water in an offshore direction. The surface waters are dragged away and replaced by deeper, colder water that rises to the surface. This upwelling of cold waters reinforces the strength and the effects of the California, Humboldt (Peru), Canary, and Benguela currents.

El Niño

As you can see in Fig. 5.25, the cold Humboldt Current flows equatorward along the coasts of Ecuador and Peru. When the current approaches the equator, the westward-flowing trade winds cause upwelling of nutrient-rich cold water along the coast. Fishing, especially for anchovies, is a major local industry in this region.

Every few years a warm countercurrent replaces the normally cold coastal waters, an occurrence that usually peaks during the month of December. Without the upwelling of nutrients from below to feed the fish, fishing comes to a standstill. Fishermen in this region have known this phenomenon for hundreds of years. They traditionally set aside this time to tend to their equipment and await the return of cold water. The people of the region have named this occurrence **El Niño**, a Spanish reference to the Christ child ("the boy") because this phenomenon tends to be strongest around Christmastime.

The warm-water countercurrent usually lasts a few months, but occasionally it can last for a year or two. During an El Niño, water temperatures are raised not just along the coast but also for thousands of miles offshore (● Fig. 5.26). Over the past decades, the term El Niño has come to describe these exceptionally strong episodes, and not just the annual event. During the past 65 years,

approximately 18 years experienced El Niño conditions (with sea surface temperatures 0.5°C warmer than normal for 6 consecutive months). El Niño events not only affect the temperature of the equatorial Pacific, but the strongest of them also have an impact on global weather patterns. The strongest El Niño in 65 years of scientific observations was in 1997. Although unpredictable, Pacific Ocean surface warming in October 2015 indicates another very strong, and perhaps an even stronger El Niño.

El Niño and the Southern Oscillation Understanding the processes that interact to produce an El Niño requires studying conditions across the Pacific, not just the waters off South America. In the 1920s, British scientist Sir Gilbert Walker discovered a connection between barometric pressure readings at weather stations on the eastern and western sides of the Pacific basin. He noted that a rise in pressure in the eastern Pacific is usually accompanied by a drop in pressure in the western Pacific and vice versa. Walker called this seesaw pattern the **Southern Oscillation**. The link between El Niño and the Southern Oscillation is so great that they are often referred to jointly as **ENSO (El Niño/Southern Oscillation)**. Today the atmospheric pressure values from Darwin, Australia, are compared to those recorded on the Island of Tahiti, and the relationship between these two values is used to define the Southern Oscillation.

During a typical year, the eastern Pacific has a higher pressure than the western Pacific. This east-to-west pressure gradient strengthens the trade winds over the equatorial Pacific waters. This results in a surface current that moves from east to west at the equator. The western Pacific develops a thick layer of warm water, while the cold Humboldt Current in the eastern Pacific (west coast of South America) is enhanced by upwelling.

Then, for unknown reasons, the Southern Oscillation swings in the opposite direction, dramatically changing the more typical conditions described earlier, with pressure increasing in the western Pacific and decreasing in the eastern Pacific. This pressure gradient change causes the trade winds to weaken; in some cases, they may even reverse. Under these conditions warm water in the western Pacific flows eastward, increasing sea surface temperatures in the central and eastern Pacific. This eastward shift signals the beginning of El Niño.

In contrast, at times and for reasons not fully understood, the trade winds intensify. These more powerful trade winds will cause even stronger upwelling than usual to occur. As a result, sea surface temperatures will be colder than normal off the South American west coast. This condition is known as **La Niña** (in Spanish, "little girl," but scientifically simply the opposite of El Niño). La Niña episodes will at times, but not always, bring about the opposite effects of an El Niño episode (see Fig. 5.26).

El Niño and Global Weather
Cold ocean waters impede cloud formation, except for coastal fogs. Thus, under normal conditions, clouds tend to develop over the warm waters of the western Pacific but not over the colder waters of the eastern Pacific ocean basin (● Fig. 5.27a). However, during an El Niño, when warm water migrates eastward, clouds develop over the equatorial region of the Pacific (Fig. 5.27b). These clouds can build to heights of 18,000 meters (59,000 ft). Clouds of this magnitude can disrupt the high-altitude wind flow above the equator. As we have seen, a change in the upper air winds in one portion of the atmosphere will trigger wind flow changes in other portions of the atmosphere. Alterations in the upper air winds result in changes to surface weather.

Scientists have tried to document as many past El Niño events as possible by piecing together historical evidence, such as sea surface temperature records, daily observations of atmospheric pressure and rainfall, fishery records from South America, and the writings of Spanish colonists living along coastal Peru and Ecuador dating back to the fifteenth century. Additional evidence comes from the growth patterns of coral and trees.

Based on historical evidence, El Niños seem to have occurred as far back as records were kept. One disturbing fact is that they appear to be occurring more often. Records indicate that during the sixteenth century an El Niño occurred, on average, every 6 years. Evidence gathered over the past few decades indicates that El

Niño conditions are now occurring, on average, every 2.2 years, although in recent years none has matched the strength of the 1997–1998. El Niño.

The record 1997–1998 El Niño brought copious and damaging rainfall to the southern United States, from California to Florida. Snowstorms in the northeast portion of the United States were more frequent and stronger than in most years. The warm El Niño waters fueled Hurricane Linda, which devastated the western coast of Mexico. Linda was the strongest hurricane ever recorded in the eastern Pacific Ocean.

Scientists now have the capability to better monitor and forecast El Niño and La Niña events. A network of ocean-anchored weather buoys and satellite observations provide an enormous amount of data that can be analyzed by computer to help predict the formation and strength of El Niño and La Niña events. The varying seasonal temperature and precipitation characteristics that tend to occur during times of El Niño or La Niña can now be mapped in terms of their general regional influence. Spatial patterns of unusual warmth or cold, storminess, precipitation probabilities, and droughts are associated with either El

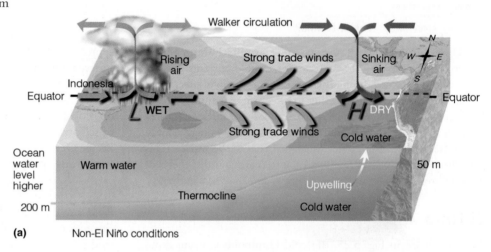

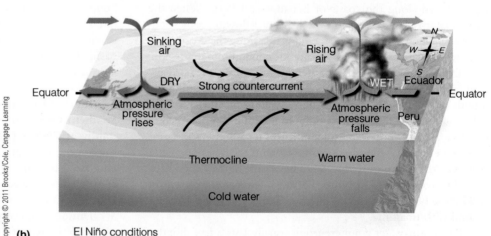

● **FIGURE 5.27** (a) In a non-El Niño (normal) year, upwelling of cold sea water and the trade winds are both strong along the west coast of South America and bring rains to Southeast Asia. (b) During El Niño, the easterly trade winds weaken, allowing the central Pacific to warm and the rainy area to migrate eastward.

Near what country or countries does El Niño begin?

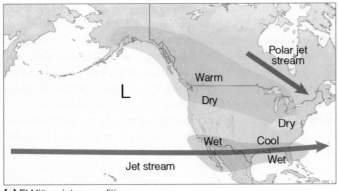

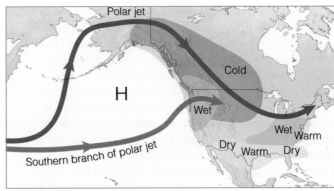

(a) El Niño winter conditions

(b) La Niña winter conditions

● **FIGURE 5.28** Associated with shifts in the jet streams, seasonal weather patterns can be mapped to illustrate how regional climates are affected during the times of (a) an El Niño and (b) a La Niña. Differing winter weather conditions tend to occur in North America during times of El Niño and La Niña.

What kind of temperature and precipitation changes tend to occur where you live during an El Niño compared to a La Niña?

Niño or La Niña, depending on the location (● Fig. 5.28a and b). The knowledge gained from scientific efforts and these generalized maps can help us prepare for weather events that are associated with El Niño and La Niña conditions.

North Atlantic Oscillation

Improved observational skills have led to the discovery of the **North Atlantic Oscillation (NAO)**—a relationship between the Azores (subtropical) High and the Icelandic (subpolar) Low.

The east-to-west, seesaw motion of the Icelandic Low and the Azores High controls the strength of the westerly winds and the direction of storm tracks across the North Atlantic. There are two recognizable phases associated with the established NAO index.

A *positive NAO* index phase is identified by higher than average pressures in the Azores High and lower than average pressures in the Icelandic Low. The increased pressure difference between the two systems results in stronger winter storms, occurring more often and following a more northerly track (● Fig. 5.29a). This promotes warm and wet winters in Europe

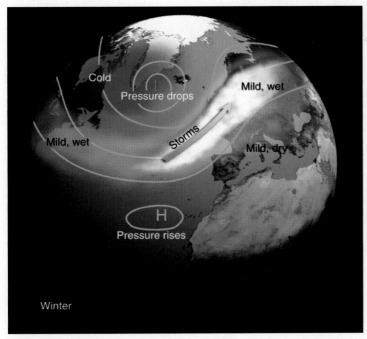

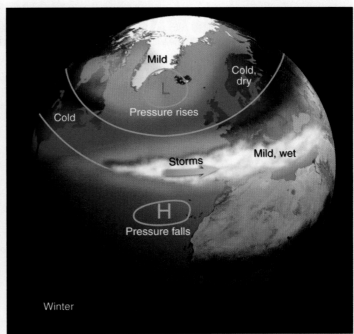

(a) Positive phase

(b) Negative phase

● **FIGURE 5.29** Positions of the pressure systems and winds involved with the (a) positive and (b) negative phases of the North Atlantic Oscillation (NAO).

Which two pressure systems are used to establish the NAO phases?

but cold, dry winters in Canada and Greenland. The eastern United States will generally experience a mild and wet winter.

A *negative NAO* index phase occurs with a weak Azores High and higher pressure in the Icelandic Low. The smaller pressure gradient between these two systems will weaken the westerlies, resulting in fewer and weaker winter storms (Fig. 5.29b). Northern Europe will experience cold air with moist air moving into the Mediterranean. The East Coast of the United States will experience more cold air and snowy winters. This index varies from year to year but it also has a tendency to stay in one phase for periods lasting several years in a row.

The NAO is not as well understood as ENSO. Truly, both oscillations require more research in the future if scientists are to better understand how these ocean phenomena affect weather and climate. Will scientists ever be able to predict the occurrence of such phenomena as ENSO or the NAO? No one can answer that question, but as technology continues to develop, the forecasting ability of scientists and other professionals will also increase. Tremendous progress has been made thus far. In the past few decades, we have come to recognize the close association between the atmosphere and hydrosphere and to understand the complex relationships between these Earth systems.

This chapter began with an examination of the behavior of atmospheric gases as they interact with solar radiation. The information enabled a definition and detailed discussion of global pressure systems and their accompanying winds. This discussion in turn permitted a description of atmospheric circulation patterns on global, regional, and local scales. Once again, we can recognize the workings of interactions among Earth's systems. Earth's energy budget influences movements in the atmosphere, which in turn affect ocean circulation, and the ocean currents influence processes and conditions in the atmosphere.

Solar radiation → Atmosphere → Hydrosphere →
Back to the atmosphere

The many interconnections between the atmosphere and the oceans directly influence conditions of weather and climate. In the following chapters, we will examine the role of atmospheric systems in controlling variations in weather and climate and, later, weather and climate systems as they affect biotic environments and surface landforms.

CHAPTER 5 ACTIVITIES

■ TERMS FOR REVIEW

mercury barometer
barometric pressure
aneroid barometer
standard sea level pressure
altitude
elevation
cyclone (low)
anticyclone (high)
convergent wind circulation
divergent wind circulation
isobar
pressure gradient
wind
Coriolis effect
wind friction
geostrophic wind
windward
leeward
prevailing wind
trough

equatorial low (equatorial trough)
subtropical highs
subpolar lows
polar highs
Siberian High
Canadian High
Icelandic Low
Aleutian Low
Pacific High
Bermuda High (Azores High)
polar easterlies
westerlies
trade winds
northeast trades
southeast trades
tropical easterlies
Intertropical Convergence Zone (ITCZ)
doldrums
"horse latitudes"
polar front

jet stream
polar jet stream
subtropical jet stream
Rossby wave
monsoon
land breeze–sea breeze
mountain breeze–valley breeze
Chinook
Foehn
Santa Ana wind
drainage wind (katabatic wind)
surface ocean current
gyre
upwelling
El Niño
Southern Oscillation
El Niño/Southern Oscillation (ENSO)
La Niña
North Atlantic Oscillation (NAO)

QUESTIONS FOR REVIEW

1. What is the standard atmospheric pressure at sea level? Explain how Earth's gravity is related to atmospheric pressure.
2. Horizontal variations in air pressure are caused by thermal or dynamic factors. How do these two factors differ? How are they related?
3. What kind of pressure (high or low) would you expect to find in the center of an anticyclone? Describe and diagram the wind patterns of anticyclones in the Northern and Southern Hemispheres.
4. What is the circulation pattern around a center of low pressure (cyclone) in the Northern Hemisphere? In the Southern Hemisphere? Draw diagrams to illustrate these circulation patterns.
5. Explain how water and land surfaces affect the pressure overhead during summer and winter. How does this relate to the afternoon sea breeze?
6. How do landmasses affect the development of belts of atmospheric pressure over Earth's surface?
7. Why do the global wind systems and pressure belts migrate with the seasons?
8. How are the land breeze–sea breeze and monsoon circulations similar? How are they different?
9. What effect on valley farms could a strong drainage wind have?
10. What causes monsoons? Name some nations that are concerned with the arrival of the "wet monsoon."
11. What is the relationship between ocean currents and global surface wind systems? How does the gyre in the Northern Hemisphere differ from the one in the Southern Hemisphere?
12. Where are the major warm and cool ocean currents located in respect to the coastal areas of continents? Which currents have the greatest effect on North America?
13. What is an El Niño? What are some of its impacts on global weather?

CONSIDER AND RESPOND

1. Look at the January (Fig. 5.10a) and July (Fig. 5.10b) maps of average sea level pressure. Answer the following questions:
   a. Why is the subtropical high pressure belt more continuous (linear, not cellular) in the Southern Hemisphere than in the Northern Hemisphere in July?
   b. During July what area of the United States exhibits the lowest average pressure? Why?
2. How did the trade winds of the Atlantic Ocean affect the sailing routes of early traders?
3. Describe the movements of the upper air winds. How have pilots applied their experience of the upper air currents to their flying patterns?
4. Is the polar front jet stream stronger in the summer or the winter? Why?
5. What effect would Chinook-type winds have on farming, forestry, and ski resorts?

PRACTICAL APPLICATIONS

1. The amount of power that can be generated by wind is determined by the equation:

$$P = \tfrac{1}{2} D \times S^3$$

   where P is the power in watts, D is the density, and S is the wind speed in meters per second (m/sec). Because $D = 1.293$ kg/m^3, we can rewrite the equation as

$$P = 0.65 \times S^3$$

   a. How much power (in watts) is generated by the following wind speeds: 2 meters per second, 6 meters per second, 10 meters per second, 12 meters per second?
   b. Because wind power increases significantly with increased wind speed, very windy areas are ideal locations for "wind farms." Cities A and B both have average wind speeds of 6 meters per second. However, city A tends to have very consistent winds; in city B, half of its winds tend to be at 2 meters per second and the other half is at 10 meters per second. Which site would be the better location for a wind generation plant?
2. Atmospheric pressure decreases at the rate of 0.036 millibar per foot as one ascends through the lower portion of the atmosphere.
   a. The Willis Tower (formerly the Sears Tower) in Chicago, Illinois, is one of the world's tallest buildings at 1450 feet. If the street-level pressure is 1020.4 millibars, what is the pressure at the top of the Willis Tower?
   b. If the difference in atmospheric pressure between the top and ground floor of an office building is 13.5 millibars, how tall is the building in feet?
   c. A single story of a building is 12 feet. You enter an elevator on the top floor of the building and want to descend five floors. The elevator has no floor markings—only a barometer. If the initial reading was 1003.2 millibars, at what pressure reading would you want to get off?

 MindTap—Make the most of your study time by accessing everything you need to succeed in one place. Read your textbook, take notes, review flashcards, watch videos, complete activities, take practice quizzes, and more online with MindTap. Log in at **www.cengagebrain.com**.

HUMIDITY, CONDENSATION, AND PRECIPITATION

OBJECTIVES

WHEN YOU COMPLETE THIS CHAPTER YOU SHOULD BE ABLE TO:

- 6.1 Explain why Earth's surface is dominated by water but freshwater remains a limited and precious resource.
- 6.2 Outline the processes in the hydrologic cycle, including how water circulates among and interacts with the lithosphere, atmosphere, hydrosphere, and biosphere.
- 6.3 Apply adiabatic lapse rates to determine temperature changes in air that rises, expands, and cools compared with air that sinks, compresses, and warms.
- 6.4 Understand that relative humidity is a percentage of moisture saturation in the air, and know why it is dependent on air temperature and moisture content.
- 6.5 Explain why if no change in moisture content occurs, as air warms the relative humidity will fall and when air cools, the relative humidity will rise.
- 6.6 Determine what processes cause the air to reach the dew point temperature and attain a relative humidity of 100%—a condition that can produce condensation and precipitation.
- 6.7 Describe the temperature, humidity, pressure, and winds that influence the potential for precipitation and the kinds of precipitation that may result.
- 6.8 Provide examples of why precipitation, evaporation, and water availability all vary in terms of their geographic distribution on Earth.

EARTH IS OFTEN REFERRED TO as "the Blue Planet." The abundance of water on our planet is unique in the solar system and vital to life on Earth. About 73% of Earth's surface is covered by water, with the largest proportion in the world's oceans (● Fig. 6.1).

The physical and chemical characteristics of water are fundamentally responsible for supporting life, but water also supports and maintains our planet's environments. Although some living organisms can survive without air, no living thing can survive without water. Water is necessary for photosynthesis, cell growth, protein formation, soil development, and the absorption of nutrients by plants and animals.

Water is the only substance on Earth that occurs naturally as a liquid, a solid, and a gas. The structure of water molecules is such that water can dissolve an enormous number of substances—so many, in fact, that it has been called the *universal solvent*. As a result of its solvent properties, water is almost never found in a pure state. Even rainwater contains impurities picked up in the atmosphere, many of which facilitate the development of clouds and precipitation. Rain often absorbs some carbon dioxide from the air, which makes most rainwater a very weak form of carbonic acid. The action of rainwater dissolves and washes away minerals from the rocks and landforms, and although this process typically acts slowly, in certain areas its influence on landforms and the landscape is significant. The weak acidity of rainwater should not be confused with the environmentally damaging *acid rain*, which is at least 10 times more acidic.

Not only can water dissolve and transport many minerals but it also can transport solid particles in suspension. These characteristics make moving water an important transportation process in many components of the Earth system. Water supplies nutrients that would not otherwise be available to plants. Water also carries minerals and nutrients down rivers and streams, through the soil, through openings in subsurface rocks, and through living organisms. It deposits solid materials on stream floodplains, in river deltas, and on the ocean floor.

As water molecules draw together, surface tension causes **capillary action**, which gives water the ability to pull itself up through small openings. Capillary action transports dissolved material in an upward direction through rock and soil against the pull of gravity. This process also moves water into the stems and leaves of plants—even to the uppermost needles of the great California redwoods and to the tops of tall rainforest trees. Capillary action is also important in the movement of blood through our bodies. Without it, many of our cells could not receive the necessary nutrients carried by the blood.

An unusual and important property of water is that its volume expands when it freezes into a solid. In contrast, most substances contract when cooled and expand when heated. Water does contract slightly until it cools below 4°C (39°F), but then it expands as the crystal structure of ice develops. Expansion from the crystallization process during freezing makes ice about 9% less dense than liquid water, which is why ice floats on water. This is also why about 90% of ice cubes, sea ice, and icebergs extend below the water surface and only about 10% of floating ice is exposed above the water level (● Fig. 6.2).

Water is also slow to heat and slow to cool. Bodies of water are reservoirs that store heat and can moderate the cold of winter, yet the same water body will have a cooling effect in

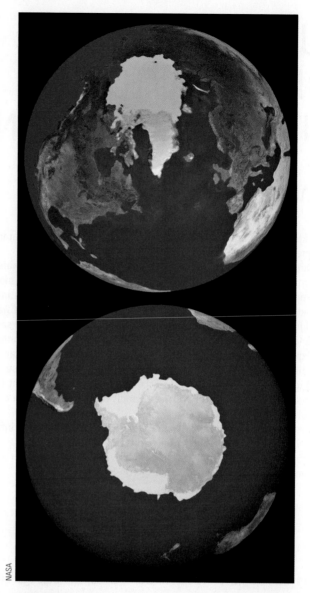

NASA

● **FIGURE 6.1** Oceans cover most of Earth's surface area. The planet's dominant freshwater source is stored in the glaciers of the polar regions, which can be seen in these images.

Can you distinguish between the Greenland and Antarctic glaciers compared to the ice shelves and seasonal sea ice that float on the ocean surface?

summer. Water's moderating effect on temperature is typically experienced in the vicinity of lakes and along seacoasts.

Earth's water—the *hydrosphere* (from Latin: *hydros* means water)—exists in all major subsystems of the planet, occurring naturally in all three states: as a liquid in rivers, lakes, oceans, and rain; as a solid as snow and ice; and as water vapor as an atmospheric gas. Even the water temporarily stored in living things could be considered part of the hydrosphere—the human body is about 60% water.

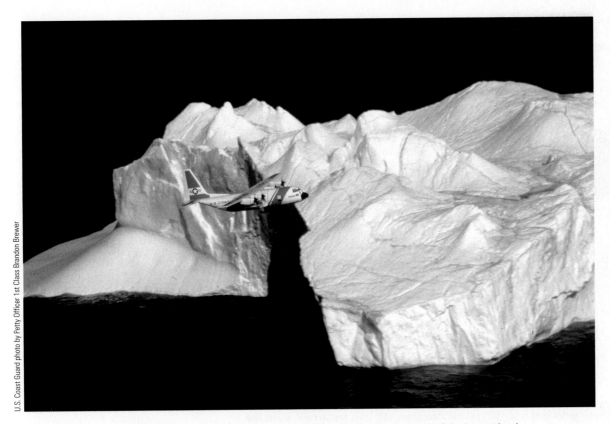

● **FIGURE 6.2** A U.S. Coast Guard airplane monitors a huge iceberg in the North Atlantic as part of the International Ice Patrol. The potential hazard to shipping makes it important to track icebergs as they drift with the currents. Because ice is about 9% less dense than water, only about 10% of the iceberg is above the surface. It is hard to imagine how much water is contained in this one iceberg.

If ice floating in a beverage glass melts completely before you can drink from it, will the liquid level rise, fall, or remain the same as before? Why?

The Hydrologic Cycle

Water circulates from one part of the Earth system to another in a system known as the **hydrologic cycle** (● Fig. 6.3). Air contains water vapor that has entered the atmosphere through evaporation from water bodies and land surfaces. In the hydrologic cycle, water is continually transferred from one state to another—as a vapor, liquid, or solid. When water vapor condenses into a liquid and falls as precipitation, several things may happen. First, precipitation can fall directly on a body of water, where it is immediately available for evaporation back into the atmosphere. Alternatively, it might fall on the land, where it runs off into rivers, streams, ponds, and lakes. Rainwater can also be absorbed into the ground, where it can either be stored in the soil or flow through open spaces, called *interstices;* in loose surface materials such as sand, gravel, and silt; or through voids in solid rock. Most of the water in the ground or on the surface will ultimately reach the oceans, returning the water to its original source. Some of the water that falls as snow can accumulate to become part of the massive ice cover in the polar regions or in mountain glaciers, whereas some snow will thaw, feeding streams with snowmelt waters. Water temporarily becomes a part of living things when it is used and stored by plants and animals. In short, the hydrologic cycle contains six major water storage areas: the atmosphere, the oceans, bodies of freshwater, plants and animals, snow and glacial ice, and the groundwater beneath the surface.

Evaporation converts liquid water into water vapor, returning this gas to the atmosphere. Ice can also revert directly from a solid to water vapor (gas) through sublimation. Evaporation occurs from bodies of water, soils, animals, and plants. Water can also evaporate from falling precipitation, and if it descends through very dry air below, rainfall can completely evaporate before it hits the surface. Evaporation returns liquid water to the atmosphere as gaseous water vapor, and the cycle repeats through condensation and precipitation.

The hydrologic cycle is a continuously operating system of evaporation, sublimation, condensation, precipitation, and transportation of water over the land, within water bodies, and in the ground. The hydrologic cycle for *the Earth system as a whole* can be considered a *closed system* because energy flows into and out of the system but there is no gain or loss of water (matter). Although our planet is a closed system, only needing energy from the sun to operate, its subsystems (such as the hydrologic cycle of a region or water body) function as open systems with both energy and matter flowing in and out. Earth's hydrosphere and its subsystems operate through many processes and interactions, changing water from one state to another and transporting water from the atmosphere to the surface and back.

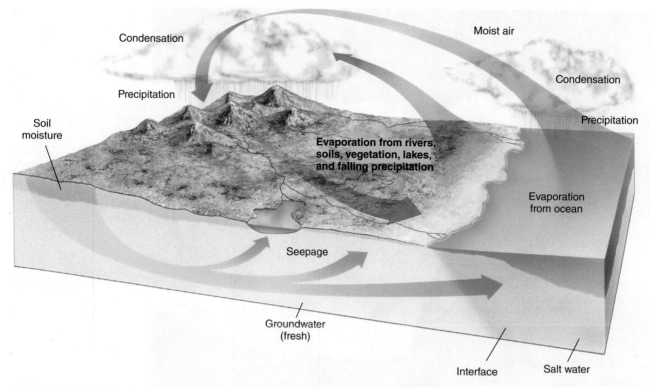

● **FIGURE 6.3** The hydrologic cycle circulates water from one part of the Earth system to another. Through condensation, precipitation, and evaporation, water cycles continually between the atmosphere, soil, lakes and streams, plants and animals, glacial ice, subsurface storage, and back to the principal reservoir—the oceans.

Is the hydrologic cycle a closed system or an open system? Why?

The hydrologic cycle is an essential, subsystem of our planet. Water and hydrologic processes are linked to numerous other subsystems that rely on water as an agent of movement, and to sustain life. The movement of water and its transfer from one state to another play major roles in redistributing energy in the atmosphere, hydrosphere, biosphere, and lithosphere.

Water in the Atmosphere

We are most familiar with water in its liquid form, as it pours from a tap, falls as rain, or falls as fine droplets in clouds or fog. Yet, water in the atmosphere exists in all three states—as ice (snow, hail), as tiny liquid droplets that form clouds and fog, and as water vapor. Water vapor is a tasteless, odorless, and transparent gas that is mixed with other atmospheric gases in varying proportions. The troposphere contains 99% of atmospheric water vapor, mainly within an altitude zone from the surface to about 5500 meters (18,000 ft). Water vapor makes up a small but highly variable percentage of the atmosphere by volume, yet atmospheric water is the source of all condensation and precipitation. Through the exchange of water among its states through evaporation, condensation, precipitation, melting, and freezing, water plays a significant role in regulating and modifying temperatures locally and globally. In addition, atmospheric water vapor reflects and absorbs significant portions of incoming solar energy and outgoing terrestrial radiation, and it re-radiates some of that absorbed energy from the atmosphere

back to Earth, warming the surface. Also, the insulating qualities of water vapor reduce the heat loss from the surface, which helps to maintain the moderate temperature ranges found on Earth.

The Water Budget

The water content of the Earth's hydrologic system is about 1.33 billion cubic kilometers (326 million mi^3), including all forms of water on our planet. The largest proportion of water on Earth is saltwater in the oceans (97%). The massive glaciers of the polar regions on Greenland and Antarctica where the average ice thickness is more than 2000 m (1.25 mi) store most of the fresh water on Earth (● Fig. 6.4).

Because our planet's hydrosphere is basically a *closed system,* virtually no water is received from outside the Earth system nor lost from it. Although water cycles in and out of the atmosphere, lithosphere, and biosphere, the total amount of water on Earth remains constant. It is important to understand that water, this essential substance for sustaining life, is a limited resource. Despite the apparent vastness of the oceans, large lakes, and huge glaciers, the comparative volume of all of Earth's waters is relatively small compared with the planet's size (● Fig. 6.5).

The Earth operates on a **water budget** in which the total supply of water remains basically the same so that any deficits should be balanced by gains elsewhere within the entire system. An increase in water within one component of the hydrologic subsystem is accounted for by a loss in another. The water budget's

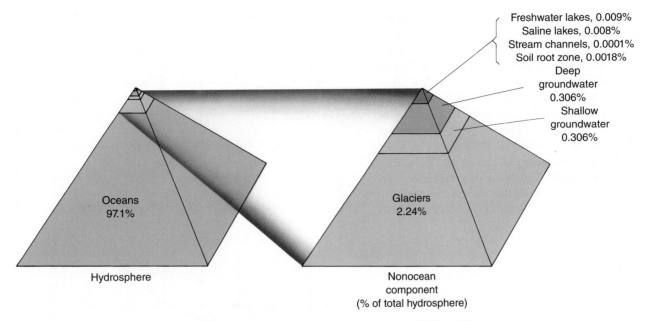

Freshwater lakes, 0.009%
Saline lakes, 0.008%
Stream channels, 0.0001%
Soil root zone, 0.0018%
Deep groundwater 0.306%
Shallow groundwater 0.306%

Oceans 97.1%

Glaciers 2.24%

Hydrosphere

Nonocean component (% of total hydrosphere)

● **FIGURE 6.4** Earth's water resources. The vast majority of water in the hydrosphere is seawater in the world's oceans. The freshwater supply stored in the geographically isolated polar ice sheets is relatively unavailable for use.

How might global warming or cooling alter the data shown in this figure?

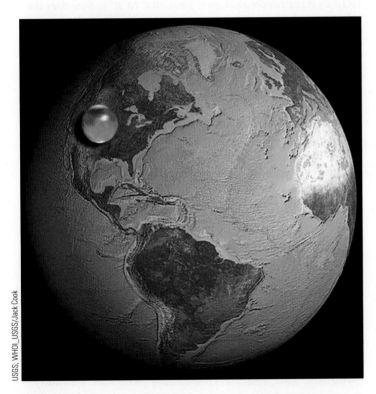

USGS; WHOI_USGS/Jack Cook

● **FIGURE 6.5** Compared to Earth's size, the amount of water in the oceans, lakes, rivers, as well as in the atmosphere and the ground, would fit into the blue sphere shown here. If Earth were the size of a basketball, all the planet's water would fit into the size of a ping-pong ball.

What does this relationship mean in terms of freshwater as a resource on Earth?

components fluctuate in this dynamic system, as the amount of water associated with any component of the hydrosphere changes continuously over time and place in terms of where water is stored. For example, about 18,600 years ago, during the last major Ice Age, glaciers expanded, causing sea level to drop, and evaporation and precipitation decreased. Less water was stored in the oceans, but this was balanced by increased water storage on the land as glacial ice. Changes in the water budget can also be human induced—for example, when rivers are dammed to form reservoirs, more water is stored on the land surface. Further, in an arid location, high evaporation rates from an artificial reservoir will cause water loss to the atmosphere and diminish river flow.

The atmosphere gives up a great deal of water by condensation into fog, dew, frost, and various forms of precipitation (rain, snow, hail, sleet). Because the quantity of water in Earth's atmosphere remains at the same level, this means that the atmosphere must be absorbing an equal amount of water to that which it gave up from other parts of the system. In a single minute, through precipitation or condensation, the atmosphere gives up more than a billion tons of water, and an equal amount of water evaporates and sublimates back to the atmosphere as water vapor.

Water and the Energy Budget

By reviewing our discussion of the energy budget in Chapter 4, you can see that a part of that budget is latent heat exchange, particularly release of latent heat through condensation. This energy, of course, is originally derived from the sun. Solar energy is used in evaporation, stored in water vapor, and released during condensation. Although the energy transfers involved in evaporation and condensation account for a small portion of the total energy budget, the actual amount of energy is significant. Imagine how much energy is released every minute when a billion tons of water condense out of the atmosphere. This vast storehouse of energy, the latent heat of condensation, is a major energy source that powers storms, particularly hurricanes and thunderstorms.

The amount of water vapor that can be held by a parcel of air is limited, and air temperature is a very important determinant of that amount. Warmer air can hold a greater quantity of water vapor compared with what can be held in colder air. Therefore, we can make a generalization that air in the cold polar regions can hold far less water vapor (approximately 0.2% by volume) than the air of the hot tropical and equatorial regions, which can contain as much as 5% by volume.

Saturation and the Dew Point Temperature

When air of a given temperature holds all the water vapor that it possibly can, it is said to be in a state of **saturation** and to have reached its **moisture capacity**. If the air maintains a constant temperature as more water vapor is added, there will be a point when the air will be saturated and unable to hold any more water vapor. This situation often occurs when taking a shower. Evaporation of water from the shower spray makes the air increasingly humid until a point is reached at which it cannot contain more water. The excess water vapor will then condense onto the colder mirrors and walls.

The air's capacity to hold water vapor, as seen in ● Figure 6.6, increases with rising temperatures. Closely examining examples on this graph will illustrate the relationship between air temperature and water vapor capacity. If a parcel of air at 30°C is saturated, it will contain 30 grams of water vapor in each cubic meter of air (30 g/m³), a relationship demonstrated in Figure 6.6. Now, suppose the air temperature increases to 40°C *without* increasing the water vapor content. The parcel will no longer be saturated because air at 40°C can hold more than 30 g/m³ of water vapor (actually 50 g/m³). Conversely, if the temperature of saturated air decreases from 30°C (containing 30 g/m³ of water vapor) to 20°C, which at that temperature has the capacity of only 17 g/m³, some 13 grams of water vapor will condense out of the air because of the reduced moisture capacity.

If unsaturated air is continually cooled, it will eventually reach a temperature at which the air will become saturated. This critical temperature is known as the **dew point temperature**—the temperature at which the air cannot hold any more water vapor and condensation takes place. For example, we know that if a parcel of air at 30°C (86°F) contains 20 grams per cubic meter of water vapor (8.7 gr/ft³), it is not saturated because it can hold 30 g/m³ (8.7 gr/ft³). However, if that air parcel cools to 21°C (69°F), it would become saturated because the capacity of air at 21°C is 20 grams per cubic meter. Thus, that air parcel at 30°C has a dew point temperature of 21°C. The cooling of air to below its *dew point temperature* brings about the condensation that precedes precipitation. Although the shortened term, "dew point" is frequently used, it is important to know that it is a temperature, and not a location.

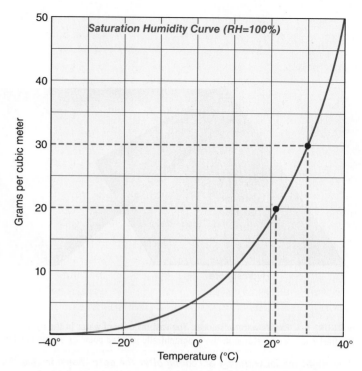

● **FIGURE 6.6** Maximum absolute humidity (saturation) increases with temperature. This graph shows the maximum amount of water vapor that can be contained in a cubic meter of air and how it varies over a range of temperatures.

Compare the change in capacity if the temperature rises from 0°C to 10°C; if the temperature rises from 20°C to 30°C. How does the relationship between temperature and capacity change?

Because the water vapor capacity of air increases with rising temperatures, warm air in the equatorial regions has a higher dew point temperature than the cold air in the polar regions. Thus, because the atmosphere in equatorial regions can hold more water, there is a greater potential for large quantities of precipitation in the tropics compared with the precipitation of the polar regions. Also, although there are some exceptions in places that have a dry summer, midlatitude locations generally have a greater potential for heavy precipitation during the warmer summer months rather than the colder winter months.

Humidity

The amount of water vapor in the air at a time and place is called **humidity**. There are three common ways to express the humidity content of the air: absolute humidity, specific humidity, and relative humidity. Each method provides information that contributes to our understanding of weather and climate.

Absolute and Specific Humidity **Absolute humidity** is the measure of the mass of water vapor that exists within a given volume of air. It is expressed in the metric system as the number of grams per cubic meter (g/m³) and in the English system as grains per cubic foot (gr/ft³). **Specific humidity** is the mass of water vapor (given in grams) per mass of air (given in kilograms). Obviously, both are measures of the actual amount of water vapor in the air, although they are expressed in different units. Because most water vapor gets into the air by the evaporation of water from the surface, both absolute and specific humidity tend to decrease with altitude and elevation. This relationship also occurs because the air is colder aloft and cannot hold as much moisture as the warmer air at lower altitudes.

We have learned that air compresses as it sinks and expands as it rises. Thus, a parcel of air changes in volume as it moves vertically, but there may be no change in the amount of water vapor in that quantity of air. What this means is that absolute humidity, which measures the actual amount of water vapor by volume of air, can vary as the volume of the air parcel also changes when it moves up or down.

Specific humidity, which compares the mass of water vapor to the mass of the air, changes *only* as the quantity of the water vapor changes. For this reason, when assessing changes in water vapor content for large masses of air, which often undergo vertical movement, specific humidity is the preferred measurement among climatologists and meteorologists. Further, metric units—grams of water vapor per kilograms of air (gm/kg)—are used as the international standard for expressing specific humidity.

Relative Humidity The best known expression of the water vapor in the air is **relative humidity**—the one commonly given on television, radio, and Internet weather reports. Relative humidity is the ratio between the actual amount of water vapor in the air and the maximum amount of water vapor that the air could hold at that temperature; it is expressed as a percentage. The stated percentage expresses how close the air is to being fully saturated with water vapor. Air at its moisture holding capacity, saturated with water vapor, has a relative humidity of 100%. Using this method to describe humidity, however, has both strengths and weaknesses. Its strength lies in how easily it is communicated and understood by the public. Most people understand the concept of "percent" but may not have a clear concept of the significance of "grams per cubic meter."

Relative humidity percentages can vary widely through the day, even when the water vapor content of the air remains the same. Two basic factors can cause the relative humidity to rise or fall—changes in air temperature and changes in the amount of water vapor that the air contains.

If the temperature and absolute humidity of an air parcel are known, its relative humidity can be determined by using Figure 6.6. For instance, if a parcel of air has a temperature of 30°C and an absolute humidity of 20 grams per cubic meter, the graph indicates that if it were saturated, its absolute humidity would be 30 grams per cubic meter. To determine relative humidity (RH), divide 20 grams (actual content) by 30 grams (content at capacity) and multiply by 100 (to get an answer in percent):

(actual content in air ÷ maximum air can hold at that temperature) × 100 = RH%

(20 grams ÷ 30 grams) × 100 = 67%

The relative humidity in this case is 67%. The air is holding only two-thirds (0.67) of the water vapor it could contain at 30°C; the air is at 67% of saturation, its maximum moisture-holding capacity.

Two important geographic factors are involved in the spatial distribution and variation of relative humidity. One of these is moisture availability at a location. For example, the air above a water body is apt to contain more moisture than air over land (of similar temperature) because there is more water available for evaporation. Conversely, the air over a region like the central Sahara Desert is usually very dry in part because it is far from the oceans and little water is available to be evaporated. The second factor that causes spatial variations in relative humidity is local temperature. In regions of higher temperature, relative humidity for air containing the same amount of water vapor will be lower than it would be in a cooler region.

The relative humidity will change if the amount of water vapor increases as a result of evaporation *or* if the temperature increases or decreases. Thus, the quantity of water vapor may not change during the day or overnight, but the relative humidity will vary along with the daily temperature cycle (● Fig. 6.7). As air temperature increases from sunrise to its maximum in mid-afternoon, the relative humidity will decrease as the air warms and becomes capable of holding greater quantities of water vapor. When the air cools, decreasing toward its minimum temperature around sunrise, the relative humidity increases with the declining nighttime temperature.

Relative humidity affects our comfort levels on hot days because of its relationship to evaporation rates. Evaporation is a cooling process because the thermal energy used to change perspiration to water vapor becomes stored in the water vapor as latent heat and is removed from our skin. On a hot day when it is 35°C (95°F), it is more uncomfortable in Atlanta, Georgia, with the relative humidity at 90%, than it is in Tucson, Arizona, where the relative humidity is only 15% at the same temperature. At a relative humidity of 15%, perspiration would evaporate faster, so you would feel much cooler. When the relative humidity is 90%, the air is nearly saturated, much less evaporation would take place, and less heat would be drawn from our skin.

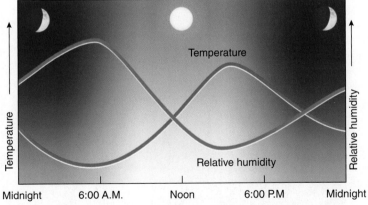

● **FIGURE 6.7** Relative humidity changes during a 24-hour day, responding to air temperature trends through the day and night. Even with no change in the moisture content in the air, as the day warms, the relative humidity drops. At night when it is cooler, the relative humidity rises.

How can the relationship between air temperature and relative humidity be applied when using a hair dryer?

The Wettest and Driest Places in the World

There is considerable debate about what place on Earth receives the greatest annual rainfall, averaged over many years of recordkeeping. One of these places is Mount Waialeale (a Hawaiian word meaning "for rippling waters"), which stands at 1569 meters (5148 ft) above sea level on the island of Kauai in the Hawaiian Island chain. Mount Waialeale averages 11.68 meters (460 in.) of rain per year, and because it receives rain all year long, it certainly may be the world's *wettest* place. Two locations in India, however, report mean annual rainfalls that are slightly higher. Mawsynram receives a yearly average of 11.87 meters (467 in.), and nearby Cherrapunji receives 11.78 meters (464 in.) of annual rainfall. These are locations that receive extremely heavy rains during the Asian monsoon that annually flood this region of India during the summer months. In fact, Cherrapunji is the record holder for the world's highest rainfall received in a single year 26.46 meters (1042 in., or 86 ft). That year, Cherrapunji, like other monsoon locations in India, also experienced the dry monsoon season during winter, but it was extremely dry with no precipitation recorded for 3 months. So, perhaps Mount Waialeale is the wettest place, but Mawsynram has the world's highest recorded annual rainfall average.

There is little debate about the world's driest region—it is the Atacama Desert of northern Chile in South America. Two locations there can claim some recognition. Arica, Chile, claims the world's lowest average annual precipitation at 0.5 mm (2/100ths of an inch—considered a *trace*). In fact, Arica once went 14.5 years without rain. However, a valley in the Andes Mountains, located between Arica and Iquique, Chile, has not received any rain according to anyone's living memory.

Northern Chile is dominated by two mechanisms that keep it arid. The first is that the stabilizing effect of the subtropical high pressure system dominates this location for long periods. Second, the Atacama Desert is on the rain shadow side of the Andes. This area does not receive much precipitation because of these two processes working together.

Ironically, in March of 2015, 24 mm (0.97 in.) of rain fell in the region, causing devastating floods that destroyed homes and took the lives of some people. Rainwater gushed off the bare rock of the desert, without soil or vegetation to absorb some of the water, and this small amount of rain caused large amounts of runoff and a serious flood.

Densely vegetated, rain-soaked area near Mawsynram, India.

⊕ In Google Earth fly to: 25.20°N, 91.70°E.

Hyperarid Atacama landscape bare of any vegetation.

⊕ In Google Earth fly to: 18.56°S, 69.96°W.

Sources of Atmospheric Moisture

Although water evaporates into the atmosphere from many different places and environments, the most important are bodies of water. Water also evaporates from wet ground surfaces and soils; from droplets of moisture on vegetation; from city pavements, building roofs, cars, and other surfaces; and even from falling precipitation.

Vegetation provides another source of water vapor to the atmosphere. Plants give up water in a process called **transpiration**, which can be a significant source of atmospheric moisture. A mature oak tree, for instance, can transpire 400 liters (105 gal) of water per day, and a cornfield may add 11,000 to 15,000 liters (2900 to 4000 gal) of water to the atmosphere per day for each acre under cultivation. In some parts of the world—notably the tropical rainforests with their heavy, lush vegetation—transpiration accounts for a significant amount of atmospheric humidity. The combined effects of evaporation and transpiration are referred to as **evapotranspiration**, which accounts for the majority of the water vapor in the atmosphere.

Evaporation Rates

Several factors have an impact on evaporation rates. First, evaporation is affected by the amount of accessible water and the water temperature. As Table 6.1 shows, evaporation rates tend to be greater over the oceans when compared with the evapotranspiration rates over the continents, even though the data for the continents includes transpiration by plants. The only regions where this generalization does not fit is in the equatorial areas between 0° and 10°N and S. Here, the land vegetation is so lush that transpiration provides the air with huge quantities of moisture.

Second is relative humidity, the degree to which the air is saturated with water vapor. The drier the air and the lower the relative humidity, the greater the evaporation rate will be. Most people have had some direct experience with this relationship. Consider the length of time it takes a swimsuit and wet towel to dry on a hot day when the air is dry compared with how long it takes on a day when the air is hot and humid.

Third is the wind, which affects the rate of evaporation. If there is no wind, the air that overlies a water surface will approach saturation as more and more molecules of water change to water vapor. When saturation is reached, however, evaporation will cease. If the conditions are windy, the wind will blow the saturated or nearly saturated air away from the evaporating water source, replacing it with lower humidity air. This allows evaporation to continue as long as the wind keeps blowing saturated air away and bringing in drier air. Anyone who has gone swimming on a windy day has experienced the chilling effects of rapid evaporation.

Air temperature also strongly affects evaporation rates by influencing the first and second factors listed earlier. As air temperature increases, so does the water temperature at the evaporation source. Temperature increases ensure that more energy is available to the water molecules, enhancing the transition from a liquid to a gas, and more evaporation takes place. As the temperature of the air increases, its moisture holding capacity also increases. As air becomes warmer, its molecules are increasingly energized, moving farther apart, and the air density decreases. With more energy and wider spacing between air molecules, additional water molecules can enter the atmosphere, thus increasing evaporation.

Potential Evapotranspiration

So far we have discussed actual evaporation and transpiration (evapotranspiration). However, geographers and meteorologists are also concerned with **potential evapotranspiration** (● Fig. 6.8),

TABLE 6.1
Distribution of Actual Mean Evapotranspiration

Zone	Latitude					
	60°−50°	50°−40°	40°−30°	30°−20°	20°−10°	10°−0°
Northern Hemisphere						
Continents	36.6 cm (14.2 in.)	33.0 (13.0)	38.0 (15.0)	50.0 (19.7)	79.0 (31.1)	115.0 (45.3)
Oceans	40.0 (15.7)	70.0 (27.6)	96.0 (37.8)	115.0 (45.3)	120.0 (47.2)	100.0 (39.4)
Mean	38.0 (15.0)	51.0 (20.1)	71.0 (28.0)	91.0 (35.8)	109.0 (42.9)	103.0 (40.6)
Southern Hemisphere						
Continents	20.0 cm (7.9 in.)	NA	51.0 (20.1)	41.0 (16.1)	90.0 (35.4)	122.0 (48.0)
Oceans	23.0 (9.1)	58.0 (22.8)	89.0 (35.0)	112.0 (44.1)	119.0 (47.2)	114.0 (44.9)
Mean	22.5 (8.8)	NA	NA	99.0 (39.0)	113.0 (44.5)	116.0 (45.7)

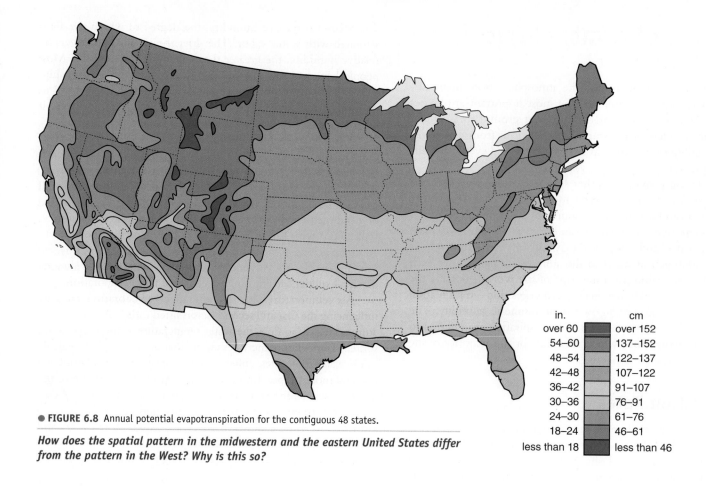

● **FIGURE 6.8** Annual potential evapotranspiration for the contiguous 48 states.

How does the spatial pattern in the midwestern and the eastern United States differ from the pattern in the West? Why is this so?

in.		cm
over 60		over 152
54–60		137–152
48–54		122–137
42–48		107–122
36–42		91–107
30–36		76–91
24–30		61–76
18–24		46–61
less than 18		less than 46

which is the maximum evapotranspiration that would take place with an unlimited supply of water available. In a desert region with hot dry air, for example, the potential for evapotranspiration may greatly exceed the water available to evaporate or transpire. Various formulas have been derived for estimating the potential evapotranspiration at a location because it is difficult to measure directly. These formulas commonly use temperature, latitude, vegetation, and soil character (permeability, water-retention ability) as factors that could affect the potential evapotranspiration.

In locations where precipitation exceeds potential evapotranspiration, a surplus of water exists that can be stored in the ground and in water bodies, allowing water to flow away from those areas in streams and rivers. Water can also be exported from wetter places to drier locations if systems of canals or pipelines are feasible. When potential evapotranspiration exceeds precipitation, as it does during the dry summer months in California and in the arid West, no water becomes available for storage. In fact, the water stored during previous rainy months evaporates into the warm, dry air or is transpired into the atmosphere by vegetation (● Fig. 6.9). During the dry season, soil dries out and the vegetation turns brown as the available water is absorbed into air that is able to hold much more moisture. This is a major reason why wildfires are a potential and serious hazard during the late summer and early fall in California and can occur almost any time in desert areas.

An understanding of potential evapotranspiration can be used to determine how much water will be lost from crops or a reservoir in environments that experience dry seasons. By assessing the daily, weekly, or seasonal relationships between potential evapotranspiration and precipitation, agriculturalists can determine when and how much irrigation water is needed for their crops.

Condensation, Fog, and Clouds

Condensation is the process by which a gas is changed to a liquid. In our present discussion of atmospheric moisture and precipitation, condensation refers to the change of water vapor to liquid water.

The process of condensation occurs when air saturated with water vapor is cooled. Viewed in another way, we can say that if we lower the temperature of air until it has a relative humidity of 100% (the air has reached the dew point), condensation will occur with additional cooling. It follows, then, that condensation depends on (1) the relative humidity of the air and (2) the degree of cooling. In the arid conditions of Death Valley, California, a great amount of cooling must take place before the dew point is reached. In contrast, on a humid summer afternoon in Biloxi, Mississippi, a minimal amount of cooling will bring on condensation.

A similar process occurs when droplets of water form on the side of a glass containing a cold drink on a warm afternoon. The temperature of the air is lowered when it comes in contact with the cold glass. Consequently, the air's capacity to hold water

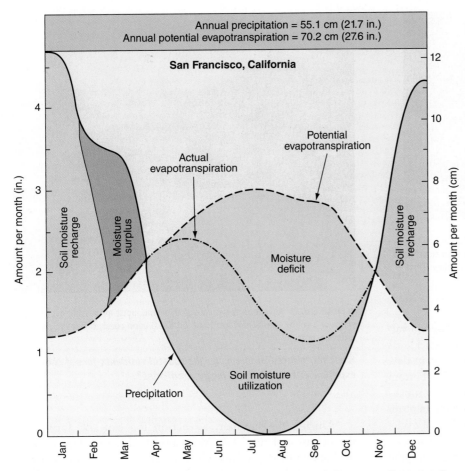

Annual precipitation = 55.1 cm (21.7 in.)
Annual potential evapotranspiration = 70.2 cm (27.6 in.)

San Francisco, California

● **FIGURE 6.9** The water budget for San Francisco, California. The water budget system "keeps score" of the balance between water input by precipitation and water loss to evaporation and transpiration, permitting month-by-month estimates of both runoff and soil moisture.

When would irrigation be necessary in areas with this kind of a seasonal water budget?

vapor is diminished. If air touching the glass is cooled sufficiently, its relative humidity will reach 100%. Any cooling beyond that point will result in condensation in the form of water droplets on the glass.

Condensation Nuclei

An atmospheric factor that reinforces the condensation process is the presence of **condensation nuclei**. These are minute particles in the atmosphere that provide a surface on which condensation can take place. Sea salt particles in the air are common condensation nuclei that come from the evaporation of saltwater spray from ocean waves. Other common nuclei include dust, smoke, pollen, and volcanic material. Some particles are more *hygroscopic* (having the property to attract moisture) than others. Typically, these are chemical particles that are the by-products of industrialization. The condensation that takes place on certain chemical nuclei can be corrosive and hazardous to human health; when it is, we know it as *smog* (a term that was derived by combining the words *smoke* and *fog*).

Theoretically, if all particles were removed from a volume of air, that air could cool below its dew point without condensation

occurring. Conversely, if there is a superabundance of particles, condensation may take place at a relative humidity that is less than 100%. For example, ocean fogs, an accumulation of condensation droplets formed on particles of sea salt, can form when the relative humidity is as low as 92%.

In nature, condensation appears in a number of forms. Fog, clouds, and dew are all the results of condensation of water vapor in the atmosphere. The type of condensation produced depends on a number of factors, including the cooling process itself. The cooling that produces condensation in one form or another can occur as a result of radiational cooling, through advection, through convection, or through a combination of these processes.

Fog

Fogs and clouds appear when condensation occurs, and forms a large number of tiny water droplets. Because these droplets are not transparent to light in the way that water vapor is, these masses of condensed water droplets appear to us as fog or clouds, in a great variety of shapes and forms, usually in shades of white or gray.

On a worldwide basis, fog is a minor form of condensation, but in certain regions it has important climatic effects. The "drip factor" from fog helps sustain vegetation along some desert coastlines where fog occurs. Fog also plays havoc with transportation systems. Navigation on the seas is more difficult in fog. Air travel can be greatly impeded by fog; dense fog can cause airports to delay flights for hours or days until visibility improves. Highway travel can also be greatly hampered by heavy fogs, which can lead to huge, chain reaction vehicle pileups.

Radiation Fog Cold ground surfaces that have cooled from the loss of thermal energy to space can produce **radiation fog**, also called **ground fog**, or **temperature inversion fog**. This kind of fog typically occurs on cold, clear nights with calm winds and it generally lasts until morning. Nighttime conditions of clear skies allow massive amounts of terrestrial radiation to be lost from the surface. With no incoming radiation at night, the ground becomes cold as it gives up much of the heat that it received during the day. The air directly above the surface is cooled by conduction through contact with the cold ground. Because the cold surface can cool only the lower few meters of the atmosphere, a temperature inversion is created as air near the surface becomes colder than the warmer air above. If this cold layer of air at the surface cools to below its dew point temperature, condensation will occur, typically as a low-lying fog. However, winds strong enough to disturb the inversion layer can prevent fog from developing by not allowing the air to stay at the surface long enough to become cooled below its dew point.

● **FIGURE 6.10** Radiation fog forms when intense nighttime cooling by loss of terrestrial radiation causes cold temperatures near the surface on clear, calm nights. These fogs can be reinforced when cold air drains into a valley. This inversion-related fog is in Reno, Nevada.

● **FIGURE 6.11** Advection fog is caused by warm, moist air passing over colder water or a colder coastal surface. Here, on the Oregon coast, the sea fog also drifts onshore.

What fog-related problems might coastal residents face if they experience these kinds of foggy conditions?

The chances of a radiation fog occurring are increased in valleys and depressions, where cold air drains down from higher areas into lowland areas. During a cold night, air cooled below its dew point can result in a fog that forms like a pond in the valley bottom (● Fig. 6.10). It is common in mountain areas to see early morning radiation fog in the valleys while snow-capped mountaintops shine against a clear blue sky. Radiation fog has a diurnal cycle, forming during the night and becoming densest around sunrise when temperatures are lowest. It typically "burns off" during the day as solar energy slowly penetrates the fog, warming the ground surface. As the ground warms, it also warms the air directly above it, increasing its temperature and its water vapor-holding capacity, which causes the fog to evaporate.

Radiation fog can become very dense in industrial areas where high concentrations of particles in the air provide abundant condensation nuclei. These fogs are usually thicker than "natural" radiation fogs and less easily dissipated by winds or the daytime sun.

Advection Fog A common type of fog, called **advection fog**, occurs when warmer, moist air moves over a colder land or water surface. The overriding warm air loses heat by conduction to the colder surface below, cooling below its dew point temperature, and condensation occurs to produce fog. Advection fog usually covers wider areas compared to radiation fog. It is also less likely to have a diurnal cycle of forming and burning off, although if the fog is not too thick, it can burn off during the day to perhaps return again in the early evening. Typically however, advection fog is persistent and spreads over a large area for days at a time.

During the winter, advection fog forms in many land areas of the middle and high latitudes. In the United States, for example, these fogs occur when warmer, moist air from the Gulf of Mexico flows northward over the cold, frozen, and snow-covered upper Mississippi Valley. Advection fog can also develop over a lake when warm air from the land flows over colder water. These fogs commonly occur when a warm air mass passes over the cool surface waters of Lake Michigan or another one of the Great Lakes.

During the summer, advection fog can also form over the oceans or large lakes. Widespread advection fogs develop offshore and along coasts when a warm air mass moves over a cold ocean current, cooling the air sufficiently to bring about condensation. These advection fogs are also known as *coastal fogs* or *sea fogs*. These are common fogs in the summertime along the West Coast of the United States and are related to cold ocean currents. In the summer months, the Pacific subtropical high moves north and air drifting toward the coast passes over the cold California Current. Condensation occurs and forms fog that can flow into coastal areas, pushed from behind by the eastward air movement and pulled inland by the low pressure of the warmer land (● Fig. 6.11). Advection fogs also occur in New England, especially along the coasts of Maine and the Canadian Maritime Provinces, when warm, moist air over the Gulf Stream flows north over cold waters of the Labrador Current. Advection fog over the Grand Banks off Newfoundland has long been a hazard for marine navigation. This region has been referred to as the foggiest area in the world.

Upslope Fog Another type of fog, **upslope fog**, clings to windward sides of mountains. Its appearance has also been the source of geographic place names—for example, the Great Smoky Mountains, where this type of fog often occurs. During early-morning hours when a moist breeze ascends a slope, if the air cools to the dew point, a blanket of fog develops (● Fig. 6.12). In wet, tropical areas, mountain slopes may be covered in a misty fog at any time of day. Because of the very humid air in those regions, reaching the dew point can happen with a very minor drop in temperature.

J. Petersen

● **FIGURE 6.12** Upslope fog is caused by moist air adiabatically cooling as it rises up a mountain slope, in this case in the German Alps.

Steve Schilling/USGS

● **FIGURE 6.13** This automatic camera was placed to monitor volcanic activity at Mt. St. Helens in Washington state. A large accumulation of rime, however, has put the camera temporarily out of commission.

How is rime related to aircraft safety?

Dew and Frost

A cover of tiny water droplets that have condensed on cool surfaces, such as plants, buildings, and metal objects, is called **dew**. Dew collects on surfaces that are good radiators of heat (such as vehicles or blades of grass) because they lose large amounts of heat during the night. If the air cools to the dew point temperature as it contacts these cold surfaces, water droplets will form in beads on the surface. If the air temperature is below freezing, 0°C (32°F), **frost** forms. It is important to note that frost is not frozen dew but results from a *sublimation* process—water vapor changing directly from the gaseous to the frozen state.

Sometimes, although air temperatures may be below 0°C (32°F), the liquid droplets that make up clouds and fogs may not freeze into solid particles. These are *supercooled* water droplets. When they come in contact with a surface such as a tree branch, a window, a vehicle, or an airplane wing, ice crystals develop on that surface in a formation known as **rime** (● Fig. 6.13). Icing on the wings, nose, and tail of an airplane is extremely hazardous and can cause aviation disasters. The airlines use de-icing treatments to avoid this problem if the weather conditions are conducive to ice forming on aircraft during flight and before takeoff.

Clouds

Clouds consist of billions of tiny water droplets and/or ice crystals so small (some measured in thousandths of a millimeter) that they remain suspended in the atmosphere. Clouds are an intriguing part of our environment. Their great variety of shapes and hues, and their impact on the colors of the sky provide us with an ever-changing backdrop to the scenery on Earth. Clouds can appear white, or as shades of gray, or even deep gray approaching black. The apparent color of clouds depends on how thick or dense they are and if the sun is shining on the cloud surface that we can see. Cloud tops reflect solar radiation away and thick clouds absorb a great deal of the sunlight that strikes them, which blocks light from our view, making the clouds appear dark. Clouds also appear dark when seen from their shaded side instead of their sunlit side.

Clouds are the most widespread form of condensation and are important for many reasons. First, they are the source of precipitation. **Precipitation** refers to both liquid water and solid ice particles that fall from the sky. Obviously, not all clouds produce precipitation, but precipitation will not occur without cloud formation. Clouds serve an important function in the energy budget because they absorb some of the incoming solar energy. They also reflect some of that energy back to space and scatter or diffuse other wavelengths of incoming energy before it strikes the surface. In addition, clouds absorb some of Earth's thermal radiation that is not lost to space and then re-radiate it back to the surface.

Cloud Forms Clouds result from various air movements in our atmosphere. In 1803, Sir Luke Howard, an Englishman, proposed the first method of cloud classification, which has been modified through time into the system that is in use today. Cloud names generally (but not always) consist of two parts. The first part refers to the cloud's height: low-level clouds, below 2000 meters (6500 ft), are called **strato**; middle-level clouds, from 2000 to 6000 meters (6500–19,700 ft), are named **alto**; and high-level clouds, above 6000 meters (19,700 ft), are termed **cirro**.

The second part of the name concerns the morphology, or shape, of the clouds. The three basic shapes are termed *cirrus, cumulus,* and *stratus*. Classification systems categorize these cloud formations into many subtypes; however, most subtypes are variations of these three basic shapes. ● Figure 6.14 illustrates the appearance and the general heights of common clouds, and ● Figure 6.15 provides images to help you recognize the major cloud types.

Cirrus clouds (from Latin: *cirrus* means "a lock or wisp of hair") form at very high altitudes, normally 6000 to 10,000 meters (19,800 to 36,300 ft), and consist of ice crystals rather than water droplets. They are thin, wispy, stringy, white clouds that move across the sky like feathers. When associated with fair weather, cirrus clouds are feathery white patches scattered in a clear blue sky.

Cumulus clouds (from Latin: *cumulus* means "heap or pile") develop vertically rather than forming the more horizontal structures of the cirrus and stratus forms. Cumulus clouds are massive piles of clouds, rounded in appearance, usually with a flat base, the altitude of which varies according to the relative humidity. Their base is the altitude where condensation begins in a rising column of air. If the air is humid, the base of cumulus clouds will be lower in altitude, and higher if the relative humidity is lower. The cloud base is the altitude where the dew point temperature is reached, and it ranges widely, generally between

500 to 3000 meters (1650 to 9850 ft) above sea level. From this base, they develop into great rounded structures, often with tops like cauliflowers. Cumulus clouds provide visible evidence of an unstable atmosphere.

Stratus clouds (from Latin: *stratus* means "layer") can appear at a variety of altitudes from near the surface to about 2000 meters (6000 ft) and the varieties of stratus clouds are based in part on their altitude. The basic characteristic of stratus clouds is their horizontal sheetlike appearance, lying in layers with fairly uniform thickness. This horizontal configuration indicates that they form in relatively stable atmospheric conditions, which minimize vertical development.

Often, stratus clouds cover the entire sky with a gray cloud layer. It is stratus clouds that make up the dull, gray, overcast sky common to winter days in much of the midwestern and eastern United States. A formation of stratus clouds may overlie an area for days and any precipitation will be light but steady and persistent.

Examine Figures 6.14 and 6.15 and become familiar with the basic cloud types and their names. Keep in mind that some cloud shapes exist in all three levels—for example, *stratocumulus* (*strato* = low level + *cumulus* = a rounded shape); *altocumulus*, and *cirrocumulus*. These three share the similar rounded or cauliflower appearance of cumulus clouds, which do exist at all three levels. You may notice that *altostratus* (*alto* = middle level + *stratus* = layered shape) and *cirrostratus* have two-part names, but low-level layered clouds are just called *stratus*. Thin, stringy *cirrus* clouds are found only as high-level clouds, so the term *cirro* (meaning "high level cloud") is not necessary here.

Other terms used in describing clouds are **nimbo**, or **nimbus**, meaning "precipitation" ("rain is falling"). Thus, the *nimbostratus*

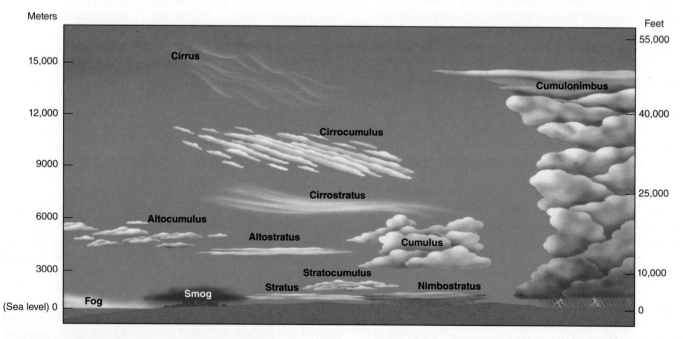

● **FIGURE 6.14** Clouds are named based on their height and their shape.

Observe this figure and the images in Figure 6.15; what cloud type is present in your area today?

● FIGURE 6.15 Types of clouds.

(a) Cirrus

(b) Cirrocumulus

(c) Cumulus

(d) Altocumulus

(e) Stratus

(f) Cumulonimbus

clouds may bring a long-lasting drizzle, and the cumulonimbus are the thunderstorm clouds. Cumulonimbus clouds develop from rapidly rising columns of air that can ascend to great heights, even as high as the tropopause, about 10,000 m (33,000 ft) in the middle latitudes. Reaching those altitudes, the highest part of a cumulonimbus cloud flattens out, forming a distinctive flat top called an *anvil head*. These clouds also become darker as they grow higher and thicker because they block the incoming sunlight. Cumulonimbus clouds are associated with storms and many atmospheric concerns, including high velocity winds, torrential rain, flash flooding, thunder, lightning, hail, and possibly tornadoes.

Adiabatic Heating and Cooling

Clouds typically develop from the cooling that results as air rises to an altitude where it reaches the dew point temperature (and relative humidity = 100%). Rising parcels of air expand as they encounter decreasing atmospheric pressure with altitude. Expansion of air allows the gas molecules to spread out, causing the parcel's air temperature to decrease. This is known as **adiabatic cooling**, a temperature decrease that occurs at a lapse rate of approximately 10°C per 1000 meters (5.6°F/1000 ft). Air descending through the atmosphere is compressed by increasing pressure and undergoes **adiabatic heating**, which increases air temperature by the same rate.

However, a rising parcel of air will eventually cool to its dew point temperature and reach its saturation capacity (relative humidity = 100%), where water vapor begins to condense, forming cloud droplets. After condensation occurs, rising will cool at a lower rate because of the release of latent heat of condensation into the air. To differentiate between these two adiabatic cooling rates, we refer to the precondensation rate (10°C/1000 m) as the **dry adiabatic lapse rate** and the lower rate postcondensation rate as the **wet adiabatic lapse rate**. The latter rate averages 5°C per 1000 meters (3.2°F/1000 ft) but varies according to the amount of water vapor that condenses out of the air.

Rising parcels of air will cool at either the dry or the wet adiabatic rate. If condensation is occurring, the wet adiabatic rate will operate; if no condensation is occurring, the dry adiabatic rate applies. The temperature of descending air will warm by compression, increasing its capacity to hold water vapor and preventing condensation. Thus, the temperature of descending air that is being compressed and warmed always increases at the dry adiabatic rate. It is important to note that adiabatic temperature changes are the result of changes in air volume and density as air moves vertically, either up or down. Adiabatic heating and cooling do not involve the addition or subtraction of heat from external sources.

It is very important to differentiate between the normal lapse rate, *where a temperature measuring device is moving up or down,* and the adiabatic lapse rates, *where the air is moving up or down.* In Chapter 4, we learned that, in general, atmospheric temperatures decrease with increasing altitude. This is the *normal lapse rate* (also called the *environmental lapse rate*), which is variable but averages 6.5°C per 1000 meters (3.6°F/1000 ft) and is measured by meteorological instruments sent aloft. The normal lapse rate reflects the vertical temperature structure of the atmosphere. Adiabatic lapse rates indicate temperature changes that result when air moves up or down and whether or not condensation is occurring (● Fig. 6.16).

Stability and Instability

Although adiabatic cooling results in cloud development, most of the varying cloud forms are related to differing degrees of vertical air movement. Some clouds are associated with air that is rapidly rising and buoyant. Other cloud forms result when air resists vertical movement.

An air parcel will be buoyant and rise as long as it is warmer than the surrounding air. When it reaches a part of the atmosphere that has the same temperature, it will stop rising. Air parcels that

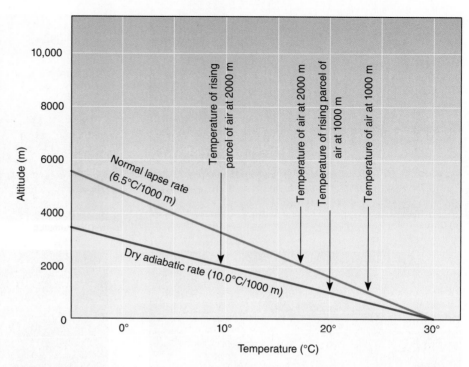

● **FIGURE 6.16** Comparison of the dry adiabatic lapse rate and the environmental, or normal, lapse rate. The normal lapse rate is the average vertical change in temperature. Air displaced upward will cool (at the dry adiabatic rate) because of expansion.

In this example, using the normal lapse rate, determine the temperature of the layer of air at 2000 meters.

rise because they are warmer than the surrounding atmosphere are said to be *unstable* (or *instable*). In contrast, air that is colder than the surrounding atmosphere tends to sink to lower levels. Sinking air is said to be *stable*.

Determining the stability or instability of an air parcel involves answering a fairly simple question. If an air parcel rises to a specific altitude (cooling at an *adiabatic* lapse rate), would it be warmer, colder, or the same temperature as the surrounding air (as determined by the normal lapse rate) at that same altitude?

If the air parcel is warmer than the air at the selected elevation, then the parcel will be unstable and would continue to rise because warmer air is less dense and therefore buoyant. Thus, under conditions of **instability**, the normal lapse rate must be *greater than* the adiabatic lapse rate in operation. For example, if the normal lapse rate is 12°C per 1000 meters and the ground temperature is 30°C, then the atmospheric air temperature at 2000 meters would be 6°C. In comparison, an air parcel (assuming that no condensation occurs) lifted to 2000 meters would have a temperature of 10°C. Because the air parcel is warmer than the atmospheric air around it, it is unstable and will continue to rise (● Fig. 6.17).

Now, assume that it is another day and all the conditions are the same except that measurements indicate the normal lapse rate on this day is 2°C per 1000 meters. Consequently, although our air parcel, if lifted to 2000 meters, would still have a temperature of 10°C; the temperature of the atmosphere at 2000 meters would now be 26°C. Thus, the air parcel would be colder and would sink back toward Earth as a result of its greater density (see Fig. 6.17).

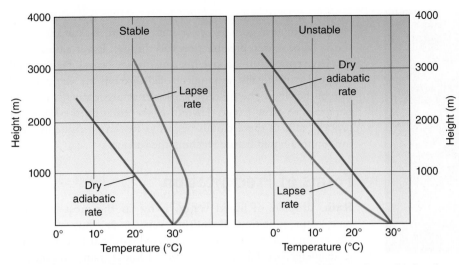

● FIGURE 6.17 Relationship between lapse rates and air mass stability. When an air parcel is forced to rise, it cools adiabatically. Whether it continues to rise or resists vertical motion depends on whether adiabatic cooling is greater or less than the prevailing vertical temperature lapse rate. (a) If the adiabatic cooling rate exceeds the lapse rate, the lifted air will be colder than its surroundings and will tend to sink when the lifting force is removed. (b) If the adiabatic cooling rate is less than the lapse rate, the lifted air will be warmer than its surroundings and will be buoyant, continuing to rise even after the original lifting force is removed.

In these examples, what would be the temperature of the lifted air if it rose to 2000 meters?

Under conditions of **stability** the normal lapse rate is *less than* the adiabatic lapse rate in operation. If an air parcel, lifted to a specific elevation, has the same temperature as the atmospheric air surrounding it, it is neither stable nor unstable. Instead, it is considered *neutral;* it will neither rise nor sink but will remain at that elevation.

Whether an air parcel will be stable or unstable is related to the amount of cooling and heating of air at Earth's surface. As air cools through radiation and conduction on a clear, cool night, air near the surface will be relatively close in temperature to that aloft, and the normal lapse rate will be low, enhancing stability. With the rapid heating of the surface on a hot summer day, there will be a very steep normal lapse rate because the air near the surface is so much warmer than that above, and instability will be enhanced. Pressure zones can also be related to atmospheric stability. In areas of high pressure, stability is maintained by air slowly subsiding from aloft, and in low pressure regions, instability is promoted by the tendency for air to rise.

Precipitation Processes

The condensed water droplets that float around within clouds do not fall to Earth because they are so tiny (0.02 mm, or less than 1000th of an inch) that gravity does not overcome the buoyant effects of air and the currents and updrafts that exist in clouds. ● Figure 6.18 shows the relative sizes of a condensation nucleus, a cloud droplet, and a raindrop. It takes about a million tiny cloud droplets to form one raindrop.

Precipitation occurs when water droplets or ice crystals become too large and heavy to be held aloft, so they fall as rain, snow, sleet, or hail. The type of precipitation depends largely on

how it formed, the temperature in the cloud, and the air temperature below as it falls to Earth. The two most widely accepted theories for how precipitation develops include the **collision–coalescence process** for warm clouds and the **Bergeron** (or **ice crystal**) **process** for cold clouds.

Collision–Coalescence Process Precipitation in the tropics and in warm clouds elsewhere is likely to form by collision–coalescence, a process that is well described by its name. Water is cohesive (able to stick to itself), so as water droplets collide while circulating in a cloud, they tend to coalesce (or grow together) until they become large and heavy enough to fall. In falling, larger droplets overtake smaller, more buoyant droplets and capture them to form larger raindrops. The mass of these growing raindrops eventually overcomes the updrafts of the cloud and fall to Earth under the pull of gravity. This process occurs in the warm section of clouds where all the moisture exists as liquid water (● Fig. 6.19).

Bergeron Process At higher latitudes, storm clouds can possess three distinctive layers. The lowest is a warm layer where the temperatures are above the freezing point of 0°C (32°F) and water droplets are liquid. Higher up, the second layer is composed of some ice crystals but mainly **supercooled water** (liquid water cooler than 0°C). In the uppermost layer of these tall clouds, if temperatures are lower than or equal to −40°C (−40°F), ice crystals will dominate (● Fig. 6.20). It is in relation to these layered clouds that Scandinavian meteorologist Tor Bergeron presented his explanation.

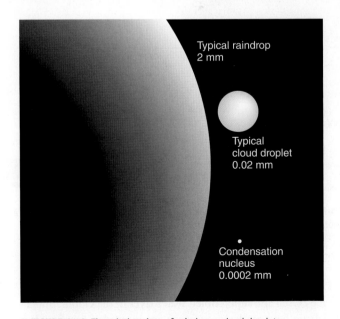

● FIGURE 6.18 The relative sizes of raindrops, cloud droplets, and condensation nuclei.

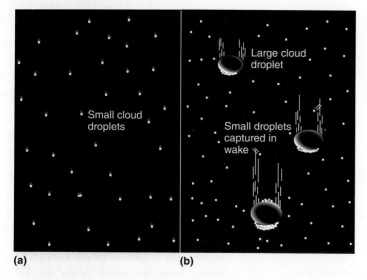

● **FIGURE 6.19** Collision and coalescence. (a) Tiny cloud droplets falling at about the same speed are unlikely to collide and coalesce, and if they do collide, they tend to bounce off of each other because of the surface tension of water. (b) Large droplets falling more rapidly can capture some of the smaller droplets.

Why do these tiny droplets fall at different speeds?

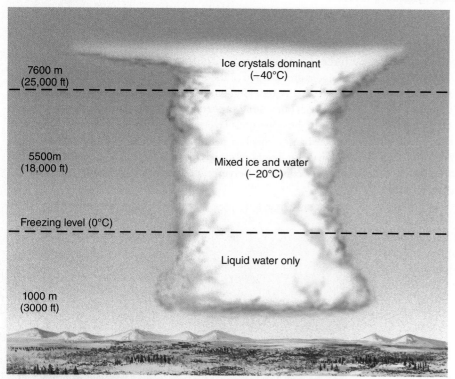

● **FIGURE 6.20** The distribution of water, supercooled water, and ice crystals in a towering cumulonimbus storm cloud according to the Bergeron process theory.

What is the difference between water and supercooled water?

The Bergeron (or ice crystal) process begins at great heights in the ice crystal and supercooled water layer of the clouds. Supercooled water has a tendency to freeze on any available surface. It is for this reason that aircraft flying through middle- to high-latitude storms run the risk of severe icing and potential disaster. The ice crystals can become *freezing nuclei* on which the supercooled water can freeze to form growing ice crystals. This process can also create snow. If the frozen precipitation falls through lower cloud layers that have temperatures above freezing, the ice crystals melt and fall as liquid rain. Finally, as raindrops fall through the warmer section of the cloud, the collision–coalescence process may cause the raindrops to grow larger. Therefore, according to the Bergeron process, rain in these clouds begins as frozen precipitation and melts into a liquid before reaching the surface.

Forms of Precipitation

Rain, droplets of liquid water, is by far the most common type of precipitation. Raindrops vary in size but are generally about 2 to 5 millimeters (approximately 0.1 to 0.25 in.) in diameter (see Fig. 6.18). As we all know, rain can come in many ways: as a brief shower, a steady rainfall, or the deluge of a strong rainstorm. When the temperature of an air mass is only slightly below the dew point, the raindrops may be very small (about 0.5 mm or smaller in diameter) and close together. The result is a fine mist called **drizzle**. Drizzle is so light that it is greatly affected by the direction of air currents and the variability of winds. Consequently, drizzle seldom falls vertically.

Snow is the second most common form of precipitation. When water vapor is frozen directly into a solid without first passing through a stage as liquid water, it forms minute ice crystals around the freezing nuclei (of the Bergeron process). These crystals develop into hexagonal ice crystals that make up the intricate six-sided form of snowflakes. Snow will reach the ground only if the cloud and the air below maintain sub-freezing temperatures (below 0°C or 32°F).

Sleet is rain that freezes as it falls through a thick layer of sub-freezing air near the surface. The cold air causes raindrops to freeze into small solid particles of clear or milky ice. In English-speaking countries outside the United States, sleet refers not to this phenomenon of frozen rain but rather to a mixture of rain and snow.

Hail is a less common form of precipitation than rain, snow, or sleet; it occurs most often during the spring and summer months as a result of thunderstorm activity. Hail forms as lumps of ice, *hailstones*, which can vary in size from 5 millimeters (0.2 in.) in diameter to larger than a baseball. The U.S. record for hailstone size fell in Vivian, South Dakota, in July 2010 (● Fig. 6.21). The world record is a hailstone 30 centimeters (12 in.) in diameter that fell in Australia. Hail forms when ice crystals are lifted by strong updrafts in a cumulonimbus (thunderstorm) cloud. As these ice crystals circulate within the storm cloud, supercooled water droplets attach themselves as increasing thicknesses of frozen layers. Sometimes these pellets are lifted up into the cold layer of air and then dropped again and again. The resulting hailstone, made up of concentric layers of ice, has a frosty, opaque appearance when

● **FIGURE 6.21** A severe thunderstorm in South Dakota produced the largest hailstone (by diameter) ever recorded in the United States in July 2010. Shown here, its diameter was 20.3 cm (8 in.) and it weighed 0.89 kilograms (1.94 pounds).

How do hailstones grow in size while up in a storm cloud?

● **FIGURE 6.22** A thick glaze of ice accumulated from freezing rain during an ice storm coats weather monitoring equipment. A glazing of ice like this also makes roads and sidewalks extremely slippery and hazardous. Tree branches often break under the extra weight of this kind of ice cover.

Why are power failures a common occurrence with ice storms?

it finally breaks out of the strong updrafts of the cloud formation and falls to Earth. The larger the hailstone, the more times it is cycled through the freezing process and accumulated additional frozen layers. Dropping from the sky, large hailstones can be highly destructive to livestock, crops and other vegetation, vehicles, and buildings. Although primarily destructive of property, hailstones have been known to injure or kill people and animals.

On occasion, raindrops can also have a temperature below freezing and maintain liquid form. These supercooled raindrops will instantly freeze if they fall onto any surface that is also at a sub-freezing temperature. The resulting icy covering on trees, plants, and utility lines is known as **freezing rain** (or **glaze**). People usually call this kind of precipitation an "**ice storm**" (● Fig. 6.22). Because of the weight of ice, glazing can break tree branches that bring down telephone and power lines. Surface ice accumulations cause extremely slippery conditions that make driving dangerous and some roads impassable. Yet, ice storms can also produce a beautiful natural landscape. Sunlight glitters on the ice, reflecting and making a diamond-like surface that covers trees, plants, buildings, and vehicles.

Factors Necessary for Precipitation

Three factors are necessary for precipitation to develop. The first is the presence of *moist air*, which provides a moisture source (for precipitation) and

energy (as latent heat of condensation). Second are the *condensation nuclei* around which the water vapor can condense. Third is an **uplift mechanism** that forces the air to rise enough to cool (by the dry adiabatic rate) to the dew point temperature and then continue to cool at the wet adiabatic lapse rate. Precipitation results from one of four major uplift mechanisms that force parcels of air to rise (● Fig. 6.23).

● **FIGURE 6.23** Four mechanisms that cause uplifting of air. The principal cause of precipitation is upward movement of moist air resulting from convectional, frontal, cyclonic, or orographic lifting.

What kind of air movement is common to all four diagrams?

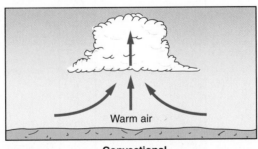

Convectional

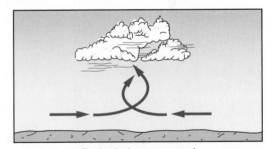

Cyclonic (convergence)

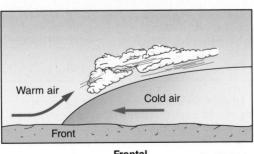

Frontal

Orographic

- **Convectional precipitation** results from the vertical displacement of warm air upward in a convectional system.
- **Frontal precipitation** takes place when a warmer air mass rises after encountering a colder, denser air mass.
- **Cyclonic** (or **convergence**) **precipitation** occurs when air converges on and is lifted upward into a low pressure system.
- **Orographic precipitation** results when a moving air mass encounters a land barrier, usually a mountain, and is forced to rise.

Convectional Precipitation Convection occurs as air, heated near the surface, expands, becomes lighter, and rises. **Convectional precipitation** is most common in the hot, humid tropical and equatorial areas and during the summer in many midlatitude locations. When condensation begins in a convectional column of air, further lifting will be encouraged by additional energy from the release of latent heat of condensation.

To reinforce our understanding of convectional precipitation, let's apply what we have learned about instability and stability. ● Figure 6.24 illustrates two different cases where air rises as a result of convection. In both, the lapse rate in the free atmosphere is the same; it is especially high during the first few thousand meters but slows after that (as on a hot summer day).

In the first case (Fig. 6.24a), the air parcel is not very humid and thus the dry adiabatic rate applies throughout its ascent. By the time the air reaches 3000 meters (9900 ft), its temperature and density are the same as those of the surrounding atmospheric air. At this point, convectional uplift stops.

In the second case (Fig. 6.24b), the latent heat of condensation is introduced. Here again the unsaturated rising column of air cools at the dry adiabatic rate of 10°C per 1000 meters (5.6°F/1000 ft) for the first 1000 meters (3300 ft). However, because the air parcel is humid, the rising air column soon reaches the dew point, condensation takes place, and cumulus clouds begin to form. As condensation occurs, the latent heat locked up in the water vapor is released, heating the moving parcel of air and retarding the adiabatic cooling so that the rising air now cools at the wet adiabatic rate 5°C per 1000 meters (3.2°F/1000 ft). Here, the temperature of the rising air parcel remains warmer than that of the atmospheric layer it is passing through and the air parcel will continue to rise. In this case, which incorporates the latent heat of condensation, massive amounts of condensation occur, reinforcing the towering cumulus clouds and their thunderstorm potential.

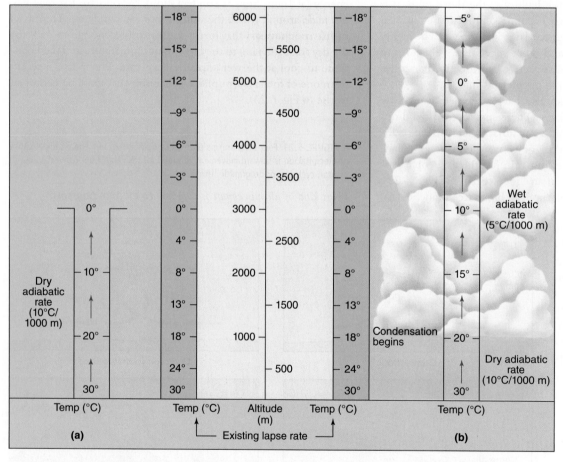

● **FIGURE 6.24** How humidity affects the stability of an air mass. (a) Warm, dry air rises and cools at the dry adiabatic rate, and when it becomes the same temperature as the surrounding air uplift stops. Because the rising dry air did not cool to its dew point temperature by the time lifting ended, no cloud formed. (b) Rising warm, moist air cools to its dew point temperature at the condensation level. The upward moving air subsequently cools at the wet adiabatic rate, keeping the air warmer than the surrounding atmosphere and uplift continues.

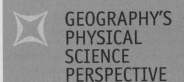
The Lifting Condensation Level

When we look at clouds in the atmosphere, it is often easy to see that many have relatively flat bases. Cloud tops may appear irregular, but the bases of clouds are often flat. Even if the cloud bases do not seem flat, it will be obvious that the clouds you see seem to have formed at the same level above the surface. This level represents the altitude to which the air must be lifted (and cooled at the dry adiabatic rate) in order to reach saturation. Basically, this is the altitude where rising air reaches its dew point temperature and condensation begins. Additional lifting will form clouds that will build upward. The altitude at which clouds form from lifting is called the **lifting condensation level (LCL)**, which will vary widely depending on the local atmospheric conditions. The following equation can be applied to estimate the lifting condensation level:

$$\text{LCL (in m)} = 125 \text{ meters} \times$$
$$(\text{Celsius temperature} - \text{Celsius dew point})$$

For example, if the surface temperature is 7.2°C (45°F) and the dew point temperature is 4.4°C (40°F), then the LCL is estimated at 350 meters (1148 ft) above the surface.

Note: Keep in mind that different layers of clouds may exist at the same time. Low, middle, and high clouds as defined in this chapter may all appear on the same afternoon. These clouds may have formed in other regions and be only passing overhead. The formula presented here is best used with the lowest level of cloud cover that appears overhead.

A cumulus cloud with a flat bottom indicating the altitude of the lifting condensation level.

J. Petersen

Convectional uplift can cause the heavy precipitation, thunder, lightning, and tornadoes that can accompany thunderstorms, especially during warm or hot, humid afternoons. The strong convectional updrafts that occur in towering cumulonimbus clouds frequently produce hail along with the thundershowers.

Frontal Precipitation The zone of contact between relatively warm and relatively cold bodies of air is a **front**. As was mentioned in the previous chapter, the concept of a weather *front* comes from the line of contact between opposing armies. When two large bodies of air that differ in temperature, humidity, and density collide, the warmer, less dense air mass is lifted above the colder, denser body of air. Collision causes uplift and uplift results in cooling—producing condensation and precipitation. *Frontal precipitation* develops as warm, moisture-laden air collides with cold air and rises above the front. Depending on many factors, especially the temperature and moisture content of the clashing *air masses*, frontal precipitation can produce many kinds of weather, from cloudy overcast conditions to rain, snow, or ice storms.

Understanding fronts requires knowing what causes unlike bodies of air to collide as well as being familiar with the weather conditions associated with different kinds of fronts. This will be discussed in Chapter 7, where we will take a more detailed look at frontal disturbances and precipitation.

Cyclonic (Convergence) Precipitation Cyclonic uplift (convergence) was introduced in Chapter 5 and involves air interacting with a cell of low pressure *(cyclone)*. Air flows into a low pressure system that is roughly circular, moving in a counter-clockwise direction in the Northern Hemisphere. When air *converges* on a cyclone, it is pulled into the rising air of the low pressure cell. Therefore, cloudy conditions and precipitation are common around the center of a cyclone. Hurricanes and the rain associated with these storms *(cyclonic precipitation)* are fed by air flowing into the convergent uplift around a circular low pressure system and the energy resulting from the release of latent heat of condensation.

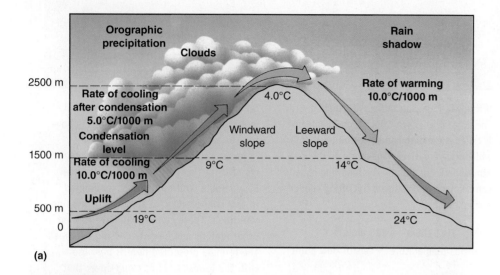

(a)

(b)

(c)

● **FIGURE 6.25** Orographic precipitation and the rain shadow effect. (a) Orographic uplift over the windward (western) slope of the Sierra Nevada Mountains produces condensation, cloud formation, and precipitation, resulting in (b) dense stands of forest. (c) Semi arid or rain shadow conditions occur on the leeward (eastern) slope of the Sierras.

Can you identify a mountain range in Eurasia in which the leeward side of that range is in the rain shadow?

Orographic Precipitation When land barriers—such as a mountain, a hilly region, or the escarpment (steep edge) of a plateau or mountain range—lie in the path of prevailing winds, air is forced to rise above these barriers. The air cools by expansion while ascending over a topographic barrier and condensation takes place. The resultant precipitation is called *orographic precipitation* (from Greek: *oros* means "mountains").

A potentially unstable air parcel may need only the initial lift provided by an orographic barrier to set it in motion. In this case, it will continue to rise of its own accord (no longer forced) as it seeks air of its own temperature and density. When a land barrier provides the initial thrust, it has performed its function as a lifting mechanism.

As orographic precipitation falls on the windward side of a mountain, the air parcel and the clouds lose some of their moisture content (the *absolute* or *specific humidity* declines). However, continued cooling as the air rises can maintain the relative humidity at 100%, so precipitation continues so long as the air contains adequate moisture and continues to rise. As the air descends the leeward slope, its temperature warms (at the dry adiabatic rate) and condensation ceases. The leeward side of an orographic barrier is called the **rain shadow** (● Fig. 6.25a). Just as being in a shadow means that you are not receiving any direct sunlight, being in a rain shadow location means that the area does not receive much rain (or other precipitation). The orographic precipitation and the rain shadow effect have a strong impact on the vegetation patterns (Fig. 6.25b and c) observed on a mountain range. The windward side of mountains (for example, the Sierra Nevada in California) will be heavily forested. The opposite slopes in the rain shadow will be drier, usually with a sparse cover of vegetation.

Distribution of Precipitation
Distribution over Time

To understand the nature of a location's precipitation, consider *average annual precipitation* to get an impression of the moisture that a region gets during a year. Also examine the annual or monthly number of *rain days*—days when 1.0 millimeter (0.01 in.) or more of rain were received during a 24-hour period. Less than this amount is known as a **trace** of rain.

By dividing the number of rain days in a month or year by the total number of days in that period, the resulting figure represents the probability of rain. Such a measure is important to farmers and to ski or summer resort owners whose incomes may depend on precipitation or the lack of it.

Another factor to be considered is the *average monthly precipitation*. By examining precipitation data for all 12 months, the seasonal variations in precipitation for a location can be determined (● Fig. 6.26). For instance, in describing the climate of California, average annual precipitation would not give the full story because an annual average would not show the distinct wet and dry seasons that characterize this region, but monthly averages would give us that information.

● **FIGURE 6.26** Average monthly precipitation in San Francisco, California, is represented by colored bars. A graph of monthly precipitation gives a good impression of how a location's weather varies over the seasons. If we had only the annual precipitation total, we would not know that nearly all the precipitation occurs in only half of the year.

How would this rainfall pattern affect agriculture?

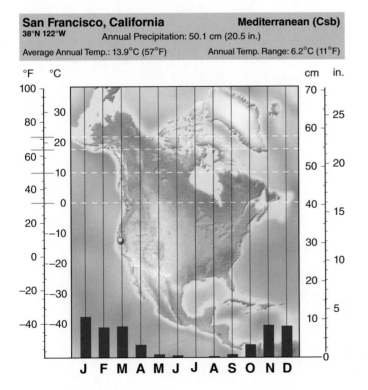

San Francisco, California **Mediterranean (Csb)**
38°N 122°W Annual Precipitation: 50.1 cm (20.5 in.)
Average Annual Temp.: 13.9°C (57°F) Annual Temp. Range: 6.2°C (11°F)

Spatial Distribution of Precipitation

Great geographic variability exists in the global distribution of precipitation. The map in Figure 6.27 shows the average annual precipitation of Earth's land areas. Although there is a latitudinal distribution pattern of precipitation, latitude is not the only factor involved in the amount of precipitation an area receives.

Two main factors affect the potential for receiving precipitation and the amount of precipitation received. First, precipitation depends on the degree of lifting that occurs in the air of a particular region. This lifting may be a result of frontal activity, air converging into a low pressure system (cyclonic), convective uplift, or orographic uplift (rising over a topographic barrier). The second factor affecting the likelihood of precipitation depends on the internal characteristics of the air itself, including its degree of instability, its temperature, and its humidity.

The rainfall amounts depicted on the map in Figure 6.27 are annual averages. However, in many parts of the world there are significant variations in precipitation, both within any one year and between years. For example, areas such as California, Chile, South Africa, Western Australia, and the Mediterranean region, located on the west sides of their respective continents, roughly between 30° and 40° latitude, receive much more rain in the winter than in the summer. Also, some areas between 10° and 20° latitude receive much more of their precipitation in the summer (high–sun season) than in the winter (low–sun season).

Latitude has a strong impact on precipitation distribution because the occurrence, absence, and variations of many weather and climate factors are related to latitude. For example, because warmer air can hold more water vapor and colder air can hold less, there is a general decrease of precipitation from the equator to the poles. Further study of the map in Figure 6.27 will reveal a great deal of spatial variability in average annual precipitation beyond the general pattern of a decrease with increased latitude.

The equatorial zone is generally an area of high precipitation—typically more than 200 cm (79 in.) annually. High temperatures and instability in the equatorial region lead to a general pattern of rising air, which generates precipitation. This tendency is reinforced by the convergence of the trade winds as they flow toward the equator from opposite hemispheres. In fact, the *intertropical convergence zone* is one of the two great zones where air masses converge. The other is along the *polar front*, between air from the subtropical highs and from the polar regions.

In general, the air of the trade wind zones is stable compared to the instability in the equatorial region. Under the influence of the steady trade winds, few atmospheric disturbances would lead to convergent or convectional lifting. However, because the trade winds have an easterly flow, when they move onshore along east coasts or islands with high elevations they carry moisture from the oceans. Thus, within the trade wind belt, east coasts tend to be wetter than west coasts. In fact, where the air of the equatorial and trade wind regions—with its high temperatures and vast amounts of moisture—moves onshore from the ocean and meets a landform barrier, record rainfalls can be measured. The windward slope of Mount Waialeale on Kauai, Hawaii, at approximately

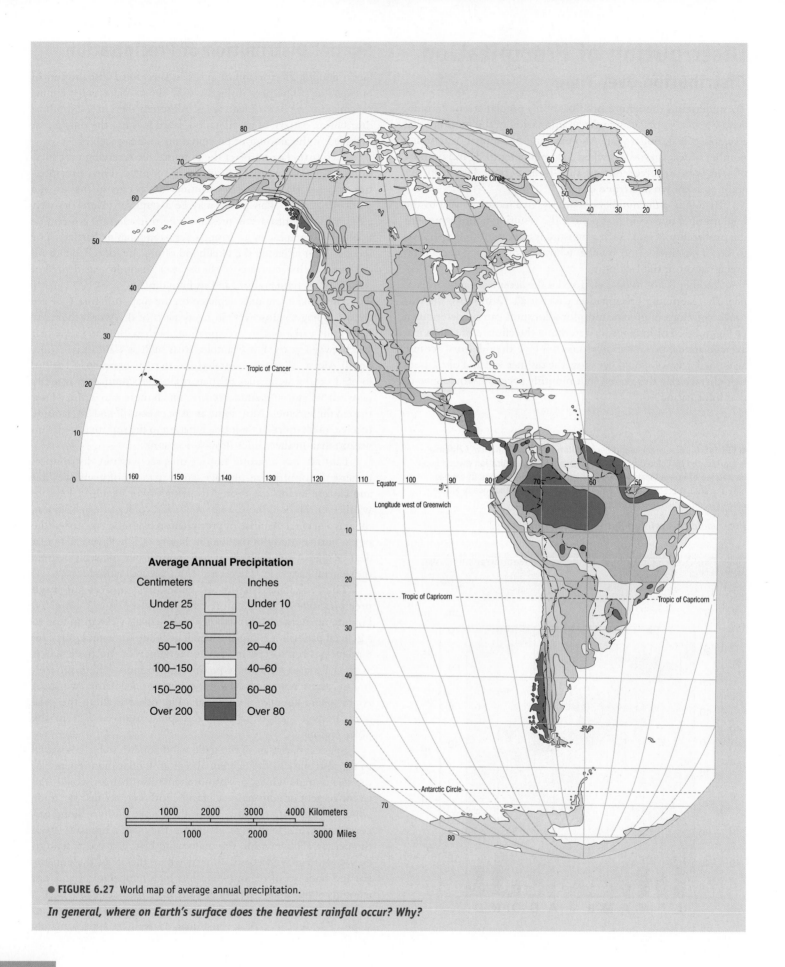

Average Annual Precipitation

Centimeters		Inches
Under 25		Under 10
25–50		10–20
50–100		20–40
100–150		40–60
150–200		60–80
Over 200		Over 80

● **FIGURE 6.27** World map of average annual precipitation.

In general, where on Earth's surface does the heaviest rainfall occur? Why?

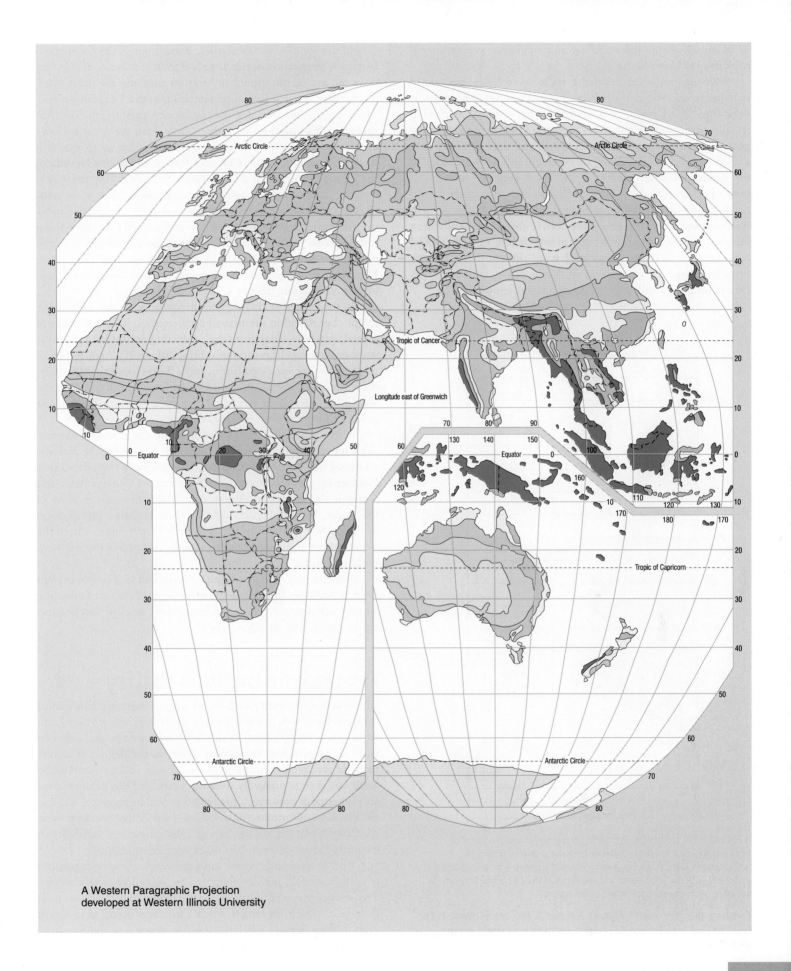

A Western Paragraphic Projection
developed at Western Illinois University

22°N latitude, claims the world's record for greatest average annual rainfall—1198 cm (471 in.).

Moving poleward from the trade wind belts, we would encounter the subtropical high pressure zones where air is subsiding. As it sinks, it is warmed adiabatically, increasing its moisture holding capacity and reducing the precipitation potential in these regions. In fact, ● Figure 6.28 indicates that a drop in precipitation amounts corresponds to the latitudes of the subtropical high pressure cells. These zones of subtropical high pressure are where most of the great deserts of the world are located—in northern and southern Africa, Arabia, North America, and Australia. The exceptions to this subtropical aridity occur along the east sides of continents, where the subtropical high pressure cells are weak and wind direction is onshore, bringing moisture from areas of warm ocean currents. This exception is especially true of the monsoon regions.

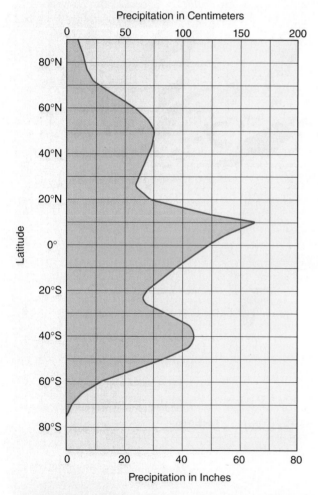

● **FIGURE 6.28** Latitudinal distribution of average annual precipitation. Earth has four distinctive precipitation zones: the tropics, with high precipitation caused by air converging; the midlatitudes, with precipitation associated with the polar front; and two zones of low precipitation caused by subsiding air in the subtropical and polar regions.

Compare this figure with Figures 5.9 and 5.10a and b. What is the relationship between world rainfall patterns and world pressure distribution?

In the zones of the westerlies, from about 35° to 65°N and S latitude, precipitation occurs largely from the collision of cold, dry polar air and warm, humid subtropical air along the polar front. Thus, there is much frontal precipitation in this latitudinal zone.

In the middle latitudes, the continental interiors are drier than the coasts because they are farther away from the oceans. However, where air in the prevailing westerlies is forced to rise, as it does when it crosses the Cascades Range and Sierra Nevada of the Pacific Northwest and California, especially during the winter months, there is heavy orographic precipitation. Thus, in the middle latitudes, continental west coasts tend to be wet and precipitation decreases eastward toward the continental interiors. Along eastern coasts within the westerlies, precipitation usually increases once again because of proximity to humid air from the oceans. Here, convection and the convergence associated with hurricanes bring precipitation, mainly in the summer months.

In the United States, the interior lowlands are not as dry as we might expect within the prevailing westerlies. This is because of frontal activity resulting from the conflicting northward and southward movements of polar and subtropical air. If a high east–west mountain range extended from central Texas to northern Florida, the lowlands of the continental United States north of that range would be much drier because the mountains would block the moist air from the Gulf of Mexico.

Also characteristic of the westerlies are desert areas in the rain shadows of mountain ranges. This is one reason for the extreme aridity of California's Death Valley; the deserts of Nevada; eastern California; the mountain-ringed deserts of eastern Asia; and Argentina's Patagonian Desert, which is in the rain shadow of the Andes. Note in Figure 6.28 that there is greater precipitation in the middle latitudes of the Southern Hemisphere, where the oceans cover more area than the continents, unlike the northern middle latitudes.

Moving poleward, low temperatures lead to low evaporation rates. In addition, the polar regions are generally areas of subsiding air and high pressure. These factors combine to cause low precipitation amounts in the polar zones.

Precipitation Variability

The rainfall amounts depicted on the map in Figure 6.27 are annual averages. However, in many parts of the world there are significant variations in precipitation, both within any one year and between years. **Precipitation variability** refers to the degree of uncertainty of precipitation amounts from one year to the next. Rainfall totals can change markedly over the years, and can be unpredictable in terms of how much precipitation will be received, if any at all. This is an unfortunate situation for many of the world's people. The drier a place is, on the average, the greater the variability in its precipitation (compare Fig. 6.29 with Fig. 6.27).

Arid or semi arid regions can experience a year with particularly high rainfall, perhaps followed by several years of drought. This makes the availability of critical water resources uncertain from year to year, or longer. Such a situation occurred in recent years in West Africa's Sahel, the Russian steppe, and the American Great Plains. There can be years of drought and years of flood,

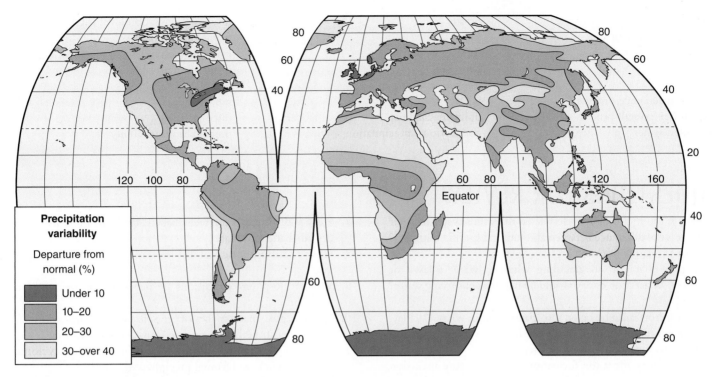

● FIGURE 6.29 World map of precipitation variability. The greatest variability in annual precipitation totals occurs in the dry regions, accentuating the critical problem of moisture supply in those parts of the world.

Compare this figure with the map in Figure 6.27. What are some of the similarities and differences?

each bringing its own kind of disaster. Farmers, construction workers, people who work in resorts, and others whose economic well-being depends in one way or another on weather are at the mercy of variable probabilities of rainfall on an annual, monthly, or even seasonal basis.

We all know from watching the weather reports that rainfall cannot always be forecast with 100% accuracy. This inability results from the interaction of many factors involved in producing precipitation—temperature, moisture, atmospheric disturbances,

landform barriers, fronts, air mass movement, upper air winds, and differential surface heating, among others. Variations in precipitation types, amounts, reliability, seasonality, and spatial distributions not only affect patterns of weather, climate, and vegetation but also the soils and landforms of our planet. These weather factors also impact our daily lives. By using our knowledge of why certain precipitation types develop and where they are likely to occur, the patterns of environmental diversity that exist on Earth can be more easily understood.

CHAPTER 6 ACTIVITIES

■ TERMS FOR REVIEW

capillary action
hydrologic cycle
water budget
saturation
moisture capacity
dew point temperature
humidity
absolute humidity
specific humidity
relative humidity

transpiration
evapotranspiration
potential evapotranspiration
condensation nuclei
radiation fog (ground or
temperature inversion fog)
advection fog
upslope fog
dew
frost

rime
precipitation
strato
alto
cirro
cirrus
cumulus
stratus
nimbus (nimbo)
adiabatic cooling

adiabatic heating
dry adiabatic lapse rate
wet adiabatic lapse rate
instability
stability
collision–coalescence process
Bergeron (ice crystal) process
supercooled water
rain

drizzle
snow
sleet
hail
freezing rain (glaze)
ice storm
uplift mechanism
convectional precipitation

lifting condensation level (LCL)
frontal precipitation
cyclonic (convergence) precipitation
orographic precipitation
front
rain shadow
trace (of rain)
precipitation variability

QUESTIONS FOR REVIEW

1. How is the hydrologic cycle related to Earth's water budget?
2. What is the difference between absolute and specific humidity? What is relative humidity?
3. Imagine that you are deciding when to water a lawn or garden. What time of day would be best for conserving water? Why?
4. What factors must be taken into consideration when calculating potential evapotranspiration? How is evapotranspiration related to the water budget of a region?
5. What factors affect the formation of temperature inversion fogs?
6. What causes adiabatic cooling? Differentiate between the normal lapse rate and the adiabatic lapse rates.

7. How and why does the wet adiabatic lapse rate differ from the dry adiabatic lapse rate?
8. How is atmospheric stability related to the adiabatic lapse rates?
9. What atmospheric conditions are necessary for precipitation to occur?
10. Find out how many inches of precipitation have fallen in your area this year. Is that average or unusually high or low?
11. Compare and contrast convectional, orographic, cyclonic (convergence), and frontal precipitation.
12. How is rainfall variability related to total annual rainfall? How might this relationship be considered a double problem for people?

PRACTICAL APPLICATIONS

1. Refer to Figure 6.6.
   a. What is the water vapor capacity of air at 0°C? 20°C? 30°C?
   b. If a parcel of air at 30°C has an absolute humidity (actual water vapor content) of 20.5 grams per cubic meter, what is the parcel's relative humidity?
   c. If the relative humidity of a parcel of air is 33% and the air temperature is 15°C, what is the absolute humidity of the air in grams per cubic meter?
   d. A major concern of residents in climate regions that experience freezing winter weather is the low relative humidity within their homes during that season. Low relative humidity is not healthy, and it has an adverse effect on home furnishings. The problem results when cold air, which can hold little water vapor, is brought indoors and heated up. The following example will illustrate the problem: Assume that the air outside is 5°C and has a relative humidity of 60%. What is the actual water vapor content of this air? If it is brought indoors (through the doors, windows, and cracks in the home) and heated to 20°C, with no increase in water vapor content, what is the new relative humidity?
2. Recall that as a parcel of air rises, it expands and cools. The rate of cooling, termed the *dry adiabatic lapse rate,* is

10°C per 1000 meters. (A descending parcel of air will always warm at this rate.) In addition, the dew point temperature decreases about 2°C per 1000 meters within a rising parcel of air. At the height at which the dew point temperature is reached, condensation begins and, as discussed in the text, the wet adiabatic lapse rate of 5°C per 1000 meters becomes operational. When the wet adiabatic lapse rate is in operation, the dew point temperature will be the same as the air temperature. When an air parcel descends through the atmosphere, its dew point temperature increases 2°C per 1000 meters. The height at which condensation begins, referred to as the *lifting condensation level (LCL),* can be determined by using the formula found in this chapter's second box, titled "The Lifting Condensation Level."

   a. A parcel of air has a temperature of 25°C and a dew point temperature of 14°C. What is the height of the LCL? If that parcel were to rise to 4000 meters, what would be its temperature?
   b. A parcel of air at 6000 meters has a temperature of 25°C and a dew point of 210°C. If it descended to 2000 meters, what would be its temperature and dew point temperature?

■ PRACTICAL APPLICATIONS

1. Using the following data set, for each month of the year calculate: (a) a running total of precipitation month to month, (b) the departure from the mean value (in surplus or deficit) for each month, and (c) the annual departure from the mean value (in surplus or deficit) for the whole year.
How does the year begin with respect to surplus or deficit? How does the year end?
Which month accumulated the greatest deficit for the year? Which month is the first to show a surplus?

Month	Recorded Rainfall (in.)	Mean Rainfall (in.)
January	4.94	4.94
February	2.40	4.10
March	3.02	5.26
April	3.21	4.36
May	3.69	4.22
June	2.69	4.17
July	5.08	4.28
August	5.41	3.62
September	7.89	3.19
October	4.10	2.68
November	4.27	3.72
December	2.77	4.36

2. Using the following data set, answer the following questions: (a) Following the procedure described in the section titled "Distribution of Precipitation," calculate the probability of precipitation for each month of the year. (b) Which month has the highest probability? Which has the lowest probability? (c) What is the average probability for rainfall for the year?

Month	Number of Days with >1 mm (0.01 in.) of Rain
January	12
February	10
March	7
April	11
May	16
June	11
July	8
August	9
September	10
October	10
November	11
December	9

■ LOCATE AND EXPLORE

1. Using Google Earth, fly to the Aral Sea (45.2°N, 59.9°E) on the Uzbekistan–Kazakhstan border. Once you arrive at your coordinates, zoom out to view the extent of the Aral Sea, which was once one of the largest lakes in the world. As a result of irrigation projects and stream diversions, much of the water flowing into the lake was cut off and the lake has been significantly reduced in size. You can see the outline of the lake before it was reduced in size. Assuming that the lake can be characterized as a rectangle (area = length × width), what has been the change in the lake's area (in square miles and as a percentage of the original lake area)? *Tip:* Use the ruler tool to measure the width and length.

MindTap—Make the most of your study time by accessing everything you need to succeed in one place. Read your textbook, take notes, review flashcards, watch videos, complete activities, take practice quizzes, and more online with MindTap. Log in at **www.cengagebrain.com**.

AIR MASSES AND WEATHER SYSTEMS

OBJECTIVES

WHEN YOU COMPLETE THIS CHAPTER YOU SHOULD BE ABLE TO:

- 7.1 Outline and explain the major air mass types, their characteristics, and their source regions.
- 7.2 Describe all four types of fronts and the weather conditions that occur with their passage or presence.
- 7.3 Contrast the general atmospheric conditions associated with anticyclones and cyclones.
- 7.4 Discuss the characteristics of a midlatitude cyclone, the factors that influence its movement, and its stages of development.
- 7.5 Review the potentially serious weather conditions, possible damage, and hazards associated with hurricanes, thunderstorms, and tornadoes.
- 7.6 Explain the difference between weather and climate, and be aware of the factors that make weather forecasting a complex process.
- 7.7 Interpret a weather map and understand the symbols that are used to show atmospheric conditions, fronts, precipitation patterns, pressure cells, air masses, and winds.

IN THIS CHAPTER WE APPLY what we have learned about insolation, heat, temperature, pressure, wind, and moisture conditions as we examine weather systems and the kinds of storms (atmospheric disturbances) that accompany them. Understanding Earth's varied weather systems—their general characteristics, when they can occur, how they form, and their influence on the regions they affect—is important in physical geography. In addition to involving temperature change and producing essential precipitation, weather systems are a major means of energy exchange, but they can also present hazards, such as floods, damaging winds, lightning, and violent thunderstorms. We begin the chapter with a detailed study of air masses and fronts, two atmospheric elements that not only affect weather systems but also strongly influence regional climates and environmental diversity, two important topics to be considered in forthcoming chapters.

Hurricane Sandy in October 2012, after it lashed several Caribbean islands with devastating force. The third largest Atlantic hurricane in history, it was over 1600 km (1000 mi.) in diameter. Later, designated as a *superstorm*, Sandy affected affected most of the eastern United States, seriously damaging areas of New York and New Jersey. NASA

Air Masses

An **air mass** is a large body of air, subcontinental in size, which is relatively homogeneous in terms of temperature and humidity. Air masses can extend through 20° to 30° of latitude, so temperature and humidity variations exist from their poleward to their equatorward edges. The characteristics, locations, and movements of an air mass have a considerable impact on the weather, and contact with differing land and ocean surfaces can modify the temperature and humidity of an air mass.

The temperature and humidity characteristics of an air mass are determined by the nature of its **source region**—the area where an air mass originates. Only certain areas on Earth make good source regions, because they require a nearly homogeneous surface for an air mass to develop similar temperature and humidity characteristics. A source region can be a desert, an ocean area, or a large landmass with relatively small elevational variations, but not a combination of surfaces. In addition, an air mass must have sufficient time to acquire the atmospheric characteristics of the source region.

On weather maps, air masses are identified by a two-letter code that refers to their source region. The first letter, always written in lowercase, is either *m* or *c*. The letter *m*, for *maritime*, means the air mass originated over water and therefore is relatively moist. The letter *c*, for *continental*, means the air mass originated over land and is relatively dry. The second letter, always capitalized, refers to the source region's latitudinal zone. *E* stands for *equatorial*, and this air is very warm. The letter *T* identifies a *tropical* origin and is also warm air. A *P* represents *polar*, air that can be quite cold; and an *A* identifies *Arctic* air, which is *very* cold (*AA* is also used for *Antarctic air*). These letters give us the classification symbols for seven air mass types: **maritime equatorial** *(mE)*, **maritime tropical** *(mT)*, **continental tropical** *(cT)*, **continental polar** *(cP)*, **maritime polar** *(mP)*, **continental arctic** *(cA)*, and **continental antarctic**

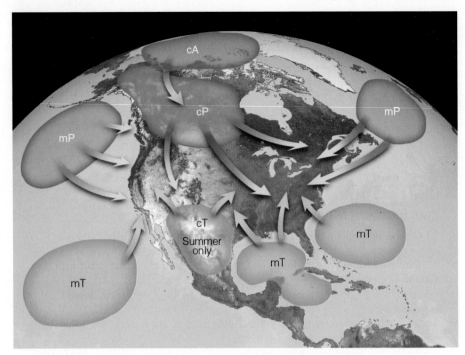

● **FIGURE 7.1** Source regions of North American air masses. Air mass movements import the temperature and moisture characteristics of these source regions into distant areas.

Use Table 7.1 and this figure to determine which air masses affect your location. Are there seasonal variations?

(cAA). The origin and typical movement of the five air mass types that affect North America are shown in ● Figure 7.1. Characteristics of all seven air masses are described in Table 7.1. From now on, the symbols rather than the full names will often be used when discussing each air mass type.

Air Mass Modification and Stability

Air masses must have sufficient time to develop under stable conditions, basically acquiring the temperature and humidity characteristics of their source region. Yet air masses do not remain stationary over their source regions indefinitely as a result of the general atmospheric circulation. As an air mass travels over the

TABLE 7.1
Types of Air Masses

Source	Region	Usual Characteristics at Source	Accompanying Weather
Maritime Equatorial *(mE)*	Equatorial oceans	Ascending air, to very high altitudes	High temperatures, humidity, moisture content, and rainfall; never reaches the United States
Maritime Tropical *(mT)*	Tropical and subtropical oceans	Subsiding air; fairly stable, but instability on western side of oceans; warm and humid	High temperatures and humidity, cumulus clouds, convectional rain in summer; mild temperatures in winter; overcast skies, fog, drizzle, and potential for heavy precipitation along *mT/cP* fronts
Continental Tropical *(cT)*	Deserts and dry plateaus of subtropical latitudes	Subsiding air aloft; generally stable, but some local instability at surface; hot and very dry	High temperatures, low humidity, clear skies, rare precipitation

surface, it generally retains its distinct characteristics. Temperature and humidity modifications, however, occur as the air mass gains or loses thermal energy or moisture by interacting with landmasses or water bodies. This gain or loss of thermal energy, humidity, or both can either make an air mass more stable or cause it to become unstable. If an air mass is colder than the surface it passes over, heat will flow from the land or water surface to the air mass. For example, an *mT* air mass that originates over the Gulf of Mexico and moves onshore over a hot land surface during the summer will warm further, possibly becoming unstable and producing heavy convective precipitation. In contrast, a similar *mT* air mass moving onshore in winter would be warmer than the land surface and lose heat to the land. Consequently, the air mass would be cooled, and possibly produce fog, stratus clouds, or light precipitation.

Air mass modification can also involve a relatively dry air mass that picks up moisture by moving over a body of water. During the early winter to midwinter seasons, cold, dry *cP* or *cA* air from Canada can move southeastward across the Great Lakes. While passing over the lakes, this air mass can pick up moisture, increasing its humidity, and it will rise slightly as it moves over the relatively warmer lake. When this modified *cP* or *cA* air reaches frigid land on the leeward shores of the Great Lakes, it can produce large amounts of *lake-effect snow*. On satellite imagery these snowy areas can be clearly seen downwind from the lakes (● Fig. 7.2). Lake-effect snows diminish in late winter as the lakes freeze, which cuts off the moisture supply to air masses that flow across them.

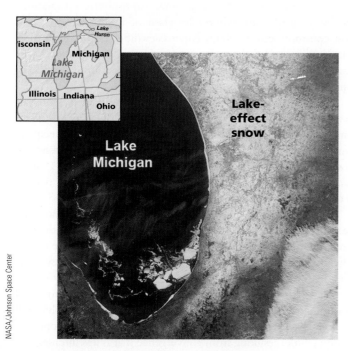

● **FIGURE 7.2** This lake-effect snow accumulated on the eastern shore of Lake Michigan as a storm moved west to east, picking up moisture as it passed over the lake.

What two main factors contribute to the increased precipitation caused by the lake effect?

North American Air Masses

Five air mass types influence the weather of North America (*cA*, *cP*, *mP*, *mT*, and *cT*). The midlatitudes are located between several source regions, where unlike air masses meet and collide, generating a great many storms and precipitation events in that latitudinal zone (see Fig. 7.1). The influence of specific air masses on a continent, as well as their characteristics, varies seasonally as source regions and wind patterns change with changing insolation.

Most of us are familiar with the weather of the United States or Canada; therefore, this chapter concentrates on the air masses of North America and their effects on weather conditions. Generally, the processes involving North American air masses are also applicable worldwide and are important factors for understanding the global climate regions that are addressed in the following chapters.

Continental Arctic Air Masses (cA)
These air masses are extremely cold, very dry, and very stable. The frozen Arctic Ocean in winter and the frigid land surface in far northern Canada and Alaska serve as source regions for this air mass type. Continental Arctic *air* masses affect parts of Canada during the winter, but if these air masses become extremely well developed, they can also travel far enough south to affect the United States. When continental Arctic air extends down into the midwestern or even the southeastern United States from Texas to Florida, these regions can experience record-setting cold temperatures. If a *cA* air mass extends into in a region that is not accustomed to extreme cold, vegetation can be severely damaged or killed. Water pipes can also freeze and break because they often are not properly insulated, unlike those in the regions that expect hard freezes in the winter.

Continental Polar Air Masses (cP)
At their source region in northern-central North America, *cP* air masses are cold, dry, and stable, resulting in clear, cold weather. Because North America has no major mountain barriers that trend east-west, *cP* air can migrate to southern Canada and as far south as the Gulf of Mexico or southern Florida. The movement of a continental polar air mass into the midwestern or the southern United States brings a cold wave characterized by clear, dry air and colder-than-average temperatures. Although the temperature in these air masses warm as they move south, continental polar air masses can bring freezing conditions as far south as Texas and Florida. The westerly circulation in the midlatitudes, along with mountain barriers, rarely allows a *cP* air mass to move westward to affect the West Coast. When a *continental polar* air mass does reach Washington, Oregon, or California, it brings with it unusual freezing temperatures that can cause significant damage to agriculture.

Maritime Polar Air Masses (mP)
Maritime polar air masses develop in the northern Pacific Ocean and move with the westerly circulation to affect the weather of southwestern Canada and the western United States, particularly in the Northwest. During winter months, oceans tend to be warmer than the land,

and *maritime polar* air masses, although damp and cool, tend to be warmer than their land counterparts (*cP* air masses). When *mP* air undergoes uplift by overriding a mass of colder, denser air, or a mountain range, cloudy weather and precipitation usually yields snow during the winter in inland regions and higher elevation areas. Maritime polar air masses may also continue eastward, becoming the moisture source for snowstorms after crossing the western mountain ranges.

Most *mP* air masses that develop over the northern Atlantic Ocean do not affect the weather in the United States because those air masses are pushed by the westerlies toward Europe. On some occasions, however, a strong low pressure cell can stall off the northern Atlantic Coast. Cyclonic winds from the poleward side of the low pressure storm system cause an onshore flow of cool or cold, damp winds from the northeast accompanied by rain or heavy snows. Known as **nor'easters** because these storms come from the northeast, they move onshore from the Atlantic Ocean to New England and can bring serious winter storm conditions to the New England states. The weather during a nor'easter typically includes strong winds, heavy snowfall, and coastal damage from high waves that are driven by winds blowing onshore.

Maritime Tropical Air Masses (*mT*) The Gulf of Mexico and the subtropical areas of the Atlantic and Pacific Oceans are source regions for *mT* air masses. Long days and intense insolation during summer produce warm ocean waters, and air masses in the *maritime tropical* source regions become warm and humid. During the summer, however, the land in the subtropics is even warmer than the maritime tropical air masses. When *mT* air masses move inland, the high temperature of the land results in convective uplift that causes precipitation and strong thunderstorms on hot, humid days. Maritime tropical air masses are responsible for much of the hot, humid summertime weather in wide areas of the United States that are east of the Rocky Mountains, particularly in the southeastern and eastern United States.

During the winter, the tropical and subtropical oceans and the Gulf of Mexico remain warm with warm and humid air aloft. As this warm, moist air moves northward into the south-central United States, it travels over increasingly cooler land surfaces. The lower layers of air are chilled and the drop in temperature increases the relative humidity, often resulting in fog. If a tropical air mass encounters *continental polar air* migrating southward from Canada, the warmer *mT* air will be forced to rise over the colder, drier *cP* air. Frontal precipitation will occur if the clash between these unlike air masses is strong enough.

Maritime tropical air masses also form over the Pacific Ocean in the subtropical latitudes. These air masses tend to be slightly cooler than those over the Gulf of Mexico and the Atlantic, partly because they pass over the cool waters of the California Current. Pacific *mT* air masses also tend to be more stable because of the strong subsidence associated with the eastern portion of the Pacific subtropical high along the West Coast. Pacific *mT* air masses contribute to the dry summers of California, but can also bring orographic moisture in winter as they rise over the mountains of the Pacific Coast region.

Continental Tropical Air Masses (*cT*) A fifth type of air mass affects North America, but it is mainly important to the weather of the southwestern United States. This is the *continental tropical* air mass that develops over large, homogeneous land surfaces in the subtropics. The typical weather associated with a *cT* air mass is very hot and dry, with clear skies and strong solar heating during the daytime.

The Sahara Desert of North Africa is a prime example of a source region for this type of air mass. In North America, source regions for a *cT* air mass of great proportion are quite limited compared to source regions for other air masses that affect the continent. During summer, continental tropical air masses form over the desert regions of the southwestern United States and northern Mexico. Areas dominated by continental tropical air masses experience clear weather that is hot and dry (often very hot and very dry). When *cT* air masses move eastward or to the north, however, they can be modified by contact with larger, stronger air masses with lower temperature and higher humidity or by passing over water bodies. Continental tropical air can also intrude relatively unmodified into the southern and central Great Plains, causing summer drought conditions in those regions. At times, *cT* air from Mexico and Texas meets with *mT* air from the Gulf of Mexico. This boundary is known as a *dry line*. Here, the denser drier air lifts the moist *mT* air over it. This mechanism of uplift may act as a trigger for precipitation episodes and can generate thunderstorm activity.

Fronts

Air masses migrate with the general atmospheric circulation. The midlatitudes are where the clash of unlike air masses is most common and frequent. Air masses differ primarily in their temperature and in their moisture content. When different air masses come together, they do not mix easily but instead come in contact along sloping boundaries called *fronts*.

The sloping surface of a front develops as a warmer, lighter air mass is lifted and rises above a cooler, denser air mass. This forced rising of air, known as *frontal uplift*, is a major source of precipitation in the midlatitude regions, where contrasting air masses are most likely to converge. The United States and southern Canada are located between the source regions for five different air masses, all of which migrate seasonally.

The steepness of the frontal surface is governed primarily by the degree of difference and relative advancement rate of the two converging air masses. When two strongly contrasting air masses converge—for example, when a warm and humid *mT* air mass meets a rapidly moving cold, dry *cP* air mass—the frontal surface tends to be steep, with strong frontal uplift. Given similar temperature and moisture conditions, a steep slope, with stronger frontal uplift, will produce heavier precipitation compared to a front with a slope that is more gentle, resulting from convergence that is less intense.

Fronts are differentiated based on whether a colder air mass is moving in on a warmer one or vice versa. The weather that occurs along a front also depends on which air mass is the "aggressor." Clashing air masses form a frontal zone that can cover an area from 2 to 3 kilometers (1 to 2 mi) wide to as wide as 150 kilometers (90 mi) or more. Although weather maps use one-dimensional line symbols to separate two different air masses along a front, fronts are actually three-dimensional surfaces with length, width, and height. Generally, it is more accurate to speak of a frontal *zone* rather than a frontal *line*.

Cold Fronts

A **cold front** occurs when a cold air mass actively moves in on a warmer air mass, pushing warm air upward. Because the colder air is denser than the warm air it displaces, it stays at the surface and forces the warmer air to rise. ● Figure 7.3 shows that cold fronts typically have a relatively steep slope. For example, the warm air may rise 1 meter vertically for every 40 to 80 meters of horizontal distance. If a warm unstable air mass with a high moisture content experiences uplift, heavy precipitation and violent thunderstorms

can develop. Cold fronts are generally associated with strong weather disturbances and sharp changes in temperature, air pressure, and wind. **Squall lines** consist of several zones of disturbance and occur in association with a cold front, often ahead and parallel to the front. A squall line can bring strong damaging winds, heavy rain, sometimes hail, and possibly tornadoes.

Warm Fronts

When a warmer air mass is the aggressor, pushing into an area occupied by a colder air mass, a **warm front** forms. At a warm front, warmer air slowly pushes against the cold air and rises over the colder, denser air mass at the surface. The slope of a warm front is much gentler than that of a cold front as shown in ● Figure 7.4. For example, the warm air may rise only 1 meter vertically for

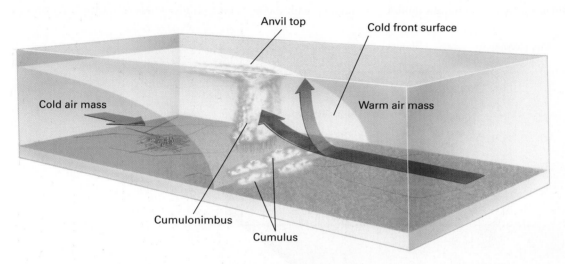

● **FIGURE 7.3** Typical structure of a cold front. Cold fronts generally move rapidly, with a blunt forward edge that drives adjacent warmer air upward. This rapid uplift can produce intense precipitation from moisture supplied by the warmer air.

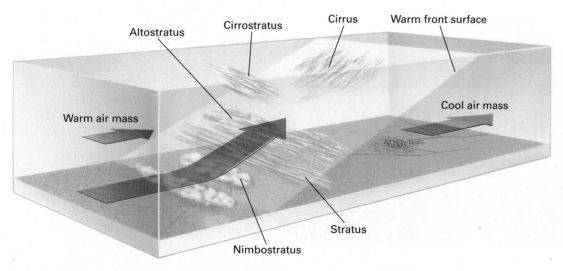

● **FIGURE 7.4** Typical structure of a warm front. Warm fronts advance more slowly compared with cold fronts, as the warm air slides upward over a wedge of colder air. The gentle rise of the warm air produces stratus clouds and gentle rain.

Compare Figures 7.3 and 7.4. How are they different? How are they similar?

every 100 or even 200 meters of horizontal distance. Thus, the uplift along a warm front will not be as strong as that occurring along a cold front. The result is that the weather associated with the passage of a warm front tends to be less violent and the changes are less abrupt than those associated with cold fronts.

Figure 7.4 shows why an advancing warm front affects the weather of areas well in advance of the surface location of the front. The cloud types that precede a front typically indicate the weather changes that can be expected as a front approaches.

Stationary and Occluded Fronts

When a frontal boundary fails to move in any appreciable direction as air masses converge, a **stationary front** develops. Under the influence of a stationary front, locations may experience clouds, drizzle, and rain (or possible thunderstorms), sometimes for several days. A stationary front and its accompanying weather will remain until it dissipates as the contrasts between the two air masses diminish or the atmospheric circulation finally causes one of the air masses to move.

An **occluded front** occurs when a faster moving cold front overtakes a warm front, pushing the warmer air aloft. This frontal situation usually occurs in the later stages of a storm and on the poleward side of a midlatitude cyclone, a storm type that will be discussed next. Areas under an occluded front tend to experience gray, overcast skies and perhaps light precipitation (see ● Fig. 7.5). Map symbols for the four frontal types are shown in ● Figure 7.6.

Atmospheric Disturbances
Anticyclones and Cyclones

In addition to the semi-permanent high and low pressure systems explained previously, secondary high and low pressure areas migrate with the prevailing winds. These secondary circulations,

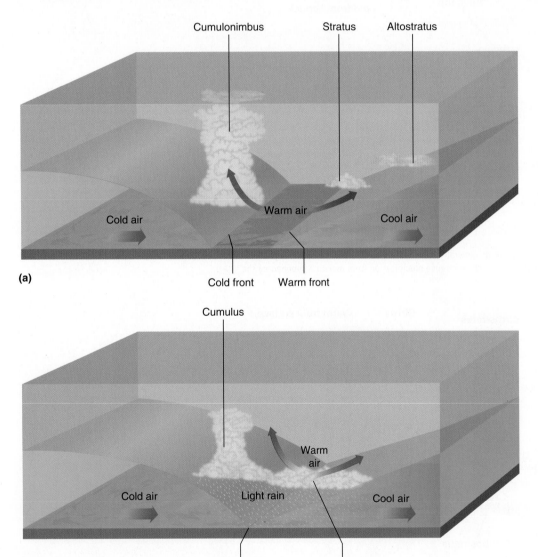

● **FIGURE 7.5** The development of an occluded front. Occluded fronts form when the faster-moving cold front approaches (a) and then catches up with a warm front (b). This forces the warmer air aloft producing light rainfall and overcast skies.

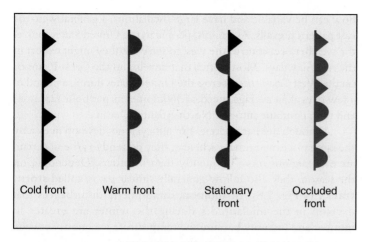

● **FIGURE 7.6** The four major symbols used to show the locations of different fronts on weather maps. The "teeth" point in the direction that an air mass on the other side of the line is moving or attempting to move.

storms, and other atmospheric disturbances are embedded within the wind belts of the general atmospheric circulation. The term **atmospheric disturbance** is used because it is a more general term than *storm* and it includes atmospheric conditions that cannot be classified as storms. Now, when referring to midlatitude atmospheric disturbances, we can use the terms *anticyclone* and *cyclone* to refer to moving cells of high and low pressure, which change with the seasons and drift along the path of the prevailing winds. It is important to remember that in a cyclone, pressure decreases toward the center, and in an anticyclone, pressure increases toward the center.

We should note here that the atmospheric pressures in these two systems (highs and lows) are relative. As systems of higher pressure, anticyclones are usually characterized by clear skies, gentle winds, and a general lack of precipitation. Cyclones are centers of converging, rising air, often with cloudy, windy conditions, with good chances for precipitation. Cyclones also include hurricanes, tornadoes, and the storms of the midlatitudes with their associated fronts of various types. The wind intensities involved in cyclonic or anticyclonic systems will depend on the steepness of the *pressure gradient*, which is determined by how much change in air pressure occurs over a horizontal distance.

Mapping Pressure Systems Cells of high and low pressure are easy to visualize if we imagine these pressure systems as if they were land surfaces. A cyclone is shaped like a basin (see ● Fig. 7.7). Winds converging toward the center of a cyclone will move rapidly if the pressure gradient is great, just as

water will flow faster into a natural basin if the sides are steep and the depression is deep. If we visualize an anticyclone as a hill or mountain, then we can see that air diverging from an anticyclone will flow at speeds directly related to how high the pressure is at the center of the cell, in a manner similar to water flowing down mountain slopes at speeds related to the landform's height and slope. Thus, if there is a steep pressure gradient in a cyclone, with the pressure much lower at the center than at the outer edges of the system, the winds will converge toward the center with considerable velocity.

On a surface weather map, cyclones and anticyclones are depicted by roughly concentric isobars of increasing pressure toward the center of a high and of decreasing pressure toward the center of a low. High pressure cells typically cover a larger area compared to low pressure cells, but both pressure systems are capable of covering and affecting extensive areas. There are times, for example, when nearly the entire midwestern United States is under the influence of the same system. The average diameter of an anticyclone is about 1500 kilometers (900 mi) and a cyclone averages about 1000 kilometers (600 mi).

Anticyclones An anticyclone is a high pressure area with subsiding air in the center, which displaces surface air with winds blowing outward, away from the center of the system. Hence, an anticyclone has diverging winds, with atmospheric pressure decreasing toward the outer limits of the system. Anticyclones tend

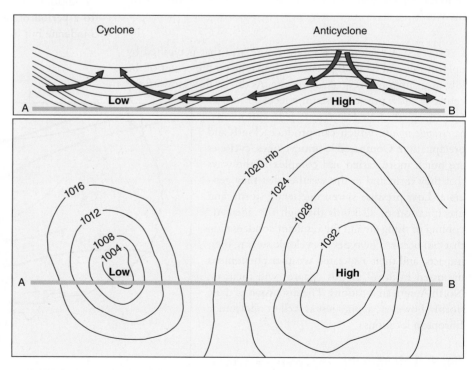

● **FIGURE 7.7** The horizontal and vertical structure of pressure systems. Close spacing of isobars around a cyclone or anticyclone indicates a steep pressure gradient that will produce strong winds. Wide spacing of isobars indicates a weaker system.

Where would be the strongest winds in this figure? Where would be the weakest winds?

to be fair-weather systems because of the temperature and stability increase that occurs in subsiding air, reducing the possibility of condensation.

Air subsidence in the center of an anticyclone encourages stability because the air is warmed adiabatically, increasing its capacity to hold moisture. Although the weather resulting from the high pressure of an anticyclone is often clear, with no rainfall, there are certain conditions under which precipitation can occur within a high pressure system. When an anticyclone is near or crosses a large body of water, the resulting evaporation can cause variations in humidity significant enough to result in some precipitation.

The high pressures associated with anticyclones in the midlatitudes of North America have two main sources. Some anticyclones move into the midlatitudes from northern Canada and the Arctic Ocean in what are called **polar outbreaks** of frigid continental polar or even continental arctic air. These outbreaks can be extensive, covering much of the midwestern and eastern United States, and they occasionally push into the subtropical Gulf Coast areas. The temperatures in an anticyclone associated with a cP or cA air mass can be markedly lower than those expected for any given time of year, dipping far below freezing in the winter. Such an outbreak would typically be preceded by squalls, clouds, and rain or snow associated with a cold front. A period of cold or cool, clear, and fair weather follows the frontal passage as the influences of modified polar air are felt. Other anticyclones are generated in the subtropical high pressure regions. When they move across the United States toward the north and northeast, they bring waves of hot, clear weather in summer and unseasonably warm days in the winter.

Cyclones A cyclone is a low pressure area that is typified by uplifting air and winds that tend to converge on the center of the low in an attempt to equalize pressure. As this air flows toward the center of a low pressure system, it feeds air into an upward spiral of air (*convergent uplift*, also known as *cyclonic uplift*), which can produce clouds and precipitation. Compared to anticyclones, cyclones are much more varied and complex in the ways that they form and in the weather that they generate. Low pressure systems generate storms and precipitation of all kinds through the adiabatic cooling of rising air and the resultant condensation that can occur. The types of cyclonic storms, their impacts, and their associated weather phenomena discussed in detail in this chapter will focus on North American locations. The discussion of these storms, however, applies just as well to midlatitude European locations.

General Movement The cyclones and anticyclones of the midlatitudes are steered, or guided, along paths influenced by the upper air westerlies (or the jet stream). Although the upper air wind flow can be variable and have large oscillations, a general west-to-east pattern prevails. As a result, people in the United States look at the weather occurring to the west to see what they might expect in the next few days. Most storms that develop in the Great Plains or on the West Coast move across the United States during a period of a few days at an average speed of 36 kilometers per hour (23 mph) and then continue into the North Atlantic Ocean.

Although neither cyclones nor anticyclones develop in exactly the same places and times each year, they do tend to arise in certain areas or regions more frequently than in others. Depending on the season, they also follow generally similar paths, called **storm tracks** (● Fig. 7.8). In addition, atmospheric disturbances that develop in the midlatitudes during the winter are greater in number and intensity because the temperature variations between air masses are stronger during the winter months.

Midlatitude Cyclones

Because they are so important to the weather of North America, we will concentrate on examining **midlatitude cyclones**, also known as **extratropical cyclones** ("extratropical" means outside the tropics). These migrating storms, typically formed of opposing cold, dry polar air and warm, humid tropical air, can cause significant variation in the day-to-day weather of the areas they affect. It is not unusual in some parts of the United States and Canada for people to experience a beautiful warm day in early spring followed by snowfall the next morning. Variability is common for weather in the midlatitudes, especially during fall and spring when conditions can also change from a period of cold, clear, dry days to a period of snow, only to be followed by one or two more moderate but humid days.

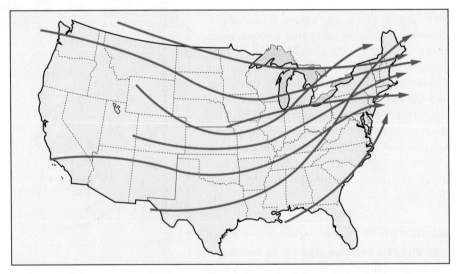

● **FIGURE 7.8** Common storm tracks for the United States. Virtually all midlatitude cyclonic storms move from west to east with the prevailing westerlies and swing northeastward across the Atlantic coast.

What storm tracks influence your location?

The weather associated with midlatitude cyclones can vary widely with the seasons and with air mass conditions, and no two of these storms are ever identical. Storms vary in intensity, longevity, speed of travel, wind strength, amount and type of cloud cover, the quantity and type of precipitation, and the area they affect. Yet there is a useful model that describes and generalizes the typical characteristics of a midlatitude cyclonic storm.

Shortly after World War I, Norwegian meteorologists Jacob Bjerknes and Halvor Solberg proposed the *polar front theory*, concerning the development, movement, and dissipation of midlatitude storms. They recognized the midlatitudes as a region where different air masses, such as cold polar air and warm subtropical air, commonly meet at a boundary called the *polar front*. Although the polar front may be a continuous boundary that circles Earth, it is typically fragmented into several frontal segments. The polar front moves north and south with the seasons and is stronger in winter than in summer. The upper air westerlies (see Figs. 5.16 to 5.18), also known as the *polar jet stream*, develop and flow along the wavy track of the polar front.

Most midlatitude cyclones develop along the polar front where warm and cold air masses meet. These two contrasting air masses do not merge but may initially move in opposite directions along the frontal zone—basically along a stationary front. Although there might be some slight uplift of the warmer air along the edge of the denser, colder air, uplift will not be significant. There may be some cloudiness and precipitation along such a frontal zone, but the conditions are not yet what we would refer to as a storm.

The line of convergence along the polar front can then develop a wave form (in map view). The wave form represents an initial step in the development of a midlatitude cyclone (● Fig. 7.9). At this bend in the polar front, warm air is pushing poleward (a warm front) and cold air is pushing equatorward (a cold front), with a low pressure center located where the two fronts are joined.

As the contrasting air masses compete for position, clouds and precipitation increase along the fronts, spreading over a larger area. Precipitation along the cold front will be less widespread than that along the warm front, but it will be more intense. One factor that affects the kind of precipitation that occurs at the warm front is the stability of the warm air mass. If it is relatively stable, then its movement over the cold air mass may cause only a fine drizzle or a light, powdery snow if the temperatures are cold enough. In contrast, if the warm air mass is moist and unstable, uplift may set off heavier precipitation. As you can see by referring to Figure 7.4, the precipitation that falls at a warm front may *appear* to be coming from the colder air. Although the weather may feel cool and damp, the precipitation is actually coming from the overriding warmer air mass above and then falling through the colder air mass to reach Earth's surface. Depending on the temperature of the colder air mass near the surface, and how far the precipitation falls through the colder air, the precipitation that results may be rain, or sleet, or freezing rain.

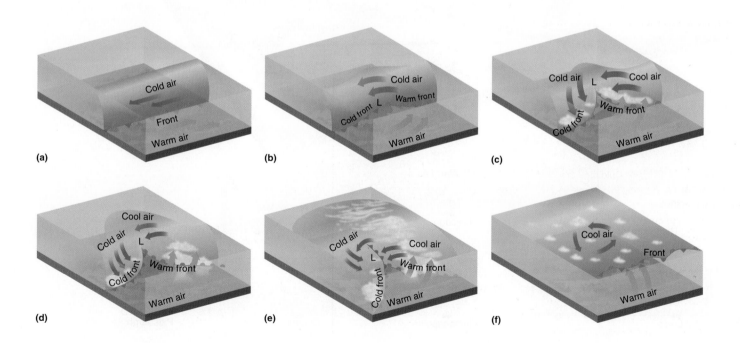

● **FIGURE 7.9** The development of a midlatitude cyclone. Each view represents a development stage of a storm as its fronts, wind directions, and air masses generally move eastward as a unit. (a) Initial contact of cold and warm air masses. (b) Cold front-warm front clash of air masses begins. (c & d) Continued development of a classic midlatitude cyclonic storm system. (e) Occlusion by air mass convergence on the poleward side of the system. (f) Dissipation of the storm.

In (c), where would you expect the strongest rain to develop? Why?

Because a cold front typically moves faster, it will eventually overtake the warm front. This produces an *occluded front*. Once this occurs, the system will soon die because the temperature, pressure, and humidity differences that powered the storm diminish at the occluded front. Occlusions are usually accompanied by cloudy overcast conditions and light rain (or snow). Occlusions represent the dissipation of a midlatitude cyclone, at least in the poleward section of the storm.

Midlatitude Cyclones and Weather The various sections of a midlatitude cyclone generate different weather conditions. Therefore the weather that a location experiences depends on which portion of the midlatitude cyclone is over the location. Because the entire cyclonic system tends to travel as a unit from west to east, a specific sequence of weather can be expected at a given location as the cyclone passes.

Let's examine the typical passage of a midlatitude cyclone during the late spring by following a westerly storm track that will take it across Illinois, Indiana, Ohio, and Pennsylvania, and finally out over the Atlantic Ocean. A view of this storm on a weather map, at a specific time in its journey, is presented in ● Figure 7.10a. Figure 7.10b is a cross-sectional view of the storm, north of the cyclone's center, and Figure 7.10c is a cross-sectional view south of the cyclone's center. As the storm continues eastward, moving at 33 to 50 kilometers per hour (20 to 30 mph), the sequence of weather will be different for Detroit, where the warm and cold fronts will pass to the south, compared to Pittsburgh, which will experience the passage of both fronts. To illustrate these differing weather sequences, we will examine the changing weather in Pittsburgh, in contrast to that which Detroit experiences as the cyclonic system moves east. We will also examine the weather of other cities that are affected by the passage of this storm system.

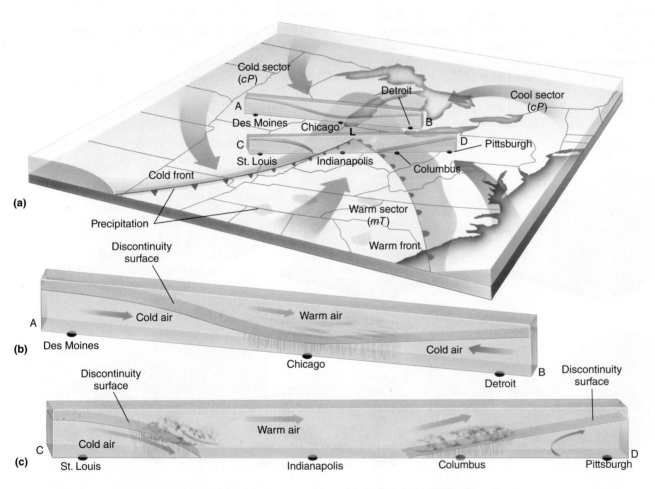

● **FIGURE 7.10** A classic midlatitude cyclonic system. This diagram models a midlatitude cyclonic storm positioned over the Midwest as the system moves eastward: (a) a map view of the weather system; (b) a cross section along line A-B north of the center of low pressure; and (c) a cross section along line C-D south of the center of low pressure.

What do the cross-sectional views show that is important to understanding the different fronts and the weather associated with them?

A midlatitude cyclonic storm is composed of two dissimilar air masses, usually with significant temperature contrasts. The sector of warm, humid *mT* air between the two fronts of the cyclone is usually considerably warmer than the cold *cP* air surrounding it. The temperature contrast between these two air masses is accentuated during the winter when the source region for *cP* air is the very cold cell of high pressure in Canada. During the summer, the contrast between these air masses is greatly reduced.

Due to the temperature difference, the atmospheric pressure in the warm sector is considerably lower than the atmospheric pressure in the cold sector behind the cold front. In advance of the warm front, the pressure is also high, but as the warm front approaches Pittsburgh, the pressure will decrease because of rising air along the front. After the warm front passes through Pittsburgh, the pressure will stop falling and the temperature will rise as warm *mT* air invades the area from the south.

At this time, Indianapolis has already experienced the warm front's passage and is now awaiting the cold front. After the cold front passes Indianapolis, the pressure will rise rapidly, the clouds will clear, and the temperature will drop. Detroit, which is to the north of the cyclone's center, will miss the warm air sector entirely and experience a slight increase in pressure and a temperature change from cool to colder as the cyclone moves to the east.

Changes in wind direction are one signal of the approach and passing of a cyclonic storm and its associated fronts. Because a midlatitude cyclone is a low pressure system, the winds tend to flow counterclockwise (in the Northern Hemisphere) and toward its center. Winds associated with a cyclonic storm tend to be stronger in winter when the pressure and temperature differences between air masses are considerable.

Indianapolis, now located in the warm sector south of the center of the low, is receiving winds from the south. Pittsburgh, located to the east and ahead of the warm front, has southeast winds. As the cyclonic system moves eastward and the warm front passes, the winds in Pittsburgh will shift to the south-southwest. After the cold front passes, the winds in Pittsburgh will be from the north-northwest. St. Louis has already experienced the passage of the cold front with cold weather and winds from the northwest.

Detroit is also experiencing winds from the southeast but will undergo a completely different sequence of directional wind changes as the cyclonic storm passes. Detroit's winds will shift, coming from the northeast as the center of the storm passes to the south. Chicago has just undergone this shift. Finally, after the storm has passed, the winds will generally blow from the northwest. Des Moines, to the west of the storm, currently has northwest winds.

The type and intensity of precipitation and cloud cover also vary as a cyclonic storm moves through a location. In Pittsburgh, the first sign of the approaching warm front will be high cirrus clouds; although the warm front is approaching, the front may still be 150 to 300 kilometers (90 to 180 mi) away. As the warm front gets closer, the clouds will thicken and lower, and in Pittsburgh, light rain and drizzle may begin (or snow in winter) and stratus clouds will blanket the sky. After the warm front has passed, precipitation will stop and the skies will clear.

As the cold front passes, warm air in its path will be forced to move aloft rapidly. This may mean that there will be a cold, hard rain (or heavy snowfall in winter), but the band of precipitation normally will not be very wide because of the steeper angle of the surface along a cold front compared to the wedge-like warm front. In our springtime example, the cold front and the band of precipitation have just passed St. Louis. During the winter, however, a cold front is likely to bring snow, followed by the cold, clear conditions of the *cP* air mass, and increasing atmospheric pressure.

Located in the latitudinal center of the cyclonic system, Pittsburgh can expect three zones of precipitation as it passes over its location: (1) a broad area of overcast and drizzle in advance of the warm front (or light snow in winter); (2) a zone within the warm sector where clearing occurs; and (3) a narrow band of heavy precipitation associated with the cold front (rain or snow, depending on the season and temperatures) (see Fig. 7.10c). However, locations to the north of the center of the cyclonic storm, such as Detroit, will usually experience light precipitation and overcast cloudy conditions resulting from warm air being lifted above cold air from the north (see Fig. 7.10b).

As the figure shows, the various fronts and sectors of a midlatitude cyclone are accompanied by different weather. If we know where the cyclone will pass relative to our location, we can make a fairly accurate forecast of what our weather will be like as the storm moves to the east along its track.

Surface Weather and Upper Air Flow

Steering surface storm systems is only one of the ways that the upper air winds influence our surface weather. A less obvious influence is related to the undulating, wave-like flow so often exhibited by the upper air winds. As the air passes through these waves, it undergoes divergence or convergence because of the atmospheric dynamics associated with curved flow of winds. These flow dynamics produce alternating pressure areas of ridges (high pressure with divergent winds) and troughs (low pressure with convergent winds).

The region between a ridge and an adjacent trough (A–B in ● Fig. 7.11) is an area of upper-level convergence. Because any action taken in one part of the atmosphere is countered by an opposite reaction somewhere else, upper-air convergence is compensated for by divergence at the surface. Specifically, upper-level convergence causes air to "pile up" and the air is pushed downward, causing high pressure at the surface. This pattern will either inhibit the formation of a midlatitude cyclone altogether, or cause an existing storm to weaken or dissipate. In contrast, is an area of upper-level divergence, where air is drawn upward, dropping the surface pressure (Fig. 7.11 B–C). This leads to cyclonic circulation at the surface that will enhance the prospects for storm development or strengthen an existing storm.

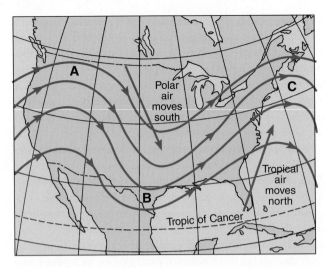

● **FIGURE 7.11** Wave forms in the jet stream. An upper air wind pattern, long waves, such as those depicted here, can significantly influence the distribution of temperatures and precipitation at the surface.

On which side of a wave (east or west) would you expect storms to develop?

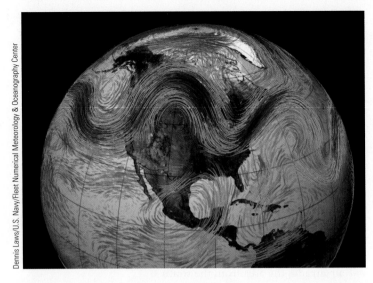

● **FIGURE 7.12** Polar front jet stream analysis. This image shows the winds at an altitude where the atmospheric pressure is 300 mb (approximately 10,000 m or 33,000 feet above sea level). At this altitude the patterns in the long waves that the jet stream follows are very apparent.

With this type of pattern, where would be the coldest region of the United States?

Upper air flow also affects temperatures in addition to the development or dissipation of storms. If the "average" flow of upper air is from west to east, then any deviation from that pattern will cause either colder air from the north or warmer air from the south to be advected into an area. For example, after the atmosphere has been in a wave-like pattern for a few days, the areas in the vicinity of a trough (area B in Fig. 7.11) will be colder than normal as polar air from higher latitudes dips into that area. This intrusion of cold dry air is a *polar outbreak*, which occurs when continental polar air or arctic air is pushed by a major expansion of the polar high and drives a strong cold front into the lower midlatitudes of North America and sometimes to the subtropics.

Just the opposite effect on temperature and weather conditions will occur at locations near a ridge (area C in Fig. 7.11). Here, warmer air from more southerly latitudes will be drawn up toward the top of the ridge. As ● Figure 7.12 shows, the jet stream actually curves with less regularity than in our model. Compare Figures 7.11 and 7.12 to see the difference between the theoretical and the real waves in the polar jet stream.

Hurricanes

A **hurricane** is a circular cyclonic system with strong low pressure, wind speeds in excess of 118 kilometers per hour (74 mph) and a diameter of 160 to 640 kilometers (100–400 mi). Hurricanes form and develop over the tropical oceans, but steered by pressure and wind systems, they are often directed toward the midlatitudes. Extending upward to heights of 12 to 14 kilometers or more (40,000–45,000 ft.), a hurricane is a towering column of spiraling air (● Fig. 7.13). Although its diameter may be less than that of a *midlatitude cyclone* including its extended *fronts*, a hurricane is essentially the Earth's largest storm, and among the most powerful.

At its base, air is sucked in by low pressure in the hurricane's eye, its center, and rises rapidly to the top where the winds spiral air outward. The rapid upward movement of moisture-laden air produces enormous amounts of rain. The power that drives the storm is increased by the massive release of latent heat energy from condensation.

The extremely low pressure at the center of a hurricane, combined with very strong pressure gradients, produce powerful high-velocity winds. Unlike midlatitude cyclones, hurricanes form from a single air mass and do not have the different sectors of strongly contrasting temperature that power a frontal system. In contrast, at the surface, hurricanes have a circular distribution of warm temperatures. The eye, at the center of the hurricane, is an area of calm, clear, and usually warm but rainless air. Sailors traveling through the eye of a hurricane have been surprised to see birds flying there. Unable to leave the eye because of the strong winds surrounding it, these birds often alight on a passing ship as a resting spot.

Hurricanes are severe *tropical cyclones* that receive a great deal of attention when they make **landfall**, primarily because of their tremendous destructive powers. Hurricanes are characterized by abundant, even torrential rains, and bring winds often exceeding 160 kilometers per hour (100 mph).

Although a great deal of time, effort, and money has been spent on studying the development, growth, and tracks of hurricanes, there is still much to learn. It is not yet possible to predict a hurricane's path with great certainty days in advance, even though it can be tracked with radar and studied from planes and from weather satellite images. There are, however, areas where hurricanes are most likely to develop and strike (● Fig. 7.14). For North America, the most susceptible areas are the Atlantic and Gulf Coastal regions.

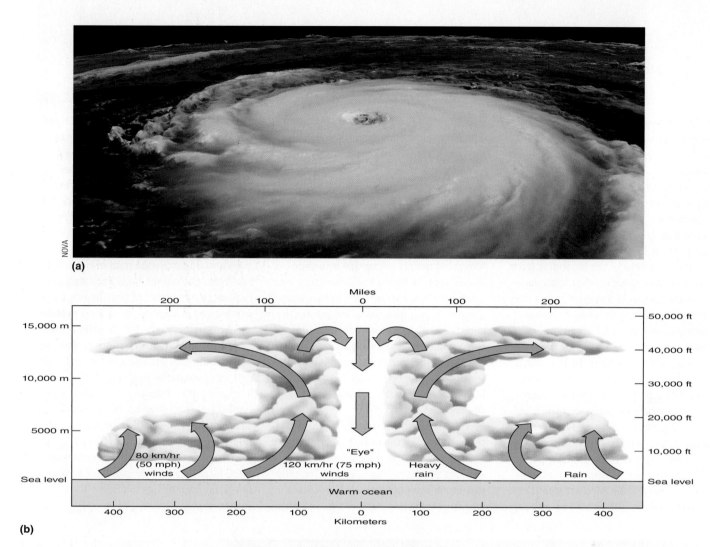

(a)

(b)

● **FIGURE 7.13** (a) Hurricane Katrina, as seen from space, can be directly compared with the diagram of a hurricane below. (b) Cross section of a hurricane (with much vertical exaggeration) showing its circulation pattern: the inflow of air in the spiraling arms of the cyclonic system, rising air in the towering circular wall cloud, and outflow in the upper atmosphere. Subsidence of air in the storm's center produces the distinctive calm, cloudless eye of the hurricane.

Which features in the Katrina image can you match with the cross-sectional diagram?

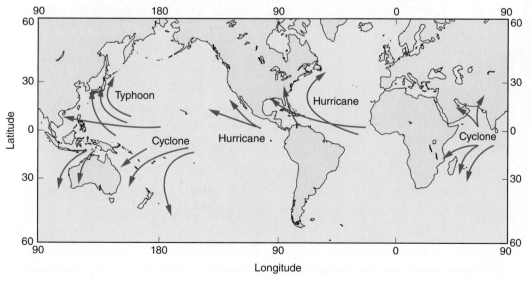

● **FIGURE 7.14** A world map showing major "Hurricane Alleys" and their regional names for intense tropical cyclones.

Which coastlines seem unaffected by these tracks?

Hurricane Landfall Probability Maps

Hurricanes are generated over tropical or subtropical oceans and build strength as they move over regions of warm seawater. Ships and aircraft regularly avoid hurricane paths by navigating away from these violent storms. People living in the path of an oncoming hurricane try to prepare their homes, other structures, and belongings, as they may have to evacuate if the hazard potential of the impending storm is great enough. *Landfall* refers to the location where the eye of the storm strikes the coastline. A *storm surge*, where the ocean violently washes over and floods low-lying coastal areas, is the most devastating hazard associated with a hurricane. In 1900, 8000 residents of Galveston Island, Texas, were killed when a hurricane pushed a 7-meter (23-foot)-high wall of water over the island. Much of the city was destroyed by this storm, the worst natural disaster ever to occur in the United States in terms of lives lost.

Today, we have sophisticated technology for tracking and evaluating storms. Computer models, developed from mapping the behavior of past storms, are used to indicate a hurricane's potential wind speed and most likely landfall location. Nature is unpredictable, and those aspects of the storm shown on a map are estimates based on probability. A hurricane's probable path is mapped with positions plotted to show where the hurricane is likely to be centered over the next few days. On the map, the forecasted path is surrounded by a cone-shaped area, widening in the direction the storm is moving, because the farther away a hurricane is from a coast, the greater the uncertainty about the hurricane's future path. Maps are revised daily to incorporate changes in the storm's path and intensity, and as a storm moves closer to the coast, the prediction about a potential landfall site becomes more certain.

In 2012, Hurricane Sandy, approaching the East Coast of the United States, collided with a frontal storm, intensifying its impact and creating "Superstorm Sandy." The term *superstorm* was used because Sandy was very large, affecting much of the east coast,

and unusual, as it merged with a cold front, making it a "hybrid" storm (see the image in the chapter opener). Sandy continued to have hurricane velocity winds (Category 1), but its form became more like storms that form outside the tropics. Sandy caused much damage in 14 states, particularly with storm surges in New Jersey and New York, no different than those generated by a hurricane. The projected path of Hurricane Sandy was determined fairly accurately days in advance of its landfall (see figure). Early warning and evacuation saved lives because many coastal residents left for safer areas.

The National Weather Service continually works to develop and improve technologies that will yield better predictions of hurricane paths, intensities, possible storm surges, and landfall areas. If you live in a coastal area affected by hurricanes, tropical disturbances, or extratropical storms, understanding the information on these maps may be very important to your safety and your ability to prepare for a coming storm.

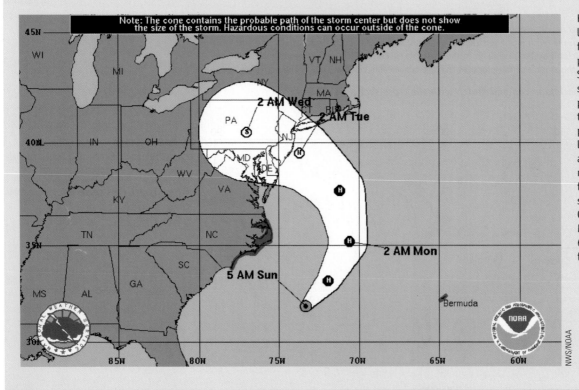

Hurricane path and landfall probability map for Hurricane Sandy produced at 5:00 a.m. Sunday, October 28, 2012, showing the most likely place for landfall at the time the map was produced. The probable landfall location was fairly accurate, but the storm moved north-northwest faster than predicted, striking the New Jersey Coast at 8:00 p.m. on Monday, October 29, 2012, 6 hours earlier than forecasted.

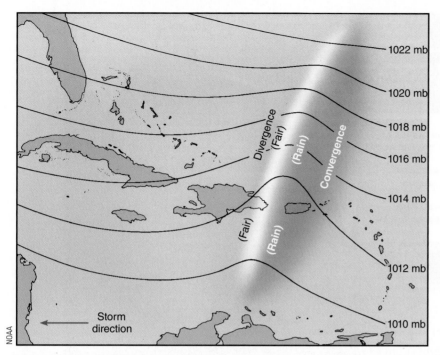

● FIGURE 7.15 A typical easterly wave in the tropics. Note that the isobars (and resulting winds) do not close in a circle but merely make a poleward "kink," indicating a low pressure trough rather than a closed cell. The resulting weather is a consequence of convergence of air coming into the trough, producing rains, and divergence of air coming out of the trough, producing clear skies.

Why do easterly waves move toward the west?

Hurricanes can occur over most subtropical and tropical oceans and seas. The South Atlantic was once the exception. However, on March 26, 2004, tropical Cyclone Catarina was the first known hurricane to attack the southern coast of Brazil, much to the amazement of atmospheric scientists. In the South Pacific, Australia, and South Asia hurricanes are called *cyclones*. Near the Philippines they are known as *bagyos*, but in most of East Asia they are called **typhoons**.

The development of a hurricane requires a warm ocean surface of about 27°C (80°F) or higher and warm, moist overlying air. These two factors explain why hurricanes occur most often in late summer and early fall when maritime air masses have maximum humidity and ocean surface temperatures are highest. Hurricane season in the Atlantic Basin officially runs from June 1 through November 30.

Hurricanes begin as weak tropical disturbances over the ocean, such as **easterly waves**. An easterly wave is a trough-shaped, weak, low pressure trough, generally aligned on an approximately north-south axis (● Fig. 7.15). Traveling slowly as the converging trade winds transport them from east to west, they are preceded by fair, dry weather and followed by cloudy, showery weather. If the pressure becomes lower and the winds strengthen, an easterly wave can develop into a tropical storm, and with further cyclonic development, it can reach hurricane strength.

Names are assigned to these storms when they reach *tropical storm* status, with wind speeds between 62 and 118 kilometers per hour (39 to 74 mph). Each year hurricane names are selected from a different alphabetical list of alternating female and male

names—one list for the North Atlantic and one for the North Pacific. The names of hurricanes that were especially destructive and of historic interest are retired, and will not be used again. Andrew, Camille, Hugo, Katrina, Ike, and more recently, Sandy, are just some of more than 80 names that have been retired since the naming of storms began in the 1950s.

Hurricanes do not last long over land because their moisture source (and consequently their source of energy) is cut off, and friction with the land surface reduces wind speeds. North Atlantic hurricanes move first toward the west with the trade winds and then move north and northeast. A hurricane makes landfall as a violent storm when it strikes a coast, bringing high winds and waves with destructive force. As hurricanes move inland they dissipate and weaken to become cyclonic storms, continuing to cause storm damage and sometimes they spawn tornadoes.

Hurricane Intensities and Impacts A hurricane landfall can result in devastating destruction of property and loss of life. The **Saffir-Simpson Hurricane Scale** provides a means of classifying hurricane intensity and potential damage by assigning a category from 1 to 5 based on sustained wind speeds (Table 7.2). All hurricanes are potentially dangerous, but intensities of 3 or more are considered major hurricanes, because they have great potential for destructive damage and significant loss of life. From year to year, the number and severity of tropical cyclones can vary dramatically. The continental United States usually averages three hurricane landfalls a year, but none occurred in 2006, 2009, 2010, 2013, or 2015. Hurricane Arthur struck the coast of North Carolina, briefly in 2014, before moving back into the Atlantic Ocean (● Fig. 7.16).

The year 2004 was a record-breaking year for tropical disturbances. Typhoon Tokage struck the Japanese Coast near Tokyo, with significant loss of life. In all, 10 tropical cyclones pounded Japan in 2004. In the Caribbean and Gulf of Mexico, three hurricanes—Charley, Frances, and Jeanne—hit Florida directly.

TABLE 7.2
Saffir-Simpson Hurricane Wind Scale (2102 revision)

Scale	Wind Speed	
(CATEGORY)	(KPH)	(MPH)
1	119–153	74–95
2	154–177	96–110
3	178–208	111–129
4	209–251	130–156
5	>250	>157

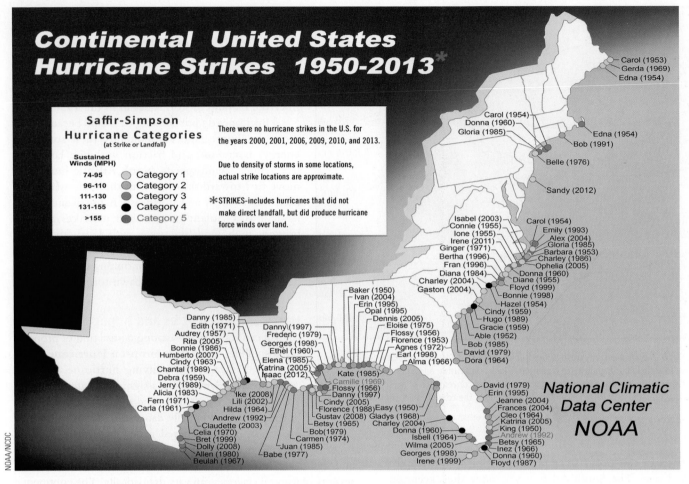

Continental United States Hurricane Strikes 1950-2013*

Saffir-Simpson Hurricane Categories
(at Strike or Landfall)

Sustained Winds (MPH)	
74-95	Category 1
96-110	Category 2
111-130	Category 3
131-155	Category 4
>155	Category 5

There were no hurricane strikes in the U.S. for the years 2000, 2001, 2006, 2009, 2010, and 2013.

Due to density of storms in some locations, actual strike locations are approximate.

*STRIKES-includes hurricanes that did not make direct landfall, but did produce hurricane force winds over land.

National Climatic Data Center **NOAA**

NOAA/NCDC

- **FIGURE 7.16** This map shows the hurricane landfall sites along the U.S. Atlantic and Gulf Coasts from 1950 to 2013, Hurricane Arthur grazed the North Carolina coast in 2014, but no hurricane made landfall in this region in 2015. The sites are labeled by the storm name, the year, and the Saffir-Simpson category.

A fourth, Ivan, struck Gulf Shores, Mississippi, and caused devastation in Florida's western panhandle. The damages from these storms were estimated at $23 billion. This amount exceeds even the $20 billion cost of damage by *Hurricane Andrew*, which, when it hit in 1992, was the costliest natural disaster in U.S. history. Among their most serious potential hazards, however, are the high seas pushed onshore by the strong winds. These **storm surges** can flood, and sometimes destroy, entire coastal communities (● Fig. 7.17a–b).

Hurricane Katrina, with 225-kilometer-per-hour (140 mph) winds and storm surges of more than 16 feet, struck the Louisiana, Mississippi, and Alabama coasts in August 2005. The storm surges breached the levee systems designed to protect the city of New Orleans, much of which is below sea level. The subsequent flooding caused massive destruction. Responsible for the deaths of more than 2000 people, the winds and floodwaters of Katrina caused damages estimated to be in excess of $125 billion. Hurricane Rita, even more powerful, followed in September with landfall near the Texas-Louisiana state line, devastating coastal areas in that region. Because Rita's eye missed the heavily populated Houston area, the estimated $10 billion damage was much less than for Katrina.

In September 2008, Hurricane Ike, a powerful storm with 230-kilometer-per-hour (145 mph) winds, made landfall on the Gulf Coast, with a storm surge that caused widespread damage to the coastal region near Galveston, Texas, and in some cases complete destruction of coastal communities (● Fig. 7.18). The storm's path continued directly to Houston, causing billions of dollars of damage. Also in 2008, major hurricanes struck Haiti, Cuba, Jamaica, and other Caribbean islands, causing many deaths and serious destruction.

In February 2011, Cyclone Yasi, a massive Category 5 hurricane, struck the coast of Queensland, Australia (● Fig. 7.19) with wind gusts as high as 290 kph (180 mph). The storm caused major flooding and considerable damage to structures and croplands in a wide area of northeastern Australia.

In October 2012, Hurricane Sandy battered the Caribbean islands before sweeping northward along the East Coast to make landfall near Atlantic City, New Jersey, with Category 1 strength and sustained winds of 129 kph (80 mph). The storm's tremendous impact serves as a reminder of the danger and destructive potential of even a Category 1 hurricane. Sandy was gigantic, and affected most of the highly populated eastern United States. Storm surges devastated coastal New Jersey and New York

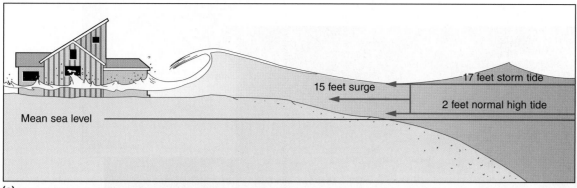

(a)

(b)

US Department of Defense

● **FIGURE 7.17** (a) Landfalling hurricanes can generate a storm surge. These surges are mounds of water, topped by battering waves that flow over low-elevation coastal areas with tremendous destructive force. (b) This aerial photograph shows what a powerful storm surge looks like as it strikes the coast, destroying an amusement park in New Jersey.

What can people who live in such regions do to protect themselves when a serious storm surge is threatening?

National Weather Service Houston/Galveston and the Galveston County Office of Emergency Management

● **FIGURE 7.18** Extensive destruction and damage to a community along the Texas Gulf Coast caused by the storm surge generated by Hurricane Ike in 2008.

(● Fig. 7.20) and affected areas from Virginia to Rhode Island. In New York City, the surge was 4.25 meters (14 ft.), flooding buildings and subway tunnels. More than eight million homes were without electricity due to damaged power lines, and some had no power for 2 weeks. Dozens of people were killed in Hurricane Sandy—the second-most costly U.S. hurricane (after Hurricane Katrina) at an estimated $65 billion in damage.

Despite the damage they cause, hurricanes also have beneficial effects. They are an important source of much-needed rainfall in the southeastern regions of the United States, and in other areas of the world as well. Hurricanes are a natural and important means of moderating latitudinal temperatures by transferring surplus heat away from the tropics and into the cooler latitudes.

Thunderstorms

Thunderstorms are common storms of the middle and lower latitudes, particularly in summer, and they are typically localized (rather than broad regional) storms. Very simply, thunderstorms

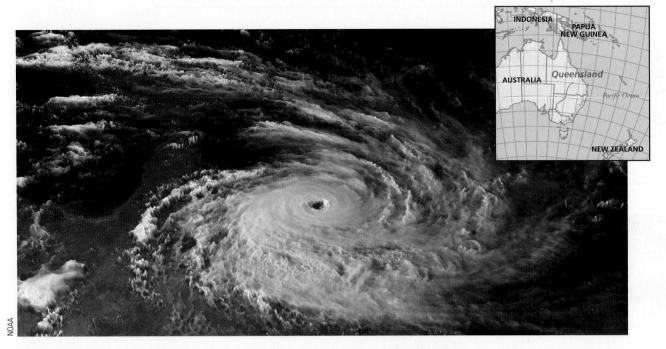

● **FIGURE 7.19** This massive hurricane, Cyclone Yasi, hit the coast of Queensland, Australia in February 2011. A Category 5 storm, Yasi had winds of nearly 300 kilometers per hour (186 mph), which demolished homes, caused severe crop damage, toppled or stripped large trees, and brought half a meter of rain.

Compare the circulation of Cyclone Yasi here with the image of Hurricane Katrina in Figure 7.13a. Explain any difference that you see.

● **FIGURE 7.20** These before and after photographs show the incredible coastal damage caused by the storm surge generated by Hurricane Sandy in October 2012. This is an area of the New Jersey coast, close to where the storm made landfall.

are often intense atmospheric disturbances that are accompanied by thunder and lightning. Lightning is a massive discharge of electricity within the atmosphere or between the atmosphere and the ground (including structures on the ground). Lightning is produced when strong positive and negative electrical charges are generated within a cloud. It is believed that collisions between moving ice particles and hail contributes to the charge separation within a cumulonimbus cloud. By colliding, ice crystals are stripped of electrons and become positively charged. Because they are lighter, they tend to bring a positive charge to the upper portion of the cloud, while heavier and negatively charged hailstones dominate the lower levels of the cloud. When the difference between these charges becomes large enough to overcome the natural insulating effects of the air, an electrical discharge takes place, producing a lightning flash. These discharges, often involving more than 1 million volts of electricity, can occur within the cloud, between two clouds, or from cloud to ground. The air immediately around the discharge is momentarily heated to temperatures in excess of 25,000°C (45,000°F), which is about four times hotter than the surface of the sun! The heated air expands nearly instantaneously and explosively, creating the sonic shock wave we call thunder (● Fig. 7.21).

● **FIGURE 7.21** A cross-sectional view of a thunderstorm showing the distribution of electrical charges.

Where would you place a lightning bolt in this diagram?

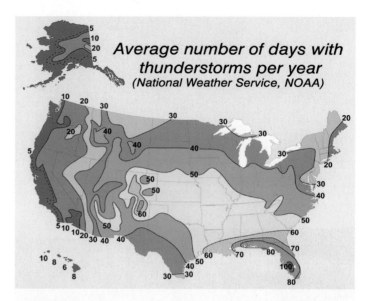

● **FIGURE 7.22** The geography of thunderstorm frequency in the United States.

Can you explain why the highest and lowest frequencies occur in terms of weather conditions?

Thunderstorms usually cover relatively small areas of a few square kilometers, although there may be a series of related thunderstorms covering a larger region. The intensity of a thunderstorm depends on the degree of instability of the air and the amount of water vapor it holds. A thunderstorm will die out when most of its water vapor has condensed and there will no longer be enough energy available for continued vertical movement; in fact, most last an hour or less. Thunderstorms result from the very strong uplift of moist air. They are most common during warm months of spring and summer and occur most often during the warmest part of the day (afternoon), but they can also occur day or night and year-round because of the different atmospheric conditions that trigger these storms. The trigger mechanism causing that uplift can be thermal convection (warm unstable air rising on a warm afternoon); orographic uplift (warm moist air ramping up a mountainside), or frontal uplift. The Great Plains area, the Southeast, and Florida are the areas of the United States that receive the most thunderstorms (● Fig. 7.22).

Convective thunderstorms, on a worldwide basis, are most common in the lower latitudes during the warmer months and during the warmer times of the day. Solar heating encourages thunderstorm development because intense surface heating steepens the environmental lapse rate, which leads to increased instability of the air and allows for greater moisture-holding capacity, adding to the buoyancy of the air.

Orographic thunderstorms occur when air is forced to rise over land barriers, providing the necessary initial trigger action that leads to the development of thunderstorm cells.

Thunderstorms of orographic origin play a large role in the tremendous precipitation that occurs in some areas of South and Southeast Asia. In North America, they occur over the mountains of the West (the Rockies and the Sierra Nevada), especially during summer afternoons when heating of south-facing slopes increases the air's instability. For this reason, pilots of small planes try to avoid flying in the mountains during summer afternoons for fear of getting caught in the turbulence generated during a thunderstorm.

Frontal thunderstorms are often associated with cold fronts where a cooler air mass forces a warm, humid air mass to rise. This kind of frontal activity can bring about the strong, vertical updrafts necessary for heavy precipitation. At times, cold fronts can be immediately preceded by a line of thunderstorms (a squall line) resulting from uplift of the edge of a warm air mass.

As was mentioned in the discussion of precipitation types, hail is a product of thunderstorms when the vertical updrafts of the cells are strong enough to repeatedly carry water droplets into a freezing layer of air that supports pellets of ice. Fortunately, only a very small percentage of storms around the world produce hail. In fact, thunderstorms only occasionally produce hail in the lower latitudes. In areas very close to the U.S. Gulf of Mexico coast hailstorms are infrequent even though thunderstorms are quite common.

Tornadoes

A **tornado** is a small, intense cyclonic storm with extremely low pressure, violent updrafts, and powerful converging winds that are most often associated with strong thunderstorms (● Fig. 7.23). Tornadoes can appear milky white or even black,

● **FIGURE 7.23** A powerful tornado crosses farmlands in Osceola, Nebraska, one of many during a June 2011 outbreak.

depending on the brightness and direction of sunlight and the amount of dust and debris picked up by the storm as it travels across the land. Although only 1% of all thunderstorms produce a tornado, 80% of all tornadoes are associated with thunderstorms and midlatitude cyclones. The remaining 20% of tornadoes are spawned by hurricanes that make landfall. The United States experiences the most tornadoes in the world, and Canada is second. Since the year 2000, more than 1000 tornadoes have occurred each year in the United States, most of them from March to July in the late afternoon or early evening and in the central part of the country.

Government records of tornado activity began in 1875. Accounts of tornadoes occurring prior to that time must be investigated by examining other sources. These early historic accounts are sometimes unverifiable and vague, but they do offer insights about early perceptions of tornadoes in North America. The following accounts describe a tornado that killed several people as it swept across several counties in western Illinois on May 21, 1859.

It was "a violent storm or hurricane [which] did immense damage to houses, barns, fences, and also caused some destruction of life." It was described as having a "frightful . . . balloon or funnel shape, and appeared . . . peculiarly bright and luminous, not at all black or dark in any of its parts, except its base." A vivid account of what must be related to the output of static electricity is given in this account of the same tornado as it swept Morgan county: "Mr. Cowell was plowing his field . . . He saw the frightful cloud approaching . . . and at once attempted to drive his horses and plow to the house.. . . The horses suddenly took fright . . . their manes and tails and all their hair 'stood right out straight' as he expressed it, and . . . the iron in the harness . . . and plow, in his language 'seemed all covered with fire.' He felt a violent pulling of his hair which left 'his head sore for some days' and

the hair itself rigid and inflexible." In addition, Mr. Cowell was one of the few individuals to apparently have had a tornado pass directly over him and live to tell about it. He described the light in the tornado's center as "so brilliant that he could not endure it with his eyes open, and for the most part kept them shut . . . Yet [inside the tornado] there was no wind, no thunder and no noise whatever. . . . ."

—Transactions of the Illinois Natural History Society
Phillips Bros., 1861

The interior of North America experiences more tornadoes than anywhere else in the world, particularly a region often referred to as "Tornado Alley" that is centered on Oklahoma and Kansas, where tornadoes are most likely to occur (● Fig. 7.24). Every state in the country, however, has experienced tornado activity. In 1976, a tornado even touched down near the village of Kina, Alaska, 47 km (29 mi) north of the Arctic Circle (the highest-known latitude for a tornado). Fortunately, tornadoes are relatively small and short-lived. Even in Tornado Alley, a tornado is likely to strike a given locale only once in 250 years.

Because of their relatively small size and limited life span, tornadoes are difficult to detect and forecast. However, a sophisticated radar technology, **Doppler radar**, improves tornado detection and forecasting significantly. Doppler radar concentrates more power in a narrower beam than previous weather radar units. This allows meteorologists to assess the patterns of storms in much greater detail (● Fig. 7.25). An important capability of Doppler radar is its ability to measure wind speeds flowing toward the radar site and wind speeds blowing away from the radar site (by measuring the Doppler effect). When the energy emitted by radar strikes precipitation, a small portion is scattered back to the radar antenna. Depending on whether the precipitation is moving toward or away from the radar site, the wavelength of the returned energy is either compressed or elongated. The faster the winds flow, the greater the change in wavelength. Previous weather radars could not measure this change. Doppler radar can estimate the wind circulation and cyclonic rotation within the storm. This technology lets

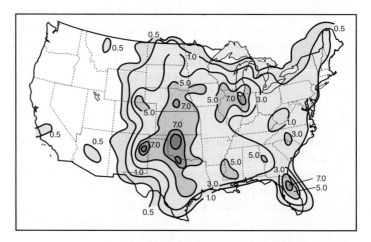

● **FIGURE 7.24** The geographic distribution of tornado frequency. Average annual number of tornadoes per 26,000 square kilometers (10,000 sq mi).

How do tornadoes affect your geographic area?

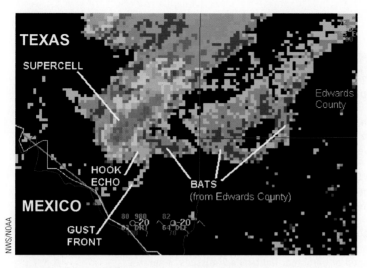

● **FIGURE 7.25** This NEXRAD Doppler radar image of thunderstorms shows details of a severe storm with a hook-shaped pattern that is associated with the development of tornadoes. Colors show rainfall intensity: green—light rainfall, yellow—moderate, and orange/red—heavy. This radar image also picked up reflections from huge groups of flying bats, among the millions that live in this region.

How do patterns on Doppler radar images help us understand weather conditions as they occur?

● **FIGURE 7.26** The destructive track of a powerful tornado is visible on this aerial photo image as a linear swath of destruction across the landscape near Tuscaloosa, Alabama, in 2011.

meteorologists see the formation of a tornado, thus increasing the warning time to the public. The U.S. government, through the NEXRAD program (NEXt-generation weather-RADar), has installed more than 150 Weather Service Doppler Radar (WSR-88D) sites across the country.

Based on studies from Doppler radar, most tornadoes (75%) are fairly weak, with wind speeds of 180 kilometers per hour (112 mph) or less. The remaining 25%, whose wind speeds reach up to 265 kilometers per hour (165 mph) are classified as strong. Nearly 70% of all tornado fatalities result from these more violent tornadoes. Although they are very rare, such killer storms can have wind speeds exceeding 320 kilometers per hour (200 mph). Before Doppler radar, tornado wind speeds could not be measured directly, and tornado intensity was estimated from the damage produced by the storm.

A tornado first appears as a swirling, twisting funnel cloud that moves across the landscape at about 35 to 51 kilometers per hour (22 to 32 mph). Its narrow end may be only 100 meters (330 ft.) across. The funnel cloud becomes a tornado when its lower, narrow end is in contact with the ground. The greatest damage is done often along a relatively linear track where the base of the tornado was in contact with the surface (● Fig. 7.26). When the base of a tornado is above the ground, the end can swirl and twist, but little or nothing is done to the surface below. Although most tornado damage is caused by the violent winds, most tornado injuries and deaths result from flying debris. The relatively small size and short duration of a tornado typically limit the number of deaths caused by tornadoes. In fact, more people die from lightning strikes each year than from tornadoes. At times, however, severe storms may spawn a **tornado outbreak**, where multiple tornadoes are produced by the same weather system. The worst outbreak in recorded history occurred on April 3 and 4, 1974, when 148 tornadoes touched down in 13 states, injuring almost 5500 people and killing 330 others.

In 1971, the late T. Theodore Fujita, a professor at the University of Chicago, originated the Fujita Scale (the F-scale), a scale of tornado intensity with levels from F-0 to F-5 based on the damage that a tornado produced, which is generally a function of wind velocity. Since Fujita's time, Doppler radar has provided the capability to accurately estimate the actual wind speed of a particular tornado. In 2007, the National Weather Service adopted a refined and modified version of the original F-scale, based on new data and observations that were not available to Fujita: the **Enhanced Fujita Scale** (the EF-scale). The main difference is in how wind speeds are now estimated based on damage observations. Also, changes in the wind speeds for intensity levels were integrated for the new EF-scale, as shown in Table 7.3. The most powerful tornadoes are EF-5, with winds that can blow houses completely away, throw vehicles through the air, and cause complete destruction in an area they strike (● Fig. 7.27).

Snowstorms and Blizzards

Snow-producing events mainly occur in the middle and higher latitudes and at high elevations, because they are associated with colder winter temperatures. However, the areas affected by these storms may be extensive. Snowstorms are also triggered by the same uplift mechanisms that cause other types

TABLE 7.3
The Enhanced Fujita Scale Tornado Intensity Scale

EF-SCALE	KPH	MPH	EXPECTED DAMAGE
	Wind Speed 3 Second Gust		
EF-0	104–137	65-85	**Light Damage** Damage to chimneys and billboards; broken branches; shallow-rooted trees pushed over
EF-1	138–177	86–110	**Moderate Damage** Surfaces peeled off roofs; mobile homes pushed off foundations or overturned; exterior doors blown off; windows broken; moving autos pushed off the road
EF-2	178–217	111–135	**Considerable Damage** Roofs torn off houses; mobile homes demolished; boxcars pushed over; large trees snapped or uprooted; light-object missiles generated
EF-3	218–265	136–165	**Severe Damage** Roofs and some walls torn off well-constructed houses; trains overturned; most trees in forest uprooted; heavy cars lifted off ground and thrown
EF-4	266–322	166–200	**Devastating Damage** Well-constructed houses leveled; structures with weak foundations blown some distance; cars thrown and large missiles generated
EF-5	>322	>200	**Incredible Damage** Strong frame houses lifted off foundations and carried considerable distance to disintegrate; automobile-sized missiles fly through the air farther than 100 meters; trees debarked; incredible phenomena occur

● **FIGURE 7.27** Destruction caused by an F5 tornado at Greensburg, Kansas, on May 16, 2007.

Greg Henshall/FEMA

of precipitation events—orographic, frontal, and convergence (cyclonic). Convection is more of a warm–weather mechanism and is less likely to be involved in producing snowfall. The mid- to high-latitude winters experience snowfall events of varying severity. They can come as a heavy snowfall or as a *snow flurry*, a brief, light snowfall that may not produce any accumulation of snow on the ground. A **snowstorm** is a weather event where frozen precipitation falls as snow, typically with much accumulation, and the storm may be severe. Some snowstorms create enough turbulence to create lightning discharges.

Receiving Warnings About Tornadoes and other Natural Hazards

The National Weather Service (NWS) is responsible for detecting and monitoring hazardous weather conditions, such as hurricanes and tornadoes. When these hazards are occurring or imminent, the NWS issues alerts and warnings depending on the seriousness of the weather event. Hurricanes develop over a span of several days or more, which allows time for monitoring the storm, tracking its movements, studying the storm's characteristics, and assessing potential threats. Usually there is ample time to give advance warnings to communities at risk. Information about a hurricane's strength, potential track, and likely landfall site is broadcast widely, well before the storm's arrival. The public generally receives storm information and warnings with adequate time to do what they can to protect their lives and property. Having emergency information before experiencing a serious natural hazard is extremely important and has helped save many lives.

Compared to hurricanes, tornadoes present a much different situation. Weather conditions that might spawn tornadoes can be reported in advance, but the warning time frame is more like hours or minutes, rather than days, and a tornado may or may not develop. Tornadoes can form in minutes or less, and then travel destructively across the land, with very little time for issuing a warning. This means that to be effective, information about an existing or imminent tornado danger must be received immediately by the weather service and communicated right away to the public.

Doppler radar images provide good information about approaching severe weather, but it may not provide a clear picture of what is happening on the ground. Doppler radar may only show where a tornado might form, and some tornadoes begin before a radar "tornado signature" appears. Tornado spotters, trained by the National Weather Service, provide information from the ground in real time by reporting severe weather problems, including pinpointing a tornado touchdown and tracking the storm from a safe distance. A spotter can communicate information to the NWS about weather conditions as they happen, so that warnings can be issued immediately. Broadcast weather reports can be accurate and up to date, but people will not receive that information unless they are listening to radio or watching television, which may not be available because electrical power can be lost during intense thunderstorms.

To minimize this problem of people not receiving warnings, a new way of issuing hazard warnings was needed in order to get critical information to the public more efficiently. Cell phones provided the solution. The National Weather Service developed a new communication system to provide crucial information and warnings directly to individual cell phones, much like a text message. The system is called WEA—Wireless Emergency Alert—and is a free service.

If a serious hazard exists in the area where the cell phone is located, short informational messages from WEA will be received with a special tone or phone vibration, and will be repeated twice (see figure). Although severe weather hazards, such as tornadoes, were the intended application, the National Weather Service is cooperating with other public safety agencies to address other kinds of environmental hazards. These include warnings and information about threats in a cell phone's location from tornadoes, hurricanes, extreme wind conditions, tsunamis, flash floods, and serious dust storms. Also, WEA includes any other emergency situations requiring immediate action or evacuation. This system has been active only for a couple of years, yet WEA messages have already saved lives and helped avoid property damage in several states.

To find out if your cell phone model is WEA capable, and if the alert system is now available in your area, contact your wireless carrier. For more information about the Wireless Emergency Alerts system, search its website address is as follows. http://www.nws.noaa.gov/com/weatherreadynation/wea.html

An example of a Weather Emergency Alert received from the National Weather Service. The system functions much like receiving text messages, but the messages are received at no cost to the cell phone user.

A **blizzard** is the most severe weather event involving snowfall. Blizzards are characterized as heavy snowstorms accompanied by very strong winds. At wind speeds of about 55 kilometers per hour (35 mph) or greater, falling and blowing snow can reduce visibility to zero, creating conditions that are known as *whiteouts*. Visibility can be reduced so much that all a person can see is white, and an individual can completely lose track of distance and direction. This is especially dangerous for people using any mode of transportation. Airport closings and traffic accidents are common during blizzards (● Fig. 7.28).

● **FIGURE 7.28** A weather station in central Illinois during a blizzard in February 2000. With winds gusting to 45 miles per hour, this blizzard greatly reduced visibility.

What other hazards are associated with blizzards?

Weather Forecasting

Weather forecasting, at least the process of it, is fairly straightforward. Meteorological observations are made, collected, and mapped to depict the current state of the atmosphere. From this information and data, the probable movement and any anticipated growth, movement, or decay of current weather systems is projected for a specific length of time into the future.

When a weather forecast goes wrong—which we all know occurs—it is usually because limited or erroneous information has been collected and processed or because of high potential variability in the anticipated path or growth of weather systems. Small errors can compound over time. For example, when recording the path of a storm, an error of a few degrees may cause the projected location of a storm to be off by several miles even in a 2-hour forecast. In a 48-hour forecast, the projected location could be off by hundreds of miles. Because of the complex uncertainties that are inherent in weather systems and atmospheric conditions, the farther into the future one tries to forecast, the greater the chance of error.

Weather forecasts are not perfect, but they are much better today than they were in the past. Much of this improvement can be attributed to increasingly sophisticated technology and equipment. Increased knowledge and surveillance of the upper atmosphere have also improved the accuracy of weather prediction. Weather satellites have helped tremendously by providing continuing images and data in real time of weather and weather systems locally, regionally, and globally (● Fig. 7.29). Satellite data and images have been of particular value to forecasters in coastal areas of the United States. Before the advent of weather satellites, forecasters had to rely on information relayed from

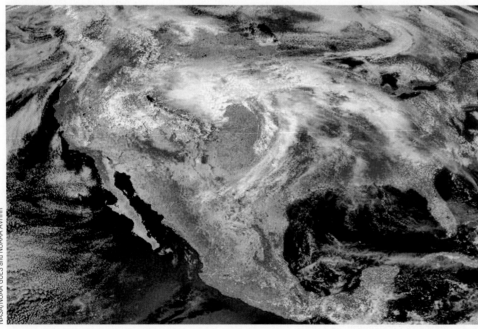

● **FIGURE 7.29** The cloud patterns shown in this weather satellite image can be combined with other weather information and data to determine regional weather conditions. Images like these are also helpful for making forecasts of the weather days in advance.

A classic midlatitude cyclonic storm. Can you identify the midlatitude cyclone's cold, warm, and cool sectors, and the cold, warm, and occluded fronts? In which general direction is this storm system moving?

ships, which left enormous areas of the oceans unobserved. Thus, forecasters could be caught off guard by unexpected weather events.

In addition, computers allow rapid processing and mapping of observed weather conditions. Computers can also be used to process numerical data and generate forecasts based on the physical laws and equations that govern our atmosphere. Numerical and long-term forecasts that are based on solving statistical relationships and complex equations would not be possible without computers, at least not within the short time span allotted to weather forecasters. Today, some of the world's largest and fastest computers are used to forecast the weather and model climatic conditions.

Although forecasters now possess a great deal of knowledge and access to a variety of highly sophisticated technologies that were previously unavailable, weather forecasting is not foolproof. If we are more aware of some of the problems that weather forecasters face, we should also be more understanding when a forecast fails. Weather forecasters combine science with art, fact with interpretation, and data with intuition to come up with some probabilities about future weather conditions.

This chapter has combined some of the atmospheric elements that we learned in previous chapters to explain major weather systems in some detail. Temperature and moisture differences characterize air masses, and fronts develop along the convergence of unlike air masses. With their characteristics of temperature, humidity, pressure, and wind, air masses and fronts form the weather systems that we deal with over the seasons. Certain weather systems may be relatively small, such as tornadoes, or they may cover large areas, such as midlatitude cyclones. Weather systems affect our lives in many different ways, from the joys of a perfect day, to the inconvenience of an incorrect forecast, to understanding how to prepare for a potentially disastrous storm. Knowledge of how weather systems vary and operate can enrich our lives as well as provide some of the background needed to understand the world's climates, which are addressed in the next chapters.

CHAPTER 7 ACTIVITIES

■ TERMS FOR REVIEW

air mass
source region
maritime equatorial *(mE)*
maritime tropical *(mT)*
continental tropical *(cT)*
continental polar *(cP)*
maritime polar *(mP)*
continental arctic *(cA)*
continental Antarctic *(cAA)*
lake-effect snow
nor'easters
cold front

squall line
warm front
stationary front
occluded front
atmospheric disturbance
polar outbreak
storm tracks
midlatitude cyclone (extratropical cyclone)
hurricane
typhoon
landfall

easterly wave
Saffir-Simpson Hurricane Scale
storm surge
convective thunderstorm
orographic thunderstorm
frontal thunderstorm
tornado
Doppler radar
tornado outbreak
Enhanced Fujita Scale
snowstorm
blizzard

QUESTIONS FOR REVIEW

1. What is an air mass?
2. Do all areas on Earth produce air masses? Why or why not?
3. What letter symbols are used to identify air masses? How are these combined? What air masses influence the weather of North America? Where and at what time of the year are they most effective?
4. Use Table 7.1 and Figure 7.1 to find out what kinds of air masses are most likely to affect your local area. How do they affect weather in your area?
5. Why are air masses classified by whether they develop over water or over land?
6. What kind of air mass forms over the southwestern United States in summer? What weather conditions are associated with this air mass?
7. What is a front? How does a front develop?
8. Compare warm and cold fronts. How do they differ in duration and precipitation characteristics?
9. What kind of weather often results from a stationary front? What kinds of forces tend to break up stationary fronts?
10. How does the westerly circulation affect air masses in your area? What kinds of weather result?
11. Draw a diagram of a fully developed midlatitude cyclone that includes the center of the low with several isobars, the warm and cold fronts, wind direction arrows, appropriate labeling of warm and cold air masses, and zones of precipitation.
12. If the wind changes to a clockwise direction, what is the shift called? Where is your location in relation to the center of a low pressure system? Explain why this happens.
13. How does the configuration of the upper air winds play a role in the surface weather conditions?
14. Describe the sequence of weather events over a 48-hour period in St. Louis, Missouri, if a typical low pressure system (cyclone) passes 300 kilometers (180 mi) north of that location in the spring.
15. List three major causes of thunderstorms. How might the storms that develop from each of these causes differ?

CONSIDER AND RESPOND

1. Collect a 3-day series of weather maps from your local newspaper. Based on the migration of high and low pressure systems during that period, discuss the likely pattern of the upper air winds.
2. Look at Figure 7.10a. Assume you are driving from point A to point B. Describe the changes in weather (temperature, wind speed and direction, barometric pressure, precipitation, and cloud cover) you would encounter on your trip. Do the same analysis for a trip from point C to point D.
3. The location of the polar front changes with seasons. Why?
4. List the ideal conditions for the development of a hurricane.

PRACTICAL APPLICATIONS

1. Wind speeds are sometimes given in knots (nautical miles per hour) instead of statute miles per hour, as most people use. A nautical mile (used mainly by sailors and pilots) is equal to 6080 feet, slightly longer than the statute mile at 5280 feet; therefore, a knot is a little faster than 1 statute mile per hour (mph). The conversion from miles per hour to knots is:

$$KNOTS = MPH \times 1.15$$

Using Table 7.2, convert the ranges of wind speeds for hurricane Categories 1 through 5 from miles per hour to knots.

2. A cold front squall line moving 35 miles per hour has spawned tornadoes at 9:00 a.m. near Memphis, Tennessee. The remainder of Tennessee is covered by an unstable *mT* air mass. Make a statement about what we might expect to happen in Nashville, Tennessee (some 200 miles away) and when.

LOCATE AND EXPLORE

1. Using Google Earth and the weather layer provided by Google Earth (in the Layers window), track the daily variation in weather (temperature, pressure, wind speed, wind direction, dew point, precipitation, and cloud type) at any weather station in the following cities. Can you see any relationship in the weather among these cities? Can you use the weather in Omaha to predict the weather later in the week in Louisville or Washington?
   Portland, Oregon (45.53°N, 122.69°W)
   Casper, Wyoming (42.86°N, 106.29°W)
   Omaha, Nebraska (41.25°N, 95.88°W)
   Louisville, Kentucky (38.27°N, 85.74°W)
   Washington, D.C. (38.93°N, 77.03°W).

2. Using Google Earth and the weather layer provided by Google Earth (in the Layers window), look at the different weather systems across the United States and their associated cloud type. Over the next week, go outside every day at the same time and look at the sky. Record the cloud type in your area using the examples in Figures 6.14 and 6.15 from this book as a guide. How do the cloud type and general weather conditions for your area (rain, sun, wind, and so on) relate to the regional weather systems (low pressures, high pressures, fronts, and so on) that are currently tracking across the United States?

3. Using Google Earth and the wind vector layer for Hurricane Katrina (download from CourseMate at www.cengagebrain.com), describe how the wind speed and direction changed in New Orleans, Louisiana (30.0°N, 90.1°W), and Biloxi, Mississippi (30.4°N, 88.9°W), as the storm passed. These layers show the speed of the wind (background color) and the direction (small arrows) every 2 hours as the storm passed. Provide an explanation for the observed wind speed and direction based on distance of each city from the center of the storm and whether the city was east or west of the storm. Tip: Create an x-y scatter plot in a digital spreadsheet (such as Excel), using time on the x-axis and wind speed on the y-axis.

 MindTap—Make the most of your study time by accessing everything you need to succeed in one place. Read your textbook, take notes, review flashcards, watch videos, complete activities, take practice quizzes, and more online with MindTap. Log in at **www.cengagebrain.com**.

Map Interpretation

WEATHER MAPS

Weather maps that portray meteorological conditions over a large area at a given moment in time are important for current weather descriptions and forecasting. Simultaneous observations of meteorological data are recorded at weather stations across the United States (and worldwide). This information is electronically relayed to the National Centers for Environmental Prediction near Washington, DC, where the data are analyzed and mapped and a U.S. map is published daily. A different map from the same day (or any recent day) can be found here at http://www.hpc.ncep.noaa.gov/dailywxmap/

Meteorologists then use the information and data to depict a general weather picture over a larger area. For example, isobars (lines of equal atmospheric pressure) are drawn to reveal the locations of cyclones (L) and anticyclones (H) and to indicate frontal boundaries. Areas that were receiving precipitation at the time the map depicts are highlighted with shading. The result is a map of weather conditions that can be used to forecast changes in weather patterns. The following map is accompanied by a satellite image that was taken on the same date.

1. What is the pressure interval (in millibars) between adjacent isobars on this weather map?

2. What kind of front is stretching across eastern Canada to Maine?

3. What kind of air mass is dominating the midwestern part of the United States?

4. What are the highest and lowest values for isobars on this map in millibars. How did you know where to look for these maximum and minimum values? Where is the pressure gradient the steepest?

5. What kind of front is approaching the Pacific Northwest?

6. Do the fronts and cyclonic storms on the map accurately depict the cloud cover shown on the satellite image? (Disregard the fact that the map and the image are the same day but not the same time of day.)

7. Find the low pressure systems on the satellite image. Does the circulation look correct for the Northern Hemisphere?

8. List the three kinds of fronts on this map with a general statement of their location. The fourth kind of front is not occurring on this day; what kind is it?

9. Do the locations of the fronts and areas of precipitation depicted on the map agree with the idealized relationships represented in Figure 7.10?

10. What kind of storm symbol lies on the map off the coast of Virginia and North Carolina?

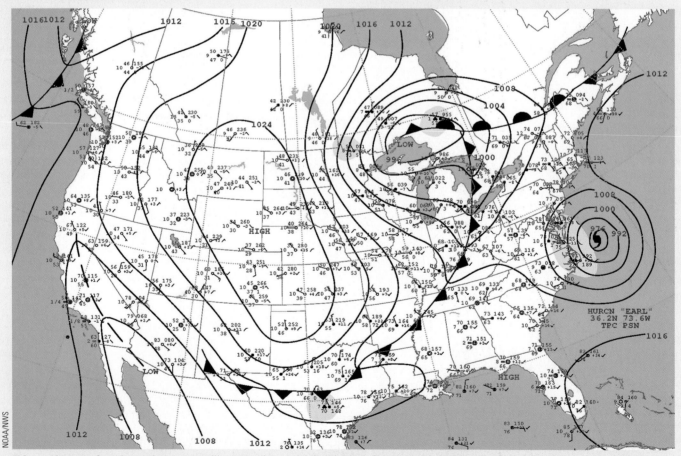

This weather map from September 3, 2010, illustrates the spatial distribution of quantitative weather elements (air pressure, temperatures, wind speed, wind directions) and the locations of fronts and areas of precipitation. Isobars define high and low pressure cells and the kinds of fronts are also identified. The weather conditions at meteorological stations are shown by numerical values and symbols.

A weather satellite image of the atmospheric conditions shown on the accompanying weather map.
NASA/GOES/GSFC

5 years later

GLOBAL CLIMATES
AND CLIMATE CHANGE

■ OBJECTIVES

WHEN YOU COMPLETE THIS CHAPTER YOU SHOULD BE ABLE TO:

- ■ 8.1 Compare and contrast the advantages and limitations of the Thornthwaite and Köppen climate classification systems.
- ■ 8.2 Apply temperature and precipitation statistics to classify climates using the Köppen system.
- ■ 8.3 Recognize how the information on climographs is used to identify the climate of a place.
- ■ 8.4 Understand why vegetation types are so closely related to climate regions in the Köppen system.
- ■ 8.5 Explain why global climates have changed over the past few million years and how Earth scientists have documented these changes.
- ■ 8.6 Appreciate why it is so difficult to determine the cause of an Ice Age and recall what hypotheses have been suggested.
- ■ 8.7 Discuss the nature of recent climate change and the impact it would likely have on Earth environments.
- ■ 8.8 Understand the results of extensive research into the role that human activities have played in climate change.

WEATHER AND CLIMATE ARE VERY different methods of referring to atmospheric conditions and their impacts, despite being strongly related. Mark Twain recognized this contrast, by saying, "Climate is what we expect and weather is what we get." The preceding chapters explained how temperature, humidity, winds, precipitation, and other atmospheric elements combine to produce the weather and briefly introduced the concept of climate. In this chapter we examine how long-term records of weather data and atmospheric conditions are used to determine the nature of climate and of climate change, both of which strongly influence the physical environments of our planet.

Climate is sometimes described as the average weather for a place that occurs at a place over the seasons, but this is a great oversimplification. Climate and the science of **climatology** are concerned with generalized atmospheric conditions of the past and present, and in forecasting future climates. Climatologists study atmospheric conditions that affect large areas and changes in climate that occur from year to year or over periods of many years. Averages of weather data are

The Bear Glacier in Alaska, like many glaciers worldwide, has been shrinking in size under current climatic conditions. These photos show the dramatic change in a 5-year period.   USGS

commonly used but it is also important to consider the kinds of meteorological extremes that can occur, their potential severity, and patterns of change over time. Climatic conditions from times that were before weather data were collected can often be reconstructed by studying natural evidence that records the influence of climatic conditions during the early historic or the ancient past. *Regional climatology* seeks to use classifications of climate types to map their geographic distribution, and to explain the spatial patterns of climates on Earth. Although climate types are primarily classified by statistical averages of weather elements such as temperature and precipitation, the description of a climate type also includes the potential for infrequent weather conditions such as severe storms, frosts, and droughts, as well as unusual precipitation, temperature, and weather events.

Understanding the world's climates and their characteristics has direct relevance to all of us. We are exposed to a great variety of natural landscapes and physical environments in person, or in books, television, movies, and on the Internet. Being aware of how climates and natural landscapes are interrelated increases appreciation of our surroundings and enriches our experience, wherever we live or travel.

For more than a century, scientists have realized that there have been major shifts in climate during Earth's history. Climates in the past were not the same as they are today, and there is every reason to believe that in the future, climates will also be different. Understanding climate change requires detailed study of past climate variability, the processes involved, and the magnitude and rates of prehistoric climate change. In prehistoric times our planet experienced both Ice Ages and periods that were warmer than today. Originally, it was thought that these climate shifts were gradual and perhaps not detectable during a human lifetime. However, recent research reveals that climate has shifted repeatedly and sometimes considerably over rather short intervals of time—fluctuations that are indicators of the natural variability of climate even in the absence of human activities. Today, it is widely recognized that human activities are affecting Earth's climates, and there is much concern about their impacts on climate change. Why this is so, how climates are currently changing, the impacts of climate change on people and on environments, and some of their potential impacts will also be discussed in this chapter.

Classifying Climates

Since ancient times people have documented their awareness that climates vary from place to place, and are related to their geographic locations. The early Greeks (e.g., Aristotle, circa 350 B.C.) classified the known world into three climate zones, called Torrid, Temperate, and Frigid, based on their relative warmth. It was recognized that these zones corresponded with latitude and that the flora and fauna reflected the climatic conditions of these zones. As world exploration expanded, naturalists observed that the geographic locations of climate types could be explained by factors in addition to latitude, such as seasonal variations in sun angles, prevailing winds, elevations, and proximity to large bodies of water.

The two most common indicators for describing the climate of a location or a region are temperature and precipitation. The invention of an instrument to reliably measure temperature—the thermometer—dates to Galileo in the early 1600s. European settlement of distant colonies, and sporadic collection of temperature and precipitation data from those areas began in the 1700s, but records were not routinely kept until the mid-1800s or later. With the availability of long-term weather records, some of the first attempts to classify global climates using actual temperature and precipitation data followed in the early twentieth century.

Climatologists have worked to classify different climates into a comprehensible number of climate types based on similar weather statistics, without having to directly consider the causes for their differences (such as the frequency of air mass movements). By applying a **climate classification system** based on numerical weather data combined with seasonal and annual similarities of precipitation and temperature, climatologists then created maps to show the geographic distribution of climate regions (● Fig. 8.1).

Despite their usefulness, climate classification systems are not without problems. Climate conditions and regions are generalizations about recorded data and observations based on averages and the occurrence probabilities of different kinds of weather. Each climate type represents a specified composite of weather conditions over the seasons. Within such a generalization, it is impossible to include the many variations that exist. Two different climate classification systems are described here; one that is usually applied on a more localized scale and another that is widely used to describe and map climates on a large regional and global scale.

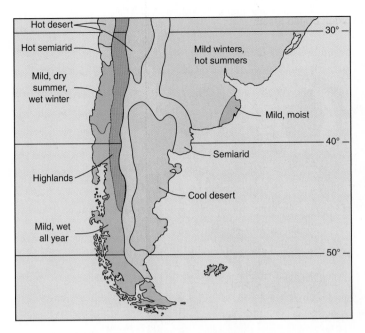

● **FIGURE 8.1** This map shows the diversity of climates possible in a relatively small area, including portions of Chile, Argentina, Uruguay, and Brazil. The climates range from dry to wet and from hot to cold, with many possible combinations of temperature and moisture characteristics.

What can you suggest as the causes for the major changes in climate types as you follow the 40°S latitude line from west to east across South America?

The Thornthwaite System

The Thornthwaite climate classification is a system that concentrates on a local scale, and is most useful for water resources specialists, soil scientists, and agriculturalists. For example, for a farmer interested in growing a certain crop in a particular area, a global system for classifying large global regions is too generalized. From an environmental or an agricultural perspective at a more local level, a major concern is whether adequate moisture will be available in the growing season, whether it comes in the form of precipitation or from the soil. Changes in soil moisture and the seasonal timing of moisture surpluses or deficits are extremely important climatic variables that have a strong impact on native vegetation and on agriculture.

Developed by an American climatologist, C. Warren Thornthwaite, the **Thornthwaite system** determines moisture availability (or shortages) at a subregional scale and is preferred for applications dealing with climates on a local scale. A detailed climate classification such as the Thornthwaite system only became possible after temperature and precipitation data were widely collected beginning in the latter half of the nineteenth century.

The Thornthwaite classification is based on the concept of **potential evapotranspiration (potential ET or PET)**, which approximates the use of water by plants if an unlimited water supply were available. (Evapotranspiration, discussed in Chapter 6, is a combination of evaporation and transpiration, the water loss to the atmosphere via vegetation.) Potential ET is a theoretical value that increases with increasing temperature, winds, and length of daylight hours, but decreases with increasing humidity. **Actual evapotranspiration (actual ET)** reflects the real (actual) evaporation loss and transpiration by plants at a location. During a dry season, water for actual ET can be supplied by soil moisture if the soil is not completely dry, and the available water supply may last through a dry season if the climate is relatively cool, and/or the daylight lengths are short. Measurements of actual ET relative to potential ET and available soil moisture are strong determining factors for most vegetation and crop growth.

The Thornthwaite system recognizes climate zones based on the relationship between the precipitation a location receives, and the potential evapotranspiration at that location. The Thornthwaite moisture index (MI) for a location is determined by this simple equation:

$$MI = 100 \times \frac{P - \text{Potential ET}}{\text{Potential ET}}$$

If precipitation (P) exceeds potential ET, the index is positive for humid or wet climates. If potential ET exceeds precipitation, the index is negative for semiarid or arid locations. Thornthwaite climate regions in the contiguous United States based on the relationship between precipitation (P) and potential evapotranspiration (ET) are shown on the map in Understanding Map Content ● Figure 8.2.

Refer to Figure 6.9, the graph shows a visual representation of the Thornthwaite system as it applies to the San Francisco, California, area. You should quickly note the striking difference between the annual potential evapotranspiration in January and that of July. It is obvious that the Thornthwaite system can help California farmers determine the best planting, growing, and harvesting seasons for their crops. The moisture index also helps them decide when irrigation might be required and how much water might be needed.

The Thornthwaite system recognizes three climate zones based on potential ET values: low-latitude climates, with potential ET greater than 130 centimeters (51 in.); midlatitude climates, with potential ET fewer than 130 but greater than 52.5 centimeters (20.5 in.); and high-latitude climates, with potential ET fewer than 52.5 centimeters. Climate zones may be subdivided based on how long and by how much actual ET is below potential ET. Moist climates have either a surplus or a minor deficit of less than 15 centimeters (6 in.). Dry climates have an annual deficit greater than 15 centimeters.

Thornthwaite's original equations for potential ET were based on analyses of data collected in the midwestern and eastern United States. The method was subsequently used with less success in other parts of the world. Over the past few decades, many attempts have been made to improve the accuracy of the Thornthwaite system for regions outside the United States.

The Köppen System

The most widely used climate classification is based on regional temperature and precipitation patterns. It is referred to as the **Köppen system** after its developer, a German botanist and climatologist. Wladimir Köppen recognized that major vegetation associations strongly reflect the area's climate. He formulated

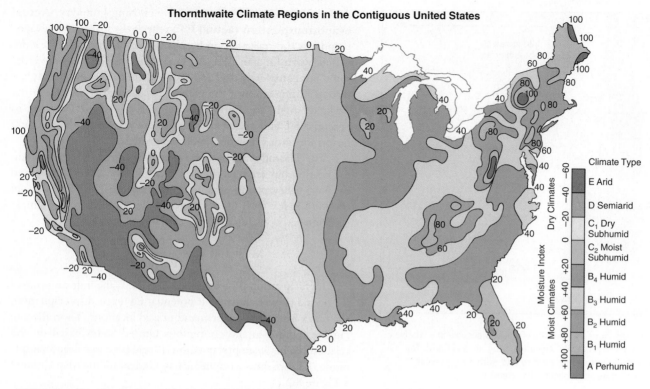

Thornthwaite Climate Regions in the Contiguous United States

Climate Type

- E Arid
- D Semiarid
- C₁ Dry Subhumid
- C₂ Moist Subhumid
- B₄ Humid
- B₃ Humid
- B₂ Humid
- B₁ Humid
- A Perhumid

● **FIGURE 8.2** Thornthwaite climate regions in the contiguous United States are divided by the Moisture Index based on the relationship between precipitation (P) and potential evapotranspiration (ET). Where precipitation exceeds potential ET, the index is positive; where potential ET exceeds precipitation, the index is negative.

***UNDERSTANDING MAP CONTENT** In some regions the pattern of climates is fairly regular, in others it is complex, what are some of the reasons for complex patterns and regular patterns? Through what region of the country does the general boundary run between humid and dry climates? That boundary roughly follows a line of longitude. Referring to a map that shows longitude, approximately what longitude is it?*

his climate regions as they were related to well-defined vegetation regions, and generally described the climate regions by the natural vegetation most often found there. Evidence of the influence of Köppen's system is seen in the wide usage of his climatic terminology, even in nonscientific literature (for example, Mediterranean climate, tundra climate, rainforest climate).

Advantages and Limitations of the Köppen System

Temperature and precipitation are two of the easiest weather elements to measure, and they are also measured more often and in more world locations than any other variables. Furthermore, these are the weather elements that most directly affect humans, animals, vegetation, soils, and many other elements of the natural landscape. Using temperature and precipitation statistics to define climatic boundaries, Köppen derived precise numerical definitions for each climate region. Variations in climate types are very apparent when examining the seasonal and annual temperature and precipitation statistics for a location.

The climate boundaries in Köppen's classification were designed to correspond closely to global vegetation regions. Thus, Köppen's climate boundaries often reflect "vegetation lines." For example, the Köppen classification map uses the 10°C (50°F) mean

monthly isotherm because of its relevance to the treeline—the line beyond which it is too cold and/or the growing season is too short for trees to thrive. For this reason, Köppen defined the treeless polar climates as those areas where the mean temperature of the warmest month is below 10°C (50°F). The climates are classified by the temperature and precipitation, and they correspond well with the vegetation types associated with particular climate types. Köppen's climate classification system and how it relates to visible natural landscapes is one of its most appealing features.

There are limitations, of course, to Köppen's system. For example, Köppen mainly considered average (mean) monthly temperature and precipitation values in his climate classifications. In some instances, the average annual precipitation total is also used along with the mean annual temperature. These monthly or annual indicators of precipitation and temperature permit estimates of precipitation effectiveness (P/PET relationships) but do not measure it directly, or with enough precision to permit detailed comparisons of different locations. For the purposes of generalization and because of very little available data concerning many other weather factors, Köppen did not consider winds, cloud cover, precipitation intensity, and humidity, although these factors exert important influences on local weather and climate.

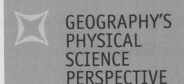

Using Climographs

The variables that Köppen based his climate classification on were those that were most available at that time. These include temperature and precipitation averages for each month of the year, and annual totals for precipitation. Records of annual and monthly averages of temperature and precipitation for the last 30 years or more are typically used for classifying climate types. A 30-year time period is considered adequate for obtaining representative values for monthly, seasonal, and annual precipitation and temperature averages.

Plotting these variables for a particular location on a graph presents an easily recognizable, visual image of the temperature and precipitation variations throughout the year. Graphs like these are called **climographs**, an example of which is shown in the figure. The plotted averages for many years of mean monthly temperature and mean monthly rainfall data, allow quick comparison of the seasonal climate regimes for the place that the data represents. Temperature is a continuous variable, because there is always a temperature value, and each month's mean temperature connects to the adjacent months. The mean monthly temperatures over the year are plotted for each month, and displayed as a continuous and smooth annual temperature curve. Mean monthly precipitation amounts are discrete values because the average accumulation for one month may have little or nothing to do with the value of adjacent months. Therefore, monthly precipitation averages are plotted with individual bars for each month of the year.

Climographs clearly display the graphed data for a location's climate. To read the graph, relate the temperature curve to the values along the left side and the precipitation amounts to the scale on the right. Other information may also be displayed, depending on the type of climograph used. A particularly useful method of analyzing the climates of two or more places is to compare the seasonal and annual differences that appear on side-by-side climographs of those locations.

The climograph shown here is an example of the type that will be used in the chapters on climate. A climograph can be used to determine the Köppen classification of the station as well as show its specific temperature and rainfall measures. The abbreviations for climate types on all climographs are found in Table 8.1.

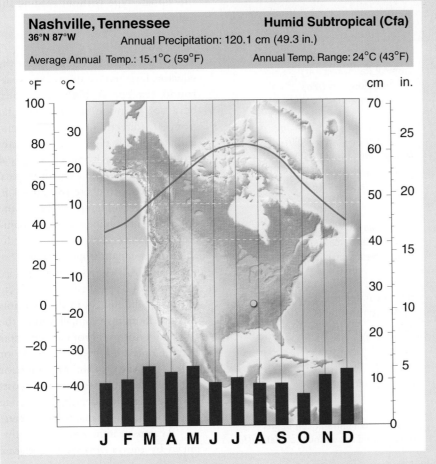

Nashville, Tennessee 36°N 87°W
Humid Subtropical (Cfa)
Annual Precipitation: 120.1 cm (49.3 in.)
Average Annual Temp.: 15.1°C (59°F)
Annual Temp. Range: 24°C (43°F)

A standard climograph showing average monthly temperature (curve) and rainfall (bars). The horizontal index lines at 0°C (32°F), 10°C (50°F), 18°C (64.4°F), and 22°C (71.6°F) are the Köppen temperature parameters by which climate stations are classified.

The Köppen system, as modified by later climatologists, divides the world into six major climate categories (noted with capital letter symbols). The first four are based on temperature characteristics and having adequate annual moisture: (A) humid tropical climates, (C) humid mesothermal climates (mild winter), (D) humid microthermal climates (cold winter), and (E) polar climates. Another category, arid climates (BW and BS) includes *desert climates* (extremely arid) and *steppe climates* (semi-arid), characteristically dry regions identified by comparing the low yearly precipitation they receive to their annual temperature characteristics. It should be noted, however, that the arid and semiarid climates include regions where the temperatures range from cold to very hot. The final category, (H) highland climates, denotes the world's mountainous regions where vegetation and climate vary rapidly because of changes in elevation and exposure. The climate types will generally be referred to by their names as shown on Table 8.1 These letter symbols for the climate types will also appear on graphs to identify specific climate types and subtypes and on maps that indicate their locations.

Within each of the first five major categories (except the highland category), individual climate types and subtypes are differentiated from one another by specific parameters of temperature and precipitation. Table 1 at the end of this chapter in the "Graph Interpretation" exercise outlines the letter designations and

procedures for determining the types and subtypes of the Köppen classification system. This table can be used with any Köppen climate type presented in this chapter and in Chapters 9 and 10.

Climate Types and Their Distribution

Five of the six major climate categories of the Köppen classification include enough differences in the ranges, total amounts, and seasonality of temperature and precipitation to produce the 13 distinctive climate types listed in Table 8.1. The tropical and arid climate types are discussed in some detail in the next chapter, and the mesothermal, microthermal, and polar climates are presented in Chapter 10, along with highland climates.

Before you continue reading, take some time to study the classification tables (in the Graph Interpretation feature at the end of the chapter). Note how many terms are already familiar to you and that they suggest a mental picture: *humid, desert, rainforest, ice sheet, continental, marine,* and so on. You may instinctively know more about climate than you realize. Then examine the photographs (● Figs. 8.3 through 8.5 and 8.7 through 8.16) associated with the brief introduction to the Köppen climate system that are shown in the next several pages. Discover how closely your mental image of each climate type matches the photograph of an environment that exists within each climate type. Even more important, what can you identify by studying each photograph that might help you determine the climate type without reading the figure caption?

TABLE 8.1
Simplified Köppen Climate Classes

Climates	Climograph and Map Abbreviations
Humid Tropical Climates (A)	
Tropical Rainforest Climate	Tropical Rainforest (Af)
Tropical Monsoon Climate	Tropical Monsoon (Am)
Tropical Savanna Climate	Tropical Savanna (Aw)
Arid Climates (B)	
Steppe (Semiarid) Climate	Low-Latitude Steppe (BSh)
	Midlatitude Steppe (BSk)
Desert Climate	Low-Latitude Desert (BWh)
	Midlatitude Desert (BWk)
Humid Mesothermal (Mild Winter) Climates (C)	
Mediterranean Climate	Mediterranean (Csa/Csb)
Humid Subtropical Climate	Humid Subtropical (Cfa)
Marine West Coast Climate	Marine West Coast (Cfb/Cfc)
Humid Microthermal (Severe Winter) Climates (D)	
Humid Continental, Hot-Summer Climate	Humid Continental H.S. (Dfa/Dwa)
Humid Continental, Mild-Summer Climate	Humid Continental M.S. (Dfb/Dwb)
Subarctic Climate	Subarctic (Dfc/Dfd/Dwc/Dwd)
Polar Climates (E)	
Tundra Climate	Tundra (ET)
Ice Sheet Climate	Ice sheet (EF)
Highland Climates (H)	
Various climates based on elevation differences.	No single climograph can depict these varied (or various) climates (H)

Tropical (A) Climates Near the equator, high temperatures exist year-round (except at high elevations) because the noon sun is never far from 90° (directly overhead). Humid climates of this type with no winter (cold) season are Köppen's **tropical climates**. As his boundary for tropical climates, Köppen chose 18°C (64.4°F) for the average temperature of the coldest month because it closely coincides with the geographic and latitudinal limit of certain tropical palms.

Table 8.1 shows that there are three humid tropical climates, reflecting major differences in the amounts and distribution of rainfall within the tropical regions. Tropical climates extend poleward to 30° latitude or higher in the continent's interior but to lower latitudes near the coasts because of the moderating influence of the oceans on coastal temperatures.

Regions near the equator are influenced by the *Intertropical Convergence*

● **FIGURE 8.3** Tropical rainforest climate: Dense vegetation cover in El Yunque National Forest in Puerto Rico.

● **FIGURE 8.4** Tropical savanna climate: East African high plains with scattered trees.

Zone (ITCZ). However, the convergent and rising air of the ITCZ, which brings rain to the tropics, is not anchored in one place; instead, it migrates with the seasons following the 90° sun angle, the most intense solar radiation (see The Analemma in Chapter 3). Within 5° to 10° latitude of the equator, rainfall occurs year-round because the ITCZ moves through those latitudes twice a year and is never far away, and heavy rainfall supports tropical rainforests (● Fig. 8.3). Poleward of this zone the stabilizing influence of *subtropical high pressure systems* causes precipitation to become seasonal. When the ITCZ is over the region during the high-sun period (summer), there is adequate rainfall. However, during the low-sun period (winter), the subtropical highs dominate the area, bringing clear, dry weather.

The tropical rainforest climate *(Af)* in the equatorial region is flanked both north and south by the dry-winter tropical savanna *(Aw)* climate (● Fig. 8.4). Finally, coasts facing the strong, moisture-laden inflow of air associated with the summer monsoon have a tropical monsoon *(Am)* climate (● Fig. 8.5).

Examine the map of world climates in ● Fig. 8.6 Note that it shows the progression of the humid tropical climates from the

● **FIGURE 8.5** Tropical monsoon climate: Himalayan foothills, West Bengal, India.

equator toward the poles in the equatorial regions of Africa and South America and the major regions associated with the tropical monsoons of Asia. The atmospheric processes that produce the various tropical *(A)* climates are discussed in Chapter 9.

Polar *(E)* Climates Just as the tropical climates lack winters (cold periods), the polar climates—at least statistically—lack summers. **Polar climates**, as defined by Köppen, are areas that do not have a month with an average temperature exceeding 10°C (50°F). Poleward of this temperature boundary, trees generally cannot survive. The 10°C isotherm for the warmest month more or less coincides with the Arctic Circle; poleward of there the sun does not rise above the horizon in midwinter and, although the length of day increases during polar summers, the insolation strikes at a low angle.

The polar climates are subdivided into tundra and ice sheet climates. The ice sheet *(EF)* climate (● Fig. 8.7) has no month with an average temperature above 0°C (32°F). The tundra *(ET)* climate (● Fig. 8.8) occurs where at least 1 month averages above 0°C (32°F). Look at the far northern regions of Eurasia, North America, and Antarctica on the world climate map. The processes creating the polar *(E)* climates are explained in Chapter 10.

Mesothermal *(C)* and Microthermal *(D)* Climates Except where arid climates intervene, the lands between the tropical and polar climates are occupied by the transitional midlatitude mesothermal and microthermal climates. Because they are neither tropical nor polar, the mild and severe winter climates must have at least 1 month averaging below 18°C (64.4°F) and 1 month averaging above 10°C (50°F). Although both midlatitude climate categories have distinct temperature seasons, the **microthermal climates** have severe winters with at least 1 month averaging below freezing. Once again, vegetation reflects the climate dropping their leaves naturally during the winter (most needle-leaf trees do not defoliate in winter). Much of the natural vegetation of the mild-winter **mesothermal climates** retains their foliage throughout the year. The line separating mild from severe winters usually lies in the vicinity of the 40th parallel.

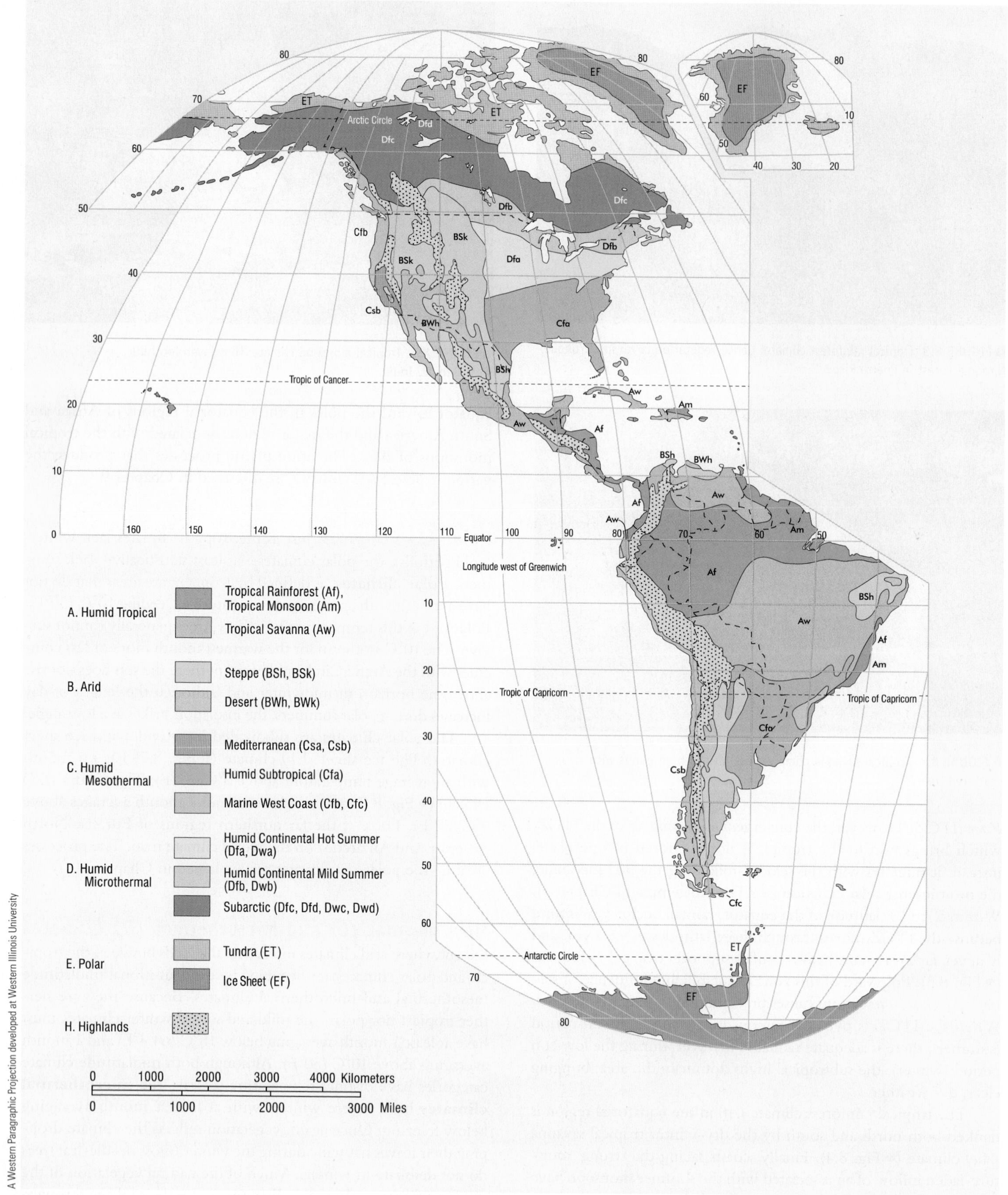

● **FIGURE 8.6** World climate map following the modified Köppen classification system.

A Western Paragraphic Projection developed at Western Illinois University

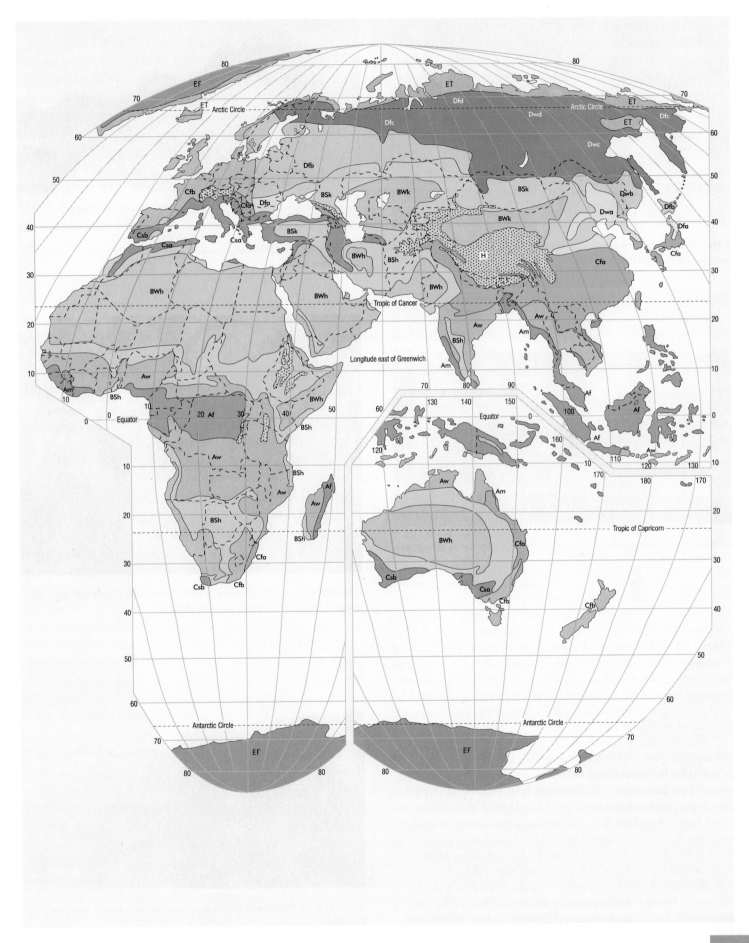

● **FIGURE 8.7** Polar ice sheet climate: Ice sheet glaciers surround and nearly bury mountains in Antarctica.

NASA/Michael Studinger

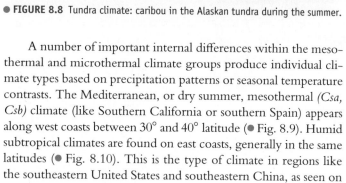

Alaska Image Library/USFWS

● **FIGURE 8.8** Tundra climate: caribou in the Alaskan tundra during the summer.

Jim Petersen

● **FIGURE 8.9** Mediterranean climate: hillside in Northern California during the dry summer season.

USGS

● **FIGURE 8.10** Humid subtropical climate: dense, mixed forest with pine, oak, and palm species on a coastal island in Georgia.

A number of important internal differences within the mesothermal and microthermal climate groups produce individual climate types based on precipitation patterns or seasonal temperature contrasts. The Mediterranean, or dry summer, mesothermal *(Csa, Csb)* climate (like Southern California or southern Spain) appears along west coasts between 30° and 40° latitude (● Fig. 8.9). Humid subtropical climates are found on east coasts, generally in the same latitudes (● Fig. 8.10). This is the type of climate in regions like the southeastern United States and southeastern China, as seen on the map of world climates.

The distinction between the humid subtropical *(Cfa)* and marine west coast *(Cfb, Cfc)* climates illustrates a second important criterion for the subdivisions of midlatitude climates: seasonal contrasts. These two mesothermal climates have year-round precipitation, but summer temperatures in the humid subtropical climate are much higher than those in the marine west coast climate. Therefore, summers are hot. In contrast, the mild and often cool summers of the marine west coast climate, located poleward of the Mediterranean climate along continental west coasts, often extend beyond 60° latitude (● Fig. 8.11). Some marine west coast climates, as shown in the world climate map are along the northwest coasts of the United States and Canada, the west coast of Europe, and the British Isles.

Jim Petersen

● **FIGURE 8.11** Marine west coast climate: heavily tree-covered and foggy coast of Oregon.

● **FIGURE 8.12** Humid continental climate: Fall color awaits a snowy winter in Minnesota.

● **FIGURE 8.13** Subarctic climate: These trees endure a long winter in Alaska.

Another example of internal differences is found among microthermal *(D)* climates, which usually receive year-round precipitation associated with midlatitude cyclones traveling along the polar front. Internal subdivision into climate types is based on summers that become shorter and cooler and winters that become longer and more severe with increasing latitude and *continentality*. Microthermal climates are found exclusively in the Northern Hemisphere (● Fig. 8.12) because there is no land in the Southern Hemisphere latitudes that would normally be occupied by these climate types. In the Northern Hemisphere these climates progress poleward through the humid continental, hot-summer *(Dfa, Dwa)* climate to the humid continental, mild-summer *(Dfb, Dwb)* climate and, finally, to the subarctic *(Dfc, Dfd, Dwc, and Dwd)* climate (● Fig. 8.13). Microthermal regions can be seen on the climate maps in the eastern United States and Canada, and northward through eastern Europe.

Arid (B) Climates
Climates dominated by considerable year-round moisture deficiency are the **arid climates**. These climates penetrate deep into the continents, interrupting the latitudinal

● **FIGURE 8.14** Desert climate: the Chihuahuan Desert in New Mexico.

zonation of climates that would otherwise exist. The definition of *climatic aridity* is that precipitation received is significantly less than potential ET. Aridity does not depend solely on the amount of precipitation received; potential ET rates and temperature must also be taken into account. In a low-latitude climate with relatively high temperatures, the potential ET rate is greater than in a colder, higher-latitude climate. As a result, more rain must fall in the lower latitudes to produce the same effects (on vegetation) that smaller amounts of precipitation produce in areas with lower temperatures and, consequently, lower potential ET rates. Potential ET rates also decrease with elevation, which helps to explain why higher elevation (highland) climates are distinguished separately.

Arid climates are concentrated in a zone from about 15°N and S to about 30°N and S latitude along the western coasts, expanding much farther poleward over the heart of each landmass. The correspondence between the arid climates and the belt of subtropical high pressure systems is unmistakable (like in the southwestern United States, central Australia, and north Africa in Fig. 8.6), and the poleward expansion is a consequence of remoteness from the oceanic moisture supply.

In desert *(BW)* climates, the annual amount of precipitation is less than half of the annual potential ET (● Fig. 8.14). Bordering the deserts are steppe *(BS)* climates—semi-arid climates that are transitional between the extreme aridity of the deserts and the moisture surplus of the humid climates (● Fig. 8.15). The definition of the *steppe climate* is an area where annual precipitation is less than potential ET but more than half the potential ET. B climates and the processes that create them are discussed in more detail in Chapter 9.

Highland (H) Climates
The pattern of climates and extent of aridity are affected by irregularities in Earth's surface, such as mountain ranges, or other significant highlands. Highlands can channel air mass movements and create abrupt climatic divides. They have microclimates that form intricate patterns related to elevation, cloud cover, and exposure (● Fig. 8.16). A significant effect of highlands that are aligned at right angles to the prevailing

● **FIGURE 8.15** Steppe climate: semi-arid prairie grasslands in South Dakota.

● **FIGURE 8.16** Highland climate: Lassen National Park, California.

wind direction is rain shadows creating arid regions that extend tens to hundreds of kilometers downwind. Look at the mountain ranges in on the climate map; the Rockies, the Andes, the Alps, and the Himalayas show *H* climates. These undifferentiated **highland climates** are discussed in more detail in Chapter 10.

Climate Regions Each of the modified Köppen climate types is defined by specific parameters for monthly averages of temperature and precipitation; thus, it is possible to draw boundaries between these types on a world map. The areas within these boundaries are examples of one type of world region. A region, as you recall, is an area that has recognizably similar internal characteristics that are distinct from those of other areas. A region may be described on any basis that unifies it and differentiates it from others.

To understand the distribution of world climate regions in the following chapters, it will be helpful to refer to the map of world climates (Fig. 8.6). It shows the geographic patterns of Earth's climates as they are distributed over each continent. However, on a map of **climate regions**, distinct lines separate one region from

another, yet the lines do not mark locations of abrupt changes in temperature or precipitation conditions. Rather, the lines generally signify **zones of transition** between different climate regions. The zones or boundaries between climate regions are basically "cutoff" locations based on monthly and annual averages and may shift as temperature and moisture statistics change over the years.

The actual transition from one climate region to another is gradual, except in cases in which the change is brought about by an unusual climate control such as a mountain barrier. It would be more accurate to depict climate regions and their zones of transition on a map by showing one color fading into another. An important concept to keep in mind is that it is the core areas of climate regions that best exhibit the characteristics that distinguish one climate type from another.

When you look closely at the world climate map you will immediately notice how climates change with latitude. This is especially apparent in North America on the East Coast of the United States moving north into Canada. Here, the seasonal differences in sun angles and length of daylight play important roles. We can also see that similar climates usually appear in latitudes and/or in locations that are similar with respect to landmasses, ocean currents, or topography. These climate patterns emphasize the geographic relationships among the weather elements, and the factors that influence climate. There is an order to Earth's atmospheric conditions and so also to its climate regions.

A striking variation in these global climate patterns becomes apparent when comparing the Northern and Southern Hemispheres. The Southern Hemisphere lacks large mid- to high-latitude landmasses like those of the Northern Hemisphere. No areas in the higher latitudes of the Southern Hemisphere can be classified having a humid microthermal climate.

Scale and Climate

Climate can be measured at different spatial scales (macro, meso, or micro). The climate of a large (macro) region, such as the Sahara, may be described correctly as hot and dry. Climate can also be described at mesoscale levels; for example, the climate of coastal Southern California is sunny and warm, with dry summers and wet winters. Finally, climate can be described at local scales, such as on the slopes of a single hill. This is termed a **microclimate**.

At the microclimate level, many factors can cause the climate to differ from nearby areas. For example, in the United States and other regions north of the Tropic of Cancer, south-facing slopes tend to be warmer and drier than north-facing slopes because they receive more sunlight (● Fig. 8.17). This variable is referred to as *slope aspect*—the direction a mountain slope faces in respect to the sun's rays. Microclimatic differences such as slope aspect can cause significant differences in vegetation and soil moisture. In what is sometimes called *topoclimates,* tall mountains often possess vertical zones of vegetation that reflect changes in the microclimates as one ascends from the base of the mountain (which may be surrounded by a tropical-type vegetation) to higher slopes with midlatitude vegetation types to the summit covered with ice and snow.

Human activities also influence microclimates. Recent research indicates that the construction of a large reservoir leads to

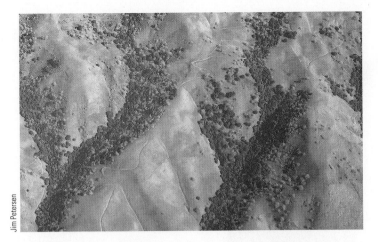

● **FIGURE 8.17** This aerial photograph, facing eastward over valleys in the Coast Ranges of California, illustrates the impact of slope aspect. South-facing slopes receive direct rays of the sun and are hotter and drier than the more shaded north-facing slopes. The south-facing slopes support grasses and only a few trees, whereas the shaded, north-facing slopes are tree covered.

Why do south-facing slopes in the Northern Hemisphere receive solar radiation that is more than the insolation received on north-facing slopes?

greater humidity and increased annual precipitation immediately downwind of this impounded water. This occurs because the lake supplies additional water vapor to the air and to passing storms, which intensifies the rainfall or snows immediately downwind of the lake. These microclimatic effects are similar to the *lake-effect snows* that occur downwind of the Great Lakes in the early winter when the lakes are not frozen (discussed in Chapter 7). Another example of human impact on microclimates is the urban heat-island effect (discussed in Chapter 4), which leads to changes in temperature (urban centers tend to be warmer than their outlying rural areas), rainfall, wind speeds, and many other phenomena.

Climate Change

Important questions are being asked today about climate change and its impacts, related to the recent and ongoing rise in Earth's average atmospheric temperatures that has been termed **global warming**. Temperature increases over the last century are well documented, but what are the causes? What role are human activities playing in these rising temperatures and to what extent are they responsible for this climate change? What causes global climate to change, and how quickly could the world's climate shift from one extreme to another? It well known that Earth's systems are dynamic, very complex, and changing constantly but at rates that range from very gradual to rapid. How has climate varied in the past and what were the causes of climate change in the past? An important reason for studying past climates is to develop an understanding of the factors that influence climates and how they interact to either stabilize climatic conditions or cause them to change. Ultimately, a question that concerns us today is, How will the effects of ongoing and expected climate change affect human populations and natural environments?

The Last Ice Ages: The Pleistocene

Scientific investigation of the premise that global climates had experienced major changes during Earth history began in the nineteenth century. In 1837, Louis Agassiz, a European naturalist, proposed that Earth had experienced extensive periods of *glaciation*—intervals that he referred to as **Ice Ages**. These were times when glaciers (masses of flowing ice) were widespread in midlatitude mountains and large areas of the Northern Hemisphere continents were covered by huge sheets of ice. Agassiz presented evidence that glaciers had once covered Great Britain, much of northern Europe and Asia, and valleys in the Alps. Arriving in the United States in 1846, he found similar evidence of widespread glaciation in North America. The most striking evidence consisted of widespread deposits of sediments and rock debris that originated in very distant sources, which could only have been transported to lower latitudes by powerful moving glaciers. Subsequent studies of evidence in sediments and in rock structures indicate that there were several major glacial advances hundreds of millions of years ago in addition to the more recent ones that Agassiz described.

Today, it is recognized that the most recent Ice Age began about 2.6 million years ago and ended about 12,000 years ago, an interval of geologic time called the **Pleistocene Epoch**. Until the 1960s, it was widely believed that during the Pleistocene, Earth experienced four major periods of global cooling significant enough to result in the expansion of glaciers from the polar regions into the midlatitudes of North America and Europe. Today, these times of glacial advance are termed **glaciations** (or *glacial periods*), which were separated by warmer times of glacial retreat, called **interglacials**. Based on the southern limits of the advance of the massive ice sheet glaciers in North America, these four colder intervals were named the Nebraskan (oldest), Kansan, Illinoian, and Wisconsinan (most recent) glaciations. Later research has shown that glacier expansion and retreat in response to climate change during the Pleistocene was actually more complex than is implied with a four-part model.

A major problem with studying the advance and retreat of glaciers on land is that each subsequent advance of the glaciers tends to destroy, bury, or greatly disrupt the sedimentary evidence of a previous glacial period. Evidence for a fourfold record of glacial advances was largely recognized on the basis of sediments related to glacial processes that were deposited beyond the limit of more recent glaciations. Evidence for earlier, but less extensive, glacial advances were rarely recognized because subsequent glacial advances overrode them.

Radiocarbon dating, developed in the 1950s, began to be widely used to measure the absolute ages of organic material found in association with glacial landforms and deposits. This research helped to establish when the glaciers retreated for the last time. Radiocarbon dating, however, can determine how old an object (containing carbon) may be, but it does not indicate the climate conditions of the past. It is also only applicable back to about 50,000 years, so its main application is in climate studies of relatively recent geologic time.

The last ice advance peaked in North America about 18,000 years ago, according to radiocarbon dates. At its maximum North

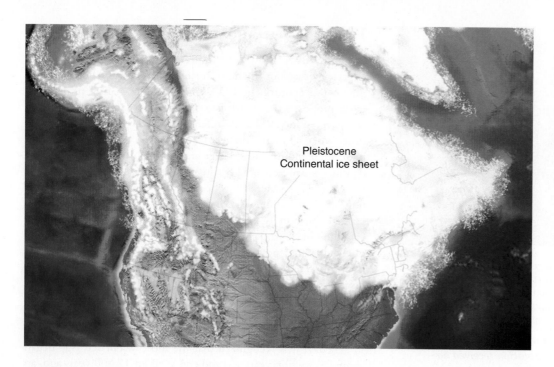

Pleistocene
Continental ice sheet

● **FIGURE 8.18** This map identifies the extensive areas of Canada, the northern United States that were covered by moving sheets of ice about 18,000 years ago. The locations of glaciers in the mountains of the West were also extensive in high mountain valleys at that time.

What are some of the U.S. states that were completely buried by the continental ice sheet at the time of this map?

American extent, a massive glacial ice sheet covered nearly all of Canada and the northern United States (● Fig. 8.18), down to the Ohio and Missouri Rivers. It flowed over the modern-day locations of cities such as Boston, New York, Chicago, Minneapolis, Indianapolis, and Des Moines. Another massive complex of glaciers covered the mountains of western Canada. Further, glaciers flowed from high elevations into the valleys of mountains in the western United States.

Methods for Revealing Past Climates

A great variety of methods are used to uncover evidence and clues about past climates. They are too numerous to mention in this one chapter. However, there are a few reliable methods that have been used for many years to reconstruct **paleoclimates** (ancient climates) and to tie climate change to a time frame. Two very useful techniques are dendrochronology and palynology.

Dendrochronology (or *tree-ring dating*) has been used for decades, and involves an examination of annual tree rings exposed by cores drilled through the middle of certain tree species. Tree cores are generally thin enough so that they do not harm the tree, and they can produce an annual tree ring sequence back to the age of the tree. Some trees are hundreds of years old or more, and the age of the tree is the limit of the age sequence. Annual rings are counted back to establish a time scale for analysis. Each ring, by its thickness, color, and texture, can reveal information about the climate conditions (temperature, precipitation, and moisture availability) during that particular year of the tree's growth. Thus, a short-term climate record can be determined through careful examination of tree rings (● Fig. 8.19). This type of analysis is often referred to as *dendroclimatology*.

Palynology (pollen analysis) is another well-established way of reconstructing past environments in order to understand the

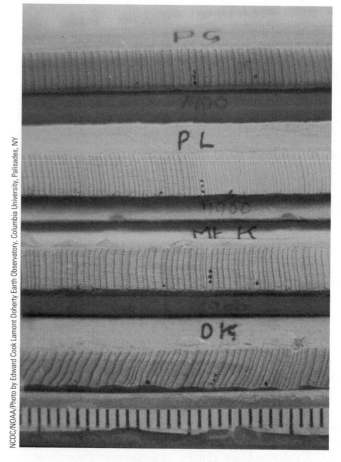

NCDC/NOAA/Photo by Edward Cook Lamont Doherty Earth Observatory, Columbia University, Palisades, NY

● **FIGURE 8.19** The thickness, color, and texture of tree rings indicate the type of climate in existence during that particular growing season.

In what part of the tree trunk are the oldest tree rings found?

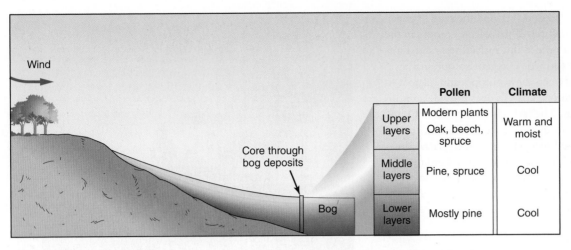

		Pollen	Climate
	Upper layers	Modern plants Oak, beech, spruce	Warm and moist
	Middle layers	Pine, spruce	Cool
	Lower layers	Mostly pine	Cool

● **FIGURE 8.20** This simple diagram shows the use of palynology to reconstruct paleoclimates from sediment cores collected from lakes and bogs.

What problems might occur with analyzing windblown pollen samples?

climate associated with the vegetation that existed during a certain time in the past. Lakes and organic bogs are among the best places to extract pollen samples for analysis. Pollen in the air settling on the water surface of a lake or bog will be deposited at the bottom along with the sediment layer being deposited at that time. A sediment core is drilled and removed to show the layers of sediment and organic material all the way to the bottom of the lake or bog, which can represent accumulations of hundreds or thousands of years (● Fig. 8.20). The age of these layers can be identified by radiocarbon dating of the organic material they contain. Pollen is removed from each sediment layer of the core and analyzed to identify the various trees and plant types that were present when the layer was deposited. Pollen analysis can identify the numbers, types, relative commonness, and distributions of trees and plants that were the source of the pollen (● Fig. 8.21). Experts in palynology and paleoclimatology then work to determine the climatic conditions that would be required to sustain a forest or other environment of the type represented by the pollen contents of layers in the core, and any environmental changes over time.

Drilled cores of sediment taken from the deep ocean bottoms led to the discovery that evidence for millions of years of climate change has been recorded in sediments deposited on the ocean floors. Unlike glacial deposits on the continents, accumulations of deep-sea sediments were not disrupted by subsequent glacial advances. Rather, the slow, continuous accumulation of sediments on the seafloor provides a fairly complete history of climate during the past several million years. An important discovery based on the deep-sea record is that Earth experienced numerous glacial advances during the Pleistocene, not just the four that had been identified previously on land. Today, the names of only two of the North American glacial advances, the Illinoian and Wisconsinan, have been retained.

Oxygen Isotope Analysis
The deep-sea sedimentary record is very important to climate-change studies. The layers of mud deposited on the deep-sea floors contain the remains of tiny

surface-dwelling marine animals (*plankton*) that built shells for protection. When they died, their microscopic shells sank and were deposited in the mud layer on the seafloor. Different species of these organisms only thrive in certain surface water temperatures; therefore, the sequence of thin mud layers that contain these fossils can produce a history of climate related fluctuations in ocean temperature.

The tiny seashells of some plankton species are composed of calcium carbonate ($CaCO_3$). Seawater (H_2O + dissolved minerals)

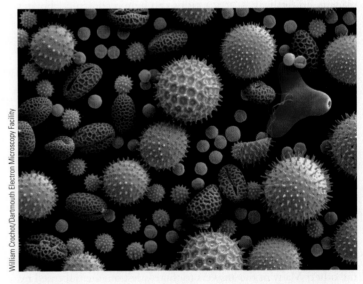

William Crochot/Dartmouth Electron Microscopy Facility

● **FIGURE 8.21** Pollen varies by species, and the mix of pollen types differs over time with climate-related environmental changes. Pollen species counts within sediment layers identify the types of vegetation that existed at the time of deposition. An electron microscope was used to take this image, and later colorized artificially, making it easy to see different species. The larger green-colored pollen grain form a morning glory flower is 100 micrometers in diameter (four one-thousands of an inch).

How can botanists (plant experts) help paleoclimatologists reconstruct climates of the past?

contains two oxygen isotopes (^{18}O and ^{16}O) and the ratio between them is related to ocean temperatures. The oxygen isotopes in the calcium carbonate fossil shells reflect the surface seawater temperature at the time they were formed. Oxygen-isotope analysis is used to determine the ratio between these isotopes and relate it to ocean temperature. Oxygen isotope analysis does provide a means of reconstructing paleoclimates. The ratios between $^{18}O/^{16}O$ isotopes in mud layers of different ages indicate changes in ocean temperatures that can be related to cycles of colder glacial times and warmer periods. Studies of the oxygen-isotope record that the last glacial advance was only one of many major glacial advances during the past 2.6 million years. Evidence suggests there may have been as many as 28 glacial-type climatic episodes.

Glacial Ice Cores

Oxygen isotope analysis can also be used to study climate history from glacial ice. Glaciers record yearly snowfall accumulations in ice layers that can provide short-term evidence of climate changes. Oxygen-isotope analysis is used along with other techniques to analyze the glacial ice cores, and these analyses have revealed a detailed record of variations in climate in ice layers deposited during the past 250,000 years. When water evaporates from the ocean, slightly more ^{16}O than ^{18}O evaporates because water containing the lighter-weight oxygen (^{16}O) evaporates more readily. During an Ice Age, water evaporated from the oceans brings snow that is stored as annual layers of glacial ice to feed the glacier. When the glaciers are large, more water is stored as ice on land, and less water is returned to the oceans. The $^{18}O/^{16}O$ ratio in the ocean changes slightly, which affects the amount of ^{16}O-enriched water being stored in the glaciers. In this way $^{18}O/^{16}O$ ratios can help reconstruct climates of the past from deep cores of glacial ice layers (● Fig. 8.22).

The graph in Figure 8.23 shows a record of average global temperatures for the past 110,000 years, based on **oxygen-isotope analysis** of ice cores. Note that temperatures have varied as much as 10°C (18°F) or more during this relatively brief period of Earth history. The modern climate epoch, the last 12,0000 years, known as the *Holocene*, is a time of relatively stable, warm temperatures when compared with most of the last 2.6 million years, which experienced intermittent glaciations and interglacials. Based on the deep-sea record, it appears that global climates tend to rest at one of two extremes: (1) a very cold interval characterized by major glaciers and lower sea levels and (2) shorter intervals between the glacial advances marked by warm temperatures and high sea levels. With the realization that global climates have changed dramatically numerous times, two obvious questions arise: What causes global climate to change, and how quickly can global climate change from one extreme to the other? Many of the answers to these questions are discerned by studying the climate changes that have occurred in the past.

Rates of Climate Change

During the peak of the last Pleistocene glaciation, freshwater lakes more than 150 meters (500 ft) deep covered many areas of comparatively low elevation in the American Intermountain West, maintained by the same general climatic trends that led to snow

● **FIGURE 8.22** Paleoclimates can be reconstructed using $^{18}O/^{16}O$ ratios from ice cores taken in Greenland and Antarctica.

Why have thousands of annual ice layers accumulated over time in Greenland and Antarctica?

and ice accumulation at higher latitudes and higher elevations. However, by about 9000 years ago, the United States (except for Alaska and the highest mountains of the West) was mostly glacier free and most of the western lake basins were dry. Abundant evidence exists that about 7000 years ago, the climate in the Northern Hemisphere (a time known as the **Altithermal**) was warmer than today (see ● Fig. 8.23). For massive glaciers to melt completely and for deep lakes to evaporate, a substantial change of climate is required over a few thousand years.

Answering questions about such rapid rates of climate change requires a more detailed record of climate than the deep-sea sediments can provide. This is because only a few centimeters of mud accumulates on the deep ocean floors in a thousand years. Rapid shifts in climate over periods of a few hundred years are not recorded clearly in the seafloor sediments. Studying the glacial ice cores of Antarctica and Greenland, where the ice is more than two miles thick, has solved this problem.

A surprising discovery of the ice sheet analyses is the relatively short time frame over which climate can change. Rather than changing gradually from glacial to interglacial conditions over

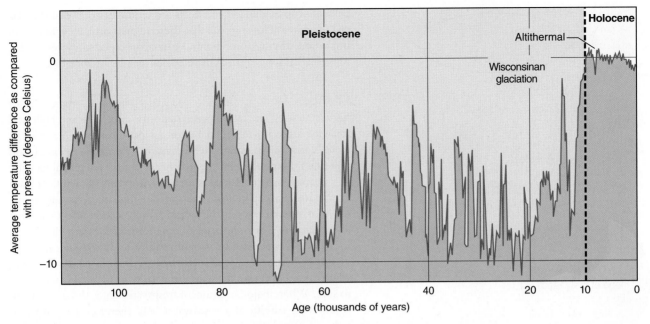

● **FIGURE 8.23** The oxygen isotope ratio record from ice cores were used to produce this graph of temperatures for the last 110,000 years. The record provides evidence of significant climate change over surprisingly short periods of time.

thousands of years, the ice record indicates that the shifts can occur in a few years or decades. Thus, whatever is most responsible for major climate changes can develop rapidly. This probably requires *positive feedback*, which means, as explained in Chapter 1, that a change in one variable will cause changes in other variables that magnify the amount of original change. For example, most glaciers have high albedos (reflective power), reflecting significant amounts of sunlight back to space. However, if the ice sheets retreat for whatever reason, low-albedo land begins to absorb more insolation, increasing the amount of energy available to melt the ice. Thus, the more ice that melts, the more energy is available to melt the ice further, magnifying the initial glacial retreat.

In contrast, a *negative feedback*, where changes in one of the variables induce the system to remain stable, can also affect the likelihood or rate of climate change. For example, increasing global temperatures cause evaporation rates to increase. When more water evaporates from the ocean, more clouds will form. With increased cloud cover, more insolation is reflected back to space from the cloud tops, cooling Earth's surface. A counterargument to this effect, however, is that clouds also operate as a greenhouse blanket, trapping heat in the lower atmosphere. Thus, for climate changes to occur rapidly, negative feed-back cycles such as this one must be overwhelmed for a period of time by positive feedback cycles.

Natural Causes of Climate Change

Ultimately, climate change occurs because of variations in three factors: the amount of solar energy entering the Earth system, the amount of energy the Earth system stores, and the amount of energy Earth loses to space. Theories about the causes of both past and future climate change are numerous, but common ones can be organized into five broad categories: (1) variations in Earth's

orbital relation to the sun; (2) changes in oceanic circulation; (3) changes in landmasses and ocean basins; (4) asteroid and comet impacts; and (5) changes in Earth's atmosphere.

Orbital Variations Astronomers have detected slow changes in Earth's orbit that affect the distance between the sun and Earth and that also occur in the tilt of Earth's axis relative to the plane of the ecliptic. These changes are cyclic and produce regular variations in the amount of solar energy that reaches Earth. Mathematician Milutin Milankovitch described these cycles and then showed how these changes in Earth's orbit and its axis could have an impact on climate. This is the **Milankovitch theory**, or **astronomical theory**, of climate variation. These orbital cycles produce regular variations in the amount of solar energy that reaches Earth that may favor either the Northern or Southern Hemisphere. The longest is the **eccentricity cycle**, a 100,000-year modification of the shape of Earth's orbit around the sun, which shifts from an elliptical (slightly oval), to a more circular orbit, and back (● Fig. 8.24a). These orbital variations affect relationships between Earth and the sun. The eccentricity cycle's impact also depends on the influence of the other two factors in this theory. In general, the 100,000-year eccentricity cycle seems to fit well with the record of glaciations and interglacials during the Pleistocene. Orbits that are more elliptical seem to be associated with warm periods, and more circular orbits may correspond to Ice Ages.

A second cyclic change, the **obliquity cycle**, represents a 41,000-year variation in the tilt of Earth's axis from a maximum 24.5° to a minimum of 22.0° and back (Fig. 8.24b). When Earth's axis has a greater tilt, seasonal differences are greater in the middle and high latitudes. Less tilt should bring cooler summers to the polar regions and less melting of ice sheets, which may promote Ice Age conditions.

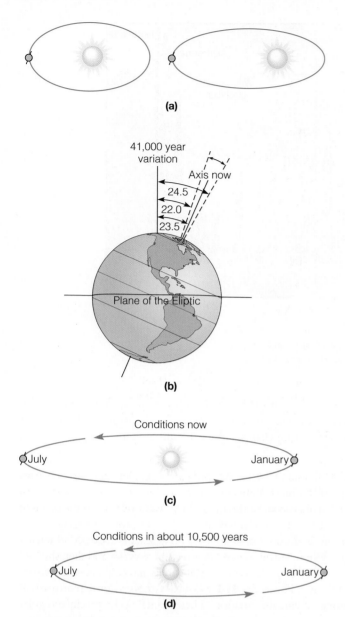

FIGURE 8.24 Milankovitch calculated the periodicity for (a) eccentricity, (b) obliquity, and (c and d) precession.

What effect should changes in these factors have on global climates?

Finally, a **precession cycle** with a periodicity of 21,000 years determines the time of year when Earth is closest to the sun (perihelion). Currently, Earth is closest to the sun in early January and, as a result, January receives about 3.5% greater than average insolation. After the time of year when Earth is farthest from the sun (aphelion, currently in July) shifts to January in about 10,500 years, the Northern Hemisphere winters should be somewhat colder (Fig. 8.24c and d).

These three cycles operate collectively, and Milankovitch's theory of their combined effects indicated that numerous glacial cycles should occur during 1 million-year intervals. Most scientists who study paleoclimates (ancient climates) are convinced that a good correlation exists between Milankovitch's predictions and paleoclimate evidence from deep-sea sediment cores and the glacial ice cores. This suggests that these regular orbital cycles are an important driving force behind glacial-interglacial cycles, and it suggests that long-term climate cycles may be predictable.

Changes in the Ocean Oceans cover more than 70% of Earth's surface. The global ocean's enormous volume and high heat capacity make it the largest buffer against changes in Earth's climate. Variations in oceanic temperatures, chemistry, or circulation typically reflect and reinforce significant changes in global climate.

Surface oceanic currents are driven mostly by winds. However, currents deep beneath the surface circulate tremendous volumes of water between the ocean basins, which affect the heat distribution on Earth. A major driving force of the deep circulation in the ocean appears to be variations in water buoyancy caused by salinity differences. Where surface evaporation is rapid, the rising salinity content causes the seawater density to increase, inducing it to sink. When major amounts of freshwater enter the ocean from landmasses, salinity of the seawater, and therefore its density, is reduced, thereby slowing its circulation to deep water. An influx of freshwater to the ocean is generally followed by a major flow of warm surface currents. This is especially important to the North Atlantic, as it contributes to warming in the Atlantic region of the Northern Hemisphere. Ocean currents are also affected by water temperature. Extremely cold Arctic and Antarctic waters are dense and tend to sink, whereas tropical water is warmer and may tend to flow near the surface. Salinity and temperature taken together bring about complex subsurface flows deep within the ocean basins that influence Earth's surface climate.

Short-term changes in oceanic circulation in the Pacific are primarily responsible for El Niño and La Niña events (discussed in Chapter 5), which are known to be associated with climatic deviations in many regions. Fluctuations in ocean water circulation may also have contributed to rapid climate change during the last 2.6 million years.

Changes in Landmasses Alterations in configurations of Earth's landmasses have also been used to explain climatic cooling and warming, and are directly related to changes in the ocean. A number of Ice Ages, some with multiple glacial advances, have occurred during Earth's history, and scientists have looked for geologic and geographical commonalities among those intervals of Earth history. For example, one characteristic that earlier glacial periods have in common with the Pleistocene is the presence of a continent in polar latitudes. Because of their rapid loss of heat, the presence of extensive landmasses at high and polar latitudes encourages the accumulation of snow and ice, and ultimately can lead to the development of massive ice sheet glaciers. As greater volumes of water become stored as ice of in the huge glacial ice sheets, sea level drops. Because the amount of water in the Earth system is fixed, an increase in the glaciers on Earth and of the volume of glacial ice that they store, the lower sea level will be, and vice versa.

Another geologic factor influence on climate is the formation, disappearance, or movement of a landmass that alters oceanic circulation, atmospheric circulation, or the composition of the

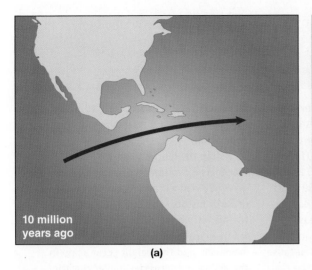 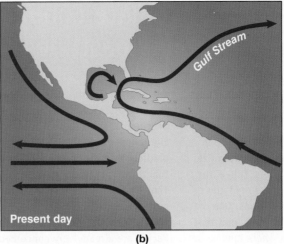

(a)   **(b)**

● **FIGURE 8.25** Volcanic eruptions constructed the Isthmus of Panama between about 10 million and 4 million years ago, blocking a connection between the Atlantic and Pacific Ocean basins and resulting in a major reorientation of ocean currents. This event led to the circular ocean current patterns that exist today in both oceans and had a great impact on climate.

How do ocean currents along east and west coasts outside the tropics affect climate?

atmosphere. Volcanic eruptions and the formation of the Isthmus of Panama blocked a connection between the Atlantic and Pacific Oceans, greatly changing ocean circulation (● Fig. 8.25). The ensuing redirection of ocean water created the Gulf Stream current. Uplift of the Himalayas altered atmospheric winds and the Asian monsoon. Creation of the Isthmus of Panama, the Himalayan uplift, and other significant changes in landmasses immediately predate the onset of the Pleistocene series of glaciations.

Another type of surface alteration involves changes in albedo. A significant increase in the spatial distribution of snow cover and the growth of ice sheets greatly increases reflectivity from the surface. A positive feedback situation is established in which cooling temperatures allow greater ice and snow cover, increasing the albedo, which leads to further cooling and accumulation of more frozen water. This positive feedback cycle may end when the polar oceans freeze, triggering negative feedback by shutting off the primary moisture source for the ice sheets, which then will begin to shrink and retreat.

Impact Events Rocky or metallic solar system bodies, called **asteroids**, usually less than 800 kilometers (500 mi) in diameter, may break into smaller pieces termed **meteoroids**. These objects orbit the sun, along with comets, which are small objects of rocky or iron materials held together by ice. Through time, some of this material has struck Earth, occasionally with devastating impact.

Most meteoroids are so small that they burn up in the atmosphere before hitting the ground. Observed at night, they are popularly known as *shooting stars*. Objects smaller than 40 meters (130 ft) in diameter incinerate because of friction with the atmosphere. Objects ranging from 40 meters to about 1 kilometer (0.6 mi) in diameter can do tremendous damage if they reach Earth's surface (● Fig. 8.26). On average, an object of this size strikes Earth approximately every 100 years. The last impact of this size occurred in 1908 near Tunguska, Siberia, and devastated

● **FIGURE 8.26** Barringer Crater, Arizona, shows the results of an impact with an iron-nickel meteorite of about 50 meters (165 ft) in diameter about 50,000 years ago. Can you imagine the damage that would occur if a similar meteorite impacted in an urban area?

⊕ **In Google Earth fly to: 35°1′38″N 111°1′21″W (Barringer Crater)**

Climate Change and Its Impact on Coastlines

When we look at a map or a globe, one aspect of the geographic information that we see is so obvious and basic that we often take it for granted, perhaps failing to recognize that it is spatial information. This is the location of coastlines—the boundaries between land and ocean regions. One of the most basic aspects of our planet that a world map shows us is where landmasses exist, where the oceans are located, and the generally familiar shape of these major Earth features. But maps of the world today only show where the coastline is currently located. Since 1880, sea level has risen about 20 centimeters (8 in.), after being at about the same level for the previous 2000 years. The total amount of water (as a gas, liquid, and solid) on our planet is fixed. When more of Earth's water is stored on the land in glaciers, there will be less water in the oceans and vice versa. As world climates experience a warming tendency, glaciers shrink and sea level rises. If global warming trends continue at their present rate, the U.S. Environmental Protection Agency (EPA) predicts that sea level could rise 31 centimeters (1 ft) in the next 25 to 50 years. That amount of sea level rise will cause problems for low-lying coastal areas. Populations of some coral islands in the Pacific are already very concerned because their homelands are barely above the high-tide level.

A map of the world's population distribution shows a strong link between settlement density and coastal areas. For low-lying coastal regions, sea-level rise is a major concern, and on coasts with a gentle slope, the inundation would be farther inland with every increment of sea-level rise. Scientists at the United States Geological Survey (USGS) have estimated that if all the glaciers on Earth were to melt, sea level would rise 80 meters (262 ft) and that a 10-meter (33 ft) rise would displace 25% of the U.S. population.

Before the glacial times of the Pleistocene, worldwide climates were generally warmer; and throughout most of Earth history, no glaciers existed on the planet. During times when Earth was ice free, sea level would have been at a maximum, and that might occur again in the distant future under similar climatic conditions. When Pleistocene glaciers were at their maximum, sea level dropped to about 130 meters (426 ft) below today's level. Maps that illustrate the positions of coastlines and the shape of continents during times of major environmental change show how temporary and vulnerable coastal areas can be. It may seem odd to think of the coastlines as being in temporary locations, but because coasts can shift over time, a future world map could look quite different from the map we know today.

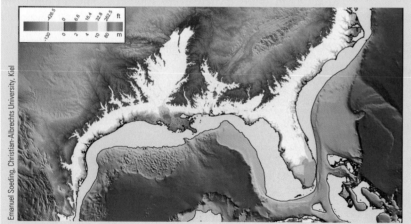

Emanuel Soeding, Christian-Albrechts University, Kiel

This map shows how changes in sea level can affect the coastline in the southeastern United States. Orange with a dark line shows our coastlines today. The green to yellow boundary shows where the coastline would be if all glaciers on Earth were to melt, as the ocean would rise to elevations on the land that are 80 meters (262 ft) above the current sea level. The tan color shows land, submerged today, that was exposed when glaciers were at their maximum extent during the Pleistocene, and the position of the shoreline at that time. Sea level had dropped 130 meters (426 ft) from where it is today.

What are some of the impacts to coastal areas of a return to a glacial maximum occurred, or if all glacial ice on Earth melted?

a huge area in the Siberian wilderness, with a blast estimated at 15 megatons (equivalent to 15 million tons of TNT). In February 2013, a large meteor traveling about 20 km per second exploded into pieces over Chelyabinsk in central Russia, with a brilliant flash, a fireball, and a tremendous shock wave from the explosive blast. Its size was smaller, estimated at 17 m (56 ft) in diameter, weighing about 10,000 tons, and the blast was the equivalent of 0.5 megaton. Buildings were damaged, and more than 1000 people were injured, many by flying glass fragments from windows that were shattered by the shock wave. These meteor strikes, as devastating as they were, did not have an appreciable effect on climate because they were not large enough.

Approximately every few hundred thousand years, an object greater than 1.6 kilometers (1 mi) in diameter has struck Earth, producing severe environmental damage and climate change on a global scale. The power of blasts from such impacts could equal a million megatons of energy. One likely effect would be an "impact winter," in which dust ejected into the atmosphere by the impact blocks insolation and causes a drastic drop in temperature. Firestorms would result from heated debris raining back down on Earth in the vicinity of the impact site. A catastrophe of this kind would result in loss of crops and starvation worldwide. The largest known meteorite impacts in Earth's history, larger than the one that may have contributed to the extinction of the dinosaurs 65 million years ago, are estimated to be about 15 kilometers (9.3 mi) in diameter, and may have exploded with a force of 100 million megatons.

A vigilant group of professional and amateur astronomers are constantly watching the night skies for **Near Earth Objects (NEOs)** or objects whose trajectories may bring them into a collision course with our planet. If we know early enough in advance that an NEO could collide with Earth, we might be able to avert a disaster by modifying the object's course to avoid a collision. At this point, however, whether anything can be done about such a collision remains speculative and a topic of science fiction.

Volcanic Activity The climatic cooling effect of volcanic activity is unquestioned; all of the coldest years on record over the past two centuries have occurred a year or so after a major eruption (● Fig. 8.27). Following the massive eruption of

Tambora (in Indonesia) in 1815, 1816 was known as "the year without a summer." Killing frosts in July ruined crops in New England and Europe, resulting in famines. Several decades later, following the massive eruption of Krakatoa (also in Indonesia) in 1883, temperatures decreased significantly during 1884. Although no eruptions have approached the magnitude of these two since the 1800s, the 1991 eruption of Mount Pinatubo (in the Philippine Islands) produced cool conditions, which continued for about 4 years in an otherwise continuous series of record warm years (● Fig. 8.28).

Atmospheric Conditions Many theories attribute climate change to variations in atmospheric constituents—some of which enter the atmosphere through natural processes and others that are a product of human activities. A characteristic closely correlated with average global temperatures is the atmosphere's composition

● **FIGURE 8.28** In 1991, the massive eruptions of Mount Pinatubo in the Philippines expelled gases and particulates 35 km (22 mi) into the atmosphere. Satellites tracked the cloud of volcanic emissions as it traveled around the world several times. Dust in the atmosphere from this eruption did cause some cooling for two to three years afterward.

In addition to affecting the climate, what other hazards result from volcanic explosions?

⊕ In Google Earth fly to: 15.145°N, 120.349°E
(Mt. Pinatubo, Philippines). Examine the
landform from multiple angles.

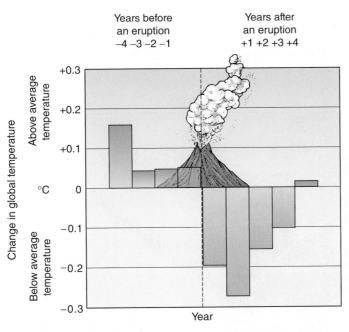

● **FIGURE 8.27** Examination of global temperatures within 4 years before and after major volcanic eruptions provides compelling evidence that volcanic activity can have a direct effect on the amounts of insolation reaching Earth's surface.

At what period after an eruption year does the effect seem the greatest?

of gases, aerosols, and particulates. Solid particles, such as dust as well as volcanic emissions, block solar radiation, and large amounts in the atmosphere tend to cause cooling.

Scientists have known for many years that carbon dioxide (CO_2) acts as a **greenhouse gas**. Atmospheric CO_2 is transparent to incoming shortwave radiation, yet it impedes outgoing long-wave radiation. Thus, as the atmospheric content of greenhouse gases rises, so will the amount of heat trapped in the lower atmosphere, some of which is returned back to the surface by radiation of thermal infrared energy.

Air bubbles containing minor samples of the atmosphere that existed at the time that the ice formed are captured in the glacial ice of Antarctica and Greenland. One of the important discoveries of the ice-core projects is that prehistoric atmospheric CO_2 levels increased during interglacial periods and decreased during major glacial advances.

The fact that average global temperatures and CO_2 levels are so closely correlated suggests that Earth will experience record warmth as the atmospheric level of CO_2 increases. The present level of approximately 400 parts per million of CO_2 is already higher than at any time in the past million years, and has been steadily rising for more than the last 50 years (● Fig. 8.29).

Carbon dioxide is not the only greenhouse gas. Methane (CH_4) is more than 20 times more effective than CO_2 as a greenhouse gas but is considered less important because atmospheric concentrations of methane are lower. Also, the time that molecules of methane remain in the atmosphere (residence time) is much shorter. Garbage dump emissions and termite mounds both produce substantial quantities of CH_4. But a much more important source of atmospheric methane may come from the tundra regions or the deep sea. If tundra areas or the ocean release large amounts of methane as theorized, the reinforcement of warming could be enormous. Other greenhouse gases include CFCs (chlorofluorocarbons) and N_2O (nitrous oxide). The relative greenhouse contribution of common gases and their average residence times in the atmosphere are presented in ● Figure 8.30.

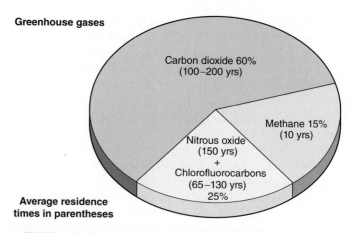

Greenhouse gases

Carbon dioxide 60%
(100–200 yrs)

Methane 15%
(10 yrs)

Nitrous oxide
(150 yrs)
+
Chlorofluorocarbons
(65–130 yrs)
25%

Average residence times in parentheses

● **FIGURE 8.30** Gases other than carbon dioxide released to the atmosphere by human activity contribute approximately 40% to the greenhouse effect. The figures in parentheses indicate the average number of years that the different gases remain in the atmosphere and contribute to temperature change.

Which gas has the longest residence time?

Recent Climate Change

Earth's climate system is complex, and made more so by human activities that can contribute to climate change; thus, it is difficult to make reliable forecasts of future climates. The world's climates are a function of both human and natural factors that interact, and can cause variability in climatic conditions. The frequency and magnitudes of climate changes that have occurred naturally over the past 150,000 years is shown in ● Figure 8.31. Although the Holocene has been the most stable interval during period of time, an examination of the Holocene record reveals a range of climates. For example, a climatic interval, warmer than the climate of today, occurred during the *Altithermal*, although the warm climate conditions that occurred at that time mainly affected the Northern Hemisphere. This interval was characterized by the dominance of grasslands in the Sahara and severe droughts on the Great Plains.

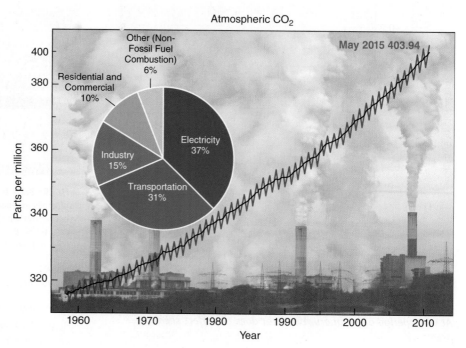

Atmospheric CO_2

Other (Non-Fossil Fuel Combustion) 6%

Residential and Commercial 10%

Electricity 37%

Industry 15%

Transportation 31%

May 2015 403.94

● **FIGURE 8.29** The steady increase since 1958 in the concentration of carbon dioxide in the atmosphere is apparent on this graph. Power generation, heavy industry, motor vehicles, and burning of forests are major contributors of CO_2 to the atmosphere.

What can you do personally to help reduce the amount of greenhouse gases going into the atmosphere?

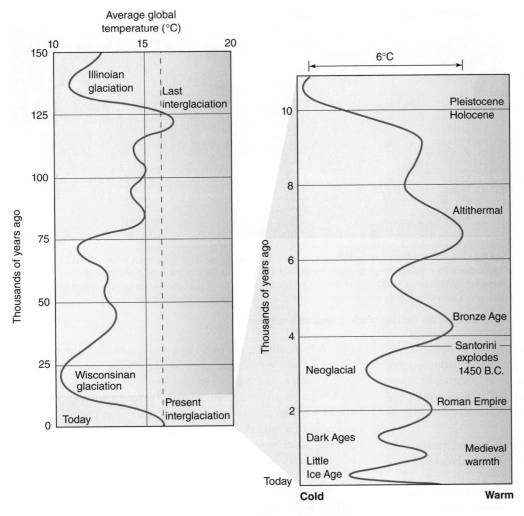

● **FIGURE 8.31** This figure shows the broad climate trends of the past 150,000 years, with significant details for the Holocene. Climatologists have been remarkably successful in dating recent climate change, but forecasting future climates remains difficult.

Why is this so?

Other, shorter warm intervals occurred during the Bronze Age, during the second half of the Roman Empire, and in medieval times. An unusually cold interval began with the massive eruption of Santorini in the Aegean Sea (a volcano that was the site of a civilization that some believe was the basis for the Atlantis myth).

Cold periods occurred during the early Middle Ages and again beginning about 1150 to 1460 in the North Atlantic region and 1560 to 1880 in continental Europe and North America. These last episodes collectively have been termed the **Little Ice Age**. The Little Ice Age had major impacts on civilizations—from the isolation of the Greenland settlements established during the medieval warm period to the abandonment of the Colorado Plateau region by the Anasazi cultures. An important point to remember is that, with the exception of the cold interval that began with the eruption of Santorini, climatologists do not know what variables changed to cause these fluctuations in climate. Furthermore, the climatic shifts of the last 9,000 years that are shown on the graph, are today thought to have mainly, and perhaps only, affected the Northern Hemisphere, particularly Europe, and were not global changes in climate.

Climate scientists are working to simulate and map the variables that affect climate, assessing their current and their future impacts on climate. **Global Circulation Models (GCMs)** are sophisticated computer simulations of climatic conditions and trends based on the relationships among multiple variables that can affect Earth's climate system—sun angles, temperature, evaporation rates, effects of land versus water, energy transfers, and so on. Some of the variables, and relationships between them, are at best difficult to model. The complexity and the usefulness of GCMs, however, are both increasing rapidly. Although improvement on GCMs continues, they are not infallible. However, these circulation models appear to do a good job in forecasting how conditions should change in specific regions as the Earth warms or cools, and they have added new insights into how some climatic variables interact. Simulations from different GCMs often vary and they can provide only painstakingly documented estimates of future climates, and the forecasting is most reliable when dealing with the next few years or decades, as opposed to far in the future.

Efforts to forecast future climates are complicated further by the operation of many feedback cycles. Simply increasing the amount of heat that is trapped by gases in the lower atmosphere may or may not result in long-term warming. Negative feedback processes such as increased cloud formation or major volcanic eruptions, and increased plant uptake of CO_2 may operate to counteract the warming.

Based on the record of climate changes in the past, one thing can be certain about the climate of the future—it will change. Looking into the distant future, the Milankovitch cycles suggest that another glacial period could be on the way, with rapid cooling between 3000 and 7000 years from now. In the near term, however, continued global warming is a virtual certainty. The increasing release of greenhouse gases such as carbon dioxide and methane into the atmosphere, and the widespread destruction of vegetation are bound to increase the average global temperature for the foreseeable future. An average global temperature increase of 1°C (nearly 2°F) would be equivalent to the change that has occurred since the end of the Little Ice Age in about 1880. A 2°C warming would be greater than anything that has happened in the Holocene, including the Altithermal. A 3°C warming would exceed anything that has happened in the past million years. Current estimates and the most reliable GCMs predict a 1°C to 3.5°C (2°F to 6°F) warming in the twenty-first century.

It is clear that not all areas of the world will be affected equally. One of the most important effects is expected to be a more vigorous hydrologic cycle, fueled largely by increases in evaporation from the ocean. Intense rainfalls will be more likely in many regions, as will droughts in other regions such as the Great Plains of North America.

Some scientists link the rising temperatures with recent increases in the frequency and severity of damaging thunderstorms, disastrous floods, tornadoes, and hurricanes. Temperatures will increase the most in polar and high-latitude regions, mainly during the winter months. As a result of the warming, sea levels will continue to rise, mostly because of melting ice sheets. By 2100, sea levels are estimated to be between 15 and 95 centimeters (0.5 to 3.1 ft) higher than today. In addition, the ranges of tropical diseases may expand into higher latitudes, tree lines will be higher, and glaciers will continue to shrink or disappear in alpine and polar regions (● Fig. 8.32). Warming of the oceans and tundra regions may release additional greenhouse gases and accelerate global warming.

However, the effects of global warming may be beneficial to some regions because of geographic differences in climate. Growing seasons in the high latitudes should increase in length. The increase in atmospheric CO_2 may help some crops grow larger or faster. In the United States, a 1°C increase in average temperature could decrease heating bills by about 11%; however, air conditioning costs would increase. The benefits and the detriments of a changing climate are still being studied. One thing is predictable, however, any changes from the status quo will be both challenging and expensive, as societies attempt to adapt to new atmospheric conditions and related environmental changes.

Anthropogenic Influences on Climate

Most Earth scientists today are convinced that the current and unusual increase in the rate of rising global temperatures can be directly linked to human activities. They identify expanding industrialization and growing emissions of greenhouse gases in both developed and developing nations as the major cause. There is ample evidence of this temperature trend. The impacts of human activities on environmental and climate changes are referred to as **anthropogenic change** (human-induced).

Incredibly, 13 of the 15 hottest years recorded have occurred since 2000, with 2014 being the hottest year ever recorded (global temperature). Average annual global temperatures have risen between 0.3°C and 0.7°C (0.5°F and 1.24°F) and sea level has risen between 12 and 23 centimeters

● **FIGURE 8.32** With few exceptions, mountain glaciers worldwide are retreating. This glacier in Alaska's Glacier Bay National Park has continued to shrink for the last 200 years. In fact, in the last 70 years the glacier has retreated by 610 meters (2000 ft) and thinned by 244 meters (800 ft). Note the line on the valley wall just to the left, which marks the glacier extent thickness, and the height where it was in the 1940s.

In what ways do melting glaciers become a problem?

(4.8 to 9 in.) during the past 100 years. Given the long residence times of many greenhouse gases (see Fig. 8.30), the heat capacity of the oceans, and current levels of human consumption, it will be difficult to prevent global warming from continuing. Also, scientists are concerned that strong evidence from multiple sources indicates that human activities that influence global warming are playing an increasing role in recent climate change, driven by industrialization, global population growth, and the massive use of natural resources.

Since the last major Ice Age ended 12,000 years or so ago, and the lesser "Little Ice Age," which ended about a century ago, the general trend of climate has been warming (with some fluctuations into colder intervals). There is little controversy about that trend, and temperatures correlate well with increases in atmospheric carbon dioxide that have occurred since the industrial revolution (● Fig. 8.33). According to extensive research and data sets produced by climate researchers worldwide, the most recent decades show global temperatures warming at an accelerated rate. A controversy comes about when trying to establish to what degree each

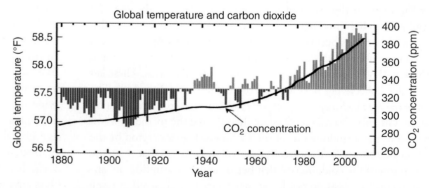

● **FIGURE 8.33** Global average annual temperatures measured over the land and the oceans from 1880 to 2009 are shown with bars that indicate temperatures above *(red)* and below *(blue)* the average global temperature for this time period. The *black line* indicates atmospheric CO_2 over time. Despite annual temperature fluctuations, this graph shows the close correspondence between the rising trends in temperature and in atmospheric CO_2.

of the major causes of global warming is contributing to rising temperatures. What are the relationships between natural processes and human activities in this upward trend of global temperatures? Any controversy about global warming is not about whether human activities have an impact on the atmosphere, the climate, and most other aspects of the physical environment. It is about agreeing to what extent human activities are responsible for global warming and to what extent humans can slow the trend of rising temperatures. A few scientists (and many more nonscientists) feel that it is incorrect to assume that human activities are responsible for global warming. They believe that natural processes drive all major climate changes.

In a truly international effort to understand global warming, including its influencing factors and its current and future impacts, the United Nations and the Intergovernmental Panel on Climate Change (IPCC) have cooperated to gather as much relevant information and data as possible. The IPCC is a worldwide group of distinguished atmospheric scientists. In 2007, after many years of research and study, the IPCC released a series of comprehensive reports on global warming that involved input from more than 800 climate scientists from 130 countries. This has been followed up with other reports to keep current with new information and data, including several in 2014. These scientists studied multiple lines of evidence worldwide, from tree-ring and ice-core data, to glacial retreat and sea-level rise, to changes in the atmosphere, to changes in weather phenomena,

and they strongly considered the many potential influences of both natural processes and human activities. The conclusion of the IPCC was that it is "very likely" (90% probability) that emissions of greenhouse gases from *anthropogenic* activities have caused "most of the observed increase in globally-averaged temperatures since the mid-twentieth century." They also state that in the past 50 years, without human activities the influence of Earth–sun relationships and volcanic activity would "likely" have caused a cooling trend. The results of computer models generated by the IPCC show how observed temperatures have increased in the past 100 years, compared to the predicted impact of natural influences alone and compared to a combination of human and natural factors. The best fit is the one that includes human influence in global warming (● Fig. 8.34).

The IPCC has summarized its findings: "Today, the time for doubt has passed. The IPCC has unequivocally affirmed the warming of our climate system, and linked it directly to human activities." The IPCC goes on to state that dealing with the environmental changes associated with global warming and working to minimize human impacts on climate change is an immediate concern worldwide that must increase in the coming years. Enacting an appropriate course of action based on these findings is complicated. The impact of global warming and of all major climate change will always vary among different geographic locations and climate regions. With the wide variety of environments on Earth, some geographic regions would benefit from a warmer climate and other areas will bear significant negative impacts (for example, the world's heavily populated coastal regions as sea level rises). There is also evidence that links rising temperatures with recent increases in the frequency and severity of damaging thunderstorms, disastrous floods, powerful tornadoes, hurricanes, droughts, and wildfires, many of which have been unusual in severity and impact. These traumatic events have caused evermore property damage and loss of life as global temperatures

● **FIGURE 8.34** Using the best computer models available to evaluate global climate change, the Intergovernmental Panel on Climate Change has found that only the models that include increasing human releases of greenhouse gases *(pink areas)* fit the temperature trends that have been observed in the last century *(black lines)*. The blue tones estimate the ranges of what the temperatures would be without human impacts on global warming.

On what continent has the observed temperature fluctuated the most during this time period, and which one the least?

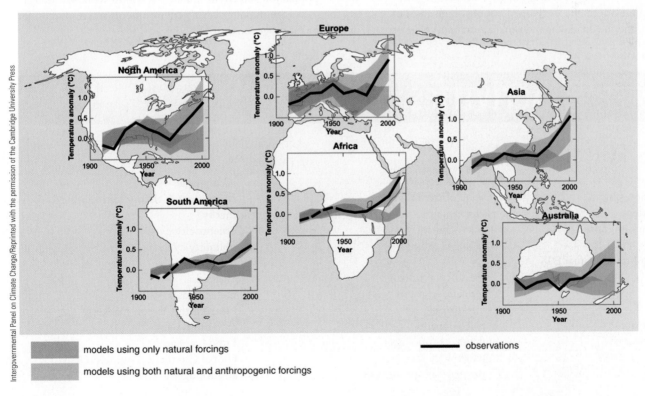

have continued to increase. We are not able to adjust our global atmosphere as easily as we can set a thermostat in our homes. The question then becomes, How should we approach this issue?

Recommendations for the Future

There is little doubt that rising levels of CO_2 and other greenhouse gases in the atmosphere are a factor in global warming and a direct product of modern industrialized economies. The rather unrestricted pollution of the atmosphere is unfortunately just one of the many ways that people's activities are negatively affecting Earth's physical environments. As cities expand, arable land gives way to highways and urbanization. As demands on fresh water supplies for agricultural, industrial, and residential uses continue to grow, serious water shortages appear. As human numbers increase and climates change in marginal areas, more plant and animal species are threatened or become extinct. The issue of global warming should be considered as part of the problem of expanding human populations. Earth's physical environments must be protected and preserved for people today and sustained for the generations of tomorrow. Much can and must be done by the nations and peoples of the world.

On a Global Scale To the extent possible, the nations of the world should devote serious research efforts and monetary resources, to the following:

1. *Developing alternative sources of energy.* In addition to the effects of burning fossil fuels on global temperatures, it also pollutes the air we breathe, making it dangerous to human health. Energy from solar radiation, wind, geothermal heat, ocean tides, biofuels, hydroelectric generation, and even nuclear reactors helps to keep the atmosphere cleaner for generations to come.
2. *Curtailing, or better managing, our energy usage.* With Earth's growing human population, the developed nations' high rates of consumption, and the increasing industrialization in developing

nations, energy demands are increasing and will continue to grow in the foreseeable future. How much energy we use and how we can conserve energy must be major considerations.

3. *Recycling our waste materials.* Currently, human populations are consuming our natural resources at rates that cannot be sustained. Recycled resources will ease the strain on those that are vanishing at such a rapid rate and will save the energy needed to create new resources.
4. *Curtailing deforestation.* This very destructive process should be restricted everywhere on Earth. Forest vegetation is a primary agent in the removal of CO_2 from our atmosphere through the process of photosynthesis.
5. *Desalinizing ocean water (removing salt from seawater).* Making this process easier and less costly should be a major research effort all over the world. Arid climates create deserts, but irrigation can turn desert regions into productive lands.

On a Personal Scale We, as individuals living on Earth, can do simple things every day to help reduce our negative impact on Earth's fragile environment. The following are just a few suggestions: (1) Use carpools and mass transportation; drive smaller cars; drive less often and at reduced speeds. (2) Use more energy-efficient lighting and appliances and turn them off when not in use. (3) Set thermostats to use less energy for air conditioning in summer and heating in the winter. (4) Recycle materials (metals, glass, plastic, paper, and others) as often as possible.

One of the few things all humans have in common, regardless of age, sex, race, religion, or nationality, is that we occupy and share Earth together. It is our responsibility to care for the planet that sustains us. If global warming continues at today's rate, or at that which well-respected groups of scientists forecast, the impacts of this climate change will accelerate as well as its impacts on the world's population and environments. We must work toward caring for and sustaining our planet's environments and resources, for us and for future generations.

CHAPTER 8 ACTIVITIES

■ TERMS FOR REVIEW

climatology	climate region	Altithermal
climate classification system	zone of transition	Milankovitch theory (astronomical
Thornthwaite system	climograph	theory)
potential evapotranspiration	microclimate	eccentricity cycle
(potential ET)	global warming	obliquity cycle
actual evapotranspiration (actual ET)	Ice Ages	precession cycle
Köppen system	Pleistocene Epoch	asteroids
tropical climate	glaciation (glacial period)	meteoroids
polar climate	interglacial	Near Earth Objects (NEOs)
microthermal climate	paleoclimates	greenhouse gas
mesothermal climate	dendrochronology	Little Ice Age
arid climate	palynology	Global Circulation models (GCMs)
highland climate	oxygen-isotope analysis	anthropogenic change

QUESTIONS FOR REVIEW

1. Why is it important to study the nature and possible causes of past climates when attempting to forecast future climate change?
2. Why are temperature and precipitation most widely used as the sources of statistics for climate classification? How are these two elements used in the Köppen system to identify six major climate categories?
3. What are the advantages and disadvantages of the Köppen system for geographers? Why are the Köppen climate boundaries often referred to as "vegetation lines"?
4. What is a climograph? What seasonal and annual patterns of climate does it reflect?
5. How does the Thornthwaite system of climate classification differ from the Köppen system? What are the advantages of the Thornthwaite system?
6. Why is the occurrence, frequency, and dating of glacial advances and retreats so important to the study of past climates? How has modern research changed earlier theories of glacial coverage and associated climate change during the Pleistocene?
7. How have scientists been able to document the rapid shifts of climates that have occurred during the latter part of the Pleistocene?
8. What are the major causes of global climate change? What contribution did the mathematician Milankovitch make to theories regarding glaciation?
9. What is the evidence that volcanic activity can affect global temperatures? How does this occur?
10. How might changes in Earth's oceans and landmasses affect global climates?
11. What effects can changes in the amounts of CO_2 and other greenhouse gases in the atmosphere have on global temperatures? How can past changes in amounts of CO_2 be determined?
12. What is the primary difficulty for any climatologist who attempts to predict future climates?
13. What changes are likely to occur in Earth's major subsystems if global warming continues for the near term, as most scientists believe?

CONSIDER AND RESPOND

1. Study the definitions for the individual Köppen climate types described in the "Graph Interpretation" exercise. Why do you think Köppen and later climatologists who modified the system selected the particular temperature and precipitation parameters that separate the individual types from one another?
2. Examine the climograph in the "Using Climographs" box earlier in the chapter. During what month does Nashville, Tennessee, experience the greatest precipitation? What major change would immediately identify this graph as representing a Southern Hemisphere location?
3. Review Figures 8.23 and 8.31. State in your own words the general conclusions you would draw from a study of these two figures.
4. After studying Chapter 7 and this chapter, which do you believe is more important to you now and in the future—the subject of weather or of climate? Defend your answer.

PRACTICAL APPLICATIONS

1. Using the climograph for Nashville, Tennessee, found in the "Using Climographs" box, find an average temperature reading for every month of the year on the line graph. Then calculate the mean temperature for the year and the annual temperature range. How close are the data you calculated compared to those given at the top of the climograph?
2. Using the precipitation data presented in the bar graph on the Nashville climograph, derive a precipitation value for every month of the year, and then calculate an annual average precipitation value and an annual precipitation range. How do your calculations compare with those printed on the climograph? What can you say about the distribution of Nashville's precipitation through the year?

LOCATE AND EXPLORE

1. Using Google Earth, fly to Lake Chad (13.42°N, 14.01°E). When you arrive at your coordinates, zoom out to view the extent of the lake. Lake Chad was once the largest lake in Africa, but ongoing drought has significantly reduced the lake in area. Because the lake is shallow, small changes in the discharge of the Chari River lead to large changes in lake area. Assuming that the lake can be characterized as rectangular (area = length × width), what has been the change in area (in square miles and as a percentage) from the original lake boundary to the lake today?

   Tip: Use the ruler tool to measure the width and length.

 MindTap—Make the most of your study time by accessing everything you need to succeed in one place. Read your textbook, take notes, review flashcards, watch videos, complete activities, take practice quizzes, and more online with MindTap. Log in at **www.cengagebrain.com**.

Graph Interpretation

THE KÖPPEN CLIMATE CLASSIFICATION SYSTEM

The key to understanding any classification system is to practice using the system. This is one reason why the review sections in Chapters 9 and 10 will provide climate data from sites around the world for you to classify, and to help you learn about these climate types. Correctly classifying these locations using the modified Köppen system may seem a bit complicated at first, but you will find that after determining the climates of a few locations and assigning the correct letter symbols it should become easy.

Before you begin, take the time to familiarize yourself with Table 1. Note that there are precise definitions of temperature or precipitation that define each of the five major Köppen climate categories (A, tropical; B, arid; C, mesothermal; D, microthermal; E, polar). Additional letters that are needed to identify the climate type more specifically, also have precise value relationships that can be determined by using the provided graphs. Table 1 in this section is all you need to classify a site if the monthly and annual precipitation and temperature averages are available. Table 1 should first be used to determine the major climate category. Once that is established, continue with the second letter symbol, and on to a third letter if it is needed to complete the classification.

TABLE 1
Simplified Köppen Classification of Climates

First Letter	Second Letter	Third Letter
E Warmest month less than 10°C (50°F) POLAR CLIMATES *ET*–Tundra *EF*–Ice Sheet	*T* Warmest month between 10°C (50°F) and 0°C (32°F) *F* Warmest month below 0°C (32°F)	NO THIRD LETTER (with polar climates) SUMMERLESS
B Arid or semiarid climates ARID CLIMATES *BS*–Steppe *BW*–Desert	*S* Semiarid climate (see Graph 1) *W* Arid climate (see Graph 1)	*h* Mean annual temperature greater than 18°C (64.4°F) *k* Mean annual temperature less than 18°C (64.4°F)

Graph 1 Humid/Dry Climate Boundaries

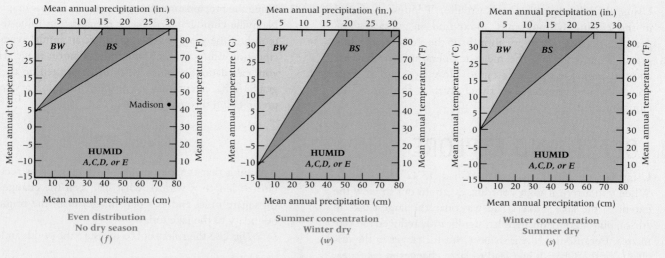

TABLE 1
Simplified Köppen Classification of Climates (continued)

First Letter	Second Letter	Third Letter
A Coolest month greater than 18°C (64.4°F) TROPICAL CLIMATES *Am*–Tropical monsoon *Aw*–Tropical savanna *Af*–Tropical rainforest	*f* Driest month has at least 6 cm (2.4 in.) of precipitation *m* Seasonally, excessively moist (see Graph 2) *w* Dry winter, wet summer (see Graph 2)	NO THIRD LETTER (with tropical climates) WINTERLESS
C Coldest month between 18°C (64.4°F) and 0°C (32°F) at least 1 month over 10°C (50°F) MESOTHERMAL CLIMATES *Csa, Csb*—Mediterranean *Cfa, Cwa*—Humid subtropical *Cfb, Cfc*—Marine west coast	*s* (DRY SUMMER) Driest month in the summer half of the year, with less than 3 cm (1.2 in.) of precipitation and less than one third of the wettest winter month	*a* Warmest month above 22°C (71.6°F) *b* Warmest month below 22°C (71.6°F), with at least 4 months above 10°C (50°F)
D Coldest month less than 0°C (32°F); at least 1 month over 10°C (50°F) MICROTHERMAL CLIMATES *Dfa, Dwa*–Humid continental, hot summer *Dfb, Dwb*–Humid continental, mild summer *Dfc, Dwc, Dfd, Dwd*–Subarctic	*w* (DRY WINTER) Driest month in the winter half of the year, with less than one tenth the precipitation of the wettest summer month *f* (ALWAYS MOIST) Does not meet conditions for *s* or *w* above	*c* Warmest month below 22°C (71.6°F), with 1 to 3 months above 10°C (50°F) *d* Same as c, but coldest month is below −38°C (−36.4°F)

Graph 2 Tropical Climate Boundaries

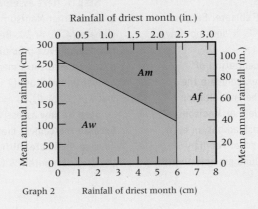

Graph 2

As you begin to classify climates, it is strongly recommended that you use the following procedure. (After a few examples you may find that you can omit some steps with a glance at the statistics.)

Step 1. Ask: Is this a polar climate *(E)*? Is the warmest month less than 10°C (50°F)? If so, is the warmest month between 10°C (50°F) and 0°C (32°F) *(ET)* or below 0°C (32°F) *(EF)*? If not, move on to Step 2.

Step 2. Ask: Is there a seasonal concentration of precipitation? Examine the monthly precipitation data for the driest and wettest summer and winter months for the site. Also, take careful note of the seasonal temperature data because you must determine whether the site is located in the Northern or Southern Hemisphere. (April to September are considered summer months in the Northern Hemisphere, but winter months in the Southern Hemisphere. Similarly, the Northern Hemisphere winter months of October to March are summer south of the equator.) As Table 1 indicates, a site has a dry summer *(s)* if the driest month in summer has less than 3 centimeters (1.2 in.) of precipitation and less than one third of the precipitation of the wettest winter month. It has a dry winter *(w)* if the driest month in winter has less than one tenth the precipitation of the wettest summer month. If the site has neither a dry summer nor a dry winter, it is classified as having an even distribution of precipitation *(f)*. Move on to Step 3.

Step 3. Ask: Is this an arid climate *(B)*? Use one of the three graphs (included in Graph 1) to decide. Based on your answer in Step 2, select one of the small graphs and compare mean annual temperature with mean annual precipitation. The graph will indicate whether the site is an arid *(B)* climate or not, and if it is, also which

type *(BW or BS)*. You should further classify the site by adding *h* if the mean annual temperature is above 18°C (64.4°F) and *k* if it is below. If the site is neither *BW* nor *BS*, it is a humid climate (A, C, or D). Move on to Step 4a.

Step 4a. Ask: Is this a tropical climate *(A)*? The site has a tropical climate if the temperature of the coolest month is higher than 18°C (64.4°F). If so, use Graph 2 to determine which tropical climate the site represents. (Note: tropical climates do not use additional lowercase letters.) If not, move on to Step 4b.

Step 4b. Ask: Which major midlatitude climate group does that site represent, mesothermal *(C)* or microthermal *(D)*? If the temperature of the coldest month is between 18°C (64.4°F) and 0°C (32°F), the site has a mesothermal climate. If it is below 0°C (32°F), it has a microthermal climate. Once you have answered that question, move on to Step 5.

Step 5. Ask: What is the precipitation distribution? This was determined back in Step 2. Add *s*, *w*, or *f* for a *C* climate or *w* or *f* for a *D* climate to the letter symbol for the climate. Then, move on to Step 6.

Step 6. Ask: What is needed to express the details of seasonal temperature for the site? Refer again to Table 1 and the definitions for the letter symbols. Add *a*, *b*, or *c* for the mesothermal *(C)* climates or *a*, *b*, *c*, or *d* for the microthermal *(D)* climates, and you have completed the classification. However, you may have already completed your classification at Steps 1, 3, or 4a.

You should now be ready use Table 1, following the recommended steps, using climate data for Madison, Wisconsin, in the following example.

	J	F	M	A	M	J	J	A	S	O	N	D	Year
T (°C)	−8	−7	−1	7	13	19	21	21	16	10	2	−6	7
P (cm)	3.3	2.5	4.8	6.9	8.6	11	9.6	7.9	8.6	5.6	4.8	3.8	77

The correct climate type is derived as follows:

Step 1. Determine whether this site has an *E* climate. Because Madison has several months averaging above 10°C, it does not have an *E* climate.

Step 2. Determine if there is a seasonal concentration of precipitation. Because Madison is driest in winter, compare the 2.5 centimeters of February precipitation with the precipitation of June (1/10 of 11.0 cm, or 1.1 cm) to conclude that Madison has neither a dry summer nor a dry winter but instead has an even distribution of precipitation *(f)*. (*Note:* The 2.5 cm of February precipitation is not less than 1/10 [1.1 cm] of June precipitation.)

Step 3. Next assess, through the use of Graphs 1*(f)*, 1*(w)*, or 1*(s)*, whether Madison has an arid climate *(BW, BS)* or a humid climate *(A, C, or D)*. Because Madison was previously determined to have an even precipitation distribution, you will use Graph 1*(f)*. Based on Madison's mean annual precipitation (77.0 cm) and mean annual temperature (7°C), you conclude that Madison is a humid climate *(A, C, or D)*.

Step 4. Now assess which humid climate type Madison falls under. Because the coldest month (28°C) is below 18°C, Madison does *not* have an *A* climate. Although the warmest month (21°C) is above 10°C, the coldest month (28°C) is not between 0°C and −8°C, so Madison does *not* have a *C* climate. Because the warmest month (21°C) is above −0°C and the coldest month is below 0°C, Madison *does* have a *D* climate.

Step 5. Madison has a *D* climate, so the second letter will be either *w* or *f*. Because precipitation in the driest month of winter (2.5 cm) is not less than one tenth of the amount of the wettest summer month (1 / 10 × 11.0 cm = 1.1 cm), Madison does *not* have a *Dw* climate. Madison therefore has a *Df* climate.

Step 6. Because Madison is a *Df* climate, the third letter will be *a*, *b*, *c*, or *d*. The average temperature of the warmest month (21°C) is not above 22°C, which means that Madison does *not* have a *Dfa* climate. Instead, the average temperature of the warmest month is below 22°C, with at least 4 months above 10°C, which means that Madison has a *Dfb* climate.

LOW-LATITUDE AND ARID CLIMATE REGIONS

OBJECTIVES

WHEN YOU COMPLETE THIS CHAPTER YOU SHOULD BE ABLE TO:

- 9.1 Explain the temperature basis for designating a climate as tropical, the precipitation characteristics for classifying the different humid tropical climates, and the criteria for designating a climate as arid or semiarid (desert or steppe).
- 9.2 Know the general locations of each tropical (rainforest, monsoon, and savanna) and arid climate (desert or steppe), and explain the factors influencing their global distribution.
- 9.3 Outline the major characteristics of each of the tropical and arid climates.
- 9.4 Describe typical vegetation types related to each of the humid tropical and arid climates.
- 9.5 Recognize a specific humid tropical or arid climate in the Köppen classification system by examining a climograph or a set of data from a representative location.
- 9.6 Explain some of the adaptations people have made to living in the humid tropical and arid climate regions.
- 9.7 Discuss the environmental challenges, hazards, and benefits to living in the humid tropical and arid climate regions.

DURING YOUR LIFETIME YOU WILL very likely travel to destinations far from where you live today, or perhaps you have done so already. You might move to a new location, or you may travel on business or for pleasure. Whatever the reason, it is likely that you will ask the question almost every other traveler asks: "What will the weather be like at my destination?"

A more relevant question would be, "What is the climate like?" Daily variability makes weather difficult to predict in advance of a trip. However, the long-term weather averages and ranges used to classify the different climates can provide a general idea of the weather to expect at a given location.

There are many reasons why an understanding of climates and how they vary around the world is useful knowledge. Being familiar with the world's climates helps us understand the adaptations that the people in various regions have made to the seasonal weather they experience over the year. This knowledge helps us to appreciate some of their economic activities and certain aspects of their cultures. Climate is also a controlling factor on native plant and animal life around the world. Soil characteristics are influenced by climatic conditions, which also affect vegetation and agricultural potential. The processes that influence the development

◀ A desert stream flowing from the High Atlas Mountains of Morocco supports a narrow ribbon of vegetation and crops on its banks. This small river, in an extremely arid region, disappears downstream into the sands of the Saharan Desert. cdrin/Shutterstock.com

and modification of landforms and physical landscapes are also related to climate. Understanding climate not only helps us understand the seasonal weather conditions of a place but also many other aspects of the world's environments.

This chapter and the next provide a broad survey of world climates: their locations, distributions, general characteristics, processes, associated features, and related human activities. The information in these two chapters can serve as a valuable knowledge base as you prepare for the future. Throughout these chapters the version of the Köppen climate classification (introduced in Chapter 8) is used. It is interesting to note that all of these climates can be found in North America, so you will find that your knowledge of climates is useful as you prepare for traveling, vacationing, or moving to a new location, even if you never journey beyond this continent.

Humid Tropical Climate Regions

The tropical (A) climate regions of the humid tropics were introduced in a preliminary discussion of these climate types in Chapter 8. Throughout this chapter and the next it will be helpful to review where these climates are located and their relation to other climate regions by examining the world climate map (see Fig. 8.6). This section provides an introduction to the major characteristics of the humid tropical climates and their associated world regions.

The discussion of Köppen's five major climate categories described in this chapter and the next provides two aids that will help you identify and recall the important characteristics of each climate type, as well as their regional locations. The first is a map of the world that clearly locates the climates, with one map for each of the five main climate types. The second is a set of tables that summarize important information about each climate. As you study humid tropical climates, for example, begin by examining the map in ● Fig 9.1 and Table 9.1. Together they provide the locations and preview significant facts associated with each tropical climate. Table 9.1 serves as a reminder that, although all humid tropical climates have high average temperatures all year, the differences among them are based on the annual amounts and seasonal distributions of precipitation. When studying a climate category, it is useful to refer to both map and table before and after reading the text.

● **FIGURE 9.1** Index map of humid tropical climates that allows you to focus on the humid tropical climates. Study the location and estimate the latitudinal extent of each tropical climate region.

UNDERSTANDING MAP CONTENT ***Which continent has the largest area of tropical rainforest? Where in the Northern Hemisphere does the tropical savanna extend to the Tropic of Cancer? Where are most of the areas with a monsoon climate found? Which tropical climate has the greatest latitudinal range?***

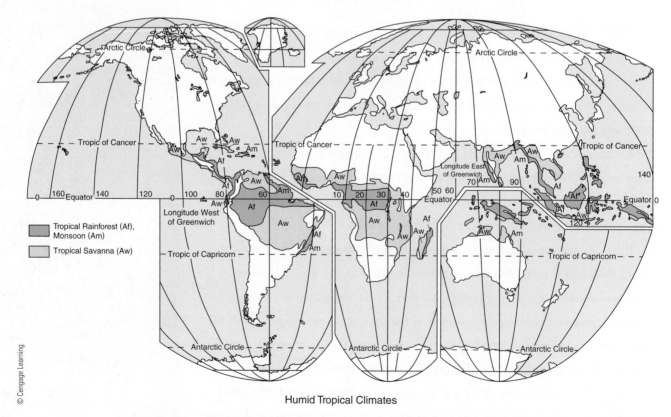

Humid Tropical Climates

TABLE 9.1
The Humid Tropical Climates

Name and Description	Controlling Factors	Geographic Distribution	Distinguishing Characteristics	Related Features
Tropical Rainforest				
Coolest month above 18°C (64.4°F); driest month with at least 6 cm (2.4 in.) of precipitation	High year-round insolation and precipitation of doldrums (ITCZ); rising air along trade wind coasts	Amazon R. Basin, Congo R. Basin, east coast of Central America, east coast of Brazil, east coast of Madagascar, Malaysia, Indonesia, and Philippines	Constant high temperatures; equal length of day and night; lowest (2°C–3°C/3°F–5°F) annual temperature ranges; evenly distributed heavy precipitation; high amount of cloud cover and humidity	Tropical rainforest vegetation (selva); jungle where light penetrates; tropical iron-rich soils; climbing and flying animals, reptiles, and insects; slash-and-burn agriculture
Tropical Monsoon				
Coolest month above 18°C (64.4°F); one or more months with less than 6 cm (2.4 in.) of precipitation; excessively wet during rainy season	Summer onshore and winter offshore air movement related to shifting ITCZ and changing pressure conditions over large landmasses; also transitional between rainforest and savanna	Coastal areas of southwest India, Sri Lanka, Bangladesh, Myanmar, southwest Africa, Guyana, Surinam, French Guiana, northeast and southeast Brazil	Heavy high-sun rainfall (especially with orographic lifting), short low-sun drought; 2°C–6°C (3°F–10°F) annual temperature range, highest temperature just prior to rainy season	Forest vegetation with fewer species than tropical rainforest; grading to jungle and thorn forest in drier margins; iron-rich soils; rainforest animals with larger leaf-eaters and carnivores near savannas; paddy rice agriculture
Tropical Savanna				
Coolest month above 18°C (64.4°F); wet during high-sun season, dry during lower-sun season	Alternation between high-sun doldrums (ITCZ) and low-sun subtropical highs and trades caused by shifting winds and pressure belts	Northern and eastern India, interior Myanmar and Indo-Chinese Peninsula; north Australia; Congo River borderlands; south central Africa; llanos of Venezuela, campos of Brazil; western Central America, south Florida, and Caribbean Islands	Distinct high-sun wet and low-sun dry seasons; rainfall averaging 75–150 cm (30–60 in.); highest temperature ranges for humid tropical climates	Grasslands with scattered, drought-resistant trees, scrub, and thorn bushes; poor soils for farming, grazing more common; large herbivores, carnivores, and scavengers

Tropical Rainforest Climate

The **tropical rainforest climate (Af)** often comes to mind when someone says the word *tropical*. Hot and wet throughout the year, tropical rainforest climates have been the environmental setting for many stories of both fact and fiction. One reason for this is that tropical rainforests are challenging environments for human occupation. People visiting this type of climate region experience high temperatures and humidity accompanied by the frequent heavy rains that sustain the massive vegetative growth for which it is known (● Fig. 9.2).

Constant Heat and Humidity To obtain an understanding of the weather conditions within a specific climate type, the first step should be to examine the representative climographs for each climate region. The climographs in ● Figure 9.3 present monthly and annual data from weather stations at two tropical rainforest locations—Akassa, Nigeria, and Nauru, a Micronesian island country in the Pacific Ocean. Note that the average monthly temperatures recorded for both stations are all above 25°C (77°F), which is about the average for most rainforest regions. Because these places are generally located within 5° of the equator, the sun's rays at noon are always close to being directly overhead.

● **FIGURE 9.2** The typical vegetation in a tropical rainforest climate forms a dense cover of trees, growing at different heights in a multilayered canopy of treetops. This rainforest area is in the Amazon region of Brazil.

How many tree layers can you see?

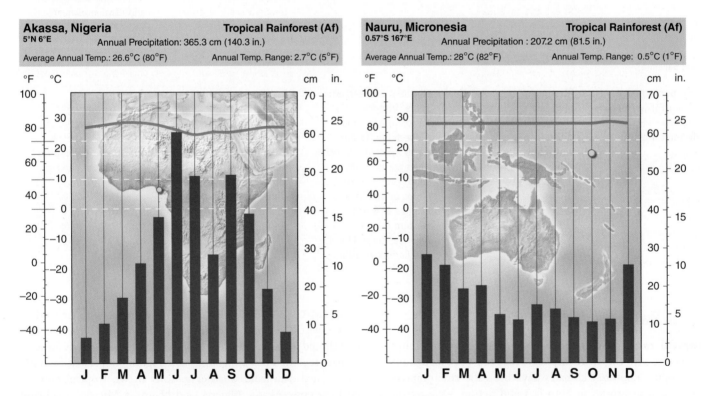

Akassa, Nigeria   Tropical Rainforest (Af)
5°N 6°E   Annual Precipitation: 365.3 cm (140.3 in.)

Average Annual Temp.: 26.6°C (80°F)   Annual Temp. Range: 2.7°C (5°F)

Nauru, Micronesia   Tropical Rainforest (Af)
0.57°S 167°E   Annual Precipitation : 207.2 cm (81.5 in.)

Average Annual Temp.: 28°C (82°F)   Annual Temp. Range: 0.5°C (1°F)

● **FIGURE 9.3** Climographs for tropical rainforest climate stations.

If you only look at the temperature curves, why is it difficult to determine whether each station is located in the Northern or Southern Hemisphere?

For most rainforest regions, hours of daylight and nighttime darkness are almost equal, and the solar radiation received is nearly constant throughout the year. Consequently, no appreciable temperature variations can be linked to the sun angle and therefore be considered seasonal. We are accustomed to regarding summer and winter as hot and cold seasons, but these seasonal temperature differences do not exist in tropical rainforest climate regions.

The **annual temperature range** is the difference between the average temperatures of the warmest and coolest months of the year. As the climographs in Figure 9.3 indicate, the annual range of temperature is seldom more than 2°C or 3°C (3.6°F or 5°F). In fact, Nauru in the central Pacific experiences an annual temperature range of only 1°C (1.8°F) because of the additional moderating influence of the ocean along with the nearly uniform insolation all year. Equatorial islands have the lowest annual temperature ranges in the world, with almost no differences in temperature from month to month over the whole year.

An interesting feature of tropical rainforest climate regions is that the **daily (diurnal) temperature ranges**—the differences between the highest and lowest temperatures during the day—are usually far greater than the annual range. Highs of 30°C to 35°C (86°F to 95°F) and lows of 20°C to 24°C (68°F to 75°F) produce daily (day to night) ranges of 10°C to 15°C (18°F to 27°F). However, the nighttime drop in temperature is small comfort. High humidity in these regions causes even the cooler evenings to seem oppressive. (Recall that water vapor and high relative humidity conditions increases the retention of heat.)

The climographs in Figure 9.3 also illustrate that significant variations in precipitation can occur even within rainforest regions. Although most tropical rainforest locations receive more than 200 centimeters (80 in.) of yearly precipitation and the average is about 250 centimeters (100 in.), some locations receive more than 500 centimeters (200 in.) of an annual precipitation. Oceanic locations, particularly if they are mountainous and near a great source of moisture, tend to receive very high rainfall amounts. Mount Waialeale in the Hawaiian Islands receives a yearly average of 1168 centimeters (460 in.), making it one of the wettest places on Earth. As a group, climate stations in the humid tropics experience much higher annual rainfall totals than typical humid midlatitude stations. Compare, for example, Akassa in Nigeria, which averages 365 centimeters (144 in.) a year, with Portland, Oregon's annual average of 112 centimeters (44 in.), or the 61 centimeters (24 in.) received annually in London, England.

Recall that the heavy precipitation received in tropical rainforest climates is associated with the warm or hot, humid air of the equatorial low and the unstable conditions along the ITCZ (Intertropical Convergence Zone). Both convection and convergence serve as uplift mechanisms, causing humid air to rise and condense, resulting in the heavy rains that are characteristic of this climate. These processes are enhanced on the east coasts of continents where warm ocean currents allow humid tropical climates to extend farther poleward. In addition, trees in the dense cover of tropical rainforests transpire great amounts of moisture back into the air, estimated to be 760 liters (200 gallons) annually per each tree in the rainforest canopy. Increasing the humidity in this way reinforces the potential for rain, thus rainforests contribute to their own climate.

Tropical rainforest areas tend to have heavy cloud cover during the warmer, daylight hours when convection is at its peak, and thundershowers are common during the hottest part of the day. The nights and early mornings when it is cooler can be quite clear. Seasonal variations in rainfall can usually be traced to the ITCZ and its low pressure cells of varying strength. Many tropical rainforest locations (Akassa, for example) experience two periods of maximum precipitation during the year, with each return of the ITCZ as it follows the migration of the sun's 90° rays (recall that the sun crosses the equator on the equinox days around March 21 and September 22). In addition, although no season in tropical rainforest regions can be called dry, during some months there may be only 15 or 20 days of rain.

Cloud Forests Highland areas near a seacoast both in the tropical regions (such as American Samoa; ● Fig. 9.4) and in the middle latitudes (coastal Washington State) may contain **cloud forests**. Here, moisture-laden maritime air is lifted up on the windward slopes of mountains. This orographic precipitation may fall as heavy rain showers or from thunderstorms but these regions tend to experience almost constant misty fogs. The cloudy environment in these forest areas can receive enough moisture to qualify as a rainforest but without the oppressive heat found in the rainforests of the low-elevation tropics. An advantage of visiting the cloud forests is a scarcity of flying insects that cannot survive in the cooler temperatures experienced at higher elevations in the tropics.

A Delicate Balance The most common vegetation association in the tropical rainforest regions are multistoried, broadleaf evergreen forests made up of many tree species whose tops form a thick, almost continuous canopy cover that blocks out much

National Park Service

● **FIGURE 9.4** A cloud forest in the highlands of American Samoa, in the South Pacific.

Can a constant misty rain drop as much water as would fall in less frequent rainstorms?

⊕ In Google Earth fly to: *14.26°S 170.69°W*

Saharan Dust Feeds the Amazon Rainforest

The systems approach to understanding Earth, outlined in Chapter 1, stresses the spatial interconnectedness of subsystems on our planet. Processes that invoke environmental change in one feature or area can have an impact on other parts of the Earth system. Many connections of this sort are described throughout this book. Interactions between processes and places often occur where the components involved are in the same vicinity; however, other interconnections have broad regional, or even global impacts. El Nino, for example, is linked to weather changes in many parts of the world, other than just the west coastal region of South America.

The environments of the Amazon rainforest and the Sahara could hardly be more different. From a continuous multilayered canopy of rainforest trees in a humid, continuously wet tropical environment, to a huge expanse of hot, extremely arid desert, with a sparse cover of shrubs in places and wide areas of sand dunes, bare rock, and stone. Separated by 2850 km (1775 mi)

across the Atlantic Ocean, and as different as they are, they share an interconnection. For many years, it has been known that clouds of dust from the Sahara are blown across the Atlantic during times of strong easterly trade winds. Depending on wind conditions, African dust has also reached the United States, mainly in the South, from Florida to Texas, but also to the East Coast up to New England. Along the way, some of the dust falls on the Atlantic Ocean, the Caribbean Sea, and islands of the Caribbean. A tremendous amount of dust—the estimated average by NASA is 182 million tons—is blown away from Africa every year. About 27 million tons reaches the Amazonian rainforest, washed out of the air by rainfall.

The nature of soils in the tropical rainforests has already been described as being deficient in nutrients that would benefit the growth of trees and other plants in these tropical regions. The input of nutrients, however, is very important to the health of the rainforest. One nutrient that is essential to vegetation growth is phosphorous, which

is commonly used in commercial fertilizers. Rainforest soils, however, are deficient in phosphates.

Recent studies have shown that African dust is important to vegetation growth in the Amazon Rainforest. The mineral content of this dust is apparently a significant contributor to nutrients that benefit rainforest vegetation. Dust from the now-dry basin of a huge ancient lake in Chad, Africa, contains phosphorous from the tiny remains of algae and other microorganisms that lived in the lake. According to NASA estimates, 22,000 tons of phosphorus make the trans-Atlantic trip to fertilize the Amazon Rainforest each year. It may seem strange that minerals that can fertilize plants come from a barren desert area, in earlier times a lake, to benefit one of the most heavily vegetated locations on Earth. It is climate that makes the difference, and the latitudinal position of these two regions, which interconnects them through transport by the tradewinds, blowing dust easterly from Africa to South America.

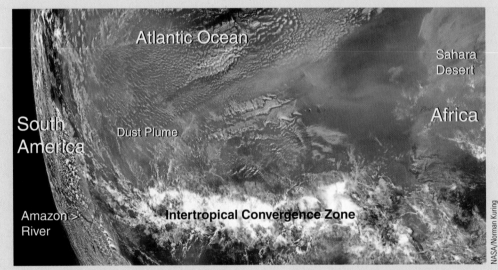

A giant plume of dust blows across the Atlantic, from Africa to South America. Driven westward by the easterly trade winds, some of the dust contains phosphorous, a fertilizing nutrient beneficial to the Amazonian Rainforest.

of the sun's light. This type of rainforest is sometimes called a **selva**, particularly in South America. Within the selva, there is usually little undergrowth on the forest floor because not enough sunlight penetrates the rainforest canopy to support very much

low-growing vegetation (● Fig. 9.5). In addition, a dense tree canopy can intercept much of the rainfall, reducing the amount that reaches the soil. When a rainforest tree dies, the new opening in the canopy allows sunlight to enter, and another tree will soon fill that void to make use of the insolation made available by the forest opening. In a rainforest, availability of sunlight can be more important to vegetative growth than rainfall.

The relationship between the rainforest vegetation and the soil that supports plant growth is so close that a strong ecological balance exists between the two, threatened only by people's efforts to earn a living from the soil. Rainforest trees supply the tropical soils with most of the nutrients that the trees themselves need for growth. As leaves, flowers, and branches fall to the ground or roots die, numerous soil-dwelling animals and bacteria act on them, transforming forest litter into organic matter with vital nutrients. However, if the trees are removed, there is almost no replenishment of these nutrients and no natural barrier (forest litter and root systems) to prevent large amounts of rainwater from percolating through the soil. This percolating water dissolves and removes nutrients and minerals from the topsoil. Intense activities of microorganisms, worms, termites, ants, and other insects cause rapid deterioration of the remaining organic debris, and soon all that remains is an infertile soil with a mixture of insoluble aluminum, iron, and manganese compounds. In essence, without the vegetation to protect and nurture the soils, rainforest soils tend to be quite infertile. In recent years, it has been discovered that the Amazon rainforest receives nutrients from dust that is wind-blown across the Atlantic Ocean from the Sahara.

For decades now, there has been large-scale harvesting of the tropical rainforests by the lumber industry, and land has been cleared for agriculture and livestock production, especially in the Amazon River basin. Such deforestation can have a significant and potentially permanent impact on the delicate balance that exists among Earth's systems. Rainforest clearing contributes to the increase of carbon dioxide in the atmosphere in two ways. Rainforest trees take in huge amounts of CO_2, and after forest clearing burning, the leftover debris puts more CO_2 into the atmosphere.

Environmental conditions vary from place to place within climate regions; therefore, the typical rainforest vegetation does not exist everywhere in the tropical rainforest climate regions. Certain areas are covered by true jungle, a term often misused when describing the rainforest. **Jungle** is a dense tangle of vines and small trees that develops where direct sunlight reaches the ground, as in clearings cut though the forest for roads and along the banks of rivers and streams (● Fig. 9.6). Some regions have soils that remain fertile through the deposition of sediments by flooding streams, and others provide nutrients through natural processes of

● **FIGURE 9.5** The tropical rainforest floor has open space between trees and is not covered over with low vegetation, in part because the rainforest canopy intercepts much of the rainfall.

What other factors are involved in the openness of the rainforest floor?

● **FIGURE 9.6** Jungle vegetation along the Rio Madre de Dios in Peru, a tributary to the Amazon River.

Why is the vegetation so dense here when it is more open inside the rainforest at ground level?

soil development from bedrock. Examples of the former region are found along river floodplains; examples of the latter are volcanic regions such as in Indonesia or in the limestone areas of Malaysia and Vietnam. Only in such regions of continuous soil fertility can agriculture be intensive and continuous enough to support population centers in a tropical rainforest region.

Human Activities In most tropical rainforest climate regions, humans are vastly outnumbered by other forms of animal life. Although there are few large animal species, a great variety of smaller tree-dwelling and aquatic species live in the rainforest. Small predatory cats, birds, monkeys, bats, alligators, crocodiles, snakes, and amphibians such as frogs of many varieties abound. Animals that can fly or climb into the food-rich leaf canopy have become the dominant animals in this world of trees.

Most common of all, however, are the insects. Mosquitoes, ants, termites, flies, beetles, grasshoppers, butterflies, and bees live everywhere in the rainforest. Insects can breed continuously in this climate without danger from cold or drought.

In addition to the insects, there are genuine health hazards for human inhabitants of the tropical rainforest. The humid, sultry weather imposes uncomfortable living conditions, and open wounds heal very slowly in this steamy environment. The tropical rainforest climate supports a variety of parasites and disease-carrying insects that threaten human survival. Malaria, yellow fever, dengue fever, and sleeping sickness are insect-borne (sometimes fatal) diseases of the tropics, but they are uncommon in the middle latitudes.

Whenever native populations have existed in the rainforest, subsistence hunting and gathering of fruits, berries, small animals, and fish have been important. As agriculture was introduced to these regions, land was cleared and crops such as manioc, yams, beans, maize (corn), bananas, and sugarcane were grown. A traditional practice has been to cut down the trees in small plots of land, burn the resulting debris, and plant crops. With the forest gone, farming is possible for only 2 or 3 years before the soil is completely exhausted of its small supply of nutrients. The native population will then move to another area of forest to begin the practice over again. This kind of subsistence agriculture is known as **slash-and-burn** or simply **shifting cultivation**. Its impact on the close ecological balance between soil and forest is obvious in many rainforest regions. Sometimes the damage done to the system is irreparable, and only jungle, thorn bushes, or scrub vegetation will return to cleared areas (● Fig. 9.7). Widespread commercial forest clearing adds greatly to this problem. It is estimated that in a hundred years, the tropical rainforests could be gone, if the current rate of rainforest clearing continues.

The **tropical monsoon climate** experiences a shift over the year from a very wet season, to a short dry season, a regime that is in sharp contrast to the consistent precipitation of tropical rainforest regions, although annual rainfall totals are somewhat similar. Tropical monsoon regions are most closely associated with South and Southeast Asia. The index map shown in Figure 9.1 also

● **FIGURE 9.7** This opening in the tropical rainforest is land cleared for shifting (slash-and-burn) cultivation. After being cleared, nutrients become depleted in the barren soil, which causes people to move on and clear a new area for their agriculture.

identifies other regions that meet the definition for this climate, including small coastal regions of South America, Africa, Australia, and some Caribbean islands. In those tropical monsoon regions, the alternating air circulation (from sea to land and from land to sea) is related to the shifting ITCZ. The climographs in ● Figure 9.8 illustrate the strikingly uneven distribution of monthly precipitation in tropical monsoon regions.

During the Northern Hemisphere summer in South Asia, the ITCZ moves north onto the Indian subcontinent and adjoins lands to latitudes of 20° to 25°N. This is due in part to humid air flowing into a strong low pressure system on the Asian continent in summer. Several months later, the moisture-laden summer monsoon is replaced by an outflow of dry air from the Siberian high pressure system that develops over central Asia in winter. The ITCZ shifts to its southernmost position in the winter (see Fig. 5.19). During the summer monsoon, uplift associated with the ITCZ is enhanced by orographic uplift as the air is forced to rise over the windward side of India's mountains, particularly the Himalayas. This orographic effect produces extremely heavy rainfalls and flooding during the summer monsoon.

Check the climograph in Figure 9.8 for Freetown, Sierra Leone in Africa. It confirms that climate regions outside of Asia also fit the Köppen classification of a tropical monsoon climate.

Distinctions between Rainforest and Monsoon Climates Despite the climatic factors that produce a tropical monsoon climate, these regions have strong similarities to those classified as tropical rainforest. In fact, although their core regions are distinctly different, the two climates are intermixed *zones of transition*. A major reason for the environmental similarity between monsoon and rainforest climates is that monsoon regions receive enough precipitation to support continuous vegetative growth

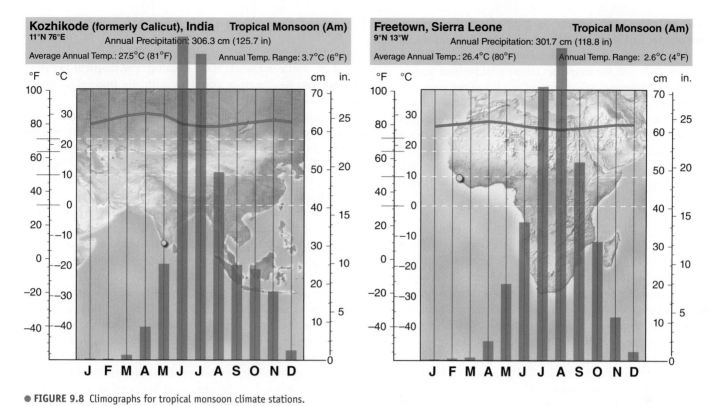

Kozhikode (formerly Calicut), India   Tropical Monsoon (Am)
11°N 76°E   Annual Precipitation: 306.3 cm (125.7 in)
Average Annual Temp.: 27.5°C (81°F)   Annual Temp. Range: 3.7°C (6°F)

Freetown, Sierra Leone   Tropical Monsoon (Am)
9°N 13°W   Annual Precipitation: 301.7 cm (118.8 in)
Average Annual Temp.: 26.4°C (80°F)   Annual Temp. Range: 2.6°C (4°F)

● **FIGURE 9.8** Climographs for tropical monsoon climate stations.

What is the approximate range of precipitation between the wettest and driest months?

without a dormant period during the year. Rains are so abundant and intense during the summer monsoon season and the dry season is so short that the soils usually do not dry out completely. As a result, this climate and its soils support a plant cover much like that of the tropical rainforests.

However, there are clear distinctions between rainforest and monsoon climate regions. The most important distinctions concern the seasonal distributions and amounts of precipitation. Monsoon climates have a short, dry season and the rainforest does not. An interesting aspect of monsoon regions is that the average rainfall varies widely from place to place. It usually totals between 150 and 400 centimeters (60 to 150 in.) and may be massive where the onshore monsoon winds rise over mountain barriers. Mahabaleshwar, India, at an elevation of 1362 meters (4467 ft), on the windward side of the Western Ghats mountain range, averages more than 630 centimeters (250 in.) of rain during the 5 months of the summer monsoon. The tremendous amounts of rain that are received during the wet season result in severe flooding in monsoon areas (● Fig. 9.9).

The annual march of temperature in monsoon regions differs from the monotony that exists in most tropical rainforest regions. Heavy cloud cover during the rainy summer monsoon reduces insolation and temperatures during that time of year. Higher temperatures typically occur prior to the period of extremely heavy rain, when the skies are clearer, and the sun angle has increased

after the spring equinox. As a result, the annual temperature range in a monsoon climate is about 2°C to 6°C (3.6°F to 10.8°F), exceeding the approximately 2°C to 3°C (3.6°F to 5.4°F) in tropical rainforest regions.

● **FIGURE 9.9** Heavy summer monsoon rains flood this area of rice fields in Kayin, Mayanmar (formerly Burma) in South Asia. August, the wettest month, averages almost 1.27 meters of rainfall (50 in).

⊕ In Google Earth fly to: 16.88°N 97.76°E

Some additional distinctions between tropical monsoon and rainforest regions exist based on their plant and animal life. Toward the wetter margins, the monsoon forests resembles the tropical rainforests, but fewer species are present because certain species become dominant. Seasonality of rainfall in the monsoon regions narrows the number of different species that will prosper. Toward the drier margins of the climate, trees grow farther apart and monsoon forest often gives way to jungle or a dwarfed thorn forest. The animal life here also changes. Climbing and flying species that dominate the forest are joined by larger, hooved leaf eaters and by larger carnivores such as the famous tigers of Bengal (northeast India and Bangladesh).

Effects of Seasonal Change The seasonal precipitation of the tropical monsoon climate is important for economic reasons, especially to the people of Southeast Asia and India. Most of the people living in those areas are farmers, and their major crop is rice, the staple food for millions of Asians. Rice is generally an irrigated crop, so the monsoon rains are very important to its growth. Harvesting is done during the dry season.

Each year an adequate food supply for much of South and Southeast Asia depends on the arrival and departure of monsoon rains. The difference between famine and survival for many people in these regions is associated with the potential for significant rainfall variation from year to year, which can be typical in monsoon climate regions.

Tropical Savanna Climate

Located well within the tropics (typically between latitudes 5° and 20° north and south of the equator), the **tropical savanna climate** also has much in common with the tropical rainforest and monsoon.

As you can see in Figure 9.1 and Table 9.1, the greatest areas of savanna climate are on the poleward sides of rainforest climates in Central and South America and in Africa. Other savanna regions occur in India, peninsular Southeast Asia, and Australia. In some instances, the climate extends poleward of the tropics, as it does in the southernmost part of Florida.

Savanna climates receive strong insolation because the sun's rays at noon are never far from overhead and temperatures remain constantly high. Daylight and nighttime hours are of nearly equal length throughout the year, as they are in other tropical regions.

A distinct seasonal precipitation pattern identifies the tropical savanna. This is caused by seasonal shifts in the wind and pressure belts that follow the latitudinal shift in sun angle. Savanna regions are under the influence of the rain-producing intertropical convergence zone (equatorial low) for part of the year and dominated by rain-suppressing subtropical highs for the other part. In fact, the poleward limits of savanna climate regions are approximately the poleward migration limit of the ITCZ, and the equatorward limits of this climate are the equatorward limits of movement by the subtropical high pressure

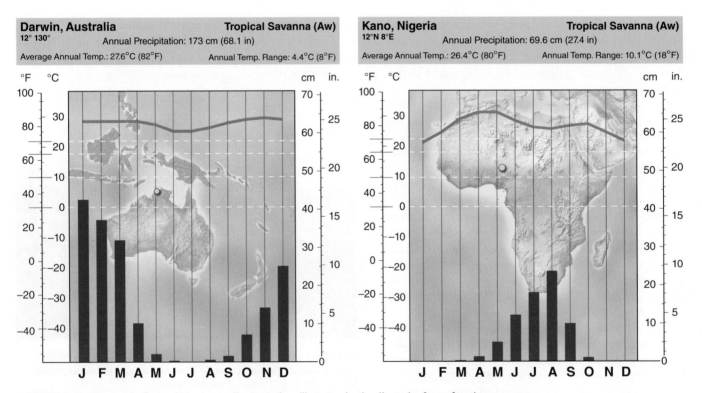

● **FIGURE 9.10** Climographs for tropical savanna climate stations illustrate why the climate is often referred to as tropical wet and dry.

Which season would be a better time of year for a photo safari—high-sun or low-sun season?

systems. Savanna climates are often referred to as *tropical wet and dry climates*, and it is easy to see why on a climograph with data from one of these areas (● Fig. 9.10).

Transitional Features of the Savanna

The geographically transitional nature of the tropical savanna is important. Situated between the humid tropical rainforest climate on one side and the rain-deficient, semiarid steppe climate on the other, the savanna alternates between some of the characteristics of both. Weather in the rainy, high-sun season resembles that of the rainforest, but in the low-sun season (winter), these regions can be as dry as the nearby arid lands are all year. The gradational nature of the climate causes precipitation patterns to vary considerably. Savanna areas close to the rainforest may have rain every month, and total annual precipitation can exceed 180 centimeters (70 in.). In contrast, locations on the drier margins of the savanna, as in Kano, Nigeria, experience longer and more intensive times of drought and lower annual rainfalls of less than 100 centimeters (40 in.).

Other characteristics of the savanna climate demonstrate its transitional nature. The higher temperatures just prior to the arrival of the ITCZ produce annual temperature ranges 3°C to 6°C (5°F 11°F) wider than those of the rainforest but still not as wide as those of the semiarid steppe regions and the deserts. Savanna environments are also transitional because they usually lie between the tropical forests and the grasslands of the semiarid regions. Typical savanna (known as *llanos* in Venezuela and *campos* in Brazil) is grassland with scattered trees (● Fig. 9.11). This is implied in the term *savanna*. Savanna grasses tend to be tall and coarse, and the trees generally are low growing with wide crowned forms.

Vegetation has developed special adaptations to the alternating wet–dry seasons of the savanna. During the wet high-sun period (summer months), the grasslands are green and the trees are covered with foliage. During the dry low-sun period (winter months), the grass turns brown, dry, and lifeless. Here, *summer* and *winter* refer to the times of year in the hemisphere where the savanna region is located. Many of the trees are deciduous and lose their leaves in the dry low-sun season to reduce moisture loss through transpiration. *Deciduous* refers to trees and other vegetation that drop their leaves in either a dry period or a cold period. Certain trees in the savanna develop deep roots that can reach down to water in the soil during the dry season. Trees in savanna regions are also fire resistant, an advantage for survival where dry grasses may burn during the drought of the low-sun season.

Vegetation in savanna regions, however, can also be a mixture of grassland with trees, scrub, and thorn bushes, and there can be considerable variation. Near the equatorward margins of savanna regions where they grade toward the rainforests, grasses are taller and trees, where they exist, grow fairly close together. Toward the drier, poleward margins where they grade toward arid climates, trees are smaller, more widely scattered, and the grasses are shorter. Soils also are affected by the climatic transition as the iron-rich, reddish soils of the wetter sections are replaced by darker-colored, more organic-rich soils in the drier regions.

Savanna Potential

Environmental conditions within tropical savanna regions are not well suited to agriculture, although many of our domesticated grasses (grains) are presumed to have grown wild there. Rainfall is much less predictable than it is in the rainforest or in the monsoon climates. For example, Nairobi, Kenya, has an average rainfall of 86 centimeters (34 in), yet from year to year the rain received can vary from 50 to 150 centimeters (20 to 60 in). In general, the unreliability of rainfall increases with increasingly drier savanna locations. However, the rains are essential for the survival of animals and people in savanna regions. When rains are late or deficient, as they have been in Africa in recent years, severe droughts and famine can result. But when the rains last longer than usual or are excessive, they can cause floods, often followed by outbreaks of serious illnesses from polluted floodwaters and disease-carrying insects.

Savanna soils (except in areas of stream deposits) also tend to limit productivity. During the wet season rains, they can become sticky wet clays and during the dry season, they are hard and almost impenetrable. Consequently, the savannas are better for grazing than for farming. However, even animal grazing has its problems. Savanna regions are poor pasture-lands during the dry season.

Courtesy of Norm Bettis

● **FIGURE 9.11** Giraffes are a majestic sight in the savanna climate regions of East Africa.

How is a giraffe's height so well adapted to the savanna environment?

● **FIGURE 9.12** In savanna climate regions, water holes are important resources for sustaining the wildlife population during the lengthy dry season.

The best known of the world's savanna regions are in Africa. The African savannas have been veritable zoological gardens for the larger tropical animals, and photo safaris have made these areas a major tourist destination. The grasslands support many different herbivores (plant eaters), such as elephants, rhinoceros, giraffes, zebras, and wildebeests. Large carnivores (flesh eaters), such as lions, leopards, and cheetahs, hunt and feed on the herbivores. Scavengers, such as hyenas, jackals, and vultures, devour what remains of a carnivore's kill.

During the dry season, the herbivores find grasses and water along stream banks, the forest margins, and at isolated water holes (● Fig. 9.12). The carnivores follow the herbivores to the water and unfortunately even in the game preserves, illegal hunters and poachers follow them both.

Arid Climate Regions

The basis for determining whether a climate is desert, semiarid, or humid is *precipitation effectiveness*, the amount of precipitation actually available for use by plants and animals. Precipitation effectiveness is related to temperature as well as rainfall totals. An area is classified as having a **desert climate** if the amount of annual precipitation received is less than half the potential evapotranspiration (ET). A **steppe climate**, often referred to as a **semiarid climate**, receives an annual precipitation that is more than half but less than the total potential ET. At higher temperatures, it takes more precipitation to have the same effect on vegetation and soils than at lower temperatures. The result is that areas with higher temperatures that promote greater evapotranspiration can receive more

precipitation than cooler regions and yet have an arid climate. The wide distribution of the world's arid climate regions is shown on the map in ● Figure 9.13 and in the information in Table 9.2.

Because of the influence of temperature, precipitation effectiveness also depends on the season when an arid region's meager precipitation is concentrated. Precipitation received during the low-sun (cooler) period will be more effective than that received during the high-sun period when temperatures are higher and more moisture will be lost through evapotranspiration. The Köppen classification graphs based on the precipitation effectiveness are included in the "Graph Interpretation" exercise (at the end of Chapter 8) and can be used to determine whether a location has a desert, steppe, or humid climate.

The map of arid climates in Figure 9.13 shows that the world's deserts are core areas of aridity, usually surrounded by regions of semiarid steppe. This is why once the locations of deserts are known; it is easy to explain where the semiarid regions are located. Most semiarid climates border desert regions because they are transitional between arid and humid climate types. Exceptions are the midlatitude areas of North America, which are arid because of their location in the rain shadows of the western mountains. These regions include an extension of the Great Basin in the rain shadow of the Cascades Range of the Pacific Northwest, and the western Great Plains, in the rain shadow of the Rocky Mountains.

Unfortunately, deserts are sometimes erroneously referred to as *wastelands*. Although vegetation may be sparse, most deserts have highly specialized ecosystems adapted to the dryness. Many deserts are also regions of spectacular natural beauty. The rocks, mountains, canyons, and other landforms in desert regions present

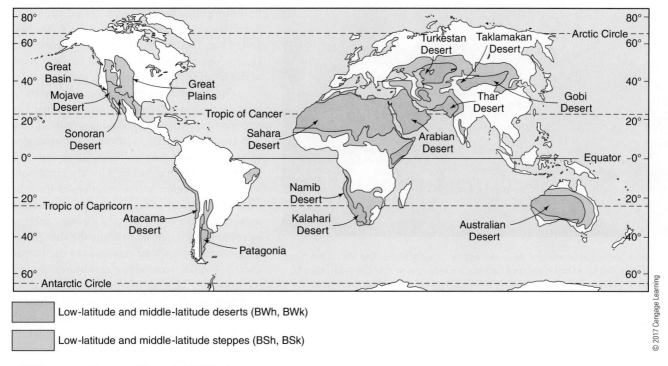

Low-latitude and middle-latitude deserts (BWh, BWk)

Low-latitude and middle-latitude steppes (BSh, BSk)

© 2017 Cengage Learning

● **FIGURE 9.13** Index map of the world's arid lands.

UNDERSTANDING MAP CONTENT   What climate factors affecting South America cause the Atacama Desert to be on the west coast, yet the Patagonian Desert is on the east coast? Where are the northernmost desert areas and the northernmost semiarid (steppe) areas? What factors influence their relatively extreme latitudinal positions for arid climates?

TABLE 9.2
The Arid Climates

Name and Description	Controlling Factors	Geographic Distribution	Distinguishing Characteristics	Related Features
Desert				
Precipitation less than half of potential evapotranspiration; mean annual temperature above 18°C (64.4°F) (low-lat.), below (mid-lat.)	Descending, diverging circulation of subtropical highs; continentality often linked with rain shadow location	Coastal Chile and Peru, southern Argentina, southwest Africa, central Australia, Baja California and interior Mexico, North Africa, Arabia, Iran, Pakistan and western India (low-lat.); inner Asia and western United States (mid-lat.)	Aridity; low relative humidity; irregular and unreliable rainfall; highest percentage of sunshine; highest diurnal temperature range; highest day-time temperatures; windy conditions	Xerophytic vegetation; often barren, rocky, or sandy surface; desert soils; excessive salinity; usually small, nocturnal burrowing animals; nomadic herding
Steppe				
Precipitation more than half but less than potential evapotrans-piration mean annual temperature above 18°C (64.4°F) (low-lat.), below (mid-lat.)	Same as deserts; usually transitional between deserts and humid climates	Peripheral to des-erts, especially in Argentina, northern and southern Africa, Australia, central and southwest Asia, and western United States	Semiarid conditions, annual rainfall distribution similar to nearest humid climate; temperatures vary with latitude, elevation, and continentality	Dry savanna (tropics) or short grass vegetation; highly fertile black and brown soils; grazing animals in vast herds; predators and small animals; ranching and dry farming

● **FIGURE 9.14** Desert landscapes accentuate the terrain and can be quite scenic. The setting sun lights up the mountains here in Red Rock Canyon National Conservation Area, just west of Las Vegas, Nevada. The area offers miles of hiking trails and a scenic drive for observing and experiencing the natural desert environment.

⊕ In Google Earth fly to: 36.136°N, 115.427°W

an entirely different landscape appearance compared to those in the densely vegetated humid regions. The National Parks in the desert areas of the western United States attract millions of visitors from around the world because of their interesting environments and scenic beauty (● Figure 9.14).

Arid climates are found in two types of locations, and each illustrates a climatic factor that causes dry conditions. The first location type is generally centered on both the Tropics of Cancer and Capricorn (23½°N and 23½°S) and extends 10° to 15° poleward and equatorward. These regions contain the world's most extensive areas of arid climates. The second type of location for arid climates is at higher latitudes and occupies continental interiors, particularly in the Northern Hemisphere. Arid regions typically have a central core of desert climate, bounded on the edges by transition zones of semiarid steppe climates.

The concentration of deserts in the vicinity of the two lines of the tropics is directly related to the subtropical high pressure systems. Although the boundaries of the subtropical highs migrate north and south with the rays of the sun, their influence remains strong in these latitudes. The subsidence and divergence of air associated with these systems is strongest along west coasts where cold ocean currents help stabilize the high pressure systems. The clear weather and dry conditions of the subtropical highs extend inland from the western coasts of continents in the subtropical latitudes. The regional extents of the some deserts, such as the Atacama, Namib, and Kalahari Deserts of Africa, and the desert of Baja California, Mexico, are restricted by the small size of the landmass or by landform barriers toward the interior. North Africa and the Middle East comprise the greatest regional stretch of desert climate that includes the Sahara, Arabian, and Thar Deserts. Similarly, the Australian Desert (often called the Outback) occupies most of the interior of the Australian continent.

The second concentration of deserts is located within continental interiors that are remote from moisture-carrying winds. These arid lands include the vast cold-winter deserts of inner central Asia and the Great Basin of the western United States. The dry conditions of the latter region extend northward into the Columbia Plateau of the Pacific Northwest and southward into the Colorado Plateau, affected by mountain barriers that restrict the movement of rain-bearing (mP) air masses from the Pacific Ocean. Similar *rain-shadow* conditions explain the Patagonia Desert's location on the east coast of Argentina, the arid lands of western China.

Wind directions and ocean currents both can accentuate aridity in coastal regions. When prevailing winds blow parallel to a coastline instead of onshore, desert conditions are likely to occur because little moisture is brought inland. This seems to be the case in eastern Africa and in northeastern Brazil, which are unusual locations for arid lands. Where a cold current flows along a coastal desert, fog can develop, when moist oceanic cools to its dew point as it passes over the current. Cool air from the coast can also produce a temperature inversion, increasing stability and calm conditions, by preventing the upward movement of air required for precipitation. The unique and barren, fog-shrouded coastal deserts such as the Atacama in Chile, the Namib in southwest Africa, and Mexico's Baja California desert have the lowest precipitation of any regions on Earth (● Fig. 9.15).

● **FIGURE 9.15** The Atacama Desert, mainly in Chile, continues into the highlands of the Andes Mountains of South America. Even at higher elevations, the Atacama is a barren landscape. Many areas of this desert region are completely free of vegetation.

⊕ In Google Earth fly to: 24.50°S, 69.25°

Desert Climates

The world's deserts extend through such a wide range of latitudes that the Köppen classification system recognizes two major subdivisions. The first consists of low-latitude deserts where temperatures are relatively high year-round and frost is absent or infrequent even along poleward margins. The second includes midlatitude deserts, which have distinct seasons, including below-freezing temperatures during winter. However, the most significant characteristic of all deserts is their aridity, which is dramatically illustrated by their climographs (● Fig. 9.16).

Land of Extremes Deserts, by definition, are associated with low precipitation, and they also include some extremes in other atmospheric conditions. As much as 90% of insolation reaches the surface in desert regions because of clear skies and very low relative humidity. This is why very high insolation levels and the world's highest temperatures are recorded in desert areas and not in the humid tropical climates that are closer to the equator. Another effect of low relative humidity in desert air and little or no cloud cover means that much of the energy received during the day is radiated away at night. Consequently, nighttime temperatures in the desert drop considerably below their daytime highs. This extreme cycle of heating and cooling between day and night gives low latitude deserts the world's greatest diurnal temperature ranges, and midlatitude deserts are not far behind. In the spring and fall, the daily temperature ranges may be as great as 40°C (72°F) in a day. More commonly, diurnal temperature ranges in deserts are about 22°C to 28°C (40°F to 50°F).

The sun's rays are so intense in the dry desert air that temperatures in shade are much lower than those a few steps away in direct sunlight. Note that all temperatures for official meteorological statistics are recorded with instruments located in the shade. Khartoum, Sudan, in the Sahara, has an *average* annual temperature of 29.5°C (85°F), which is a *shade* temperature that includes daytime high and nighttime low temperatures averaged over the year. Temperatures in the intense desert sun under cloudless skies at Khartoum are often 43°C (110°F) or more. Soil temperatures have risen close to 95°C (203°F) in midsummer in the Mojave Desert of Southern California.

During the low-sun or winter months, most desert areas experience colder temperatures than humid areas at the same latitude, and in summer, desert areas experience hotter temperatures. Just as with the high diurnal ranges in deserts, these high annual temperature ranges can be attributed to the low relative humidity.

Annual temperature ranges are usually greater in midlatitude deserts, such as the Gobi in Asia, than in the low-latitude deserts because of the colder winters experienced at higher latitudes. Compare, for example, the climograph for Aswan in south central Egypt—at 24°N, a low-latitude desert location, with the climograph for Turtkul, Uzbekistan—a midlatitude desert at 41°N (see Fig. 9.16). The annual range for Aswan is 18°C (33°F); in Turtkul, it is 31.5°C (61.2°F).

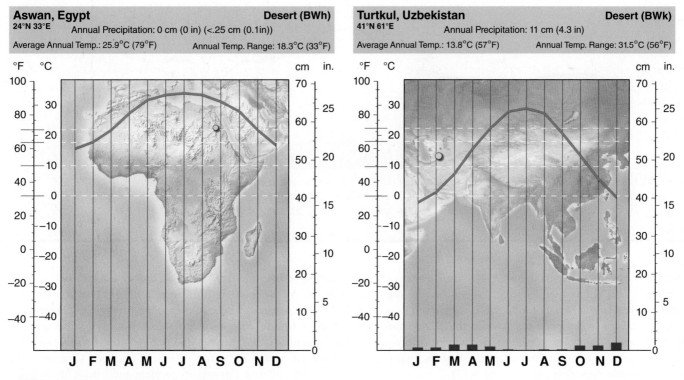

Aswan, Egypt — **Desert (BWh)**
24°N 33°E — Annual Precipitation: 0 cm (0 in) (<.25 cm (0.1in))
Average Annual Temp.: 25.9°C (79°F) — Annual Temp. Range: 18.3°C (33°F)

Turtkul, Uzbekistan — **Desert (BWk)**
41°N 61°E — Annual Precipitation: 11 cm (4.3 in)
Average Annual Temp.: 13.8°C (57°F) — Annual Temp. Range: 31.5°C (56°F)

● **FIGURE 9.16** Climographs for desert climate stations.

Considering the serious limitations of desert climates, why do some people choose to live in desert regions?

Desertification

The United Nations defined the term **desertification**. as "the land degradation in arid, semiarid and sub-humid areas resulting from various factors, including climatic variations and human activities." Although desertification is sometimes confused with drought, these are distinctly different processes. Droughts are times of little or no rainfall that last longer than normal and are naturally recurring events common in arid and semiarid regions that also can occur in humid regions. Desertification expands desertlike landscapes, generally as a result of human activities in marginal lands that cannot sustain them. The encroachment of sand dunes on former agricultural lands is one type of land degradation, but desertification does not necessarily refer to the advance of deserts. Rather, it is the persistent degradation of dryland ecosystems by unsustainable overcultivation, overgrazing, clear-cutting of vegetation, and salt accumulation in the soil.

Land degradation creates desertlike landscapes where people are trying to live in locations that are experiencing a serious lack of rain. Sparse vegetation is stripped by grazing animals, and cut down by people needing firewood. This leads to a die-off of the vegetative cover and exposes the bare ground. Under these conditions, particularly in marginal semiarid or desert environments, there is a die-off of the vegetative cover, exposing the ground to increased erosion, and there is a loss of soil. Wind erosion increases, causing dust storms, and in some instances sand dunes move into settled areas, grasslands, and farmlands. The real problem is that there are growing populations of poverty-stricken people living in environments that cannot sustain them, given their current use of the land. Unfortunately, when the resources there are exhausted, starvation can ensue. Along with the threat to human populations, desertification endangers habitats for wildlife.

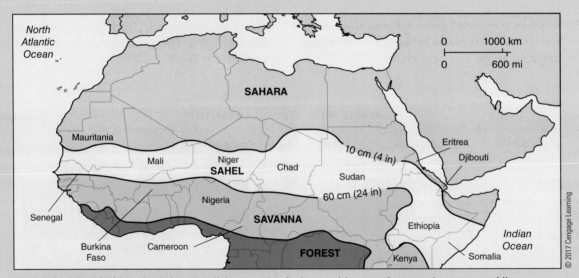

The Sahel region of Africa, shown here in a light-tan color, is the transition zone between the extreme aridity of the Sahara and the tropical humid areas of Africa (in green tones). In recent years, the Sahel has experienced serious desertification through climate change and overuse of this marginal land by human activities.

Convection currents set up by the intense daytime heating of the land can make the desert a windy place. In addition, the vegetation sparseness and the absence of topographic barriers in some deserts allow winds to sweep unimpeded across these arid lands. Sand and dust carried by desert winds lowers visibility, irritates eyes and throats, and causes soil erosion because there is little or no vegetation to hold the soil in place. (● Fig. 9.17).

Hot daytime temperatures increase the atmosphere's capacity to hold moisture so that the relative humidity during daylight hours is very low (10% to 30%). Cooling occurs during the desert night from the rapid radiation loss of thermal energy in the clear air. As the temperature drops, relative humidity increases and the formation of dew in the cool hours of early morning is common. In some regions where measurements have been made, the amount of dew formed has sometimes considerably exceeded the annual rainfall for that location. It has been suggested that dew may be of great importance to plant and animal life in many desert regions. Studies are now being carried out on the use of dew as a moisture source for certain crops, to minimize the need for irrigation.

It was not until about 40 years ago that desertification became well known, as the media revealed starving people in the African Sahel, bordering the southern margin of the Sahara. Television showed bone-thin cattle trying to find a blade of grass in a barren landscape and villages invaded by sand dunes. Desertification is also occurring in areas of Spain, in northwest India, and throughout much of the Middle East, northern China, and northern Africa. Today, according to the United Nations, 2.1 billion people live in the world's deserts and drylands, and many of these areas are experiencing desertification.

The United Nations has declared 2010 to 2020 as the decade for fighting against desertification and for promoting action to protect drylands and the people who live there. The aim is to reduce large-scale environmental deterioration and the resultant human suffering. This effort includes promoting sustainable agricultural practices given local conditions and by planting trees and vegetation to protect the soil. This will require the application of land uses that foster a balance between the local ecology and human consumption of resources to meet their needs.

These two images separated by 10 years show landscape degradation in Senegal, West Africa, just south of the Sahara. Note the difference in the trees along the horizon, between the upper image, which is older and the more recent one below. Overgrazing, soil loss and human impacts, combined with drought to make this land unproductive.

Desert climates experience irregular and unreliable precipitation. Although the actual amount of water vapor (specific humidity) in the air may be high in many desert regions, the daytime relative humidity is low, and high pressure tends to dominate. When an occasional rainfall comes, it may bring more precipitation than has been recorded in years. A single rainfall can arrive as an enormous cloudburst and may cause flash flooding (● Fig. 9.18a/b).

Antofagasta, Chile, in the Atacama Desert—perhaps the world's driest location—averages 1.8 mm (0.07 in.) of annual precipitation. In March 2015, a freak storm dumped 24.4 mm (0.96 in.) of rain in 24 hours, flooding normally dry stream channels, with a rainfall accumulation that was the equivalent of 14 years of precipitation for that location.

Adaptations by Plants and Animals Vegetation in deserts tends be sparse, and large areas may be barren bedrock, sand, or gravel. The plants that survive naturally in arid lands are **xerophytic**, which means they are adapted to extreme drought. They may have thick bark, little foliage, small waxy

● **FIGURE 9.17** A giant dust storm invades Phoenix, Arizona, in July of 2011. Visibility was reduced to 1.5 m (5 ft), and this wall of dust, 1.6 kilometer high (1 mile) moved as fast as 80 kilometers per hour (50 mph). The dust storm traveled at least 240 km (150 miles) across this desert region. Although dust storms are not rare here, a dust storm of this size is expected only about once in 100 years.

How might storms like these affect human health? What about machinery and vehicles?

(a)   **(b)**

● **FIGURE 9.18** Intense rainfall in arid regions can cause flash floods like this one in the Chihuahuan Desert of west Texas. (a) The leading edge of the flash flood going down a previously dry channel. (b) Looking upstream about two minutes later.

What factors could cause floodwaters to rise so rapidly in a desert environment?

leaves, and grow in a compact shape close to the ground, all of which reduce loss of water by transpiration. Sharp thorns on xerophytic plants, and often a bitter taste as well, tend to keep browsing animals away, which can injure plants and make survival even more difficult. Another adaptation is storing moisture in stems or leaf cells, as in the many types of cactus (saguaro, barrel, and prickly pear cacti) (● Fig. 9.19). Some trees and shrubs have deep root systems to reach water, and others spread their roots widely near the surface for gathering their moisture supply.

Even humans, the most adaptable of animals, find desert environments to be a lasting challenge. For the most part, the people of the desert have been hunters and gatherers, nomadic herders, and subsistence farmers wherever there was a water supply from wells, oases, or exotic streams (desert streams carrying water from humid regions), such as the Nile, Tigris, Euphrates, Indus, and Colorado Rivers. Desert people have adjusted their ways of living to the arid environment. For example, they wear loose clothing to protect themselves from the intense rays of the sun and to prevent moisture loss from the skin by evaporation. At night, when the temperatures drop, their clothing keeps them warm by insulating and minimizing the loss of body heat.

Agriculture has been established in desert regions around the world wherever river or well water is available. Some produce mainly subsistence crops, but others have become producers of commercial crops for export.

● **FIGURE 9.19** Desert vegetation is well adapted to arid conditions, yet it also responds well to occasional precipitation. After a recent rainfall, the desert plants here in Big Bend National Park, Texas have become greener and will come into bloom.

What physical characteristics of cacti help them to survive the heat, drought, and evaporation rates of the desert?

⊕ In Google Earth fly to: 29.294°N, 103.494°W

Steppe Climates

The map shown in Figure 9.13 provides a reminder that the geographic distribution of the world's semiarid lands is closely related to the location of deserts. Like deserts, steppe regions include both low-latitude and midlatitude varieties and are identified by their mean annual temperatures (see Table 9.2). These moisture-deficient climate types share the controlling factors of the subtropical high pressure systems, rain shadow location, continentality, or some combination of these three. Their transitional nature of semiarid regions may make steppe climates seem like better-watered deserts in some locations or like slightly subhumid versions of their neighboring humid climates at another. This notion leads to a major problem of steppe regions: To what extent should people use the resources of these regions of variable and unpredictable climate?

Similarities to Deserts

A comparison of the climographs in ● Figure 9.20 with those in Figure 9.16 illustrates the basic criteria that distinguish between deserts and steppes. Steppe regions are differentiated from deserts by their greater precipitation (compared, of course, with evapotranspiration). Most low-latitude desert locations receive fewer than 25 centimeters (10 in.) of rain annually, but low-latitude steppe regions typically receive between 25 and 50 centimeters (10 to 20 in.). However, deserts and semiarid locations share many similarities. In these climates, the potential ET exceeds the precipitation. Precipitation in steppe regions, like in deserts, is unpredictable and varies widely in total amounts from year to year. Annual rainfall differs significantly from place to place within both desert and steppe regions, and vegetation varies accordingly.

The general precipitation pattern and the nature of the vegetation of a steppe region are usually closely related to the more humid climate immediately adjacent to it. Thus, when a semiarid region is located between desert and tropical savanna, the steppe's rains come with the high-sun season. Where they are adjacent to a Mediterranean climate, steppe climate regions receive primarily winter precipitation. Similarly, the short, shallow-rooted grasses most commonly associated with the steppe climate occur in the areas of transition from mesothermal and microthermal climates to desert. However, transitional areas from tropical savanna to desert, the vegetation is the dry savanna type, including scrub tree and bush growth, which becomes more stunted and sparse toward the drier margins until the typical desert shrub–vegetation type is dominant. The area shown in ● Figure 9.21 can be described as a *tropical scrub* or *tropical thorn forest, vegetative* regions with environmental conditions that can be severely limiting for human activities.

The low-latitude and midlatitude steppe climates are divided by mean annual temperature (see Table 9.2). As in the desert, temperatures in the steppe climate regions vary

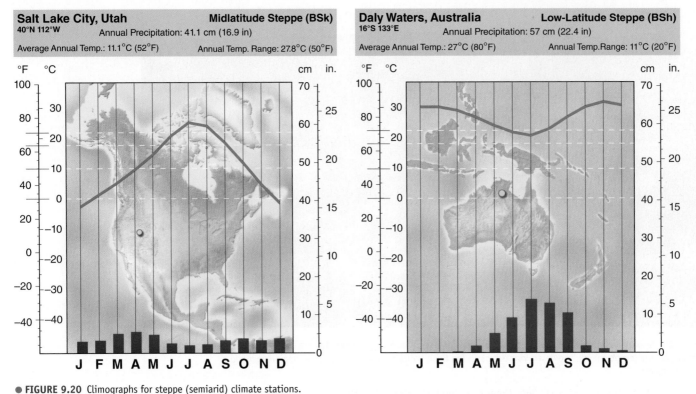

Salt Lake City, Utah	Midlatitude Steppe (BSk)
40°N 112°W	Annual Precipitation: 41.1 cm (16.9 in)
Average Annual Temp.: 11.1°C (52°F)	Annual Temp. Range: 27.8°C (50°F)

Daly Waters, Australia	Low-Latitude Steppe (BSh)
16°S 133°E	Annual Precipitation: 57 cm (22.4 in)
Average Annual Temp.: 27°C (80°F)	Annual Temp. Range: 11°C (20°F)

● **FIGURE 9.20** Climographs for steppe (semiarid) climate stations.

What are the major differences between the climograph data for Salt Lake City and Daly Waters? What are the causes of these differences?

throughout the climate type with latitude, distance from the sea, and elevation. The climographs of Figure 9.20 demonstrate that although summer temperatures are high in all steppe regions, the differences in winter temperatures can produce annual ranges in midlatitude semiarid regions that are two or three times as great as those in the low-latitudes. The environments of semiarid climates include steppe, originally the name for the grasslands of Eurasia, the prairies of the Midwestern North America, and the shrublands of higher elevation areas such as the Great Basin of the Western United States (● Fig. 9.22a, b, c).

A Dangerous Appeal Many midlatitude steppe regions support grasses that have been excellent for grazing of large herbivores. In North America, the semiarid steppe region was the realm of bison and antelope; in Africa, it is the domain of wildebeests and zebras. The soils of these regions are usually high in organic matter and soluble minerals. Attributes such as these have attracted farmers and herders alike to the rich grasslands—but not without penalty to both humans and land.

The semiarid steppe climate is problematic for agriculture, even in the midlatitude areas, and people assume a sizable risk when they attempt to farm. During dry cycles, crops can fail year

● **FIGURE 9.21** The scrub or thorn forest environment in a tropical steppe climate can be a hostile environment for humans and many animals, but these African warthogs thrive there.

Why is it important not to wander off the well-worn paths and trails in this environment?

(a)

(b)

(c)

Jim Petersen

Wind Cave National Park, NPS

Bureau of Land Management

after year, and as the land is stripped of its natural sod for agriculture, the soil is exposed to wind erosion. Grazing domesticated animals is not always a solution because overgrazing can create "Dust Bowl" conditions. Crop failure, great dust storms, and much soil loss occurred during the extreme droughts of the 1930s in the southern Great Plains of the United States.

The difficulties in making steppe regions more productive once again point out a sensitive ecological balance. Natural rains in the steppe regions are usually sufficient to support a cover of short grasses that can feed roaming herbivores that graze on them. Herbivore populations are kept in check by preying carnivores. However, when people introduce more grazing animals, plow the land, and kill off the predators, the ecological balance is tipped.

The humid tropical and arid climates represent some of the greatest worldwide extremes in weather conditions and climatic characteristics. These extremes vary from the heavy precipitation in tropical rainforest and tropical monsoon climates to the extreme dryness in subtropical desert regions. Daily temperatures vary the greatest in midlatitude desert regions, and precipitation variability is greatest in the extremely arid desert regions. Ironically, these climates may be harsh and delicate at the same time. Survival is difficult for the human populations that inhabit them, and intervention by human activities can have devastating effects on these environments.

● **FIGURE 9.22** Vegetation associations vary in midlatitude semiarid climate regions: (a) The short grass steppe of the Anatolian Plateau in central Turkey during the dry season. (b) Bison graze on the prairie grasslands of the Great Plains of Wind Cave National Park, South Dakota. (c) Sagebrush shrublands in the Great Basin region of the western United States.

CHAPTER 9 ACTIVITIES

■ TERMS FOR REVIEW

tropical rainforest climate
annual temperature range
daily (diurnal) temperature range
cloud forest
selva

jungle
slash-and-burn (shifting cultivation)
tropical monsoon climate
tropical savanna climate
desert climate

desertification
steppe climate (semiarid climate)
xerophytic

QUESTIONS FOR REVIEW

1. Describe the three humid tropical (*A*) climates, and the main climatic factors that influence each of them.
2. How do ocean currents affect the humid tropical and arid climates?
3. Describe the delicate balance between vegetative growth and soil fertility in a tropical rainforest environment.
4. What is the difference between a rainforest and a jungle? What factor differentiates these two forest types?
5. Where do you find the greatest difference in temperature between direct sunlight and shade, the rainforest or the desert?

6. Do each of the following for the listed climates: tropical rainforest, tropical monsoon, tropical savanna, desert, and steppe.
   a. Identify each climate type from a set of data or a climograph provided in the chapter indicating average monthly temperature and precipitation.
   b. Cite a geographic locational example for each of the tropical and arid climates, stated in terms of physical or political location.
   c. Distinguish between the important subtypes (if any) of each climate by identifying the characteristics that separate them from one another.

CONSIDER AND RESPOND

1. Explain the factors that affect the seasonal precipitation pattern of a tropical savanna climate. How have vegetation and soils adapted to the regime of alternating wet and dry seasons?
2. What are the three main conditions that give rise to desert climates?
3. How do steppes differ from deserts? Why might semiarid regions in some ways be more hazardous than the use of deserts?

4. What exceptions are there to the general rule that Earth–sun relationships produce the geographic pattern of climate regions?
5. What are some of the characteristics of tropical and arid climates that make them attractive places to visit?
6. Why are climographs an effective means of understanding the characteristics of the different types of climates, and making comparisons among them?

PRACTICAL APPLICATIONS

1. Based on the Köppen classification in the "Graph Interpretation" exercise (found at the end of Chapter 8), classify the following climate stations from the data provided.

		J	F	M	A	M	J	J	A	S	O	N	D	Yr.
a.	Temp. (°C)	20	23	27	31	34	34	34	33	33	31	26	21	29
	Precip. (cm)	0.0	0.0	0.0	0.0	0.2	0.1	1.1	0.2	0.0	0.0	0.0	0.0	1.6
b.	Temp. (°C)	21	24	28	31	30	28	26	25	26	27	25	22	26
	Precip. (cm)	0.0	0.0	0.2	0.8	7.1	11.9	20.9	31.1	13.7	1.4	0.0	0.0	87.2
c.	Temp. (°C)	19	20	21	23	26	27	28	28	27	26	22	20	24
	Precip. (cm)	5.1	4.8	5.8	9.9	16.3	18.0	17.0	17.0	24.0	20.0	7.1	3.0	149.0
d.	Temp. (°C)	2	4	8	13	18	24	26	24	21	14	7	3	14
	Precip. (cm)	1.0	1.0	1.0	1.3	2.0	1.5	3.0	3.3	2.3	2.0	1.0	1.3	20.6
e.	Temp. (°C)	27	26	27	27	27	27	27	27	27	27	27	27	27
	Precip. (cm)	31.8	35.8	35.8	32.0	25.9	17.0	15.0	11.2	8.9	8.4	6.6	15.5	243.8
f.	Temp. (°C)	13	17	17	19	22	24	26	26	26	24	19	15	20
	Precip. (cm)	6.6	2.0	2.0	0.5	0.3	0.0	0.0	0.0	0.3	1.8	4.6	6.6	26.7

2. The preceding table presents the climate data for the following six locations, but *not* in the order listed: Albuquerque, New Mexico; Belém, Brazil; Benghazi, Libya; Faya, Chad; Kano, Nigeria; Miami, Florida. Use this chapter and your knowledge of climates to match the climatic data with each of those locations.

3. Benghazi, Libya, and Albuquerque, New Mexico, both exhibit semiarid climates, but the cause (or control) of their dry conditions is quite different. Describe the primary factor responsible for the dry conditions at each location.

4. Tropical *B* climates are located farther away from the equator compared with *A* climates, yet their daytime high temperatures are often higher. Why?

5. Kano, Nigeria, receives great amounts of rainfall during the summer season. What are the two main sources, or causes, of this rainfall?

6. The preceding data set shows measurements collected from Death Valley, California, between 1961 and 2014. For this data set, calculate the mean annual maximum temperature and annual maximum temperature range; the mean annual minimum temperature and annual minimum temperature range; and the mean annual precipitation values and the annual precipitation range.

7. The highest temperature ever recorded in the United States was set in Death Valley, California, in July 1913, at 134°F. How much higher was that record from the mean maximum temperature for July in the data set? Is July typically the month with the highest maximum temperatures? Note: The measurements here are left in Fahrenheit and inches to match the original data.

	Jan	Feb	Mar	Apr	May	Jun	Jul	Aug	Sep	Oct	Nov	Dec
Average Max. Temperature (°F)	66.3	73.4	81.4	89.7	99.9	109.5	115.9	114.1	105.9	92.9	76.6	65.2
Average Min. Temperature (°F)	39.3	46.0	53.8	61.6	71.5	80.9	87.8	85.4	75.5	61.7	47.8	38.2
Average Total Precipitation (in.)	0.31	0.50	0.28	0.13	0.07	0.04	0.13	0.14	0.19	0.10	0.19	0.16

 MindTap—Make the most of your study time by accessing everything you need to succeed in one place. Read your textbook, take notes, review flashcards, watch videos, complete activities, take practice quizzes, and more online with MindTap. Log in at **www.cengagebrain.com.**

MIDLATITUDE, POLAR, AND HIGHLAND CLIMATE REGIONS

■ OBJECTIVES

WHEN YOU COMPLETE THIS CHAPTER YOU SHOULD BE ABLE TO:

■ 10.1 Locate the general areas of each mesothermal, microthermal, and polar climate on a world map, and explain the major factors that control their global distribution.

■ 10.2 Outline the major characteristics of each mesothermal (moderate winter), microthermal (cold winter), and polar climate.

■ 10.3 Describe the important vegetation types and human adaptations related to each mesothermal, microthermal, and polar climate.

■ 10.4 State the significant differences between the Mediterranean climate and the humid subtropical climate, and explain why the two are found on opposite coasts.

■ 10.5 Identify the differences between the humid continental hot-summer and mild-summer climate subtypes in North America.

■ 10.6 Recognize a specific mesothermal, microthermal, or polar climate according to the Köppen classification system by examining a climograph or a data set from a representative station.

■ 10.7 Explain why highland climates are so variable and what factors control the nature of a highland climate at a given location and time of year.

OUR EXAMINATION OF THE WORLD'S climates continues in this chapter, which focuses on the climates of midlatitude, polar, and highland regions. The tropical, arid, and polar climates each have a dominant and representative climatic characteristic—hot, dry, and cold, respectively. Even young children, if asked to draw a desert, a polar, or a tropical scene, can produce a picture that illustrates some characteristic of that climate's environment. The climates of the middle latitudes, however, do not have such strong defining criteria, which is one reason why they are so diverse and interesting. Variability over the year is the constant characteristic of the mesothermal and microthermal midlatitude climates. The changing of the seasons dominates these climates as they experience the varying influences of migrating air masses and seasonal shifts in cyclonic activity. In the midlatitudes, summers can vary from hot to cool and winters vary from mild to extremely cold. Precipitation regimes also exhibit a wide range because some climates receive adequate precipitation throughout the year; others experience a dry winter, and some experience dry months during the summer growing season, which can be challenging to the local populations. The midlatitude climate regions are

California redwoods (*Sequoia sempervirens*) can reach heights of 100 meters (330 ft) and live for thousands of years. They grow in cool, shady, moist, and often foggy areas. Feargus Cooney/Getty Images

home to the majority of the world's people and they include some of Earth's most productive physical environments. Understanding the climates of mesothermal, microthermal, polar and highland regions discussed in this chapter is best accomplished by studying the related index maps, the world climate map (see Fig. 8.6) and the tables that summarize the these climates.

Mesothermal Climate Regions

When the term **mesothermal** (from Greek: mesos, middle) is used to describe climates, it refers to the moderate temperatures that characterize such regions, particularly in winter. However, *mesothermal* could also refer to their middle position between climates that have high temperature throughout the year and the climates that experience severe cold (● Fig. 10.1). Mesothermal climates experience seasonality, with distinct summers and winters that distinguish them from the humid tropics. Their long summers and mild winters separate them from the seasonal regime of the higher latitude microthermal climates.

The three distinct mesothermal climates introduced in Chapter 8 are outlined in Table 10.1. All three receive an annual precipitation that exceeds the annual potential evapotranspiration. The Mediterranean climate, however, has a lengthy dry summer season with a precipitation deficit that distinguishes it from the humid subtropical and the marine west coast climates. The latter two are further differentiated by the fact that the humid subtropical regions have hot summers, whereas the marine west coast regions experience mild summers.

Mediterranean Climate

The **Mediterranean climate** *(Csa, Csb)* is an excellent example for understanding the division of world regions based on climatic characteristics and classification. Figure 10.1 shows that the Mediterranean climate is found with remarkable regularity in the vicinity of 30° to 40° N and S latitudes along the west coasts of continents. The alternating controls of subtropical high pressure in summer and the delivery of storms from westerly winds in winter are so predictable that the world's Mediterranean climate regions have similar and easily recognizable precipitation characteristics, however, summer temperatures vary considerably between cooler coastal areas and warmer inland locations. The special appearance, combinations, and climatic adaptations of Mediterranean vegetation not only are unusual but also are distinguishable from those of other climate regions. Agricultural practices, crops, recreational activities, and architectural styles all exhibit strong similarities among the world's Mediterranean climate regions.

Characteristics The major characteristics of the Mediterranean climate are a dry summer; a mild, moist winter; and abundant sunshine (90% of possible sunshine in summer and as much as

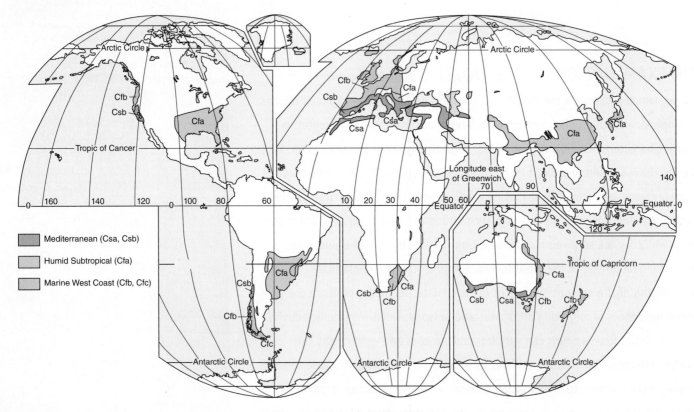

Humid Mesothermal Climates

● **FIGURE 10.1** Index map of humid mesothermal climates.

TABLE 10.1
The Mesothermal Climates

Name and Description	Controlling Factors	Geographic Distribution	Distinguishing Characteristics	Related Features
Mediterranean				
Warmest month above 10°C (50°F); coldest month between 18°C (64.4°F) and 0°C (32°F); summer drought; hot summers (inland), mild summers (coastal)	West coast location between 30° and 40°N and S latitudes; alternation between subtropical highs in summer and westerlies in winter	Central California; central Chile; Mediterranean Sea borderlands, Iranian highlands; Cape Town area of South Africa; southern and southwestern Australia	Mild, moist winters and hot, dry summers inland with cooler, often foggy coasts; high percentage of sunshine; high summer diurnal temperature range; frost danger	Sclerophyllous vegetation; low, tough brush (chaparral); scrub woodlands; varied soils, erosion in Old World regions; winter-sown grains, olives, grapes, vegetables, citrus, irrigation
Humid Subtropical				
Warmest month above 10°C (50°F); coldest month between 18°C (64.4°F) and 0°C (32°F); hot summers; generally year-round precipitation, winter drought (Asia)	East coast location between 20° and 40°N and S latitudes; humid onshore (monsoonal) air movement in summer, cyclonic storms in winter	Southeastern United States; southeastern South America; coastal southeast South Africa and eastern Australia; eastern Asia from northern India through south China to southern Japan	High humidity; summers like humid tropics; frost with polar air masses in winter; precipitation 62–250 cm (25–100 in.), decreasing inland; monsoon influence in Asia	Mixed forests, some grasslands, pines in sandy areas; soils productive with regular fertilization; rice, wheat, corn, cotton, tobacco, sugarcane, citrus
Marine West Coast				
Warmest month above 10°C (50°F); coldest month between 18°C (64.4°F) and 0°C (32°F); year-round precipitation; mild to cool summers	West coast location under the year-round influence of the westerlies; warm ocean currents along some coasts	Coastal Oregon, Washington, British Columbia, and southern Alaska; southern Chile; interior South Africa; southeast Australia and New Zealand; northwest Europe	Mild winters, mild summers, low annual temperature range; heavy cloud cover, high humidity; frequent cyclonic storms, with prolonged rain, drizzle, or fog; 3- to 4-month frost period	Naturally forested, green year-round; soils require fertilization; root crops, deciduous fruits, winter wheat, rye, pasture and grazing animals; coastal fisheries

50 to 60% during the winter rainy season). Summers are warm to hot, but there are enough differences between the monthly temperatures in coastal and interior locations to recognize two distinct subtypes of Mediterranean climates. The moderate-summer subtype *(Csb)* has lower summer temperatures associated with coastal fogs and a strong maritime influence. The hot-summer subtype *(Csa)* is located farther inland and reflects an increased influence of *continentality*. The inland version has higher summer and daytime temperatures along with cooler winter and nighttime temperatures compared with its coastal counterpart, and it also has greater annual and diurnal temperature ranges. Compare, for example, the annual temperature range for Red Bluff, California, an inland location, with that of San Francisco, directly on the coast, and about 240 kilometers (150 mi) farther south (● Fig. 10.2).

All Mediterranean climate subtypes have summers that reflect a strong influence of the subtropical highs. Weeks go by without rain, and evapotranspiration rates are high, except in very foggy areas right along the coasts. Effective precipitation is lower than actual precipitation, and the summer drought condition can be as intense as that of the desert. Days are warm to hot, skies are blue and clear, and sunshine is abundant. The high percentage of insolation coupled with the sun's high-angle rays can drive daytime temperatures as high as 30°C to 40°C+ (86°F to 104°F+), except in locations that are moderated by ocean breezes or coastal fogs.

Fog is common in coastal locations and is especially noticeable in summer. As moist maritime air moves onshore, it passes over the cold ocean currents that flow along west coasts in Mediterranean latitudes. The air cools and condenses into fogs that creep onshore from the ocean, and generally "burns off" during the daytime as the land warms. As in the desert, radiation loss is rapid at night, and even in summer, nighttime temperatures are commonly only 10°C to 15°C (50° to 60°F), especially at locations close to the coasts.

Winter is the rainy season in Mediterranean climate regions. Average annual rainfall in these areas is generally between 35 and 75 centimeters (14 and 30 in.), with 75% or more of the annual

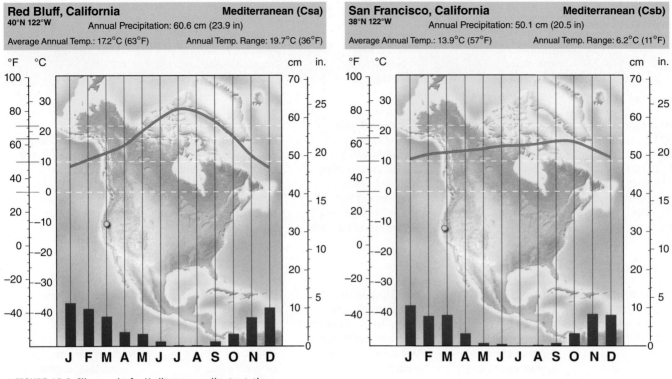

Red Bluff, California	Mediterranean (Csa)
40°N 122°W	Annual Precipitation: 60.6 cm (23.9 in)
Average Annual Temp.: 17.2°C (63°F)	Annual Temp. Range: 19.7°C (36°F)

San Francisco, California	Mediterranean (Csb)
38°N 122°W	Annual Precipitation: 50.1 cm (20.5 in)
Average Annual Temp.: 13.9°C (57°F)	Annual Temp. Range: 6.2°C (11°F)

● **FIGURE 10.2** Climographs for Mediterranean climate stations.

What is responsible for the temperature differences between these two locations?

rainfall occurring during winter. The precipitation results primarily from frontal storms associated with the westerlies during the late fall and winter. Precipitation amounts increase with elevation and generally decrease with increased distance from the ocean. Despite the long, dry summer, Mediterranean climates are classified among the humid climates, because the rain comes during cooler months, when evapotranspiration rates are lower, and annual precipitation is sufficient to qualify as a humid climate.

The rain comes mainly during the winter season, yet there are also many days of fine, mild weather, and the average temperature of the coldest month rarely falls below 4°C to 10°C (40°F to 50°F). Frost is uncommon, which is why many less hardy tropical varieties of fruits and vegetables are grown in these regions. However, when frost does occur, it can do great damage.

Special Adaptations The summer drought, not frost, is the challenge to vegetation growth in Mediterranean regions, but the natural vegetation is well adapted to the wet-dry seasonal pattern. During the rainy season, the land is covered with lush, green grasses that turn golden brown under the summer drought. When winter returns and brings rain, the landscape becomes green again. Much of the natural vegetation in Mediterranean regions is **sclerophyllous** (hard-leafed) and drought resistant. Like xerophytes, these plants have tough surfaces; shiny, thick leaves, and compact shapes that resist moisture loss; and deep roots to combat dry season aridity.

One of the most familiar plant communities is made up of many low, scrubby bushes that grow together in a thick tangle. In the western United States, this is called **chaparral** (● Fig. 10.3).

Chaparral areas have the potential for frequent wildfires typically during the summer and fall when the vegetation is especially dry. The fires help perpetuate the chaparral because the associated heat opens seedpods, which allows many chaparral species to reproduce. People living in chaparral environments often remove the vegetation as a preventive measure against fires, but this can cause other problems. When the chaparral vegetation is removed or burned off by wildfires, the soil can wash down hillsides causing mudflows that unfortunately sometimes destroy homes.

The shrub-brush vegetation association similar to the chaparral of California is called *mallee* in Australia, *matorral* in Chile, and *maquis* in France. In fact, the *Maquis* French Underground that fought against Nazi occupation in World War II literally meant "underbrush."

Trees in Mediterranean climate regions are also adapted to the wet and dry seasons. Because of their drought-resistant qualities, pines and oaks are among the more common species. Groves of deciduous and evergreen oaks appear in depressions where moisture collects and on the shady north sides of hills where evapotranspiration rates are lower. Woodlands of grasses and scattered oak trees is another common vegetation association that somewhat resembles the tropical savannas, again responding to a wet-dry seasonal regime (● Fig. 10.4). The great redwood forests of northern California exist in areas that tend to be cool and wet all year where heavy fogs regularly invade the coastal lands during summer.

In response to the dry summers, most soils in Mediterranean regions are high in soluble nutrients, but the potential for agriculture differs widely from one region to another. The lime-rich soils around the Mediterranean Sea originally were highly productive,

● **FIGURE 10.3** Typical chaparral vegetation forms a dense cover of shrubs and low trees near San Diego, California.

(a)

(b)

● **FIGURE 10.4** Seasonal change in a Mediterranean climate. (a) The summer dry season in an area of oak woodland in Northern California. (b) A nearby location in spring after the wet winter season.

How are the seasonal changes in the Mediterranean climates different from those of the Savanna climates?

but destructive agricultural practices and overgrazing over thousands of years have caused serious erosion problems. Today, bare white limestone is widely visible on hillsides in Spain, Italy, Greece, Crete, Syria, Lebanon, Israel, and Jordan.

In all Mediterranean regions, the most productive areas are lowlands covered by stream deposits. There is sufficient rainfall in the cool season to permit fall planting and spring harvesting of winter wheat and barley. These grasses originally grew wild in the eastern Mediterranean region. Grapevines, fig and olive trees, and cork oaks are especially well adapted to the dry summers because of their deep roots and thick, well-insulated stems or bark. Where water for irrigation is available, an incredible diversity of crops may be grown. These include, in addition to those already mentioned, cirrus fruits, melons, dates, rice, cotton, deciduous fruits, nuts, and vegetables. California, blessed both with fertile valleys and with snow meltwater for irrigation, is probably the most agriculturally productive of the Mediterranean regions.

The houses of Mediterranean climate regions typically reflect people's adaptations to the climate. Usually white or pastel in color, they gleam in the brilliant sunshine against clear blue skies. Many have shuttered windows to reduce incoming sunlight and to keep the houses cool in summer. However, less attention is paid to heating homes during the cooler winter months. This is true even in the United States, where many Midwesterners and easterners are surprised at the modest insulation in California homes and the small number and size of heating devices.

Humid Subtropical Climate

The **humid subtropical climate** *(Cfa)* extends inland from continental east coasts between 15° to 20° and 40° N and S latitudes (see Fig. 10.1 and Table 10.1). Thus, these climate regions are located within approximately the same latitudes and in a similar transitional position as the Mediterranean climate but on the eastern instead of the western continental margins. There is ample evidence of this climate's transitional nature. Summers in the humid subtropics are similar to the tropical climates located farther equatorward. When the noon sun is nearly overhead, these regions are subject to the inflow of warm and moist tropical air masses. High temperatures, high relative humidity, and frequent convectional showers are all characteristics that they share with the tropical climates. In contrast, during the winter months, when the pressure and wind belts have shifted equatorward, the humid subtropical regions are more commonly under the influence of the cyclonic systems of the continental midlatitudes. Polar air masses can bring colder temperatures and occasional frost.

Comparison with the Mediterranean Climate The major similarities and differences between humid subtropical and Mediterranean climates are seen by comparing the climographs of

humid subtropical New Orleans, Louisiana, and Brisbane, Australia (● Fig. 10.5) with that of Red Bluff, California (see Fig. 10.2), an inland Mediterranean climate. The two climographs indicate that both climates receive adequate precipitation, with mild winters, and hot summers—but New Orleans has no dry season. The Mediterranean regions, represented by Red Bluff in summer, are under the stable, dry, drought-producing subtropical high pressure system. In contrast, the humid subtropical regions, represented by New Orleans, near the warm ocean water of the Gulf of Mexico, receive maritime tropical air in summer, which brings humidity and rain. Furthermore, a modified monsoon effect affects many humid subtropical climates by increasing summer precipitation as moist, unstable tropical air is drawn in over the land. The climographs for New Orleans, Louisiana, and Brisbane, Australia, illustrate these effects (see Fig. 10.5).

As might be expected, the year-round precipitation for humid subtropical locations usually exceeds that for the seasonally wet Mediterranean locations. Humid subtropical regions receive anywhere from 60 to 250 centimeters (25 to 100 in.) a year. Precipitation generally decreases inland toward continental interiors and away from oceanic moisture sources. It is not surprising that these regions become increasingly drier the closer they are to inland semiarid regions along their western margins.

Both the Mediterranean and humid subtropical climates receive winter moisture from cyclonic storms, which travel with the westerlies along the polar front. As we have noted, the great contrast occurs in the summer when the humid subtropics receive substantial precipitation from convectional showers. In addition, because of the shift of the subsolar point and the wind belts during

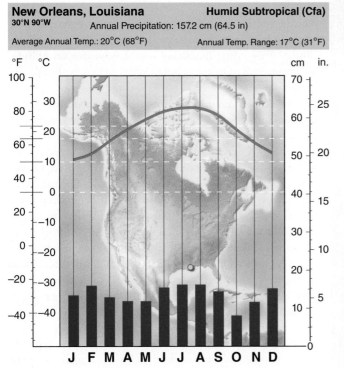

New Orleans, Louisiana   **Humid Subtropical (Cfa)**
30°N 90°W   Annual Precipitation: 157.2 cm (64.5 in)

Average Annual Temp.: 20°C (68°F)   Annual Temp. Range: 17°C (31°F)

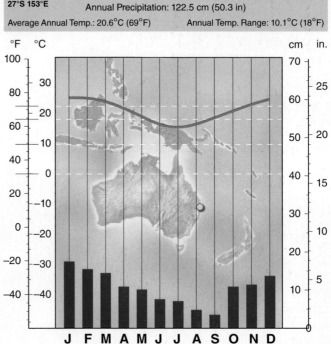

Brisbane, Australia   **Humid Subtropical (Cfa)**
27°S 153°E   Annual Precipitation: 122.5 cm (50.3 in)

Average Annual Temp.: 20.6°C (69°F)   Annual Temp. Range: 10.1°C (18°F)

● **FIGURE 10.5** Climographs for humid subtropical climate stations.

What hemispheric characteristics are shown in these graphs?

the summer months, most humid subtropical climate regions are subject to tropical storms, some of which develop into hurricanes (or typhoons), especially in later summer. Convectional activity and tropical storms combine in most of these regions to produce a precipitation maximum in late summer.

Temperatures in the humid subtropics are much like those of the Mediterranean regions. Annual ranges are similar, despite a greater variation among climate stations in the humid subtropical climate, primarily because the climate covers a much larger inland area. Mediterranean stations record higher summer daytime temperatures, but summer months in both climates average around 25°C (77°F), increasing to as much as 32°C (90°F) as maritime influence decreases inland. Winter months in both climates average around 7°C to 14°C (45°F to 57°F). Frost is a similar problem. The long growing season in the warmest humid subtropical regions enables farmers to grow delicate crops as oranges, grapefruit, and lemons, but, as in the Mediterranean climates, farmers must be prepared with various means to protect sensitive crops from the danger of freezing.

Variation in humidity greatly affects the effective temperatures—the temperatures we feel—in the humid subtropical and Mediterranean climates. Summer temperatures in humid subtropical regions can feel far warmer than they are because of the high relative humidity. In fact, summers in these climate regions can be oppressively hot, sultry, and uncomfortable. There is little or no relief from lower night temperatures, as there is in the Mediterranean regions. The high humidity of the humid subtropical climate retards the radiative loss of heat at night. Consequently, the air remains hot and sticky. Diurnal temperature ranges, in winter and in summer, are far smaller in the humid subtropical than in the drier Mediterranean regions. Despite the relatively mild temperatures, humid subtropical winters seem cold and damp, again because of the high humidity.

A Productive Climate Vegetation generally thrives in humid subtropical regions, which experience abundant rainfall, high temperatures, and long growing seasons. The wetter portions support forests of broad-leaf trees, pine forests on sandy soils, and mixed forests (● Fig. 10.6). In the drier interiors near the semiarid regions, forests give way to grasslands, which require less moisture. There are many and varied fauna. A few of the common species are deer, bears, foxes, rabbits, squirrels, opossums, raccoons, skunks, and birds of many sizes and species. Alligators inhabit the coastal swamps and wetlands of the southeastern United States from North Carolina to Texas.

Where forests still form the major natural vegetation, forest products, are important commercially (● Fig. 10.7). The long-leaf and slash pines of the southeastern United States are a source of lumber, as well as the resinous products of the pine tree (pitch, tar, resin, and turpentine). The absence of temperature and moisture limitations strongly favors forest growth. In Georgia, for example, trees may grow two to four times faster than in colder regions such as New England. This means that trees can be planted and harvested in much less time than in cooler forested regions, offering distinct commercial advantages.

National Scenic Byways / Tourism Council of Frederick County

(a)

Jereme Phillips/U.S. Fish and Wildlife Service

(b)

● **FIGURE 10.6** Variations in humid subtropical vegetation: (a) Mixed forest in the fall along a creek in Maryland, near the northern limit of this climate type in the United States. (b) Forest vegetation in Bon Secour National Wildlife Refuge, Alabama, a coastal environment on the Gulf of Mexico.

Why do these vegetation scenes look so different, yet they are both in humid subtropical climates?

⊕ In Google Earth fly to: 29.294°N, 103.494°W

Farm Service Agency/USDA

● **FIGURE 10.7** A commercial tree farm in northern Florida. Note that the pine trees are planted in orderly rows to expedite cultivation and harvest.

Why are tree farms common in the U.S. Southeast and Pacific Northwest but not in New England or the upper Midwest?

As in the tropics, soils tend to have limited fertility because of rapid removal of soluble nutrients. However, there are exceptions in drier grassland areas, such as the pampas of Argentina and Uruguay, which constitute South America's "bread basket." Whatever the soil resource, the humid subtropical regions are of enormous agricultural value because of their favorable temperature and moisture characteristics. They have been used intensively for both subsistence crops, such as rice and wheat in Asia, and commercial crops, such as cotton and tobacco in the United States. When we consider that this climate (with its monsoon phase) is characteristic of south China as well as of the most densely populated portions of both India and Japan, we realize that this climate regime contains and feeds far more human beings than any other type (● Fig. 10.8). The fact that living things thrive in the humid subtropical climate also presents certain problems because parasites

and disease-carrying insects also thrive there. Africanized "killer" bees, for example, have migrated from South America through Mexico and pose a threat to humans and livestock in the southern United States.

Agriculture and lumbering are not the only important industries in the humid subtropical climates. In the southeastern United States, livestock raising has been greatly increasing in importance. Despite the commercial advantages of this climate, people often find it an uncomfortable one in which to live. The development and spread of air conditioning helps mitigate this problem. Where the ocean offers relief from the summer heat, as in Florida, the humid subtropical climate is an attractive recreation and retirement region. The beauty of its more unusual features, such as its cypress swamps and forests draped with Spanish moss, has to be experienced to be fully appreciated.

Marine West Coast Climate

To begin this study of the **marine west coast climate** *(Cfb, Cfc)*, examine the climate's global locations. A quick review of Figure 10.1 and 8.6 reveals that marine west coast regions are found in latitudes that seem unusually high for a mesothermal climate. In some places, the northern hemisphere extent of marine west coast climates even reaches beyond the Arctic Circle. In those latitudes, this climate type is located in coastal regions adjacent to subarctic climates, the coldest of the microthermal climates. Mesothermal marine west coast regions serve as one exception to the trend of decreasing temperatures as latitudes increase toward the poles. Proximity to the sea and prevailing onshore winds make the marine west coast climate one of the most temperate in the world. Mainly found in coastal regions between 40 and 65° north and south that are continuously influenced by the westerlies, and cold water offshore, the climate receives ample precipitation and no monthly temperatures average below freezing, 0°C (32°F). Summers are mild to cool. The climograph for Bordeaux, in ● Fig. 10.9, is representative of a mild-summer marine west coast climate *(Cfb)*, and the climograph for Reykjavik, Iceland represents the cool-summer variety *(Cfc)*.

Oceanic Influences As the westerlies travel onshore, they carry with them the moderating marine influence on temperature, as well as much moisture. In addition, warm ocean currents, such as the North Atlantic Drift, bathe some of the coastal lands in the higher latitude locations of marine west coast climates, further moderating climatic conditions and accentuating humidity. This latter influence is particularly noticeable in Europe where the marine west coast climate extends along the coast of Norway to beyond the Arctic Circle (see Fig. 10.1).

In this climate zone, the marine influence is so strong that temperatures decrease relatively little in a poleward direction. The oceanic influence is even stronger than latitude in determining these temperatures. This is obvious when we examine isotherms in the areas of marine west coast climates on a map of world

Robert Essel NYC/CORBIS

● **FIGURE 10.8** The terraced fields in this humid subtropical climate region near Wakayama, Japan, are ideally suited for rice production, but level land for growing crops is limited.

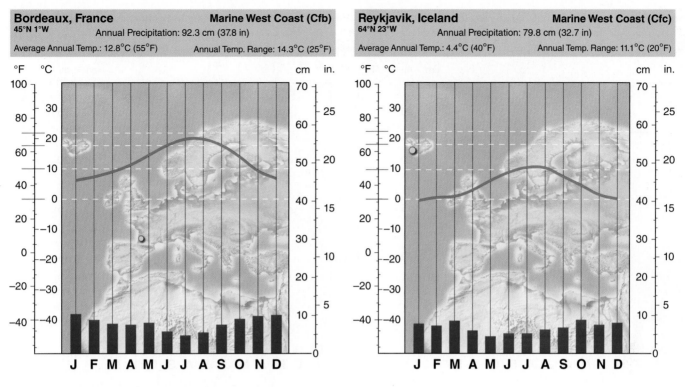

Bordeaux, France **Marine West Coast (Cfb)**
45°N 1°W Annual Precipitation: 92.3 cm (37.8 in)
Average Annual Temp.: 12.8°C (55°F) Annual Temp. Range: 14.3°C (25°F)

Reykjavik, Iceland **Marine West Coast (Cfc)**
64°N 23°W Annual Precipitation: 79.8 cm (32.7 in)
Average Annual Temp.: 4.4°C (40°F) Annual Temp. Range: 11.1°C (20°F)

● **FIGURE 10.9** Climographs for marine west coast climate stations.

How do you explain why Reykjavik has a lower temperature range than Bordeaux?

temperatures (see Fig. 4.27). Wherever a marine west coast climate prevails, the isotherms swing poleward, parallel to the coast, clearly demonstrating the dominant *marine influence.*

Another result of the ocean's moderating effect is that the annual temperature ranges in the marine west coast climates are relatively small, considering the latitude. For an illustration of this, compare the monthly temperature graphs for Portland, Oregon and Eau Claire, Wisconsin (● Fig. 10.10). Although these two cities are at about the same latitude, the annual temperature range in Portland is 15.5°C (28°F), whereas at Eau Claire, it is 31.5°C (57°F). The moderating effect of the ocean on Portland's temperatures is in contrast to the effect of *continentality* on temperatures in Eau Claire.

Diurnal temperature ranges are also smaller than they are in other climate regions at similar latitudes and in more arid climates. Heavy cloud cover and high humidity in both summer and winter diminish daytime heating and prevent much radiational cooling at night. Consequently, the difference between the daily maximum and minimum temperatures is small.

Of course, climographs are based on climate statistics, averages derived from many years of weather data, which can mask occasional occurrences of unusually high or low temperatures or precipitation totals. Marine west coast climate regions experience the unpredictable weather conditions associated with the polar front. Occasional invasions of a tropical air mass in summer or a polar air mass in winter can move against the general westerly flow of air in these latitudes and produce surprising results. For example, under just such weather conditions, temperatures in Seattle, Washington, have reached a high of 39°C (103°F) and a low of −18°C (0°F).

The insulating effect of cloudy skies and high-moisture content in the air slows heat loss at night, yet, frost is a significant factor in the marine west coast climate. It occurs more often and may last longer than in other mesothermal regions. The growing season is limited to about 8 (or fewer) months, but even during the months when freezing temperatures can occur, only up to half of the nights might experience a frost. The possibility of frost and the frequency of its occurrence increase inland more rapidly than it does poleward, once more illustrating the strong marine influence.

As further evidence of oceanic influences, study the distribution of marine west coast climates in Figure 10.1. Where mountain barriers prevent the inland movement of maritime air, marine west coast climates are often restricted to a narrow coastal strip, as in the Pacific Northwest of North America and the west coast of Chile. Where the land is surrounded by water, as in New Zealand, or where the air masses can move across broad plains, as in much of northwestern Europe, the climate extends well into the interior.

Clouds and Precipitation Marine west coast climates have a deserved reputation as one of world's cloudiest, foggiest, rainiest, and stormiest climates. This is particularly true during the winter. Rain or drizzle may occur off and on for days, although the amount of rain received is small for the number of rainy days. Even when rain is not falling, the weather is apt to be cloudy or foggy. Advection fog can be especially common and long lasting in the winter months when air masses pass over cool ocean currents and pick up considerable moisture, which condenses as a fog when the air masses move over colder land. Cyclonic storms

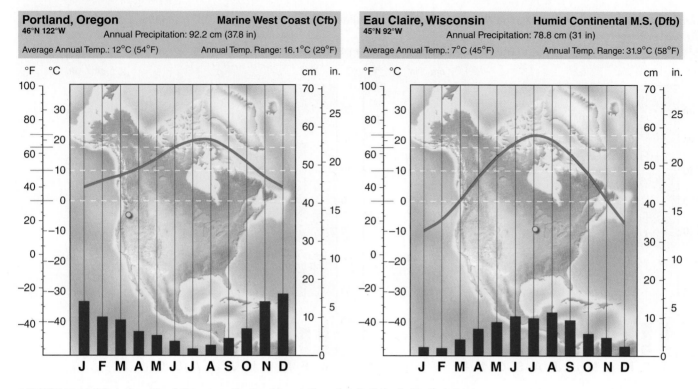

Portland, Oregon 46°N 122°W   **Marine West Coast (Cfb)**
Annual Precipitation: 92.2 cm (37.8 in)
Average Annual Temp.: 12°C (54°F)   Annual Temp. Range: 16.1°C (29°F)

Eau Claire, Wisconsin 45°N 92°W   **Humid Continental M.S. (Dfb)**
Annual Precipitation: 78.8 cm (31 in)
Average Annual Temp.: 7°C (45°F)   Annual Temp. Range: 31.9°C (58°F)

● **FIGURE 10.10** Effect of maritime influence on climates of two stations of similar latitude. Portland, Oregon, exemplifies the maritime influences of marine west coast climates. Eau Claire, Wisconsin, shows the effect of continentality. The difference in temperature ranges for the two stations is significant, but note also the interesting differences in precipitation distribution.

How do you explain these differences?

and frontal systems are also strongest in winter, a time when the subtropical highs have shifted equatorward. Conspicuous heavy winter rainfalls occur near the coasts and near the boundaries with Mediterranean climates (see again the climograph for Portland, Oregon, in Fig. 10.10).

All marine west coast climate types receive ample precipitation, although there is greater site-to-site variation in precipitation averages than in temperature statistics. Precipitation tends to decrease gradually inland, and away from the oceanic moisture source. Precipitation also decreases equatorward, especially during summer months, because the influence of the subtropical highs increases and the influence of the westerlies decreases. This can bring about periods of beautiful, clear weather; something not usually associated with this climate yet are not uncommon in the Pacific Northwest.

The most important factor affecting the amount of precipitation in this region received is local topography. When a mountain barrier such as the Cascades Range in the Pacific Northwest or the Andes in Chile parallels the coast, abundant precipitation, both cyclonic and orographic, falls on the windward side of the mountains. Valdivia, Chile, located on the windward side of the Andes, receives an average of 267 centimeters (105 in.) of precipitation a year. A similar location in Canada, Henderson Lake, British Columbia, averages 666 centimeters (262 in.) of rain a year, the highest figure for the entire North American continent. During the colder times of the Pleistocene epoch, high precipitation amounts, falling largely as winter snow, produced large glaciers in the mountains. In many cases, the glaciers in coastal areas flowed

down to the ocean, excavating deep troughs that today are elongated inlets or *fjords*, flooded by the sea. Examples of fjord coasts include Norway, British Columbia, Chile, and New Zealand—all areas of marine west coast climates today (● Fig. 10.11). In contrast, where there are lowlands and no mountains, precipitation is spread more evenly over a wide area, and the amount received at individual weather stations is more moderate, around 50 to 75 centimeters (20 to 30 in.) annually. This is the situation in much of the Northern European plain, which extends from western France to eastern Poland.

Two aspects of precipitation are directly related to the moderate temperatures of this climate regimen. Snow falls relatively infrequently and tends to melt or turns to slush without covering the ground for very long. Snow is unusual in lowland regions of this climate zone but occurs more frequently in the highest latitude areas. Paris averages only 14 snow days a year; London, 13 days; and Seattle, 10 days. In addition, thunderstorms and convectional showers are uncommon, although they occur occasionally. Even in summer, surface heating is rarely sufficient to produce the towering cumulonimbus clouds.

Resource Potential The marine west coast climate offers certain advantages for agriculture. The small annual temperature ranges, mild winters, long growing seasons, and abundant precipitation all favor plant growth. Many crops, such as wheat, barley, and rye can be grown farther poleward than they can

● **FIGURE 10.11** The scenic fjords of coastal Norway, shown here, were produced by glacial erosion during the Pleistocene ice advance.

In what other areas of the world are fjords common?

in more continental regions. Root crops (such as potatoes and beets), deciduous fruits (such as apples and pears), berries, and grapes join grains as important agricultural products. Grass in particular requires little sunshine, and pastures are always lush. The greenness of Ireland—the Emerald Isle—is an example of these

favorable conditions, as is the abundance of beef and dairy cattle. The magnificent forests that form the natural vegetation of the marine west coast regions have been a readily available resource (● Fig. 10.12). Some of the finest stands of commercial timber in the world are found along the Pacific coast of North America

● **FIGURE 10.12** A dense stand of needle-leaf coniferous trees in the marine west coast climate of Washington state in the Pacific Northwest covers the mountains along the coastline and extend well inland.

where pines, firs, and spruces commonly exceed 30 meters (100 ft) in height. Europe and the British Isles were once heavily forested, but most of those forests (even Sherwood Forest of Robin Hood fame) were cut down for building material or firewood and have been replaced by agricultural lands and urbanization.

Microthermal Climate Regions

Although a mesothermal climate type (marine west coast) extends inland from western coasts, the middle latitudes of the Northern Hemisphere are primarily the realms of **microthermal** climates (● Fig. 10.13). The definition of a *microthermal climate* includes temperatures high enough during part of the year to have a recognizable summer with at least one month averaging over 10°C (50°F) and a distinct winter with the coldest month averaging less than 0°C (32°F). It is the cold winters that distinguish microthermal climates from the moderate winter mesothermal climates. As Fig. 10.13 indicates, these climates are generally located between 35°N and 75°N on the North American and Eurasian landmasses. The dominance of continentality is well demonstrated because these climate types exist only in the Northern Hemisphere. There are no large landmasses at similar latitudes in the Southern Hemisphere that could support microthermal climates.

Three different microthermal climates are recognized based mainly on differences in the length and severity of the seasons mainly related to their latitudinal location. All of the microthermal climates experience four readily identifiable seasons. Winters tend to be longer and colder toward the northern margins because of

latitude and toward interiors because of continentality. Summers in the inland regions are inclined to be hotter, but become progressively shorter because the winter season becomes longer at higher latitudes. The three microthermal climates are *humid continental, hot summer; humid continental, mild summer;* and *subarctic,* which has a cool summer and in extreme cases, a long, bitterly cold winter. Information about these climate types is summarized in Table 10.2.

The microthermal climates share several common characteristics. As humid climates, they all experience a surplus of precipitation over potential evapotranspiration and most receive precipitation all year. The exceptions are some of the subarctic areas, which experience bitterly cold, dry air that causes winter droughts. Intrusion of maritime tropical air masses during the summer, combined with convection caused by warm temperatures, produce a precipitation maximum in the summer. Although the length of time that snow remains on the ground increases poleward and toward continental interiors, all three microthermal climates experience lasting snow cover (● Fig. 10.14). The high albedo of snow decreases the effectiveness of warming by solar radiation, which contributes to the cold winter temperatures experienced at these locations. Finally, unpredictable and variable weather is especially apparent in the humid microthermal climates.

Considering these generalizations, compare the microthermal climates with the mesothermal climates we previously examined. Regions with microthermal climates have more severe winters, a lasting snow cover in winter, shorter summers, shorter growing seasons, shorter frost-free seasons, a year with four distinct seasons, greater average annual temperature ranges, and much more variable weather compared to the mesothermal climate regions.

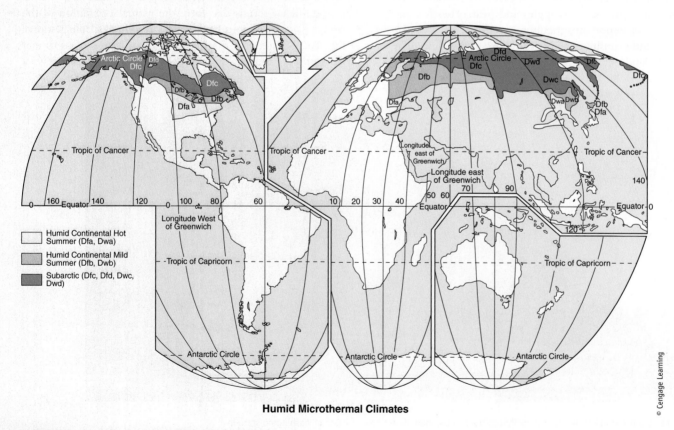

Humid Microthermal Climates

Humid Continental Hot Summer (Dfa, Dwa)

Humid Continental Mild Summer (Dfb, Dwb)

Subarctic (Dfc, Dfd, Dwc, Dwd)

© Cengage Learning

● **FIGURE 10.13** Index map of humid microthermal climates.

TABLE 10.2
The Microthermal Climates

Name and Description	Controlling Factors	Geographic Distribution	Distinguishing Characteristics	Related Features
Humid Continental, Hot Summer				
Warmest month above 10°C (50°F); coldest month below 0°C (32°F); hot summers; usually year-round precipitation, winter drought (Asia)	Location in the midlatitudes (35°–45°N); cyclonic storms along the polar front; prevailing westerlies; continentality; polar anticyclone in winter (Asia)	Eastern and Midwestern U.S. from Atlantic coast to 100°W longitude; east central Europe; northern and northeast (Manchuria) China, northern Korea, and Honshu (Japan)	Hot, often humid summers; occasional winter cold waves; rather large annual temperature ranges; weather variability; precipitation 50–115 cm (20–45 in.), decreasing inland and poleward; 140 to 200-day growing season	Broad-leaf deciduous and mixed forest; moderately fertile soils with fertilization in wetter areas; highly fertile grassland and prairie soils in drier areas; "corn belt," soybeans, hay, oats, winter wheat
Humid Continental, Mild Summer				
Warmest month above 10°C (50°F); coldest month below 0°C (32°F); mild summers; usually year-round precipitation, winter drought (Asia)	Location in the middle latitudes (45°–55°N); cyclonic storms along the polar front; prevailing westerlies; continentality; polar anticyclone in winter (Asia)	New England, the Great Lakes region, and south central Canada; southeastern Scandinavia; eastern Europe, west central Asia; eastern Manchuria (China) and Hokkaido (Japan)	Moderate summers; long winters with frequent spells of clear, cold weather; large annual temperature ranges; variable weather with less total precipitation than farther south; 90- to 130-day growing season	Mixed or coniferous forest; moderately fertile soils with fertilization in wetter areas; highly fertile grassland and prairie soils in drier areas; spring wheat, corn for fodder, root crops, hay, dairying
Subarctic				
Warmest month above 10°C (50°F); coldest month below 0°C (32°F); cool summers, cold winters poleward; usually year-round precipitation, winter drought (Asia)	Location in the higher midlatitudes with part in the polar regions (50°–70°N); westerlies in summer, strong polar anticyclone in winter (Asia); occasional cyclonic storms; extreme continentality	Northern North America from Newfoundland to Alaska; northern Eurasia from Scandinavia through most of Siberia to the Bering Sea and the Sea of Okhotsk	Brief, cool summers; long, bitterly cold winters; largest annual temperature ranges; lowest temperatures outside Antarctica; low precipitation, 20–50 cm (10–20 in.); unreliable 50- to 80-day growing season; permafrost common	Northern coniferous forest (taiga); strongly acidic soils; poor drainage and swampy conditions in warm season; experimental vegetables and root crops

Humid Continental, Hot-Summer Climate

The **humid continental, hot-summer climate** (Dfa, Dwa) is relatively limited in its distribution in Eurasia (see Table 10.2). In the United States, this climate is distributed over a wide area that begins with the eastern seaboard of New York, New Jersey, and southern New England and stretches across the heartland of the eastern United States to encompass much of the American Midwest.

The environmental conditions in the hot-summer variety of humid continental climates have some advantages over their higher latitude counterparts. Higher summer temperatures and a longer growing season permit a wide variety of agriculture. Areas within the hot-summer regions have soils that are generally more fertile, especially under forest cover. Of course, some advantages are matched by liabilities. The lower fuel bills of winter in the humid continental, hot-summer climate are often more than offset by the cost of air conditioning during the long, hot summers that are not experienced in other microthermal climates.

Internal Variations From place to place there are significant differences in temperature characteristics within humid continental, hot-summer climate regions. The length of the growing season

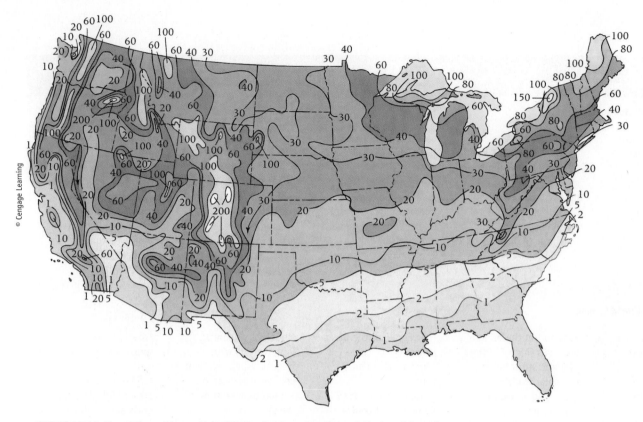

● **FIGURE 10.14** Map of the contiguous United States showing average annual number of days of snow cover.

What areas of the United States average the greatest number of days of snow cover?

is closely related to latitude, varying from 200 days equatorward to as little as 140 days along poleward margins of this climate type.

Continentality has the greatest impact on temperature differences. The degree of continentality and the size of a continent significantly affect both summer and winter temperatures and, as a result, on the annual temperature range. The climographs in ● Fig. 10.15 display the differences in summer and winter temperatures recorded at two weather stations for comparing the temperature ranges. Galesburg, Illinois, a typical humid continental (Dfa) station, has a summer that is less hot, less frigid winters, and a lower annual temperature range than Shenyang, in northeast China, which is located at almost the same latitude but experiences the greater seasonal contrasts of the Eurasian landmass.

The humid continental, hot-summer climate regions, experience yearly temperature ranges that are consistently large and they become larger toward continental interiors. Coastal locations in this climate region experience temperatures that are modified by a slight *maritime* influence and have summers that are less hot, and the winters are not as cold, as those at inland locations at comparable latitudes. Nearness to large lakes can cause a similar moderating effect, mainly in the summer. Precipitation is also variable from station to station, as the annual precipitation decreases both poleward and inland as those locations become farther away from the source regions of warm maritime air masses that provide moisture for storms and convectional showers. This decrease can be seen in the annual precipitation figures for the following cities, all at a latitude

of about 40°N: New York (longitude 74°W), 115 centimeters (45 in.); Indianapolis, Indiana (86°W), 100 centimeters (40 in.); Hannibal, Missouri (92°W), 90 centimeters (35 in.); and Grand Island, Nebraska (98°W), 60 centimeters (24 in.). Most of these locations have a precipitation maximum in summer when the warm, moist air masses dominate. Certain regions of Asia also receive a summer maximum of precipitation, but have a winter drought because the monsoon circulation inhibits precipitation during that season (see again Shenyang in Figure 10.15).

Vegetation and soils vary in response to precipitation amounts and seasonality. In the wetter regions, forests and forest soils predominate. At one time in sections of the American Midwest, tall prairie grasses grew where precipitation was too low to support forests. In the drier areas of this climate type, grasslands are the natural vegetation, and the soils that develop here are among the richest in the world.

Seasonal Changes All four seasons are highly developed in the humid continental, hot-summer climate regions, and each has a its own distinct character. The winter is cold and often snowy; the spring is warmer, with showers that produce flowers, budding leaves, and green grasses; the hot, humid summer brings occasional thunderstorms; and the fall brings both clear and rainy weather, with mild days and frosty nights as the green trees turn to colorful reds, oranges, yellows, and browns before dropping their leaves (● Fig. 10.16).

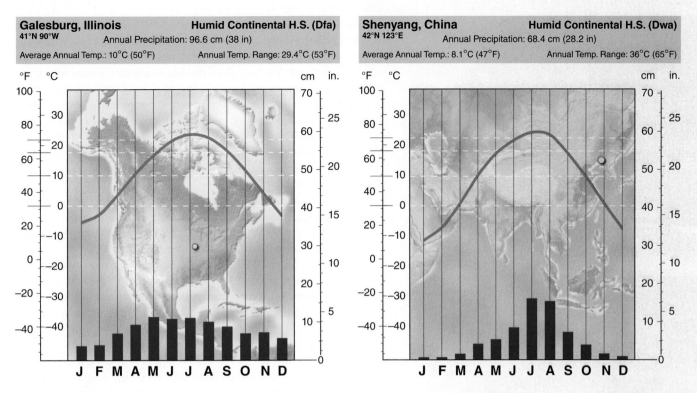

Galesburg, Illinois	Humid Continental H.S. (Dfa)
41°N 90°W	Annual Precipitation: 96.6 cm (38 in)
Average Annual Temp.: 10°C (50°F)	Annual Temp. Range: 29.4°C (53°F)

Shenyang, China	Humid Continental H.S. (Dwa)
42°N 123°E	Annual Precipitation: 68.4 cm (28.2 in)
Average Annual Temp.: 8.1°C (47°F)	Annual Temp. Range: 36°C (65°F)

● **FIGURE 10.15** Climographs for humid continental, hot-summer climate stations.

What are the reasons for the differences in temperature and precipitation between the two stations?

The summers are long, humid, and hot in these humid continental climates. The centers of migrating low pressure systems usually pass to the north, and in summer maritime tropical air dominates. Hot spells can continue day and night for a week or more, with only temporary relief available from convectional thunderstorms or an occasional cold (cool) front. Asia, in particular, experiences the heavy summer precipitation associated with the monsoon of that season. The summer heat and humidity in the humid continental, hot-summer climates encourage vigorous vegetative growth and are ideal for insects—mosquitoes, flies, gnats, and bugs of all kinds.

Winters are not as severe as those of the cool summer variety, located farther poleward, but average January temperatures are usually between −5°C and 0°C (23°F to 32°F) or lower. The averages tell only part of the story, because there invariably is an invasion of very cold, dry, arctic air once or twice during the winter. This often occurs just after a cyclonic storm has passed and the ground is covered with snow. The sky remains clear and blue for days at a time; the temperatures will be freezing −18°C (0°F) or colder and may occasionally dip to below zero °F at night with typical temperatures of −30°C or −35°C (−22°F or −31°F). The ground remains frozen for long periods and snow cover may be present for several days, or even weeks, at a time. However, these characteristics do not last all winter because sudden weather contrasts are common. Cold air precedes warmer air, and a thaw follows a freeze. Vegetation remains dormant throughout the winter season but bursts into life again in spring. During its early growth period, vegetation is in constant danger of a late-spring frost.

The atmospheric changes within seasons are just as significant as those between seasons. The humid continental, hot-summer climate is the classic example of variable middle-latitude weather. Cyclonic storms are born as tropical air masses move northward and confront polar air masses migrating to the south. Stormy frontal activity is followed by the clear conditions of a following anticyclone. The general atmospheric circulation in these latitudes continuously carries cyclones and anticyclones toward the east along the polar front. When the polar front is most directly over these regions, as it is in winter and spring, one storm and its associated fronts seem to follow another with such regularity that the only safe weather prediction is that the weather will change.

Humid Continental, Mild-Summer Climate

If you review the geographic distribution of the humid continental, hot-summer, and mild-summer climates in Figure 10.13, their close relationship is unmistakable. Where one is found, the other also is found; in each situation, the mild-summer climates invariably lie adjacent to and north of the humid continental hot-summer climates.

In most instances, the **humid continental, mild-summer climate** *(Dfb, Dwb)* is a more continental or severe-winter version of its equatorward counterpart, characterized by distinct seasonality. There is significant variation, particularly with respect to precipitation. Variable weather is the rule, and storms along the polar front provide most of the precipitation within this climate type.

(a)

(b)

(c)

R. Gabler

● **FIGURE 10.16** Deciduous forests in humid microthermal regions change dramatically with the seasons. (a) The green leaves of summer (b) change to reds, golds, and browns in fall and (c) drop to the forest floor in winter. Leaf dropping in cold winter climates, such as here in western Illinois, is a means of minimizing transpiration and moisture loss when the soil water is frozen.

Mild-Summer–Hot-Summer Comparison In all three microthermal climates, precipitation tends to decrease poleward; therefore, the humid continental, mild-summer climate tends to have less precipitation than the hot-summer regions, which are closer to sources of humid air. Precipitation continues to decrease throughout this climate type toward the poleward margins and from the coasts toward the semiarid continental interiors. Just as in its hot-summer counterpart, the monsoon effect in the Asian mild-summer climate regions is strong enough to produce a dry-winter season as seen on the climograph for Vladivostok in ● Fig. 10.17.

Other significant differences between the mild-summer and hot-summer types of humid continental climates become readily apparent by comparing the climographs in Figure 10.17 with those of Figure 10.15. The climographs of typical hot-summer stations in Figure 10.15 indicate both higher temperatures and summer warmth that lasts longer compared with those of mild-summer stations. Note that in humid continental, hot-summer climates, the temperatures average 10°C (50°F) or more for 1 to 2 months longer than temperatures in the mild-summer climates. In addition, this comparison also shows that the winters of mild-summer stations are colder and average sub-freezing temperatures for 1 to 2 months longer than winters of hot-summer stations. The combination of more severe winters and shorter summers makes for a growing season of between 90 and 130 days, which is 1 to 3 months shorter than in hot-summer regions. In addition, although overall precipitation totals of 50 to 100 centimeters (20 to 39 in.) are generally lower, snowfall is greater, and snow cover is thicker and lasts longer (● Fig. 10.18).

The humid continental, mild-summer regions exhibit seasonal changes just as clearly as the hot-summer regions. Annual temperature ranges are generally larger. Vigorous interaction between polar and tropical air masses makes for frequent changes in the weather. However, the more poleward positions of the mild-summer climate experience a greater dominance of colder air masses, and temperature variability is not as great as it is farther south. Under normal conditions, tropical air is strongly modified by the time it reaches the higher latitude mild-summer regions and even in summer, intrusions of warm, humid air rarely last more than a few days. By contrast, winter invasions of very cold arctic air periodically bring several successive days or weeks of clear skies and frigid temperatures.

As in the hot-summer climate, the wetter regions of the mild-summer climate are associated with natural forest vegetation. However, many common trees in the hot-summer climate, such as oaks, hickories, and maples, find it difficult to compete with firs, pines, and spruces toward the colder, polar margins of these regions.

Human Activity in the Humid Continental Climates

Perhaps the greatest contrast between the hot-summer and mild-summer humid continental regions is exhibited in agriculture. Despite the unpredictability of the weather, humid continental, hot-summer agricultural regions are among the world's finest. The favorable combination of long hot summers, ample rainfall, and

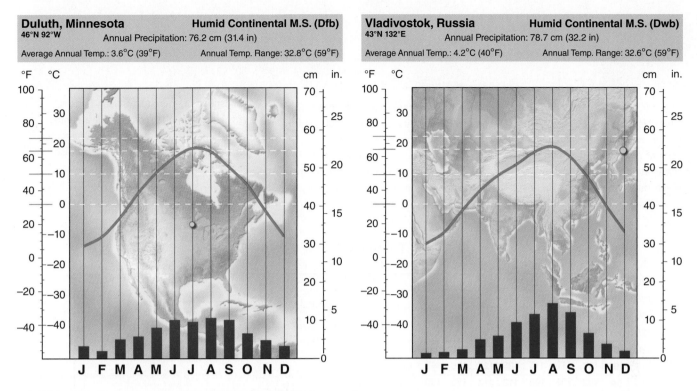

Duluth, Minnesota Humid Continental M.S. (Dfb)
46°N 92°W
Annual Precipitation: 76.2 cm (31.4 in)
Average Annual Temp.: 3.6°C (39°F)   Annual Temp. Range: 32.8°C (59°F)

Vladivostok, Russia Humid Continental M.S. (Dwb)
43°N 132°E
Annual Precipitation: 78.7 cm (32.2 in)
Average Annual Temp.: 4.2°C (40°F)   Annual Temp. Range: 32.6°C (59°F)

● **FIGURE 10.17** Climographs for humid continental, mild-summer climate stations.

What characteristics of these climographs distinguish them from the climographs of Figure 10.15?

● **FIGURE 10.18** The humid continental, mild-summer regions can receive abundant snow, which may be present continuously for long periods. The snowfalls are occasionally extremely heavy, particularly when enhanced by the lake effect, such as here in Buffalo, New York, in 2014.

What other problems does snow cause for people living in these regions?

highly fertile soils has made the American Midwest a leading agricultural producer. Soybeans, which are native to similar climate regions in northern China, are now second to corn throughout the Midwest. Wheat, barley, and other grains are especially important. In the mild-summer climate, however, a shorter growing season restricts the crops that can be grown. Farmers in the more northerly regions rely more on quick-ripening varieties, grazing animals, orchards, root crops, and dairy farming.

The length of the growing season is one reason for the differences in agriculture between the two humid continental climates, but there is another climate-related factor. The great ice sheets of the Pleistocene epoch had different effects on the mild-summer and hot-summer regions. In the hot-summer regions, the ice sheets thinned and receded, releasing an enormous load of soil and solid rock debris and sediments. The material was deposited in a blanket hundreds of feet thick in the areas of maximum glacial advance. As the ice retreated northward, less and less debris was deposited, much of it flushed away by meltwater streams. Consequently, the more southerly, hot-summer microthermal regions have an undulating topography underlain by thick masses of glacial debris. The soils formed on this debris are well developed and fertile. The more northerly, mild-summer region, however, mainly shows the effects of glacial erosion, rather than deposition. Rockbound lakes and marshy lowlands alternate with ice-scoured bedrock hills. Soils are either thin and stony or waterlogged. Because of its lower agricultural potential, large sections of this area remain in forest. However, because of its wilderness character and the abundance of lakes in basins produced by Ice Age glaciers, recreational possibilities in the mild-summer regions far exceed those of a more subdued hot-summer region. Minnesota calls itself the "Land of

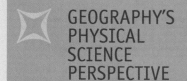

Effective Temperatures

Effective temperatures (formerly known as *sensible temperatures*) are temperatures as they might be experienced by a person at rest, in ordinary clothing, in a motionless atmosphere. In other words, at any given time, how comfortable does a temperature feel to an individual? This temperature value cannot be obtained by simply reading a thermometer.

Several factors come into play when considering effective temperatures. These factors can be divided into two categories: atmospheric factors and human factors. Of the atmospheric factors, four are most important. First is the actual air temperature; thermometers will help distinguish between cold and warm days. Second is humidity; because the evaporation of sweat

Relative humidity (%)

Air temperature (°F)	0	5	10	15	20	25	30	35	40	45	50	55	60	65	70	75	80	85	90	95	100
140	125																				
135	120	128																			
130	117	122	131																		
125	111	116	123	131	141																
120	107	111	116	123	130	139	148														
115	103	107	111	115	120	127	135	143	151												
110	99	102	105	108	112	117	123	130	137	143	150										
105	95	97	100	102	105	109	113	118	123	129	135	142	149								
100	91	93	95	97	99	101	104	107	110	115	120	126	132	138	144						
95	87	88	90	91	93	94	96	98	101	104	107	110	114	119	124	130	136				
90	83	84	85	86	87	88	90	91	93	95	96	98	100	102	106	109	113	117	122		
85	78	79	80	81	82	83	84	85	86	87	88	89	90	91	93	95	97	99	102	105	108
80	73	74	75	76	77	77	78	79	79	80	81	81	82	83	85	86	86	87	88	89	91
75	69	69	70	71	72	72	73	73	74	74	75	75	76	76	77	77	78	78	79	79	80
70	64	64	65	65	66	66	67	67	68	68	69	69	70	70	70	70	71	71	71	71	72

Heat index (or apparent temperature)

A table of effective temperatures can be used to determine the heat index, which combines conditions of temperature and humidity to find an apparent temperature.

10,000 Lakes," and in New York and New England, lakes, rough mountains, and forests combine to produce some of the most spectacular scenery east of the Rocky Mountains (● Fig. 10.19).

Subarctic Climate

The **subarctic climate** *(Dfc, Dfd, Dwc, Dwd)* is the farthest poleward and most extreme of the microthermal climates. By definition, it has at least 1 month with an average temperature warmer than 10°C (50°F), so the poleward limit of this climate roughly coincides with the 10°C isotherm for the warmest month of the year. Forests cannot survive where at least 1 month does not have an average temperature over 10°C, so the poleward boundary of subarctic climates is also the latitudinal limit of forest growth.

As the map in Figure 10.13 indicates, subarctic climates, like the other microthermal climates, are found only in the Northern Hemisphere. They cover vast areas of subpolar Eurasia and North America, and conditions vary widely over these great distances. Extremely severe winter regions are located along the polar margins or deep in the interior of the Asia, and climate subtypes with winter drought are found in association with the Siberian high with its clear skies, and bitter cold over interior Asia during winter. Other subarctic regions experience less severe winters and year-round precipitation.

Ocean currents have an influence on the geographical distribution of subarctic climates. Along the west coasts of continents, especially in North America, the ocean modifies temperatures sufficiently to extend marine west coast climates into latitudes that normally would be subarctic and the subarctic to extend

is a cooling process for the human body, humid days feel warmer than dry days. Third is wind speed; winds not only carry heat from the body but can also accelerate the evaporation of sweat. Fourth is the percentage of clear sky; shady areas are cooler than sunny areas.

Of the many human factors, the following stand out as important. First is respiration; breathing in a lungful of cold air will make one feel colder. Second is perspiration; the evaporative cooling process is quite efficient for the human body, but it differs from one individual to another. Third is the amount of activity involved; physical work or playing a physical sport can heat the body rapidly. Fourth is the amount of exposed skin; tank tops versus sweatshirts can make a world of difference.

Effective temperatures are established by considering the interplay between these two sets of factors. For example, the well-known heat index takes into account the temperature and humidity of a summer day and calculates how it might feel. The equally well-known wind chill index considers both temperature and wind speed to establish how cold one might feel on a winter day. Keep in mind that these are only theoretical values. No one can predict exactly how comfortable a particular person will feel on a given day. However, these temperature indices can help guide us when dealing with seasonally extreme days.

Wind Chill Chart

Temperature (°F)

Wind (mph) \ Calm	40	35	30	25	20	15	10	5	0	-5	-10	-15	-20	-25	-30	-35	-40	-45
5	36	31	25	19	13	7	1	-5	-11	-16	-22	-28	-34	-40	-46	-52	-57	-63
10	34	27	21	15	9	3	-4	-10	-16	-22	-28	-35	-41	-47	-53	-59	-66	-72
15	32	25	19	13	6	0	-7	-13	-19	-26	-32	-39	-45	-51	-58	-64	-71	-77
20	30	24	17	11	4	-2	-9	-15	-22	-29	-35	-42	-48	-55	-61	-68	-74	-81
25	29	23	16	9	3	-4	-11	-17	-24	-31	-37	-44	-51	-58	-64	-71	-78	-84
30	28	22	15	8	1	-5	-12	-19	-26	-33	-39	-46	-53	-60	-67	-73	-80	-87
35	28	21	14	7	0	-7	-14	-21	-27	-34	-41	-48	-55	-62	-69	-76	-82	-89
40	27	20	13	6	-1	-8	-15	-22	-29	-36	-43	-50	-57	-64	-71	-78	-84	-91
45	26	19	12	5	-2	-9	-16	-23	-30	-37	-44	-51	-58	-65	-72	-79	-86	-93
50	26	19	12	4	-3	-10	-17	-24	-31	-38	-45	-52	-60	-67	-74	-81	-88	-95
55	25	18	11	4	-3	-11	-18	-25	-32	-39	-46	-54	-61	-68	-75	-82	-89	-97
60	25	17	10	3	-4	-11	-19	-26	-33	-40	-48	-55	-62	-69	-76	-84	-91	-98

Frostbite occurs in 15 minutes or less

A table of effective temperatures can be used to determine wind chill, which shows what the combination of cold temperatures and wind speed will make the temperature outside feel like.

well beyond the Arctic Circle. Along east coasts, with a stronger influence of continentality, subarctic climates reach farther south. Subarctic climates are also not as extensive in North America as they are in Eurasia. This is because the huge Eurasian landmass increases the effect of continentality, and the large water surface of Hudson Bay in Canada provides a modifying influence, which partly counters the effect of continentality there.

Effects of High Latitude and Continentality Subarctic regions experience short, cool summers and long, bitterly cold winters, as seen in the temperature curves of the subarctic stations in ● Figure 10.20. Both Eagle, Alaska, and Verkhoyansk, Russia, experience more months with average temperatures below freezing than above. Note how steeply the temperatures rise and fall between summer and winter. The rapid heating and cooling associated with continental interiors in the higher latitudes allow little time for the in-between seasons of spring and fall. At Eagle, Alaska, a station in the Klondike region of the Yukon River Valley, the temperature climbs 8°C to 11°C (15°F to 20°F) per month as summer approaches and drops just as rapidly prior to the next winter season. At Verkhoyansk in Siberia, the change between the seasonal extremes is even more rapid, averaging 17°C to 22°C (30°F to 40°F) per month.

Because these regions are in the high latitudes, summer daylight hours are quite long and nights are short. The noon sun is as high in the sky during a subarctic summer as during a subtropical winter. The combination of a moderately high sun angle and many hours of daylight means that some subarctic locations receive as much solar

● **FIGURE 10.19** Although the effect of glacial action farther south was to deposit material, here in New Hampshire we see an area of glacial erosion—Newfound Lake, a former glacial valley.

How might this area be considered an economic resource?

recorded there and are officially −68°C (−90°F) at both Verkhoyansk and Oymyakon, Russia, and unofficially −78°C (−108°F) at Oymyakon. In addition, the winter nights, with an average 18 to 20 hours of darkness extending well into one's working hours can be mentally depressing and increase the impression of climatic severity.

As a direct result of the intense seasonal heating and cooling of the land, the subarctic has the largest annual temperature ranges of any climate. Annual ranges in the equatorward margins of the climate vary from near 40°C (72°F) to more than 45°C (80°F) between summer and winter. The exceptions are near western coasts, where warm ocean currents and the marine influence may significantly modify winter temperatures. Annual temperature ranges for poleward stations are even greater. The climograph for Verkhoyansk indicates an extreme example with a summer-to-winter range of 64°C (115°F).

Latitude and continentality also influence the precipitation in subarctic regions. These climate controls combine to limit annual precipitation amounts to less than 50 centimeters (20 in.) for most

energy during the summer solstice as the equator does. As a result, temperatures during the 1 to 3 months of the subarctic summer usually average 10°C to 15°C (50°F to 60°F), and on some days they may approach 30°C (86°F). Thus, the brief summer in the subarctic climate is often pleasantly warm, and occasionally even hot.

The winter season in the subarctic is bitter, intense, and lasts for as long as 8 months. Eagle, Alaska, has 8 months with average temperatures below freezing. In the Siberian subarctic, the January temperatures regularly *average* −40°C to −51°C (−40°F to −60°F). The coldest temperatures in the Northern Hemisphere have been

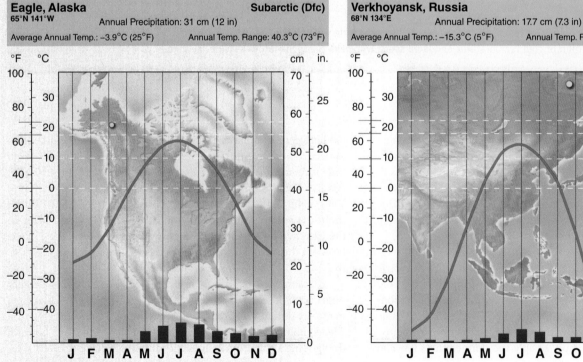

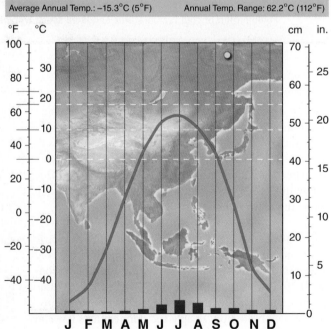

Eagle, Alaska — 65°N 141°W — **Subarctic (Dfc)**
Annual Precipitation: 31 cm (12 in)
Average Annual Temp.: −3.9°C (25°F) — Annual Temp. Range: 40.3°C (73°F)

Verkhoyansk, Russia — 68°N 134°E — **Subarctic (Dwd)**
Annual Precipitation: 17.7 cm (7.3 in)
Average Annual Temp.: −15.3°C (5°F) — Annual Temp. Range: 62.2°C (112°F)

● **FIGURE 10.20** Climographs of the subarctic climate stations.

Why would people settle in such severe-winter climate regions?

regions and to 25 centimeters (10 in.) or less in northern and interior locations. The low temperatures of subarctic climates reduce the moisture-holding capacity of the air, and minimize precipitation received from cyclonic storms. A location toward the center of large landmasses increases the distance from oceanic sources of moisture. The higher latitude subarctic climates are also dominated by the polar high, especially in the winter season, which limits opportunities for precipitation. In addition, this high pressure system also blocks the entry of moist air from warmer areas to the south.

Subarctic precipitation is cyclonic or frontal because the polar high is weaker and has moved farther north along with the polar front during the warmer summer months, so more precipitation comes during that season. Winter precipitation is meager and falls as fine, dry snow, and it is so cold for so long that the snow cover lasts for as much as 7 or 8 months. During this time, there is almost no melting of snow, especially in the shaded areas of the forest.

A Limiting Environment The climatic restrictions of subarctic regions place distinct limitations on plant and animal life and on human activity. The characteristic vegetation is coniferous forest, adapted to the severe temperatures; the physiologic drought associated with frozen soil water; and the infertile soils. Seemingly endless **boreal forests** of spruce and fir thrive over enormous areas (● Fig. 10.21). In Russia, this forest type is called the *taiga*.

The brief summers and long, cold winters severely limit the growth of vegetation in subarctic regions. The trees are shorter and more slender than comparable species in less severe climate regimens. There is little hope for agriculture, as the growing season averages only 50 to 75 days, and frost can occur even during June, July, or August. In some years, a subarctic location might not have a truly frost-free season.

A particularly vexing problem to people in subarctic (and tundra) regions is **permafrost**, a permanently frozen layer of subsoil and underlying rock. Permafrost is present in wide areas of subarctic climates, but it varies greatly in thickness and is often discontinuous. Where permafrost exists, both the ground surface and the subsurface are completely frozen in the winter. Spring and summer bring warm temperatures, causing the top few feet to thaw out, but because the land beneath this thawed top layer remains frozen, water cannot percolate downward. The thawed soil becomes saturated with moisture, especially in spring, when there is an abundant supply of water from melting snow. The seasonal freeze and thaw cycle causes repeated expansion and contraction, heaving the surface up and then letting it sag down. The effects of this cycle break up roads, force buried pipelines out of the ground, cause walls and bridge piers to collapse, and even cause buildings to tilt or even sink a few feet. Landscapes with surfaces called **patterned ground (frost polygons)** are commonly found in subarctic and tundra regions of seasonal freezing and thawing (● Fig. 10.22).

There are few economic incentives to draw people to subarctic regions. Logging is unimportant because trees are often crooked, small, and thin in the boreal forests. Even use of the vast forests for paper, pulp, and wood products is restricted by their distant interior location in difficult terrain. Miners work ore deposits, and many people who are native to the subarctic regions pursue hunting, trapping, and fishing.

Melissa Gabrielson/US Fish and Wildlife Service, Alaska

● **FIGURE 10.21** Boreal forest is typical of the vegetation throughout much of the American and Canadian subarctic. This scene is in Alaska's Yukon Delta National Wildlife Refuge.

Why are these kinds of virgin forests currently of little economic value?

⊕ **In Google Earth fly to: 60.83° N, 161.65° W**

Polar Climate Regions

The polar climate is the last of Köppen's humid climate subdivisions that are differentiated on the basis of temperature. These climate regions are situated at the greatest distance from the equator. All of the average monthly temperatures in polar climate areas are below 10°C (50°F), so they do not experience a warm summer (Table 10.3). Trees cannot survive under these conditions. In the polar climate regions where at least 1 month averages above freezing 0°C (32°F), forests are replaced by treeless tundra vegetation. Elsewhere, the surface is covered by great expanses of glaciers. There are two polar climate types: tundra *(ET)* and ice sheet *(EF)*.

Two significant characteristics of polar climate regions are important to know. First, these climates experience a large net annual radiation loss; that is, they radiate away much more energy than they receive from the sun during a year. The heat transfer from lower to higher latitudes that helps make up this deficiency is a driving force of the global atmospheric circulation. Without this

● **FIGURE 10.22** Ice freezing in the subsurface pushed up these gravel mounds, here in this tundra of the Yukon Delta National Wildlife Refuge, Alaska. Repeated freezing and thawing cause the surface to produce mounds, some with polygonal shapes. When the ground freezes, it expands, and it shrinks when it thaws.

Why are the edges heaved upward?

compensating transfer of heat from the lower latitudes, the polar regions would become too cold to permit any form of life and the equatorial regions would heat to temperatures that no organisms could survive.

A second and equally important characteristic of polar climates is the unique pattern of day and night. The North and South Poles experience 6 months of relative darkness when the sun is positioned below the horizon, alternating with 6 months of

TABLE 10.3
The Polar Climates

Name and Description	Controlling Factors	Geographic Distribution	Distinguishing Characteristics	Related Features
Tundra				
Warmest month between 0°C 32°F and 10°C (50°F); precipitation exceeds potential evapotranspiration	Location in the high latitudes; subsidence and divergence of the polar anticyclone; proximity to coasts	Arctic Ocean borderlands of North America, Greenland, and Eurasia; Antarctic Peninsula; some polar islands	At least 9 months average below freezing; low evaporation; precipitation usually below 25.5 cm (10 in.); coastal fog; strong winds	Tundra vegetation; tundra soils; permafrost; swamps and bogs during melting period; life most common in nearby seas; Inuit; mineral and oil resources; defense industry
Ice Sheet				
Warmest month below 0°C (32°F); precipitation exceeds potential evaporation	Location in the high latitudes and interior of landmasses; year-round influence of the polar anticyclone; ice cover; elevation	Antarctica; interior Greenland; permanently frozen portions of the Arctic Ocean and associated islands	Summerless; all months average below freezing; world's coldest temperature; extremely meager precipitation in the form of snow, evaporation even less; gale-force winds	Ice- and snow-covered surface; no vegetation; no exposed soils; only sea life or aquatic birds; scientific exploration

daylight when the sun stays above the horizon. Even when the sun is above the horizon, however, the solar radiation arrives at a sharply oblique angle, and little insolation is received, considering the number of daylight hours. Moving outward from the poles, the lengths of periods of continuous winter night and continuous summer daylight decrease rapidly from 6 months at the poles to 24 hours at the Arctic and Antarctic Circles (66 1/2°N and S). In each hemisphere during the winter solstice there will be 24-hours of darkness and on the summer solstice, 24-hours of daylight everywhere poleward of 66 1/2°N and S.

Tundra Climate

Compare the locations of the **tundra climate** *(ET)* with that of the subarctic climate on the world climate map in Figure 8.6. You can see that the tundra climate regions are situated closer to the poles. Now compare the winter temperatures and temperature ranges of two tundra climates—Barrow, Alaska, and Vaygach, Russia, in ● Fig. 10.23 Figure 8.6—with those of subarctic climates at Eagle, Alaska, and Verkhoyansk, Russia, in Figure 10.20. Note that even though the annual temperature ranges in tundra climates are large, they are not as large as those of subarctic climates, and winter temperatures are not as severe. Why is this so, when the tundra climate is closer to the poles? If you noticed that the tundra climates are situated along the periphery of landmasses and, with the exception of the Antarctic Peninsula, are adjacent to the Arctic Ocean, you know the answer. Tundra temperatures are influenced by the tundra's maritime location.

It seems inappropriate to call the generally chilly and damp conditions of the tundra's warmer season "summer." Temperatures average around 4°C (40°F) to just below 10°C (<50°F) for the warmest month, and frosts occur regularly. The summer warms sufficiently to melt the thin snow cover and the ice on small bodies of water, forming marshes, swamps, and bogs across the land because permafrost does not allow surface water drainage (● Fig. 10.24). Clouds of black flies, mosquitoes, and gnats swarm in this soggy landscape, known as **muskeg** in Canada and Alaska. One bright note on the land-scape is the wildflower bloom that occurs in spring when the conditions are right. Another is provided by the enormous number of migratory birds that nest in the arctic regions in summer and feed on the insects. However, as soon as the shrinking daylight hours of autumn approach, the birds depart for warmer climates.

Winters are cold and seem to last forever, especially in tundra locations where the sun may be below the horizon for days at a time. The climograph for Barrow, Alaska, illustrates the low temperatures of this climate. Note that average monthly temperatures are *below freezing* 9 months of the year. The average annual temperature is −12°C (10°F). The low-growing tundra vegetation survives despite the forbidding environment. It consists of lichens, mosses, sedges, flowering herbaceous plants, small shrubs, and grasses, plants are adapted to the conditions of permafrost (● Fig. 10.25).

The tundra regions exhibit several other significant climatic characteristics. Diurnal temperature ranges are small because insolation is uniformly high during the long summer days and uniformly low during the long winter nights. Precipitation is generally low because of exceedingly low absolute humidity and the influence of the polar high, except in eastern Canada and Greenland. Frigid, dry winds that sweep across the land are an additional factor in eliminating tree growth. Coastal fog is characteristic in marine locations, where cool polar maritime air drifts onshore and is chilled below the dew point by contact with the colder land.

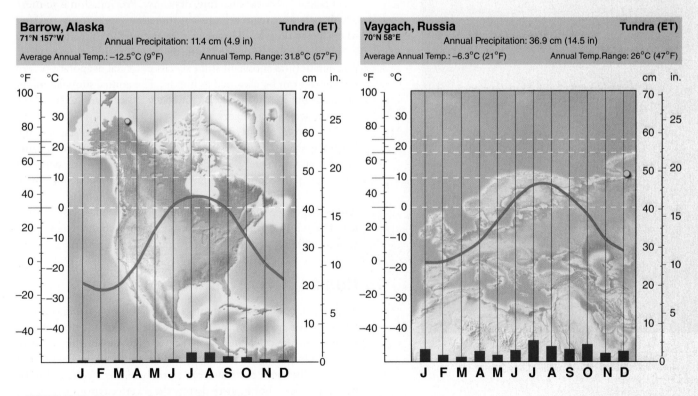

● **FIGURE 10.23** Climographs for tundra climate stations.

Why is it not surprising that both stations are located in the Northern Hemisphere?

● **FIGURE 10.24** Permafrost regions, such as this area at the base of the Alaska Range, become almost impenetrable swamplands during the brief Alaskan summer. Travel over land is feasible only in the winter season.

What is the preferred means of travel in the summer?

● **FIGURE 10.25** Autumn in the tundra at Denali National Park and Preserve in Alaska makes a colorful landscape of low-growing plants in this treeless arctic region.

What climate controls help to form a tundra landscape.

⊕ In Google Earth fly to: 63.41° N, 150.30° W

Ice-Sheet Climate

The **ice-sheet climate** *(EF)* is the most severe and restrictive climate on Earth. As Table 10.3 indicates, it covers large areas in both the Northern and Southern Hemispheres, a total of about 16 million square kilometers (6 million sq mi)—nearly the same area as the United States and Canada combined. Every average monthly temperature is below freezing, and surfaces are covered with glacial ice.

Antarctica is the coldest place on Earth (although Siberia has more severe periods of cold in winter). The world's coldest temperature, −89°C (−128°F), was recorded at Vostok, Antarctica. The climographs for the Amundsen-Scott South Pole Station, and Eismitte, Greenland, present a detailed picture of the cold temperature regimes of ice-sheet climates (● Fig. 10.26).

The primary reason for continually low temperatures of ice-sheet climates is the minimal effective solar radiation received in these regions. Not only is little or no insolation received during half of the year, but also the sun's radiant energy arrives at sharply oblique angles. In addition, solar energy is reflected away by the high albedo of the perpetual snow and ice cover existing in these climates. An additional factor, in both Greenland and Antarctica, is elevation. The ice sheets covering both regions rise more than 3000 meters (10,000 ft) above sea level (● Fig. 10.27). This high elevation further contributes to the cold temperatures.

The polar high pressure severely limits precipitation in the ice-sheet climate to occasional fine, dry snow. Precipitation is so meager that these climate regions are sometimes incorrectly referred to as "polar deserts." However, because of the exceedingly low evaporation rates associated with the severely cold temperatures, precipitation (as snowfall) still exceeds potential evaporation, and the climate can be classified as humid. The annual snowfall might be small, but it continually accumulates rather than melts, which feeds the ice-sheet glaciers.

The strong and persistent polar winds are another staple of the harsh ice-sheet climate. Mawson Station, Antarctica, for example, has approximately 340 days a year with gale-force winds of 15 meters per second (33 mph) or more. *Katabatic winds*, which are caused by the downslope drainage of heavy cold air over the dome-shaped ice sheets, are common along the edges of the polar ice. The winds of these regions can result in whiteouts—periods of zero visibility as a result of blowing fine snow and ice crystals.

Human Activity in Polar Regions

The climatic severity that limits animal life in polar regions to a few scattered species in the tundra is just as restrictive on human settlement. The celebrated Lapps of northern Europe migrate with their reindeer to the tundra from the adjacent forest during warmer months. They join the musk ox, arctic hare, fox, wolf, and polar bear that make a home there despite the severe environment. Only the Inuit (Eskimos) of Alaska, northern Canada, and Greenland have in the past succeeded in developing a year-round lifestyle

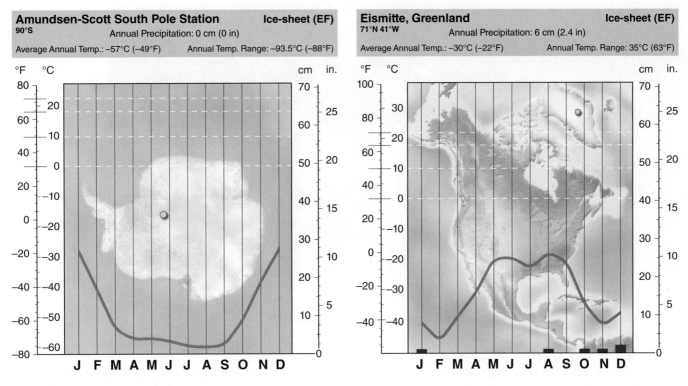

Amundsen-Scott South Pole Station   Ice-sheet (EF)
90°S   Annual Precipitation: 0 cm (0 in)

Average Annual Temp.: −57°C (−49°F)   Annual Temp. Range: −93.5°C (−88°F)

Eismitte, Greenland   Ice-sheet (EF)
71°N 41°W   Annual Precipitation: 6 cm (2.4 in)

Average Annual Temp.: −30°C (−22°F)   Annual Temp. Range: 35°C (63°F)

● **FIGURE 10.26** Climographs for ice-sheet climate stations.

If you were to accept an offer for an all-expense-paid trip to visit either Greenland or Antarctica, which would you choose, and why would you go?

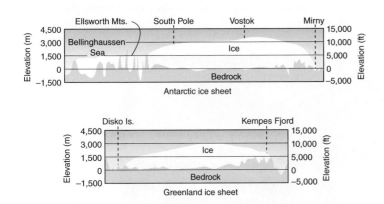

● **FIGURE 10.27** The ice-sheet climate is named after two huge sheet-like glaciers in the polar regions. In Antarctica, the glacial ice is as much as 4000 meters (13,200 ft) thick. Where it flows beyond the coast, the ice floats on seawater, forming an ice shelf. The smaller Greenland ice sheet is about 3000 meters (10,000 ft) thick.

What reasons might be given for the fact that more land in Greenland than in Antarctica is free of glacial ice?

adapted to the tundra climate, yet even this group relies less on the resources of the tundra than on the large variety of fish and sea mammals that occupy the adjacent seas.

The life for those living in the tundra includes working at defense installations or in the acquisition of mineral or energy resources. However, population centers based on the construction and maintenance of defense stations or, as in the case of Alaska's North Slope, on the production and transportation of oil, are not self-sustaining. Residents depend on other regions for support; thus, many workers often inhabit this region only temporarily. The barren ice sheets cannot serve as a natural home for people or animals. Inuit communities exist in Greenland, but these are located in tundra areas along the coast. Even the penguins, gulls, seals, and polar bears are coastal inhabitants. The ice-sheet climates are the harshest, most restrictive, most nearly lifeless areas on Earth (● Fig. 10.28). Antarctica and Greenland, however, are of great scientific interest. Scientists analyze the oxygen-isotope ratios of cores from the ice sheets and the gas bubbles trapped within them to reconstruct climate history. Antarctica's environment is widely recognized as important for cooperative scientific research and exploration.

Highland Climate Regions

The general relationship between temperature and increasing elevation was discussed in Chapter 4, as an average of −6.5°C per 1000 meters (−3.6°F per 1000 ft) temperature drop as one ascends in elevation. Knowing this, you might suspect that highland regions exhibit broad climate zones based on temperature changes with elevation. Furthermore, these zones might roughly correspond to Köppen's climate zones based on declining temperature with increased latitude. This might be generally true; however, the seasons that occur in highland areas reflect the seasons in the latitude of nearby lowlands. For example, at very high elevations on the equator, cold environments resembling tundra, and even perennial glacial ice can exist, and although colder than the lowlands, the seasonal changes will not be like those of the

● **FIGURE 10.28** Greenland's massive glacial ice sheet covers about 85 percent of it's surface.

What kind of activities might bring individuals from other regions to an ice-sheet climate?

latitudinal tundra regions or the ice-sheet climates. One reason for this is that the equatorial regions at either high or low elevations have a regime of relatively even temperatures all year, because the sun angle and daylight hours do not appreciably vary with the seasons as they do in higher latitude locations.

Complexity is the hallmark of highland climates (● Fig. 10.29). Every mountain range of significance is composed of a mosaic of climates far too intricate to differentiate on a world map or even on a map of a single continent. Highland climates according to Köppen's system are undifferentiated, signifying that much variation in microclimates occurs in relatively small areas. Highland climates are indicated on the world climate map wherever there are considerable local variations in climate as a consequence of elevation, and exposure. Highlands are distributed widely but are concentrated in Asia, central Europe, and western North and South America.

● **FIGURE 10.29** Highland climate areas experience many climatic conditions in a small area. The influence of elevation and exposure on changes in weather (and therefore also in climate) can clearly be seen here in Great Basin National Park, Nevada. How many different weather conditions do you see here in this area?

In what ways might high elevations be similar to high latitudes?

⊕ In Google Earth fly to: 36.557° N, 105.417° W

Elevation is only one of several controls of highland climates; **exposure** is another. Just as some continental coasts face the prevailing winds, so do some mountain slopes; others are downwind slopes or are sheltered behind higher topography. The nature of the wind, its temperature, and its moisture content depends on whether the mountain is (1) in a coastal location or deep in a continental interior and (2) at a high or low latitude within or beyond the reaches of cyclonic storms and monsoon circulation. In the middle and high latitudes, mountain slopes and valley walls that face the equator receive the direct rays of the sun and are warm; poleward-facing slopes are shadowed and cool. West-facing slopes feel the hot afternoon sun, whereas east-facing slopes are sunlit only in the cool of the morning. The higher one rises in the mountains, the more important direct sunlight is as a source of warmth and energy for plant and animal life processes.

Areas of highland climate are cooler islands in the midst of the major climate zones that dominate the areas around them. Consequently, highland areas are also biotic islands, supporting a flora and fauna adapted to cooler and wetter conditions that could not survive in the surrounding lowlands. This coolness is part of the attractions that highlands offer, particularly where forested mountains rise above arid and semiarid regions, as do the Rocky Mountains, California's Sierra Nevada, the Andes, and many other mountain ranges.

Orographic precipitation is stimulated as air masses are forced to rise over highlands (● Fig. 10.30). Where mountain slopes are rocky and not forested, their surfaces can grow warm during the day, causing upward convection, which can also produce afternoon thundershowers. Mountains generally receive abundant precipitation and are the source for multitudes of streams that join to form the world's great rivers.

● **FIGURE 10.30** Variations in precipitation is caused by air uplifting as it crosses the Sierra Nevada range of California from west to east. The heaviest precipitation occurs on the windward slope because air in the summit region is too cool to retain a large supply of moisture. Note the strong rain shadow downwind that gives Reno, Nevada a desert climate.

Taking into consideration the locations of the recording stations, during what season of the year does the maximum precipitation occur on the windward slope?

Characteristics of Mountain Climates

The weather in mountains can be highly variable, changing from hour to hour and from place to place. Strong orographic flow over mountains can cause clouds to form quickly, leading to thunderstorms and rains that often do not affect the surrounding cloud-free lowlands. When the skies are clear, the diurnal temperature range in high elevations is far greater than those of nearby lowlands. Because mountains rise to elevations where the atmosphere is thinner, the air above a mountain does not greatly impede solar radiation, and surfaces can warm dramatically during the daytime. At night the thin air does little to impede loss of thermal infrared radiation. Consequently, overnight air temperatures are cooler than the elevation alone would indicate. Because the atmospheric is thin at high elevations, plants, animals, and humans receive proportionately more ultraviolet radiation at high altitudes. Without protection, severe sunburn is a real hazard of a day in the high country.

Mountains that are high enough rise into zones of tundra and areas of perennial snow cover. The lower slopes of mountains are commonly forested with trees that become more stunted at higher elevations, until the last dwarfed tree is passed at the **treeline**. This is the elevation above which low winter temperatures and severe wind stress eliminate vegetation that does not grow low to the ground, where they can be protected from winds by a blanket of snow during the coldest months (● Fig. 10.31). Where mountains are high enough, snow or ice can permanently cover the land surface. The elevation above which summer temperatures are insufficient to melt all of the preceding winter's snowfall is called the **snow line**, although glaciologists call this the *equilibrium line*, to be explained in the discussion of glaciers in Chapter 19. On tropical mountains, both the treeline and snow line occur at higher elevations and lower elevations at higher latitudes. Seasonal changes are mainly variations in rainfall amounts; in the tropics, especially near the equator, temperatures are stable year-round, regardless of elevation.

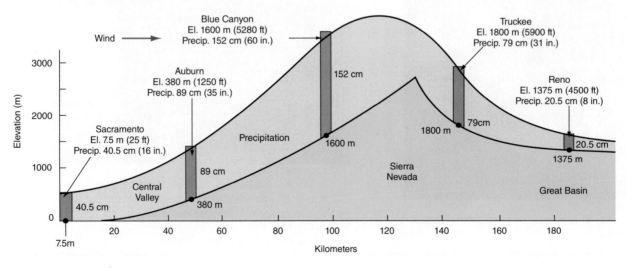

The Effects of Elevation on the Human Body

Many people are strongly attracted to mountain environments for living; for recreational activities such as hiking, skiing, and mountain climbing; or for just enjoying cool weather, spectacular landscapes, and the ecologies at high elevations. However, the environmental conditions at high elevations have several impacts on the human body. In addition to food and water, a person's metabolism also depends on the consumption of oxygen. A lowering of the normal oxygen intake can have profound physiological and psychological effects. One expression of the amount of oxygen available for human respiration is called the *partial pressure of oxygen gas* (pO_2). Partial pressure in this case refers to the portion of total atmospheric pressure attributed to oxygen alone. At sea level, atmospheric pressure is 1013.2 millibars, and the pO_2 is about 212 millibars. In other words, 212 of the 1013.2 millibars result from oxygen gas. At an altitude of 10 kilometers (6.2 mi) the pO_2 drops to only

55 millibars! Even at moderate elevations, the effects of *hypoxia* (oxygen starvation) can cause headaches and nausea. Above 6 kilometers (3.7 mi), this lack of oxygen can seriously affect the brain. Because the body's need for oxygen does not alter with major changes in altitude, any significant drop in pO_2 with height can cause severe body stress. Pressure-controlled cabins of high-flying aircraft and the use of oxygen by mountain climbers provide dramatic examples of ways to meet the vital need for oxygen.

Fortunately, the human body can acclimatize to moderate changes in elevation, but it takes time (days to sometimes weeks, depending on the altitude). During this time, however, one may experience sleeplessness, headaches, loss of weight, thyroid deficiency, increased excitability, muscle pain, gastrointestinal problems, swelling of the lungs, severe infections, psychological disturbances, decreased mental abilities, and other problems. At the very least, generally

when visiting significantly higher elevation locations, most people can expect headaches (perhaps leading to nausea), a drop in physical endurance (climbing stairs, hiking and other physical activities will become more difficult), and hampered mental function (solutions to simple problems may be more difficult to grasp). There is little need to worry, however, as most people do acclimatize and these symptoms will pass, unless they are at extremely high elevations.

It is also important to remember that, in addition to hypoxia, higher altitudes may cause increased susceptibility to severe sunburn. If you hike or drive to a higher elevation, this means that more of Earth's atmosphere is below you. Correspondingly, there is less protective atmosphere above you to filter harmful ultraviolet radiation. Keeping latitude constant, sunburn of exposed skin is more likely on a high-elevation mountainside than on a beach at sea level.

Mountain climbers carry an oxygen supply to help them function in very high elevations.

Mt. Evans in the Colorado Rocky Mountains has the highest paved road in North America, with the summit at 4350 meters (14,271 ft). Many visitors who have driven there are not used to such a high elevation, and this sign provides important information about potential health problems with the thin air.

R. Gabler

● **FIGURE 10.31** Here at treeline in the Colorado Rocky Mountains, the last tree species found at the treeline are stunted, prostrate forms, which often produce an elfin forest. Where the trees are especially gnarled and misshapen by wind stress, the vegetation is called *krummholz* (crooked wood).

What do you see in the photograph that indicates prevailing wind direction?

Highland Climates and Human Activity

In midlatitude highlands, soils are poor; the growing season is short; and the winter snow cover is heavy in the conifer zone, which dominates the lower and middle mountain slopes. Midlatitude highlands serve mainly as sources of timber and minerals, and as recreation areas in both summer and winter. As the winter snow melts off the high ground below bare rocky peaks, grass springs into life and herds of cattle and flocks of sheep and goats are driven up from the warmer valleys. The high pastures are green and lush in summer, but in early fall they are again vacated by the animals and their keepers. This seasonal movement of herds and herders between alpine pastures and villages in the valleys, termed *transhumance*, was once common in the European highlands (the Alps, Pyrenees, Carpathian Mountains, and mountains throughout Scandinavia) and is still practiced there on a reduced scale.

In contrast to higher-latitude mountain regions, tropical highlands may actually experience more favorable climatic conditions and have attracted human settlement (from ancient to modern times) from adjacent lowlands. In tropical South America, four climate zones that change with elevation have been recognized for hundreds of years: *tierra caliente* (hot lands), *tierra templada* (temperate lands), *tierra fría* (cool lands), and *tierra helada* (frozen lands). People living in these areas have long interacted with and adapted to these altitudinal environments (● Fig. 10.32).

● **FIGURE 10.32** The vegetation and agricultural products of vertical life zones in tropical mountains. This example extends from lowland tropical zones to the zone of permanent snow and ice. Little temperature change occurs in tropical mountains over the year, which allows life forms to survive at relatively high elevations without experiencing seasonal extremes.

When Europeans first settled in the highlands of tropical South America, in which vertical climate zone did they prefer to live?

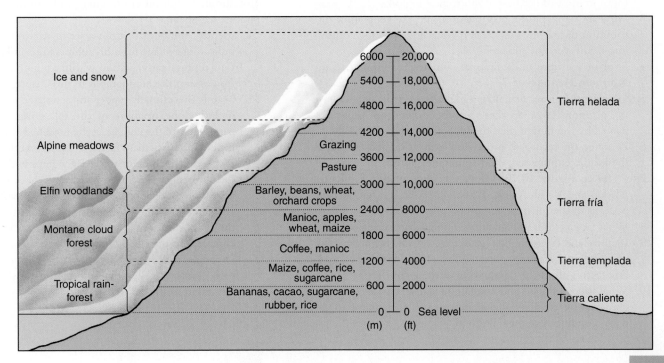

Large permanent populations are supported in other tropical highland areas where the topography and soils favor agriculture. In steep mountain regions, slopes have been extensively terraced to produce level land for agriculture. Spectacular agricultural terraces can be seen in Peru, the Philippines, and many other tropical highlands.

This chapter has discussed the major climates that dominate the middle and high latitudes, and the variable climatic conditions of highland regions. These climates range from some of the most productive to some of the harshest climates on the planet. Vegetation, animal life, and human activities have all had to adapt to the climatic conditions they experience in their particular region.

CHAPTER 10 ACTIVITIES

■ TERMS FOR REVIEW

mesothermal
Mediterranean climate
sclerophyllous
chaparral
humid subtropical climate
marine west coast climate
microthermal

humid continental, hot-summer climate
humid continental, mild-summer climate
subarctic climate
boreal forests
permafrost
patterned ground (frost polygons)
tundra climate

muskeg
ice-sheet climate
exposure
treeline
snow line

■ QUESTIONS FOR REVIEW

Summarize the special adaptations of vegetation and soils in Mediterranean regions.

1. Compare the humid subtropical and Mediterranean climates. What are their most obvious similarities and differences?
2. What factors combine to cause a summer precipitation maximum in most of the humid subtropical regions?
3. How are temperature, precipitation, and the geographic distribution of marine west coast regions linked to the controlling factors for this climate?
4. Explain why the microthermal climates are limited to the Northern Hemisphere.
5. List several features that all humid microthermal climates have in common. How do these features differ from those displayed by the humid mesothermal climates?
6. Describe the relationship between vegetation and climate in the humid continental, mild-summer regions.
7. Refer to Figure 10.20. Using the climographs for Eagle, Alaska, and Verkhoyansk, Russia, describe the temperature patterns of the subarctic regions.
8. What factors limit precipitation in the subarctic regions?
9. Identify and compare the climate factors that strongly influence the tundra and ice-sheet regions. How do these controlling factors affect the distribution of these climates?
10. What kinds of plant and animal life can survive in the polar climates? What special adaptations must they make to the harsh conditions of these regions?

11. How do elevation and exposure affect the microclimates of highland regions? What are the major climatic differences between highland regions and nearby lowlands?
12. Do each of the following for these climates: Mediterranean; humid subtropical; marine west coast; humid continental, hot-summer; humid continental, mild-summer; subarctic; tundra; ice-sheet.
    a. Identify the climate from a set of data or a climograph indicating average monthly temperature and precipitation for a representative station within a region of that climate (use one of the climographs in this chapter).
    b. Match the climate type with a written statement that includes one or more of the following: the statistical parameters of the climate in the modified Köppen classification; the particular climate controls (controlling factors) that produce the climate; the geographic distribution of the climate as stated in terms of physical or political location; the unique climate characteristics or combination of characteristics that distinguishes the climate from others; the types of plants, animals, and soils associated with the climate; and the human utilization typical of the climate.
    c. Distinguish between the important subtypes (if any) of each climate by identifying the characteristics that separate them from one another.

CONSIDER AND RESPOND

1. Based on the classification scheme presented in the "Graph Interpretation" exercise at the end of Chapter 8, classify the following climate stations from the data provided.

		J	F	M	A	M	J	J	A	S	O	N	D	Yr
a.	Temp. (°C)	−42	−47	−40	−31	−20	15	−11	−18	−22	−36	−43	−39	−30
	Precip. (cm)	0.3	0.3	0.5	0.3	0.5	0.8	2.0	1.8	0.8	0.3	0.8	0.5	8.6
b.	Temp. (°C)	3	3	5	8	19	13	15	14	13	10	7	5	9
	Precip. (cm)	4.8	3.6	3.3	3.3	4.8	4.6	8.9	9.1	4.8	5.1	6.1	7.4	65.8
c.	Temp. (°C)	23	23	22	19	16	14	13	13	14	17	19	22	18
	Precip. (cm)	0.8	1.0	2.0	4.3	13.0	18.0	17.0	14.5	8.6	5.6	2.0	1.3	88.1
d.	Temp. (°C)	−27	−28	−26	−18	−8	1	4	3	−1	−8	−18	−24	−12
	Precip. (cm)	0.5	0.5	0.3	0.3	0.3	1.0	2.0	2.3	1.5	1.3	0.5	0.5	10.9
e.	Temp. (°C)	−4	−2	5	14	20	24	26	25	20	13	3	−2	12
	Precip. (cm)	0.5	0.5	0.8	1.8	3.6	7.9	24.4	14.2	5.8	1.5	1.0	0.3	62.2
f.	Temp. (°C)	9	9	9	10	12	13	14	14	14	12	11	9	11
	Precip. (cm)	17.0	14.8	13.3	6.8	5.5	1.9	0.3	0.3	1.6	8.1	11.7	17.0	97.6
g.	Temp. (°C)	−3	−2	2	9	16	21	24	23	19	13	4	−2	11
	Precip. (cm)	4.8	4.1	6.9	7.6	9.4	10.4	8.6	8.1	6.9	7.1	5.6	4.8	84.8
h.	Temp. (°C)	0	0	4	9	16	21	24	23	20	14	8	2	12
	Precip. (cm)	8.1	7.4	10.7	8.9	9.4	8.6	10.2	12.7	10.7	8.1	8.9	8.1	111.5

2. The data in the previous table represent the following eight locations, although not necessarily in this order: Beijing, China; Point Barrow, Alaska; Chicago, Illinois; Eismitte, Greenland; Eureka, California; Edinburgh, Scotland; New York City, New York; and Perth, Australia. Use an internet mapping site or an atlas and your knowledge of climates to match the climatic data with the locations.

3. Eureka, California; Chicago; and New York City are located within a few degrees latitude of one another, yet they represent three distinctly different climate types. Discuss these differences and identify the primary cause, or source, of the differences.

4. The precipitation recorded at Albuquerque, New Mexico (see Consider and Respond in Chapter 9) is almost twice that recorded at Point Barrow, Alaska, yet Albuquerque is considered a dry climate and Point Barrow is a humid climate. Why?

5. Why is the *Dw* climate type found only in Asia? The precipitation recorded at Albuquerque, New Mexico (see Consider and Respond in Chapter 9) is almost twice that recorded at Point Barrow, Alaska, yet Albuquerque is considered a dry climate and Point Barrow is a humid climate. Why?

6. *Csa* climates are dry during the summer and *Cfa* climates are wet. Why?

PRACTICAL APPLICATIONS

Use the data charts in the box titled "Geography's Physical Science Perspective: Effective Temperatures":

1. Determine the heat index values for the following data:
   a. 95°F and 70% relative humidity
   b. 85°F and 90% relative humidity
   c. 80°F and 80% relative humidity
   d. 75°F and 100% relative humidity
   e. 70°F and 50% relative humidity
   f. 100°F and 0% relative humidity

2. Determine the wind chill index values for the following data:
   a. 40°F and 35 mph winds
   b. 20°F and 20 mph winds
   c. 15°F and 35 mph winds
   d. −15°F and 40 mph winds
   e. −30°F and 30 mph winds
   f. −40°F and calm winds

 MindTap—Make the most of your study time by accessing everything you need to succeed in one place. Read your textbook, take notes, review flashcards, watch videos, complete activities, take practice quizzes, and more online with MindTap. Log in at **www.cengagebrain.com**.

BIOGEOGRAPHY

OBJECTIVES

WHEN YOU COMPLETE THIS CHAPTER YOU SHOULD BE ABLE TO:

- 11.1 Define the four major components of an ecosystem, and explain their inter-dependence.
- 11.2 Recognize that other environmental controls may be more important on a local scale but that climate has the greatest influence over ecosystems on a worldwide basis.
- 11.3 Explain how vegetation becomes established on barren or devastated areas, and cite an example of the steps in plant succession.
- 11.4 Provide examples of how plants, animals, and the environments in which they live are interdependent, each affecting the others.
- 11.5 Outline the climatic factors that have the greatest effect on plants and animals, and summarize the nature of those climatic impacts.
- 11.6 Describe Earth's major terrestrial ecosystems (biomes) and the dominant vegetation types that occupy the ecosystems.
- 11.7 Cite a few reasons why humans affect ecosystems more than all other life forms, and provide some examples of major impacts.

BIOGEOGRAPHY IS THE STUDY OF how environmental factors affect the locations, distributions, and life processes of plants and animals. Basically, this discipline seeks explanations for the geography of life forms. Biogeographers delineate the spatial boundaries of natural environments and investigate how and why environmental characteristics change spatially and over time. Along with ecologists and scientists from other disciplines, biogeographers often study *ecosystems*. An *ecosystem* is a community of organisms that occupy a given area, along with their interdependent relationships with each other and with characteristics of the nonliving environment. Biogeography is also concerned with both human and natural impacts on organisms and their ecosystems. It also focuses on how those external influences affect the environmental conditions that support life.

Ecology is an old science. The voyages of discovery beginning in the fifteenth century carried colonists and explorers to uncharted lands with exotic environments. Scientists eventually accompanied exploratory voyages travelling to distant places and recording descriptions and illustrations of the previously unknown flora and fauna they encountered. It was apparent that certain plants and animals that were found together bore direct relationships to the climate and environment in which they lived. As widespread information about global environments became more complete and

◄ Autumn vegetation patterns in this mountain area of Alaska are caused by differences in elevation, temperature, winds, aspect, soil, moisture, and slope steepness. These factors produce varying microenvironments, each suitable for different vegetation species.   Tim Rains/NPS Photo

detailed, early biologists studied plant communities and classified vegetation types for much of the world. As the relationships of animals to these plant communities were recognized, early twentieth century naturalists began dividing Earth's life forms into *biotic associations*. Recently, the functional relationships of plants and animals with their physical environment have been a primary focus of attention, and the ecosystem concept has become widely applied. The study of biogeography, ecology, and ecosystems provides excellent opportunities to demonstrate the environmentally integrative nature of physical geography. Relationships and interactions among the different climate types, their associated vegetative biomes, and certain soils were introduced in Chapters 9 and 10. This chapter takes a closer look at biogeography, ecology, and the nature of certain environments.

● **FIGURE 11.1** This area in Grand Tetons National Park, Wyoming, demonstrates the close relationship between living organisms and their nonliving environment. Beavers have dammed a stream to make a pond. Here, there are high mountain, forest, stream, pond, and meadow environments.

What are some of the basic ways that the ecosystems might differ among these varied environments?

Ecosystems

Our definition of an **ecosystem** is broad and flexible. The term can be used in reference to the Earth system in its entirety or to any group of organisms that occupy a given area and function together with their nonliving environment. An ecosystem may be large or small, marine or terrestrial (on land), short-lived or long lasting (● Fig. 11.1). There are also artificial ecosystems, created by human activities such as an agricultural field. When farmers plant crops, spread fertilizer, practice weed control, and spray insecticides, they create new ecosystems, but plants and animals still live together in interdependent relationships with the soil, precipitation, temperatures, sunshine, and other factors that constitute the physical environment.

Ecosystems are *open systems*. There is movement of both energy and materials into and out of these systems, and storage of both within them. Ecosystems are not isolated elements of nature; rather, they are usually closely related to nearby ecosystems and integrated with the larger ecosystems of which they are a part. The ecosystem concept is a valuable model for examining the structure and function of living communities and their life forms.

Major Components

Despite their great variety on Earth, ecosystems typically have four basic components (● Fig. 11.2). The first is the nonliving, or **abiotic**, parts of the ecosystem in which the plants and animals live. In a terrestrial ecosystem, the abiotic component provides life-supporting elements and compounds from the soil, the groundwater, and the atmosphere.

The abiotic setting is the physical environment in which the plants and animals of an ecosystem live. In an aquatic ecosystem

(a pond, for example), the abiotic component would include inorganic substances as calcium, mineral salts, oxygen, carbon dioxide, and water. Some of these would be dissolved in the water, and others lie at the bottom as sediments—a natural reservoir of nutrients for both plants and animals. In a terrestrial ecosystem, the abiotic component provides life-supporting elements and compounds in the soil, groundwater, and atmosphere.

The second component of an ecosystem consists of the basic **producers**, or **autotrophs** (meaning "self-nourished"). Plants are important autotrophs because through photosynthesis they can use solar energy to convert water and carbon dioxide into organic molecules. The sugars, fats, and the proteins plants produce during photosynthesis supply food to support other forms of life. Some bacteria are also capable of photosynthesis, and the sulfur-dependent organisms that exist at thermal vents on the sea floor are also classified as autotrophs.

A third component of most ecosystems consists of **consumers**, or **heterotrophs** (meaning "other-nourished"). These are animals that survive by eating plants or other animals. Herbivores eat only plant material, carnivores eat other animals, and omnivores feed on both plants and animals. Animals contribute to our planet's ecosystem in many ways. They use oxygen in respiration and exhale carbon dioxide that is required for photosynthesis by plants. Animals also influence soil development through digging and trampling, and these activities also affect local plant distributions.

We might assume that plants, animals, and a supporting environment are all that is needed for a functioning ecosystem, but this is not the case. Without a fourth component in ecosystems, plant growth would come to a halt—these are **decomposers**, or **detritivores** that feed on dead plant and animal material and waste products. They promote decay, which returns mineral nutrients to the soil and to bodies of water in a form that plants can use.

Erin Himmel/National Park Service

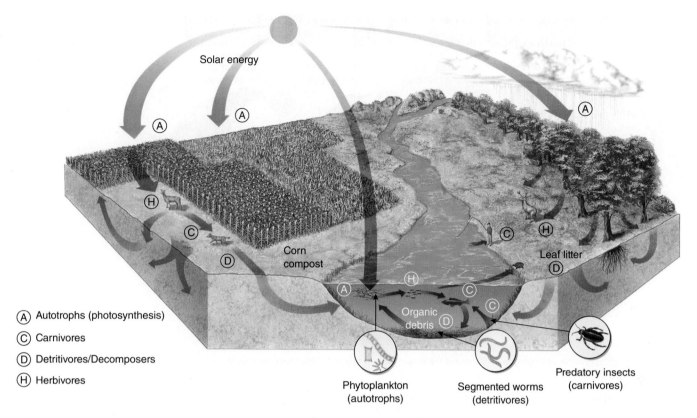

(A) Autotrophs (photosynthesis)

(C) Carnivores

(D) Detritivores/Decomposers

(H) Herbivores

Solar energy

Corn compost

Leaf litter

Organic debris

Phytoplankton (autotrophs)

Segmented worms (detritivores)

Predatory insects (carnivores)

● **FIGURE 11.2** Ecosystems illustrate the interdependence of variables in environmental systems. A focus is on the close relationships between living organisms (of the biosphere) and the nonliving (abiotic) components in an environmental system (the atmosphere, hydrosphere, and lithosphere).

Can you trace a trophic structure through this diagram?

Trophic Structure

The living components of an ecosystem are organized in a sequence defined by their eating habits. *Herbivores* eat plants, *carnivores* may eat herbivores or other carnivores, and *decomposers* feed on dead plants and animals and their waste products. This sequence of feeding levels is referred to as a **food chain**, and organisms are identified by their **trophic level**, the number of steps that they are removed from the producers. Plants occupy the first trophic level, **herbivores** the second, **carnivores** feeding on other animals the third, and so forth until the last level, the decomposers, is reached (● Fig. 11.3). **Omnivores** may belong to several trophic levels because they eat both plants and animals. The simplest food chain, however, would include only plants and decomposers. More complex food chains may include six or more levels as carnivores feed on other carnivores—for example, zooplankton eat plants, small fish eat zooplankton, larger fish eat small fish, bears eat larger fish, and decomposers consume the bear after it dies.

Actually, most food chains do not operate in a simple linear sequence; instead they overlap and interact to form a feeding network within an ecosystem called a *food web*. Food chains and food webs can be used to trace the movement of food and energy from one level to another in an ecosystem.

Nutrient Cycles Some biologists and ecologists find it helpful to separate the trophic structure into specific **nutrient cycles**. There are several such cycles that help explain the routing of nutrients through ecosystems. Particularly important cycles have been recognized for water, carbon, nitrogen, and oxygen, and knowledge of these nutrient cycles is essential to understanding energy flows in ecosystems. Parts of the oxygen and carbon cycles as well as the water (hydrologic) cycle were discussed in previous chapters. ● Figure 11.4 illustrates the major processes involved in these cycles. Knowledge of chemical nutrient cycles is essential to an understanding of energy flow in ecosystems.

Energy Flow and Biomass

Ecosystems are studied by tracing flows of energy through the system much like analyzing the energy flow in other natural systems, such as streams or glaciers. As in other systems, the laws of thermodynamics also apply to ecosystems. The first law of thermodynamics states that energy cannot be created or destroyed; it can only be changed from one form to another. Sunlight provides energy to an ecosystem, which is used by plants in photosynthesis, and the energy is stored in the organic materials of plants and animals. Energy flows through the system along food chains and webs from one trophic level to the next. It is finally released from the system when oxygen combines with

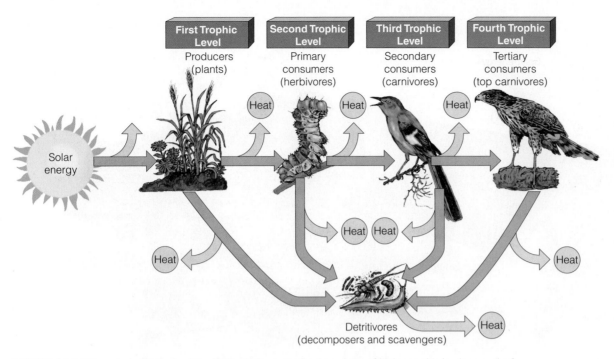

● **FIGURE 11.3** This ecological example of trophic levels illustrates the dependence of higher trophic levels on each of the lower levels in their food chain.

Can you outline a trophic structure that exists where you live through four tropic levels?

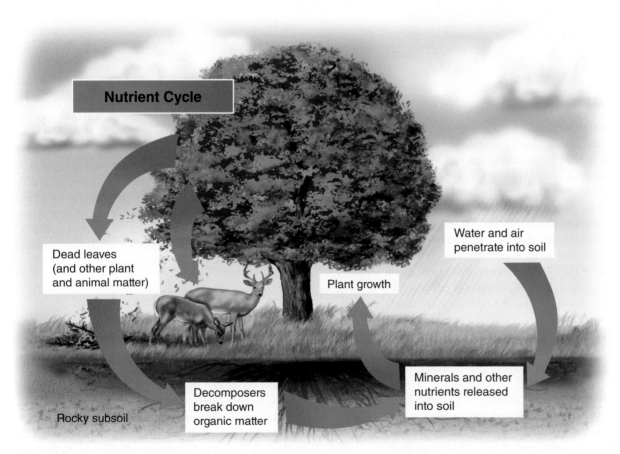

● **FIGURE 11.4** The processes involved in a nutrient cycle as they travel through an ecosystem, shown in a simplified diagram.

How are aspects of the biosphere linked with the atmosphere, lithosphere, and hydrosphere in this scene?

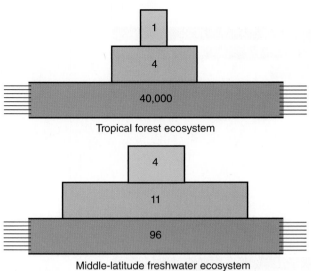

● FIGURE 11.5 Trophic pyramids showing biomass of organisms at various trophic levels in two contrasting ecosystems. Dry weight is used to measure biomass because the proportion of water to total mass differs from one ecosystem to another.

Why is there an exceptionally large reduction in biomass between the first and second trophic levels of a tropical forest ecosystem?

the chemical compounds of organic materials through the process of oxidation. Respiration, the combination of oxygen with chemical compounds in living cells that can occur at any trophic level, is the major form of oxidation. Fire is yet another oxidation process.

The total amount of living material in an ecosystem is its **biomass**. Because the energy of an ecosystem is stored in the biomass, scientists measure the biomass of each trophic level to trace energy flow through an ecosystem. They usually find that the biomass decreases with each successive trophic level (● Fig. 11.5). There are a number of explanations for this, each involving a loss of energy. The first instance occurs between trophic levels. The second law of thermodynamics states that whenever energy is transformed from one state to another, there will be a loss (dissipation) of energy through heat. When any living thing feeds on another organism, not all of the food energy is used; some is lost from the system. Additional energy is expended through respiration and movement. At each successive trophic level, the amount of energy required becomes greater. A deer may graze in a limited area, but the wolf that preys on the deer must hunt over a much larger territory. As the flow of energy decreases with each successive trophic level, the biomass also decreases—a principle that also applies to agriculture. A great deal more biomass (and food energy) is available in a field of corn than there is in the cattle that eat the corn.

Primary and Secondary Productivity

Productivity is the rate at which new organic material is created at a particular trophic level. **Primary productivity** refers to the formation rate of new organic matter through photosynthesis by producers. **Secondary productivity** refers to the rate of formation of new organic material at the consumer level.

Primary Productivity How efficient are plants at producing new organic matter through photosynthesis? The answer to this question depends on a number of variables. Photosynthesis requires sunlight, the amount of which is related to the length of daylight hours and the angle of the solar radiation—two factors that vary widely with latitude and season. Photosynthesis is also affected by soil moisture, temperature, the availability of mineral nutrients, the atmosphere's carbon dioxide content, and the age and species of the individual plants.

Most studies of ecosystem productivity have been concerned with measuring the net biomass at the autotroph level. Wherever studies have compiled data about the efficiency of photosynthesis, efficiency levels have been surprisingly low. In general, these studies indicate that less than 5% of the available sunlight is used to produce new biomass in ecosystems. For planet Earth as a whole, the figure is probably less than 1%. Nonetheless, the annual net primary productivity of the biosphere is enormous, and estimated to be about 170 billion metric tonnes (a metric *tonne* is about 10% greater than a U.S. ton) of organic matter. Even though the oceans cover about 70% of the planet's surface, slightly more than two-thirds of net annual productivity is from terrestrial ecosystems and less than one-third is from marine ecosystems. Even more surprising is the fact that humans consume less than 1% of Earth's primary productivity as plant food. However, we also use biomass in a variety of other ways—for example, lumber for construction and paper production and using biomass energy for feedstock and as fodder for range animals.

● Figure 11.6 illustrates the wide range of net primary productivity that exists among various ecosystems. Latitudinal influence on insolation and photosynthesis results in a noticeable decrease in terrestrial productivity from tropical ecosystems to those in middle and higher latitudes. Even the tropical savannas, which are dominated by grasses, produce more biomass in a year than the boreal forests of colder climates in high latitudes. Today, monitoring by satellites and the maps produced from those images provide us with information about global ecosystems and their distributions on our planet (● Fig. 11.7).

Some agricultural (artificial) ecosystems can be fairly productive when compared with the natural ecosystems they have replaced. This is especially true in warmer regions where farmers can raise two or more crops in a year, or in arid lands where irrigation supplies the water for growth. However, mean productivity for cultivated land does not approach that of forested land and is just about the same as that of midlatitude grasslands. Most agricultural ecosystems are significantly less productive than the natural systems that exist in the same environment.

The reasons for differences among aquatic, or water-controlled, ecosystems are not quite as apparent. Swamps and marshes are well supplied with plant nutrients that support a relatively large biomass at the first trophic level. In contrast, water depth has the greatest impact on ocean ecosystems. Most nutrients in the open ocean sink to the bottom, beyond the depth of sunlight penetration that makes photosynthesis possible. This is why the most productive marine ecosystems are found in areas where ocean upwelling carries nutrients nearer to the surface, and in the sunlit, shallow waters of continental shelves, coral reefs, or estuaries (● Fig. 11.8).

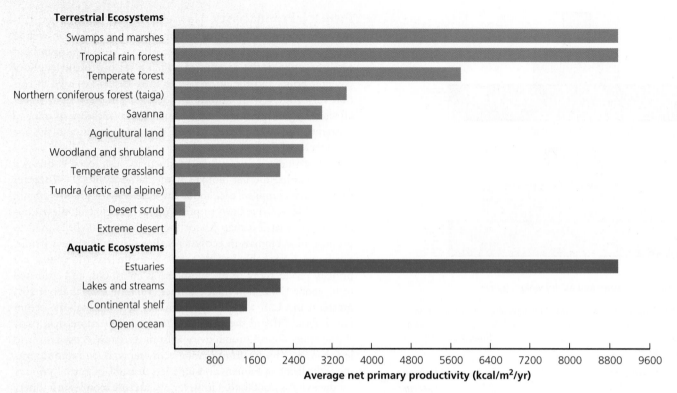

Terrestrial Ecosystems

Swamps and marshes
Tropical rain forest
Temperate forest
Northern coniferous forest (taiga)
Savanna
Agricultural land
Woodland and shrubland
Temperate grassland
Tundra (arctic and alpine)
Desert scrub
Extreme desert

Aquatic Ecosystems

Estuaries
Lakes and streams
Continental shelf
Open ocean

800 1600 2400 3200 4000 4800 5600 6400 7200 8000 8800 9600

Average net primary productivity (kcal/m²/yr)

● **FIGURE 11.6** This graph shows the variability in annual net primary productivity among different terrestrial and aquatic ecosystems in kilocalories per square meter per year.
Source: Data From R.H. Whittaker, Communities and Ecosystems, 2nd ed., Newyork: Macmillan, 1975.

What climatic or environmental characteristics seem to be most responsible for highest and lowest net productivity?

Secondary Productivity Secondary productivity results from the conversion of plant materials to animal substances. We have also noted that the ecological efficiency, or the rate of energy transfer from one trophic level to the next, is low (● Fig. 11.9). A huge biomass is required at the producer level to support one animal that eats only meat. The efficiency of transfer from autotrophs to heterotrophs varies widely from one ecosystem to another. The amount of net primary production actually eaten by herbivores may range from as high as 15% in some grassland areas to as low as 1% or 2% in certain forested regions. In ocean ecosystems, the figure may be much higher, but there is a greater loss during the digestion process. Once the food is eaten, energy loss through respiration or body movement reduces secondary productivity to a small fraction of the biomass available as net primary productivity.

Most authorities consider 10% to be a reasonable estimate of ecological efficiency for both herbivores and carnivores. If both herbivores and carnivores have ecological efficiencies of only 10%, the ratio of biomass at the first trophic level to biomass of carnivores at the third trophic level is several thousand to one. It obviously requires a huge biomass at the autotroph level to support one animal that eats only meat. As human populations grow at increasing rates and agricultural production lags behind, it is fortunate that human beings are omnivores and can adopt a more vegetarian diet (● Fig. 11.10).

Ecological Niche

A surprising number of species exist in nearly every ecosystem— the exception being species that are severely restricted by adverse environmental conditions. Yet each organism performs a specific role in the system and lives in a locational setting with certain characteristics, described as its **habitat**. A combination of the role and habitat for a particular species is referred to as its **ecological niche**. Many factors influence an organism's ecological niche. Some species are **generalists** and can survive on a wide variety of food. The North American brown, or grizzly, bear, an omnivore, will eat berries, honey, and fish. In comparison, the koala of Australia is a **specialist** that eats only the leaves of certain eucalyptus trees. Specialists do well when their particular food supply is abundant, but they cannot adapt to changing environmental conditions that affect their food source. Generalists are in the majority in most ecosystems because their broader ecological niche allows survival on alternative food supplies. Raccoons are an excellent generalist example, having adapted well to suburban and urban areas, in addition to their original wooded environments.

Some generalist species occupy an ecological niche in one ecosystem that is quite different from the niche they occupy in another. As food supply varies with habitat, so varies the ecological niche. Humans are also generalists. In some regions, they are carnivores, and in others, herbivores; and in most locations, omnivores. It is also true that different species may occupy the same ecological niche in habitats that are similar but located in separate ecosystems.

Succession and Climax Communities

Vegetative associations are called **plant communities**, aggregations of vegetation species that have adapted to existing

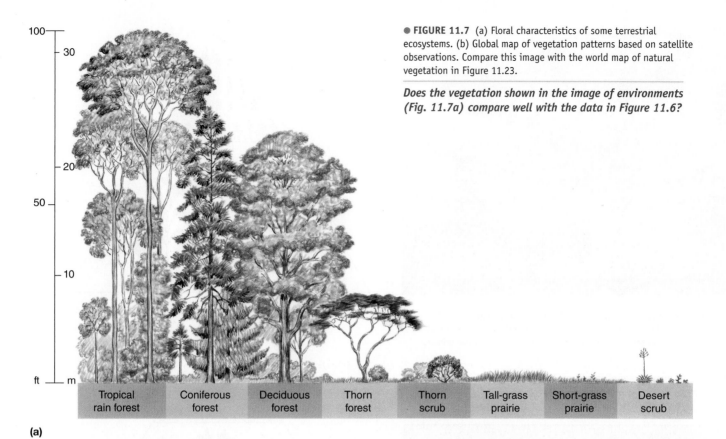

● FIGURE 11.7 (a) Floral characteristics of some terrestrial ecosystems. (b) Global map of vegetation patterns based on satellite observations. Compare this image with the world map of natural vegetation in Figure 11.23.

Does the vegetation shown in the image of environments (Fig. 11.7a) compare well with the data in Figure 11.6?

100 —
30
20
50 —
10
ft — m

| Tropical rain forest | Coniferous forest | Deciduous forest | Thorn forest | Thorn scrub | Tall-grass prairie | Short-grass prairie | Desert scrub |

(a)

Evergreen Needleleaf Forest
Evergreen Broadleaf Forest
Deciduous Needleleaf Forest
Deciduous Broadleaf Forest
Mixed Forests
Closed Shrublands
Open Shrublands
Woody Savannas
Savannas
Grasslands
Permanent Wetlands
Croplands
Urban and Built-Up
Cropland/Natural Vegetation Mosaic
Snow and Ice
Barren or Sparsely Vegetated

(b)

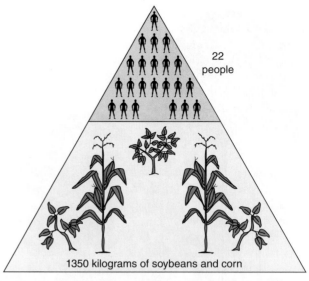

22 people

1350 kilograms of soybeans and corn

● **FIGURE 11.8** Estuaries, where rivers flow into the sea, create a variety of nutrient-rich environments, including associated wetlands. These coastal ecosystems are among the most productive on Earth in terms of net primary productivity.

Why are very productive ecosystems found in the shallow waters bordering the world's continents?

1 m² of field for 1 yr

d. Net carnivore productivity:
0.4 g (0.15% of a; 6.7% of c)

c. Net herbivore productivity:
6 g (2.2% of a; 20.7% of b)

b. Every year herbivores eat:
29 g (10.7% of a)

a. Net primary productivity
(plants): 270 g (dry weight)

● **FIGURE 11.9** Productivity at the autotroph, herbivore, and carnivore trophic levels in a Tennessee field. The figures represent productivity for 1 square meter of field in 1 year. Note the extremely small proportion of primary productivity that reaches the carnivore level of the food chain.

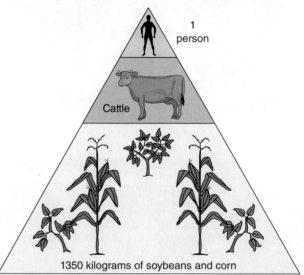

1 person

Cattle

1350 kilograms of soybeans and corn

● **FIGURE 11.10** These triangles illustrate the advantages of a vegetarian diet as the world experiences another century of rapid population growth. People are fortunate that they are omnivores and can choose to eat grain products. The same 1350 kilograms of grain, if converted to meat, supports only one person, but it will support 22 people if cattle or other animals are omitted from the human food chain.

What does this relationship mean if the world's population continues to grow unimpeded?

environmental conditions. At least for terrestrial ecosystems, it is the autotrophs, the plant species at the first trophic level, that most easily distinguish one ecosystem from another. All other species in an ecosystem depend on the autotrophs for food, and the association of all living organisms determines the energy flow and the trophic structure of the ecosystem. It should be noted that the species occupying the first trophic level are strongly influenced by climate, which is another example of the interconnections between the major Earth subsystems.

If the vegetation that comprise the biomass of the first trophic level is allowed to develop naturally without obvious interference from or modification by humans, the resulting association of plants is called **natural vegetation**. These plant associations, or

Roy W. Lowe/USFWS

plant communities, are compatible because each species within the community has different requirements in relation to major environmental factors such as light, moisture, and mineral nutrients. If two species within a community were to compete for the same resources, one would eventually eliminate the other. The species forming a community at any specific place and time will be an aggregation of those that can adapt to the prevailing environmental conditions.

Plant Succession

As natural vegetation becomes established, it often undergoes a progressive sequence of different plant communities over time. This developmental process, known as **succession**, usually begins with a relatively simple plant community. There are two major types of succession, primary and secondary. In primary succession, a bare substrate is the beginning point. No soil or seedbed exists yet. A **pioneer community** invades the bare substrate (whether it be volcanic lava, glacially deposited sediment, or a bare beach, among others) and begins to alter the environment. As a result, the species structure of the ecosystem does not remain constant. In time, the environmental alteration becomes sufficient to allow a new plant community (that could not have survived under the original conditions) to appear and eventually to dominate the original community. The process continues with each succeeding community rendering further changes to the environment. Primary succession can take place in a few hundred or even a few thousand years because of the barren conditions under which the pioneer vegetation began this process.

Secondary succession begins when some natural process, such as a forest fire, tornado, or landslide, has destroyed or damaged a great deal of the existing vegetation. Ecologists refer to this process as **gap** creation. Even with such damage, however, soil still typically exists and seeds may be lying dormant in that soil, ready to invade the newly opened gap. Secondary succession, therefore, can occur much more quickly than primary succession.

A common form of secondary succession associated with abandoned agricultural fields in the southeastern United States is depicted in ● Figure 11.11. If agriculture ceases, pioneer plants such as weeds and grasses colonize the bare fields. These plants stabilize the topsoil, add organic matter, and produce favorable conditions for the growth of shrubs and brush such as sassafras, persimmon, and sweet gum. During this stage, the soil is enriched with nutrients and organic matter, increasing its ability to retain moisture. These conditions encourage the development of pine forests, the next stage of vegetative succession in this region. As pine forests thrive and dominate in this newly created environment, the forest cover will shade out the weeds, grasses, and brush.

Ironically, the growth of these pine forests can also lead to their demise. Pine trees require much sunlight for their seeds to germinate. When competing with scattered brush, grasses, and weeds, there is adequate sunlight for germination. Once a pine forest develops, however, shade and litter from the pine trees will not allow pine seeds to germinate. Hardwood trees, such as oak and hickory, whose seeds can germinate in shady conditions, begin as an understory, and eventually can replace the pines. In this southeastern United States example, a complete succession

● **FIGURE 11.11** A common plant succession in the southeastern United States. Successional changes over time are shown from left to right. Until succession is complete each succeeding vegetation type alters the environment in ways that favor their replacement, by being out competed by new species.

What are some reasons why plant succession would differ in another region of the United States?

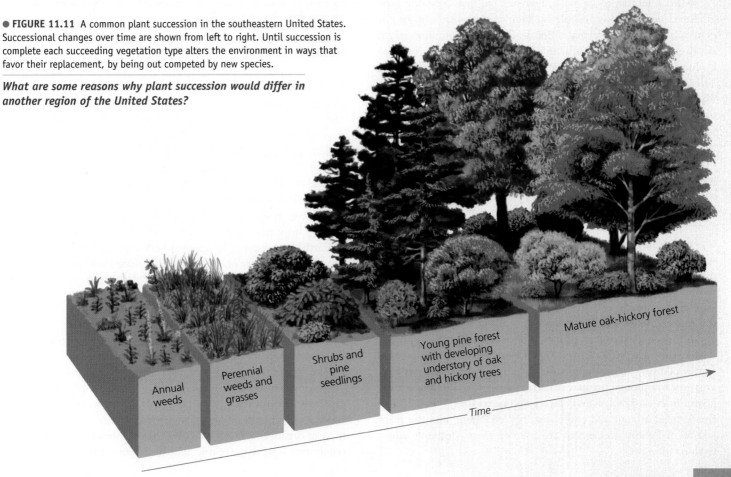

Annual weeds

Perennial weeds and grasses

Shrubs and pine seedlings

Young pine forest with developing understory of oak and hickory trees

Mature oak-hickory forest

Time

from field to oak-hickory forest takes about 100 to 200 years if it continues unimpeded. Through succession, the area returns to a natural oak-hickory forest, the plant community that existed prior to agricultural clearing. In other ecosystems, such as tropical rainforests, succession from deforestation back to a natural forest can take many centuries.

The concept of plant succession was introduced early in the twentieth century. At that time, succession was considered to be an orderly process of vegetation change in steps or phases that ended with a dominant and stable vegetative cover. This implied that the vegetation association would remain unless disturbed by human activity—or major environmental change (for example, climate change). The final hypothesized result in a vegetative succession is called a **climax community**. It was generally agreed that such a community was self-perpetuating and had reached a state of equilibrium or stability with the environment. In our illustration of plant succession (Fig. 11.11) in the southeastern United States, the oak-hickory forest would be the climax community. Tropical rainforests have been considered to be an approximate example of a climax community because except for human interference they have been relatively stable for thousands of years.

Succession remains as a useful model for studying ecosystems, but some of the original ideas have undergone considerable modification. For one thing, early proponents emphasized a *predictable* sequence of successional steps. One plant community would follow another in regular order as the ecosystem changed over time. Today, it is known that changes in ecosystems do not follow a rigid or completely predictable sequence.

Most scientists today no longer believe that only one type of climax vegetation is possible for each of the world's major climate regions. One of several different climax communities might develop in a given area, influenced not only by climate but also by local conditions of drainage, nutrients, soil, or topography. The dynamic nature of climate is better understood today than it was when the original theories of succession and climax vegetation were developed. For example, in the time it takes the species structure of a plant community to adjust to climatic conditions, the climate may change again. Also, because every habitat has a dynamic nature, no climax community can exist in equilibrium with an environment indefinitely.

Today, many biogeographers and ecologists view plant communities and their ecosystems as a *landscape* that is the interactive expression of all of its various environmental factors functioning together. They view an area's landscape as a vegetative **mosaic** of interlocking parts, much like tiles in a mosaic artwork. In a pine forest, for example, other plants also exist, and within the mosaic some areas might not support pine trees. The dominant area of the mosaic—in this example, the pine forest—is called the **matrix**. Gaps within the matrix, resulting from different soil conditions or from human or natural processes, are called **patches** within the matrix (● Fig. 11.12). Relatively continuous linear features cutting across the mosaic, including natural features such as rivers and human-created structures such as roads, fence lines, power lines, and hedgerows, are termed **corridors** (● Fig. 11.13). Every habitat is unique and constantly changing, and the plant and animal communities that are affected must constantly adjust to these changes.

USGS/Mark Dornblaser

● **FIGURE 11.12** This patch in a coniferous forest matrix is the result of a wildfire that burned the tree cover. The reddish area consists of fireweed, a wildflower that has colonized the patch. This is an example of the first stage of secondary succession—eventually shrubs will follow to shade out the fireweed, and later, conifers may reforest the area. Other patches here are most likely older fire scars.

What human activities might be responsible for patches in the vegetation matrix of a midlatitude forest?

Climate is a dominant environmental influence, but it is changing today as it has throughout Earth history. Global climates have changed over relatively short time periods, such as over decades and centuries, and also over time frames measured in millennia. Plant and animal communities have to adapt to environmental changes, or they will not survive. Many modern biogeographers are deeply involved in the reconstruction of the vegetation communities from past climate periods by examining evidence such as tree rings, pollen, insect fossils, and plant fossils. By determining how past climate changes affected and induced change in Earth's ecosystems, biogeographers hope to be able to suggest how future changes may develop as climate continues to change.

Environmental Controls

The plants and animals that exist in a particular ecosystem are those that have been successful in adjusting to their habitat's environmental conditions. Every living organism requires a certain range of environmental conditions to survive. For example, some plants can exist under a wide range of temperatures, whereas others can survive only under a narrow temperature range. This is referred to as an organism's **range of tolerance** for various environmental conditions. The range of tolerance determines where a species may

● **FIGURE 11.13** An aerial view of Taskinas Creek, Virginia, shows riparian corridors passing through a forest matrix on the Atlantic coastal plain.

How does a corridor differ from a patch?

exist, and species with wide ranges of tolerance will be the most widely distributed geographically. The *ecological optimum* refers to the environmental conditions under which a particular species will thrive (● Fig. 11.14). The farther away a species is from its ecological optimum or from the geographic core of its plant or animal community, the more difficult conditions will become for that species or the community to survive. However, those same conditions may be more amenable for another species or community. An **ecotone** is the overlap, or zone of transition, between two plant or animal communities (see Fig. 11.14).

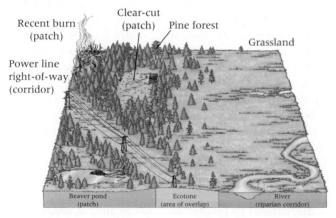

● **FIGURE 11.14** The concepts of *ecotone, ecological optimum, range of tolerance, mosaic, matrix, patch,* and *corridor* are each illustrated in the diagram.

How would a change to more arid conditions affect the relative sizes of the pine forest and the grassland ecosystems?

Climate has the greatest influence over natural vegetation when we consider plant communities on a worldwide basis. The major types of terrestrial ecosystems, or **biomes**, are associated with certain climatic conditions, such as temperature ranges, critical annual or seasonal precipitation, and evaporation characteristics. Climatic conditions have influenced the sizes and shapes of tree leaves and determined whether trees can exist in a region. At the local scale, however, other environmental factors can be just as important. A plant's range of tolerance for acidity, salinity, soil moisture, or other environmental characteristics may be the critical factor in determining whether a plant will grow, flourish, or die. The discussion that follows illustrates how major environmental factors influence the organization and structure of ecosystems.

Natural Factors

Climatic Factors Sunlight is one of the most critical factors related to climate that influence an ecosystem. Solar radiation provides the energy for plant photosynthesis, and it strongly influences the behavior of both plants and animals. Competition for light can make forest trees grow taller and limit plant growth on the forest floor to shade-tolerant species such as ferns. The sizes, shapes, and colors of leaves may result from variations in light reception, with large leaves developing in areas of limited light. The *quality* of light is also important, especially in mountain areas, where plant growth may be retarded by strong ultraviolet radiation. In the thin air at higher elevations ultraviolet radiation is more intense, compared to locations at lower elevations where it is more effectively screened out by the denser atmosphere. Light *intensity* affects both the rate of photosynthesis and the rate of primary productivity in an ecosystem. Intense sunlight in the low latitudes produces a greater biomass in the tropical forests compared to the much lower light intensity of the high latitude Arctic regions. *Duration* of daylight, which varies seasonally and with latitude, has a profound effect on the flowering of plants and on animal mating and migration.

Virtually all organisms require water. Plants need water for germination, growth, and reproduction, and most plant nutrients must be dissolved in soil water to be absorbed by plants. Marine and aquatic plants are adapted to living in water. Mangroves grow in coastal tidal zones, and survive by being able to extract freshwater from seawater (● Fig. 11.15). Bald cypress trees (● Fig. 11.16) rise from the waters of marshes and swamps and also along the banks of streams. Certain tropical plants become dormant during dry seasons, dropping their leaves, and others store water received during wet periods for surviving the drought seasons. Desert plants, such as cacti, are well adapted to storing water when it is available while minimizing water loss from transpiration.

Organisms are affected less by temperature variations than by the availability of sunlight and water. Many plants can tolerate a wide range of temperatures, although each species has optimum conditions for germination, growth, and reproduction. These functions, however, can be impeded by temperature extremes. Temperatures may also have indirect effects on vegetation. For example, high temperatures will lower the relative humidity, and increase transpiration. If a plant's root system cannot extract enough moisture from the soil to meet this increase in transpiration, the plant will wilt and eventually die. Tundra areas are

FIGURE 11.15 A mangrove tree with underwater roots on a coast in the U.S. Virgin Islands. Mangrove roots are able to expel the salt from seawater to use fresh water for growth.

USGS/Caroline Rogers

NPS/Everglades/G. Garner

● FIGURE 11.16 Bald cypress trees in Everglades National Park, Florida. Extensive cypress forests exist in swampy areas of the southeastern United States. They are called *bald cypress* because they are deciduous needleleaf trees that change color and lose their leaves during the fall and winter. Note the alligator.

generally free of tree growth because the growing season is too short and the temperatures too cool to sustain trees.

Animals, because of their mobility, are not as dependent as plants are on climatic conditions, although animals are also subject to climatic stresses. The geographic distribution of some groups of animals reflects this sensitivity to climate. Cold-blooded animals are, for example, more widespread in warmer climates and more restricted in colder climates. Some warm-blooded animals develop a layer of fat or fur and are able to shiver to protect themselves against the cold. In hot periods, they may sweat, shed fur, or lick their fur in an attempt to stay cool. In regions that have extremely cold or arid seasons, some animals hibernate. During hibernation, an animal's body temperature changes roughly in response to the temperatures outside and of the ground. Cold-blooded animals in hot climates—rattlesnakes, for example—move in and out of shade in response to temperature changes. Warm-blooded animals migrate great distances into amenable areas and out of environmentally harsh regions. Others, like the camel, can travel for great distances and live for extended periods without a water supply.

Some warm-blooded animals exhibit interesting relationships between body shape and size variations and the average environmental temperature of their habitats. These adaptations have been described by biologists as Bergmann's Rule and Allen's Rule.

Bergmann's Rule states that, within a warm-blooded species, the body size of the subspecies usually increases with the decreasing mean annual temperature of its habitat; **Allen's Rule** notes that, in warm-blooded species, the relative size of exposed portions of an animal's body decreases with decreasing mean annual temperatures. This means that members of the same species living in colder climates eventually evolve shorter or smaller appendages (ears, noses, arms, legs, and so on) than their relatives in warmer climates (Allen's Rule). Also in cold climates animals will be larger with more mass to provide the body heat needed for survival and to protect the main trunk of the body where vital organs are located (Bergmann's Rule). In cold climates, small appendages are advantageous because they reduce the amount of exposed area subject to temperature loss, frostbite, and cellular disruption (● Fig. 11.17). In warm climates, large body sizes are not necessary for the protection of internal organs, but long limbs, noses, and ears allow for heat dissipation in addition to that provided by panting or fur licking.

The influence of wind on vegetation is most significant in deserts, polar regions, coastal zones, and highlands. Wind can directly injure vegetation and have an indirect effect by increasing evapotranspiration. To prevent water loss in areas of severe wind stress, plants twist and grow close to the ground, minimizing their wind exposure (● Fig. 11.18). Trees that have taken on this low, twisted form because of harsh conditions near the limit of tree growth are referred to as **krummholz**, a German word for *crooked*

● **FIGURE 11.18** Krummholz (stunted trees) along the upper edge of the treeline in the Colorado Rocky Mountains. The low vegetation is covered by snow much of the year, protecting it from bitterly cold temperatures and gale-force winter winds. Note the flag-shaped trees, with no limbs on one side, which indicates the prevailing wind direction.

What biome type would be found at elevations higher than the one depicted in this photograph?

● **FIGURE 11.17** The Arctic polar bear is an excellent example of both Allen's and Bergmann's Rules.

What warm-climate animal can you suggest as a good example of Allen's Rule?

wood. During severe winters, trees are better off being buried by snow than being exposed to bitterly cold gale-force winds. In some windy coastal regions, the shoreline may be devoid of trees or other tall plants. In windy coastal and mountain regions where the trees do grow, they are often misshapen or swept bare of leaves and branches on their windward sides.

Soil and Topography

In terrestrial ecosystems, soils supply much of the moisture and minerals for plant growth. Soil variations influence plant distributions and can produce sharp boundaries between vegetation types. This is partly a consequence of varying chemical requirements of different plant species and partly a result of factors such as soil texture. Clay soils can retain too much moisture for some plants, whereas sandy soils retain too little. Pines generally thrive in sandy soils, grasses in clays, cranberries in acid soils, and chili peppers in alkaline soils. The subject of soils and their influence on vegetation will be explored more thoroughly in Chapter 12.

Topography, particularly in highlands, influences ecosystems by providing diverse microclimates within a relatively small area. Plant communities vary from place to place in highland areas in response to differing microclimatic conditions. **Slope aspect** has a direct effect on vegetation patterns in areas outside of the equatorial tropics. North-facing slopes in the middle and high latitudes of the Northern

Hemisphere have microclimates that are cooler and wetter than those on South-facing exposures. Some plants thrive on the sunny south-facing slopes in the Northern Hemisphere; others survive on the colder, shaded, north-facing slopes (● Fig. 11.19). The steepness and shape of slopes also affect the amount of time water is present there before draining downslope.

● **FIGURE 11.19** The strong impact of slope aspect is visible here in Washington State. The cooler, wetter North-facing slopes (left) support coniferous evergreen forest, but the South-facing slopes receive more direct sunlight so they are warmer, drier, and covered with annual grasses that have dried out.

Are there good examples of the influence of slope aspect on vegetation in the area where you live?

The Theory of Island Biogeography

Biogeographers are intrigued by the life forms and the diversity of species that exist on isolated islands, separated from larger landmasses by extensive areas of ocean. How can land plants and animals be found living on an island surrounded by a wide expanse of sea? How do flora and fauna expand their spatial distribution to become established and flourish on these distant, often geologically recent, and originally barren terrains (volcanic islands, for example)? The farther an island is from the nearest landmass, the more difficult it is for species to migrate to the island and establish a viable population there. Winds, birds, or ocean currents can carry some seeds to islands, where they germinate to develop the vegetative environments on isolated landmasses. But why did the species adapt and survive? Humans have also introduced many species into island environments.

The theory of island biogeography offers an explanation for how natural factors interact to affect either the successful colonization or the extinction of species that come to live on an island. The theory considers an island's locational isolation (the distance from a mainland source of migrating species), the island's size, and the number of species living on an island. Generally, the biological diversity on islands is low compared with mainland areas with

Laura Beauregard/USFWS

Palmyra Atoll, at one of the most remote locations in the Pacific Ocean, is an example of how palm trees become established on tropical islands. Coconuts are the seeds for coconut palm trees. Washed away by surf from their original location, they float and drift hundreds of kilometers from one island to another. Waves deposit and bury coconuts on the beach, and palms spout and grow. Relatively few plant species grow on this very small island. Greater species diversities tend to exist on larger islands and on islands that are closer to major landmasses.

⊕ In Google Earth fly to: 5.878°N, 162.073°W

similar climates and other environmental characteristics. Low species diversity typically means that the floral and faunal populations exist in an environmentally challenging location. Many extinctions have occurred on islands because some factor was introduced that made the habitat no longer viable for certain species to survive.

Several factors affect the species diversity on islands (as long as other environmental conditions such as climate are comparable):

1. The farther an island is from the area from which species must migrate, the lower the species diversity. Islands nearer to large landmasses tend to have higher diversity than those that are more distant.

2. The larger the island, the greater the species diversity. This is partly because larger islands tend to offer a wider variety of environments to colonizing organisms than smaller islands do. Larger islands also offer more space for species to occupy.

3. An island's species diversity results from a balance between the rates of extinction of species on the island and the colonization of species. If the island's extinction rate is higher, only a few hardy species will live there; if the extinction rate is lower compared to the colonization rate, more species will thrive and the diversity will be higher.

The theory of island biogeography has also been useful in understanding the ecology and species diversity of other areas surrounded by environments where their biota would not survive. For example, high mountain areas that stand, much like islands, above extensive desert areas. In those regions, plants and animals adapted to cool, wet environments live in isolation from similar populations on nearby mountains, separated by inhospitable arid environments.

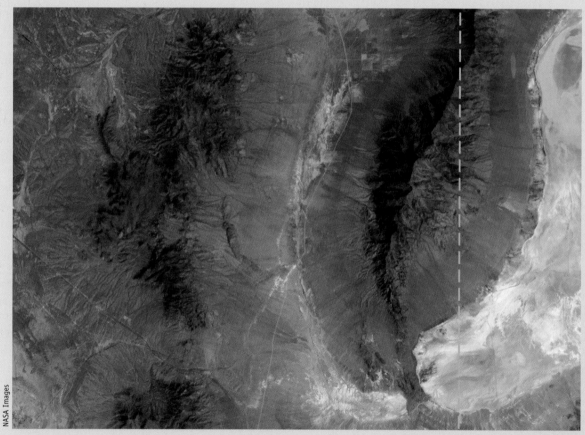

NASA Images

A mountain range along the Nevada/Utah state line supports green forests, alpine zones, and animal species associated with those environments. The blue (flooded) and white (dry) areas to the east *(left)* are the edge of the Bonneville Salt Flats. Flora and fauna that could not survive in the surrounding desert environments flourish in the green, cooler, wetter, higher elevation zones. Mountains surrounded by desert environments tend to fit the concept of island biogeography, as they are isolated by lower elevation deserts rather than ocean waters.

⊕ In Google Earth fly to: 41.04°N, 114.06°W

Luxuriant forests tower above the well-watered windward sides of mountain ranges such as the Sierra Nevada and Cascades, but semi-arid grasslands and sparse forests cover the leeward sides. Spatial variations in precipitation drainage and resulting vegetation differences would not exist were it not for the presence of the topographic barrier-inducing orographic uplift. Each major increase in elevation produces a different mixture of plant species that can tolerate the lower ranges of temperature found at the higher elevation.

Natural Catastrophes

The distributions of plants and animals are also affected by a diversity of natural processes that are sometimes referred to as *catastrophes*. It should be noted, however, that this term is applied from a strictly human perspective. What may be catastrophic to a human—such as a hurricane, forest fire, landslide, tsunami, or avalanche (● Fig. 11.20)—is simply a natural process operating to produce openings (gaps) in the prevailing vegetative mosaic of a region. The resulting successional processes, whether primary or secondary, produce a variety of patch habitats within the broader regional matrix of vegetation. The study of natural catastrophes and the resulting patch dynamics among the plant and animal residents of an area is a topic of strong interest and ongoing research in modern landscape biogeography.

Biotic Factors

Although their influence on a particular species might tend to overlooked, other plants and animals may also affect whether a given organism exists as part of an ecosystem.

USGS/P. Carrara

● **FIGURE 11.20** The vegetation mosaic here in Glacier National Park, Montana is coniferous forest, but snow avalanches keep conifer trees from invading this patch of open, low shrubs and grasses.

Why are there so many broken tree stumps in the foreground?

Some interactions between organisms are beneficial to both species involved; this is called a **symbiotic relationship**. Other relationships may be directly *competitive* and have an adverse effect on one or both species. Because most ecosystems are suitable to a wide variety of plants and animals, there is always competition between species and among members of a given species to determine which organisms will survive. The greatest competition occurs between species that occupy the same ecological niche, especially during the earliest stages of life cycles when organisms are most vulnerable. Depending on the climate, the greatest competition among plants is for light or water. The trees that become dominant in a forest are those that grow the tallest and partially shade the plants growing beneath them. Other competition occurs underground, where the roots compete for soil water and plant nutrients. In sunny arid locales, competition for water may be the greatest determinant for plant survival.

Interactions between animals and plants as well as competition both within and among animal species also can significantly affect an ecosystem. Many animals are helpful to plants through pollination or by the dispersal of seeds, and plants are the basic food supply for many animals. Grazing can also influence the species that make up a plant community. During dry periods, herbivores may be forced to graze an area very closely, grazing out the taller plants. Plants that are unpalatable, have thorns, or have the strongest root development are the ones that tend to survive. Grazing is a part of the natural selection process, yet serious overgrazing rarely results under natural conditions because wild animal populations increase or decrease with the available food supply. For most animals, predators are another control of population numbers. Fortunately, when a predator's favorite prey is scarce, it will seek an alternative species for its food supply.

Human Impacts on Ecosystems

Throughout history, humans have modified ecosystems. Except in regions too remote to be altered significantly by civilization, people have eliminated or had a significant impact on much of Earth's natural vegetation. Farming, fire, domesticated animal grazing, deforestation, road building, urbanization, dam building and irrigation, impacts on water resources, mining, and the draining or infilling of wetlands are a few examples of how humans have modified plant communities. Overgrazing by domesticated animals can seriously harm marginal environments in arid and semi-arid climates. Trampling and soil compaction by grazing herbivores can reduce a soil's ability to absorb moisture, leading to increased surface runoff of precipitation. In turn, the decreased absorption and increased runoff may lead to *land degradation* and gully erosion.

Human alteration of ecosystems and natural environments can often have a negative impact on the people themselves. The desertification of large semi-arid sections of East Africa has resulted in widespread famines in countries such as Ethiopia and Somalia (● Fig. 11.21). Elsewhere, the continuing destruction of wetlands not only eliminates valuable plant and animal communities but also often seriously threatens the quality and reliability of the water supplies for the people who drained the land.

Invasive Exotic Species—Burmese Pythons

Exotic species are plants or animals that people have introduced to a place that is not their native environment. Some exotics are regarded as beneficial; many landscaping plants were originally native to distant regions. Many problems occur with exotics, whether they were introduced to a new area purposefully or inadvertently. A significant ecological problem occurs when the exotics thrive in the new environment to the detriment of native plants or animals. Introduced exotic species that are unable to adapt will die off. Exotics that outcompete native species and reproduce in great numbers are invasive exotic species, which can seriously affect the ecology and biodiversity of a region. Invasive exotics are a major cause of endangered or threatened native flora and fauna worldwide. Hundreds of invasive exotic plants and animals are causing problems in the aquatic and terrestrial environments of the United States.

One example is the Burmese python, a constrictor native to South and South-east Asia, originally brought to the United States by the exotic pet trade. Acquired as a pet when they are young and small, these snakes can grow as large 6 meters (20 ft) long and can weigh more than 90 kilograms (200 lb). Burmese pythons are semi-aquatic, very powerful, and excellent swimmers, and they have proliferated in the Florida Everglades. Although it is not exactly known how they were introduced to the Everglades, some owners have released pets that had become too large, and some snakes escaped their keepers after damage by Hurricane Andrew in 1992.

Park rangers in Everglades National Park first collected two Burmese pythons in 2000 and more than 2000 have been removed since that time. Their population is exploding, and is very difficult to estimate although it is thought that there may be about 10,000 Burmese pythons in the Everglades (earlier estimates were as high as 10 times that). These snakes are voracious carnivores that are disrupting the ecology by preying on native wildlife in great numbers, including mammals, birds, and other reptiles. Some of the pythons' prey are considered to be threatened or endangered species. The American alligator, once an endangered species, might also be threatened. These pythons have been known to consume alligators that are 1.5 to 1.8 meters (5–6 feet) long.

The National Park Service is working to control the population of Burmese pythons in the Everglades. This process is not only important to the ecology of South Florida, but there is also concern that these pythons could spread to other wetland regions along the Gulf of Mexico and the south Atlantic coast.

A Park Ranger in Everglades National Park holds the tail of a large invasive Burmese python.

NPS/Bob DeGross

● **FIGURE 11.21** Overgrazing in a marginal environment is a major cause of desertification here in Somalia, Africa. The area normally would have been a grassy savanna region with scattered trees.

What are some other causes of desertification?

Classification of Terrestrial Ecosystems

Earth's terrestrial ecosystems can be categorized into one of four basic types: forest, grassland, desert, and tundra (● Fig. 11.22). To include vegetation adapted to cold climates that occurs not only in high-latitude regions but also at high elevations at any latitude, biogeographers often refer to the last of these types as either arctic or alpine tundra. The forests in equatorial lowlands, however, are entirely different from those of Siberia or of New England, and the original grasslands of Kansas bore little resemblance to those in Kenya or the Sudan. Hence, the four major ecosystem types can be subdivided into distinctive biomes, with each having a different association of plants and animals. Pure stands of certain trees, shrubs, or even grasses are extremely rare and are limited to small areas having peculiar soil or drainage conditions.

Plant communities are highly visible natural environments, and they can be categorized on the basis of vegetative form, structure, or major physical characteristics. The natural vegetation changes from place to place in a transitional manner, of course, just as temperature and rainfall do, and although distinct biomes are apparent, there may be broad transition zones (ecotones) between them. Nevertheless, certain recurring plant communities have distinctive similarities, resulting from a botanical response to environmental controls that are strongly climatic. It is the dominant vegetation of these plant communities that we recognize when we classify the world's major terrestrial ecosystems.

The world's major biomes are mapped in ● Fig. 11.23 on the basis of natural vegetation associations that give each its distinctive character. The influence of climate on the distribution of biomes is immediately apparent on the map.

Latitudinal and/or elevational effects on temperature and solar radiation and the availability of moisture are key factors in the location of major biomes on a world regional scale. There is some similarity of between biomes influenced by the conditions associated with latitudinal zones to biomes in zones affected by

elevation (● Fig. 11.24). Because climate is strongly related to vegetation associations, general descriptions of major vegetation types were discussed with their corresponding climate types in Chapters 9 and 10.

Forest Biomes

Forests are associations of large, woody, perennial tree species, generally several times the height of a person and with a more or less closed canopy of leaves overhead. They vary widely in density and physical appearance. The kinds of forest trees also vary—they may be needleleaf or broadleaf, and either evergreen or deciduous, which means they drop their leaves to reduce moisture loss during seasons that are dry or cold. Forests exist only where the annual moisture balance is positive—where moisture availability exceeds potential evapotranspiration, particularly during the growing season. Thus, they occur in the tropics, where either the intertropical convergence zone (ITCZ) or monsoonal circulation brings plentiful rainfall, and in the middle latitudes, where precipitation is associated with midlatitude cyclones, summer convectional rainfall, or orographic uplift.

Tropical and midlatitude forests have different characteristics that have evolved in response to any physical limitations that exist. In general, tropical forests have developed in less climatically restrictive environments. Temperatures in the humid tropics are high but not extreme, and there is ample moisture to encourage rapid and luxuriant growth. Midlatitude forests, however, must adapt to either seasonal cold (ranging from occasional frosts to subzero temperatures) or seasonal drought, which may occur at the worst possible time for vegetative growth.

Tropical Forests

The forests of the tropics are far from uniform in appearance and composition. They grade poleward from the equatorial rainforests, which support the world's greatest biomass, to the last scattering

● **FIGURE 11.22** The four major types of Earth biomes: (a) forest biome near Dornbirn, Austria; (b) pronghorn antelope on a grassland biome in the Dakotas; (c) desert biome in Arizona; and (d) tundra biome in Alaska.

(a) R. Gabler

(b) USDA/Dakota Prairie National Grasslands

(c) BLM/Bob Wick

(d) Jason Neely/Polar Field Service

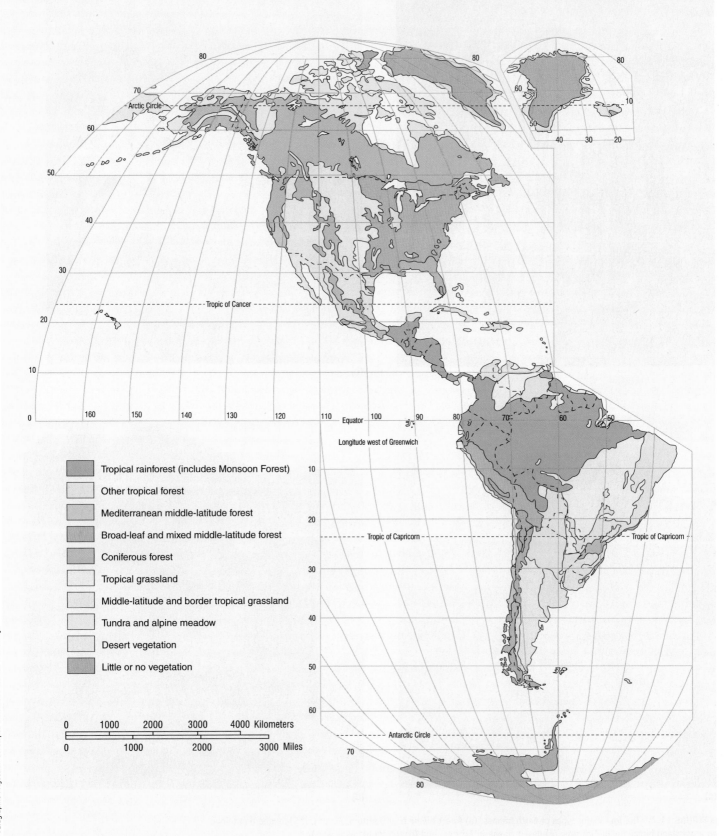

Tropical rainforest (includes Monsoon Forest)

Other tropical forest

Mediterranean middle-latitude forest

Broad-leaf and mixed middle-latitude forest

Coniferous forest

Tropical grassland

Middle-latitude and border tropical grassland

Tundra and alpine meadow

Desert vegetation

Little or no vegetation

Arctic Circle

Tropic of Cancer

Equator

Longitude west of Greenwich

Tropic of Capricorn

Antarctic Circle

0   1000   2000   3000   4000 Kilometers

0   1000   2000   3000 Miles

● **FIGURE 11.23** World map of natural vegetation.

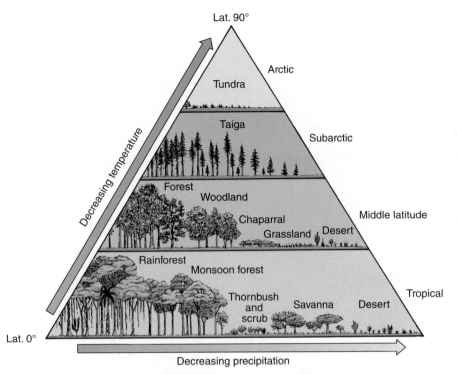

Lat. 90°

Arctic

Tundra

Decreasing temperature

Taiga

Subarctic

Forest

Woodland

Chaparral

Grassland   Desert

Middle latitude

Rainforest

Monsoon forest

Thornbush and scrub

Savanna   Desert

Tropical

Lat. 0°

Decreasing precipitation

Influence of latitude and moisture on distribution of biomes

● FIGURE 11.24 This schematic diagram shows distribution of major biomes as they relate to temperature (latitude) and to moisture availability.

What major biome dominates the wetter margins of all latitudes but the Arctic?

of low trees that overlook endless expanses of tall grass or desert shrubs on the tropical margins. The tropical forests are divided into three distinct biomes: the tropical rainforest, the monsoon rainforest, and other tropical forest types (primarily thornbush and scrub). Of course, gradations exist (ecotones) between the different types and distinctive variations found only in certain localities.

Tropical Rainforest

In the equatorial lowlands dominated by Köppen's tropical rainforest climate, the limitation for vegetation growth is competition among adjacent species for light. Temperatures are high enough to promote constant growth, and water is always sufficient. These forests consist of an amazing number of broadleaf evergreen tree species. A cross-section of the rainforest typically reveals multiple levels of leaf canopies. The trees composing the distinctive individual tiers have similar light requirements—lower than those of the higher tiers but higher than those of the lower tiers. Relatively little sunlight reaches the forest floor, which may support ferns but is often rather sparsely vegetated. The forest is bound together by woody vines, called **lianas** (● Fig. 11.25), which climb the tree trunks and intertwine toward the canopy in a search for light.

Epiphytes (● Fig. 11.26a), sometimes called *air plants*, grow on the limbs of the forest trees, deriving nutrients from the water and plant debris that falls from higher levels. Large colorful fruits and flowers attract animals that carry out seed and the dispersal of pollen.

Rainforest trees commonly depend on **buttressed trunks** (● Fig. 11.26b) for support because their root systems are shallow. This is a consequence of the surface soil's richness and its nutrient deficiency at lower levels. The forest litter quickly decomposes, releasing its nutrients, which are soon reabsorbed by the forest root systems Tropical soils that maintain the tremendous rainforest biomass are fertile only as long as the forest remains undisturbed. Forest clearing interrupts the cycling of nutrients between the vegetation and the soil; the copious amounts of water percolating through the soil leach away its soluble constituents, leaving behind iron and aluminum oxides that cannot support forest growth. The present rate of clearing threatens the worldwide tropical rainforests within the foreseeable future. The largest remaining areas of unmodified rainforests are in the upper Amazon Basin, where they cover hundreds of thousands of square kilometers.

Because of the darkness and extensive root systems present on the forest floor, animals of the tropical rainforest are primarily arboreal (tree-dwelling). A wide variety of species of tree-dwelling monkeys and lemurs, snakes, tree frogs, birds, and insects characterize the rainforest. Even the large herbivorous and carnivorous mammals—such as sloths, ocelots, and jaguars, respectively—are primarily arboreal.

Jim Petersen

● FIGURE 11.25 The dense forest canopy of a tropical rainforest effectively shades the forest floor. Here, lianas (woody vines) climb the trees in search of light.

How might this rainforest differ from the rainforests of the Pacific Northwest of North America?

(a)

(b)

● **FIGURE 11.26** (a) Epiphytes like this one grow on trees in humid, wet areas, such as here in a tropical rainforest. Sometimes called *air plants*, they take their nutrients from the humid air. They are not parasites, as they do not directly harm the trees they grow on. (b) Distinctive buttressed trunks with flared-out roots help support the tall and massive hardwood trees of the tropical rainforests.

Within the rainforest areas along streams, roads and other cleared areas are patches or strips of jungle. Jungle consists of an almost impenetrable tangle of vegetation contrasting strongly with the relatively open floor of the true rainforest. It is often composed of secondary growth that quickly invades where a clearing has allowed light to penetrate to the forest floor. Jungle commonly extends along the forest margins and the courses of streams. There it forms a canopy of vegetative growth closing over the water-course called **galleria forest** (● Fig. 11.27).

Monsoon Rainforest In tropical monsoon areas, the low-sun dry monsoon alternates with the high-sun wet monsoon season, which brings moist air inland from tropical coasts. Rainfall in the wet monsoon season may be very high and even hundreds of centimeters when enhanced by orographic uplift. Annual rainfall, despite a short dry season, is sufficient to produce a dense forest. Monsoon forests may have discernible tiers of vegetation related to the varying light demands of different species, and they are classified as a variety of tropical rainforest. The number of species, however, is less than that of the equatorial rainforests, and the overall height and density of forest vegetation are also somewhat less. Some of the species are evergreen, but many are also deciduous, responding to the dry monsoon.

Thornbush and Scrub Forests Where seasonal drought or soil characteristics do not support the growth of tropical rainforests, other kinds of tropical forests have developed. These tend to be found on the subtropical margins of the rainforests and in locations where the soils are especially poor in nutrients. The vegetation of this category varies greatly but is generally lower growing in comparison to rainforest, without a tiered structure, and is denser at ground level. It is commonly thorny, which is a defense against browsing animals, and resists drought by dropping leaves

● **FIGURE 11.27** Tropical rainforest trees compete for light, because of shading by the continuous cover of the forest canopy. Where streams flow through the rainforest, trees along the banks reach laterally for light and can form a cover of branches over the stream channel. Tree cover that forms a "roof" over a water channel is called *galleria* vegetation.

to conserve moisture during the dry low-sun season. Typically, grass grows beneath the trees and shrubs. Farther away from the equatorial zone, the trees are more widely spaced and the grassy areas become dominant.

Midlatitude Forests

The forest biomes of the midlatitudes differ from those of the tropics because the dominant trees are adapted to withstand periods of water deprivation resulting from low temperatures and annual variations in precipitation. Evergreen and deciduous plants are present and equipped to cope with seasonal extremes that are not encountered in tropical latitudes.

Mediterranean Sclerophyllous Woodland
Surrounding the Mediterranean Sea and on the southwest coasts of the continents between approximately 30° and 40° N and S latitude, a distinctive climate exists—Köppen's Mediterranean climate. Here, annual temperature variations are moderate, and freezing temperatures are uncommon. Little or no rainfall occurs during the warmest months, the growing season, and vegetation must be drought resistant. A distinctive vegetation association has evolved that consists of relatively low-growing shrubs and small trees, with small, hard-surfaced leaves and roots that probe for water. The leaves are capable of photosynthesis with minimal moisture transpiration. The general look of the vegetation is a thick scrub plant community, called *chaparral* in the western United States and *maquis* in the Mediterranean region (● Fig. 11.28).

Wherever moisture is concentrated in depressions or on the cooler North-facing hill slopes, deciduous and evergreen oaks occur in groves. Drought-resistant needleleaf trees, especially pines, are also part of the vegetation associations in these regions. Thus, the vegetation is a mosaic related to site characteristics and microclimate. Nevertheless, the similarity of the natural vegetative cover in such widely separated areas as Spain, Turkey, Chile, and California is astonishing. (Note the location of Mediterranean forest in Fig. 11.23.)

Broadleaf Deciduous Forest
The humid regions of the midlatitudes experience a seasonal rhythm dominated by warm tropical air during the summer and invasions of cold continental polar air in the winter. To avoid frost damage during the cold winters and to survive moisture deprivation when the ground is frozen, leaves with large transpiring surfaces drop from the trees as they become dormant, to be replaced by new leaves after the danger period has passed. A large variety of trees have evolved this mechanism; certain oaks, hickory, chestnut, beech, and maples are common examples. The seasonal rhythms produce some beautiful scenes, particularly during the periods of transition between dormancy and activity, with the sprouting of new leaves in the spring and the brilliant coloration of the fall as chemical substances draw back into the plant for winter storage (● Fig. 11.29).

Jim Petersen

● **FIGURE 11.28** Chaparral, a distinctive association of evergreen shrubs *(sclerophylls)*, grows in many Mediterranean climate regions and is adapted to a regime of dry summers alternating with rainy winters.

What are the general characteristics of sclerophyllous vegetation?

USFWS

● **FIGURE 11.29** Deciduous hardwood forests in midlatitude regions, like here in Maryland, change dramatically in the fall to beautiful colors before dropping their leaves for winter dormancy. A beaver lodge is in the lake.

Trees of the deciduous forests can be almost as tall as those of the tropical rainforests and can also produce a closed canopy of leaf-covered branches overhead or, in the cold season, an interlaced network of bare branches. However, lacking a multistoried structure and having a lower forest density, midlatitude deciduous forests allow much more light to reach ground level. Forests of this type are the natural vegetation in much of western Europe, eastern Asia, and eastern North America. To the north and south, they merge with mixed forests composed of broadleaf deciduous trees and conifers.

Broadleaf Evergreen Forest

Beyond the tropics, broadleaf trees remain evergreen (active throughout the year) in significant areas only in certain Southern Hemisphere locations. Here, the mild maritime influence is strong enough to prevent either dangerous seasonal droughts or severely low winter temperatures. Southeastern Australia and portions of New Zealand, South Africa, and southern Chile are the principal areas of this type. Limited areas occur in the United States in Florida and along the Gulf Coast as a belt of evergreen oaks and magnolia trees.

Mixed Forest

Poleward and equatorward, the broadleaf deciduous forests in North America, Europe, and Asia gradually merge into mixed forests, including needleleaf coniferous trees, mainly pines. In general, where conditions permit the growth of broadleaf deciduous trees, coniferous trees cannot compete successfully with them. Thus, in mixed forests, the conifers, which are more adaptable to soil and moisture deficiencies, can be found in less hospitable sites: in sandy areas, on acid soils, or where the soil is thin. The northern mixed forests reflect a transition to colder climates with increasing latitude, and eventually, conifers become dominant in this direction. The southern mixed forests in the United States are transitional to pine forests situated on sandy soils of the coastal plains. In Eurasia, they coincide with highlands dominated by conifers during a stage in the plant succession that began with climate change at the close of the last Ice Age, some 10,000 years ago.

Coniferous Forest

Coniferous forests occupy the frontiers of tree growth. They survive where most of broadleaf species cannot endure the climatic severity and poor soils. The hard, narrow needles of coniferous species transpire less moisture compared with broad leaves so that needleleaf species can tolerate unavailability of moisture because of high soil permeability, or a dry season, or because of frozen soil water. Pines demand little from the soil and can grow in sandy places and where the soil is acidic. As a whole, conifers are well adapted to regions having long, severe winters combined with summers warm enough for vigorous plant growth. Because all but a few exceptions retain their leaves (needles) throughout the year, they are ready to begin photosynthesis as soon as temperatures permit without having to produce a new set of leaves to do the work.

A great band of coniferous forests, the **boreal forests**, or **taiga**, dominated by spruce and fir species, sweeps the breadth of North America and Eurasia north of the 50th parallel, the region

of Köppen's subarctic climate (see Fig. 10.21). Conifers differ from other trees in that their seeds are not enclosed in a case or fruit but are carried naked on cones. All are needleleaf and drought resistant trees, but a few are not evergreen. A large area of Siberia is dominated by larch, a deciduous tree, which produces a deciduous coniferous forest. In this area, January mean temperatures may be $-35°$ C to $-51°$ C ($-30°$ F to $-60°$ F). This is the most severe winter climate that can maintain trees; even needleleaf foliage is shed for the vegetation to survive. Hardy broadleaf deciduous birch trees share this extreme climate with robust larches.

Extensive coniferous forests are not confined to high latitude areas with short summers. Higher elevations in the midlatitude mountains of the Northern Hemisphere have forests of pine, hemlock, and fir (● Fig. 11.30), with subalpine larch and specially adapted pine species characterizing the harshest and highest forest sites. The forests along the sandy coastal plain of the eastern United States are there in part because of the sandy soils.

● **FIGURE 11.30** As this scene in the Gifford Pinchot National Forest in Washington State indicates, evergreen coniferous forest is the characteristic vegetation of higher-elevation midlatitude regions. These needleleaf trees are well adapted to drought in the winter season, which is longer and more severe at higher elevations. The broadleaf trees are in fall color prior to losing their leaves.

Why are needleleaf trees better adapted to physiologic drought than broadleaf trees?

A maritime coniferous forest occupies the west coast of North America from southern Alaska to central California. These forests include redwoods, Douglas fir, cedars, hemlock, and farther north, Sitka spruce. Many of the California redwood trees are thousands of years old and more than 100 meters (330 ft) high. The southern regions experience summer drought, and farther north sandy, acidic, or coarse-textured soils dominate.

Grassland Biomes

Enormous, continuous expanses of grasslands exist on Earth. On a global scale, grassland biomes are mainly located in continental interiors where most of the precipitation falls in the summer. Two large grassland realms are generally recognized: the tropical and the midlatitude grasslands. In general, it is thought that grasses are dominant only where trees and shrubs cannot maintain themselves because of either excessive or deficient soil moisture. However, in many regions it is suspected that human interference with the natural vegetation has caused the expansion of grasslands into forested areas of both the tropical and midlatitudes.

Tropical Savanna Grasslands

The tropical grassland biome differs from those of the midlatitudes in that it ordinarily includes a scattering of trees; this is implied in the term **savanna** (● Fig. 11.31). In fact, the demarcation between tropical scrub forest and savanna is seldom clear. Savanna grasses tend to be coarse with bare ground visible between individual tufts, and the tree species are generally low growing with wide crowns. They also have both drought and fire-resisting qualities to survive fires that can sweep the savannas during the drought season. Savannas occur under a variety of temperature and rainfall conditions but generally fall within Köppen's tropical savanna climate. They commonly occur on red-colored soils, leached of all but iron and aluminum oxides, which become bricklike when dried. They likewise coincide with areas where the water table (the zone below which all subsurface pore space is water saturated) fluctuates dramatically. The up-and-down movement of the water table may inhibit forest development, and it is no doubt a factor in the peculiar chemical nature of savanna soils. Large herds of grazing animals and associated predators migrate seasonally, responding to the abundant grasses of the wet summer and the drought conditions of the low-sun season.

Midlatitude Grasslands

The midlatitude grasslands occupy the zone of transition between the deserts and forests of the middle latitudes. On their dry margins, they pass gradually into deserts in Eurasia and are cut off westward by mountains in North America. However, on their humid side, they terminate rather abruptly against the forest

● **FIGURE 11.31** A classic savanna biome landscape of grasses and scattered trees near Alice Springs, Australia. A difference from other savanna areas, however, is that these are eucalyptus trees, native to Australia.

What season is likely shown here, summer or winter, and what is a possible month?

⊕ **In Google Earth fly to: 23.75°S, 133.80°E**

margin, raising questions as to whether their limits are natural or have been created by human activities, particularly the intentional use of fire to drive game animals. The midlatitude grasslands of North America, like the African savannas, formerly supported enormous herds of grazing animals—in this case, antelope and bison, which were the principal means of support for the American Plains Indians.

Like the savannas, the midlatitude grasslands were diverse in appearance. They, too, consisted of varying associations of plant species that were never uniform in composition. In North America, the grasses were as much as 3 meters (10 ft) tall in the more humid sections, as in Iowa, Indiana, and Illinois, but only 15 centimeters (6 in.) high on the dry margins from New Mexico to western Canada. Thus, the midlatitude grassland biomes are usually divided into tall-grass and short-grass prairie, often with a zone of mixture recognized between them. In contrast to the tropical savannas, the germination and growth of midlatitude grasses are attuned to the melting of winter snows followed by summer rainfall. Whether the grasses are annuals that complete their life cycle in one growing season or perennials that grow from year to year, they are dormant in the winter season. Also, in contrast to the savannas, the soils beneath these grasslands are extremely rich in organic matter and soluble nutrients. As a consequence, most of the midlatitude grasslands have been completely transformed by agricultural activity. Their wild grasses have been replaced by domesticated varieties, such as wheat, corn, and barley. The midlatitude grasslands have become the "bread-baskets" of the world.

Tall-Grass Prairie

The **tall-grass prairies** (● Fig. 11.32a) were an impressive sight: endless seas of grass moving in the breeze, with conspicuous flowering plants adding to the effect. Unfortunately, this tall-grass prairie environment, once reaching continuously from Alberta, Canada to Texas, has been almost completely destroyed. Today, only a few tall-grass prairie areas remain in government-protected preserves. Their tough sod, formed from a dense network of roots from grasses was subdued by the steel plow in the 1830s, aided by the introduction of well-digging machinery and barbed wire. These innovations transformed the tall-grass prairie from grazing land to cropland.

In North America, the tall-grass prairie extended as far eastward as Lake Michigan. Why trees did not invade the prairie in this relatively humid area remains an unanswered question. Farther west, shallow-rooted grass cover exists, which could otherwise support

trees, but the region is too dry. In these areas, trees can survive only along streams or where depressions collect water.

In Eurasia, tall-grass prairies were found in a discontinuous belt from Hungary through the Ukraine, Russia, and central Asia to northern China. In South America, these grasslands are known as the *pampas* and in South Africa as the *veldt*. Today, all of these areas have been modified by agriculture. The main factor that supports tall-grass prairies is precipitation that is moderate but variable from year to year, and this is the principal hazard in using these regions as farmlands. This problem is even greater in the drier areas of short-grass prairie.

(a)

Photo courtesy of Tallgrass Prairie National Preserve

(b)

Photo courtesy of Tallgrass Prairie National Preserve

● **FIGURE 11.32** (a) Spring wildflowers bloom in the Tallgrass Prairie National Preserve, Kansas, one of the remaining protected areas of this prairie type. The grasses will grow taller in later spring and summer. (b) Short-grass prairie, with bison grazing in South Dakota's Wind Cave National Park.

Both kinds of prairies develop under a semiarid climate, but which grassland type receives a higher annual precipitation?

Short-Grass Prairie West of the 100th meridian in the United States and extending across Eurasia from the Black Sea to northern China, roughly coinciding with the areas of Köppen's midlatitude steppe climate, are vast, nearly level grasslands composed of a mixture of tall and short grasses, with short grass being dominant where annual precipitation is lower (● Fig. 11.32b). On the Great Plains between the Rocky Mountains and the tall-grass prairies, the **short-grass prairie** more or less coincides with the zone where moisture rarely penetrates more than 60 centimeters (2 ft) into the soil, and the subsoil is permanently dry. Moving toward the drier areas, the grassland vegetation dwindles in height. The amount of ground cover also declines toward the drier margins as the sod-forming grasses give way to bunchgrasses, which grow in isolated clumps (bunches), instead of forming a continuous grass cover. In their natural conditions, the short-grass prairies of North America and Eurasia supported high densities of grazing animals—bison and antelope in the former, wild horses in the latter. Indeed, it is suspected that the plant associations of the short-grass prairies may have been a consequence of overgrazing under natural conditions. The short-grass prairies cannot be cultivated without either irrigation or dry farming methods, so they remain primarily the domain of wide-ranging grazing animals. Today's resident animals are domesticated cattle, not the thundering self-sufficient wild herds that formerly made these plains one of world's marvels.

Desert Environments

Where the potential evapotranspiration greatly exceeds available moisture throughout the year, as in the desert climates, either special forms of plant life have evolved or the surface is bare. Desert plants adapt by probing deeply or widely for moisture, by keeping minimizing moisture losses, or by storing moisture when it is available. Other desert plants evade drought by lying dormant, perhaps for years, until enough moisture is available to ensure growth and reproduction. The desert biome is recognized by the presence of plants that are either drought resisting or drought evading, and resistant to being browsed by animals (● Fig. 11.33).

Plants with mechanisms to combat drought are known as **xerophytes**. They are perennial shrubs and other plants that have adapted mechanisms that allow them to grow in arid environments. Some xerophytes have evolved tiny leaves with waxy coatings to combat transpiration. Others leaves that are needlelike, trunks and limbs that photosynthesize like leaves, or they may have expandable tissues or accordion-like stems to store water when it is plentiful. They can also be plants that can tolerate saline water and soils, or shrubs that shed their leaves until sufficient moisture is available for new leaf growth. Nonxerophytic vegetation in deserts consists mainly of short-lived plants that germinate and complete their growth cycle—leaf production, flowering, and seed dispersal—in a matter of weeks when triggered by available moisture. These ephemeral plants require certain conditions to grow, and the month-to-month and year-to-year variation in their appearance is enormous. Other non-xerophytic plants live along water sources such as the few streams and springs that exist in desert areas.

National Park Service

● **FIGURE 11.33** Desert plants employ a variety of strategies for surviving in an arid environment, with little rain and high potential evapotranspiration. This cactus has needles to defend against browsing animals, but also has a brilliant flower that attracts and welcomes pollinators.

What other kinds of adaptations help xerophytic vegetation survive in a desert?

Animals of the deserts are primarily nocturnal to avoid the searing heat of the daytime, and many have evolved long ears, noses, legs, and tails that allow for greater blood circulation and cooling. The similar life forms and habits of the plant and animal species found in the deserts of widely separated continents are a remarkable display of adaptive evolution to ensure survival under similar arid conditions. In extremely dry deserts, only a few plants can survive and ground cover is much less common. The driest parts of the Atacama Desert of Chile are so arid that there is no vegetation at all in some locations (● Fig. 11.34).

Arctic and Alpine Tundra

Proceeding upward in elevation and poleward in latitude, we come to regions with growing seasons that are too brief to support tree growth. Nearing the poles, we enter a vast realm dominated by sub-freezing temperatures and thin snow cover much of the year, so the ground is frozen to depths of hundreds of meters. Only the top 36 to 60 centimeters (15 to 25 in.) thaw during the short summer interval. Still, vegetation survives and can form a nearly complete surface cover. Vegetation in these regions must be equipped to tolerate frozen subsoil (permafrost), icy winds, a low sun angle (except in low-latitude, high-elevation locations), summer frosts, and waterlogged during the short growing season. The result is **tundra**—a mixture of grasses, flowering herbs, sedges, mosses, lichens, and occasional low-growing shrubs. Most of the

● **FIGURE 11.34** The absence of vegetation in this moonlike landscape in the Atacama Desert in Chile is clear evidence the extremely low rainfall and high evaporation rates of this region. The Atacama Desert is the driest area in the world.

What are some reasons why people might be found in such desolate regions?

⊕ In Google Earth fly to: 22.95°S, 68.28°W

plants are perennials that produce buds close to or beneath the soil surface, protected from the wind. Many of the plants show xerophytic adaptations—such as small, hard leaves, and compact form—in response to dessication resulting from cold dry winds. The effect of wind is evident because coniferous woodlands often occupy the less exposed valleys in tundra regions.

In a zone of varying width that reaches across northern Alaska, Canada, Scandinavia, and northern Russia, several types of tundra are recognized: *bush tundra*, consisting of dwarf willow, birch, and alder, which grow along the edge of the boreal forest; *grass tundra*, which is hummocky and water soaked during the summer; and *desert tundra*, in which expanses of bare rocks may be covered by colorful lichens. In a few ice-free valleys of Antarctica, only desert tundra occurs.

Alpine conditions are not exactly like those in the Arctic latitudes. The deeper snow cover of the high mountains prevents the development of permafrost, and the summer sun results in considerably more evaporation. However, many high elevation areas are swept clean of snow by wind and are thus extremely exposed, supporting only desert tundra. Microclimatic conditions become an important influence on vegetation because of the varying exposures to sun and wind. Nevertheless, the short growing season and severe wind stress combine to produce a plant community similar to that in the Arctic regions (● Fig. 11.35).

Animals of the tundra must cope with long periods of extreme cold and darkness. Many animals hibernate through the long Arctic winters (or the cold and windy alpine winters); others, such as the caribou of Alaska and Canada, migrate into the boreal forest to escape the extreme cold exacerbated by Arctic winds. Year-round residents, such as the polar bear and musk ox, must have extremely large fat reserves around their large chests in addition to extremely efficient fur. Many insects, such as the ubiquitous mosquito, cope with the extreme climate by emerging from eggs in the spring, maturing and laying eggs, and dying within one short Arctic summer. Enough eggs survive, buried under insulating snow, to continue the cycle in the following year.

Marine Ecosystems

The living organisms of the ocean can be divided into three groups according to where or how they live in the ocean. The first group is plankton. **Plankton** is made up mostly of the ocean's smallest—mainly microscopic—plants (**phytoplankton**) and animals (**zooplankton**). These tiny plants and animals that float freely with the ocean currents are true "drifters" and form the basis of the oceanic food chain (● Fig. 11.36). The second group is composed of swimming animals. This group, called **nekton**, includes fish, squid, marine reptiles, and marine mammals such as whales and seals. The third group is composed of the plants and animals that live on the ocean floor. This group is called the **benthos**. It includes corals, sponges, and many algae; such burrowing or crawling animals as barnacles, crabs, lobsters, and oysters; and attached plants such as turtle grass and kelp.

Life in the ocean depends on the availability of solar energy and on the nutrients in the water. Phytoplankton species are the most important link in the ocean food chain.

At the base of the marine food web, phytoplankton use dissolved nutrients in the water, and through the process of photosynthesis, phytoplankton produce oxygen and foods needed by zooplankton and by the smallest nekton. Phytoplankton are the food supply for these animals, which in turn form the food source for larger carnivorous fish and marine mammals. These, in turn, are prey for still larger animals. For example, tiny shrimplike creatures

● **FIGURE 11.35** Alpine tundra vegetation growing during summer here in Rocky Mountain National Park. The tundra vegetation is a mixture of grasses, sedges, and wildflowers. Note that there is little or no vegetation in the higher elevation areas in the background.

What climatic-environmental factors cause tundra vegetation to grow so close to the ground?

● **FIGURE 11.36** A giant, and spectacular, phytoplankton bloom is shown here off the west coast of Iceland on this satellite image.

Why would we want to monitor plankton blooms from space or by other means?

⊕ **In Google Earth fly to: 65.20°N, 24.06°W**

known as *krill* are nicknamed the "power food of the Antarctic." These crustaceans feed on plankton and then become the main food source for birds, penguins, seals, and whales.

The marine food chain just described is part of a full food cycle in the ocean. Through animal excretions and the decomposition of marine plants and animals, chemical nutrients return to the ocean and are transformed by phytoplankton into usable foods. Phytoplankton also play an essential role in the production of oxygen for the atmosphere.

The uneven nutrient distribution in the oceans and the fact that sunlight can penetrate only to a depth of about 120 meters (400 ft), depending on the water conditions, greatly affects the distribution of marine organisms. Most organisms are concentrated in the upper layers of the ocean where the most solar energy is available. In deep waters, where the ocean floor lies below the level of sunlight penetration, the benthos organisms depend on whatever nutrients and organic detritus filter down to them. For this reason, benthos animals are scarce in deep and dark ocean waters. They are most common in shallow waters near coasts, such as coral reefs and tide pools, where there is sunlight, a rich supply of nutrients, and abundant phytoplankton and zooplankton (● Fig. 11.37).

The oceans over the continental shelves have the highest concentration of marine life. The chemical nutrient supply is greater near the coasts where nutrients are washed into the sea from rivers.

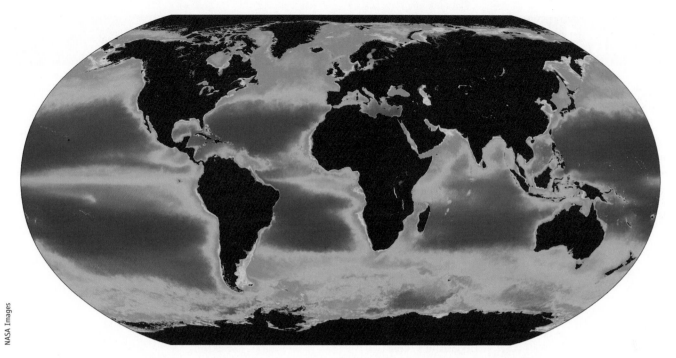

● **FIGURE 11.37** The global distribution of chlorophyll-producing marine plankton can be mapped from satellite observations. Blue to light blue and green and then to yellow to red indicates increasingly higher levels of chlorophyll concentration.

What two spatial observations can you make about geographic locations that have high plankton concentrations?

NASA Images

Marine organisms are also concentrated where deeper waters rise (upwelling) to the surface layers where sunlight is available. These vertical exchanges are generally the result of variations in salinity or density. Marine life is also abundant in areas where there is a mixing of cold and warm ocean currents, as there is off the northeastern coast of the United States.

The manned submersible vessel *Alvin* took scientists down to depths of more than 2,500 meters (8,100 ft) in the East Pacific, where for the first time they could observe sea life on the deep ocean floor. It was previously presumed that the cold waters (2°C/36°F) at that depth and the totally darkness would create a biological desert. However, undersea volcanic activity produces vents of warm, mineral-rich waters, which nourish bacteria and colonies of crabs, clams, mussels, and giant 3-meter (10 ft) red tubeworms. The discovery of this deep-ocean ecosystem, which exists without the benefit of sunlight, has caused scientists to rethink previous theories about the ocean and its chemistry. This unusual "chemosynthetic" vent community has also been observed at other oceanic ridges in the Pacific and Atlantic Oceans.

The Resilience of Life Forms

One overarching theme is apparent throughout this chapter: Earth's life forms are extremely resilient. First came the *autotrophs* (plant life), followed by the *heterotrophs* (animal life) that reside within and feed in the plant communities. Temperature, precipitation, sunlight, and winds are among the most important of the climatic factors that influence the geography of life forms, but it is also essential to note that wherever a climatic limit can be breached, or a pocket of favorable environmental conditions is present, life can be found. Living things have even found a way to exist in rather inhospitable environments. Where one life form fails, a more adaptive life form will survive.

As climates change through time. flora and fauna must adapt to the new climatic conditions, migrate to areas with favorable conditions, or cease to exist. Today, and in the future, analyzing the geographic distribution and limits of vegetation associations along with animal migrations will help us to monitor environmental shifts and their potential links to changes in climatic conditions.

TERMS FOR REVIEW

ecosystem
abiotic
producer (autotroph)
consumer (heterotroph)
decomposer (detritivore)
food chain
trophic level
herbivore
carnivore
omnivore
nutrient cycle
biomass
primary productivity
secondary productivity
habitat
ecological niche
generalist
specialist

plant community
natural vegetation
succession
pioneer community
secondary succession
gap
climax community
mosaic
matrix
patch
corridor
range of tolerance
ecotone
biome
Bergmann's Rule
Allen's Rule
krummholz
slope aspect

symbiotic relationship
liana
epiphytes
buttressed trunks
galleria forest
boreal forest (taiga)
savanna
tall-grass prairie
short-grass prairie
xerophytes
tundra
plankton
phytoplankton
zooplankton
nekton
benthos

QUESTIONS FOR REVIEW

1. Why is the study of ecosystems important in the world today? Give several examples of natural ecosystems within easy driving distance of your own residence.
2. What are the four basic components of an ecosystem?
3. What factors are most critical in affecting the net primary productivity of a terrestrial ecosystem?
4. In what ways has the original theory of succession been modified?
5. How do the terms *mosaic, matrix, patch, corridor,* and *ecotone* relate to each other and to a vegetation landscape?
6. What are the four major types of Earth biomes? What important climate characteristics are related to each of the major biomes?

7. Describe a true tropical rainforest. How does such a forest differ from jungle?
8. What are the distinctive features of chaparral vegetation? What climate conditions are associated with chaparral?
9. What conditions of climate or soil might be anticipated for each of the following in the midlatitudes: broadleaf deciduous forest, broadleaf evergreen forest, needleleaf coniferous forest?
10. How have xerophytes adapted to desert climate conditions?
11. What are the chief characteristics of tundra vegetation? What adaptations to climate does tundra vegetation make?
12. The highest concentrations of marine life (see Fig. 11.37) are found in which parts of the oceans? Why?

CONSIDER AND RESPOND

1. Refer to Figure 11.6. Based on the data, which is more important for terrestrial ecosystem productivity between the equator and the polar regions—annual temperatures or the availability of water? Use examples from the figure to explain your choice.

2. What broad groups might you use if you were to devise a classification of Earth's vegetation on the basis of latitudinal zones? Why do you suppose your textbook authors chose not to do this, electing instead to identify broad groups based on major terrestrial ecosystems (biomes)?

3. Refer to Figure 11.23, which shows considerable variation in the natural vegetation of the United States between 30° and 40° north latitude. Describe the broad changes in vegetation that occur within that latitudinal band as you move from the East Coast to the West Coast of the United States.

4. Examine the climate variation (see Fig. 8.6) within the same latitudinal band described in Question 3. Does there appear to be a relationship between natural vegetation and climate? If so, describe this relationship.

PRACTICAL APPLICATIONS

1. Kudzu is a climbing, woody, perennial vine that originates from Japan. From 1935 to 1950, farmers in the southeastern United States were encouraged to plant it to help reduce soil erosion. Without natural enemies, this vine began to grow out of control. Since 1953, it has been identified as a pest weed, and efforts to eradicate this invader continue to this day. Kudzu inhabits about $30,000\,km^2$ of the southeastern United States. From 1935 to present, what is the spread of this pest in km^2/year?

2. Research has shown that the harvest dates for vineyards in Switzerland vary systematically with the local mean temperatures between April and August. With every 1°C rise in temperature, the harvest date for the grapes moves back by 12 days. Assuming an average harvest date of September 15, what would be the expected harvest date if the April to August mean temperature was the following: 1°C above normal, 1.25°C below normal, 1.5°C above normal, and 1°F above normal?

MindTap—Make the most of your study time by accessing everything you need to succeed in one place. Read your textbook, take notes, review flashcards, watch videos, complete activities, take practice quizzes, and more online with MindTap. Log in at **www.cengagebrain.com.**

SOILS AND SOIL DEVELOPMENT

12

OBJECTIVES

WHEN YOU COMPLETE THIS CHAPTER YOU SHOULD BE ABLE TO:

- 12.1 Appreciate the interaction and integration of subsystems that influence the soil system.
- 12.2 Describe the major components of a soil and how they vary to produce different soil types.
- 12.3 Discuss the role of water in soil processes and how different amounts of available soil water can affect the vegetation growing on a soil.
- 12.4 Understand the factors that determine a soil's formation, development, and fertility, especially the roles of climate and vegetation.
- 12.5 Explain why soils are among the world's most critical resources and the need for effective soil conservation practices.
- 12.6 Cite a few examples of major human impacts on soils.

WHAT ARE THE FOUR MOST important natural constituents that permit life as we know it to exist on Earth? Most people would immediately answer, "air, water, and sunlight," but they might have to think harder and longer about their fourth answer. People generally give little attention to that fourth natural resource, but it is essential for their life on Earth, and it lies right below their feet. *Soil* is that critical resource. The soil mantle that covers most land surfaces is indispensable, but fragile, and threatened by erosion, pollution, or being covered over by the human-built environment (● Fig. 12.1). Soils provide nutrients that directly or indirectly support much of life on Earth.

Soil is a dynamic body of natural materials capable of supporting a vegetative cover. It contains chemical solutions, gases, organic refuse, flora, and fauna. The physical, chemical, and biological processes that take place among the components of a soil are integral parts of its dynamic character. Soil responds to climatic conditions (especially temperature and moisture), to the land surface configuration, to vegetative cover and composition, and to animal activity.

The characteristics of a soil are distinctive and reflect the environmental conditions under which it developed and exists today. This is a healthy soil in Hawaii.   Photo by Scott Bauer USDA/NRCS

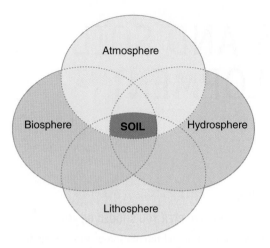

● **FIGURE 12.2** The intertwined links between soil and the major Earth subsystems.

Why is soil considered to be such an integrator of Earth systems?

of interacting factors. When viewed on a world regional scale, a strong and significant influence on soil is climate. The relationships between soils, climate types, and their associated environments were considered in Chapters 9 and 10.

Major Soil Components

Soil is an exceptional example of the interdependence and overlap among Earth's subsystems because a soil develops through long-term interactions of atmospheric, hydrologic, lithologic, and biotic conditions. The nature of a soil reflects the ancient environments under which it formed as well as current environmental conditions. Soils contain four major components: inorganic materials, soil water, soil air, and organic matter (● Fig. 12.3).

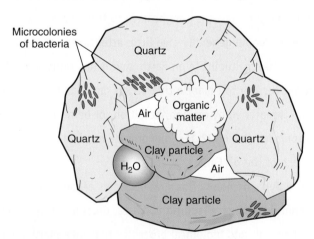

● **FIGURE 12.3** The four major components of soil. Soil contains a complex assemblage of inorganic minerals and rocks, along with water, air, and organic matter. The interactions among these components and the proportion of each are important factors in soil development.
From Purves et al., *Life: The Science of Biology*, Fourth Edition. Used with permission of Sinauer Associates, Inc.

How do each of these soil components contribute to making a soil suitable to support plant life?

● **FIGURE 12.1** Scientists work to understand and control erosional losses and other problems that threaten our precious natural resource—soil.

Soil has been called "the skin of the Earth." The condition and nature of a soil reflects both the ancient environments under which it formed and today's environmental conditions. A soil is an environmental system that adapts, reflects, and responds to a great variety of natural and human-influenced processes. Soil is an exceptional example of the integration, interdependence, and overlap among Earth's subsystems because the characteristics of a soil reflect the atmospheric, hydrologic, lithologic, and biotic conditions under which it developed (● Fig. 12.2). In fact, because soils integrate these major subsystems so well, they are sometimes considered a separate system called the *pedosphere* (from Greek: *pedon*, ground).

Soil is also home to many living organisms that affect their living environments, both above and below the ground surface. The life forms that live in or on a soil play significant roles in the development and characteristics of a soil. A soil's characteristics and development result from a great number

Inorganic Materials

Soils contain varying amounts of insoluble materials—rock fragments and minerals that will not readily dissolve in water. Soils also contain soluble minerals as dissolved chemicals held in solution. Most minerals found in soils are combinations of the common elements of surface rocks: silicon, aluminum, oxygen, and iron. Some of these constituents occur as solid chemical compounds, terrestrial ecosystems by providing a variety of necessary chemical elements and compounds to life forms. Carbon, hydrogen, nitrogen, sodium, potassium, zinc, copper, iodine, and compounds of these elements are important in soils. The chemical constituents of a soil come from multiple sources, including the breakdown (*weathering*) of underlying rocks, deposits of loose sediments, and minerals dissolved in water. Organic activities help to disintegrate rocks, create new chemical compounds, and release gases into the soil.

Plants need a variety of chemical substances for growth, and knowing a soil's mineral and chemical content is necessary for determining its productivity potential. **Soil fertilization** is the process of adding nutrients or other constituents in order to meet the soil conditions that certain plants require (● Fig. 12.4).

Soil Water

When precipitation falls on the land, the water that does not run downslope or evaporate is absorbed into the rock or soil, or by vegetation. Moving through a soil, water dissolves certain materials

USDA/NRCS/Lynn Betts

● **FIGURE 12.4** Fertilizers increase the productivity of soils. This farmer is adding nitrogen fertilizer into the soil on an Iowa farm.

Why can soil fertilizer be either useful or detrimental when it is introduced into the soil system?

and carries them through the soil. Soil water is both an ingredient and a catalyst for chemical reactions that sustain life and influence soil development. As water moves through a soil, it washes over and through various soil components, dissolving some of these materials and carrying them through the soil. The water in a soil is not pure, as it contains nutrients dissolved in liquid, a form that can be extracted by vegetation. Plants need air, water, and minerals from the soil to live and grow. Soil is a critical natural resource that functions as an *open system* (● Fig. 12.5). Matter and energy flow in and out, and a soil also holds them in storage. Understanding the flows of inputs and outputs, the components and processes involved, and how they vary among different soils are the keys to appreciating the complexities of soil.

The water in a soil is found in several different circumstances (see Fig. 12.5). Soil water adheres to soil particles and soil clumps by surface tension (the property that causes small water droplets to form rounded beads instead of spreading out in a thin film). This soil water, called **capillary water**, serves as a stored water supply for plants. Capillary water can move in all directions through soil because it migrates from areas with more water to areas with less. Thus, during dry periods, when there is no gravitational water flowing through the soil, capillary water can move upward or horizontally to supply plant roots with moisture and dissolved nutrients.

Capillary water moving upward moves minerals from the subsoil toward the surface. If this capillary water evaporates, the formerly dissolved minerals remain, generally as salt or lime (calcium carbonate) deposits in the topsoil. High concentrations of certain minerals like these can be detrimental to some plants and animals that live in the soil. Lime (calcium carbonate) deposited by evaporating soil water can build up to produce a cement-like layer, called *caliche*, which like a clay **hardpan** prevents the downward percolation of water.

Soil water is also found as a very thin film, invisible to the naked eye that is bound to the surfaces of soil particles by strong electrical forces. This is **hygroscopic water**, which does not move through the soil, nor does it supply plants with the moisture they need.

Water that percolates down through a soil under the force of gravity is called **gravitational water**. Gravitational water moves downward through voids between soil particles and toward the *water table*—the level below which all available spaces are filled with water. The quantity of gravitational water in a soil is related to several conditions, including the amount of precipitation, the time since it fell, evaporation rates, the space available for water storage, and how easily the water can move through the soil.

Gravitational water performs several functions in a soil (● Fig. 12.6). As gravitational water percolates downward, it dissolves soluble minerals and carries them into deeper levels of the soil, perhaps to the zone where all open spaces are saturated. Depleting nutrients in the soil by the through-flow of water is called **leaching**. In regions of heavy rainfall, leaching is common and can be intense, robbing a topsoil of all but the insoluble substances.

Gravitational water moving down through a soil also can wash the finer solid particles (clay and silt) away from upper soil layers. This downward removal of soil components by water is called **eluviation**. As gravitational water percolates downward,

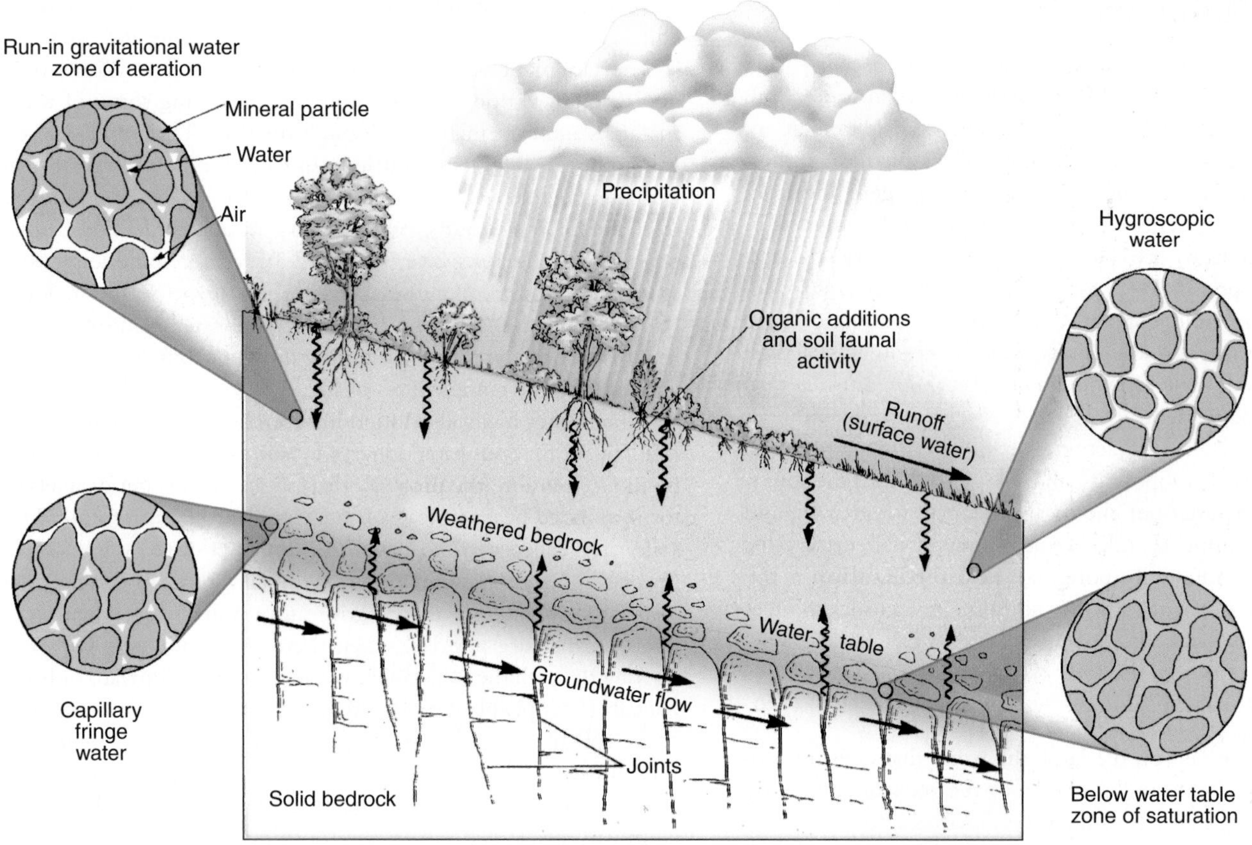

Run-in gravitational water
zone of aeration

Mineral particle

Water

Air

Precipitation

Hygroscopic
water

Organic additions
and soil faunal
activity

Runoff
(surface water)

Weathered bedrock

Water table

Groundwater flow

Joints

Capillary
fringe
water

Solid bedrock

Below water table
zone of saturation

● **FIGURE 12.5** The interrelationships between a soil and the environmental factors that influence soil development. Soil is an example of an open system because it receives inputs of matter and energy, stores part of these inputs, and outputs both matter and energy. Note the inputs and outputs on this diagram.

What are some examples of energy and matter that flow into and out of the soil system?

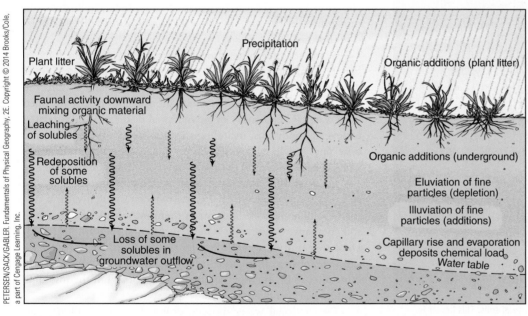

PETERSEN/SACK/GABLER. Fundamentals of Physical Geography, 2E. Copyright © 2014 Brooks/Cole, a part of Cengage Learning, Inc.

Precipitation

Plant litter

Organic additions (plant litter)

Faunal activity downward
mixing organic material

Leaching
of solubles

Redeposition
of some
solubles

Organic additions (underground)

Eluviation of fine
particles (depletion)

Illuviation of fine
particles (additions)

Loss of some
solubles in
groundwater outflow

Capillary rise and evaporation
deposits chemical load

Water table

● **FIGURE 12.6** Water plays several important roles in the processes that affect a soil. Water is important in moving nutrients and particles vertically, both up and down, in a soil.

How does deposition by capillary water differ from deposition (illuviation) by gravitational water?

it deposits the fine materials that were removed from the topsoil at a lower level in the soil. Deposition by water in the subsoil is called **illuviation**. Gravitational water also mixes soil particles as it moves them downward. One result of eluviation is that the texture of the topsoil tends to become coarser as the fine particles are removed. Consequently, the topsoil's ability to retain water is reduced. Illuviation may eventually cause the subsoil to become dense and compact, forming a clay hardpan.

Leaching and eluviation both strongly influence the characteristic layered changes with depth, or **stratification**. Fine particles and substances dissolved from the upper soil are deposited in lower levels, which become dense and may be strongly colored by accumulated iron compounds.

Soil Air

Perhaps as much as nearly 50% of a soil consists of open spaces between soil particles and *clumps* (aggregates of soil particles). Voids that are not filled with water contain air or gases. Compared to the composition of the lower atmosphere, the air in a soil is likely to have less oxygen, more carbon dioxide, and a fairly high relative humidity because of the presence of capillary and hygroscopic water.

For most microorganisms and plants that live in the ground, soil air supplies oxygen and carbon dioxide necessary for life processes. The problem with a water-saturated soil is not necessarily excess water but, if all pore spaces are filled with water, there is no air supply. The lack of air is why many plants find it difficult to survive in water-saturated soils.

Organic Matter

Soil contains organic matter in addition to minerals, gases, and water. The decayed remains of plant and animal materials, partially transformed by bacterial action, are collectively called **humus**. Humus is an important catalyst in chemical reactions that help plants extract soil nutrients. Soils that are rich in humus are workable and have a good capacity for water retention. Humus supplies a soil with nutrients and minerals and is important in chemical reactions that help plants extract nutrients. Humus also provides an abundant food source for microscopic soil organisms.

Most soils are actually microenvironments teeming with life that ranges from bacteria and fungi to earthworms, rodents, and other burrowers. Animals mix organic material deeper into the soil, and move inorganic fragments toward the surface. In addition, plants and their root systems are integral parts of the soil-forming system.

Soils vary at local, regional, and global scales. Particularly strong relationships exist between a soil and its location's vegetation and climate. For example, soils in midlatitude grasslands normally have a very high proportion of organic matter; those in deserts are thin and rich in minerals such as lime and salts left behind by evaporating water; and tropical soils typically have a high content of iron and aluminum oxides. Knowing a soil's water, mineral, and organic components and their proportions can help scientists determine its productivity and what the best use for that particular soil might be.

Soil Characteristics

Several soil properties that can be readily tested or examined are used to describe and differentiate soil types. The most important properties include color, texture, structure, acidity or alkalinity, and the capacity to hold and transmit water and air.

Color

The color of a soil is immediately visible, but it might not be its most important characteristic. Most people know that soils vary in color from place to place. For example, the well-known red clay soils of Georgia are not far from Alabama's belt of black soils. Soils vary in color from black to brown to red, yellow, gray, and near-white. A soil's color is typically related to its physical and chemical characteristics. When describing soils in the field or samples in a laboratory, soil scientists use a book of standardized colors to clearly and precisely identify this coloration (● Fig. 12.7).

If the humus content is low because of limited organic activity or loss of organics through leaching, soil colors typically are light brown or gray. Decomposed organic matter is black or brown, so soils rich in humus tend to be dark. Soils with high humus contents attend to be very fertile, so dark brown or black soils are often referred to as *rich*. However, this is not always true because some dark soils have little or no humus but are dark because of other soil characteristics.

Red or yellow soils typically indicate the presence of iron. In moist climates, a light gray or white soil indicates that iron has been leached out, leaving oxides of silicon and aluminum, but in dry climates, the same color typically indicates an accumulation of calcium or salts.

Courtesy of James P. Shroyer, Kansas State University Research and Extension

● **FIGURE 12.7** Determining soil color. A standardized classification system is used to determine precise color by comparing the soil to the color samples found in Munsell soil color books.

In general, how would you describe the color of the soils where you live?

Soil colors provide clues to the characteristics of soils and make the job of recognizing different soil types easier. But color alone does not always indicate a soil's qualities or its fertility.

Texture

Soil texture refers to the particle sizes (or distribution of sizes) that make up a soil (● Fig. 12.8). In **clayey** soils, the dominant size consists of **clay** particles; defined as having diameters of less than 0.002 millimeter (soil scientists universally use the metric system). In **silty** soils, the dominant **silt** particles are defined as being between 0.002 and 0.05 millimeter. **Sandy** soils have mostly **sand**-sized particles, with diameters between 0.05 and 2.0 millimeters. To compare relative sizes, if a clay particle was pea-sized (although they are flat), a silt particle would be the size of a ping-pong ball, and a sand grain would be the size of a basketball. Rocks larger than 2.0 millimeters (less than 1 in.) are regarded as pebbles, gravel, or rock fragments and technically are not considered to be soil particles.

The proportion of sand, silt, and clay particles in a soil is used to classify its texture. A triangular graph (● Fig. 12.9) is used to discern different classes of soil texture based on the plot of percentages for the sand, silt, and clay particles within each class. For example, a soil composed of 50% silt-sized particles, with 45% clay and 5% sand would be identified as a *silty clay*. Point A on the graph lies within the silty clay class in the example just given. A second soil sample (B) that is 20% silt, 30% clay, and 50% sand would be referred to as a *sandy clay loam*. **Loam** soils, which occupy the central areas of the triangular graph, are soils with a mix of the three grades (sizes) of soil particles without any size being greatly dominant. It is interesting to note that loam soils are generally best suited for supporting vegetation growth.

Soil texture helps determine a soil's capacity to retain moisture and air that are necessary for plant growth. Soils with a higher proportion of larger particles tend to be well aerated and allow water to **infiltrate** (seep through) the soil quickly—sometimes so quickly that plants are unable to use the water. Clayey soils retard water movement, becoming waterlogged and deficient in air. Aeration of the soil is an important process in cultivation, and plowing a soil opens its structure and increases its air content.

Structure

Scientists classify soil structures according to their form. In most soils, particles clump into masses known as **soil peds**, which give a soil a distinctive structure. Soil peds range from columns, prisms, and angular blocks to nutlike spheroids, laminated plates, crumbs, and granules (● Fig. 12.10). Soils with massive and fine structures tend to be less useful than aggregates of intermediate size and stability, which permit good drainage and aeration.

Soil structure and texture both influence a soil's **porosity**—the amount of space that may contain fluids—and they also affect **permeability**—the rate at which water can pass through. Permeability is usually greatest in sandy soils and poorest in clayey soils. Both of these factors control soil drainage and a soil's available moisture. Soils with similar textures may have different structures and vice versa.

The structure of a soil can be influenced by external factors such as moisture regime and plant nutrient cycles interchanging chemicals with the soil, keeping certain ones in the system while others are leached away. We have all seen the structural change in soils that occurs when they are wet compared to when they are dry. Human activities also affect soil structure through cultivation, irrigation, and fertilization. Fertilizers, and lime or decayed organic debris, encourage clumping of soil particles and the maintenance of clumps. Excess sodium and magnesium have the opposite effect, causing clay soils to become a sticky muck when wet and like concrete when dry. The absence of smaller particles typically hinders the development of a well-defined soil structure.

Acidity and Alkalinity

An important aspect of soil chemistry is acidity, alkalinity (baseness), or neutrality. Levels of acidity or alkalinity are measured on the **pH scale** of 0 to 14. A pH reading indicates the concentration of reactive hydrogen ions present. The pH scale is logarithmic, meaning that each change in a whole pH number represents a tenfold change. It is also an inverted scale—a lower pH means a greater amount of hydrogen ions (higher acidity). Low pH values indicate an acidic soil, and high pH indicates alkaline conditions (● Fig. 12.11).

Either acidity or alkalinity of a soil can affect the nutrients that are available for plant growth. Plants absorb nutrients that are dissolved in liquid. However, soil water that lacks some degree of acidity has little ability to dissolve these nutrients. Under those

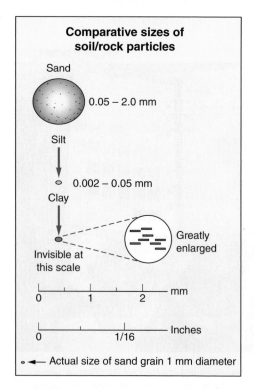

● **FIGURE 12.8** Particle sizes in soil. *Sand, silt,* and *clay* are terms that refer to the size of these particles for scientific and engineering purposes. Here, the sizes of each can be compared. Clay particles are tiny, sheet-like particles that cannot be seen.

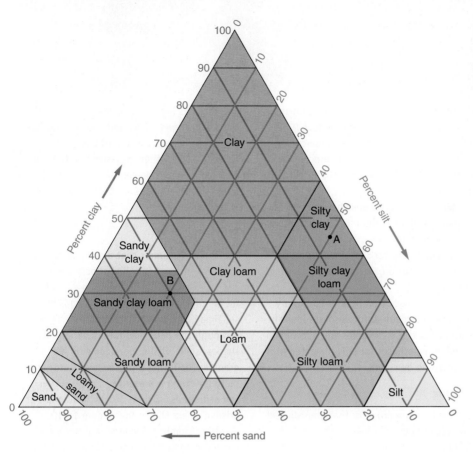

● **FIGURE 12.9** The texture of a soil can be represented by a plotting a point on this diagram. Texture is determined by sieving the soil to determine the percentage of particles falling into the size ranges for clay, silt, and sand. Note that each of the three axes of the triangle is in a different color and the line colors also correspond (clay—red, silt—blue, sand—green).

What would a soil that contains 40% sand, 40% silt, and 20% clay be classified as?

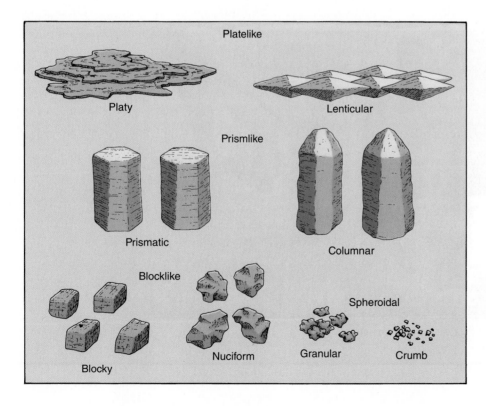

● **FIGURE 12.10** This guide to classifying the structure of a soil on the basis of soil peds can be used to help determine other characteristics of a soil.

How does soil structure affect a soil's usefulness or suitability for agriculture?

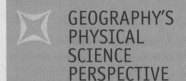

Basic Soil Analysis

After studying a single chapter on soils, no one would expect you to be able to do a detailed soil analysis. However, there are some basic and easy-to-perform observations that will help you better understand a few properties of a local soil. No equipment is required to make some simple analyses; only visual and hands-on examinations are necessary.

Soil Color

Soil color can hold clues to the composition and/or the formation processes of a soil. Figure 12.7 shows the Munsell color book, a standard guide for matching and recognizing precise colors of soil types. The book includes common soil colors, but each color can also appear in a wide variety of tones.

Red: Reddish soil usually indicates that oxidation has been an active process—oxygen has chemically reacted with the soil minerals. Red also indicates that iron is in the soil. Just like rusting iron, many iron-rich minerals turn red when oxidized. The formula for this process is $FeO + O_2 \rightarrow Fe_2O_3$ (ferrous oxide + oxygen becomes hematite, a reddish iron oxide).

Blue/Silver/Gray: These tones mean that the soil has likely been reduced; in other words, oxygen has been removed from the soil. The formula for this process is $Fe_2O_3 \rightarrow FeO + O_2$ (the previous formula in reverse).

White: Usually means that calcium carbonate ($CaCO_3$) or salts (such as NaCl) may be present in the soil.

Black: A very dark color may indicate a high amount of organic material present in the soil.

More sophisticated field or laboratory analyses are required for absolute identification, but these examples will allow good working hypotheses for soil characteristics represented by colors.

Texture

Soil texture is a property that you can feel, and your fingers can help in the analysis. Sand-sized particles can be easily recognized because they feel gritty to the touch. Wetting the soil and working it with your hands can help in this process. If the sample is not gritty but is smooth to the touch, then the soil contains silt or clay.

Silt particles are fine and feel like flour. If the sample feels sticky and you can squeeze a small soil sample into a ribbon (like with modeling clay), then clay-sized particles are abundant. Actual percentages of particle sizes in a soil sample are best established in a laboratory.

Structure

The shape of clumps that a soil makes when it is broken apart is called *structure* and can be examined by breaking up a handful of soil. The peds (or small clumps of soil) may take on some distinctive shapes. Although the peds may form a variety of shapes, some of the more common are granular (denoting a presence of sand) and platy (showing a presence of clay). Other soil structures, such as blocky, columnar, or prismatic, are shown in ● Figure 12.10.

These simple procedures will not yield a complete analysis of a soil sample, but they can certainly be the first steps in the process. It is interesting to note that pedologists (soil scientists) in the field perform many of these same procedures.

© Jeff Vanuga/USDA Natural Resources Conservation Service

Soil analysis in the field is a hands-on process.

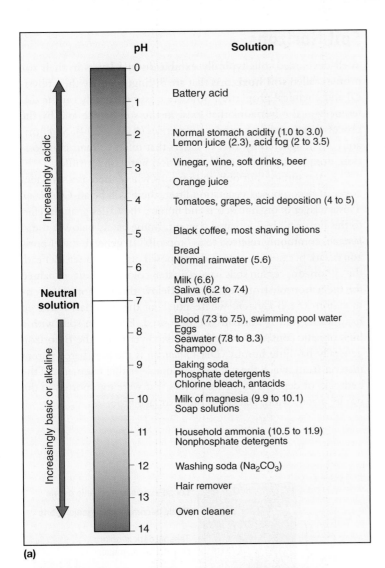

pH	Solution
0	
1	Battery acid
2	Normal stomach acidity (1.0 to 3.0) Lemon juice (2.3), acid fog (2 to 3.5)
3	Vinegar, wine, soft drinks, beer
	Orange juice
4	Tomatoes, grapes, acid deposition (4 to 5)
5	Black coffee, most shaving lotions
6	Bread Normal rainwater (5.6)
7	Milk (6.6) Saliva (6.2 to 7.4) Pure water
8	Blood (7.3 to 7.5), swimming pool water Eggs Seawater (7.8 to 8.3) Shampoo
9	Baking soda Phosphate detergents Chlorine bleach, antacids
10	Milk of magnesia (9.9 to 10.1) Soap solutions
11	Household ammonia (10.5 to 11.9) Nonphosphate detergents
12	Washing soda (Na_2CO_3)
13	Hair remover
14	Oven cleaner

Increasingly acidic

Neutral solution

Increasingly basic or alkaline

(a)

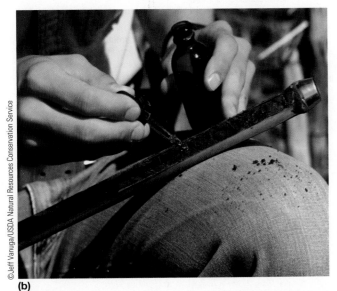

Jeff Vanuga/USDA Natural Resources Conservation Service

(b)

● **FIGURE 12.11** (a) The pH scale of acidity, neutrality, and alkalinity. The degree of acidity or of alkalinity, called pH, can be easily understood when numbers on the scale are linked to common substances. Low pH means acidic, and high pH means alkaline; a reading of 7 is neutral. (b) Alkalinity in a soil can be tested in the field with drops of a dilute acid. If the soil fizzes in the acid solution, alkalinity is typically high.

conditions, even if nutrients are present in the soil, plants may not have access to them.

Most complex plants will grow only in soils with levels between pH 4 and pH 10, although the optimum pH for vegetation growth varies with the plant species. Around the world, vegetation has evolved in and adapted to a variety of climates and soil environments, both of which can affect soil pH. Certain species tolerate alkaline soils, and others thrive under more acid conditions. The ranges of pH in soils vary with parent material, vegetation and other factors, but between pH 5 and pH 7 is a common range for humid region soils; and between pH 7 and pH 9 for soils in arid regions.

Leaching caused by high rainfall gradually replaces soil elements such as sodium (Na), potassium (K), magnesium (Mg), and calcium (Ca) with hydrogen. Falling rain picks up atmospheric carbon dioxide and becomes slightly acidic: $H_2O + CO_2 = H_2CO_3$ (carbonic acid), and this is one reason why desert soils tend to be alkaline and soils in humid regions tend to be acidic (● Fig. 12.12). The humus content in the soils of humid areas also contributes to higher soil acidity.

Alkaline soils, common in the arid regions and in areas of limestone bedrock, can be made more productive by flushing the soil with irrigation water. Strongly acidic soils are also detrimental to plant growth. In acidic soils, soil moisture dissolves nutrients, but they may be leached away before plant roots can absorb them. Soil acidity can generally be corrected by adding lime to the soil. In addition to affecting plant growth, soil acidity or alkalinity also have an impact on microorganisms in the soil. Microorganisms are highly sensitive to a soil's pH, and each species has a range of pH levels required for survival.

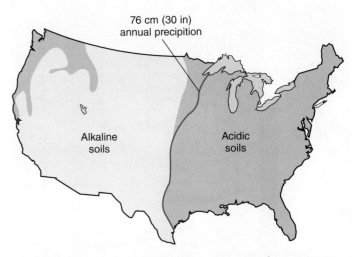

● **FIGURE 12.12** The distribution of alkaline and acidic soils in the United States is generally related to climate. Soils in the East tend to be acidic and those in the West usually are alkaline. The dividing line corresponds fairly well with the 30-inch annual precipitation isohyet.

Other than climate, what environmental factors might cause this east-west variation, and why are some places west of the 30-inch line acidic?

Development of Soil Horizons

Soil development begins when plants and animals colonize rocks or deposits of rock fragments, the **parent material** on which soil will form. Once organic processes begin among mineral particles or rock fragments, differences begin to develop from the surface down through the parent material.

Vertical differences initially result from the surface accumulation of organic litter and the removal of fine particles and dissolved minerals from upper layers by percolating water that deposits these materials at a lower level. The vertical cross-section of a soil from the surface down to the parent material is called a **soil profile** (● Fig. 12.13). Examining the vertical differences in a soil profile is important to recognizing different soil types and understanding how a soil developed. Over time, as climate, vegetation, animal life, and the land surface affect soil development, this vertical differentiation becomes increasingly apparent.

Soil Horizons

Well-developed soils typically exhibit distinct layers in their soil profiles called **soil horizons** that are distinguished by their physical and chemical properties. Soils are classified largely on the differences in the horizons that exist in the soil profile and by the processes responsible for those differences (● Fig. 12.14). Soil horizons are designated by a set of letters that refers to their composition, dominant process, and/or position in the soil profile.

At the surface, but only in locations where there is a sufficient cover of decomposed vegetation litter, there will be an *O horizon*. This is a layer of organic debris and humus; the *O* designation refers to this horizon's high organic content. Immediately below is the *A horizon*, commonly referred to as "topsoil." In general, the *A* horizon is dark because it contains decomposed organic matter. Beneath the *A* horizon, certain soils have a lighter-colored *E horizon*, named for the action of strong eluviation. Below this is a zone of accumulation, the *B horizon*, where much of the materials removed from the *A* and *E* horizons are deposited. Except in soils with a high organic content that has been mixed vertically, the *B* horizon generally has little humus. The *C horizon* is the weathered parent material from which the soil has developed—either fragments of the bedrock, or deposits of rock materials that were transported to the site by water, wind, glacial, or other surface processes.

● **FIGURE 12.13** A soil profile is examined by digging a pit with vertical walls to clearly show variations in color, structure, composition, and other characteristics that occur with depth. This soil is in a grassland region of northern Minnesota.

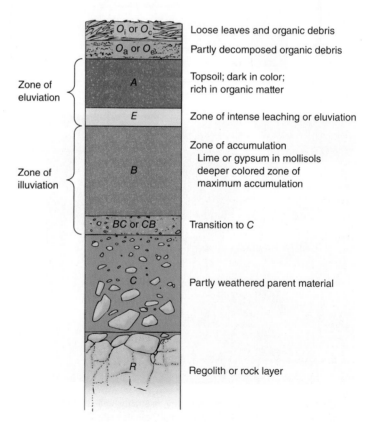

● **FIGURE 12.14** Soils are categorized by the degree of development and the physical characteristics of their horizons. *Regolith* is a generic term for broken bedrock fragments at or very near the surface.

The lowest layer, sometimes called the *R horizon*, is unchanged parent material, either bedrock or transported deposits of rock fragments. Certain horizons in some soils may not be as well developed as others, and some horizons may be missing altogether. Because soils and the processes that form them vary widely and can be transitional between horizons, the horizon boundaries may be either sharp or gradual. Variations in color and texture within a horizon are also not unusual.

Factors Affecting Soil Formation

Because of the great variety among the components of soils and the processes that affected them, no two soils are identical in all of their characteristics. One important factor is rock *weathering*, which refers to the many natural processes that break rocks down into smaller fragments (weathering will be discussed in detail in Chapter 15). Chemical reactions can cause rocks and minerals to decompose, and physical processes cause the breakup of rocks. Just as statues, monuments, and buildings become "weather-beaten" over time, rocks exposed to the elements eventually break up and decompose.

Hans Jenny, a distinguished soil scientist, observed that soil development was a function of *climate*, *organic matter*, *relief*, *parent material*, and *time*—factors that are easy to remember by their initials arranged in the following order: **Cl, O, R, P, T**. Among these factors, parent material is distinctive because it is the raw material. The other factors influence the type of soil that forms from the parent material.

Parent Material

All soils contain weathered rock fragments. If these weathered rock particles have accumulated in place—through the physical and chemical breakdown of bedrock directly beneath the soil—we refer to these fragments as **residual parent material** (● Fig. 12.15). If the rock fragments that form a soil have been carried to the site and deposited by streams, waves, winds, gravity, or glaciers, this mass of deposits is called **transported parent material** (● Fig. 12.16). It is the development and action of organic matter through the life cycles of organisms and the climatic conditions that are primarily responsible for changing fragmented rocks or other parent materials into a soil.

Parent material influences the characteristics of a soil in varying degrees. Some parent materials, such as sandstone that contains extremely hard and resistant sand-sized fragments, are less subject to weathering than others. Soils that develop from weathering-resistant rocks tend to have a high level of similarity to their parent materials. If the bedrock is easily weathered, the soils that develop tend to be somewhat similar to soils in regions that have a comparable climate.

On a global basis, climate and the associated plant communities produce greater variations in soil characteristics compared with the influence of parent materials. Soil differences that are related to differences in parent material are most visible on a local level.

● **FIGURE 12.15** Residual parent material. Grasses, yucca, and cactus grow in a soil developed on a bare granite surface. Soils developed on a bedrock surface are referred to as residual soils.

● **FIGURE 12.16** Transported parent material. These trees in the Rocky Mountains are growing in soil developed on glacial deposits transported down a valley about 10,000 years ago, which provides an approximate time when soil development began. After thousands of years of development, the soil layer is still thin, illustrating that soil development is a long term process.

What does this tell us about the need to conserve soils?

R. Gabler

What other parent materials provide the basis for continuously fertile soils in wet tropical climates?

Organic Activity

Plants and animals affect soil development in many ways. The life processes of plants growing in a soil are as important as its microorganisms—the microscopic plants and animals that live in a soil.

Generally, a dense vegetative cover protects a soil from erosion by running water or wind. The protective forest canopy and the surface litter it produces, combine to keep rain from beating directly on the soil. This reduces the amount of rainwater that would run off of the surface, and increases the moisture infiltrating the soil. Variations in vegetation species and the density of cover also affect the evapotranspiration rates. A sparse vegetative cover will allow greater evaporation of soil moisture, and dense vegetation tends to maintain soil moisture.

The characteristics of a plant community affect the nutrient cycles that are involved in soil development (● Fig. 12.18). As plants die and decompose or leaves fall to the ground, nutrients are returned to the soil. Soils, however, can become impoverished if soluble nutrients that are not used by plants are lost through leaching. The roots of plants help break up the soil structure, making it more porous, and roots also absorb water and nutrients from the soil.

Over the long term, as a soil develops, the influence of parent material on its characteristics diminishes. Given the same soil-forming conditions, recently developed soils will more strongly reflect the characteristics of their parent material, compared to soils that have developed over a long time.

Many of the chemicals and nutrients in a soil reflect the composition of its parent material (● Fig. 12.17). For example, calcium-deficient parent materials will produce soils that are low in calcium, and its natural fauna and plant cover will consist of the types that require little calcium. Likewise, a parent material with a high aluminum content will produce a soil that is rich in aluminum. In fact, the ore of aluminum is bauxite, found in tropical soils where it has been concentrated by intense leaching away of the other bases.

The particle sizes that result from the breakdown of parent material are a prime determinant of a soil's texture and structure. A rock material such as sandstone, which contains little clay and weathers into relatively coarse fragments, will produce a soil of coarse texture. Parent materials are also an important influence on the availability of air and water to a soil's living population.

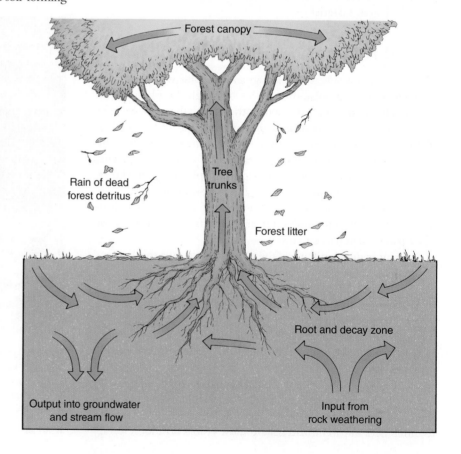

● FIGURE 12.18 The nutrient cycle in a forest. Trees take up nutrients from the soil through soil water absorbed into their root systems. Nutrients are supplied by the breakdown of rocks and minerals and by leaf and other organic litter.

Forest canopy

Rain of dead forest detritus

Tree trunks

Forest litter

Root and decay zone

Output into groundwater and stream flow

Input from rock weathering

Leaves, bark, branches, flowers, and root networks contribute to the organic composition of soil through litter and through the remains of dead plants. The organic content of a soil depends on its associated plant life. For example, a grass-covered prairie supplies much more organic matter than the thin vegetative cover of desert regions. There is some question, however, as to whether forests or grasslands (with their thick root networks and annual life cycle) furnish the soils with a higher organic content. Many of the world's grassland regions, like the North American prairies, provide some of the world's most fertile soils for cultivation in part because of the high amount of organic matter that a grass cover generates.

In terms of their contribution to soil formation, bacteria are perhaps the most important microorganisms that live in soils. Bacteria break down organic matter, humus, and the debris of living things into organic and inorganic components, allowing the formation of new organic compounds that promote plant growth. It has been suggested that the number of bacteria, fungi, and other microscopic plants and animals living in a soil may be 1 billion per gram (a fifth of a teaspoon) of soil. The activities and remains of these microorganisms, minute though they are individually, add considerably to the organic content of a soil.

Earthworms, nematodes, ants, termites, wood lice, centipedes, burrowing rodents, snails, and slugs also stir up the soil, mixing mineral components from lower levels with organic components from the upper portion. Earthworms contribute greatly to soil development because they take soil in, pass it through their digestive tracts, and excrete it in casts. The process not only helps mix the soil but also changes the texture, structure, and chemical qualities of the soil. In the late 1800s, Charles Darwin estimated that earthworm casts produced in a year would equal as much as 10 to 15 tons per acre. As for the number of earthworms, a study in New Zealand suggested that the total weight of earthworms beneath a pasture there equaled the weight of the sheep grazing above them.

Climate

On a world regional scale, climate is a very important factor in soil formation. Of course, if different soils exist within a particular climate region, other factors must be responsible for the local variation. Soil differences that are apparent at a local level tend to reflect the influence of factors such as parent materials, land surface configurations, vegetation types, and time.

Temperature directly affects soil microorganisms that influence the decomposition of organic matter. In hot equatorial regions, intense activities by soil microorganisms preclude a thick accumulation of organic debris or humus. ● Figure 12.19 shows that the amounts of organic matter and humus in a soil increase toward the middle latitudes and away from polar regions and the tropics. In the mesothermal and microthermal climates (C and D), the activity of microorganisms is slow enough to allow decaying

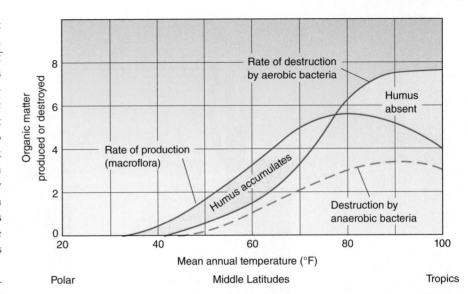

● **FIGURE 12.19** The relationship of temperature to production and destruction of organic matter in the soil.

What range of mean annual temperatures is most favorable for the accumulation of humus?

organic matter and humus to accumulate in rich layers. The climate of colder regions retards microorganism activity and limits plant growth, which tends to result in thin accumulations of decomposed organic matter.

Chemical activity increases and decreases directly with temperature, given equal availability of moisture. As a result, parent materials of soils in hot, humid equatorial regions are chemically altered to a far greater degree compared with parent materials in colder zones.

Temperature affects soil indirectly through a climate's influence on vegetation associations. Soils generally reflect the character of plant cover because of nutrient cycles that tend to keep both vegetation and soil in chemical equilibrium. The combined effects of vegetative cover and the climatic regime tend to produce soil profiles and characteristics that tend to share certain characteristics among different regions that have similar climates and vegetation associations (● Fig. 12.20).

Moisture conditions affect the development and character of soils more directly than any other climatic factor. Precipitation amounts affect plant growth, which directly influences a soils organic content and fertility. Extremely high rainfall will cause leaching of nutrients, and a relatively infertile soil. Extreme aridity may result in the absence of any soil development.

Gravitational and capillary water have pronounced effects on soil development, structure, texture, and color. Precipitation is the original source of soil water (disregarding the minor contribution of dew), and the amount of precipitation received affects leaching, eluviation, and illuviation and thereby rates of soil formation and horizon development. The evaporation rate is also an important factor. Salt and gypsum deposits from the upward migration of capillary water are more extensive in hot, dry regions—such as the southwestern United States where evaporation rates are higher—compared to those of colder, dry regions (see Fig. 12.20).

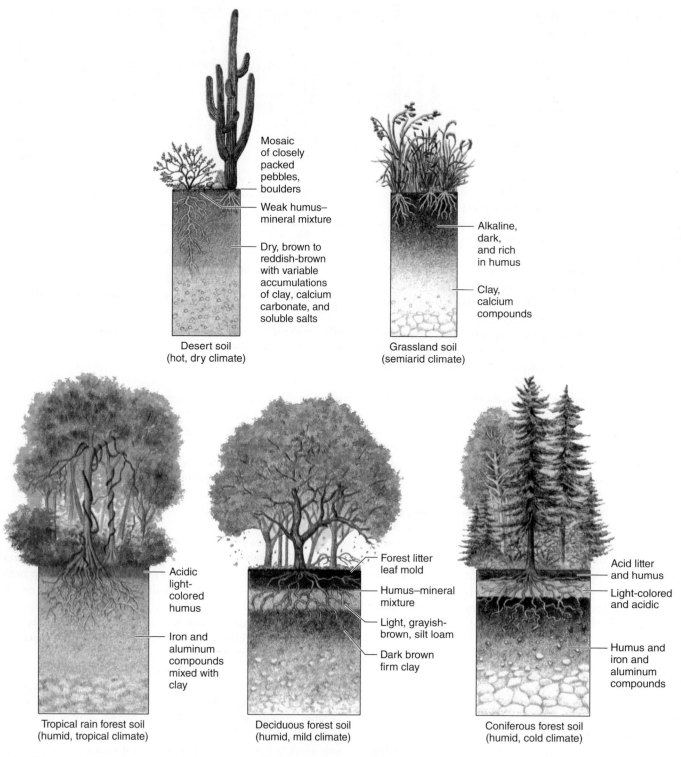

● **FIGURE 12.20** Idealized diagrams of five different soil profiles illustrate the effects of climate and vegetation on the development of soils and their horizons.

Which two environments produce the most humus, and which two produce the least?

Relief

Relief includes characteristics of the land surface, such as the topography, the slope, and its *aspect* (the direction it faces) all influence soil development. Steep slopes are generally better drained than gentler ones, and they are also subject to rapid runoff of surface water. As a consequence, there is less infiltration of water on steeper slopes, which inhibits soil development, sometimes to the extent that there might be no soil. In addition, rapid runoff on steep slopes can erode surfaces as fast or faster than the soil development processes. On gentler slopes, where there is less runoff and higher infiltration, more water is available to encourage soil development and to support vegetation growth, and erosion is not as intense. Well-developed soils typically form on land that is flat or has a gentle slope.

Slope aspect has a direct effect on microclimates and vegetative cover in areas outside of the equatorial tropics. North-facing slopes in the middle and high latitudes of the Northern Hemisphere have microclimates that are cooler and wetter than the warmer, drier south-facing exposures, which receive the sun's rays at a steeper angle. Local variations in soil depth, texture, and profile development result directly from microclimate differences.

Topography, through its influence on vegetation, indirectly affects the development of soil. Steep slopes prevent the formation of a soil that would support abundant vegetation, and a modest plant cover yields less organic debris for the soil.

Time

Soils have a tendency to develop toward a state of equilibrium with their environment. A soil is sometimes called "mature" when it has reached such an equilibrium condition. Young soils are still in the process of developing toward being in equilibrium with their environmental conditions. Mature soils have well-developed horizons that indicate the conditions under which they formed. Young or "immature" soils typically have poorly developed, or perhaps no horizons at all (● Fig. 12.21).

Another effect of time is that, as soils develop, their influence of their parent material decreases and they increasingly reflect their climate and vegetative environments. On a global scale, climate typically has the greatest influence on soils, provided sufficient time has passed for the soils to become well developed.

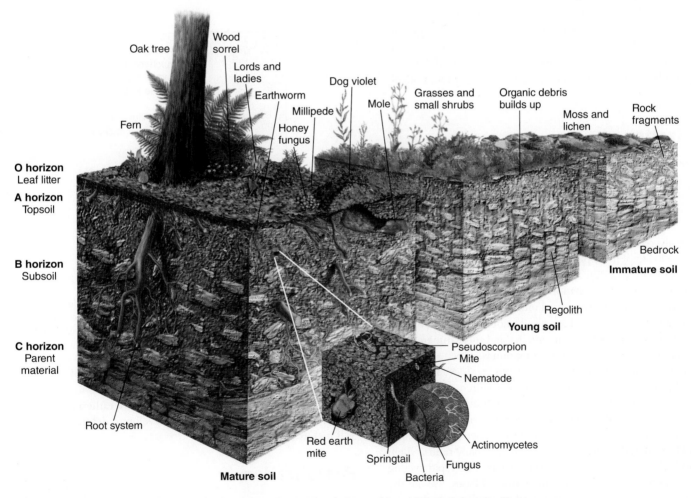

● **FIGURE 12.21** The time that a soil has been developing is important to its composition and physical character. Given enough time and the proper environmental conditions, soils will become more maturely developed with a deeper profile and stronger horizon development.

From Derek Elsom, *Earth*, 1992. Copyright © 1992 by Marshall Editions Developments Limited. New York. Macmillan. Used by permission.

What major changes occur as the soil illustrated here becomes better developed over time?

The importance of time in soil formation is especially clear in soils developed on transported parent materials. Depositional surfaces are in many cases quite recent in geologic terms and have not been exposed to weathering long enough for a mature soil to develop. Deposition occurs in a variety of settings: on river floodplains where the accumulating sediment is known as *alluvium*; downwind from dry areas where dust settles out of the atmosphere to form blankets of wind-deposited silts, called *loess*; and in volcanic regions showered by ash and covered by lava. Ten thousand years ago glaciers withdrew from vast areas, leaving behind jumbled deposits of rocks, sand, silt, and clay.

Because of the great variety of materials and processes involved in soil formation and development, there is no fixed amount of time that it takes for a soil to become mature. The Natural Resources Conservation Service, however, estimates that it takes about 500 years to develop 1 inch of soil in the agricultural regions of the United States. Generally, though, it takes thousands of years for a soil to reach maturity.

Soil-Forming Regimes

The characteristics that make major soil types distinctive and different from one another result from their **soil-forming regimes**, which vary mainly because of differences in climate and vegetation. At the broadest scale of generalization, climate differences produce three primary soil-forming regimes: laterization, podzolization, and calcification.

Laterization

Laterization is a soil-forming regime that occurs in humid tropical and subtropical climates as a result of high temperatures and abundant precipitation. These climatic environments encourage rapid breakdown of rocks and decomposition of nearly all minerals. A soil of this type is known as **laterite**, and these soils are generally reddish in color from iron oxides; the term *laterite* means "brick-like." In tropical areas, laterite is quarried for building material (● Fig. 12.22).

Despite the dense vegetation that is typical of these climate regions, little humus is incorporated into the soil because the plant litter decomposes so rapidly. Laterites do not have an *O* horizon, and the *A* horizon loses fine soil particles and most minerals and bases except for iron and aluminum compounds, which are insoluble primarily because of the absence of organic acids (● Fig. 12.23). As a result, the topsoil is reddish, is coarse textured, and tends to be porous. In contrast to the *A* horizon, the *B* horizon in a lateritic soil has a heavy concentration of illuviated materials.

In the tropical forests, soluble nutrients released by weathering are quickly absorbed by vegetation, which eventually returns them to the soil where they are reabsorbed by plants. This rapid cycling of nutrients prevents the total leaching away of bases, leaving the soil only moderately acidic. Removal of vegetation permits total leaching of bases, resulting in the formation of crusts of iron and aluminum compounds (laterites), and accelerated erosion of the *A* horizon.

● **FIGURE 12.22** Laterite cut for building stone and stacked along a village road in the state of Orissa, India.

Why is building with brick or stone rather than wood so important in heavily populated, less developed nations such as India?

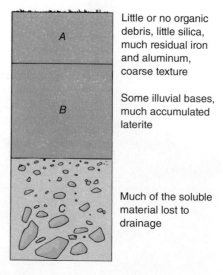

● **FIGURE 12.23** Soil profile horizons in laterite, one of three major soil-forming regimes. Laterization is a soil development process that occurs in tropical and equatorial zones that experience warm temperatures year round and wet climates.

Laterization is a year-round process because of the small seasonal variations in temperature or soil moisture in the humid tropics. This continuous activity and strong weathering of parent material cause some tropical soils to develop to depths of as much as 8 meters (25 ft) or more.

Podzolization

Podzolization occurs mainly in the high midlatitudes where the climate is moist with short, cool summers and long, severe winters. The coniferous forests of these climate regions are an integral part of the podzolization process.

How Much Good Soil Is There on Earth?

A widely used analogy for understanding the amount of soil that exists on Earth uses an apple to demonstrate that soil, which is adequate for the world's agricultural needs, is a rather limited natural resource. If the apple is cut into quarters, three of those pieces (75%) represent the water bodies on Earth and should be put away. The piece that remains represents the Earth's land area (25%) where soil *could* exist. Half of the remaining piece should be cut off and put away, representing rocky desert, polar, or high mountain regions, where soil does not exist or will not grow much because of harsh climatic conditions. Now there is an eighth (12.5%) of the apple remaining. This eighth of the apple should then be cut into four pieces, and three of them should be put away because they represent areas with local conditions that preclude agriculture (too rocky, steep, wet, and so on) and the areas that we have already covered over with cities, towns, and roads.

The remaining apple slice (about 3% of Earth) represents areas with good soils for agriculture to feed the world's population.

Soil Degradation and Soil Loss

Today, even areas with good soils are under continuing threat. It may take 200 to 1000 years or more to develop 2.5 centimeters (1 in.) of soil, but through erosion by water or wind, thousands of years of soil development may be lost in a season or in days. Land degradation by humans is a major cause of soil loss but impacts from changing climates, particularly desertification, also play a role. Overgrazing, deforestation, overuse of the land, and poor agricultural practices are the major human-related causes of soil loss. The United Nations estimates that, just through erosion, the world loses between 5 and 7 million hectares (12.36 and 17.30 million acres) of farmland every year. This is an area equal to the size of West Virginia.

In addition to the degradation of soil that is ongoing, farmlands with excellent soils are being taken out of production by urbanization/suburbanization. Many farmlands are attractive to land developers because they are already cleared, the soils are good for lawns and landscaping, and cities and suburbs are expanding into surrounding agricultural lands.

Our planet is losing its arable soil, and the world's population is growing—trends that cannot continue forever. One estimate is that soils are being lost at a rate 17 times faster than the rate at which they form. Many government and private agencies around the world are working to educate people about the critically important problem of soil degradation and to support solutions to minimize human impacts on soils. The problems associated with soil degradation are globally significant, and the necessary solutions are proper conservation and management at local levels.

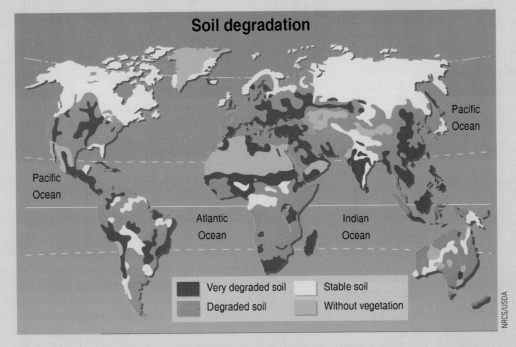

The United Nations produced this world map to show the geographic distribution of soil degradation by various degrees.
Source: UNGP, International Soil Reference and Information Centre (ISRIC), World Atlas of Desertification, 1997. Philippe Rekacewicz, UNEP/GRID-Arendal

What major geographic factors explain the areas with the highest levels of degradation, and what explains areas with the lowest impact on soils?

Where temperatures are low much of the year, microorganism activity is reduced enough that humus does accumulate; however, because of the small number of animals living in the soil, there is little mixing of humus below the surface. Leaching and eluviation by acidic solutions remove the soluble bases and aluminum and iron compounds from the *A* horizon (● Fig. 12.24). The remaining silica gives a distinctive ash-gray color to the *E* horizon (*podzol* is derived from a Russian word meaning "ashy"). The needles that coniferous trees drop are chemically acidic and contribute to the soil acidity. It is difficult to determine whether the soil is acidic because of the vegetative cover or whether the vegetative cover is adapted to the acidic soil.

Podzolization can take place outside its typical cold, moist climate regions if the parent material is highly acidic—for example, on the sandy areas common along the East Coast of the United States. The pine forests that grow in such acidic conditions return acids to the soil, promoting the process of podzolization.

Calcification

A third distinctive soil-forming regime is called **calcification**. In contrast to both laterization and podzolization, which require humid climates, calcification occurs in regions where evapotranspiration significantly exceeds precipitation. Calcification is important in the climate regions where moisture penetration is shallow. The subsoil is typically too dry to support tree growth, and shrubs or shallow-rooted grass are the primary forms of vegetation. Calcification is enhanced as grasses use calcium, drawing it up from lower soil layers and returning it to the soil when the annual grasses die. Grasses and their dense root networks provide large amounts of organic matter, which is mixed deep into the soil by burrowing

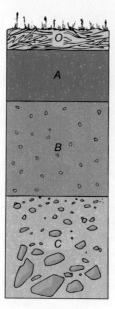

● **FIGURE 12.25** Soil profile horizons in a calcified soil, formed by the third major soil-forming regime. Calcification is a soil development process that is most prominent in cool to hot subhumid or semiarid climate regions, particularly in grassland regions but also in deserts.

animals. Middle-latitude grassland soils are rich in bases and in humus and are the world's most productive agricultural soils. The deserts of the American West generally have no humus, and the rise of capillary water can leave calcium carbonate ($CaCO_3$) and sodium chloride ($NaCl$) and other salts at the surface.

In many areas of low precipitation, the air is often loaded with alkali dusts such as calcium carbonate. When calm conditions prevail or when it rains, the dust settles and accumulates in the soil. The rainfall produces an amount of soil water that is just sufficient to translocate these materials to the *B* horizon (● Fig 12.25). Over hundreds to thousands of years, the $CaCO_3$ enriched dust concentrates in the *B* horizon, forming hard layers of *caliche*. Much thicker accumulations called *calcretes* (● Fig. 12.26) form by the upward (capillary) movement of dissolved calcium in groundwater when the water table is near the surface.

Localized Soil Regimes

Two additional localized soil-forming regimes merit attention. Both characterize areas with poor drainage, although they occur under very different climate conditions. The first, **salinization**, or the concentration of salts in the soil, is often detrimental to plant growth. Salinization occurs in stream valleys, interior basins, and other low-lying areas, particularly in arid regions with high groundwater tables. High groundwater levels can be the result of water from adjacent mountain ranges, stream flow originating in humid regions, or a wet–dry seasonal precipitation regime (● Fig. 12.27). Salinization can also occur because of intensive irrigation under arid conditions. Rapid evaporation leaves behind a high concentration of soluble salts and may destroy a soil's agricultural productivity. An extreme example of salinization exists in certain areas of the Middle East, where thousands of years of irrigated agriculture in the desert have made the soils too saline to cultivate today.

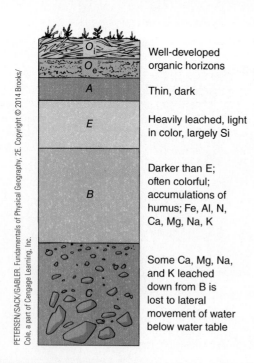

Well-developed organic horizons

Thin, dark

Heavily leached, light in color, largely Si

Darker than E; often colorful; accumulations of humus; Fe, Al, N, Ca, Mg, Na, K

Some Ca, Mg, Na, and K leached down from B is lost to lateral movement of water below water table

● **FIGURE 12.24** Soil profile horizons in a podzol soil, another of the three major soil-forming regimes. Podzolization typically occurs under cool, wet climates in regions of coniferous trees or in boggy environments and is a very acidic soil type.

● **FIGURE 12.26** This shallow-rooted grass is growing in a thin layer of topsoil over a much thicker layer of calcium carbonate called calcrete.

What climate characteristics are associated with the calcification soil-forming process?

● **FIGURE 12.27** Salinization from high evaporation loss of irrigation water is indicated by these white deposits on this furrowed field. Surface salinity results from the upward capillary movement of water and evaporation at the surface causing deposits of salt. Note that the crops grow in a narrow zone where water is moving through the soil from the water in the furrows to the crest where the salt is deposited.

What negative soil effects can result when humans practice irrigated agriculture in regions that experience high evaporation rates?

Another localized soil regime, **gleization**, occurs in poorly drained areas under cold and wet environmental conditions. Gley soils, as they are called, are typically associated with peat bogs where the soil has an accumulation of humus overlying a blue-gray layer of thick, gummy, water-saturated clay. In poorly drained regions that were formerly glaciated, such as northern Russia, Ireland, Scotland, and Scandinavia, peat has long been harvested and used as a source of energy.

Soil Classification

Soils, like climates, can be classified by their characteristics and mapped by their spatial distributions. In the United States, the Soil Survey Division of the Natural Resources Conservation Service (NRCS), a branch of the Department of Agriculture, is responsible for soil classification (termed **soil taxonomy**). As with any classification system, the methods and categories are continually being updated and refined.

Soil classifications are published in **soil surveys**, books that outline and describe the kinds of soils in a region and include maps of the distribution of soil types, usually at the county level. These documents, available for most areas of the United States, are useful references for information about local soil types, soil fertility, irrigation, and drainage.

The NRCS Soil Classification System

The NRCS soil classification system is based on the development and composition of soil horizons. The largest division in the classification of soils is the **soil order**, of which 12 are recognized by the NRCS. To provide greater detail, soil orders can also be subdivided into suborders, great groups, subgroups, families, and series. More than 10,000 soil series have been recognized in the United States.

The NRCS system of soil types uses names derived from root words of classical languages such as Latin, Arabic, and Greek to refer to the different soil categories. The names and the classification are precise in describing the distinguishing characteristics of each soil type. Some soil orders reflect regional climate conditions. Other soil orders, however, reflect their regency or type of parent material, and their distribution does not conform to climate regions.

When examining a soil for classification under the NRCS system, particular attention is paid to characteristic horizons and textures. Some of these horizons are below the surface (**subsurface horizons**); others, called **epipedons**, are surface layers that usually exhibit dark shading associated with organic material (humus). Examples of some of the more common horizons, illustrating how names were chosen to represent actual soil properties, are found in Table 12.1.

NRCS Soil Orders

The 12 soil orders are based on a variety of characteristics and processes that can be recognized by examining a soil and its profile. The soil descriptions that follow are based on the sequence shown in ● Fig. 12.28, which illustrates the links between climate,

TABLE 12.1
Common Soil Horizons (NRCS Soil Classification System)*

Oxic horizon (from *oxygen*)
Subsurface horizon, in low-elevation tropical and subtropical climates containing oxides of iron and aluminum.

Argillic horizon (from Latin: *argilla*, clay)
Layer beneath the *A* horizon by illuviation containing a high content of accumulated clays.

Ochric epipedon (from Greek: *ochros*, pale)
A light-colored surface horizon that is either very low in organic matter or very thin.

Albic horizon (from Latin: *albus*, white)
A sandy and light-colored A2 horizon above a spodic horizon, because of the removal of clay and iron oxides.

Spodic horizon (from Greek: *spodos*, wood ash)
A dark colored layer beneath an *A2* horizon formed by the illuviation of humus and oxides of aluminum and/or iron.

Mollic epipedon (from Latin: *mollis*, soft)
A dark-colored surface layer with a high content of basic substances (calcium, magnesium, and potassium).

Calcic horizon (from *calcium*)
A subsurface horizon that is rich in accumulated calcium carbonate or magnesium carbonate.

Salic horizon (from *salt*)
A soil layer, common in desert basins, that is at least 6 inches thick and contains at least 2% salt.

Gypsic horizon (from *gypsum*)
A subsurface soil horizon that is rich in accumulated gypsum.

*This table includes only some of the more common horizons.

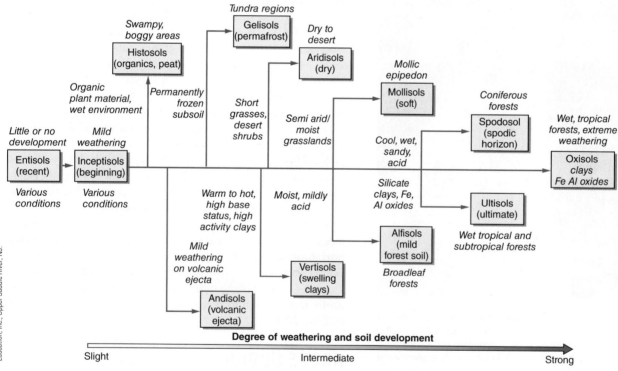

Modified from *Nature and Properties of Soils*, 12/e by Brady and Weil, © 1998. Reprinted by permission of Pearson Education, Inc., Upper Saddle River, NJ.

● **FIGURE 12.28** The Natural Resources Conservation Service (NRCS) soil orders. The soil orders of the NRCS can be linked to the parent materials, climate, and vegetation of the region in which they formed. The linkages form a treelike pattern, as seen here.

How is degree of weathering related to climatic characteristics?

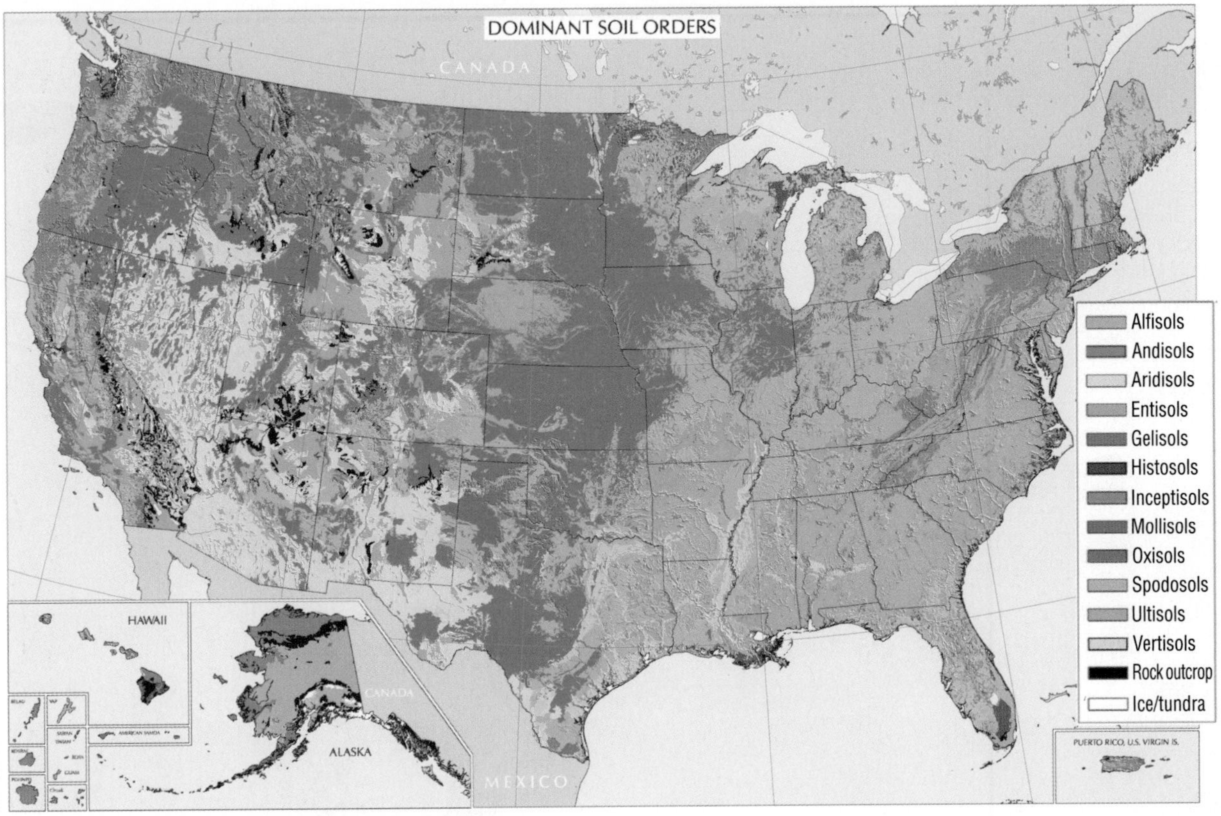
DOMINANT SOIL ORDERS
CANADA
Alfisols
Andisols
Aridisols
Entisols
Gelisols
Histosols
Inceptisols
Mollisols
Oxisols
Spodosols
Ultisols
Vertisols
Rock outcrop
Ice/tundra
HAWAII
ALASKA
CANADA
MEXICO
PUERTO RICO, U.S. VIRGIN IS.

USDA/NRCS

● **FIGURE 12.29** The distribution of soils in the United States according to the National Resource Conservation Service's classification.

What kind of soil dominates the place where you live, according to this map?

environmental characteristics, and soil orders. ● Figure 12.29 is a map showing the distribution of dominant soil orders in the United States. Global soils based on the NRCS classification of soil orders are shown in ● Figure 12.30. Frequent comparison of Fig. 12.30 with the map of world climates in Chapter 8 will illustrate the relationships between the global distributions of soil and climate.

Entisols are soils that have undergone little or no soil development and lack horizons because they have only recently begun to form (● Fig. 12.31a). They are often associated with the continuing erosion of sloping land in mountainous regions or with the frequent deposition of alluvium by flooding or in areas of windblown sand.

Inceptisols are young soils with weak horizon development (● Fig. 12.31b). The processes of *A* horizon depletion (eluviation) and *B* horizon deposition (illuviation) are just beginning, usually because of a very cold climate, repeated flood-related deposition, or a high rate of soil erosion. In the United States, inceptisols are most common in Alaska, the lower Mississippi River floodplain, and the western Appalachians. Globally, inceptisols are especially important along the lower portions of the great river systems of South Asia, such as the Ganges-Brahmaputra, the Irrawaddy, and the Mekong. In these areas, sediments associated with periodic flooding constantly enrich inceptisols and form the basis for agriculture that supports millions of people.

Histosols develop in poorly drained areas, such as swamps, meadows, or bogs, as a product of gleization (● Fig. 12.31c). They are largely composed of decomposing plant material. The waterlogged soil conditions deprive bacteria of the oxygen necessary to decompose the organic matter. Although histosols may be found in low areas with poor drainage at all latitudes, they are most common in tundra areas or in recently glaciated, high-latitude locations such as Scandinavia, Canada, Ireland, and Scotland. Histosols in the subpolar latitudes are commonly acidic and only suitable for special bog crops such as cranberries. Histosols are the primary source of peat, which is a fuel source in some regions, and also is used in landscaping.

Andisols are soils that develop on volcanic parent materials, usually *volcanic ash*, the dust-sized particles emitted by volcanoes (● Fig. 12.32a). Because many of these soils are replenished by eruptions, they are often fertile. Intensive agriculture atop andisols supports dense populations in the Philippines, Indonesia, and the West Indies. In the United States, andisols are most common on the slopes of and downwind from volcanoes in the Pacific Northwest and to a lesser extent in Hawaii and Alaska.

Gelisols are soils that experience frequent freezing and thawing of the ground, above *permafrost*, permanently frozen subsoil (● Fig. 12.32b). When soil freezes, the ice that forms takes up 9% more space than the liquid water that it replaces. To accommodate

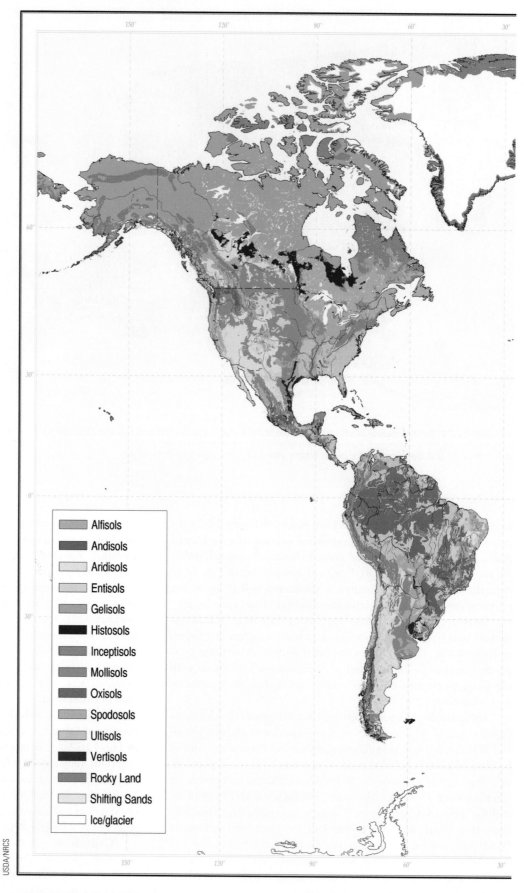

Alfisols
Andisols
Aridisols
Entisols
Gelisols
Histosols
Inceptisols
Mollisols
Oxisols
Spodosols
Ultisols
Vertisols
Rocky Land
Shifting Sands
Ice/glacier

USDA/NRCS

● **FIGURE 12.30** The global distribution of soils by Natural Resources Conservation Service soil orders.

How do these patterns resemble the spatial distribution of world climates?

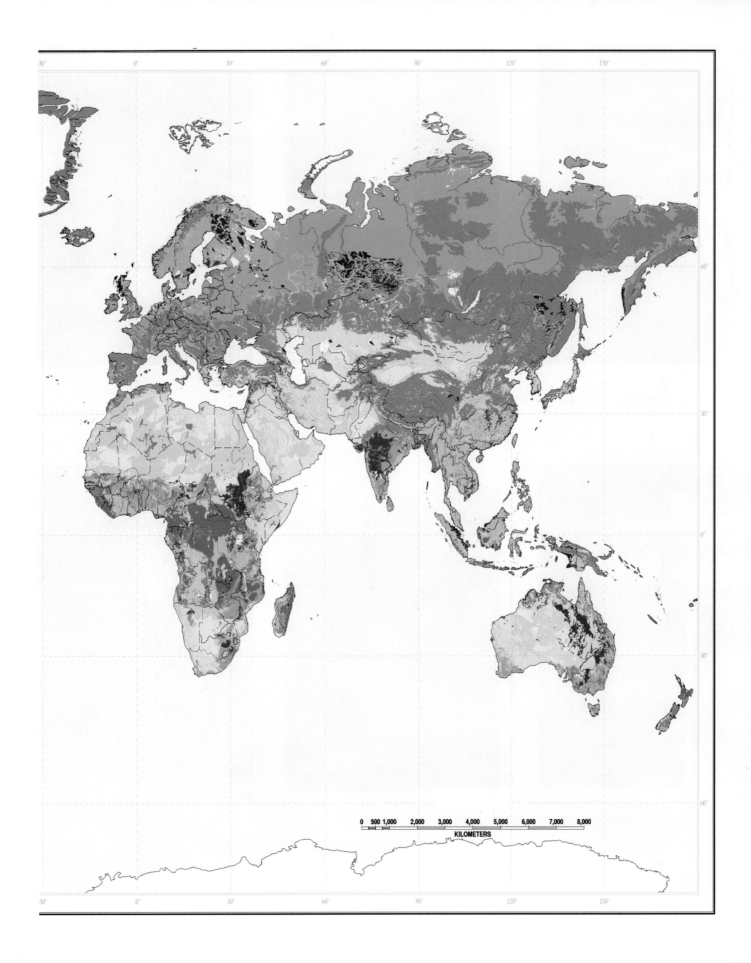

(a)

(b)

(c)

● **FIGURE 12.31** Soil profile examples: (a) entisols, (b) inceptisols, and (c) histosols.

(a)

(b)

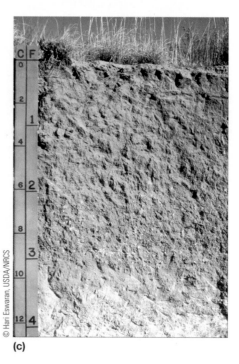

(c)

● **FIGURE 12.32** Soil profile examples: (a) andisols, (b) gelisols, and (c) aridisols.

the increased space taken up by ice, the soil and the particles in it are pushed upward and outward, away from forming ice cores. When the surface soil thaws, gravity pulls the waterlogged ground back downward. Repeated cycles of freezing and thawing mix and churn the upper soil in a process called *cryoturbation*—mixing *(turbation)* related to freezing *(cryo)*. Only the upper part of the soil undergoes freeze–thaw cycles. Permafrost does not permit soil water to percolate downward, so gelisol soils are typically water saturated when they are not frozen at the surface.

Gelisols occur in tundra and subarctic climate regions where soil development tends to be slow because chemical processes operate slowly in cold environments. This soil type is found in north and central Alaska and Canada, in Siberia, and in high-altitude tundra areas.

Aridisols are soils of desert regions that develop primarily under conditions where precipitation is much less than half of potential evaporation (● Fig. 12.32c). Consequently, most aridisols reflect the calcification process. Where groundwater tables are high, evidence of salinization may also be present.

Although aridisols tend to have weak horizon development because of limited water movement in the soil, there is often a subsurface accumulation of calcium carbonate (calcic horizon), salt (salic horizon), or calcium sulfate (gypsic horizon). Soil humus is minimal because vegetation is sparse in deserts; therefore, aridisols are often light in color. Aridisols are usually alkaline, but because few nutrients have been leached, they can support productive agriculture if irrigated to reduce the pH and salinity. Geographically, aridisols are the most common soils on Earth because deserts cover such a large portion of the land surface.

Vertisols are typically found in regions of strong seasonality of precipitation such as the tropical wet and dry climates (● Fig. 12.33a). In the United States, they are most common where

the parent materials produce clay-rich soils. The combination of clayey soils in a wet and dry climate leads to the drying of the soil and consequent shrinkage that forms deep cracks during the dry season, followed by expansion of the soil during the wet season. The constant shrink–swell process disrupts horizon formation to the point that soil scientists often describe vertisols as "self-plowing" soils. Vertical soil movement may damage highways, sidewalks, foundations, and basements that are built on shrink–swell soils. Vertisols are dark-colored, are high in bases, and contain considerable organic material derived from the grasslands or savanna vegetation with which they are normally associated. Although they harden when dry and become sticky and difficult to cultivate when swollen with moisture, vertisols can be agriculturally productive.

Mollisols are most closely associated with grassland regions and are among the best soils for sustained agriculture (● Fig. 12.33b). Because they are located in semiarid climates, mollisols are not heavily leached and they have a generous supply of bases, especially calcium. The characteristic horizon of a mollisol is a thick, dark-colored surface layer rich in organic matter from the decay of abundant root material. Grasslands and associated mollisols served as the grazing lands for countless herds of antelope, bison, and horses. Before the invention of the steel plow, the thick root material of grasses made this soil nearly uncultivable in the United States and thus led to the widespread public image of the Great Plains as a "Great American Desert."

In regions of adequate precipitation, such as the tall-grass prairies of the American Midwest, the combination of soils and climate is unexcelled for agriculture. In areas of lesser precipitation, periodic drought is a constant threat, and the temptation of fertile soils was the downfall of many farmers prior to the advent of center pivot irrigation.

(a) **(b)** **(c)**

● **FIGURE 12.33** Soil profile examples: (a) vertisols, (b) mollisols, and (c) alfisols.

Alfisols occur in a wide variety of climate settings. They are characterized by a subsurface clay horizon (argillic *B* horizon), a medium to high base supply, and a light-colored ochric epipedon (● Fig. 12.33c). The five suborders of alfisols reflect climate types and exemplify the hierarchical nature of the classification system: *aqualfs* are seasonally wet and can be found in mesothermal areas such as Louisiana, Mississippi, and Florida; *boralfs* are found in moist, microthermal climates such as Montana, Wyoming, and Minnesota; *udalfs* are common in both microthermal and mesothermal climates that are moist enough to support agriculture without irrigation, such as Wisconsin, Ohio, and Tennessee; *ustalfs* are found in mesothermal climates that are intermittently dry, such as Texas and New Mexico; and *xeralfs* are found in California's Mediterranean climate, which is characterized by wet winters and long, dry summers.

Because of their abundant bases, alfisols can be very productive agriculturally if local deficiencies are corrected: irrigation for the dry suborders and properly drained fields for the wet suborders.

Spodosols are most closely associated with the podzolization soil-forming process. They are readily identified by their strong horizon development (● Fig. 12.34a). There is often a white or light-gray *E* horizon (albic horizon) covered with a thin, black layer of partially decomposed humus and underlain by a colorful *B* horizon enriched in relocated iron and aluminum compounds (spodic horizon).

Spodosols are generally low in bases and form in porous substrates such as glacial drift or beach sands. In New England and Michigan, spodosols are also acidic. In these regions, and in similar regions in northern Russia, Scandinavia, and Poland, only a few types of agricultural plants, such as cucumbers and potatoes, can tolerate the microthermal climates and sandy, acidic soils. Consequently, the cuisine of these regions directly reflects the spodosols that dominate the areas.

Ultisols, like spodosols, are low in bases because they develop in moist or wet regions. Ultisols are characterized by a subsurface clay horizon (argillic horizon) and are often yellow or red because of residual iron and aluminum oxides in the *A* horizon (Fig. 12.34b). In North America, the ultisols are most closely associated with the southeastern United States. When first cleared of forests, these soils can be agriculturally productive for several decades, but a combination of high rainfall with the associated runoff and erosion from the fields decreases the natural fertility

(a)

(b)

(c)

● **FIGURE 12.34** Soil profile examples: (a) spodosols, (b) ultisols, and (c) oxisols.

of the soils. Ultisols remain productive only with the continuous application of fertilizers. Today, forests cover many former cotton and tobacco fields of the southeastern United States because of a reduction in soil fertility and extensive soil erosion.

Oxisols have developed over long periods in tropical regions with high temperatures and heavy annual rainfall. They are almost entirely leached of soluble bases and are characterized by a thick development of iron and aluminum oxides (Fig. 12.34c). The soil consists mainly of minerals that resist weathering (for example, quartz, clays, and hydrated oxides). Oxisols are most closely associated with the humid tropics, but they also extend into savanna and tropical thorn forest regions. In the United States, oxisols are present only in Hawaii. Oxisols are dominated by laterization and retain their natural fertility only as long as the soils and forest cover maintain their delicate equilibrium. The bases in the tropical rainforests are stored mainly in the vegetation. When a tree dies, epiphytes and insects must recycle the bases rapidly before the heavy rainfall leaches them from the system.

The burning of vegetation associated with slash-and-burn agriculture in rainforests releases the nutrients necessary for crop growth but quickly results in their loss from the ecosystem. Many tropical oxisols that once supported lush forests are now heavily dissected by erosion and only support a combination of weeds, shrubs, and grasses.

Soil as a Critical Natural Resource

Soil is an essential, yet vulnerable, natural resource. It is the responsibility of all of us to help protect our valuable soils and the world's ecosystems with which we also maintain interdependent symbiotic relationships. Soil erosion, soil depletion, and mismanagement of the land are problems that we should have great concern about today (● Fig. 12.35). Detrimental impacts on soils have negative consequences on the natural ecology as well as on the agricultural productivity that humankind depends on. These problems, however, often have reasonable solutions (● Fig. 12.36). Conserving soils and maintaining soil fertility are critical challenges, essential to natural environments, and to our planet's life-giving resources. The information and knowledge gained from studying natural environments and soils can help us learn to work in concert with nature to sustain and improve life on Earth.

R. Gabler

● **FIGURE 12.35** Gully erosion on farmlands is a significant problem that can often be avoided or overcome by proper agricultural practices. Gullying, if unchecked, can alter the landscape to the point that the original productivity of the land cannot be regained.

What could have been done to prevent the kind of soil loss shown in this example?

USDA/NRCS/Lynn Betts

● **FIGURE 12.36** Contour farming techniques, and the use of buffer zones between fields and along water courses, are excellent examples of soil conservation methods.

What other soil conservation practices are often used to preserve the soil resource?

CHAPTER 12 ACTIVITIES

■ TERMS FOR REVIEW

soil	soil ped	soil taxonomy
soil fertilization	porosity	soil survey
capillary water	permeability	soil order
hardpan	pH scale	subsurface horizon
hygroscopic water	parent material	epipedon
gravitational water	soil profile	entisol
leaching	soil horizon	inceptisol
eluviation	Cl, O, R, P, T	histosol
illuviation	residual parent material	andisol
stratification	transported parent material	gelisol
humus	soil-forming regime	aridisol
soil texture	laterization	vertisol
clay (clayey)	laterite	mollisol
silt (silty)	podzolization	alfisol
sand (sandy)	calcification	spodosol
loam	salinization	ultisol
infiltrate	gleization	oxisol

■ QUESTIONS FOR REVIEW

1. Why is soil such an outstanding example of the integration and interaction among Earth's subsystems?
2. Describe the different circumstances in which water is found in soil.
3. Under what conditions does leaching take place? What is the effect of leaching on the soil and, consequently, on the vegetation that it supports?
4. How can capillary water contribute to the formation of caliche? What is the effect of caliche on drainage?
5. How is humus formed? What relation does humus have to soil fertility?
6. How is texture used to classify soils? Describe the ways scientists have classified soil structure.
7. What pH range indicates soil suitable for most complex plants?
8. What are the general characteristics of each horizon in a soil profile? How are soil profiles important to scientists?
9. What factors are involved in the formation of soils? Which is most important on a global scale?
10. How does transported parent material differ from residual parent material? List those factors that help determine how much effect the parent material will have on the soil.
11. What are the most important effects of parent material on soil?
12. How does the presence of earthworms and other burrowing animals affect soil?
13. Describe the various ways in which temperature and precipitation are related to soil formation.
14. Describe the three major soil-forming regimes.

CONSIDER AND RESPOND

1. Refer to Figure 12.13 and associated pages in the text.
   (a) What horizons make up the zone of eluviation?
   (b) What are two processes that occur in the zone of eluviation?
   (c) The various *B* horizons are in what zone?
   (d) Weathered parent material is the major constituent of what horizon?
   (e) Partly decomposed organic debris makes up which horizon?

2. Refer to Table 12.1.
   (a) What materials accumulate in an argillic horizon?
   (b) Which would generally be better suited for agriculture— a soil with an ochric epipedon or a mollic epipedon? Why?
   (c) What name would be given to a 7-inch-thick horizon that contained at least 2% salt?

3. How important do you consider soils to be among a nation's, and the world's, environmental resources? Explain why.

4. Give your opinion of the overall value of soils in the United States and the extent to which these soils are preserved and protected.

PRACTICAL APPLICATIONS

1. Refer to Figure 12.9. Using the texture triangle, determine the textures of the following soil samples.

	Sand	Silt	Clay
Soil a.	35%	45%	20%
Soil b.	75%	15%	10%
Soil c.	10%	60%	30%
Soil d.	5%	45%	50%

What are the percentages of sand, silt, and clay of the following soil textures? (Note: Answers may vary, but they should total 100%.)
   (a) Sandy clay
   (b) Silty loam

2. Obtain a small sample of soil (a handful or so) and try to discover its texture by using the following method: Wet the soil a bit and work it in your hand.

   a. **First, if you can, form a ribbon of soil by kneading it with your fingers:**
      If you can form a ribbon that is long, relatively strong, and flexible, the soil is a clay.
      If you can form a ribbon that is weak and breaks easily, and the soil can be rolled into a coherent ball, the soil is a clay loam.
      If the clay loam looks powdery when dry, the soil is a silty clay loam.
      If the clay loam has a gritty feel with visible sand, it is a sandy clay loam.

   b. **Second, if you cannot form a ribbon because the soil breaks up:**
      If damp soil breaks up easily but is gritty yet still sticky enough to make a ball, the soil is a loam.
      If it feels gritty and sand can be seen and the ball breaks up easily in your hands, the soil is a sandy loam.

   c. **Third, if the soil is very loose with visible grains of about the same size:**
      If you can see grains in the soil and the mass in your hand breaks up easily, the soil is a sand.

 MindTap—Make the most of your study time by accessing everything you need to succeed in one place. Read your textbook, take notes, review flashcards, watch videos, complete activities, take practice quizzes, and more online with MindTap. Log in at **www.cengagebrain.com**.

EARTH MATERIALS AND PLATE TECTONICS

13

OBJECTIVES

WHEN YOU COMPLETE THIS CHAPTER YOU SHOULD BE ABLE TO:

- 13.1 Compare the relative size and material properties of Earth's core, mantle, and crust.
- 13.2 Relate the principal differences between oceanic and continental crust.
- 13.3 Contrast the lithosphere with the asthenosphere.
- 13.4 Differentiate between minerals and rocks.
- 13.5 Recall the definitions of the major categories of igneous, sedimentary, and metamorphic rocks.
- 13.6 Explain the meaning of the rock cycle.
- 13.7 Understand the relationship between the theories of continental drift and plate tectonics.
- 13.8 Provide evidence for the theory of plate tectonics.
- 13.9 Describe major Earth features associated with tectonic plate convergence, divergence, and transform motion.
- 13.10 Explain how the geologic timescale is structured and why paleogeography is important.

IF WE COULD TRAVEL BACK in time to view Earth as it was 90 million years ago, in addition to seeing now-extinct life-forms, including dinosaurs, we would notice an altered spatial distribution of land and water. The shapes of the oceans and continents, their locations, and their orientations relative to the poles and equator were very different then from what they are today. A vast inland sea cut across what is now the heartland of North America. Dinosaurs left their footprints in large trackways on floodplains of rivers that flowed out of the early Rocky Mountains. Forests grew above the present Arctic Circle, and grasses did not exist yet. The dramatic differences between then and now in the size, shape, and distribution of mountain ranges and water bodies—as well as the accompanying differences in climate, soils, and organisms—merit a scientific explanation.

Like the atmosphere, hydrosphere, and biosphere, the extensive part of the Earth system that lies beneath our feet—the lithosphere—undergoes change resulting from flows of energy and matter. Processes inside Earth help determine the present distribution of various rock types, mineral resources, and natural hazards. Over long intervals of geologic time, the flows of energy and matter within Earth significantly alter the size, shape, and location of major Earth surface features and environments.

◄ Relative age of rock material on the floor of the Atlantic Ocean. Rocks are youngest (red) along the extensive mid-Atlantic submarine mountain chain, and they are progressively older with increasing distance from that midoceanic ridge. E. Lim and J. Varner, CIRES & NOAA/NDGC

In essence, forces inside Earth build the structural foundation that other processes, those originating at Earth's surface, modify. Together, the internal and surface processes create the familiar landscapes in which we live. In this chapter we begin our study of the solid, or rock, portion of Earth—the lithosphere.

Earth's Planetary Structure

The science of physical geography predominantly focuses on that part of the Earth system that lies at the interface of the atmosphere, hydrosphere, biosphere, and lithosphere, and these come together at Earth's surface. Still, basic knowledge of our planet's internal structure, composition, properties, and processes is needed to understand many aspects of Earth's natural surface characteristics.

All of the gas, liquid, and solid matter comprising the Earth system is held within that system by mutual forces of gravitational attraction. Sir Isaac Newton determined that the degree to which particles are drawn to each other by gravity depends in part on the mass of each particle, which is commonly expressed in units of grams (g) or kilograms (kg). The gravitational force of attraction is greater for objects that have a larger mass than for those with a smaller mass. Scientists commonly use *density*, which is mass per unit volume (for example, grams per cubic centimeter, abbreviated g/cm³), to compare how equal amounts of various materials differ in mass. Those types of Earth materials that have the greatest density have the greatest gravitational force of attraction, and as a result they have tended to concentrate close together at and near the center of Earth. High-density iron and nickel occupy the center of the planet, whereas low-density gas molecules, which have a much smaller gravitational force of attraction, are held in the outer atmosphere, far from Earth's center.

Earth's interior is primarily composed of solids, the densest of the three states of matter. Liquids are not as dense as solids, thus most of Earth's liquid water lies at the planet's surface thousands of kilometers above the densest solids located deep inside Earth. Gases, with an even lower density than liquids, have the weakest gravitational attractive force and are held more loosely around Earth as the atmospheric envelope, rather than within the planet or on its surface. Traveling outward from the center of the Earth system, a *density continuum* (spectrum) extends from the densest materials at Earth's center to the least dense substances at the outer edge of the atmosphere.

Planet Earth has a radius of about 6400 kilometers (4000 mi). Through direct means by mining and drilling, we have been able to penetrate and examine only an extremely small part of that distance. In South Africa, the lure of gold has taken miners to a depth of 3.5 kilometers (2.2 mi), whereas in drilling for oil and gas our machinery has reached a depth of about 12 kilometers (7.5 mi). These explorations have been helpful in providing information about the solid Earth. For example, they have helped establish the notion of the *geothermal gradient*—that temperature increases with increasing depth inside Earth. Our direct explorations into the solid Earth, however, have just barely scratched the planet's surface. Scientists are continually working to learn more about Earth's interior. Extending scientific knowledge about the structure, composition, and processes operating within Earth enhances our understanding of such lithospheric phenomena as earthquakes, volcanic eruptions, the formation and distribution of mineral deposits, and the origins of continents. It can even help us learn more about the origin of the planet itself.

Most of what we know about Earth's internal structure and composition has not been determined directly but has instead been deduced through indirect means by various forms of remote sensing. Thus far, the most important evidence that scientists have used to gain indirect knowledge of Earth's interior is the behavior of various shock waves, known as **seismic waves**, as they travel through the planet. Scientists generate some of these shock waves artificially through controlled explosions, but they mainly use evidence derived by tracking the natural seismic waves of earthquakes as they travel through Earth. A sensitive instrument called a **seismograph** can record seismic waves from an earthquake even when the earthquake is centered thousands of kilometers away from the seismograph's location (● Fig. 13.1).

Earthquakes produce two major types of seismic waves that travel at different speeds through varying types and densities of Earth

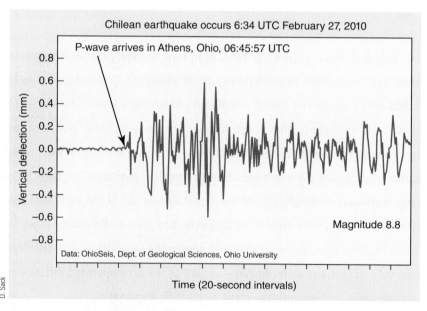

● **FIGURE 13.1** This seismograph trace, or seismogram, shows the ground shaking experienced in Athens, Ohio, when seismic waves first started arriving there from a major earthquake in Chile in 2010. Data from multiple seismographs are used to determine characteristics of Earth's interior and characteristics of individual earthquakes. Over Earth's surface, Athens lies 8420 kilometers (5230 mi) from the site of the earthquake; the straight-line distance through Earth is 7873 kilometers (4890 mi).

At what average speed did the seismic waves travel through Earth from Chile to Athens? What maximum vertical motion (vertical deflection) did Athens experience from this earthquake?

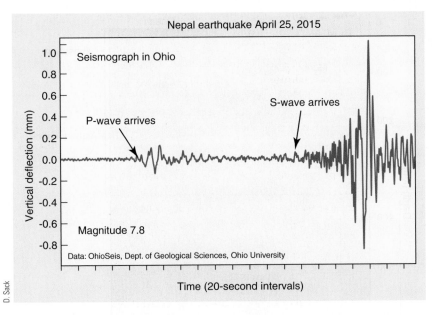

Nepal earthquake April 25, 2015

Seismograph in Ohio

P-wave arrives

S-wave arrives

Magnitude 7.8

Data: OhioSeis, Dept. of Geological Sciences, Ohio University

Time (20-second intervals)

Vertical deflection (mm)

D. Sack

● **FIGURE 13.2** Difference in arrival time of the seismic P (primary) and S (secondary) waves from the 2015 earthquake in Nepal, as recorded on the seismogram in Athens, Ohio. The farther a seismograph station is from the earthquake, the greater the lag between the P and S wave arrival times.

material. Of the two types, P (primary) waves travel faster and are the first to arrive at a recording seismograph. S (secondary) waves travel more slowly than P waves, thus they arrive at the seismograph later (● Fig. 13.2). Studies of numerous seismograph records

show that P and S waves are either refracted or reflected when they reach the zones of major density change that mark the boundaries between Earth's various interior layers. Seismic waves *refract* if they cross a boundary and continue traveling but with a shift in direction caused by the change in material density. Waves that *reflect* bounce off the boundary and do not enter the next layer. Both P and S waves speed up in denser material and slow down when passing through material that is less dense. P waves pass through all types of matter, including liquids and gases. In fact, P waves traveling through the atmosphere are responsible for the rumbling sound that we hear during an earthquake. S waves, however, can only move through solids; they do not travel through fluids, whether liquid or gas (● Fig. 13.3a).

By considering these and other properties of P and S waves and by analyzing global data on travel patterns of earthquake waves collected over decades, scientists have been able to develop a general model of the composition of Earth's interior. This information, supplemented by studies of Earth's magnetism and gravitational pull, reveals a series of layers, or zones, in Earth's internal structure distinguished on the basis of density and composition. From the center of Earth to the surface, these zones are the core, mantle, and crust (Fig. 13.3b). Using a different criterion, how rock material behaves under stress, the outer approximately 350 kilometers (220 mi) of Earth consists of two distinct layers known as the lithosphere and asthenosphere.

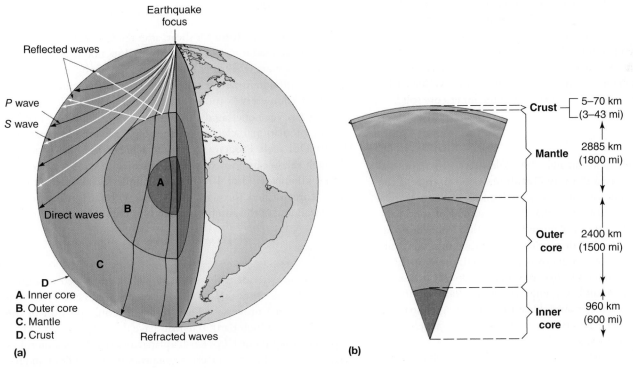

● **FIGURE 13.3** Earth's internal structure as revealed by seismic waves. (a) Seismic P (primary) waves refract (bend) where material density changes significantly, whereas seismic S (secondary) waves cannot pass through liquids, including Earth's outer core. (b) Cross section through Earth's internal structural zones.

Which of Earth's internal zones is the largest?

Core

Earth's innermost section, the **core**, contains one third of Earth's mass and has a radius of about 3360 kilometers (2100 mi), which is larger than the planet Mars. Earth's core is under enormous pressure—several million times atmospheric pressure at sea level. Scientists have deduced that the core is composed primarily of iron and nickel and consists of two distinct sections, the inner core and the outer core.

Earth's **inner core** has a radius of about 960 kilometers (600 mi). The speed of P waves traveling through the inner core shows that it is a solid with a very high material density of about 13 grams per cubic centimeter (0.5 lb/in.[3]). The **outer core** forms a 2400 kilometer (1500 mi) thick band around the inner core. Rock matter at the top of the outer core has a density of about 10 grams per cubic centimeter (0.4 lb/in.[3]). Because the outer core blocks the passage of seismic S waves by reflecting them, but allows continuation of refracted P waves, Earth scientists know that the outer core is *molten*, that is, it consists of liquid (melted) rock matter. The high density of both sections of Earth's core supports the notion that they are composed of iron and nickel.

Temperatures are estimated to be 4800°C (8600°F) at the top of the outer core, increasing to 6900°C (12,400°F) at the very center of Earth. Why, then, is Earth's inner core solid but its cooler outer core molten? The answer involves the facts that the melting temperature of rock matter increases with pressure, and pressure increases with depth beneath Earth's surface. Rock existing under greater pressure needs to achieve a higher temperature to melt than rock at a lower pressure does. The actual temperature of rock material in the outer core exceeds that material's pressure-altered melting temperature, thus rock in the outer core is molten. The extreme high pressure of the inner core, however, has elevated the melting temperature there to a value that lies above the rock's actual temperature, leaving the material of the inner core solid.

Mantle

With a thickness of approximately 2885 kilometers (1800 mi) and representing nearly two thirds of Earth's mass, the **mantle** is the largest of Earth's interior zones. Seismic P and S waves both pass through the mantle, indicating that it is composed of solid rock matter, in contrast to the molten outer core that lies beneath it. Mantle material is also less dense than that of the core, with values ranging from 3.3 to 5.5 grams per cubic centimeter (0.12 to 0.20 lb/in.[3]). Scientists agree that the mantle consists of silicate rocks (high in silicon and oxygen) that also contain significant amounts of iron and magnesium. Heat from decaying radioactive materials inside Earth drives the thermal convection currents that exist in much of the mantle.

The interface between the mantle and the overlying crust is marked by a significant change of density, called a *discontinuity*, which is indicated by an abrupt decrease in the velocity of seismic waves as they travel up through this internal boundary (● Fig. 13.4). Scientists call this zone the **Mohorovičić**

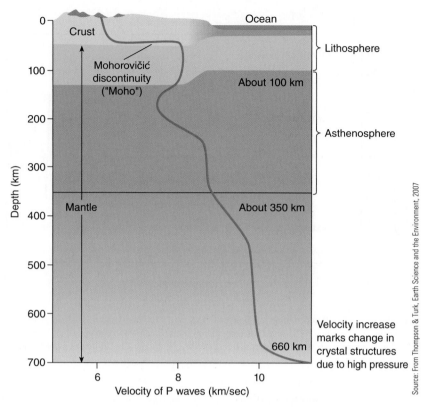

● **FIGURE 13.4** Velocity of seismic P (primary) waves increases with depth as material density increases. An abrupt increase in seismic wave velocity marks the Mohorovičić discontinuity.

Where is the most abrupt change in material density in the upper 700 kilometers of the solid Earth?

Source: From Thompson & Turk, Earth Science and the Environment, 2007

discontinuity, or **Moho** for short, after the Croatian geophysicist who first detected it in 1909. The Moho does not lie at a constant depth but generally mirrors the surface topography, being deepest under mountain ranges where the crust is thick and rising to within 8 kilometers (5 mi) of the thin ocean floor. No geologic drilling has yet penetrated through the Moho into the mantle, but an international scientific partnership, called the International Ocean Discovery Program, will try to achieve that goal by drilling through the ocean floor where the distance to the Moho is only 5 to 6 kilometers (3 to 4 mi). Rock samples eventually retrieved from cores drilled through the Moho will add greatly to our understanding of the composition and structure of Earth's interior.

Crust

Earth's solid exterior is the **crust**, which is composed of a great variety of rock types that respond in diverse ways and at varying rates to surface processes. The crust is the only portion of the lithosphere of which Earth scientists have direct knowledge, yet it represents only about 1% of Earth's planetary mass. As the outermost layer of the solid Earth, the crust comprises the ocean floor and the continents and is of primary importance in understanding surface processes and landforms. Earth's deep interior components, the core and mantle, are of concern to physical geographers primarily for how they affect the crust.

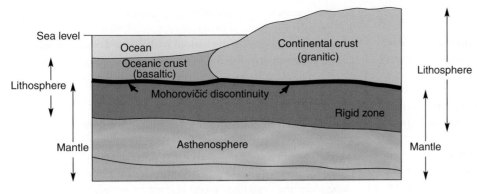

● **FIGURE 13.5** Earth has two distinct types of crust: oceanic and continental. The crust and the uppermost part of the mantle, which is rigid like the crust, make up the lithosphere. Below the lithosphere, but still in the upper part of the mantle, is the plastic asthenosphere.

The density of rock matter in Earth's crust is significantly lower than that in the core and mantle, and ranges from 2.7 to 3.0 grams per cubic centimeter (0.10 to 0.11 lb/in.^{3}). The crust is also extremely thin in comparison to the diameter of the planet. Two kinds of Earth crust, oceanic and continental, are distinguished by their location, thickness, and composition (● Fig. 13.5). Crustal thickness varies from 3 to 6 kilometers (2 to 4 mi) in the ocean basins to as much as 70 kilometers (44 mi) under some continental mountain systems. The crust is cold compared with the mantle and behaves in a more rigid and brittle fashion, especially in its upper 10 to 15 kilometers (6 to 9 mi). The crust responds to stress by fracturing, crumpling, or warping.

Oceanic crust is composed of *basalt*, a heavy, dark-colored, iron-rich rock that is also high in silicon (Si) and magnesium (Mg). Oceanic crust has a density of 3.0 grams per cubic centimeter and is only a few kilometers thick. Forming the vast, deep ocean floors and the flows of lava on all the continents, basaltic rocks are the most common rocks on Earth.

Continental crust comprises the major landmasses on Earth that are exposed to the atmosphere. At 2.7 grams per cubic centimeter, the average density of continental crust is less than that of oceanic crust (3.0 g/cm^3). Continental crust, however, is considerably thicker than oceanic crust. It ranges from about 20 to 70 kilometers (12 to 44 mi) with an average thickness of 32 to 40 kilometers (20 to 25 mi). At places where continental crust extends to high elevations in mountain ranges, it also descends to great depths below the surface. Continental crust contains more light-colored rocks than oceanic crust does and can be regarded as *granitic* in composition. Granite, basalt, and other common rocks are discussed more fully later in the chapter.

Lithosphere and Asthenosphere

The extreme uppermost part of the mantle, with a thickness of about 100 kilometers (60 mi), has a chemical composition like the rest of the mantle, but it responds to applied stress like the overlying Earth layer, the crust. The response is that of an elastic solid. **Elastic solid** materials are rigid and brittle. They do not flow, but instead withstand a certain amount of applied

stress (force per unit area) with little deformation until a threshold limit of stress is reached. At the threshold value, elastic solids fail by fracturing. Behaviorally, then, this uppermost mantle and overlying crust form a single structural unit called the **lithosphere**. The term *lithosphere* has traditionally been used to describe the entire solid part of the Earth system (as discussed in Chapter 1 and earlier in this chapter). In recent decades, however, that term has also been used in a separate, structural sense to refer to the brittle outer shell of Earth, including the crust and the rigid, uppermost mantle layer (● Fig. 13.6).

Beneath the lithosphere lies the **asthenosphere** (from Greek: *asthenias*, without strength), a 180-kilometer (110-mi) thick layer of the upper mantle that responds to stress by deforming and flowing slowly rather than by fracturing. In other words, the asthenosphere has the characteristics of a **plastic solid**. Rock in the asthenosphere can flow vertically or horizontally at rates of a few centimeters per year. As material in the asthenosphere flows, it drags segments of the overlying, rigid lithosphere along with it. This movement within the plastic asthenosphere drives **tectonic forces**, large-scale forces from inside Earth that break and deform the lithosphere, sometimes resulting in earthquakes and often responsible for mountain building. Movement in the asthenosphere, in turn, is produced by thermal convection currents originating deeper in the mantle below the asthenosphere. These currents are driven by heat from decaying radioactive materials in the planet's interior.

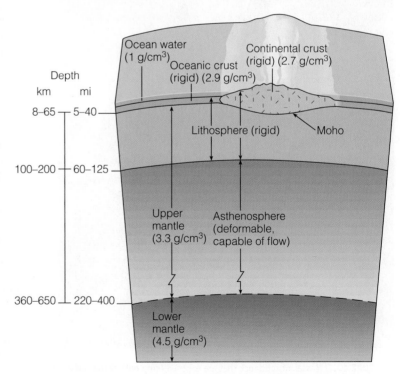

● **FIGURE 13.6** Cross section through Earth's lithosphere and asthenosphere, and extending lower into the mantle.

Which material is denser, oceanic crust or continental crust?

Earth Materials

The solid part of Earth, including the crust, is composed of rock. At many locations on Earth's surface, however, large masses of consolidated rock are not visible. Instead, the solid rock lies concealed under an accumulation of loose sediment and soil. This cover forms because when exposed to the environmental conditions at Earth's surface, rock tends to *weather*, that is, rock begins to break into smaller pieces, known as *regolith*. A mass of consolidated rock that has not been weathered, whether exposed at the surface or buried, is termed **bedrock**. Bedrock often lies buried under regolith or under rock fragments deposited by other surface processes, such as stream flow or gravity. Soil will develop on this bedrock-covering sediment if it is relatively stable. On steep slopes, including road cuts, processes such as running water or the pull of gravity can remove surface sediment to leave the bare bedrock exposed at the surface as a rock **outcrop** (● Fig. 13.7).

Rocks contain valuable resources, contribute to soil formation, are a major factor in determining Earth's surface elevation changes, and record various aspects of Earth's history. Rocks also contain evidence of the environmental conditions that existed on Earth at the time and place of their formation. For these reasons, it is important for physical geographers to understand the main characteristics and origins of the various rock types. To accomplish this, we must first discuss the difference between rocks and minerals.

Minerals versus Rocks

Minerals are the building blocks of rocks. A **mineral** is an inorganic, naturally occurring, crystalline substance represented by a specific chemical formula. A *crystalline* substance displays a specific, repeated, three-dimensional structure at the molecular level. In some cases, the crystalline molecular nature is visible only with the use of a microscope, but minerals sometimes form actual *crystals*, which are geometric forms visible to the unaided eye and consisting of smooth faces and sharp edges (● Fig. 13.8). The shape of a crystal is an expression of the mineral's molecular structure. The mineral *halite*, which is used as table salt, has the specific chemical formula NaCl and, as a crystal, adopts a cubic form. Quartz, calcite, fluorite, talc, topaz, and diamond are just a few other examples of the more than 2000 known minerals.

Every mineral has distinctive and recognizable physical characteristics that aid in its identification. Some of these characteristics include hardness, color, luster, cleavage, tendency to fracture, and specific gravity (density compared to that of water), in addition to the shape of the crystal. *Luster* specifies the shininess of a mineral. *Cleavage* describes how a mineral tends to break along preferred planes determined by its molecular structure, whereas *fracture* refers to the nature of irregular breaks not along the preferred planes.

Chemical bonds hold together the atoms and molecules that compose a mineral. The strength and nature of those chemical bonds affect the resistance and hardness of minerals and of the rocks they form. Minerals with weak internal bonds undergo chemical alteration most easily. Charged particles, or *ions*, that form part of a molecule in a mineral may leave or be traded for other substances, generally weakening the mineral structure and forming the chemical basis of rock weathering.

Out of more than 100 known elements, Table 13.1 lists the eight most common in Earth's crust and therefore in the minerals that compose the rocks that make up the crust. These eight elements account for almost 99% of Earth's crust by weight. The most abundant minerals are combinations of these eight elements.

Minerals are categorized into groups based on their chemical composition. Certain elements, particularly silicon, oxygen, and carbon, combine readily with many other elements. As a result, the most common mineral groups are silicates, oxides, and carbonates. Calcite ($CaCO_3$), for example, is a relatively soft but widespread mineral that consists of one atom of calcium (Ca) linked together with a carbonate molecule (CO_3), which is composed of one atom of carbon (C) plus three atoms of oxygen (O). The silicates, however, are by far the largest and most common mineral group, constituting 92% of Earth's crust.

The two most abundant elements in Earth's crust, oxygen and silicon (Si), frequently combine to form SiO_2, which is called *silica*. *Silicate minerals* are compounds of oxygen and silicon that also include one or more metals and/or bases. They are generally created when molten rock matter containing these elements cools

● **FIGURE 13.7** Masses of solid rock that are exposed (crop out) at the surface are referred to as *outcrops*.

What physical characteristics of this rock outcrop have caused it to protrude above the general land surface?

Jim Petersen

● **FIGURE 13.8** Crystals of the mineral calcite, which has the chemical formula $CaCO_3$. The geometric arrangement of atoms determines the crystal form of a mineral.

and solidifies, causing the crystallization of different minerals at successively lower temperatures and pressures. Dark, heavy, iron-rich silicate minerals crystallize first (at high temperatures), and light-colored, lower density, iron-poor minerals crystallize later at cooler temperatures. The order of mineral crystallization parallels their relative chemical stability in a rock, with silicate minerals that crystallize later tending to be more stable and more resistant to breakdown. In rocks composed of a variety of silicate minerals, the dark, heavy minerals are the first to decompose, whereas the iron-poor minerals decompose later. The mineral *quartz* is crystalline silica; as a crystal, it adopts a distinctive prismatic shape. Because quartz is one of the last silicate minerals to form from solidifying molten rock matter, it is a relatively hard and resistant mineral.

A **rock** is a consolidated aggregate of various types of minerals or a consolidated aggregate of multiple individual pieces (grains) of the same kind of mineral. In other words, a rock is not one single, uniform crystal. Although a few rock types are composed of many particles of a single mineral, most rocks consist of several different minerals (● Fig. 13.9). Each constituent mineral in a rock remains separate and retains its own distinctive characteristics. The properties of the rock as a whole are a composite of those of its various mineral constituents. The number of rock-forming minerals that are common is limited,

TABLE 13.1
Most Common Elements in Earth's Crust

Element	Percentage of Earth's Crust by Weight
Oxygen (O)	46.60
Silicon (Si)	27.72
Aluminum (Al)	8.13
Iron (Fe)	5.00
Calcium (Ca)	3.63
Sodium (Na)	2.83
Potassium (K)	2.70
Magnesium (Mg)	2.09
Total	98.70

Source: J. Green, "Geotechnical Table of the Elements for 1953," *Bulletin of the Geological Society of America 64* (1953).

● **FIGURE 13.9** Most rocks consist of several different minerals. This piece of cut and polished granite displays intergrown crystals of differing composition, color, and size that give the rock a distinctive appearance.

How many different kinds of minerals can you see in this sample of granite?

but they combine through a multitude of processes to produce an enormous variety of rock types. (Refer to Appendix C for further information and pictures of common rocks mentioned in the text.)

Geologists distinguish three major categories of rocks based on mode of formation. These rock types are igneous, sedimentary, and metamorphic.

Igneous Rocks

When molten rock material cools and solidifies, it becomes an **igneous rock**. Molten rock matter below Earth's surface is called **magma**, whereas molten rock material at the surface is known specifically as **lava** (● Fig. 13.10). Lava, therefore, is the only form of molten rock matter that we can see. Lava erupts from volcanoes or fissures in the crust at temperatures as high as 1090°C (2000°F). The two major categories of igneous rocks are extrusive and intrusive.

Molten material that solidifies at Earth's surface creates **extrusive igneous rock**, also called *volcanic rock*. Extrusive igneous rock, therefore, is made from lava. Very explosive eruptions of molten rock material can cause the accumulation of fragments of volcanic rock, dust-sized or larger, termed **pyroclastics** (from Greek: *pyros*, fire; *clastus*, broken), also known as **tephra**, that settle out of the air (● Fig. 13.11a). When molten rock beneath Earth's surface, that is, magma, changes to a solid (freezes), it forms **intrusive igneous rock**, also called *plutonic*

USDI/USGS

● **FIGURE 13.10** Recently erupted basaltic lava on the island of Hawaii. Molten basalt typically erupts at very high temperatures, flows easily, and cools rapidly into a fine-grained extrusive igneous rock.

What features besides the red color indicate the part of the flow that was added most recently?

rock after Pluto, Roman god of the underworld. Igneous rocks are classified in terms of their mineral composition as well as the size of constituent minerals, which is referred to as *texture*. Igneous rocks vary in texture, chemical composition, crystalline structure, tendency to fracture, and presence or absence of layering.

Rocks composed of small-sized individual minerals not visible to the unaided eye are described as having a *fine-grained* texture, whereas those with large minerals that are visible without

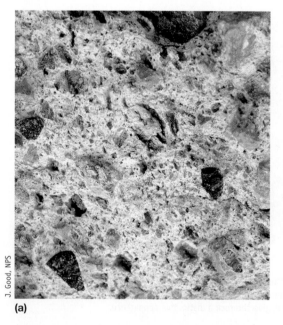

J. Good, NPS

(a)

J. Good, NPS

(b)

● **FIGURE 13.11** (a) Pyroclastic rocks are made of fragments ejected during a volcanic eruption. (b) Obsidian—volcanic glass—results when molten lava cools too quickly for crystals to form.

magnification are referred to as *coarse-grained*. Molten rock matter extruded onto the surface cools quickly—at Earth surface temperatures—and are fine-grained as a result of the little time available for crystal growth prior to solidification. An extreme example is the extrusive rock *obsidian*, which cools so rapidly that it is essentially a glass (Fig. 13.11b). Large masses of intrusive rock matter solidifying deep inside Earth cool very slowly because, like all rocks, those surrounding the magma are poor conductors and inhibit heat loss. Slow cooling allows more time for crystal formation prior to solidification. In special circumstances, however, such as if magma lies in thin stringers or close to Earth's surface, it can cool rapidly enough to produce a fine-grained, instead of a coarse-grained, intrusive rock.

The chemical composition of igneous rocks varies from *felsic*, which is rich in light-colored, lighter weight minerals, especially silicon and aluminum (*fel* for the mineral feldspar; *si* for silica), to *mafic*, which is lower in silica and rich in heavy minerals, such as compounds of magnesium and iron (*ma* for magnesium; *f* for *ferrum*, Latin for iron). Granite, a felsic, coarse-grained, intrusive rock, has the same chemical and mineral composition as *rhyolite*, a fine-grained, extrusive rock. Likewise, basalt is the dark-colored, mafic, fine-grained extrusive chemical and mineral equivalent of *gabbro*, a coarse-grained intrusive rock that cools at depth (● Fig. 13.12).

Igneous rocks also form with an intermediate composition represented by a rough balance between felsic and mafic minerals. The intrusive rock *diorite* and the extrusive rock *andesite* (named for the Andes where many volcanoes erupt this type of lava) represent this intermediate composition (see Figure 13.12).

Many igneous rocks are fractured, often by multiple cracks that may be evenly spaced or arranged in regular geometric patterns. In the Earth sciences, simple fractures or cracks in bedrock are called **joints**. Joints caused by regional stresses in the crust are common features in any type of rock, including igneous rocks. Igneous rocks, however, also develop joints because molten rock shrinks in volume as it cools and solidifies, resulting in fractures.

Sedimentary Rocks

As their name implies, **sedimentary rocks** are derived from sediment, that is, unconsolidated fragments that have accumulated together in a loose collection. After the fragments accumulate, often in horizontal layers, pressure from the addition of more material above compacts the sediment, expelling water and reducing pore space. Cementation occurs when silica, calcium carbonate, or iron oxide bonds the fragments together. The processes of compaction and cementation transform, or *lithify*, sediments into solid, coherent layers of rock. The three major categories of sedimentary rocks are clastic, organic, and chemical precipitate.

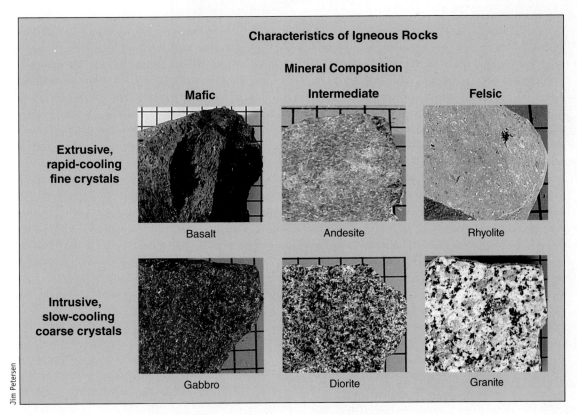

● **FIGURE 13.12** Igneous rocks are distinguished on the basis of texture (crystal size) and whether their mineral composition is mafic, felsic, or intermediate. Rocks that cool rapidly, at or near Earth's surface, have fine (small) crystals. Those that cool slowly, deep beneath the surface, have coarse crystalline texture (large crystals).

How does granite differ from basalt in composition and texture?

Broken fragments of solids are called **clasts** (from Latin: *clastus*, broken). In order of increasing size, clasts range from clay, silt, and sand, to gravel, which is a general category for any fragment larger than sand (larger than 2.0 mm) and includes granules, pebbles, cobbles, and boulders. Clasts consist of fragments of previously existing rocks, shells, or bones that were deposited on a river bed, beach, sand dune, lake bottom, the ocean floor, and other environments where fragments of solids accumulate. Sedimentary rocks that form from fragments of preexisting rocks, shells, or bones are called **clastic sedimentary rocks**.

Examples of clastic sedimentary rocks include conglomerate, sandstone, siltstone, and shale (● Fig. 13.13). *Conglomerate* is a lithified mass of cemented, roughly rounded pebbles, cobbles, and boulders and may have clay, silt, or sand filling in spaces between those larger particles. A somewhat similar rock composed of lithified fragments that are angular rather than rounded is called *breccia*. *Sandstone* consists of cemented sand-sized particles, most commonly grains of quartz. Sandstone is usually granular (has visible grains), porous, and resistant to weathering, but the cementing material influences its strength and hardness.

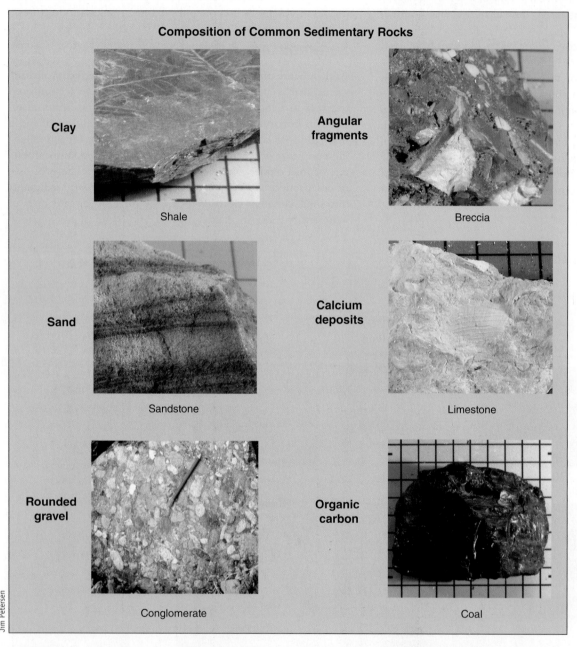

Composition of Common Sedimentary Rocks

Clay — Shale

Angular fragments — Breccia

Sand — Sandstone

Calcium deposits — Limestone

Rounded gravel — Conglomerate

Organic carbon — Coal

Jim Petersen

● **FIGURE 13.13** Sedimentary rocks are made from broken pieces of preexisting solids (clasts), organic remains, or chemical precipitates. Clastic sedimentary rocks are further classified by the size and/or shape of the sediment particles they contain.

Why are there different sizes and shapes of clastic sediments?

If cemented by silica, sandstone tends to be more resistant to weathering than if it is cemented by calcium carbonate or iron oxide. Unlike sandstone, individual grains in *siltstone*, which is composed of silt-sized particles, are not easily visible with the unaided eye. *Shale* is produced from the compaction of very fine-grained sediments, especially clays. Shale is often thinly layered with a smooth surface and a low permeability. It is, however, easily cracked, broken apart, and eroded.

Sedimentary rocks may be further classified by their origin in either a marine or terrestrial (continental) environment. Marine sandstones typically form from sediments deposited in a nearshore coastal zone; terrestrial sandstones commonly derive from sediments deposited in desert or floodplain environments on land. The nature and arrangement of sediments in a sedimentary rock provide a great deal of evidence about the environment in which they were deposited, for example, whether on a stream bed, a beach, the deep ocean floor, or elsewhere.

Organic sedimentary rocks lithify from the remains of organisms, both plants and animals. *Coal* is created by the accumulation and compaction of partially decayed vegetation in acidic, swampy environments where water-saturated ground prevents oxidation and complete decay of the organic matter. The initial transformation of such organic material produces peat, which, when subjected to deeper burial and further compaction, is lithified to produce coal. Most of the world's greatest coal deposits originated between 290 and 354 million years ago during the Mississippian and subsequent Pennsylvanian Periods of geologic time.

Other organic sedimentary rocks develop from the remains of organisms in lakes and seas. The remains of shellfish, corals, and microscopic drifting organisms called *plankton* sink to the bottom of such water bodies where they are compacted and cemented together. Rich in calcium carbonate ($CaCO_3$), they form a type of *limestone* that typically contains fossil shell and coral fragments (● Fig. 13.14).

When the amounts of dissolved minerals in ocean or lake water reach saturation they begin to precipitate and build up as a deposit on the sea or lake bottom. When lithified, these sediments become **chemical precipitate sedimentary rocks**. Some fine-grained limestone rocks form in this manner from chemical precipitates of calcium carbonate. Limestone, therefore, may vary from a jagged and cemented complex of visible shells or fossil skeletal material (an organic sedimentary rock) to a chemical precipitate sedimentary rock with a smooth surface. Where magnesium is a major constituent along with calcium carbonate, the chemical precipitate sedimentary rock is *dolomite*. Because the calcium carbonate in limestone can slowly dissolve in water, limestone tends to be a weak rock in humid environments, but in arid or semiarid climates it tends to be resistant.

Mineral salts that have reached saturation in evaporating seas or lakes will precipitate to form a variety of sedimentary deposits that are useful to humans. These include *gypsum* (used in wallboard), halite (common salt), and *borates*, which are important in hundreds of products such as fertilizer, fiberglass, detergents, and pharmaceuticals.

D. Sack

● **FIGURE 13.14** The White Cliffs of Dover, England. These striking, steep cliffs along the English Channel are made of chalky limestone from the skeletal remains of microscopic marine organisms.

Many types of sediment accumulate in distinct layers, or *strata*, that remain visible after lithification. This layering in rocks is known as **stratification**. A **bedding plane** is the boundary between two sedimentary layers that represent separate depositional events. Bedding planes denote a pause or some change in the nature of deposition, and the strata may differ in grain size, color, or composition (● Fig. 13.15). Where markedly mismatching strata meet along an irregular, eroded surface, that contact between the rocks is an **unconformity**. An unconformity indicates a gap in the section caused by removal of sediment from the top of one layer before deposition of the overlying sediment. Another type of stratification, called **cross bedding**, is characterized by multiple thin sediment layers that accumulated in inclined, rather than horizontal, beds. The compass direction toward which cross beds tilt often reflects the travel direction of the medium that deposited the cross-bedded sediment, whether

FIGURE 13.15 Stratification (layering) in sedimentary rocks exposed along the Book Cliffs, Utah.

What are some characteristics that differ among the various strata seen in this photograph?

waves along a coast, currents in streams, or winds over sand dunes (● Fig. 13.16). All types of stratification provide evidence about the environment in which the sediments were deposited, and changes from one sedimentary layer to the next reflect elements of the local geologic history. For example, if shale representing an offshore environment lies directly on top of sandstone strata deposited on an ancient beach, it indicates that the location was first a beach that the sea later covered.

Sedimentary rocks become jointed, or fractured, when they are subjected to crustal stresses after they lithify. The impressive rock "fins" of Arches National Park, Utah, owe their vertical, tabular shape to vertical joints in great beds of sandstone (● Fig. 13.17). Structures such as bedding planes and joints are important in the development of physical landscapes because they are weak points in the rock that weathering and erosion can attack with relative ease. Often, water travels preferentially along joints, causing faster rock breakdown and removal along the joints than elsewhere in the rocks.

Metamorphic Rocks

Metamorphic means "changed form." Enormous heat and pressure deep in Earth's crust can alter (metamorphose) an existing rock into a new rock type that is completely different from the original by recrystallizing the minerals without creating molten rock matter. In *recrystallization*, mineral grains or their constituent atoms and molecules are compacted and rearranged into a new form that is more stable at the higher temperature and pressure. Compared with the original rocks, the resulting **metamorphic rocks** are typically harder and more compact, have a reoriented crystalline structure, and are more resistant to weathering. There are two major categories of metamorphic rocks, foliated and nonfoliated, distinguished by the presence (foliated) or absence (nonfoliated) of platy surfaces or wavy alignments of light and dark minerals that form during metamorphism.

Metamorphism occurs most commonly where crustal rocks are subjected to great pressures by tectonic processes or deep burial, where rising magma generates heat that modifies the nearby rock, or where rocks are altered by both pressure and heat. With enough heat and pressure, metamorphism causes minerals to recrystallize perpendicular to the applied stress, forming platy flat surfaces or wavy bands known as **foliation** (● Fig. 13.18). For example, some shales change to a hard metamorphic rock *slate*, which exhibits a tendency to break apart, or *cleave*, along parallel planar surfaces that actually

FIGURE 13.16 Cross bedding in sandstone composed of sand dune sediments, Valley of Fire, Nevada. How many different dip angles and directions of cross-bed orientations do you see in this photograph?

● **FIGURE 13.17** Vertical jointing of sandstone in Arches National Park, Utah, is responsible for creating the vertical rock walls, called *fins*, that make up this outcrop. Preferential erosion occurs along the vertical joints so that only the rock that was farthest from the joints remains standing.

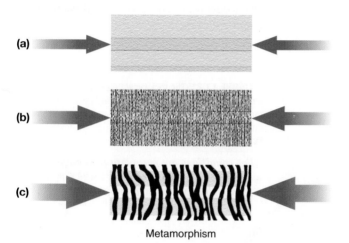

Metamorphism

● **FIGURE 13.18** During metamorphism, applied stress (arrows) can lead to an alignment of minerals that produces foliation. (a) Layered rocks under moderate pressure. (b) At greater pressure, metamorphism can realign minerals perpendicular to the applied stress, creating a platy (elongated) mineral structure and thin foliation layers. (c) With even greater pressure, broader foliation layers can develop as wavy bands of light and dark minerals.

How do foliation layers differ from bedding planes?

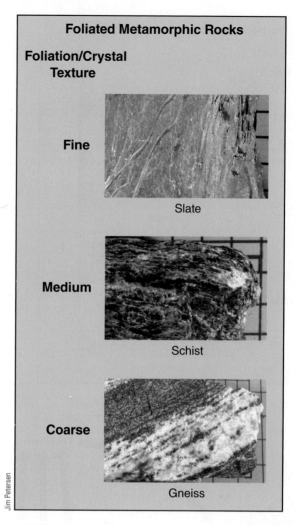

Foliated Metamorphic Rocks

Foliation/Crystal Texture

Fine — Slate

Medium — Schist

Coarse — Gneiss

Jim Petersen

● **FIGURE 13.19** Slate, schist, and gneiss are examples of foliated metamorphic rocks that represent an increase in thickness of foliation planes.

represent extremely thin foliation (● Fig. 13.19). Where foliation layers are moderately thin, individual minerals exhibit a flattened elongate structure, described as platy, and the rocks tend to flake apart along these bands. A common metamorphic rock with thin foliation layers is *schist*. Where the foliation develops into broad mineral bands, the rock is extremely hard and is known as *gneiss* (pronounced "nice"). Coarse-grained rocks, such as granite, generally metamorphose into gneiss, whereas fine-grained rocks tend to become schist.

Preexisting rocks composed of one dominant mineral do not develop foliation when metamorphosed (● Fig. 13.20). Limestone

is metamorphosed into much denser rock *marble*, and impurities in the rock can produce a beautiful variety of colors. Silica–rich sandstones fuse into *quartzite*. Quartzite is brittle, harder than steel, and almost inert chemically. It is virtually immune to chemical weathering and commonly forms cliffs or rugged mountain peaks after the surrounding, less resistant rocks have been removed by erosion.

The Rock Cycle

Many rocks do not remain in their original form indefinitely but instead, over a long time, undergo processes of transformation. The **rock cycle** is a conceptual model for understanding processes that generate, alter, transport, and deposit mineral materials to produce different kinds of rocks (● Fig. 13.21). The term *cycle* emphasizes that existing rocks supply the materials to make new and sometimes very different rocks. Whole existing rocks can be

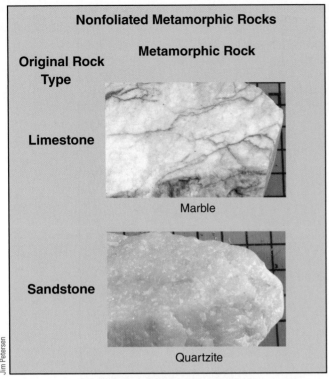

Nonfoliated Metamorphic Rocks

Original Rock Type

Metamorphic Rock

Limestone

Marble

Sandstone

Quartzite

Jim Petersen

● **FIGURE 13.20** Marble and quartzite are nonfoliated metamorphic rocks. They have a recrystallized composition that is harder than the limestone and sandstone from which they are made.

recycled to form new rocks. The geologic age of a rock is based on the time when it assumed its current state; melting, lithification, or metamorphism resets the age of origin.

A complete cycle is shown in the outer circle of Figure 13.21, but as indicated by the arrows that cut across the diagram, rock matter does not have to go through every step of the full rock cycle. For example, after igneous rocks are created by the cooling and crystallizing of magma or lava, they can weather into fragments that lithify into sedimentary rocks. Igneous rocks, however, could also be remelted and crystallized to make new igneous rocks, or they could be changed into metamorphic rocks by heat and pressure. Sedimentary rocks consist of particles derived from any of the three basic rock types. Metamorphic rocks are created by means of heat and pressure changing any preexisting rock—igneous, sedimentary, or metamorphic—into a new rock type. In addition, with sufficient heat, any rock can melt completely into magma that will eventually cool and solidify into an igneous rock. The rock cycle encompasses all the possible pathways for the recycling of rock matter over time.

● **FIGURE 13.21** The rock cycle helps illustrate how igneous, sedimentary, and metamorphic rocks are formed and how rock-forming materials can cycle through different rocks over geologic time. Note the links that bypass some parts of the cycle.

What conditions are necessary to change an igneous rock to a metamorphic rock? Can a metamorphic rock be metamorphosed?

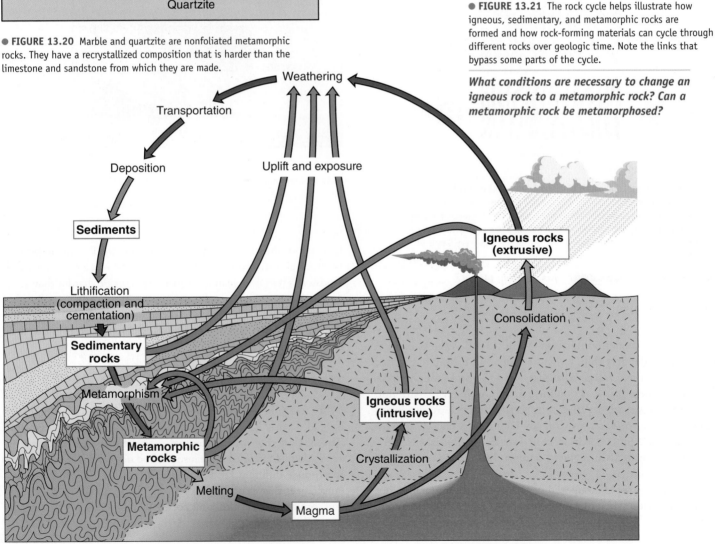

What characteristics can you determine from the terrain pictured here about the past and present physical geography of this location?

Is there one broad theory compatible with the principle of uniformitarianism that can explain how and why Earth's lithospheric processes work? Such a theory would have to explain many diverse phenomena, including the growth of continents, the movement of solid rock that causes earthquakes, the location of great mountain ranges, differing patterns of temperature in the rocks of the seafloor, and violent volcanic eruptions. One theory explains all of these phenomena; that concept, plate tectonics, involves the continual movement of segments of Earth's surface over millions of years.

Sometimes it requires many years to develop, test, and refine a scientific concept to the point where it is more fully understood and widely accepted. As data and information are gathered and analyzed, new methods and technologies can arise and contribute to the process of testing hypotheses via the scientific method, and bit by bit an acceptable explanatory framework emerges. This was the case in the twentieth century with the idea that segments of Earth's outer shell undergo changes in location and orientation over long periods of time. Over the last century, an initial theoretical framework postulating moving continents was eventually replaced by the theory of plate tectonics, which has been well tested using a great deal of evidence collected from the lithosphere. The theory of plate tectonics revolutionized the Earth sciences and our understanding of Earth's history.

Plate Tectonics

Scientists in all disciplines constantly seek broad explanations that account for the detailed facts, recurring patterns, and interrelated processes that they observe and analyze. From the seventeenth to the early nineteenth centuries, many Earth scientists believed that present landscapes were created by a previous period of great and sudden cataclysms. They might have believed, for example, that the Grand Canyon split open one violent day and has remained that way ever since or that the Rocky Mountains appeared overnight. That notion, called **catastrophism**, was largely rejected after about 1830. For almost two centuries, physical geographers, geologists, and other Earth scientists have instead accepted the concept of **uniformitarianism**, which is the idea that internal and external Earth processes operated in the geologic past not by cataclysms unparalleled in recent centuries, but slowly as they do today.

Uniformitarianism does not imply that processes have always operated at the same rate or with equal strength everywhere on Earth. In fact, the spatial variability of our planet's surface features is the result of changing intensities of internal and external processes, influenced by geographic position. These processes have varied in intensity and location throughout Earth's history. Furthermore, regular or episodic changes in the Earth system that may seem relatively small to us can dramatically alter a landscape after progressing, even on an irregular basis, for millions of years.

The Theory of Continental Drift

Most of us have probably noticed on world maps that the Atlantic coasts of South America and Africa look as if they would fit neatly together if we could move them closer to each other. In fact, several widely separated landmasses on Earth appear as though they would fit alongside each other without large gaps or overlaps if we could slide them together (● Fig. 13.22). How can this phenomenon be explained scientifically?

In the early 1900s, Alfred Wegener, a German climatologist, proposed the theory of **continental drift**, the idea that continents and other landmasses have shifted their positions during Earth's history. Wegener's evidence for continental drift included the close fit of continental coastlines on opposite sides of oceans and the trends of mountain ranges that also appear to match across oceans. Wegener also cited comparable geographic patterns of fossils and rock types found on different continents. To explain

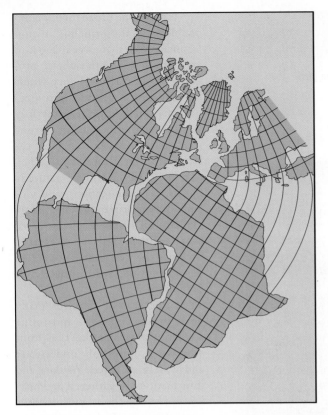

● **FIGURE 13.22** The close fit of the edges of the continents that border the Atlantic Ocean today has been used as geographic evidence supporting the notion of continental drift. The actual fit is even closer if the true continental edges, currently submerged, are used.

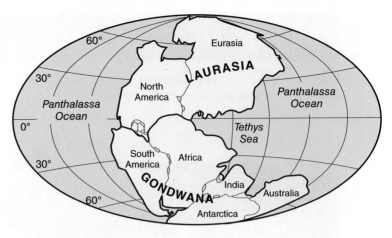

● **FIGURE 13.23** The supercontinent of Pangaea included all of the present major landmasses joined together. Pangaea later split to make Laurasia and Gondwana. Further plate motion left the continents arranged as they are today.

How would continental movement affect the climates of landmasses?

Evidence of Moving Landmasses

It was almost half a century after Wegener first presented his ideas that Earth scientists began to seriously consider the notion of slowly moving continents. Increasingly, new data were being collected and additional observations were made that seemed to require drifting landmasses for adequate explanation.

One field of study that helped initiate this reassessment of Wegener's theory involves measurement of magnetic-field orientations in rocks. Iron-bearing minerals in rock record the magnetic field of Earth as it existed when the rock solidified, which is a phenomenon known as **paleomagnetism**. By about the mid-twentieth century, scientists knew that the positions of the magnetic poles shift, or wander, through time, but they could not account for the confusing range of magnetic-field orientations indicated from iron-bearing rocks. Magnetic-field orientations of rocks of the same age but from different locations did not point toward a single spot on Earth. The position that they indicated for the north magnetic pole ranged widely, and some even pointed toward the present south magnetic pole. The observed variations were more than could be accounted for by the known magnetic polar wandering.

Scientists eventually used the paleomagnetic data in the opposite way—to reconstruct where the sampled rocks would have to have been located relative to a common north magnetic pole. Successful alignment was only possible if the continents had previously occupied different positions than they do today. Using rocks of different age, scientists reconstructed locations of the continents during past periods in geologic history (● Fig. 13.24). The paleomagnetic data revealed that the continents were grouped together about 200 million years ago, just as Wegener's hypothesized two large continents, Gondwana and Laurasia, began to split apart to form the beginnings of the modern Atlantic Ocean. The paleomagnetic data also revealed that the polarity of Earth's magnetic field has reversed many times in the

the spatial distributions of these features, he reasoned that the now-separate continents must have been joined together at some time in the past. Wegener also noted indications of great climate change, such as evidence of former glacial conditions in the Sahara Desert and tropical fossils in Antarctica, that could be explained best by large landmasses moving from one climate zone to another.

Wegener hypothesized that all the continents had once been part of a single supercontinent, which he called *Pangaea*, that later divided into two large landmasses, one in the Southern Hemisphere (Gondwana) and one in the Northern Hemisphere (Laurasia). Laurasia in the Northern Hemisphere consisted of North America, Europe, and Asia. Gondwana in the Southern Hemisphere was made up of South America, Africa, Australia, Antarctica, and India (● Fig. 13.23). Wegener suggested that Gondwana and Laurasia each broke apart later to produce the present continents, which eventually drifted to their current positions by plowing through the ocean.

The reaction of most of the scientific community to Wegener's proposal of drifting continents ranged from skepticism to ridicule. A major objection to his hypothesis was that no one could adequately explain the source of the energy that would be needed to break apart huge continental landmasses and drive them across the ocean over Earth's surface.

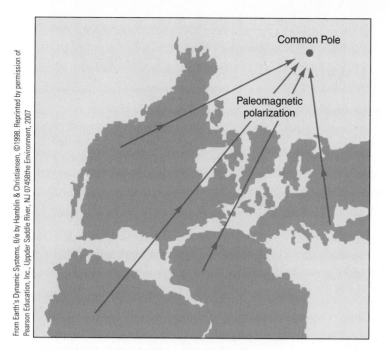

From Earth's Dynamic Systems, 8/e by Hamblin & Christiansen, ©1998. Reprinted by permission of Pearson Education, Inc., Upper Saddle River, NJ 07458the Environment, 2007

● **FIGURE 13.24** Paleomagnetic properties of rocks of the same age indicate the location of magnetic north at that time. Former positions and orientations of the continents are reconstructed by determining where the rocks would have to have originated to point to a common magnetic pole.

past. A record of these polarity reversals is imprinted within the iron-rich basaltic rocks of the seafloor.

Other supporting evidence for crustal movement eventually came from a variety of sources. The widely separated locations having similar reptile and plant fossils found in Australia, India, South Africa, South America, and Antarctica, previously noted by Wegener, received further study. Fossils within both the reptile and plant groups were considered too similar and specialized to have originated in multiple distant locations. When the positions of the continents were reassembled on a paleomap derived from paleomagnetic data and representing the time when the organisms were living, the fossil locations grouped together spatially. Other types of ancient environmental evidence, such as that left by glaciations, also formed logical geographic patterns on reconstructed paleomaps of the continents or the world (● Fig. 13.25).

Improved knowledge of the true, submerged edges of continents, which lie a few hundred meters below sea level, supported the notion of continental drift. The precision with which Earth's landmasses fit together when joined on a paleomap was even better when using the true continental edges rather than their outlines at sea level. Mountain ranges line up and rock ages and types match well on opposite sides of oceans, as Wegener had suggested, when joining landmasses along their true (submerged) continental edges. Refined understanding of the geographic distribution of Earth's environments relative to latitude and climate zones provided additional insight. Evidence that an ancient glaciation occurred simultaneously in India and South Africa while tropical forest climates existed in the northeastern United States and in Great Britain could only be explained by the latitudinal movement

of landmasses, and their locations came together well on paleogeographic reconstructions.

Despite this evidence in support of past changes in position of large landmasses, it remained difficult for many Earth scientists in the mid-twentieth century to accept Wegener's theory. The processes responsible for moving continents, and how they would slide across the ocean basins, remained unexplained.

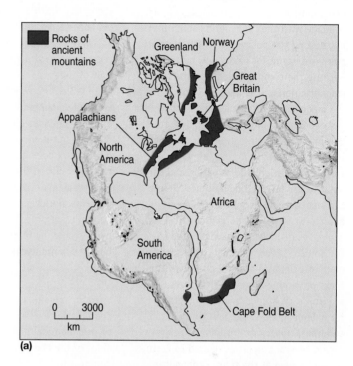

(a)

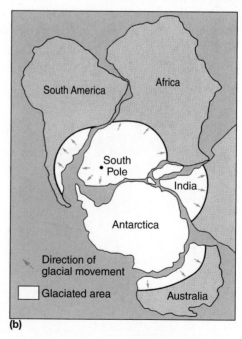

(b)

● **FIGURE 13.25** Multiple types of paleogeographic evidence help reveal the previous locations and distributions of Earth's landmasses in the geologic past. Two sources of this evidence are (a) rocks of ancient mountain ranges, and (b) evidence of ancient glaciations.

Paleomagnetism: Evidence of Earth's Ancient Geography

Earth's magnetic field encircles the globe with field lines that converge at two opposite magnetic poles. The geographic North and South Poles do not coincide with their magnetic counterparts, but beyond the polar regions the magnetic poles are useful for navigating by compass. Because the compass needle points toward magnetic north, it is necessary to adjust for the magnetic declination (angle between the geographic and magnetic poles) to determine geographic direction accurately.

Over geologic time, the intensity of Earth's magnetic field has varied and Earth's polarity has reversed many times. Before the last reversal, about 700,000 years ago (to what we call normal polarity), a compass needle would have pointed to the south. *Paleomagnetism* is the study of magnetic fields in mineral crystals within rocks of varying ages. Basaltic rocks, which are iron rich, are most commonly used for paleomagnetic research. When basalt solidifies, iron oxide crystals in the rock record several properties related to Earth's magnetic field at the time. Radiometric dating and paleomagnetic studies help scientists determine a rock's geographic location when it cooled. Paleomagnetic studies have yielded much evidence of plate tectonics, and they help scientists reconstruct the shifting geographic positions of landmasses during Earth history.

Three important characteristics that these rocks record are polarity (normal, like that of today, or reversed), declination, and inclination, which is measured with a vertically mounted compass needle. Each property provides different evidence about changes in the magnetic field and about how Earth's paleogeography varied as plate tectonics moved the landmasses. Numerous measurements of these three paleomagnetic variables worldwide have given scientists a good picture of Earth's continually changing paleogeography throughout the last several hundred million years.

Polarity

Seafloor spreading was confirmed by polarity changes discovered in stripelike patterns of oceanic basalts that match on opposite sides of the Mid-Atlantic Ridge spreading center, where basalts form. Going farther away from a midoceanic ridge, the rocks are progressively older, and each stripe has a counterpart of the same age and same magnetic polarity on the opposite side of the ridge. The basaltic seafloor has recorded the polarity history of Earth's magnetic field and the widening of the Atlantic Ocean.

Declination

Declination shows the direction to the magnetic pole. By studying basalts of the same age but on several continents, it is possible to triangulate directions to the position of the magnetic north pole at the time the rocks formed. The information provided by these paleodeclinations is the orientation of ancient landmasses—in other words, whether or not they rotated relative to north as they drifted.

Inclination

The magnetic field surrounding Earth causes a magnetic compass needle not only to point north but also to dip downward in a straight-line direction to north. This is called *magnetic dip*, and a needle's angle off of horizontal approximates its latitudinal location. Paleoinclinations recorded from ancient basalts provide the latitude of their location at the time of cooling.

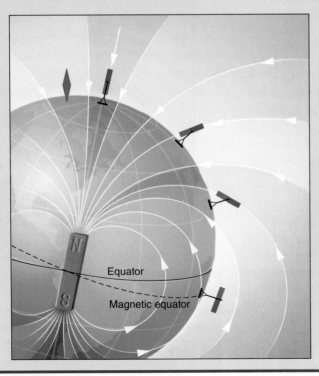

Earth's magnetic field encircles the planet and makes a magnetized dip needle point downward at an angle that equals the latitude of the needle's location. At the equator, the magnetic dip is 0° (horizontal), and at the north magnetic pole the needle points straight down (90°).

Equator

Magnetic equator

Seafloor Spreading and Subduction

In the 1960s, intensive study and mapping of the ocean floor, aided by sonar, radioactive dating of rocks, and improvements in instruments for measuring Earth's magnetism, yielded several key lines of evidence related to traveling landmasses. First, detailed mapping conducted on extensive submarine mountain chains, called **midoceanic ridges**, revealed spatial trends remarkably similar to those of the continental coastlines. Second, it was discovered in the Atlantic and Pacific Oceans that basaltic seafloor displays matching patterns of magnetic properties in rocks of the same age but on opposite sides of midoceanic ridges. Third, scientists made the surprising discovery that although some continental rocks are 3.6 billion years old, rocks on the ocean floor are all geologically young, having been in existence less than 250 million years. Fourth, researchers found that the oldest rocks of the seafloor lie in trenches beneath the deepest ocean waters or close to the continents, and they found that rocks become progressively younger toward the midoceanic ridges, where the youngest basaltic rocks exist (● Fig. 13.26). Finally, it was determined that temperatures of rocks on the ocean floor vary significantly, being hottest near the midoceanic ridges and progressively cooler farther away.

Only one logical explanation emerged to fit all of this ocean-floor evidence. It became apparent that new oceanic crust forms along the midoceanic ridges, where basaltic rocks are youngest, and that the oceanic crust moves slowly in opposite directions away from the axis of a ridge. This phenomenon is called **seafloor spreading** (● Fig. 13.27). Movement of ocean floor in both directions away from a midoceanic ridge is evidenced by the symmetrical pattern of increasing age with distance from the ridge and the symmetrical pattern of paleomagnetism measured in those rocks. As molten basalt cools and crystallizes in the seafloor, the iron minerals record the orientation of Earth's magnetic field at the time. The iron-rich basalts of the seafloor have preserved a historical record of changes in Earth's magnetic field, including **polarity reversals**, when the north and south magnetic poles switch location.

The young age of oceanic crust as a whole, its increasing age with both distance from the midoceanic ridges and proximity to the oceanic trenches, and the occurrence of deep-centered earthquakes along the trenches indicate that the oldest oceanic crust returns to the subsurface at the trenches near the ocean basin margins. **Subduction** is the process by which Earth material moves down to the subsurface in these zones.

These and other related scientific efforts discovered much new evidence that pointed to the horizontal movement of segments of the entire structural lithosphere, including the uppermost mantle, oceanic crust, and continental crust, rather than just the continents as Wegener had suggested. Clearly, a new theory more expansive than Wegener's was needed to account for all of these observations.

Elliot Lim, CIRES & NOAA/NDGC

● **FIGURE 13.26** Basaltic rocks forming the floor of the Atlantic Ocean are youngest (red) along its extensive midoceanic ridge and are progressively older with increasing distance from the ridge axis.

What is the relative age of oceanic crust along the East Coast of the United States?

The Theory of Plate Tectonics

Plate tectonics is the modern, comprehensive theory that explains seafloor spreading, subduction, and the horizontal movement of **lithospheric plates**, large segments of the structural lithosphere. By acknowledging that the entire rigid and brittle lithosphere, rather than just continental crust, is broken into multiple sections that move,

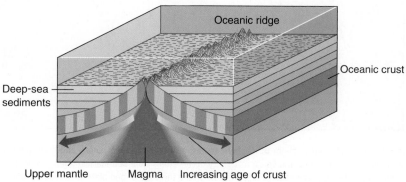

● **FIGURE 13.27** Seafloor spreading at an oceanic ridge produces new oceanic crust.

the theory of plate tectonics allows a plausible explanation, which had eluded Wegener, of the driving force for the movement—convection in the mantle. Plate tectonics is an appropriate name for the theory because the large-scale tectonic forces originating within Earth move, break, and deform the lithospheric plates.

According to the theory, thermal convection currents in Earth's mantle cause the deformable, plastic asthenosphere, near the top of the mantle, to flow. As material in the asthenosphere flows horizontally, it carries along the overlying rigid and brittle lithosphere, causing it to break into lithospheric plates (● Fig. 13.28). Seven major plates have proportions as large as or larger than continents or ocean basins. Five other plates are of minor size, although they have maintained their own identity and direction of movement for some time. Several additional plates are even smaller and exist in active zones at the boundaries between major plates. All major plates consist of both continental and oceanic crust, although the largest, the Pacific plate, is primarily oceanic. In some places, plates pull away from each other (diverge), in other places they push together (converge), and elsewhere they slide alongside each other (move laterally).

The huge thermal convection cells driving plate tectonics consist of hot mantle material traveling upward toward Earth's surface and cooler material returning downward deeper into the mantle (● Fig. 13.29). Where rising mantle material in the convection cells reaches the asthenosphere, it spreads laterally and flows plastically in opposite directions (diverges), pulling apart the rigid overlying lithosphere. This opens a midoceanic ridge, which marks the boundary between two separate plates. Basaltic magma wells up into fractures along the ridge, cooling to form new crust. In this process, the ocean becomes wider by the width of the now-sealed fracture. As the convective motion continues, recently solidified crustal material moves away from the ridges. In a time frame of up to 250 million years, this crustal material reaches the deep trench at the other boundary of the lithospheric plate, where it is recycled into Earth's interior by subduction.

Tectonic Plate Movement

The movement of lithospheric plates relative to one another provides an explanation for many of Earth's surface features. The plate tectonics theory enables physical geographers to better understand not only our planet's ancient geography but also the modern global distributions and spatial relationships among such diverse, but often related, phenomena as earthquakes, volcanic activity, zones of crustal movement, and major landform features (● Fig. 13.30).

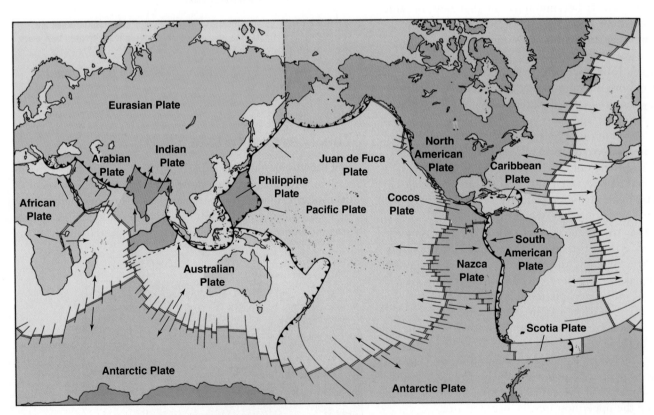

● **FIGURE 13.28** Earth's major lithospheric plates, also called *tectonic plates*, and their general directions of movement. Most tectonic and volcanic activity occurs along plate boundaries, where the large segments separate, collide, or slide past each other. Barbs indicate where the edge of one plate is overriding another.

Does every lithospheric plate include a continent?

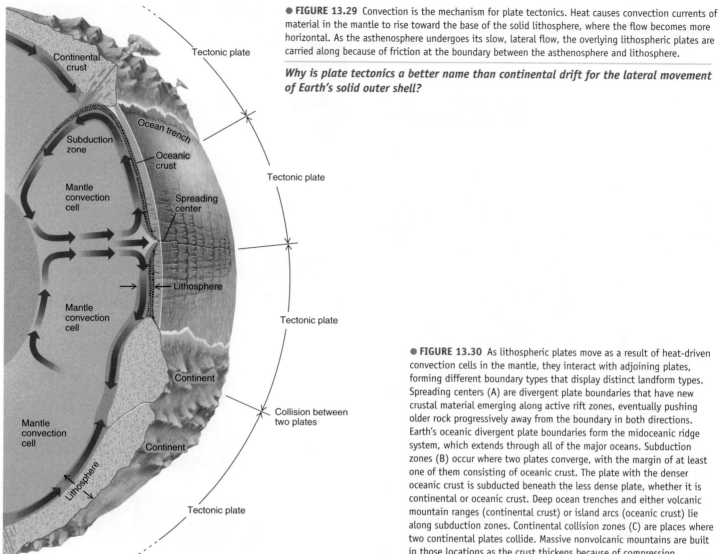

● FIGURE 13.29 Convection is the mechanism for plate tectonics. Heat causes convection currents of material in the mantle to rise toward the base of the solid lithosphere, where the flow becomes more horizontal. As the asthenosphere undergoes its slow, lateral flow, the overlying lithospheric plates are carried along because of friction at the boundary between the asthenosphere and lithosphere.

Why is plate tectonics a better name than continental drift for the lateral movement of Earth's solid outer shell?

● FIGURE 13.30 As lithospheric plates move as a result of heat-driven convection cells in the mantle, they interact with adjoining plates, forming different boundary types that display distinct landform types. Spreading centers (A) are divergent plate boundaries that have new crustal material emerging along active rift zones, eventually pushing older rock progressively away from the boundary in both directions. Earth's oceanic divergent plate boundaries form the midoceanic ridge system, which extends through all of the major oceans. Subduction zones (B) occur where two plates converge, with the margin of at least one of them consisting of oceanic crust. The plate with the denser oceanic crust is subducted beneath the less dense plate, whether it is continental or oceanic crust. Deep ocean trenches and either volcanic mountain ranges (continental crust) or island arcs (oceanic crust) lie along subduction zones. Continental collision zones (C) are places where two continental plates collide. Massive nonvolcanic mountains are built in those locations as the crust thickens because of compression.

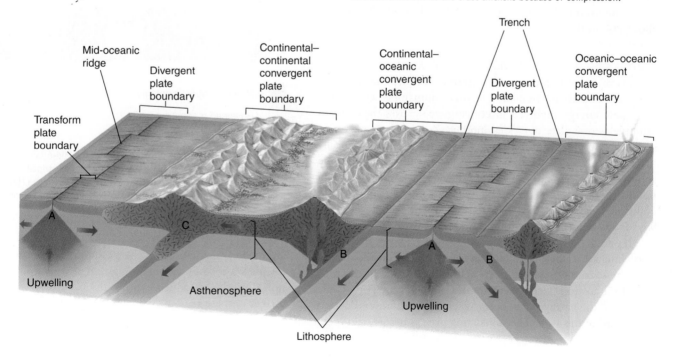

We will next examine the three ways in which lithospheric plates relate to one another along their boundaries as a result of tectonic movement: by pulling apart, pushing together, or sliding alongside each other.

Plate Divergence The pulling apart of lithospheric plates, as occurs with seafloor spreading, is tectonic **plate divergence** (see Figure 13.27). Tectonic forces that act to pull rock masses apart cause the crust to thin and weaken. Shallow earthquakes are often associated with this crustal stretching, and basaltic magma from the mantle wells up along crustal fractures. When oceanic crust is pulled apart, the process creates new ocean floor as the plates move away from each other along a spreading center. The diverging plates separate at an average rate of 2 to 5 centimeters (1 to 2 in.) per year as they are carried along with the flowing plastic asthenosphere in the mantle. The formation of new crust in these spreading centers gives the label *constructive plate margins* to these zones. In some places, volcanoes, like those of Iceland, the Azores, and Tristan da Cunha, mark such boundaries (● Fig. 13.31).

Most plate divergence occurs along oceanic ridges, but this process can also break apart continental crust, eventually reducing the size of the continents involved. The Atlantic Ocean floor formed as the continent that included South America and Africa broke up and moved apart a few centimeters per year over millions of years. The Atlantic Ocean continues to grow today at about the same rate. The best modern example of divergence on a continent is the rift valley system of East Africa, stretching from the Red Sea south to Lake Malawi (● Fig. 13.32). Crustal blocks that have moved downward with respect to the land on either side, with lakes occupying many of the depressions, characterize the entire system, including the Sinai Peninsula and the Dead Sea. Measurable widening of the Red Sea suggests that it may be the beginning of a future ocean that is forming between Africa and the Arabian Peninsula, similar to the young Atlantic between Africa and South America about 200 million years ago.

Plate Convergence A wide variety of crustal activity occurs at areas of tectonic **plate convergence**. Despite the relatively slow rates of plate movement in terms of human perception, the incredible energy involved in convergence causes the crust to crumple as one plate overrides another. Zones where plates are converging mark locations of major, and some of the tectonically more active, landforms on our planet. Deep trenches, volcanic activity, and mountain ranges can develop at convergent plate boundaries, depending on the type of crust involved in the plate collision. The distinctive spatial arrangement of these features worldwide is understood within the framework of plate tectonics.

If one or both margins of a convergent plate boundary consist of oceanic crust, the margin of one plate—always one composed of oceanic crust—is forced deep below the surface in the process of subduction. Deep ocean trenches, such as the Peru–Chile trench and the Japan trench, occur where oceanic crust is dragged downward in this way. The subducting plate is heated, and rocks

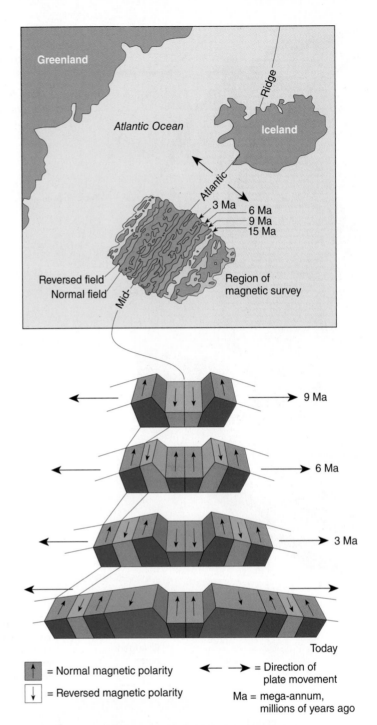

= Normal magnetic polarity

= Reversed magnetic polarity

= Direction of plate movement

Ma = mega-annum, millions of years ago

● **FIGURE 13.31** Iceland represents a part of the Mid-Atlantic Ridge that reaches above sea level to form a volcanic island. The striped pattern of polarity reversals documented in the basaltic rocks along the Mid-Atlantic Ridge helped scientists discover the process of seafloor spreading.

are melted, as the edge of the plate plunges downward into the mantle. As the subducting plate grinds downward, enormous friction is produced, which explains the occurrence of major earthquakes in these regions.

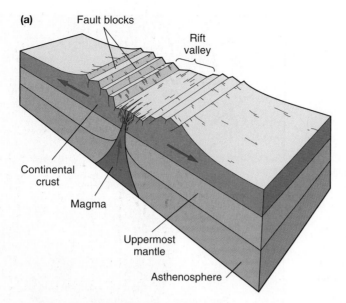

(a)

Fault blocks

Rift valley

Continental crust

Magma

Uppermost mantle

Asthenosphere

● **FIGURE 13.32** (a) A continental divergent plate boundary breaks continents into smaller landmasses. (b) The roughly triangular-shaped Sinai Peninsula, flanked by the Red Sea to the south (lower left), Gulf of Suez on the west (photo center), and Gulf of Aqaba toward the east (lower right), illustrates the breakup of a continental landmass. The Red Sea rift and the narrow Gulf of Aqaba are both zones of spreading.

Mediterranean Sea

ISRAEL

Sinai Peninsula

Gulf of Aqaba

Gulf of Suez

Red Sea

EGYPT

NASA, Johnson Space Center

(b)

Where oceanic crust collides with continental crust, the oceanic crust, which is denser, is subducted beneath the less dense continental crust (● Fig. 13.33). This is the situation along South America's Pacific coast, where the Nazca plate subducts beneath the South American plate, and in Japan, where the Pacific plate dips under the Eurasian plate. As oceanic crust, and the lithospheric plate of which it forms a part, is subducted, it descends into the asthenosphere to be partially melted and recycled into Earth's interior. Often, sediments deposited at continental margins are carried along down into the deep trenches.

As these sediments melt, the resulting magma migrates upward into the overriding plate. Where molten rock reaches the surface, it produces a series of volcanic peaks, as in the Cascade Range of the northwestern United States. Rocks can also be squeezed and contorted between colliding plates, becoming uplifted and greatly deformed or metamorphosed. The great mountain ranges, such as the Andes, are produced at convergent plate margins by these processes.

Where oceanic crust lies on both sides of a convergent plate boundary, the plate with the denser oceanic crust subducts

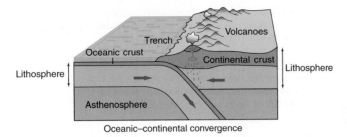

● **FIGURE 13.33** An oceanic–continental convergent plate boundary where continent and seafloor collide. An example is the west coast of South America, where collision has formed the Andes and an offshore deep-ocean trench.

below the other plate. Volcanoes can also develop at this type of boundary, creating major volcanic **island arcs** on the overriding plate. The Aleutians, the Kuriles, and the Marianas are all examples of island arcs lying near oceanic trenches that border the Pacific plate.

Continental crust converging with continental crust is termed **continental collision**, and it causes two continents or major landmasses to fuse or join, creating a new larger landmass (● Fig. 13.34). This process closes an ocean basin that once separated the colliding landmasses, and therefore it has also been called *continental suturing*. The crustal thickening that occurs along this type of plate boundary generally produces major mountain ranges resulting from massive folding and crustal block movement rather than volcanic activity. The Himalayas, the Tibetan Plateau, and other high Eurasian mountain ranges formed in this way as the plate containing the Indian subcontinent collided with Eurasia some 40 million years ago. India is still pushing into Asia today to produce the highest mountains in the world, and the 2015 earthquake in Nepal resulted from crustal block movement associated with that convergence. In a similar fashion, the Alps were created as the African plate was thrust against the Eurasian plate.

Transform Movement Lateral sliding along plate boundaries, called **transform movement**, occurs where plates neither pull apart nor converge but instead slide past each other as they move in opposite directions. Such a boundary exists along the San Andreas Fault zone in California (● Fig. 13.35). Mexico's Baja

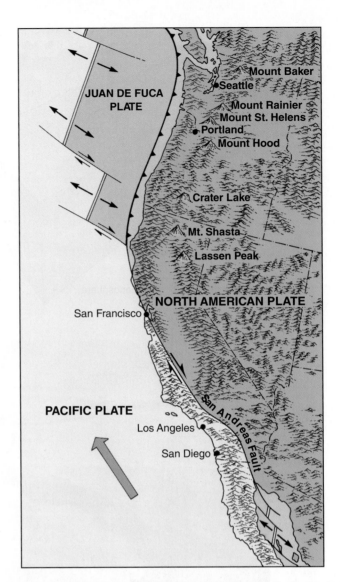

● **FIGURE 13.35** Along this lateral plate boundary, marked by the San Andreas Fault in western North America, the Pacific plate moves northwestward relative to the North American plate. Note that north of San Francisco the boundary type changes.

What boundary type is found north of San Francisco, and what types of surface features indicate this change?

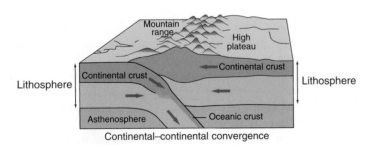

● **FIGURE 13.34** Continental collision along a convergent plate boundary fuses two landmasses together. The Himalayas, the world's highest mountains, were formed in this way when India drifted northward to collide with Asia.

Peninsula and a portion of southern and central California are west of the San Andreas Fault on the Pacific plate. San Francisco and the rest of California east of the fault zone lie on the North American plate. Along the fault zone, the Pacific plate is moving laterally northwestward in relation to the North American plate at a rate of about 8 centimeters (3 in.) a year (80 km or about 50 mi per million years). If movement continues at this rate, Los Angeles will lie alongside San Francisco, 725 kilometers (450 mi) to the northwest, in about 10 million years and eventually pass that city on its way to finally colliding with the Aleutian Islands at a subduction zone.

Another type of lateral plate movement occurs on ocean floors in areas of plate divergence. As plates pull apart, they usually do so along a series of fracture zones that tend to form at right angles to the major zone of plate contact. These crosshatched plate boundaries along which lateral movement takes place are *transform faults*. Transform faults, or fracture zones, are common along mid-oceanic ridges, but examples can also be seen elsewhere, as on the seafloor offshore from the Pacific Northwest coast between the Pacific and Juan de Fuca plates (see Figure 13.35). Transform faults result when adjacent plates travel at variable rates, causing lateral movement of one plate relative to the other. The most rapid plate motion is on the East Pacific rise where the rate of movement is more than 17 centimeters (7 in.) per year.

Hot Spots in the Mantle

The Hawaiian Islands, like many major landform features, owe their existence to processes associated with plate tectonics. As the Pacific plate in that region moves toward the northwest, it passes over a mass of molten rock in the mantle that does not move with the lithospheric plate. Called **hot spots**, these almost stationary molten masses occur in a few other places in both continental and oceanic locations. Melting of the upper mantle and oceanic crust causes undersea eruptions and the outpouring of basaltic lava on the seafloor, eventually constructing a volcanic island. This process

is responsible for building the Hawaiian Islands as well as the chain of islands and undersea volcanoes that extends for thousands of kilometers northwest of Hawaii. Today, this hot spot causes active volcanic eruptions on the island of Hawaii. The other islands in the Hawaiian chain came from a similar origin, having formed over the hot spot as well, but these volcanoes have now drifted along with the Pacific plate away from their magmatic source. Evidence of the plate motion is indicated by the fact that the youngest islands of the Hawaiian chain, Hawaii and Maui, are to the southeast, and the older islands, such as Kauai and Oahu, are located to the northwest (● Fig. 13.36). A newly forming undersea volcano, named Loihi, is now developing southeast of the island of Hawaii and will someday be the next island member of the Hawaiian chain.

Growth of Continents

The origin of continents is still being debated. It is clear that the continents tend to have a core area of very old igneous and metamorphic rocks that are interpreted as representing the deeply eroded roots of ancient mountains. These core regions have been worn down by hundreds of millions of years of erosion to create areas of relatively low relief that are located far from active plate boundaries. As a result, they have a history

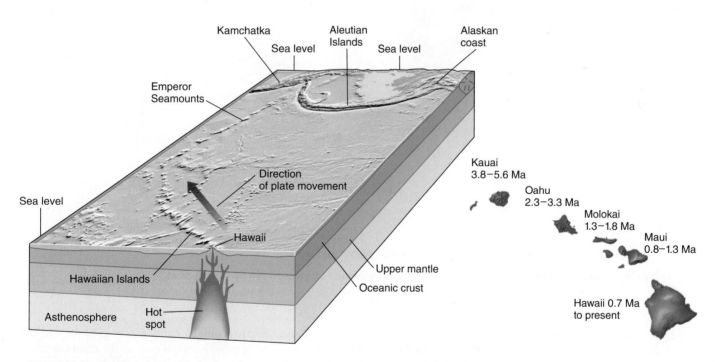

● **FIGURE 13.36** Over the last few million years, a stationary zone of molten material in the mantle—a hot spot—created each of the volcanic Hawaiian Islands in succession. Because they move to the northwest with the Pacific plate, the islands are progressively older to the northwest (ages are in millions of years, abbreviated Ma). The hot spot is currently located at the island of Hawaii, which is about 300 kilometers (185 mi) from Oahu.

Approximately how long did it take the Pacific plate to move Oahu from the hot spot to its current position?

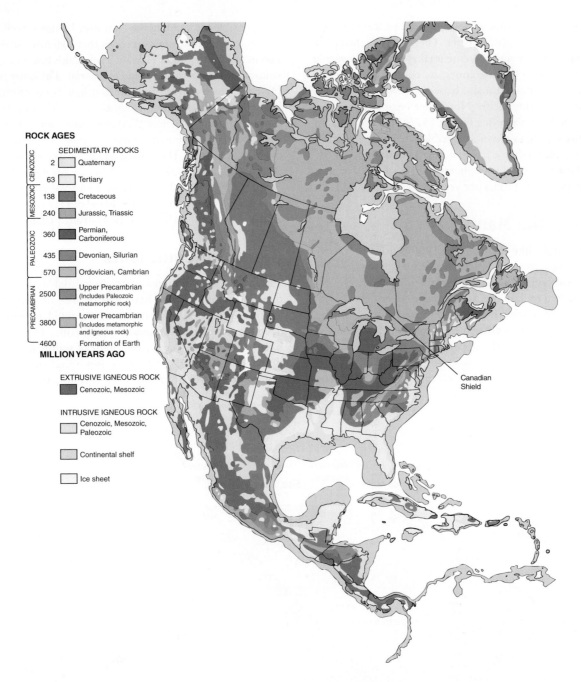

ROCK AGES

SEDIMENTARY ROCKS

CENOZOIC	2		Quaternary
	63		Tertiary
MESOZOIC	138		Cretaceous
	240		Jurassic, Triassic
PALEOZOIC	360		Permian, Carboniferous
	435		Devonian, Silurian
	570		Ordovician, Cambrian
PRECAMBRIAN	2500		Upper Precambrian (Includes Paleozoic metamorphic rock)
	3800		Lower Precambrian (Includes metamorphic and igneous rock)
	4600		Formation of Earth

MILLION YEARS AGO

EXTRUSIVE IGNEOUS ROCK
Cenozoic, Mesozoic

INTRUSIVE IGNEOUS ROCK
Cenozoic, Mesozoic, Paleozoic

Continental shelf

Ice sheet

Canadian Shield

● **FIGURE 13.37** Map of North America showing the continental shield (Canadian shield) and the general ages of rocks.

Going outward from the shield toward the coast, what generally happens to the age of the rocks?

of tectonic stability over an immense period of time. These ancient crystalline rock areas are known as **continental shields** (● Fig. 13.37). The Canadian, Scandinavian, and Siberian shields are outstanding examples. Around the peripheries of the exposed shields, flat-lying, younger sedimentary rocks at the surface indicate the presence of a stable and rigid rock mass below, as in the American Midwest, western Siberia, and much of Africa.

Most Earth scientists consider continents to grow by *accretion*, that is, by adding numerous chunks of crust to the main continent by collision. Western North America grew in this manner over the past 200 million years by adding segments of crust, known as *microplate terranes*, as it moved westward over the Pacific plate and previous oceanic plates (● Fig. 13.38). Paleomagnetic data show that parts of western North America from Alaska to California originated south of the equator and moved to join the continent.

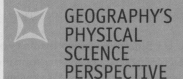
Isostasy: Balancing Earth's Lithosphere

Structurally, the solid uppermost mantle, oceanic crust, and continental crust constitute the rigid and brittle lithosphere, which rests on top of the plastic, deformable asthenosphere. Mantle material in the asthenosphere flows like a very thick fluid at a rate of about 2 to 5 centimeters (1 to 2 in.) per year. The lithosphere is broken into several plates (segments) that behave like rafts moving along with currents in the flowing asthenosphere. The plates float because material in the lithosphere is less dense than material in the asthenosphere.

Because of the principle of buoyancy, an object will float in a fluid as long as its total mass is less than that of the fluid it displaces. The volume of water displaced by a floating object is the amount that has the same total mass as the object. The difference in density (mass per unit volume) between the object and the fluid is represented by the proportion of the object that floats above the surface. An iceberg having 90% of the density of ocean water floats with 10% of the iceberg extending above the water surface. As long as the average density of a cargo ship is less than that of the water, a balance (equilibrium) will be maintained and the ship will float. If the ship takes on so much cargo that its average density exceeds that of the water, the ship will sink.

Isostasy is the term for a similar concept regarding the equalization of hydrostatic pressure (fluid balance) between the lithosphere and asthenosphere. Isostasy suggests that a column of lithosphere (and the overlying hydrosphere) anywhere on Earth weighs about the same as a column of equal diameter from anywhere else, regardless of vertical thickness. The lithosphere is thicker (taller and deeper) where it contains a high percentage of low-density materials and thinner where it contains more high-density materials. Oceanic crust is thinner than continental crust because oceanic crust has a higher density than continental crust.

If an additional load is placed on an area of Earth's surface by a massive accumulation of glacial ice, lake water, or sediments, the lithosphere there will subside in a process called *isostatic depression* until it attains a new equilibrium level. If the surface accumulation is later removed, the region will tend to rise in a process called *isostatic rebound*. Neither subsidence nor rebound of the lithosphere will be instantaneous because flow in the asthenosphere is only a few centimeters per year.

Isostasy suggests that mountains are made of relatively low-density crustal materials and thus exist in areas of very thick crust, whereas regions of low elevation have thin, high-density crust. Similarly, a tall iceberg requires a massive amount of ice below the surface to expose ice so high above sea level, and as ice above the surface melts, ice from below will rise above sea level to replace it until the iceberg has completely melted.

Isostatic balance helps explain many aspects of Earth's surface, including the following:

- Why most of the continental crust lies above sea level.

- Why wide areas of the seafloor are at a uniform depth.

- Why many mountain ranges continue to rise even though erosion removes material from them.

- Why some regions where rivers are depositing great amounts of sediments are subsiding.

- Why the crust subsided in areas that were covered by thick accumulations of ice during the last glacial age and now continues to rebound after deglaciation.

The density of ice is 90% that of water, thus icebergs (and ice cubes) float with 90% of their volume below the water surface and 10% above it.

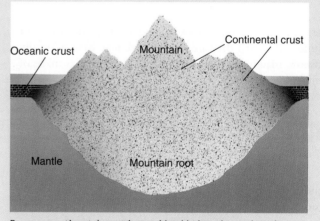

Because continental crust is considerably less dense than the material in the asthenosphere, where continental crust reaches high elevations it also extends far below the surface. Oceanic crust is also less dense than mantle material, but because it is denser than continental crust, it is thinner than continental crust.

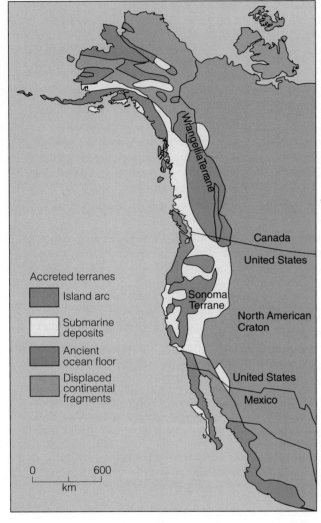

● **FIGURE 13.38** Location of various accretion terranes in western North America.

Terranes, which have their own distinct geology from that of the continent to which they are now joined, may have originally been offshore island arcs, undersea volcanoes, or islands made of continental fragments.

Geologic Time and Paleogeography

The study of past geographic environments is known as **paleogeography**. The goal of paleogeography is to try to reconstruct the past environment of a geographic region based on geologic and climatic evidence. For students of physical geography, it may seem that the present is complex enough without trying to know what the geography of ancient times was like. However,

peering into the past helps us forecast and prepare for changes in the future.

The immensity of geologic time over which major events or processes (such as plate tectonics, ice ages, or the formation and erosion of mountain ranges) have taken place is difficult to picture in a human time frame of days, months, and years. The geologic timescale is a calendar of Earth history (● Fig. 13.39). It is divided into *eras*, which are long units of time, such as the Mesozoic Era (*meso*, middle; *zoic*, life), and eras are divided into *periods*, such as the Cretaceous Period. *Epochs*—for example, the Pleistocene Epoch (recent ice ages)—are shorter time units and are used to subdivide the periods of the Cenozoic Era (*ceno*, recent), for which geologic evidence is more abundant. Today, we are in the Holocene Epoch (last 10,000 years) of the Quaternary Period (last 2.6 million years) of the Cenozoic Era (last 65 million years). In a sense, these divisions are used like we would use days, months, and years to record time.

If a 24–hour day is used to represent the approximately 4.6 billion years of Earth history, the Precambrian, an era of which we know very little, would consume the first 21 hours. The current geologic period, the Quaternary, which has lasted about 2.6 million years, would occupy less than 30 seconds, and the time of humans, beginning about 4 million years ago, would be present for about 1 minute.

Each era, period, and epoch in Earth's geologic history had a unique paleogeography with its own distribution of land and sea, climate regions, plants, and animal life. If we look at evidence for the paleogeography of the Mesozoic Era (248 million to 65 million years ago), for instance, we would find a much different physical geography than what exists now. This was a time when the large continents of Gondwana and Laurasia each gradually split apart as new ocean floors widened, creating the continents that are familiar to us today. Global and local Mesozoic climates were very different from those of today but were changing as North America drifted to the northwest. During the Cretaceous Period, much of the present United States experienced warmer climates than now. Ferns and conifer forests were common. The Mesozoic Era was the age of the dinosaurs, but other life also thrived, such as marine plants and invertebrates, reptiles, insects, mammals, and the earliest birds.

The Mesozoic Era ended with a great extinction event that included the end of the dinosaurs and virtually all large terrestrial animals. Geologists, paleontologists, and paleogeographers are still investigating the cause of this great extinction. Some of the strongest evidence points to a large meteorite striking Earth 65 million years ago, disrupting global climate and causing global environmental change. Other evidence points to plate tectonic changes leading to increased volcanic activity, which could cause climate change that might trigger or contribute to a mass extinction.

Available maps depicting Earth in early geologic times show only approximate and generalized patterns of mountains, plains, coasts, and oceans, with the addition of some environmental characteristics. These maps portray a general picture of how global geography has changed through geologic time (● Fig. 13.40). Much of the evidence and the rocks that bear this

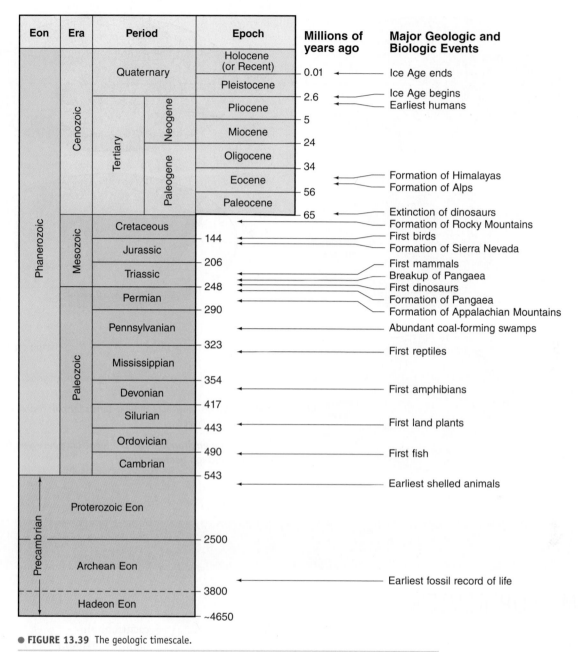

Eon	Era	Period		Epoch	Millions of years ago	Major Geologic and Biologic Events
Phanerozoic	Cenozoic	Quaternary		Holocene (or Recent)	0.01	Ice Age ends
				Pleistocene	2.6	Ice Age begins / Earliest humans
		Tertiary	Neogene	Pliocene	5	
				Miocene	24	
			Paleogene	Oligocene	34	
				Eocene	56	Formation of Himalayas / Formation of Alps
				Paleocene	65	
	Mesozoic	Cretaceous			144	Extinction of dinosaurs / Formation of Rocky Mountains / First birds / Formation of Sierra Nevada
		Jurassic			206	First mammals
		Triassic			248	Breakup of Pangaea / First dinosaurs / Formation of Pangaea / Formation of Appalachian Mountains
	Paleozoic	Permian			290	
		Pennsylvanian			323	Abundant coal-forming swamps
		Mississippian			354	First reptiles
		Devonian			417	First amphibians
		Silurian			443	First land plants
		Ordovician			490	First fish
		Cambrian			543	Earliest shelled animals
Precambrian		Proterozoic Eon			2500	
		Archean Eon			3800	Earliest fossil record of life
		Hadeon Eon			~4650	

● **FIGURE 13.39** The geologic timescale.

Geologists consider 90 million years ago to be within which geologic era and period?

information have been lost through metamorphism or erosion, buried under younger sediments or lava flows, or recycled into Earth's interior. The further back in time, the sketchier is the paleoenvironmental information presented on the map. Paleomaps, like other maps, are simplified models of the regions and times they represent.

As time passes and additional evidence is collected, paleogeographers may be able to fill in more of the empty spaces on those maps of the past that are so unfamiliar to us. These paleogeographic studies aim not only at understanding the past but also at understanding today's environments and physical landscapes, how they have developed, and how processes act to change them. By applying the theory of plate tectonics and the concept of uniformitarianism to our knowledge of how the Earth system and its subsystems function, we can gain a better understanding of our planet's geologic past, as well as its present, and this will facilitate better forecasts of its potential future.

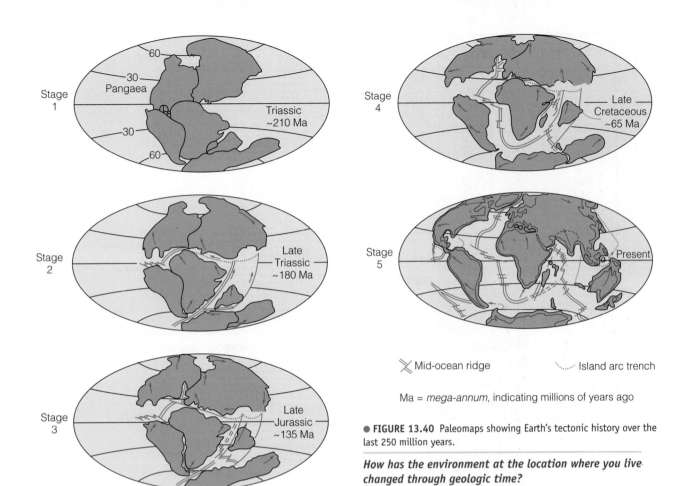

Stage 1 Pangaea 60 30 30 60 Triassic ~210 Ma

Stage 2 Late Triassic ~180 Ma

Stage 3 Late Jurassic ~135 Ma

Stage 4 Late Cretaceous ~65 Ma

Stage 5 Present

⋈ Mid-ocean ridge   ⋯⋯ Island arc trench

Ma = *mega-annum*, indicating millions of years ago

● **FIGURE 13.40** Paleomaps showing Earth's tectonic history over the last 250 million years.

How has the environment at the location where you live changed through geologic time?

CHAPTER 13 ACTIVITIES

■ TERMS FOR REVIEW

seismic wave
seismograph
core
inner core
outer core
mantle
Mohorovičić discontinuity (Moho)
crust
oceanic crust
continental crust
elastic solid
lithosphere (planetary structure)
asthenosphere
plastic solid
tectonic force
bedrock
outcrop
mineral
rock

igneous rock
magma
lava
extrusive igneous rock
pyroclastics (tephra)
intrusive igneous rock
joint
sedimentary rock
clast
clastic sedimentary rock
organic sedimentary rock
chemical precipitate sedimentary rock
stratification
bedding plane
unconformity
cross bedding
metamorphic rock
foliation
rock cycle

catastrophism
uniformitarianism
continental drift
paleomagnetism
midoceanic ridge
seafloor spreading
polarity reversal
subduction
plate tectonics
lithospheric plate
plate divergence
plate convergence
island arc
continental collision
transform movement
hot spot
continental shield
paleogeography

QUESTIONS FOR REVIEW

1. Name the major zones of Earth's interior from the center to the surface. How do these zones differ from one another?
2. Define and distinguish (a) continental crust and oceanic crust, and (b) the lithosphere and asthenosphere.
3. How do minerals differ from rocks? Provide examples of each.
4. Describe the three major categories of rock and the principal means by which each is formed. Give an example of each.
5. What is the rock cycle?
6. What evidence did Wegener rely on in the formulation of his theory of continental drift? What evidence did he lack?
7. Compare and contrast the theory of continental drift and the theory of plate tectonics.
8. What type of lithospheric plate boundary is found (a) paralleling the Andes, (b) at the San Andreas Fault, (c) in Iceland, and (d) near the Himalayas?
9. How does the formation of the Hawaiian Islands support plate tectonic theory?
10. Define paleogeography. Why are geographers interested in this topic?

CONSIDER AND RESPOND

1. What evidence has been found to support the theory that lithospheric plates move around on Earth's surface?
2. How is the concept of uniformitarianism related to plate tectonics theory and the arrangement of the continents today?
3. Explain why the eastern portion of the United States has relatively little tectonic activity compared with the western portion of the United States.

PRACTICAL APPLICATIONS

1. Two plates are diverging and both are moving at a rate of 3 centimeters per year. How long will it take for them to move 100 kilometers apart?
2. An area of oceanic crust has a density of 3.0 grams per cubic centimeter and a thickness of 4 kilometers. The same size area of continental crust has a density of 2.7 grams per cubic centimeter. How thick in kilometers would the continental crust have to be for its total mass to equal that of the oceanic crust?

 MindTap—Make the most of your study time by accessing everything you need to succeed in one place. Read your textbook, take notes, review flashcards, watch videos, complete activities, take practice quizzes, and more online with MindTap. Log in at **www.cengagebrain.com**.

TECTONISM AND VOLCANISM

14

■ OBJECTIVES

WHEN YOU COMPLETE THIS CHAPTER YOU SHOULD BE ABLE TO:

- 14.1 Distinguish the endogenic from the exogenic landforming processes.
- 14.2 Demonstrate how compressional, tensional, and shearing forces each stress rock matter.
- 14.3 Sketch examples of the different types of faults, indicating direction of motion.
- 14.4 Associate the different types of faults with the type of tectonic force responsible them.
- 14.5 Draw a cross section of a fault that shows the relationship between an earthquake's focus and epicenter.
- 14.6 Discuss the two ways in which the severity of an earthquake is measured.
- 14.7 List various factors that help determine the devastation caused by an earthquake.
- 14.8 Differentiate the various types of igneous intrusions.
- 14.9 Explain why some volcanic eruptions are explosive whereas others are effusive.
- 14.10 Summarize the principal differences among the six major types of volcanic landforms.
- 14.11 Account for the general, global distribution of active volcanoes and earthquake-prone regions.

OUR PLANET'S SURFACE **TOPOGRAPHY,** THE distribution of landscape highs and lows, is complex and intriguing. Landscapes can consist of rugged mountains, gently sloping plains, rolling hills and valleys, or elevated plateaus cut by steep canyons. These are just a few examples of the types of surface terrain features, referred to as **landforms**, that contribute to the beauty and diversity of Earth's environments. Landforms are one of the most appealing and impressive elements of Earth's surface. Each year, local, state, and national parks attract millions of visitors who seek to observe and experience firsthand spectacular examples of landforms. Landforms owe their existence and development to processes and materials that originate in Earth's interior, at its surface, or, most typically, some combination of both.

Understanding landforms and landscapes—how they originate, why they vary, and their significance in a local, regional, or global context—is the primary goal of **geomorphology**, a major subfield of physical geography devoted to the scientific study of landforms. Geomorphologists seek explanations for the origin, shape, and spatial distribution of terrain features of all kinds and for the processes that modify

◄ This small volcanic hill, called a spatter cone, on the surface of lava flows from Kilauea in Hawaii, ejects molten lava much like a miniature volcano.   USGS/HVO/Tim Orr

and destroy them. Landforms represent the interplay of one set of processes that elevate, depress, or disrupt Earth's surface, creating topographic inequalities, and another set of processes that wear down, fill in, and work to level the landscape.

Landforming *tectonic processes* (from Greek: *tekton*, carpenter or builder), which are movements of parts of the crust and upper mantle, and *igneous processes* (from Latin: *ignis*, fire), which are related to the eruption and solidification of molten rock matter, constitute the primary geomorphic mechanisms that increase the topographic irregularities on Earth's surface. These tectonic and igneous forces originate in the planet's interior, yet they contribute significantly to the nature of Earth's exterior topography, affecting the composition, shape, size, and appearance of surface terrain features in many regions. Areas of the crust can be uplifted or downdropped by tectonism. Parts of Earth's surface can also be built up or depressed because of intrusion or ejection of molten rock matter from Earth's interior by igneous processes. Tectonic and igneous processes build extensive mountain systems and a great variety of other landforms. The geographic distribution of these terrain features, moreover, is not random. The greatest tectonic forces, most dangerous earthquake zones, and largest mountain ranges lie along lithospheric plate boundaries, and plate margins are the most common locations for extensive volcanic landforms.

Tectonic and igneous processes have produced many impressively scenic landscapes, but they also present serious natural hazards to people and their property. This chapter and those that follow focus on understanding how various landforms and landscapes develop and on the potential hazards that are related to them. It is extremely important to understand how geomorphic processes work to shape Earth's surface landforms because they are active, ongoing, and often powerful processes that affect human welfare. Landforms are a fundamental, dynamic, beautiful, diverse, and sometimes dangerous aspect of the human habitat.

Introduction to Geomorphology

A fundamental characteristic of all landforms and landscapes is their relative amount of **relief**, which is the difference in elevation between the highest and lowest points within a specified area or on a particular surface feature (● Fig. 14.1). With no variations in relief, our planet would be a smooth, featureless sphere and certainly much less interesting. It is hard to imagine Earth without dramatic terrain as seen in the high-relief mountainous regions of the Himalayas, Alps, Andes, Rockies, and Appalachians or in the huge chasm that we call the Grand Canyon. Interspersed with high-relief features, large expanses of low-relief features, like the Great Plains, are equally impressive and inspiring.

Earth's landforms result from mechanisms that act to increase relief by raising or lowering the land surface and from mechanisms that work to reduce relief by removing rock matter from high places and using it to fill in depressions. Geomorphic processes that

(a)

(b)

● **FIGURE 14.1** An area of (a) low relief in Selawik National Wildlife Refuge, Alaska, and (b) high relief in the Sierra Nevada of California.

layer, the **dip**, is always measured at right angles to the strike and in degrees of angle from the horizontal (0° dip = horizontal). The direction toward which the rock dips down is expressed with the general compass direction. For example, a rock layer that strikes northeast and is inclined 11° from the horizontal down to the southeast would be described as having a dip of 11° to the southeast (see Fig. 14.5).

Earth's crust has been subjected to tectonic forces throughout its history, although the forces have been greater during some geologic periods than others and have varied widely over Earth's surface. Most of the resulting changes in the crust have occurred slowly over hundreds of thousands or millions of years, but others have been rapid and cataclysmic. The response of crustal rocks to tectonic forces yields a variety of configurations in rock structure, depending on the nature of the rocks and the nature of the applied forces.

Tectonic forces are divided into three principal types that differ in the direction of the applied forces (● Fig. 14.6). **Compressional tectonic forces** push crustal rocks together. **Tensional tectonic forces** pull parts of the crust away from each other. **Shearing tectonic forces** slide parts of Earth's crust past each other.

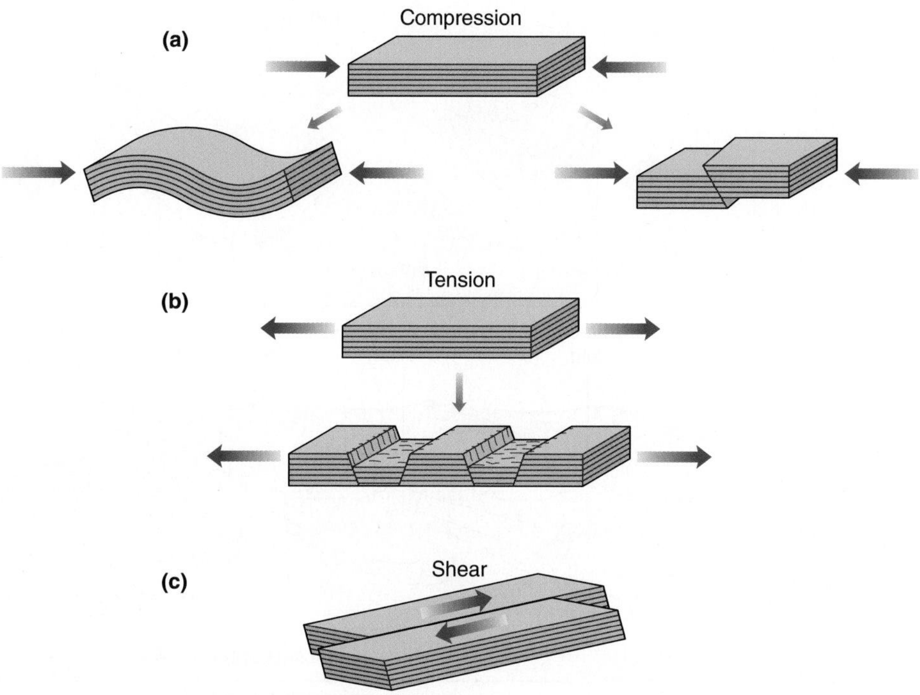

Compressional Forces

Tectonic forces that push two areas of crustal rocks together tend to shorten and thicken the crust. How the affected rocks respond to compressional forces depends on how brittle (breakable) the rocks are and the speed with which the forces are applied. **Folding**, which is a bending or crumpling of rock layers, occurs when compressional forces are applied to rocks that are *ductile* (bendable) rather than brittle. Rocks that lie deep within the crust and that are therefore under high pressure are generally ductile and particularly susceptible to behaving plastically, that is, to deforming without breaking. As a result, rocks deep within the crust typically fold rather than break in response to compressional forces (● Fig. 14.7). Folding is also more likely than fracturing when the compressional forces are applied slowly. Eventually, however, if the force per unit area, the *stress*, is great enough, the rocks can break, with one section pushed up over another.

As elements of rock structure, upfolds are called **anticlines** and downfolds are called **synclines** (● Fig. 14.8). Rock layers that form the flanks of anticlinal crests and synclinal troughs are the *fold limbs*. Folds in some rock layers are very small, occupying

● **FIGURE 14.6** Three types (directions) of tectonic force cause deformation of rock layers. (a) Compressional forces push rocks together and can bend (fold) rocks or cause rocks to break and slide along the breakage zone (fault). (b) Tensional forces pull rocks apart and may also lead to the breaking and shifting of rock masses along faults. (c) Shearing forces slide rocks past each other horizontally, and this likewise can cause movement along a fault.

When you stretch a rubber band, are you subjecting it to compressional, tensional, or shearing forces?

J. Petersen

● **FIGURE 14.7** Compressional forces have made complex folds in these layers of sedimentary rock.

How can solid rock be folded without breaking?

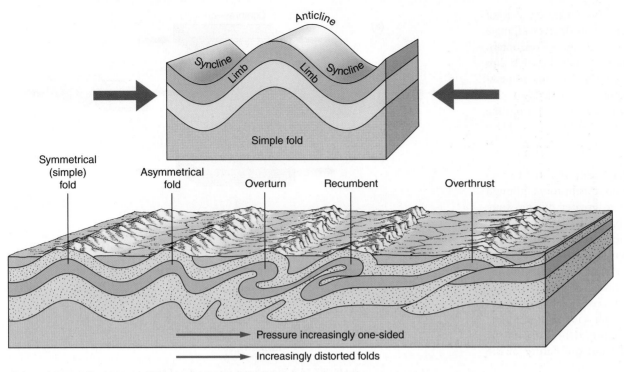

Anticline

Syncline
Limb
Limb
Syncline

Simple fold

Symmetrical
(simple)
fold

Asymmetrical
fold

Overturn

Recumbent

Overthrust

Pressure increasingly one-sided

Increasingly distorted folds

● **FIGURE 14.8** Folded rock structures become increasingly complex as the applied compressional forces become more unequal from the two directions.

a few centimeters, whereas others are enormous with vertical distances between the upfolds and downfolds measured in kilometers. Folds can be tight or broad, symmetrical or asymmetrical. Folds are symmetrical if the two limbs have approximately the same dip angle, which results from compressional forces that were relatively equal from both sides. If compressional forces were stronger from one side, the fold may be asymmetrical, with the dip of one limb being much steeper than that of the other. Eventually, asymmetrically folded rocks can become overturned and perhaps so compressed that the fold lies horizontally; these are known as *recumbent folds* (see Fig. 14.8).

Almost all mountain systems exhibit some degree of folding. Much of the Appalachian Mountain system is an example of folding on a large scale. Spectacular folds exist in the Rocky Mountains of Colorado, Wyoming, and Montana, as well as in the Canadian Rockies. Highly complex folding produced the European Alps, where folds are overturned, sheared off, and piled on top of one another.

Rock layers that are near Earth's surface and not under high confining pressures are too rigid to bend into folds when experiencing compressional forces. If the tectonic force is large enough, or if it is applied rapidly, these rocks will break rather than bend and the rock masses will move relative to each other along the fracture. *Faulting* is the slippage or displacement of rocks along a fracture surface, and the fracture along which movement has occurred is a **fault**.

When compressional forces cause faulting, either one mass of rock is pushed up relative to the other along a steep-angled fault or one mass of rock slides over the other along a shallow, low-angle fault. The steep, high-angle fault resulting from compressional forces

is termed a **reverse fault** (● Fig. 14.9a). Where compression pushes a mass of rock along a low-angle fault so that it overrides rocks on the other side of the fault, the fracture surface is called a **thrust fault**, and the shallow displacement is an **overthrust** (Fig. 14.9b). Major overthrusts exist in the northern Rocky Mountains and in the southern Appalachians, and both the 2011 earthquake in Japan and the 2015 earthquake in Nepal resulted from movement along thrust faults.

In both reverse and thrust faults, one block of crustal rocks is wedged up relative to the other. Direction of motion along faults is always given in relative terms because even though it might seem obvious that one block was pushed up along the fault, the other block might have slid down some distance as well, and it is not always possible to determine with certainty if one or both blocks moved.

In some cases, reverse or thrust faulting resulting from the rapid application of compressional forces affects rocks that have already responded to the force by folding. Where folded rocks undergo thrust faulting, the upper part of a fold breaks, sliding over the lower rock layers along the low-angle fault forming an overthrust. Together, recumbent folds and overthrusts are important rock structures that have formed in the Andes, Alps, Himalayas, and other complex mountain ranges.

Tensional Forces

Tensional tectonic forces pull in opposite directions in a way that stretches and thins the impacted part of the crust. Rocks, however, typically respond to tensional forces by faulting, rather than bending or stretching plastically. Tensional forces commonly cause the crust to break into discrete blocks, called **fault blocks**, that are separated

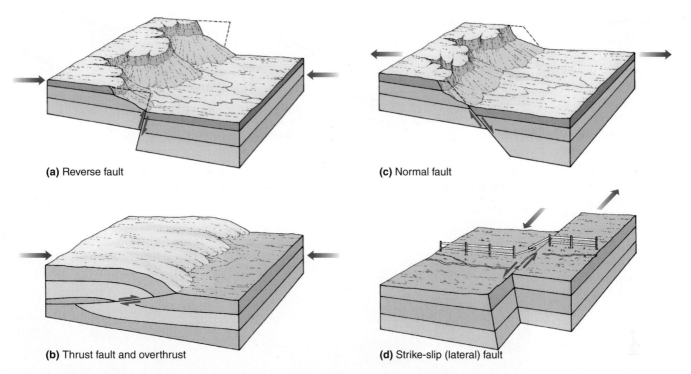

(a) Reverse fault

(c) Normal fault

(b) Thrust fault and overthrust

(d) Strike-slip (lateral) fault

● **FIGURE 14.9** The major types of faults and the tectonic forces that cause them (indicated by large arrows). Compressional forces form (a) reverse or (b) thrust faults. (c) Tensional tectonic forces break rocks along normal faults. (d) Shearing forces move rocks horizontally past each other along strike-slip (lateral) faults.

How does motion along a normal fault differ from that along a reverse fault?

from each other by **normal faults** (Fig. 14.9c). To accommodate the extension of the crust, one crustal fault block slides downward along the normal fault relative to the adjacent fault block. Notice that the direction of motion along a normal fault is opposite to that along a reverse or thrust fault (see Fig. 14.9a).

In map view, tensional forces affecting a large region produce a repeated pattern of roughly parallel normal faults, creating a series of alternating downdropped and upthrown fault blocks. Each block that slid downward between two normal faults, or that remained in place while blocks on either side slid upward along the faults, is called a **graben** (● Fig. 14.10). A fault block that moved relatively upward between two normal faults—that is, it actually moved up or it remained in place while adjacent blocks slid downward—is a **horst**. The great Ruwenzori Range of East Africa is a horst, as is the Sinai Peninsula between the fault troughs in the Gulfs of Suez and Aqaba (see again Fig. 13.32). Horsts and grabens are rock structural features that are identified by the nature of the offset of rock units along normal faults; topographically, horsts form mountain ranges and grabens form basins.

The large Basin and Range region of the western United States, which extends eastward from California to Utah and southward from Oregon to New Mexico, is undergoing tensional tectonic forces that are pulling the region apart to the west and east.

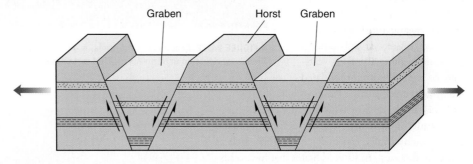

Graben   Horst   Graben

● **FIGURE 14.10** Horsts (upthrown blocks) and grabens (downdropped blocks) are bounded by normal faults.

What kind of tectonic force leads to the formation of horsts and grabens?

A transect from west to east across that region, for example from Reno, Nevada, to Salt Lake City, Utah, encounters an extensive series of alternating downdropped and upthrown fault blocks comprising the basins and ranges for which the region is named. Some of the ranges and basins are simple horsts and grabens, but others are tilted fault blocks that resulted from the uplift of one side of a fault block while the other end of the same block rotated downward (● Fig. 14.11). Death Valley, California, is a classic example of the downtilted side of a tilted fault block (● Fig. 14.12).

In some cases, large-scale tensional tectonic forces create **rift valleys**, which are long and comparatively narrow zones of crust downdropped between normal faults. Examples of rift valleys include the Rio Grande rift of New Mexico and Colorado and the Great

Rift Valley of East Africa. A rift valley in the Middle East contains the Dead Sea, the surface of which lies about 390 meters (1280 ft) below the level of the Mediterranean Sea, which is only 64 kilometers (40 mi) away. Other rift valleys extend along the crests of the oceanic ridges where diverging lithospheric plates pull apart oceanic crust.

An **escarpment**, often shortened to **scarp**, is a steep cliff, which may be tall or short. Scarps form on Earth surface terrain for many reasons and in many different settings. A cliff that results from movement along a fault is specifically a **fault scarp**. Fault scarps are commonly visible in the landscape along normal fault zones, where they may consist of impressive rock faces on fault blocks that have undergone extensive uplift over long periods. In areas of normal faults, unconsolidated sediments eroded from the uplifted block are deposited at the base of the slope near the fault zone and extending onto the down-dropped block. If subsequent movement along the fault vertically offsets those unconsolidated sediments, it produces a *piedmont fault scarp* in the sediments (● Fig. 14.13).

Some fault scarps account for spectacular mountain walls, especially in regions such as much of the western United States that have a history of recent tectonic activity. In southern Utah and northern Arizona, the northwestern part of the elevated Colorado Plateau steps down to the Great Basin in an impressive series of west-facing fault scarps. In California, the 645-kilometer-long (400 mi) Sierra Nevada Range is a great tilted fault block with the east side faulted upward and the west side tilted down (see Fig. 14.11). The east face of the Sierra Nevada essentially represents a fault scarp that rises abruptly a steep 3350 meters (11,000 ft) above the desert (● Fig. 14.14). In contrast, the west side of the Sierra (the "back slope") dips down gradually over a distance of 100 kilometers (60 mi), forming terrain that includes a band of foothills. Like the Sierra Nevada, the Grand Tetons of Wyoming rise dramatically along an east-facing fault scarp.

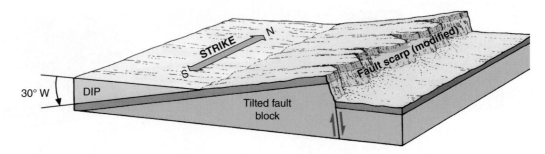

● **FIGURE 14.11** A tilted fault block similar to the kind that produced the basin occupied by Death Valley. Here, the east-facing cliff is a fault scarp that has been worn back to lower slope (modified) by erosion.

Is erosion an endogenic or an exogenic process?

Courtesy Sheila Brazier

● **FIGURE 14.12** Death Valley, California, is a fine example of a topographic basin created by a tilted fault block. The lowest elevation in North America, 86 meters (282 ft) below sea level, is found on the valley floor.

D. Sack

● **FIGURE 14.13** This piedmont fault scarp in Nevada is the topographic expression of a normal fault.

On which side of the fault does the horst lie?

● **FIGURE 14.14** The east front of the Sierra Nevada in California is essentially the steep scarp side of a tilted fault block.

● **FIGURE 14.15** The steep fault scarp of the Sierra del Carmen on the Texas-Mexico border has undergone limited modification by weathering and erosion. The Rio Grande flows through the canyon in the mountain front, which it carved during uplift of the fault block.

Another excellent example of a fault scarp lies along the fault block across which the Rio Grande has carved Santa Elena Canyon in Big Bend National Park, Texas. Other than that 500-meter-deep (1640 ft) canyon, the fault block has been so little modified by erosion in the arid climate that it retains much of its blocklike shape (● Fig. 14.15).

Major uplift of faulted mountain ranges has a strong impact on other aspects of the physical geographic system, as exemplified by the uplift of the Sierra Nevada, which began several million years ago. As the mountain range rose in response to endogenic processes, precipitation increased on its windward (western) side because of orographic lifting of air masses, and the greater elevation brought cooler temperatures. Along the steep, lee (eastern) side of the tilted Sierra fault block, the rain shadow effect led to increased aridity. These climate changes influenced the vegetation, soils, and animal life. Exogenic processes attacked the rocks as uplift progressed. Weathering, erosion of rock and soil by flowing water, and the downslope movement of rock material by gravity intensified as a result of the greater runoff, elevation, and slope. As in many high mountain ranges, glaciers formed, altering and etching the terrain further. Thus, over time, exogenic processes carved and shaped valleys in the Sierra fault block, leaving spectacular canyons and mountain peaks. High relief persists because the Sierra Nevada continues to rise rapidly, in a geologic sense, on average about a centimeter per year.

Shearing Forces

Vertical displacement along a fault occurs when the rocks on one side move up or drop down relative to rocks on the other side. Faults with this kind of movement, up or down along the dip of the fault plane extending into Earth, are **dip-slip faults**. Reverse and normal faults have dip-slip motion. There also exists, however, a completely different category of fault along which displacement of rock units is horizontal rather than vertical. In this case, the direction of slippage is parallel to the surface trace, or strike, of the fault, thus it is called a **strike-slip fault** or, because of the horizontal motion, a **lateral fault** (Fig. 14.9d). The offset along strike-slip faults is most easily seen in map view (from above), rather than in cross-sectional view. Active strike-slip faults cause horizontal offset of roads, railroad

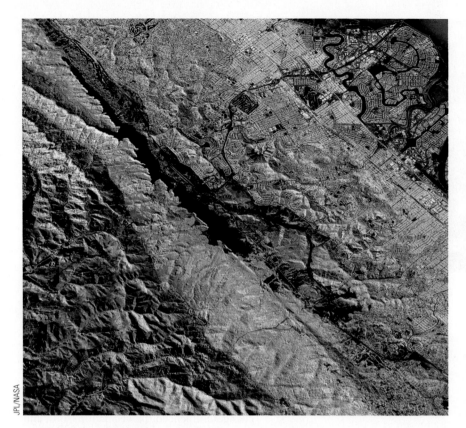

JPL/NASA

● **FIGURE 14.16** The San Andreas Fault extends from upper left to lower right across this false-color image of an area west of San Mateo, California. The lithosphere to the west (left) of the fault is moving to the northwest relative to that on the east side of the fault.

Is the San Andreas a left or right lateral fault?

tracks, fences, streambeds, and other features that extend across the fault. The direction of motion along a strike–slip fault is described as left lateral or right lateral. Direction of motion is determined by imagining yourself standing on one block looking across the strike–slip fault to assess if the other block moved to your left or right. Movement along a left-lateral fault caused the 2010 earthquake in Haiti. The San Andreas Fault, which runs through much of California, has right lateral strike–slip movement. A long and narrow, rather linear valley composed of rocks that have been crushed and weakened by faulting marks the trace of the San Andreas Fault zone (● Fig. 14.16).

The amount that Earth's surface is offset during instantaneous movement along a fault varies from fractions of a centimeter to several meters. Faulting moves rocks laterally, vertically, or both. The maximum horizontal displacement along the San Andreas Fault in California during the 1906 San Francisco earthquake was more than 6 meters (21 ft). Analysis of satellite radar imagery showed that parts of Nepal rose 1.4 meters (4.6 ft) during the

earthquake of 2015, and vertical offset of more than 10 meters (33 ft) occurred during the Alaskan earthquake of 1964. Over millions of years, the cumulative displacement along a major fault may be tens of kilometers vertically or hundreds of kilometers horizontally, although the majority of faults have offsets that are much smaller than these values.

Rock Structure and Topography

Tectonic activity produces different types of structural features that range from microscopic fractures to major folds and fault blocks. At Earth's surface, these structural features comprise various landforms and are subject to modification by weathering, erosion, transportation, and deposition. It is important to distinguish between elements of rock structure and topographic features because rock structure reflects endogenic factors, whereas landforms reflect the balance between endogenic and exogenic factors. As a result, a specific type of structural element can assume a variety of topographic expressions (● Fig. 14.17). For instance, a structural upfold is an anticline even though geomorphically it may comprise a ridge, a valley, or a plain, depending on surface erosion processes and the nature of broken or weak rocks. Nashville, Tennessee, occupies a topographic valley, yet it is sited in the remains of a structural dome (a circular domal anticline). Likewise, even though synclines are structural downfolds, topographically a syncline can comprise a valley or a ridge. Some mountaintops in the Alps are the erosional remnants of synclines.

Words such as *mountain, ridge, valley, basin,* and *fault scarp* are geomorphic terms that describe the surface topography, whereas *anticline, syncline, horst, graben,* and *normal fault* are structural terms that describe the arrangement of rock layers. Elements of rock structure might or might not be directly represented in the surface

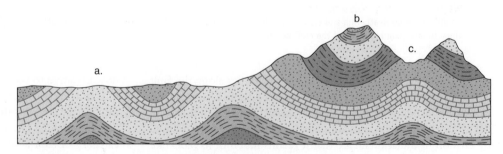

● **FIGURE 14.17** Geologic structure, the rock response to applied tectonic forces, may or may not be directly represented in the surface topography, which results from the interplay of exogenic and endogenic geomorphic processes. (a) The structure is an anticline, but the surface landform is a plain. (b) A topographic peak composed of a structural downfold (syncline). (c) A valley eroded into the crest of an anticline.

Why do some anticlines not form mountains?

This highway cut through a mountain exposes what feature of geologic structure? What type of tectonic force created this structure? Provide a plausible explanation for why these rocks form a highland.

D. Sack

topography. It is important to remember that the topographic variation on Earth's surface results from the interaction of three major factors: endogenic processes that create relief, exogenic processes that shape landforms and reduce relief, and the relative strength or resistance of different rock types to weathering and erosion.

Earthquakes

Earthquakes, evidence of present-day tectonic activity, are ground motions of Earth caused when accumulating tectonic stress is relieved by the sudden displacement of rocks along a fault. The abrupt, lurching movement of crustal blocks past one another to new positions releases energy that moves through Earth as traveling *seismic waves*. Chapter 13 described how seismic waves help us understand the characteristics of Earth's interior, but they also have a great impact on Earth's surface. Seismic waves passing along the crustal exterior or emerging at Earth's surface from below cause the damage, injury, and loss of life that we associate with major tremors. The vast majority of earthquakes are so slight that we cannot feel them and they produce no injuries or damage.

The subsurface location where rock displacement and the resulting earthquake originate is the earthquake **focus**, which may be located anywhere from near the surface to a depth of 700 kilometers (435 mi). The earthquake **epicenter** is the point on Earth's surface that lies directly above the focus, and it is where the strongest shock is normally felt (● Fig. 14.18). Most earthquakes occur at a focus deep enough that no displacement is visible at

Earth's surface. Others cause mild shaking that might do little more than rattle dishes. A small percentage of earthquakes are so strong that they topple buildings and break power lines, gas mains, and water pipes. Surface offset or ground shaking during an earthquake can also trigger potentially deadly rockfalls, landslides, and avalanches. Earthquakes that occur beneath or near the ocean, moreover, are one of the main causes of *tsunamis*, which are sets of large ocean waves generated by the sudden displacement of water. Aftershocks commonly follow a large earthquake as crustal adjustments continue to occur, and foreshocks sometimes precede large quakes.

Measuring Earthquake Size

Scientists express the severity of an earthquake in two distinct ways: the size of the event as a physical Earth process, and the degree of its impact on people. These two characteristics are sometimes related because, all other factors being equal, more powerful earthquakes should have a greater effect on humans than smaller earthquakes. However, large tremors that strike in sparsely inhabited places will have limited human impact, whereas small earthquakes that strike

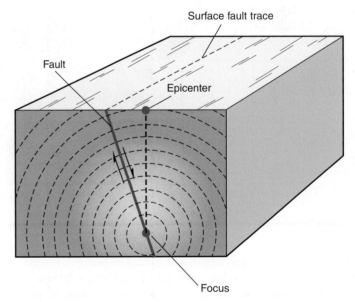

● **FIGURE 14.18** The relationship between an earthquake's focus, the location where movement along the fault began, and its epicenter, the point on Earth's surface directly above the focus.

Why is the epicenter in this example not located where the fault crosses Earth's surface?

Mapping the Distribution of Earthquake Intensity

When an earthquake strikes a populated area, one of the first pieces of scientific information released is the magnitude of the tremor. Magnitude is a numeric expression of an earthquake's size at its focus in terms of energy released. In this sense, earthquakes can be compared with explosions. For example, a magnitude 4.0 earthquake releases energy equivalent to exploding 1000 tons of TNT. Because the scale is logarithmic rather than linear, a 5.0 magnitude tremor is the equivalent of 32,000 tons of TNT, whereas energy released in a 6.0 magnitude quake approximates 1 million tons of TNT.

Because of their greater energy, earthquakes of larger magnitude have the potential to cause much more damage and human suffering than those of smaller magnitude, but the reality is much more complex than that. A moderate earthquake in a densely populated area may cause great injury and damage, whereas a very large earthquake in an isolated region may not affect humans at all. Many factors relating to physical geography influence an earthquake's impact on people and our built environment. In general, the farther a location is from the earthquake epicenter, the less the effect of shaking,

but this generalization does not apply in every case. An earthquake in 1985 caused great damage in Mexico City, including the complete collapse of buildings, even though the epicenter was 385 kilometers (240 mi) away.

The Mercalli Scale of earthquake intensity (I–XII) was devised to measure a tremor's impact on people and the structures we have built. Although every earthquake has only one magnitude, intensity often varies greatly from place to place, and a single tremor typically generates a range of intensity values. The spatial variation in Mercalli intensity for

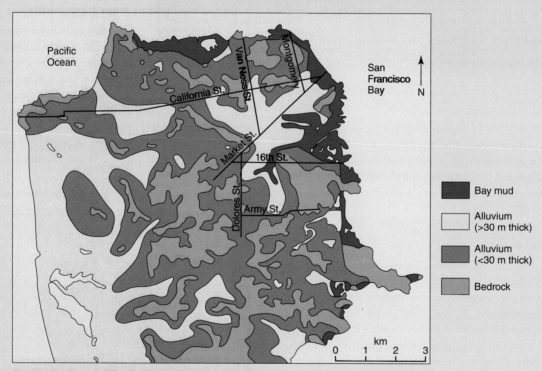

The distribution of different Earth materials at the time of the 1906 San Francisco earthquake.

densely populated areas can cause considerable damage and suffering. Many factors other than earthquake size affect the damage and loss of life resulting from a tremor. Measuring the physical size of earthquakes and, separately, their effects on people help scientists and planners understand the local and regional hazard potential of an earthquake.

In 1935, Charles F. Richter developed a numeric scale of **earthquake magnitude** to express the size of earthquakes as

recorded by a *seismograph*, a sensitive instruments that measures ground shaking. Earthquake magnitude is reported as a number, generally with one decimal place, that represents the amount of energy released in an earthquake. As established by Richter, any whole number increase in earthquake magnitude (for example, from 6.0 to 7.0) represents a 10 times increase in amplitude on the seismograph trace (see again Fig. 13.1), but the actual energy released is about 32 times greater. The extremely destructive 1906

a given earthquake is portrayed on maps by *isoseismals*, lines connecting points of equal shaking and earthquake damage expressed in Mercalli intensity levels. Patterns of isoseismals help scientists assess the local conditions that contributed to spatial variations in shaking and impact. Earthquake intensity factors vary with the nature of the substrate, construction type and quality, topography, and population density. Areas with unconsolidated Earth materials, poor construction, or high population densities generally suffer more from ground shaking and experience greater damage.

The 1906 San Francisco earthquake and ensuing fire caused the destruction of many buildings, numerous injuries, and an estimated 3000 deaths. The fire resulted from earthquake-damaged electrical and gas lines. Neither the magnitude nor the intensity scales existed in 1906. Subsequent studies suggest that the earthquake magnitude was about 8.3. Its Mercalli intensities have been reconstructed and mapped, showing the great variations in ground shaking and damage that the earthquake caused. In some cases, buildings on one side of a street were destroyed, whereas those on the other side of the street suffered little damage. Geographic patterns of the isoseismals have been analyzed and compared to the distribution of other factors to determine why some areas suffered more than others. These analyses reveal that areas of bedrock experienced lower intensities (less damage) than areas of unconsolidated Earth materials. Much of the worst damage occurred where bayfront lands and stream valleys had been artificially filled with sediment to allow construction.

Analyzing the nature of sites where intensities in an earthquake were higher or lower than expected helps us understand the reasons for spatial variations in earthquake hazards. The geographic patterns of Mercalli intensity that are generated even by small tremors help in planning for larger earthquakes in the same area. The overall patterns of ground shaking and isoseismals should be similar for a large earthquake having the same epicenter as a small one, but the amount of ground shaking, Mercalli intensity, and size of the area affected would be greater for the larger magnitude tremor.

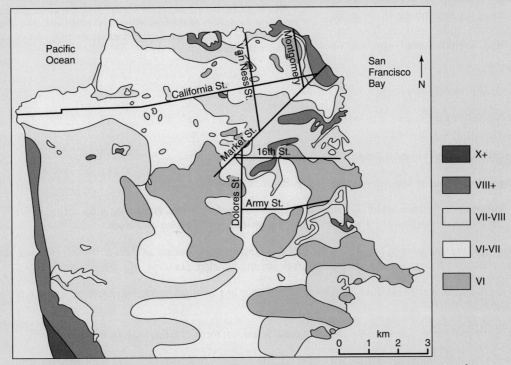

Geographic pattern of Mercalli intensity values for the 1906 San Francisco earthquake. Map areas of intensity VIII+ in the northeastern quarter of the city are located on lands that were artificially built up by infilling with sediment.

San Francisco earthquake occurred before the magnitude scale was devised but is estimated to have had a magnitude of 8.3. The 1989 Loma Prieta earthquake near San Francisco had a Richter magnitude of 7.1 but was still responsible for dozens of fatalities. The strongest earthquake in North America to date, a magnitude 8.6, occurred in Alaska in 1964, and the strongest in the world in recent history was a 9.5 magnitude earthquake that occurred off the coast of Chile in 1960.

Over the last several decades, scientists have derived alternative ways for calculating earthquake magnitude from seismograph data. Today the energy released in most moderate and major earthquakes is determined using formulas for the more accurate **moment magnitude** instead of using Richter's original procedure (Richter magnitude). Regardless of how it is calculated, earthquake magnitude is reported using the same numeric scheme devised by Richter, where a whole number increase in magnitude represents 10 times greater amplitude and approximately 32 times greater energy released. Furthermore, as with Richter magnitude, each earthquake has only one moment magnitude representing the earthquake's size in terms of energy released. Like most recent earthquakes, the 9.1 magnitude reported for the 2004 Sumatra-Andaman earthquake off the west coast of northern Sumatra, Indonesia, which generated the deadliest tsunami on record (● Fig. 14.19), the 7.0 magnitude determined for the 2010 earthquake in Haiti, the 9.0 magnitude tsunami-generating earthquake in Japan in 2011, and the 7.8 magnitude 2015 earthquake in Nepal are all moment magnitudes.

A very different type of measurement scale is that used to record and understand patterns of **earthquake intensity**, the damage caused by an earthquake and the degree of its impact on people and their property. The **modified Mercalli scale** of earthquake intensity uses categories numbered from I to XII (Table 14.1) to describe the effects of an earthquake on humans and the spatial variation of those impacts. The categories are represented with Roman numerals to

● **FIGURE 14.19** The 9.1 magnitude Sumatra-Andaman earthquake and tsunami destroyed the city of Banda Aceh on the Indonesian island of Sumatra. Here, above the lowest set of windows, the only piece of a building left standing displays dark scratches made by debris carried in the tsunami.

TABLE 14.1
Modified Mercalli Intensity Scale

I.	Not felt except by a very few under especially favorable conditions.
II.	Felt only by a few persons at rest, especially on upper floors of buildings.
III.	Felt quite noticeably by persons indoors, especially on upper floors of buildings. Many people do not recognize it as an earthquake. Standing vehicles may rock slightly. Vibrations similar to the passing of a truck.
IV.	Felt indoors by many, outdoors by few during the day. At night, some people are awakened. Dishes, windows, doors disturbed; walls make a cracking sound. Sensation like heavy truck striking building. Standing vehicles rock noticeably.
V.	Felt by nearly everyone; many people are awakened. Some dishes and windows broken. Unstable objects overturned. Pendulum clocks may stop.
VI.	Felt by all, many frightened. Some heavy furniture moved; a few instances of fallen plaster. Damage slight.
VII.	Damage negligible in buildings of good design and construction; slight to moderate in well-built ordinary structures. Considerable damage in poorly built or poorly designed structures. Some chimneys broken.
VIII.	Damage slight in specially designed structures; considerable damage in ordinary substantial buildings with partial collapse. Damage great in poorly built structures. Fall of chimneys, factory stacks, columns, monuments, walls. Heavy furniture overturned.
IX.	Damage considerable in specially designed structures; well-designed frame structures thrown out of plumb. Damage great in substantial buildings, with partial collapse. Buildings shifted off foundations.
X.	Some well-built wooden structures destroyed; most masonry and frame structures with foundations destroyed. Rails bent.
XI.	Few, if any, masonry structures remain standing. Bridges destroyed. Rails bent greatly.
XII.	Complete destruction. Lines of sight and level are distorted. Objects thrown into the air.

Abridged from J. Major/USGS

avoid confusion with earthquake magnitude values. Although a maximum intensity is determined for each earthquake, one earthquake typically produces a variety of intensity levels, depending on varying local conditions. These conditions include distance from the epicenter, how long shaking lasted, how various local surface materials responded to the shaking, population density, and the nature of building construction in the affected area. After an earthquake, observers gather information about Mercalli intensity levels by noting the damage and talking to residents about their experiences. The variation in intensity levels is then mapped so that geographic patterns of damage and shaking intensity can be analyzed. Understanding the spatial patterns of damage and ground response helps us plan and prepare so that we can reduce the hazards of future earthquakes.

● **FIGURE 14.20** Damage in Ofunato, Japan, from the 2011 earthquake and ensuing tsunami.

Earthquake Hazards

Unfortunately, there is plenty of evidence concerning the deadliness of earthquakes as illustrated by just a few recent examples. More than 8000 people perished in the earthquake in Nepal in 2015. Japan's 9.0 magnitude earthquake and tsunami in 2011 left almost 28,000 people missing or confirmed dead (● Fig. 14.20). In 2010, the 7.0 magnitude earthquake in Haiti killed approximately 225,000, injured 300,000, and displaced 1.3 million people (● Fig. 14.21), whereas the 8.8 magnitude quake in Chile that year killed hundreds, injured thousands, and displaced 800,000 people. In 2008 a magnitude 7.9 earthquake devastated extensive areas of eastern Sichuan, China, causing 68,800 fatalities. A magnitude 7.6 tremor in Pakistan in 2005 resulted in 86,000 deaths. Together with the ensuing tsunami, the 2004 Sumatra–Andaman earthquake led to almost 300,000 fatalities.

The impact of an earthquake on people, their property, and the built environment depends on many factors. Some of the most important pertain to the likelihood of structural collapse and number of people potentially affected by such failures. Factors influencing structural collapse include the location of an earthquake epicenter relative to population centers, construction materials and methods, and the stability of Earth materials on which buildings and other structures are erected. A strong tremor of magnitude 7.5 that struck the Mojave Desert of Southern California in 1992 caused little loss of life because of sparse settlement. Structures made of brick, unreinforced masonry, or other inflexible materials or without adequate structural support do not withstand ground shaking well, and this contributed greatly to the high number of injuries and fatalities in the 2010 earthquake in Haiti. Freeway spans collapsed in the San Francisco Bay area in the Loma Prieta earthquake of 1989 and in 1994 during the 6.7 magnitude

● **FIGURE 14.21** Left-lateral, strike-slip movement along the boundary between the North American and Caribbean plates caused tremendous damage and loss of life in the 2010 earthquake in Haiti. The 7.0 magnitude quake destroyed a majority of the homes in this southern Haitian city of Jacmel.

What factors contributed to the large death toll for this earthquake in Haiti?

Northridge earthquake in the San Fernando Valley area of Los Angeles (● Fig. 14.22). Wood frame houses generally fare better than brick, adobe, or block structures because the wood frame has greater flexibility during ground shaking. Buildings located on loosely deposited Earth materials (unconsolidated sediment) tend to shake violently compared to those in areas of solid bedrock. The effects of an 8.1 magnitude earthquake with an epicenter 385 kilometers (240 mi) from Mexico City in 1985 caused a death toll of more than 9000 and extensive damage to many buildings, including high-rise sections of the city. Mexico's densely populated capital city is built on an ancient lake bed made of soft sediments

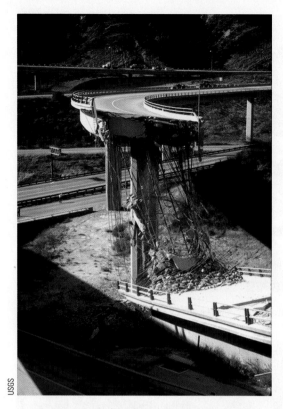

that shook greatly, causing much destruction even though the earthquake was centered in a distant mountain region.

In addition to structural collapse, loss of life in an earthquake is influenced by several other factors, including time of day, time of year, weather conditions at the time of the quake, resulting fires from downed power lines and broken gas mains, and whether ground shaking triggers rockfalls, landslides, avalanches, or tsunamis. Freeway spans that collapse from an earthquake will cause much greater loss of life if the tremor occurs at rush hour instead of in the middle of the night. Contributing to the death toll in Nepal in 2015 were numerous earthquake-induced avalanches, including one that tore through base camp at Mount Everest, as well as rockfalls and landslides that obliterated mountain villages. Even slope failures that only block roads can increase the death toll and extent of human suffering by hampering access for rescue operations.

Most major earthquakes are related to known faults and have epicenters in or near mountainous regions, but the one most widely felt in North America was not. It was one of a series of tremors that occurred during 1811 and 1812, centered near New Madrid, Missouri. It was felt from Canada to the Gulf of Mexico and from the Rocky Mountains to the Atlantic Ocean. Fortunately, the region was not densely settled at that time. Because of that huge earthquake, the St. Louis area is considered today to have a high earthquake-hazard potential, as shown on the seismic-risk map for the conterminous United States (● Fig. 14.23). Although not common in these regions, recent earthquakes have occurred in Virginia, New England, and the Mississippi Valley. Probably no area on Earth could be properly described as entirely safe from earthquakes.

● FIGURE 14.22 Elevated freeways collapsed in both the 1994 Northridge earthquake in the Los Angeles area (pictured here) and the 1989 Loma Prieta earthquake in the San Francisco Bay area. Each tremor killed close to 60 people, but additional lives were spared in 1994 because the Northridge earthquake occurred at 4:30 a.m. on a holiday. The Loma Prieta earthquake struck during a weekday evening at rush hour.

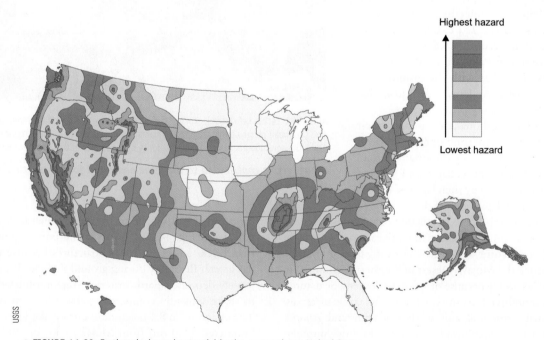

Highest hazard

Lowest hazard

● FIGURE 14.23 Earthquake hazard potential in the conterminous United States.

What is the earthquake hazard potential where you live?

Igneous Processes and Landforms

Like tectonism, processes involving molten rock matter—igneous activity—also originate inside Earth yet can have a profound influence on the topography and relief at Earth's surface. **Plutonism** refers to the slow, subsurface cooling of magma into various types of intrusive igneous rock masses. When those rock emplacements are eventually uplifted through tectonism and exposed at Earth's surface, they comprise landforms of distinctive shapes and properties. Other igneous landforms are created by **volcanism**, the eruption of molten rock material from Earth's subsurface to the exterior. *Volcanoes* are mountains or hills that form in this way, but the type of terrain feature created by volcanic activity depends on the nature and circumstances of the eruption. Both plutonism and volcanism increase surface relief.

Plutonism and Intrusions

Bodies of magma that exist beneath the surface of Earth, or masses of intrusive igneous rock that cooled and solidified beneath the surface, are **igneous intrusions**, or *plutons*. A great variety of shapes and sizes of magma bodies results from this plutonism. When they are first formed, smaller plutons have little or no effect on the surface terrain. During their formation, larger plutons may be associated with uplift of the land surface under which they are intruded.

Because they form beneath Earth's surface, typically at great depth, considerable time must pass before the igneous intrusions themselves become part of the visible landscape. It can take millions of years of regional uplift, accompanied by erosion of rocks overlying the pluton, even for small intrusions to achieve surface exposure. Those that do crop out tend to stand higher than the surrounding terrain because intrusive igneous rocks, such as granite, that constitute the plutons have a resistance to weathering and erosion exceeding that of many other kinds of rocks.

The many different kinds of igneous intrusions are classified by their size, shape, and relationship to the surrounding rocks (● Fig. 14.24). Where exposed at Earth's surface, an irregularly shaped intrusion is a **stock** if it covers an area smaller than 100 square kilometers (40 sq mi); if larger, it is known as a **batholith**. Batholiths are complex masses of solidified magma, usually granite. Batholiths represent large plutons that, during their development kilometers beneath Earth's surface, melted, metamorphosed, or pushed aside other rocks. Some batholiths are as much as several hundred kilometers across and thousands of meters thick. Because of their size and the resistance of intrusive igneous rocks to weathering and erosion, batholiths form the core of many major mountain ranges. The Sierra Nevada batholith, Idaho batholith, and the Peninsular Ranges batholith of Southern and Baja California represent hundreds of thousands of square kilometers of granite landscapes in western North America.

Magma creates other kinds of igneous intrusions by forcing its way into fractures and between rock layers without melting the surrounding rock. A **laccolith** is a mushroom-shaped intrusion that develops where molten magma pushes its way between preexisting horizontal layers of other rock, causing the overlying strata to bulge upward as the intrusion grows. Laccoliths resemble mushrooms because the upper dome-like part is usually connected to the magma source by a pipe or stem-like component. As the laccolith is forming and causing the overlying rock strata to bulge, Earth's surface arches upward like a giant blister, with magma beneath the overlying layers comparable to the fluid beneath the skin of a blister. Although laccoliths are smaller, like batholiths they form mountains or hills after erosion has worn away the overlying less resistant rocks. The La Sal, Abajo, and Henry Mountains in southern Utah are excellent examples of exposed laccoliths (● Fig. 14.25).

Landforms smaller than stocks, batholiths, and laccoliths also result when erosion of overlying rocks exposes igneous intrusions at Earth's surface. Magma sometimes intrudes between rock layers

● **FIGURE 14.24** Because intrusive igneous rocks tend to be more resistant to erosion than sedimentary rocks, when they are eventually exposed at the surface, sills, dikes, laccoliths, stocks, and batholiths generally stand higher than the surrounding rocks. Irregular, pod-shaped plutons less than 100 square kilometers (40 sq mi) in area form stocks when exposed, whereas larger ones form extensive batholiths.

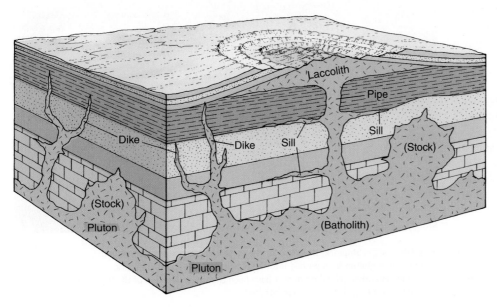

Parvinder S. Sethi

● **FIGURE 14.25** The La Sal Mountains in southeastern Utah are composed of a laccolith that was exposed at the surface by uplift and subsequent erosion of the overlying sedimentary rocks.

How do laccoliths deform the rocks they are intruded into?

without doming them upward, solidifying as a horizontal sheet of intrusive igneous rock called a **sill**. The cliff-forming Palisades, along New York's Hudson River, provide an example of a sill made of *gabbro*, the intrusive compositional equivalent of basalt, which is an extrusive igneous rock. Molten rock under pressure may also intrude into a nonhorizontal fracture that cuts across the surrounding rocks. The solidified magma in this case has a wall-like shape and is known as a **dike**. When exposed by erosion of surrounding, less resistant rocks, dikes appear as vertical or near-vertical walls of rock rising above the surrounding topography (● Fig. 14.26). At Shiprock, in New Mexico, resistant dikes many kilometers long rise vertically more than 90 meters (300 ft) above the surrounding plateau (● Fig. 14.27). Shiprock itself is a **volcanic neck**, a tall spire of igneous rock which solidified in the vertical conduit that about 30 million years ago fed an overlying volcano. Erosion has removed the bulk of the volcanic mountain, exposing the resistant dikes and neck that were once internal features of the volcano at Shiprock.

Volcanic Eruptions

Few spectacles in nature are as awesome as a volcanic eruption (● Fig. 14.28). Although large violent eruptions tend to be infrequent events, they can devastate the surrounding environment and completely change the nearby terrain. Yet volcanic eruptions are natural processes and should not be unexpected by people who live or travel in the vicinity of active volcanoes.

Eruptions vary greatly in their size and character, and the volcanic landforms that result are extremely diverse. *Explosive eruptions* violently blast pieces of molten and solid rock into the air, whereas molten rock pours less violently onto the

J. Petersen

● **FIGURE 14.26** When erosion of overlying rocks exposes them at Earth's surface, dikes, like this one in Big Bend National Park, Texas, often stand somewhat higher than the rock into which they were intruded.

How does a dike differ from a sill?

J. Petersen

● **FIGURE 14.27** Shiprock, New Mexico, is a volcanic neck, a resistant remnant of rock that in molten form pushed into the pipe of a volcano. Like Shiprock itself, the long dikes extending from it have been exposed by erosion of the less resistant rock into which the magma intruded.

● **FIGURE 14.28** A spectacular eruption of Italy's Stromboli volcano, located on an island near Sicily.

higher viscosity—for example, if it has a rhyolitic composition—expanding gases move more slowly through the molten rock and do not vent easily to the atmosphere. Gas pressure builds quickly in this type of magma and can soon exceed the strength of the overlying rock. When the overlying rock breaks, the gases surge with great speed and force from the high pressure in the magma toward the much lower pressure at Earth's surface taking molten rock out with it in an explosive eruption.

Explosive eruptions hurl into the air fragments of solidified lava, clots of molten lava that solidify in flight, or molten clots that solidify once they land. All of these represent **pyroclastic materials** (fire fragments), also known as **tephra**. These rock fragments erupt in a range of sizes from *volcanic ash*, which is sand-sized or smaller (<2 mm), to *cinders* (2–4 mm), *lapilli* (4–64 mm), and *blocks* (>64 mm). Pyroclastic materials include volcanic "bombs," large spindle-shaped clasts that solidified while flying through the air. In the most explosive eruptions, clay and silt-sized volcanic ash is hurled into the atmosphere to an altitude of 10,000 meters (32,800 ft) or more (● Fig. 14.29). The 1991 eruptions of Mount Pinatubo in the Philippines ejected a volcanic aerosol cloud that circled the globe. The suspended material caused spectacular reddish orange sunsets from increased scattering and lowered global temperatures slightly for 3 years by increasing reflection of solar energy back to space.

Volcanic Landforms

The type of landform that results from a volcanic eruption depends mostly on the explosiveness of that eruption. We consider six major kinds of volcanic landforms, beginning with those associated with the most effusive (least explosive) eruptions. Four of the six major landforms are types of volcanoes.

Lava Flows Layers of erupted rock matter that poured or oozed over the landscape when they were molten are **lava flows**. After cooling and solidifying, the rock retains the appearance of having flowed. Lava flows can be made from any composition of lava (see Appendix C), but basalt is by far the most common because its low viscosity and hot eruptive temperature allow gases to escape, greatly reducing the potential for an explosive eruption.

Solidified lava flows tend to have many fractures, known as *joints*. When basaltic lava cools and solidifies it shrinks, and this contraction can produce a network of vertical fractures that break the rock into numerous hexagonal columns. This creates columnar-jointed basalt flows (● Fig. 14.30).

Lava flows display distinctive surface characteristics. Extremely fluid lavas tend to flow rapidly and for long distances before solidifying. In this case, a thin surface layer of lava in contact with the atmosphere solidifies, while the molten lava beneath continues to move, carrying the thin, hardened crust along and wrinkling it into a ropy surface form called **pahoehoe**. Lavas of slightly greater

surface as flowing streams of lava in *effusive eruptions*. Variations in eruptive style and in the landforms produced by volcanism stem mainly from differences in the **viscosity**—that is, resistance to flowing—of the magma feeding the eruption and in its gas content. Viscosity and gas content are determined especially by the chemical properties of the magma, but temperature is also involved.

Like honey compared with water at the same room temperature, some molten rock matter is naturally more resistant to flowing (more viscous) than other molten rock matter at the same temperature because of differences in chemical composition. The primary material component that contributes to a higher viscosity in molten rock is silica (SiO_2) content. Everything else being equal, because they are rich in silica, *felsic* magmas have a higher viscosity than *mafic* magmas, which contain less silica. Temperature also affects the viscosity of a liquid—consider the viscosity difference between cooler and warmer honey. As it does with honey, an increase in temperature reduces to at least some extent the viscosity of any type of magma (makes it flow more easily).

Magma viscosity influences the explosiveness of a volcanic eruption through its effect on the gas content of the magma. Magmas contain large amounts of gases that remain dissolved when under high pressure at great depths. As molten rock rises closer to the surface, the decrease in pressure from above releases expanding gases from the magma. The gases move with comparative ease through low viscosity, mafic magma into adjacent rocks and through cracks into the atmosphere relieving the eruptive pressure of the expanding gases. The relief in gas pressure delays eruption, often allowing the magma to become hotter, further reducing its viscosity. As a result, mafic magmas, such as those with a basaltic composition, typically erupt in the less violent effusive fashion, with streams of flowing lava, rather than by exploding. If, however, the magma has a felsic composition and therefore a

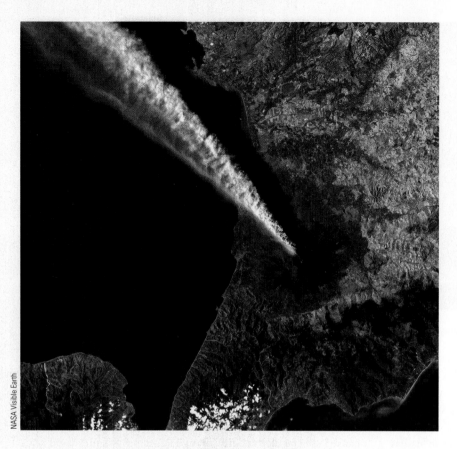

What do you think conditions were like at the time of this eruption for settlements located under the ash cloud?

● FIGURE 14.30 Columnar-jointed lava at Devil's Postpile National Monument, California.

Why are the cliffs shown in this photograph so steep?

viscosity flow more slowly, allowing a thicker surface layer to harden while the still-molten interior lava keeps on flowing. This causes the thick layer of hardened crust to break up into sharp-edged, jagged blocks, making a surface known as **aa**. The terms pahoehoe and aa originated in Hawaii, where effusive eruptions of basalt, and these surface characteristics, are common (● Fig. 14.31).

Lava flows do not have to emanate directly from volcanoes. In many cases, low viscosity lava pours out of deep fractures in the crust, called **fissures**, that can be independent of mountains or hills of volcanic origin. Very fluid basaltic lava erupting from fissures has traveled up to 150 kilometers (93 mi) before solidifying.

In some regions, multiple layers of basalt flows have constructed elevated, relatively flat-topped tablelands known as *basaltic plateaus*. The Columbia Plateau in Washington, Oregon, and Idaho, covering 520,000 square kilometers (200,000 sq mi), is a major example of a basaltic plateau (● Fig. 14.32), as is the Deccan Plateau in India. In the geologic past, huge amounts of basalt issued from some fissures, eventually burying existing landscapes under thousands of meters of lava flows. These very extensive flows are often referred to as *flood basalts*.

Shield Volcanoes

Shield Volcanoes When numerous successive basaltic lava flows occur in a given region, they may eventually pile up into the shape of a large mountain, called a **shield volcano**, which resembles a giant knight's shield resting on Earth's surface (● Fig. 14.33a).

Parvinder S. Sethi

(a)

D. Sack

(b)

● **FIGURE 14.31** Lava flow surfaces often consist of (a) ropy-textured pahoehoe, where the molten lava was extremely fluid, or (b) blocky and angular aa, where viscosity of the molten rock matter was slightly higher (less fluid).

USFWS

● **FIGURE 14.32** In southeastern Washington the Palouse River has cut an impressive gorge exposing layers of Columbia Plateau flood basalts.

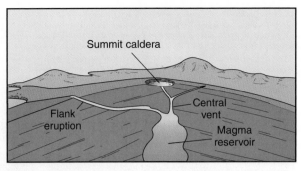

(a)

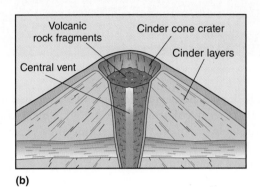

(b)

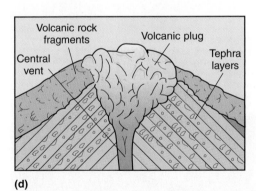

(c)

(d)

● **FIGURE 14.33** The four basic types of volcanoes are (a) shield volcano, (b) cinder cone, (c) composite cone, also known as stratovolcano, and (d) plug dome.

What are the key differences in the shapes of these volcanoes? What properties are alike or different in their internal structure?

● **FIGURE 14.34** Mauna Loa, on the island of Hawaii, clearly displays the dome, or convex, shape of a classic shield volcano. From its base on the ocean floor to its summit at 4170 meters (13,681 ft) above sea level, Mauna Loa is almost 17 kilometers (56,000 ft) tall.

Why do Hawaiian volcanoes erupt less explosively than volcanoes of the Cascades or Andes?

The gently sloping, dome-shaped cones of Hawaii best illustrate this largest type of volcano (● Fig. 14.34). Shield volcanoes erupt extremely hot mafic lava at temperatures near 1100°C (2000°F). Escape of gases and steam occasionally hurls fountains of molten lava a few hundred meters into the air (● Fig. 14.35), causing some accumulations of solidified pyroclastic materials, but the major feature of shield volcanoes is the outpouring of fluid basaltic lava flows. Compared with other volcano types, these eruptions are not very explosive, although they are still potentially dangerous and damaging. The extremely hot and fluid basalt can flow long distances before solidifying, and the accumulation of flow layers develops broad, dome-shaped volcanoes with gentle slopes. On the island of Hawaii, active shield volcanoes also erupt lava from fissures on their flanks so that living on the island's edges, away from the summit craters, does not guarantee safety from volcanic hazards. Neighborhoods in Hawaii have been destroyed or threatened by lava flows. The Hawaiian shield volcanoes form the largest volcanoes on Earth in terms of both their height—beginning at the ocean floor—and diameter.

Cinder Cones

The smallest type of volcano, typically only a couple of hundred meters high, is a **cinder cone**. Most cinder cones consist of predominantly gravel-sized pyroclastics. Gas-charged eruptions throw molten lava and solid pyroclastic fragments into the air. Falling under the influence of gravity, these particles accumulate around the almost pipelike conduit for the eruption, the *vent*, as a large pile of tephra (Fig. 14.33b). Each eruptive burst ejects more pyroclastics that fall and cascade down the sides to build an internally layered volcanic cone. Cinder cone volcanoes typically have a rhyolitic composition but can be made of basalt if conditions of temperature and viscosity keep gases from escaping easily. The form of a cinder cone is distinctive, with steep straight sides and a crater (depression) at the top of the hill (● Fig. 14.36). The steep slopes of accumulated pyroclastics lie at or near the **angle of repose**, the steepest angle that a pile of loose material can maintain without rocks sliding or rolling downslope. The angle of repose for unconsolidated rock matter generally ranges between 30° and 34°. Cinder cone examples include several in the Craters of the Moon area in Idaho, Capulin Mountain in New Mexico, and Sunset Crater, Arizona. In 1943, a remarkable cinder cone called Paricutín grew

● **FIGURE 14.35** This fountain of lava in Hawaii reached a height of 300 meters (1000 ft).

● **FIGURE 14.36** A cinder cone stands among lava flows in Lassen Volcanic National Park, California.

Why is the crater so prominent on this volcano?

from a fissure in a Mexican cornfield to a height of 92 meters (300 ft) in 5 days and to more than 336 meters (1100 ft) in a year. Eventually, the volcano began erupting basaltic lava flows, which completely buried a nearby village except for the top of a church steeple.

Composite Cones

A third kind of volcano, the **composite cone**, results when formative eruptions are sometimes effusive and sometimes explosive. Composite cones are therefore composed of a combination—that is, they represent a composite—of lava flows and pyroclastic materials (Fig. 14.33c). They are also known as **stratovolcanoes** because they are constructed of layers (strata) of pyroclastics and lava. The topographic profile of a composite cone represents what many people might consider the classic volcano shape, with concave slopes that are gentle near the base and steep near the top (● Fig. 14.37). Composite volcanoes form from andesite, which is a volcanic rock intermediate in silica content and explosiveness between basalt and rhyolite. Although andesite is only intermediate in terms of these characteristics, composite cones are dangerous. As a composite cone grows larger, the vent eventually becomes plugged with unerupted andesitic rock. When this happens, the pressure driving an eruption can build to the point where either the plug is explosively forced out or the mountain side is pushed outward until it fails, allowing the great accumulation of pressure to be relieved in a lateral explosion. Such explosive eruptions may be accompanied by **pyroclastic flows**, fast-moving density currents of airborne volcanic ash, hot gases, and steam that flow downslope close to the ground in a manner similar to some avalanches. The speed of a pyroclastic flow can reach 100 kilometers per hour (62 mi/hr) or more.

Most of the world's best-known volcanoes are composite cones. Some examples include Mounts Fuji in Japan, Cotopaxi in Ecuador, Vesuvius and Etna in Italy, Rainier and St. Helens in Washington, and Shasta in California. The highest volcano on Earth, Nevados Ojos del Salado, is an andesitic composite cone that reaches an elevation of 6887 meters (22,595 ft) on the border between Chile and Argentina in the Andes, the mountain range for which andesite was named.

● **FIGURE 14.37** Composite cones are composed of lava flows as well as pyroclastic material and have distinctive concave side slopes. This beautiful example is Mount Augustine in Cook Inlet, Alaska.

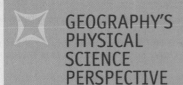

Eruption of Mount St. Helens

On the morning of May 18, 1980, residents of the American Pacific Northwest were stunned by the enormous eruption of Mount St. Helens in the Cascade Mountain Range. Mount St. Helens, a composite cone in southwestern Washington located about 80 km (50 mi) northeast of Portland, Oregon, had been venting steam and ash for two months prior to the eruption. Thousands of small and moderate earthquakes over those weeks indicated pressure changes and shifts in the magma below the surface, and a menacing bulge was growing on the north flank of the volcano. Earth scientists knew from these events and other lines of evidence that the situation could be building toward a major eruption. Warnings were issued, but the exact timing and magnitude of the blast could not be predicted. The power and effects of the eruption—the most destructive in U.S. history—surprised, shocked, and astonished the public, in part because it had been 65 years since the previous major eruption in the conterminous 48 states.

Unlike most explosive volcanic eruptions, in which the force of the blast is directed vertically upward, the 1980 eruption of Mount St. Helens blew gases and pyroclastic debris out laterally. In the weeks leading up to the eruption, pressure from the viscous magma pushing up into the mountain caused part of the volcano's north slope to bulge out, eventually by 140 meters (450 ft). On May 18, at the time of a 5.1 magnitude earthquake, the bulging slope failed, initiating a huge landslip that sent an avalanche of debris down the forested terrain. This breaking away of part of the mountainside greatly reduced the confining pressure on the trapped magma allowing intensely hot steam, noxious gases, and pyroclastic material to burst laterally through the failed side slope, the path of least resistance, at more than 320 kilometers (200 mi) per hour. The slope failure and lateral blast removed part of the mountain's summit, thereby decreasing confinement pressure from the mountain top and enabling upward eruption of volcanic ash. Dense clouds of volcanic ash built up

quickly to an altitude of 24 kilometers (15 mi) and interfered with air travel. Choking ash up to several centimeters thick blanketed the region.

The volcano's lateral explosion destroyed forests, lakes, streams, and campsites for nearly 32 kilometers (20 mi). The original landscape within a few kilometers of the initial point of eruption was obliterated and replaced with thick accumulations of bluish gray volcanic ash. Farther out, extensive tracts of trees were blown down by the force of the blast. More than 500 square kilometers (200 sq mi) of forest and recreational lands were destroyed, untold numbers of wildlife were killed, and 57 people lost their lives in the eruption.

The eruptive blast and volcanic ash were not the only hazards faced by the area's residents. Considerable amounts of snow and ice were present on the mountain in mid-May, and it melted quickly with the heat of the eruption. When this meltwater combined with volcanic ash huge mudflows formed that choked streams, buried valleys,

USDA Forest Service

Prior to the 1980 eruption, Mount St. Helens towered majestically over Spirit Lake in the foreground.

swept away buildings and vehicles, and engulfed everything in their paths.

Mount St. Helens and the other volcanoes in the Cascade Range owe their existence to plate tectonics. They mark the zone where the North American lithospheric plate is overriding the Pacific plate along the Juan de Fuca subduction zone. Comparatively low-density, felsic sediments eroded from the continent start down the subduction zone along with oceanic crust. As increasing pressure from subduction melts the low-density continental sediments, the viscous magma erupts back onto the surface.

Geologists have traced the eruptive history of Mount St. Helens as far back as about 275,000 years ago. Over that geologic time it has experienced several phases of intense activity lasting thousands of years separated by millennia of relative quiescence. The volcano is currently in an active interval that began 3900 years ago after 10,000 years of much less activity. Mount St. Helens, the most active of the Cascade Range volcanoes, experienced multiple major eruptions in the past 3900 years, and is expected to continue this trend. Its most recent major eruption before 1980 occurred in the 1850s, when the region was still sparsely populated. Future eruptions have the potential to cause greater property damage and loss of life than occurred in 1980 because of the population growth and development experienced by the Pacific Northwest since that time.

Volcanic ash billowing up into the atmosphere during the May 18, 1980, eruption of Mount St. Helens.

Steam erupting from Mount St. Helens in 2004.

● **FIGURE 14.38** Beginning in 1995 the Caribbean island of Montserrat was struck by a series of volcanic eruptions, including pyroclastic flows that devastated much of the island. This town was completely abandoned because of the amount of destruction and threat of future eruptions. Prior to the 1995 disaster, the volcano had not erupted for 400 years.

Volcanic ash is much like tiny slivers of glass. It is abrasive and causes respiratory problems in people and animals, clogging of air intakes for engines, destructive scratching and pitting of metal and glass, and the collapse of building roofs under the heavy weight of significant ash accumulations. The damaging effects of volcanic ash on airplane engines, fuel lines, windshields, and the multitude of other sensitive parts on airplanes makes flying through volcanic ash extremely hazardous. Ash erupting from an Icelandic composite cone in 2010 shut down air travel over most of Europe for 6 days, affecting millions of travelers and having major economic impacts.

Plug Domes Where extremely viscous silica-rich magma pushes up into the vent of a volcanic cone but does not flow farther, it forms a **plug dome** (Fig. 14.33d). Solidified outer portions of this blockage create the dome-shaped summit, and jagged blocks that break away from the plug, or preexisting parts of the cone, form the steep side slopes of the volcano. Great pressures build up causing more blocks to break off from the plug and producing the potential for extremely violent explosive eruptions, including pyroclastic flows. In 1903, Mount Pelée, a plug dome on the French West Indies island of Martinique, caused the deaths in a single blast of all but a few people from a town of 30,000. Lassen Peak in California is a large plug dome that erupted with great violence in 1915 (● Fig. 14.39). Other plug domes exist in Japan, Guatemala, the Caribbean, and the Aleutian Islands.

Calderas Occasionally, the eruption of a volcano expels so much material and relieves so much pressure within the magma chamber that only a large and deep depression remains in the area that previously contained the volcano's summit. A large depression made in this way is a **caldera**. The best known caldera in North America is the large depression in south-central Oregon that contains Crater Lake, a circular body of water 10 kilometers (6 mi) across, almost 600 meters (2000 ft) deep, and surrounded by near-vertical cliffs. The caldera that contains Crater Lake was formed 7700 years ago by the enormous eruption and resulting collapse of a prehistoric composite volcano that geologists have named Mount Mazama. The eruption left volcanic ash deposits in several western states and three Canadian provinces. A cinder cone, Wizard Island, subsequently built up from the floor of the caldera to a height almost 235 meters (770 ft) above the lake's surface (● Fig. 14.40). Calderas are also found in the Philippines, the Azores, Japan, Nicaragua, Tanzania, and Italy, many of them occupied by deep lakes. Island remnants remain of some calderas in coastal locations, including Krakatoa in Indonesia and Santorini (Thera) in Greece.

Some of the worst natural disasters in history have occurred in the shadows of composite cones. The oldest known written description of a volcanic eruption is Pliny the Younger's account of the 79 A.D. eruption of Mount Vesuvius, a composite cone in Italy. That eruption may have killed as many as 20,000 people in the cities of Pompeii and Herculaneum. Mount Etna, on the Italian island of Sicily, destroyed 14 cities in 1669, also killing thousands, and is still active much of the time. The greatest volcanic eruption in the last 200 years was the 1883 explosion of Krakatoa in what is now Indonesia. The explosive eruption killed more than 36,000 people, many as a result of the tsunamis generated by the eruption that swept the coasts of Java and Sumatra. In 1985, the Andean composite cone Nevado del Ruiz, in the center of Colombia's coffee-growing region, erupted and melted its snowcap sending torrents of mud and debris down its slopes, burying cities and villages, and resulting in a death toll in excess of 23,000. The 1991 eruption of Mount Pinatubo in the Philippines killed more than 800 people, and airborne ash caused climatic effects for 3 years following the eruption. In 1997, a series of violent eruptions from the Soufriere Hills volcano destroyed more than half of the Caribbean island of Montserrat with volcanic ash and pyroclastic flows (● Fig. 14.38). Mexico City, one of the world's most populous urban areas, continues to be threatened by ash falls from eruptions of a large, active composite cone 70 kilometers (45 mi) away. At this distance, ash fall from a major eruption is the most severe hazard expected.

● **FIGURE 14.39** Lassen Peak in northern California is a plug dome that was active between 1914 and 1921. Stiff silica-rich lava plugs are the darker areas protruding from the volcanic peak, and give plug domes their steep slopes.

Why are plug dome volcanoes considered dangerous?

Some calderas are remnants of gigantic eruptions much larger than have occurred in historic time. A volcano that has erupted more than 1000 cubic kilometers (240 cu mi) of material onto the landscape is known as a **supervolcano**. The area of Yellowstone National Park in Wyoming and Montana, Long Valley in California, Taupo in New Zealand, and Toba in Indonesia are among the world's supervolcanoes. Volcanic ash from an eruption of the Yellowstone supervolcano 2.1 million years ago was deposited over half the area of the conterminous 48 U.S. states and into northern Mexico.

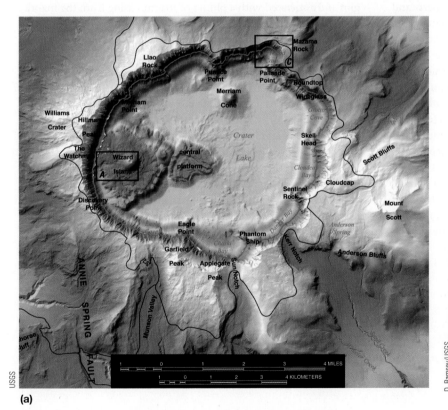

(a)

(b)

● **FIGURE 14.40** (a) Crater Lake, Oregon, formed about 7700 years ago when a violent eruption of Mount Mazama blasted out solid and molten rock matter, leaving behind a deep crater, the caldera, which later accumulated water. (b) Wizard Island is a later, secondary volcano that has risen within the caldera.

Could other Cascade volcanoes erupt to the point of destroying the volcano summit and leaving a caldera?

Distribution of Endogenic Processes

The geographic distributions of most of Earth's igneous intrusions, earthquakes, and volcanic activity are related to plate tectonics. The greatest tectonic forces, most extensive faults, most dangerous earthquakes, and most of the world's volcanic activity are concentrated in relatively narrow zones along the boundaries of lithospheric plates or at hot spots associated with plumes of magma rising from the mantle. Plutons also form along plate boundaries, but because of the time needed for the igneous intrusions to achieve surface exposure, their distribution tends to represent ancient locations of plate margins. The amount and severity of earthquakes and volcanic activity vary with the type of plate boundary.

Tensional tectonic forces associated with lithospheric plate divergence cause faulting and create fractures that provide avenues for molten rock to reach the surface. The divergent midoceanic ridges experience comparatively mild volcanic eruptions and small to moderate earthquakes that originate at a shallow depth. These hazards affect islands along the midoceanic ridges, including Iceland and the Azores. Where continental crust is diverging and breaking, as in rift valleys, earthquakes tend to be small to moderate. Compared to diverging oceanic crust, volcanic eruptions associated with diverging continental crust are more varied because of mafic magma mixing with continental crust. As a result, the continental eruptions are potentially quite violent. Mount Kilimanjaro and Mount Kenya are volcanoes in rift valleys of East Africa.

The potential severity of earthquakes and volcanic eruptions is much greater where, as around the border of the Pacific Ocean, compressional forces produce converging rather than diverging plate boundaries. Along the oceanic trenches where crust is subducted, earthquakes are common and volcanoes develop where melted rock moves upward near the edge of the overriding plate, whether it consists of oceanic or continental crust. Although most of the earthquakes are small to moderate, the largest recorded earthquakes have been related to subduction zones. Earthquake foci become deeper toward the overriding plate, indicating the subducting plate's progress downward toward where it is recycled into the mantle. Subduction along the Pacific Ocean is associated with extensive volcanoes in the Andes of South America, the Cascade Range and the Aleutian Islands of North America, the Kuril Islands and the Kamchatka Peninsula in Russia, as well as Japan, the Philippines, New Guinea, Tonga, and New Zealand. Many of these volcanoes erupt andesitic rock matter and can be dangerously explosive. Because the Pacific Ocean is bordered by active plate margins that are particularly prone to earthquakes and volcanic eruptions, that perimeter is often referred to as the Pacific Ring of Fire (● Fig. 14.41).

Another volcanic and seismic belt occupies the continental collision zone between northward-moving lithospheric plates of the Southern Hemisphere and the Eurasian plate. Volcanoes of the Mediterranean region and Indonesia are located along this collision zone, and major deadly earthquakes have plagued Italy, Turkey, Iran, Nepal, China, and Indonesia.

Transform plate boundaries, where shearing forces and lateral sliding occurs, also experience many earthquakes. The potential for major earthquakes mainly exists in places such as the Caribbean and along the San Andreas Fault zone in California where thick continental crust does not slide easily. Volcanic activity along transform plate boundaries ranges from moderate on the seafloor to slight in continental locations.

Areas far from active plate boundaries are not necessarily immune from earthquakes and volcanism. The Hawaiian Islands, the Galapagos Islands, and the Yellowstone National Park region are examples of intraplate "hot spots" located away from plate margins and associated with a plume of magma rising from the mantle. Oceanic crustal areas that lie over hot spots, such as the Hawaiian Islands, have strong volcanic activity and moderate earthquake activity. In midcontinental areas broken edges of ancient landmasses can still shift, even though today they lie far from active plate margins and are deeply buried by more recent rocks.

Many regions with earthquake or volcanic activity are incredibly scenic and offer attractive environments in which to live and vacation, so it is not surprising that some of these hazard-prone regions are densely populated. To minimize the impacts of tectonic, volcanic, and other related hazards, such as landslides and tsunamis, it is essential that people in these regions have a basic understanding of the nature of the hazards with which they live. It is equally imperative that residents and governmental agencies make detailed preparations for coping with such disasters before they occur.

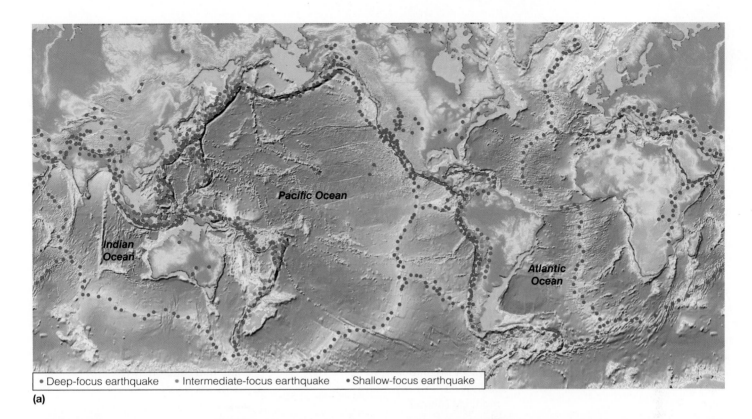

(a)

• Deep-focus earthquake • Intermediate-focus earthquake • Shallow-focus earthquake

Pacific Ocean

Indian Ocean

Atlantic Ocean

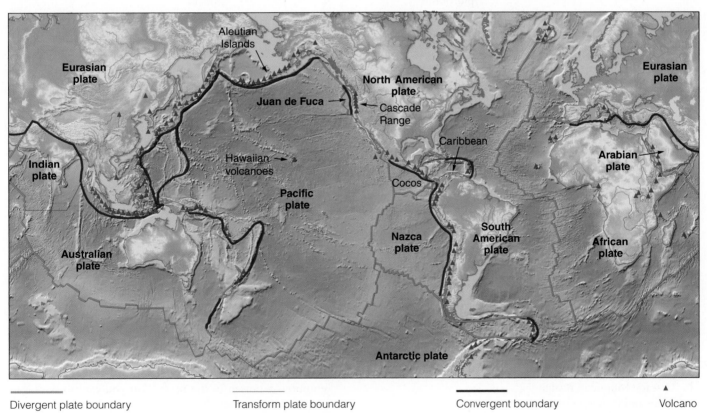

Eurasian plate

Aleutian Islands

North American plate

Eurasian plate

Juan de Fuca

Cascade Range

Caribbean

Arabian plate

Indian plate

Hawaiian volcanoes

Cocos

Pacific plate

Australian plate

Nazca plate

South American plate

African plate

Antarctic plate

Divergent plate boundary   Transform plate boundary   Convergent boundary   Volcano

(b)

● **FIGURE 14.41** The spatial correspondence of lithospheric plate boundaries, especially convergent boundaries having oceanic subduction zones, with (a) earthquake activity, and (b) volcanic eruptions. Because of the high density of earthquakes and volcanoes around the margin of the Pacific Ocean, that perimeter is often referred to as the Pacific Ring of Fire.

Besides the Pacific Ring of Fire, what is another region that shows high incidence of both earthquakes and volcanic activity?

TERMS FOR REVIEW

topography	thrust fault	stock
landform	overthrust	batholith
geomorphology	fault block	laccolith
relief	normal fault	sill
endogenic process	graben	dike
exogenic process	horst	volcanic neck
erosion	rift valley	viscosity
transportation	escarpment (scarp)	pyroclastic material (tephra)
deposition	fault scarp	lava flow
geomorphic agent	dip-slip fault	pahoehoe
rock structure	strike-slip fault (lateral fault)	aa
strike	earthquake	fissure
dip	focus	shield volcano
compressional tectonic force	epicenter	cinder cone
tensional tectonic force	earthquake magnitude	angle of repose
shearing tectonic force	moment magnitude	composite cone (stratovolcano)
folding	earthquake intensity	pyroclastic flow
anticline	modified Mercalli scale	plug dome
syncline	plutonism	caldera
fault	volcanism	supervolcano
reverse fault	igneous intrusion (pluton)	

QUESTIONS FOR REVIEW

1. How do endogenic processes differ from exogenic processes?
2. Distinguish among compressional, tensional, and shearing forces.
3. Draw a diagram to illustrate folding, showing anticlines and synclines.
4. How does a reverse fault differ from a normal fault?
5. How is a fault different from a joint?
6. What causes an earthquake?
7. What is the relationship between the focus and the epicenter of an earthquake?
8. How do earthquake magnitude and earthquake intensity differ? Why are there two systems for evaluating earthquakes?
9. How is a sill different from a dike?
10. How do plutonism and volcanism differ from each other? How are they alike?
11. What are the four basic types of volcanoes, and what are the distinguishing features of each?
12. Which is the most dangerous type of volcano? Explain why.
13. Explain the distribution of most of the world's earthquakes and volcanoes.

CONSIDER AND RESPOND

1. Name the type of tectonic or volcanic activity that is primarily responsible for the following:
   a. Sierra Nevada
   b. Cascade Mountain Range
   c. Basin and Range region
   d. Appalachian Mountains
2. Name several areas in the United States that are highly susceptible to natural hazards from earthquakes or volcanic activity.
3. List five countries bordering the Pacific Ocean that exhibit evidence of major volcanic activity.
4. Name a few countries not in the Pacific region that face tectonic hazards.
5. Recall the locations of some recent earthquakes or volcanic eruptions reported in the news. Can you explain why they occurred where they did?
6. What hazards should a community plan for if it is located in a region of active faults and volcanoes? What recommendations for land use and settlement patterns might reduce the danger from earthquake and volcanic hazards?

PRACTICAL APPLICATIONS

1. If the first seismic wave to arrive at the epicenter of an earthquake traveled 6 kilometers per second and took 4.3 seconds to reach the surface, how deep was the earthquake focus?
2. A 6.5 magnitude earthquake rocked a city on the California coast. In an older community along a bay everyone felt it, and those who were asleep were awakened by it. People were frightened and stood under doorways for protection. Ultimately, there were no fatalities and only minor injuries. Almost every house experienced at least broken dishes, and there were some reports of fallen plaster, broken windows, and damaged chimneys. Three homes experienced collapse of overhangs that had been added to cover the porch after the homes were built. What was the maximum intensity of the earthquake for this coastal community?

LOCATE AND EXPLORE

1. Using Google Earth and the USGS Earthquake Layer (download from CourseMate at www.cengagebrain.com), describe the distribution of recent earthquakes in the United States. In what areas do they tend to cluster and why? What was the largest recent earthquake in Utah within the last 7 days? Was the event large enough to reach the national media?
2. The Hawaiian hot spot is a mantle plume that has persisted for millions of years. As the Pacific plate moved to the northwest, a progression of volcanoes was created and then died as their source of magma was shut off. As shown in Figure 13.36 of your text, formation of the island of Kauai (22.05°N, 159.54°W) was over by about 3.8 million years before present, whereas the island of Hawaii (19.60°N, 155.50°W) started to form approximately 0.7 million years before present. Using the ruler tool in Google Earth, calculate the distance between the two islands and use the distance to calculate the speed of the Pacific Plate in meters per century.

3. Using Google Earth, identify the landforms or structural features at the following locations (latitude, longitude) and provide a brief explanation of why the feature is found there. Use the zoom, tilt, rotate, and elevation exaggeration tools to help view and interpret the feature and the area in which it is found.
   a. 34.5449°N, 115.7907°W
   b. 27.45°N, 54.44°E
   c. 38.876°N, 113.010°W
   d. 52.822°N, 169.947°W
   e. 27.682°N, 16.622°E
   f. 35.2705°N, 119.8264°W
   g. 19.47°N, 155.59°W
   h. 36.688°N, 108.837°W

 MindTap—Make the most of your study time by accessing everything you need to succeed in one place. Read your textbook, take notes, review flashcards, watch videos, complete activities, take practice quizzes, and more online with MindTap. Log in at **www.cengagebrain.com**.

Map Interpretation

■ VOLCANIC LANDFORMS

The Map

The Folsom, New Mexico, map area is located in the far northeastern part of New Mexico, about 300 kilometers (190 mi) northeast of Albuquerque and 80 kilometers (50 mi) west of the Oklahoma panhandle. This region is found within the high western strip of the Great Plains, adjacent to the southern Rocky Mountains. The Basin and Range province, a large region undergoing tensional forces, lies to the southwest, extends south into Mexico and Texas, and west and then north into several western states. Elevation on the entire Folsom map, only a part of which is shown here, ranges from about 1868 meters (6130 ft) to almost 2500 meters (8200 ft) above sea level.

Vegetation is transitional between the grassland of the plains and the forests of the Rocky Mountains. Grasses dominate the lower map areas, whereas higher terrain exhibits juniper and pinyon trees.

Topographically, the map area is dominated by Capulin Mountain, one of almost 100 volcanic landforms in the region's 21,000 square kilometer (8000 mi²) Raton-Clayton volcanic field. Capulin Mountain formed about 62,000 years ago near the end of an 8 million year period of activity for the Raton-Clayton volcanic field. Activity was marked by both effusive and explosive eruptions, occurring in some cases at different times from approximately the same eruptive source. Capulin Mountain has been protected since 1891. It acquired National Monument status in 1916.

Interpreting the Map

1. What specific type of volcanic landform is Capulin Mountain? On what landform characteristics is your decision based?
2. What is the maximum local relief of Capulin Mountain from its base to its peak? What is the drop in elevation from the peak to the bottom of the mountain's crater (hachured contours indicate a depression)?
3. What is Capulin Mountain's maximum diameter? What geometric shape does the mountain most closely resemble in map view?
4. Describe how the terrain adjacent to and slightly beyond the base of the mountain's east and southeast side differs from that at the base of the mountain's northwest side. What accounts for the difference?
5. Between the northwest base of Capulin Mountain and the nearby main highway lies the narrow, southern end of a terrain feature that widens substantially as it extends to the north edge of the map. Use the aerial image and the topographic map to help you determine what type of landform that is. What relationship, if any, do you think it has to Capulin Mountain?
6. A roughly north-trending secondary road runs just west of the feature you identified in question 5. Consider the landform occupying the map area west of that road. In what major way does the surface terrain of that feature differ from the surface terrain of the landform to the east of the road? What might account for the difference in surface terrain?
7. Provide a reasonable explanation for the landform labeled Mud Hill. What observations support your explanation?
8. Volcanic activity during the last several million years is not typical of either the Great Plains or the southern Rocky Mountains. What large-scale processes associated with another nearby region might have contributed to the volcanism in this map area?
9. How helpful is the aerial photograph in analyzing the terrain displayed on the map?

Black and white orthophotograph of Capulin Mountain and vicinity.

Opposite:
Folsom, New Mexico
Scale 1:24,000
Contour interval = 20 ft
USGS

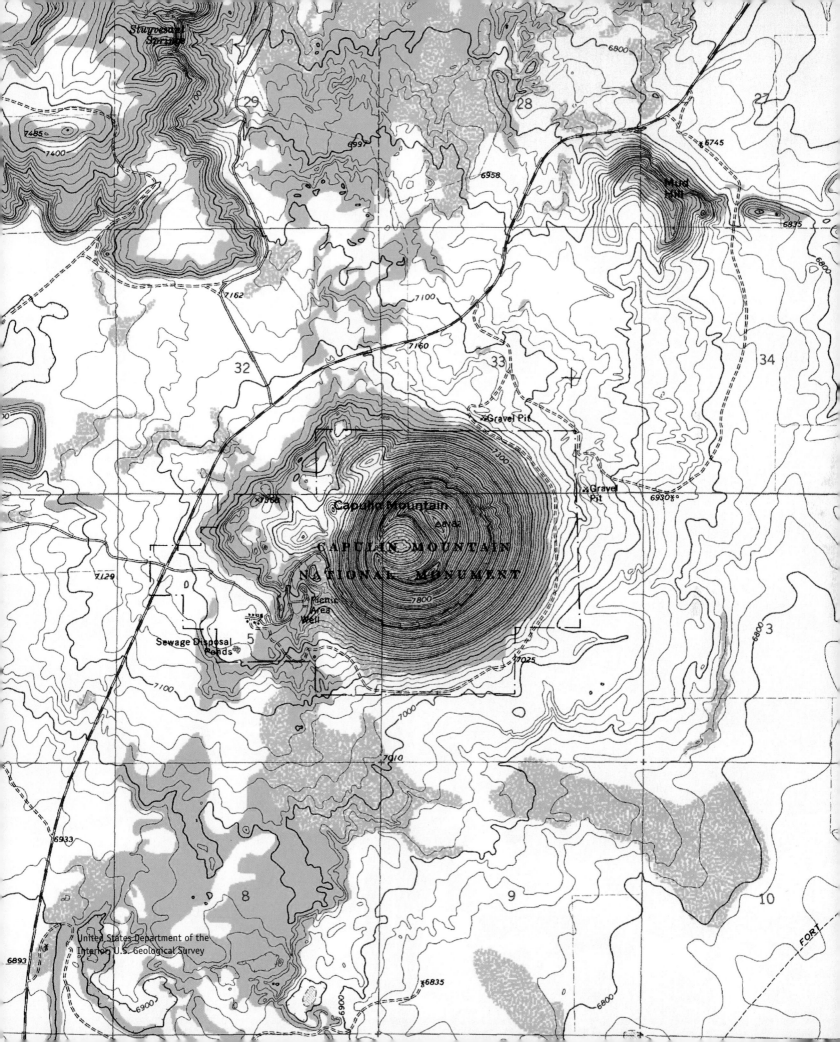

WEATHERING AND MASS WASTING

■ OBJECTIVES

WHEN YOU COMPLETE THIS CHAPTER YOU SHOULD BE ABLE TO:

■ 15.1 Explain that by disintegrating and decomposing rock, weathering prepares rock fragments for erosion, transportation, and deposition.

■ 15.2 Recount the principal differences among the various physical and chemical weathering processes.

■ 15.3 Discuss how variations in climate, rock type, and rock structure influence weathering rates.

■ 15.4 Provide examples of the topographic effects of differential weathering and erosion.

■ 15.5 Describe the role of gravity in mass wasting.

■ 15.6 Use appropriate terms to describe the types of material involved in slope failures.

■ 15.7 Categorize various mass wasting events as slow or fast.

■ 15.8 Summarize the various ways materials move downslope.

■ 15.9 Recognize some of the major landforms and landscape features created by mass wasting.

■ 15.10 Identify ways people influence weathering and mass wasting and ways weathering and mass wasting influence people.

IN THE PREVIOUS TWO CHAPTERS, we examined the materials, processes, and structures associated with the construction of topographic relief at the surface of Earth. We also noted that the relief-building endogenic geomorphic processes of tectonism and volcanism, which arise from within Earth, are opposed by the relief-reducing exogenic geomorphic processes, which originate at the surface. The exogenic processes break down rocks and erode rock fragments from sites of higher energy, transporting them to locations of lower energy. The relocation of rock fragments can be accomplished by the force of gravity alone or with the help of one of the geomorphic agents: flowing water, wind, moving ice, or waves. This chapter focuses on those exogenic processes that cause rocks to decay and on the ways that erosion, transportation, and deposition of surficial Earth materials are accomplished when gravity, rather than a geomorphic agent, is the dominant factor in relocating these materials.

Gravity constantly pulls downward on all Earth surface materials. Weakened rock and broken rock fragments are especially susceptible to downslope movement by gravity, which may be slow and barely noticeable or rapid and catastrophic.

Fortunately no one was hiking on this trail in Bryce Canyon National Park, Utah, when a large mass of rock suddenly detached from the cliff above, crashing down onto the ground below. National Park Service

Slope instability causes costly damage to buildings, roadways, pipelines, and other types of construction and is also responsible for injury and loss of life. Some gravity-induced slope movements are entirely natural in origin, but human actions contribute to the occurrence of others. Understanding the processes involved and circumstances that lead to slope movements can help people avoid these costly and often hazardous events.

Nature of Exogenic Processes

Earth's surface provides a harsh environment for rocks. Most rocks originate under much higher temperatures and pressures and in very different chemical settings than those found at Earth's surface. Surface and near-surface conditions, which have comparatively low temperature, low pressure, and extensive contact with water, cause rocks to undergo varying amounts of disintegration and decomposition (● Fig. 15.1). This breakdown of rock material at and near Earth's surface is known as **weathering**. Rocks weakened and broken by weathering become susceptible to the other exogenic processes: erosion, transportation, and deposition. A rock fragment broken (weathered) from a larger mass will be removed from that mass (eroded), moved (transported), and set down (deposited) in a new location. Together, weathering, erosion, transportation, and deposition actually represent a chain or continuum of processes that begins with the breakdown of rock.

Erosion, transportation, and deposition of weathered rock material often occur with the assistance of a geomorphic agent, such as flowing water, wind, moving ice, and waves. Sometimes,

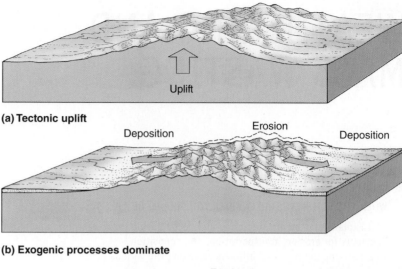

(a) Tectonic uplift

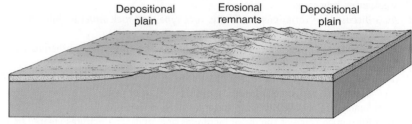

(b) Exogenic processes dominate

(c) Reduced relief

● **FIGURE 15.2** (a) Tectonic uplift, an endogenic process, is opposed by (b) the exogenic processes of weathering, mass wasting, erosion, transportation, and deposition that (c) eventually decrease relief at Earth's surface significantly if no additional uplift occurs.

however, the only factor involved is gravity. Gravity-induced downslope movement of rock material that happens without the assistance of a geomorphic agent, as in the case of a rock falling from a cliff, is known as **mass wasting**. Although gravity also plays a role in the redistribution of rock material by geomorphic agents, the term mass wasting is reserved for movement caused by gravity alone. Whether it is mass wasting or a geomorphic agent doing the work, fragments and ions of weathered rock are removed from high-energy locations and transported to positions of low energy, where they are deposited.

The variations in elevation at Earth's surface, and the shape of individual landforms, reflect the opposing tendencies of endogenic and exogenic processes. The relief created by tectonism and volcanism will decrease over time if endogenic processes cease or operate at a slow rate compared to the exogenic processes (● Fig. 15.2). Rates of the exogenic processes depend on factors such as rock resistance to weathering and erosion, the amount of relief, and climate.

● **FIGURE 15.1** This boulder, which was once hard and solid, has undergone disintegration and decomposition due to conditions at Earth's surface.

Why are some exposed parts of the boulder darker than others?

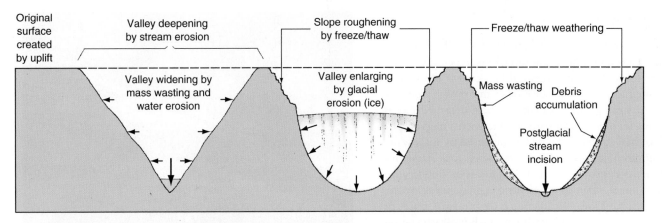

● **FIGURE 15.3** The landscape of most high mountain regions has been created by at least three distinct phases of landforming processes: stream cutting and valley formation during tectonic uplift; glacial enlargement of former stream valleys with intense freeze–thaw weathering above the level of the ice; and postglacial streamcutting, weathering, and mass wasting.

How does the cross-sectional profile of the valley change at each phase?

Different exogenic geomorphic processes impart visually distinctive features to a landform or landscape. Typically, weathering, mass wasting, or one of the various geomorphic agents do not work alone in shaping and developing a landform. More often, they work together to modify the landscape, and evidence of multiple processes can be discerned in the appearance of the resulting landform (● Fig. 15.3). For example, both the northern Rockies and the Sierra Nevada were produced by tectonic uplift (an endogenic process), but much of the spectacular terrain seen in both mountain ranges today is the result of weathering, mass wasting, running water, and glacial activity (exogenic processes) that have sculpted the mountains and valleys in distinctive ways (● Fig. 15.4).

● **FIGURE 15.4** Sculpted mountain summits of the White Cloud Peaks in the Sawtooth Range of Idaho.

Can you identify evidence in this photograph of the three phases shown in Figure 15.3?

Austin Post/USGS

Weathering

Environmental conditions at and near Earth's surface subject rocks to temperatures, pressures, and substances, especially water, that contribute to physical and chemical breakdown of exposed rock. Broken fragments of rock, called *clasts*, that detach from the original rock mass can be large or small. These detached pieces continue to weather into smaller particles. Fragments might accumulate close to their source or be widely dispersed by mass wasting and the geomorphic agents. Many weathered rock fragments become sediments deposited in such landforms as floodplains, beaches, or sand dunes, whereas others blanket hillslopes as *regolith*, the inorganic portion of soils. Weathering is the principal source of the inorganic constituents of soil, without which most vegetation could not grow. Likewise, ions chemically removed from rocks during weathering are transported in surface or subsurface water to other locations. These ions represent a major source of nutrients for terrestrial as well as aquatic ecosystems, including rivers, ponds, lakes, and the ocean.

The several types of rock weathering fall into two basic categories. **Physical weathering**, also known as *mechanical weathering*, disintegrates rocks, breaking larger blocks or outcrops of rock into smaller clasts. **Chemical weathering** decomposes rock through chemical reactions that remove ions from the original rock-forming minerals. Disintegration and decomposition of rock matter accomplished in ways influenced by organisms is sometimes referred to as **biological weathering**, even though the processes are fundamentally physical or chemical in nature. Many physical and chemical processes contribute to rock weathering, and water plays an important role in almost all of them.

Physical Weathering

The mechanical disintegration of rocks by physical weathering is especially important to landscape modification in two ways. First, the resulting smaller clasts are more easily eroded and transported than the initial larger ones. Second, the breakup of a large rock into smaller rocks encourages additional weathering because it increases the rock surface area exposed to weathering processes. Rocks can be physically weathered in several ways; even a person breaking a rock with a hammer is carrying out physical weathering. Five principal types of physical weathering are discussed here.

Unloading Most rocks originate under much higher pressure (weight per unit area) than the 1013.2 millibars (29.92 in. Hg; 15 lbs/in.²) of average atmospheric pressure that exists at Earth's surface. Intrusive igneous rocks solidify slowly deep beneath the surface under great pressure from the weight of the overlying rocks. Sedimentary rocks solidify partly or wholly as a result of compaction from the weight of overlying sediments. Metamorphic rocks form when high pressure and temperature substantially change preexisting rocks.

Many rocks that originated under conditions of high pressure from deep burial are later uplifted through mountain-building

● **FIGURE 15.5** Outward expansion as a result of unloading has caused this granite in the Sierra Nevada, California, to break, forming joints and sheets of rock that parallel the surface.

tectonic processes and are eventually exposed at Earth's surface. For example, a large pluton of the intrusive igneous rock granite can be uplifted in a fault block during tectonism. Uplift to a high elevation helps drive erosional stripping of the rocks overlying the pluton. Eventually, through this erosional removal of overlying weight—the **unloading** process—the upper part of the granite is exposed at Earth's surface, where it is subjected to the low pressure of the atmosphere. The difference between the higher pressure on the rock at depth and the lower (atmospheric) pressure on the granite's surface causes the outer few centimeters to meters of the rock mass to expand outward toward the low pressure of the atmosphere (● Fig. 15.5). This expansion causes an extensive crack to form a few centimeters or meters below the exposed rock surface, and roughly parallel to the surface. Additional cracks commonly develop a few centimeters to meters deeper within the rock approximately parallel to the first crack. These shallow expansion cracks are *joints*, not faults, and they break the rock into concentric sheets. The breakage is mechanical weathering resulting from erosional removal of all the rock that was originally on top of the pluton.

As the outer sheet of an unloaded rock outcrop continues to weather, sections of it can slide off and slough away, further reducing the load on the underlying rock and allowing additional roughly parallel joints to form. The successive removal of these outer rock sheets is known as **exfoliation**, and each concentric broken layer of rock is an *exfoliation sheet*. An **exfoliation dome** is an unloaded exfoliating outcrop of rock with a domelike surface form.

Weathering by unloading is especially common on granite, but sometimes it affects other rocks that lack internal bedding and other planes of weakness. Well-known examples of exfoliation domes are Stone Mountain in Georgia; Half Dome in Yosemite National Park; Sugar Loaf Mountain, which overlooks Rio de Janeiro, Brazil; and Enchanted Rock in central Texas (● Fig. 15.6).

(a)

(b)

● **FIGURE 15.6** (a) Enchanted Rock in central Texas appears distinctly dome-shaped from the air. From left to right the photo represents about 0.6 kilometers (0.4 mi). (b) In ground view, the sides of the large outcrop display slabs of exfoliating granite.

Why is granite so susceptible to unloading and exfoliation?

Thermal Expansion and Contraction Many early Earth scientists believed that the extreme diurnal temperature changes common in deserts physically weather rocks because of expansion and contraction as the rocks heat and cool. Those scientists cited widespread existence of split rocks in arid regions as evidence of the effectiveness of this **thermal expansion and contraction** weathering (● Fig. 15.7). Laboratory studies in the early twentieth century seemed to refute the idea of rock weathering by this process, yet the notion remained popular. Recent field studies in the

● **FIGURE 15.7** The many fractures in this split rock may be the result of physical weathering by thermal expansion and contraction.

desert of the American Southwest now lend scientific support to the concept that alternating heating and cooling can lead to the mechanical splitting of rocks. The broken rocks found after a forest or brush fire provide additional evidence for the effectiveness of thermal expansion and contraction weathering, although fire heats rocks to much higher temperatures than the atmosphere can.

Less controversial has been the notion that differential thermal expansion and contraction of individual mineral grains in coarse crystalline rocks contributes to **granular disintegration**, the breaking free of individual mineral grains from a rock (● Fig. 15.8). How much a grain expands and contracts upon heating and cooling depends not only on air temperature but also on the type of mineral. Different minerals often attain unequal maximum temperatures despite receiving the same insolation if, for example, they have dissimilar colors and therefore different albedos (percentage reflectivity of solar energy). The volume change that a mineral grain undergoes for a given increase in temperature likewise varies with the type of mineral. Such factors lead to a multitude of different stress fields in coarse-grained rocks that are composed of various mineral types. These differential stresses eventually loosen and dislodge individual mineral grains.

Freeze–Thaw In regions subject to numerous diurnal freeze–thaw cycles, water repeatedly freezing in fractures and small cracks in rocks contributes significantly to rock breakage by **freeze–thaw weathering**, sometimes referred to as *frost weathering* or *ice wedging*. When water freezes, it expands in volume up to 9%, and this can cause large pressures to be exerted on the walls and bottom of the crack, widening it and eventually leading to a piece of rock breaking off (● Fig. 15.9). The damaging effects of the expansion of freezing water is why in regions with freezing temperatures vehicles must have antifreeze instead of water in their radiator systems. Likewise, water pipes to and within buildings will burst

D. Sack

● **FIGURE 15.8** Weathering of this intrusive igneous boulder is causing granular disintegration, as evidenced by the numerous individual mineral grains that have collected around the base of the boulder.

What other evidence exists on the boulder to suggest that it has been subjected to extensive weathering?

● **FIGURE 15.9** Within rock fractures, the force of expansion of water freezing into ice is great enough to cause some rock weathering.

How important is freeze–thaw weathering where you live?

if they are not sufficiently insulated to protect the water inside them from freezing. Freeze–thaw weathering is particularly effective in the upper middle latitudes and lower high latitudes, and results of freeze–thaw weathering are especially noticeable in mountainous regions near the tree line where angular blocks of rock attributed to freeze–thaw weathering are common (● Fig. 15.10). Because of warmer temperatures, freeze–thaw weathering is not significant at lower latitudes except in areas of high elevation.

Salt Crystal Growth The development of salt crystals in cracks, fractures, and other void spaces in rocks causes physical disintegration in a way that is similar to freeze–thaw weathering. With **salt crystal growth**, water containing dissolved salts accumulates in these spaces. When the water evaporates, the salts stay behind growing into crystals that are capable of wedging pieces of rock apart. This physical weathering process is most common in arid regions and rocky coastal locations where salts are abundant, but people also contribute to salt weathering when using salt to melt ice from roads and sidewalks in winter. Salt crystal growth leads to granular disintegration in coarse crystalline (intrusive igneous) rocks and to the removal of clastic particles from sedimentary rocks, especially sandstones. Continued salt weathering, accompanied by removal of the weathered fragments, contributes to the creation of the numerous small hollows that occur along many sandstone cliffs in arid and coastal locations (● Fig. 15.11).

Hydration In weathering by **hydration**, water molecules attach to the crystalline structure of a mineral without causing a permanent change in that mineral's composition. The water molecules

M.A. Shroeder, Washington Department of Fish and Wildlife

● **FIGURE 15.10** Angular blocks of rock attributed to freeze–thaw weathering, like these on Armstrong Mountain in Washington, are common in mountainous areas that experience numerous freeze–thaw cycles every year.

Why are these rocks angular in shape rather than rounded?

J. Petersen

● **FIGURE 15.11** Physical weathering by salt crystal growth helps break apart rocks, especially along bedding planes exposed in cliffs and on rocky seacoasts. Weathered fragments are removed by gravity, wind, and water, leaving behind hollows.

Once a small hollow is formed, how might it affect further weathering at that site?

D. Sack

● **FIGURE 15.12** The orange coloration on this boulder reveals that the rock contains iron-bearing minerals and that it has been weathered by oxidation.

What is a likely chemical formula for the reddish-orange substance?

are able to join and leave the host mineral during hydration and dehydration, respectively. The mineral expands when hydrated and shrinks when dehydrated. As in weathering by freeze–thaw and salt crystal growth, when hydrating materials expand in cracks or voids within rocks, pieces of the rock can wedge apart. Clasts, mineral grains, and thin flakes are broken from rock masses because of hydration weathering. Salts and **clay minerals**, which are clay-sized materials formed during chemical weathering, commonly occupy cracks and voids in rocks and are subject to hydration and dehydration.

Chemical Weathering

Chemical reactions between rock-forming minerals and other substances at Earth's surface also work to break down rocks. In chemical weathering, ions from a rock are either released into water or recombine with other substances to form new materials, such as clay minerals. New materials made by chemical weathering are generally more stable at Earth's surface than the original rocks. The most important catalysts and reactive agents performing chemical weathering are water, oxygen, and carbon dioxide, all of which are common in soil, precipitation, surface water, groundwater, and air.

Oxidation Water that has regular contact with the atmosphere contains plenty of oxygen. When oxygen in water comes in contact with certain elements in rock-forming minerals, a chemical reaction can occur. In this reaction, the element releases its bond with the mineral, leaving behind a substance with an altered chemical formula, and establishes a new bond with the oxygen. This chemical union of oxygen atoms with another substance to create a new product is **oxidation**. Metals, particularly iron and aluminum, are commonly oxidized in rock weathering and form iron and aluminum oxides as the new products. Compared with the original rock, these oxides, including Fe_2O_3 and Al_2O_3, are chemically more stable, not as hard, and larger in volume. Iron oxides produced in this way often have a red or orange color, whereas oxidized aluminum from rock weathering typically appears yellow (● Fig. 15.12). Oxidation of iron is common, and when it affects iron and steel objects we call it rust.

Solution and Carbonation Under certain circumstances, some rock-forming minerals dissolve in water. In this process of **solution**, a chemical reaction causes mineral-forming ions to dissociate and the separated ions are carried away in the water. Rock salt, which contains the mineral halite (NaCl), is quite susceptible to solution in water. Most minerals that are insoluble or only slightly soluble in pure water will dissolve more readily if the water is acidic. Water at or near Earth's surface increases in acidity by picking up organic acids from the soil or carbon dioxide from the air or soil.

The chemical weathering process of **carbonation** is a common type of solution that involves carbon dioxide and water molecules reacting with, and thereby decomposing, rock material. Carbonation weathering is most effective on carbonate rocks (those containing CO_3), particularly limestone, which is an abundant sedimentary rock composed of calcium carbonate ($CaCO_3$). When water with sufficient carbon dioxide comes into contact with limestone, the chemical reaction decomposes some of the calcium carbonate into separate, detached ions of calcium

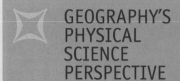
Expanding and Contracting Soils

Expansion and contraction of rocks, sediments, or materials in fractures and other void spaces in rocks and sediments is a common theme in weathering and mass wasting. Physical weathering by thermal expansion and contraction, freezing and thawing of water, and hydration and dehydration of salts and clay minerals all involve cycles of expansion and contraction, as does downslope movement of soil by creep.

Soils consist of inorganic and organic materials, soil water, and soil air. The inorganic fraction of soils—rock fragments and minerals—derives from rock weathering. These fragments are of sand, silt, and clay size. As weathering continues, clasts become smaller and smaller. In addition, the chemical weathering process of hydrolysis produces very small substances called clay minerals. Clay minerals are colloidal in size, meaning they are smaller than 0.0001 millimeter. In addition to rock fragments and clay minerals, some decaying organic matter (humus) in the soil is colloidal in size.

Having colloidal-sized materials in the soil benefits soil fertility because colloids increase the cation exchange capacity of the soil, which is essentially the ability of the soil to hold plant nutrients. Colloids

are so small that their negative electrical charge becomes an important characteristic. Plant nutrients, such as potassium, calcium, magnesium, phosphorous, copper, zinc, and several others, are positively charged ions, or cations. The negatively charged colloids, therefore, attract nutrient cations. When a plant takes up soil water, nutrient cations are also drawn up. A high cation exchange capacity means that the soil has plenty of colloids and therefore the potential to maintain a large number of nutrient cations in the soil for plants to use. The specific capacity of a colloid to hold cations is a function of its composition and structure. The cation exchange capacity of a clay mineral depends on the type of clay mineral, which, in turn, depends on the kind of parent material and the extent of weathering. Illite and vermiculite are two clay minerals associated with slightly weathered soils; montmorillonite is an important component of moderately weathered soils; and kaolinite is characteristic of highly weathered soils.

Nutrient cations are not the only substances that join and leave some colloids. Water molecules come into and leave the crystal structure of colloidal-sized clay minerals during hydration and dehydration. Clay minerals have layered, "sandwichlike"

structures. Water molecules enter and exit their layered crystalline structure and cause the clay mineral to expand when wet and contract when dry. Of the four basic clay minerals, montmorillonite expands and contracts the most upon hydration and dehydration. The expansion and contraction, or swelling and shrinking, of clay minerals applies physical pressure on surrounding soil particles and rock surfaces and contributes to physical weathering. According to some sources, the expansion in clay mineral volume can vary from a very small percentage to more than 100%. In most cases, however, the expansion in volume is less than 50%.

A high content of clay minerals advantageous for soil fertility sometimes interferes with the human-built environment. Soils with a significant amount of expansive clays swell and shrink considerably as they become wet and dry. The expansion and contraction in soil volume can shift and crack roads, sidewalks, and building foundations. In regions with typically fertile, expansive clay-rich soils, this causes an estimated $7 billion damage a year to structures in the United States alone. Special construction techniques can be used to mitigate the negative effects of shrinking and swelling on the built environment.

A concrete driveway slab has been heaved upward by expansive soils that have also damaged the adjacent railing and retaining wall.

Foundation cracks resulting from expansive soils represent serious structural damage that is expensive to repair.

J. Petersen

● **FIGURE 15.13** Limestone weathered by carbonation often takes on a fluted, pitted, or even honeycombed appearance.

Why does the rock near the bottom of the outcrop seem to be more weathered than that at the top?

D. Sack

● **FIGURE 15.14** Interfingering of tree root networks with rocks can lead to rock breakage if the tree is uprooted by strong winds or other means.

(Ca^{2+}) and bicarbonate (HCO_3^-) that are carried away in the water (● Fig. 15.13). The weathering event occurs as follows: $H_2O + CO_2 + CaCO_3 = Ca^{2+} + 2HCO_3^-$. Similar reactions take place when other types of carbonate rocks undergo carbonation weathering. Because of the role of water in carbonation, limestone in particular, but carbonate rocks overall, tend to weather extensively in humid regions but are resistant and often form cliffs in arid climates. Because water can obtain carbon dioxide by moving through the soil, carbonation operates on rocks affected by soil water or groundwater in the subsurface as well as on rocks exposed at the surface.

Hydrolysis In the weathering process of **hydrolysis**, water molecules alone, rather than oxygen or carbon dioxide in water, react with chemical components of rock-forming minerals to create new compounds, of which the H^+ and OH^- ions of water are a part. Many common minerals are susceptible to hydrolysis, particularly the silicate minerals (those containing oxygen and silica) that compose igneous rocks. As an example, hydrolysis of the feldspar mineral $KAlSi_3O_8$ can proceed as follows: $KAlSi_3O_8 + H_2O = HAlSi_3O_8 + KOH$.

Hydrolysis of silicate minerals often produces clay minerals, some of which swell and shrink considerably when hydrated and dehydrated, and thereby contribute to physical weathering by hydration. It is chemical weathering by hydrolysis, however,

that produces many of the clay minerals. The specific type of clay mineral resulting from hydrolysis depends on the composition of the preexisting mineral and other substances that may be present and active in the water. Because water is the weathering agent, hydrolysis is not limited to rocks that are exposed at the surface but can occur in the subsurface through the action of soil water or groundwater; in tropical humid climates, hydrolysis can decompose rock to a depth of 30 meters (100 ft) or more.

The chemical weathering process of hydrolysis differs from the physical weathering process of hydration, in which water molecules join and leave a substance causing it to swell and shrink without changing its inherent chemical formula. Both weathering processes, however, may involve clay minerals because those clay-sized substances are produced by hydrolysis, and many clay minerals swell and shrink substantially during hydration and dehydration.

Biological Weathering

Plants, animals, decaying organic matter, lichens, algae, fungi, and even bacteria acting as agents of physical and chemical weathering is biological weathering. For example, plant roots physically wedge into and expand cracks in rocks causing them to break further. Mechanical breakage of rocks also occurs when trees in rocky terrain are knocked over and uprooted, such as by strong winds (● Fig. 15.14). Acids from decaying organic matter contribute to chemical weathering by solution. Lichens growing on rocks can physically dislodge particles and they also secrete acidic substances that mix with water from precipitation, encouraging chemical weathering by solution (● Fig. 15.15). Some bacteria obtain nutrients directly from rock-forming minerals, and are able to oxidize metals. Clearly, organisms contribute in a variety of ways to chemical and physical weathering of rock.

Also, chemical weathering increases as more water comes into contact with rocks. Chemical weathering, then, is particularly effective and rapid in humid climates (● Fig. 15.16). Most arid regions have enough moisture to allow some chemical weathering, but it is much more restricted than in humid climates. Arid regions typically receive sufficient moisture to enable the physical weathering processes of salt crystal growth and hydration of salts. Abundant salts, high humidity, and contact with seawater make salt weathering of rocks very effective in marine coastal locations.

The other principal climatic variable, temperature, also influences dominant types and rates of weathering. Most chemical reactions proceed faster at higher temperatures. Low latitude regions with humid climates consequently experience the most intense chemical weathering. In the tropical rainforest, savanna, and monsoon climates, chemical weathering is more significant than physical weathering, soils are deep, and landforms appear rounded. Although chemical

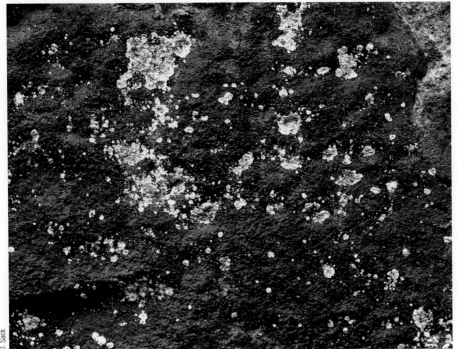

● **FIGURE 15.15** The light green patches of lichens growing on this reddish rock contribute to solution weathering because of the acids that they secrete.

How does the fact that lichens retain moisture also contribute to weathering?

Variability in Weathering

How and why types and rates of weathering vary, at both the regional and local spatial scales, are of particular interest to physical geographers. Climate, the type of rock (lithology), and the nature and amount of fractures or other weaknesses in the rock are major influences on the effectiveness of the various weathering processes. Understanding how specific rocks weather in natural settings is important for explaining the characteristics of regolith, soil, and relief, and it is also important in our cultural environment because weathering affects building stone as well as rock in its natural setting.

Climate Factors

In almost all environments, whether biological or not, physical and chemical weathering processes operate together, even though one of these two categories usually dominates. Water plays a role in most of the physical weathering processes, but it is essential for all types of chemical weathering.

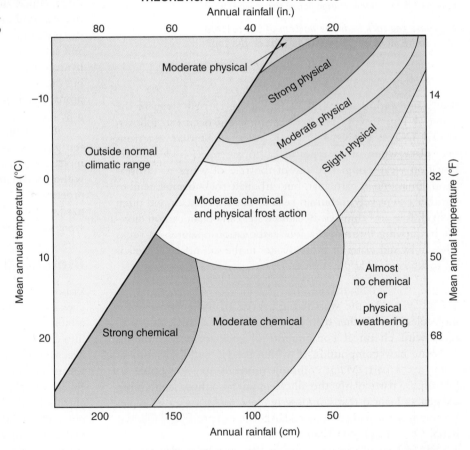

● **FIGURE 15.16** This generalized, theoretical diagram indicates that chemical and physical weathering both occur in any climate. Chemical weathering, however, is generally more intense in regions of high temperature and rainfall, whereas physical weathering tends to be more pronounced where temperature and rainfall are low.

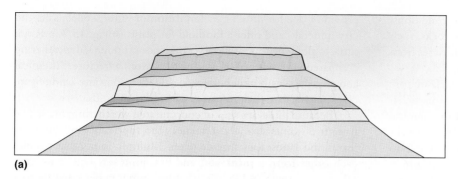

(a)

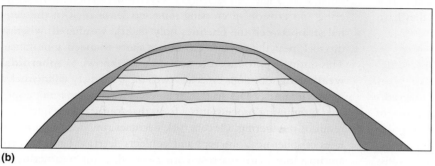

(b)

● **FIGURE 15.17** (a) Because of the dominance of physical weathering, which is comparatively slow, and the sparseness of vegetation, slopes in arid and semiarid environments tend to be bare and angular. Slope angles reflect differences in rock resistance to weathering and erosion. (b) Because chemical weathering dominates in humid regions and abundant vegetation holds the regolith in place longer, slopes in wetter climates have a deep weathered mantle and a rounded appearance.

weathering is somewhat less extreme in the midlatitude humid climates, its influence is apparent in the moderate soil depth and rounded forms of most landscapes in those regions. In contrast, the landforms and rocks of both arid and cold regions, where physical weathering dominates, tend to be sharper, angular, and jagged, but this depends to some extent on rock type, and rounded features may remain in an arid landscape as relicts from prehistoric times when the climate was wetter (● Fig. 15.17). Comparatively low rates of chemical weathering are reflected in the thin soils found in arid, subarctic, and polar climate regimes. Daily temperature ranges are the critical factor for thermal expansion and contraction weathering in arid climates and for freeze–thaw weathering in areas with cold winters.

Air pollution that contributes to the acidity of atmospheric moisture accelerates weathering rates. In some regions, extensive damage has already occurred to historic cultural artifacts made of limestone and marble (metamorphosed limestone), which both contain calcium carbonate (● Fig. 15.18). Because many of the world's great monuments and sculptures are made of limestone or marble, there is a growing concern about damage to these treasures from weathering. The Parthenon in Greece, the Taj Mahal in India, and the Great Sphinx in Egypt are just a few examples of cultural artifacts undergoing pollution-induced solution, salt crystal growth, hydration of salts, and other destructive weathering processes.

Rock Type

Some rock types are more resistant to weathering than other rock types. Commonly, a rock that is strong under certain environmental conditions is easily weathered and eroded in a different environmental setting. Rocks that are resistant in a humid climate dominated by chemical weathering may be weak in arid or cold environments where physical weathering processes dominate, and

vice versa. The metamorphic rock quartzite is a good example. It is nearly inert chemically, and even though it is harder than steel, it is brittle and can be fractured by physical weathering. Shale, a fine-grained clastic sedimentary rock, is also chemically inert, but it breaks apart easily. In humid regions, limestone is highly

● **FIGURE 15.18** These limestone tombstones have undergone extensive chemical weathering, accelerated by air pollution.

What kind of chemical weathering has affected the iron fence?

VARIABILITY IN WEATHERING   429

susceptible to solution and carbonation, but under arid conditions, limestone is much more resistant. Some sedimentary rocks, such as sandstone, are only as strong as their cement, which varies from soluble calcium carbonate to inert and resistant silica.

Rocks composed of large mineral grains or coarse clastic particles have some small gaps between individual grains or particles. When water reaches these gaps it instigates physical or chemical weathering. Outcrops of granite in an arid or semiarid region resist weathering. However, the minerals in granite are susceptible to alteration by hydration, oxidation, and hydrolysis, particularly in regions with warm, humid conditions. Accordingly, granitic areas are often covered by a deeply weathered regolith where they have been exposed to a tropical humid environment.

Structural Controls

In addition to rock type, the relative resistance of a rock to weathering depends on other characteristics, such as the presence of joints, faults, folds, and bedding planes that make rocks susceptible to enhanced weathering. In general, the more *massive* the rock, that is, the fewer joints and bedding planes it has, the more resistant it is to weathering.

The processes of volcanism, tectonism, and rock formation produce fractures in rocks that can be exploited by exogenic processes, including weathering. Joints can be found in any solid rock that has been subjected to crustal stresses, and some rocks are intensely jointed (● Fig. 15.19). Joints and other fractures that commonly develop in igneous, sedimentary, and metamorphic rocks represent zones of weakness that expose more surface area of rock,

provide space for flow or accumulation of water, collect salts and clay minerals, and offer a foothold for plants (● Fig. 15.20). Rock surfaces along fractures tend to experience pronounced weathering compared with parts of the rock farther from a fracture. Chemical and physical weathering both proceed faster along any kind of gap, crack, or fracture than in places without such voids.

Because joints are sites of concentrated weathering, the spatial pattern of joints strongly influences the appearance of the landforms and landscapes that develop. Multiple joints that parallel each other form a **joint set**, and two joint sets will cross each other at an angle (● Fig. 15.21). Over time, preferential weathering and erosion in crossing joint sets leave rock in the central area between the fractures only slightly weathered, whereas the rock near the fractures acquires a more rounded appearance. This distinctive, rounded weathered form, known as **spheroidal weathering**, develops especially well on jointed crystalline rocks, such as granite (● Fig. 15.22). Spheroidal weathering is not a specific weathering process; it is a form that some rocks acquire as a result of weathering. Once a rock becomes rounded, weathering rates of spheroidal outcrops and boulders decrease because there are no more sharp, narrow corners or edges for weathering to attack, and a sphere exposes the least amount of surface area for a given volume of rock.

Differential Weathering and Erosion

Wherever several different rock types occupy a given landscape, some will naturally be more resistant and others will be less resistant to the weathering processes operating there. Because erosion

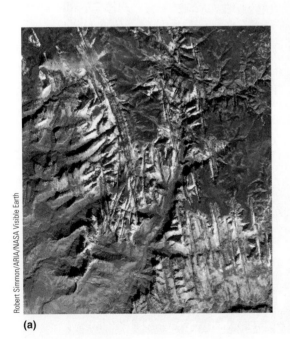

(a)   (b)

● **FIGURE 15.19** (a) Extensive jointing influences the location and shape of canyons in and around Zion National Park, Utah. To get an idea of the scale of this satellite image, note the smoke from the wildfire at upper left. (b) Farther north, in Bryce Canyon National Park, Utah, numerous closely spaced vertical joints in sedimentary rocks are sites of preferential weathering and erosion, leaving narrow rock walls between joints.

● **FIGURE 15.20** High up at a narrow spot along this vertical joint, enough soil has accumulated to allow a cactus to grow, while enhanced weathering and erosion lower down has noticeably widened the joint.

● **FIGURE 15.21** Multiple cross-cutting joint sets are visible in this aerial view of part of the Colorado Plateau.

With north at the top of this photo, what directions do the two most apparent joint sets trend?

● **FIGURE 15.22** Spheroidally weathered blocks of rock in cross-jointed granite east of the Sierra Nevada in California.

removes small, weathered rock fragments more easily than large, intact rock masses, areas of diverse rock types undergo **differential weathering and erosion**.

Variation in rock resistance to weathering exerts a strong and often highly visible influence on the appearance of landforms and landscapes. Given sufficient time, rocks that are resistant to weathering and erosion tend to stand higher than less resistant rocks. Resistant rocks stand out in the topography as cliffs, ridges, or mountains, whereas weaker rocks undergo greater weathering and erosion to create gentler slopes, valleys, or subdued hills.

An outstanding example of how differential weathering and erosion can expose rock structure and enhance its expression in the landscape is the scenery at Arizona's Grand Canyon (● Fig. 15.23). In the arid climate of this region, limestone is resistant, as are sandstones and conglomerates, but shale is relatively weak. Strong and resistant rocks are necessary to maintain steep or vertical cliffs. The stair-stepped walls of the Grand Canyon have cliffs composed of limestone, sandstone, or conglomerate, separated by gentler slopes of shale. At the canyon base, ancient resistant metamorphic rocks have produced a steep-walled inner gorge.

The topographic effects of differential weathering and erosion tend to be more prominent and obvious in landscapes of arid and semiarid climates. In dry environments, chemical weathering is minimized, so slopes and varying rock units are not generally covered under a thick mantle of soil or weathered rocks. In addition, sparse vegetation makes details of topography easy to see in arid regions.

Another fine example of differential weathering and erosion is the Appalachian Ridge and Valley region of the eastern United States (● Fig. 15.24). The rock structure here consists of sandstone, conglomerate, shale, and limestone folded into anticlines and synclines. These folds have been eroded so that the edges of steeply dipping rock layers are exposed as prominent ridges. In this humid climate region, forested ridges composed of resistant sandstones and conglomerates stand up to 700 meters (2000 ft) above agricultural lowlands that have been excavated by weathering and erosion out of weaker shales and soluble limestones.

VARIABILITY IN WEATHERING 431

● **FIGURE 15.23** Rock layers of varying thickness and resistance to weathering and erosion along the Grand Canyon create a distinctive array of cliffs composed of strong rocks and gentler slopes formed by less resistant rocks.

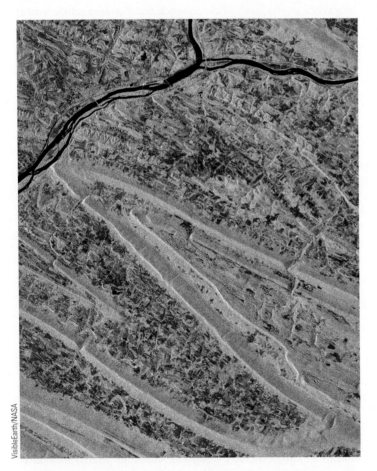

● **FIGURE 15.24** A satellite image of Pennsylvania's Ridge and Valley section of the Appalachians clearly shows the effects of weathering and erosion on folded rock layers of different resistance. Resistant rocks form ridges, and weaker rocks form valleys.

Can you see how the topography of the Ridge and Valley section influences human settlement patterns?

Mass Wasting

Mass wasting, also called *mass movement*, is a collective term for the downslope transport of surface materials in direct response to gravity. Everywhere on the planet's surface, gravity pulls objects toward Earth's center. This gravitational force is represented by the weight of each object. Heavier objects have a greater downward pull from gravity than lighter objects. The force of gravity encourages rock, sediment, and soil to move downhill on sloping surfaces.

Mass wasting operates in a wide variety of ways and at many scales. A single rock rolling and tumbling downhill is a form of this gravity-driven transfer of materials, as is an entire hillside sliding hundreds or thousands of meters downslope, burying homes, cars, and trees (● Fig. 15.25). Some mass movements act so slowly that they are imperceptible by direct observation, and their effects appear gradually over long periods of time. In these cases, tilted telephone poles, gravestones, fence posts, retaining walls, trees, or cracks in buildings can reveal how mass wasting processes are affecting the ground beneath those objects. Other types of mass wasting occur rapidly, sometimes with great violence and disastrous consequences.

The cumulative impact of all forms of mass wasting rivals the work of running water as a modifier of physical landscapes because gravitational force is always present. Wherever loose rock, regolith, or soil lies on a slope, gravity will sooner or later cause some particles to move downslope. Friction and rock strength are factors that resist this downslope movement of materials. Friction increases with the roughness and angularity of a rock fragment and the roughness of the surface on which it rests. Rock strength depends on physical and chemical properties of the rock and is reduced by any kind of break or gap in the rock. Fractures, joints, faults, bedding planes, and spaces between mineral grains or clasts all weaken the rock. Furthermore, because all of these gaps invite the accumulation of water, their bonds to the outcrop continue to weaken over time through weathering.

(a)

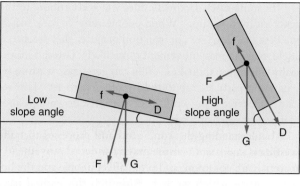

● **FIGURE 15.26** Besides slope angle and material strength, vegetation cover and soil moisture can influence the occurrence of mass wasting. G = total force of gravity (weight of the block); F = component of the block's weight resisting motion; f = frictional forces resisting motion; D = downslope component of gravity.

How might vegetative cover or moisture content affect the potential for downslope movement of soil?

(b)

● **FIGURE 15.25** (a) A massive boulder suddenly dislodged from a steep hillslope and crashed onto the residential street below, fortunately injuring no one. (b) In central Utah, gravity caused a large section of rock and soil to detach from a slope, slide, and then flow to a lower elevation.

What made the material in these two examples stop moving?

Slope angle also helps determine whether mass wasting will occur. Gravitational forces act to pull objects straight downward, toward the center of Earth. The closer a slope is to being parallel to that downward direction—in other words, the steeper the slope angle—the easier it is for the gravitational forces to overcome the resistance provided by friction and rock

strength. Gravity is more effective at pulling rock materials downslope on steep hillsides and cliffs than on gently sloping or level surfaces. The steeper the slope, the greater the friction or rock strength must be to resist downslope motion (● Fig. 15.26). Surface materials on a slope that do not have the strength or stability to resist the force of gravity will respond by creeping, falling, sliding, or flowing downslope until they reach a location with a gentle enough slope or sufficient friction to resist further movement. As a result, accumulations of soil and regolith are thinner on steep slopes and thicker on gentle slopes, and intense mass wasting is one reason why bedrock tends to be exposed in areas of steep terrain.

Gravity is the principal force responsible for mass wasting, but water commonly plays a role and can do so in several ways. Water (1) contributes to weathering, which prepares rock matter for mass movement, (2) adds weight to porous materials on a slope, (3) decreases the strength of unconsolidated slope sediments, and (4) can increase the slope angle. We have seen that water is involved in many weathering processes that break and weaken rocks, making them more susceptible to mass movement. Unconsolidated soil and regolith have a considerable volume of voids, or pore spaces, between particles. Usually some of these voids contain air and some contain water, but storms, wet seasons, broken pipelines, irrigation, and other situations can cause the voids to fill with water. Saturated conditions encourage mass wasting because water adds weight to the collection of materials, and an object's weight represents the amount of gravitational force pulling on it. As water replaces air in the voids, unconsolidated sediments and soil also experience decreasing strength as the rock fragments come into greater contact with the liquid, which tends to flow downslope. Finally, streams, and in coastal locations waves, undercut the base of slopes by erosion, thereby increasing the slope angle and facilitating mass movement.

In some cases, factors besides water contribute to the occurrence of mass wasting. When people remove rock material from the base of a slope for construction projects, they steepen the slope angle, making mass movement more likely. Ground shaking, especially during earthquakes, often triggers mass wasting by briefly separating particles supported by other particles. This separation causes a decrease in two factors that resist downslope movement: material strength and friction.

Understanding the conditions and processes that affect mass wasting is important because gravity-induced movements of Earth materials are common, and often affect people and the built environments in which we live. Although this natural hazard cannot be eradicated, we can avoid actions that aggravate the hazard potential and pay close attention to evidence of impending failure in susceptible terrain.

Materials and Motion

Physical geographers categorize mass wasting events according to the kinds of Earth materials involved and the ways in which they move. Mass wasting events are described with a specific term, for example *rockfall*, that summarizes the type of material and the type of motion.

Anything on Earth's surface that exists in or on an unstable, or potentially unstable, land surface is susceptible to gravity-induced movement and can therefore be transported downslope as a result of mass wasting. Mass wasting involves almost any kind of surface material. Rock, snow and ice, soil, earth, debris, and mud commonly experience downslope movements. In a mass wasting sense, **soil** denotes a thin layer of predominantly fine-grained, unconsolidated surface material. Soil moving downslope by gravity could involve only particles at the ground surface, or it could extend beneath the surface by no more than a couple of meters. Where mass wasting involves a thicker unit of fine-grained, unconsolidated material, often tens of meters thick, the material is called **earth**. **Debris** specifies a mass of sediment composed of a wide range of grain sizes, at least 20% of which is gravel-sized. **Mud** indicates saturated sediment composed mainly of clay and silt, which are the smallest particle sizes.

Surface materials move in response to gravity in many different ways, depending on the type of material, its water content, and characteristics of the setting. Some types of mass movement happen so slowly that no one can watch the motion occurring. With these **slow mass wasting** types we can only measure the movement and observe its effects over long periods of time. The motion of **fast mass wasting** can be witnessed directly by people. The speed of downslope movement of material varies greatly according to details of the slope, the nature of the rock matter, and whether a triggering factor is involved. In addition, slow and fast mass movements often work in combination. Mass movement that is initially slow may be a precursor to more destructive rapid motion, and most materials that have undergone rapid mass wasting continue to shift with slow movements. Specific types of motion, and common Earth materials associated with them, are discussed next and are summarized in Table 15.1.

TABLE 15.1
Common Types of Mass Wasting

Motion	Common Material	Typical Speed	Effect
Creep	Soil	Slow	
Solifluction	Soil	Slow	
Fall	Rock	Fast	
Avalanche	Ice and snow or debris or rock	Fast	
Slump (rotational slide)	Earth	Fast	
Slide (linear)	Rock or debris	Fast	
Flow	Debris or mud	Fast	

Slow Mass Wasting

Slow mass wasting has a significant, cumulative effect on Earth's surface. Landscapes dominated by slow mass wasting tend toward rounded hillcrests and an absence of sharp angular features.

Creep Most hillslopes covered with weathered rock or soil undergo **creep**, the slow migration of particles along the slope to successively lower elevations. This gradual downslope motion often occurs as *soil creep*, primarily affecting a relatively thin surface layer of weathered rock particles. Creep is so gradual that it is visually imperceptible; the rate of movement is usually less than a few centimeters per year. Yet creep is the most widespread and persistent form of mass wasting because it affects nearly all slopes that have soil or weathered rock fragments at the surface.

Creep results from some kind of **heaving** process, which causes individual soil particles or rock fragments to be first pushed upward perpendicular to the slope, and then eventually fall straight downward because of gravity. Freezing and thawing of soil water, and wetting and drying of soils or clay minerals, can lead to soil heaving. For example, when soil water freezes, it expands, pushing overlying soil particles upward relative to the surface of the slope. When that ice thaws, the soil particles move back, but not to their original positions because the force of gravity pulls downward

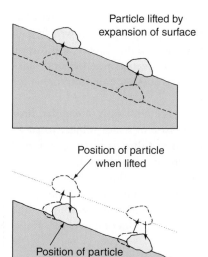

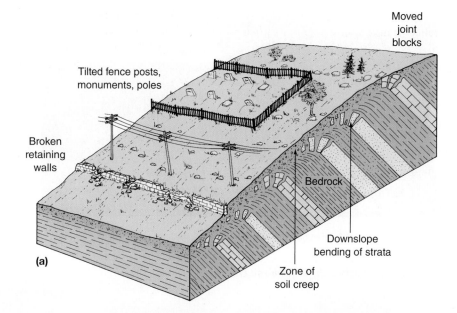

FIGURE 15.27 Repeated cycles of expansion and contraction cause soil particles to be lifted at right angles to the surface slope but to fall straight downward by the force of gravity, resulting in soil creep.

Are there places near where you live that show evidence of soil creep?

on them, as illustrated in ● Figure 15.27. Soil creep results from repeated cycles of expansion and contraction related to freezing and thawing, or wetting and drying, which cause lifting followed by the downslope movement of soil and rock particles.

Organisms contribute to soil creep as well as other kinds of mass wasting. The traversing of slopes by people and animals tends to push surface material downhill. Every step up, down, or across a steep slope shifts some soil or regolith downhill to a slightly lower position. Even small burrowing animals, such as ground squirrels and chipmunks, are effective soil movers. When they dig their tunnels on a sloping surface, the excavated material tends to fall downslope. The growth of plant roots can also move soil slowly outward and in a downward direction on a slope.

Rates of soil creep are greater near the surface and diminish into the subsurface of a slope because the factors instigating it are more frequent near the surface and because frictional resistance to movement increases with depth. As a result, telephone poles, fence posts, gravestones, retaining walls, other structures built by people, and even trees become tilted when affected by the downward movement of creep (● Fig. 15.28).

For the most part, creep does not produce discrete distinctive landforms, but it contributes to rounded, rather than angular, topography in hilly terrain. After gravitational forces deposit creeping surface sediments at the base of the slope, the sediments can eventually be carried away by one of the geomorphic agents, usually running water.

Solifluction The word **solifluction**, which literally means "soil flow," refers to the slow downslope movement of water-saturated soil or regolith. Solifluction is most common in high-latitude or

FIGURE 15.28 (a) Effects of soil creep are visible in natural and cultural landscapes. (b) Trees attempt to grow vertically, but their trunks become bent if surface creep is occurring.

What other constructed features might be damaged by creep?

high-elevation tundra regions that have **permafrost**, a subsurface layer of permanently frozen ground. Above the permafrost layer lies the **active layer**, which freezes during winter but thaws during summer. During the summer thaw of the active layer, which

● **FIGURE 15.29** Solifluction has formed these tongue-shaped masses of soil on a slope in Alaska.

How does solifluction differ from soil creep?

In steep mountainous areas, rockfalls are particularly common during the spring when snowmelt, rain and alternating freezing and thawing loosen and disturb rocks positioned on cliffs or steep slopes. Ground shaking caused by earthquakes is another common trigger for rockfalls.

Over time, a sloping accumulation of angular, broken clasts piles up at the base of a cliff subject to rockfall. This slope is known as **talus**, sometimes referred to as a *talus slope* or, where cone-shaped, a *talus cone* (● Fig. 15.31). The presence of talus is good evidence that the cliff above is undergoing rockfall. Like other accumulations of loose, unconsolidated sediment, such as the slopes of cinder cones or piles of gravel in a gravel

can be centimeters to meters in thickness, the permanently frozen ground underneath the active layer prevents downward percolation of melted soil water. As a result, the active layer becomes a heavy, water-saturated soil mass that, even on a gentle incline, sags slowly downslope by the pull of gravity until the next surface freeze arrives. Movement rates by solifluction are typically only centimeters per year.

Evidence of solifluction exists in many tundra landscapes. It consists of relatively thin, irregular, tongue-shaped lobes of soil that may partially override each other to produce hummocky terrain or low mounds (● Fig. 15.29).

Fast Mass Wasting

Four major kinds of mass wasting usually occur so quickly—from seconds to days—that people in the right place at the right time can watch the material move. The actual speed of movement varies with the situation and depends on the quantity and composition of the material, the steepness of slope, the amount of water present, the vegetative cover, and the triggering factor. The effects of fast mass wasting events on the land surface are more dramatic than those of slow mass wasting. Rapid mass movements usually leave a visible upslope scar on the landscape, revealing where material has been removed, and a definite deposit where transported Earth material has come to rest at a lower elevation.

Falls Mass wasting events that consist of Earth materials plummeting downward freely through the air are **falls**. **Rockfalls** are the most common type of fall. Rocks fall from steep bedrock cliffs, either one by one as weathering weakens the bonds between individual clasts and the rest of the cliff, or as large rock masses that fall from a cliff face or an overhanging ledge (● Fig. 15.30). Large slabs that fall typically break into angular individual clasts when they hit the ground at the base of the cliff.

● **FIGURE 15.30** Eventually this overhanging sandstone ledge will fail in the form of a rockfall. As more individual rocks fall from the damper, shaded zone beneath the ledge, the size of the overhang will increase until its weight exceeds the strength of the bonds holding the ledge in place.

What weathering processes might be acting on the sandstone beneath the overhang when it becomes wet?

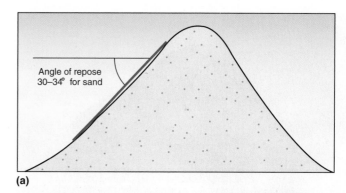

(a)

Angle of repose 30–34° for sand

● **FIGURE 15.31** Rockfall deposits in Glacier National Park, Montana, have accumulated into a cone-shaped talus, the slope of angular clasts.

USGS/P. Carrara

J. Petersen

(b)

● **FIGURE 15.32** (a) The angle of repose is the steepest natural slope angle that loose material can maintain. Particles are held at this angle by friction between clasts. (b) Nearly identical piles of sediment containing the same size and shape of particles lie at the same angle of repose.

How would the angle of repose of rounded particles differ from that of angular particles of the same size?

pit, talus slopes typically lie at or near the steepest angle they can maintain without instability, their *angle of repose* (● Fig. 15.32). The angle of repose commonly lies between 30° and 34°, varying with the size and angularity of the clasts. Large or angular clasts have a steeper angle of repose than small or more rounded rock fragments.

Falling rocks create hazardous conditions wherever bedrock cliffs are exposed, including mountainous regions and areas with steep roadcuts. Rockfall hazard mitigation remains a high priority along transportation routes in these locations. Massive rockfalls have contributed to the scenic beauty of the towering and steep granite cliffs in Yosemite National Park, California, but frequent rockfalls also represent serious hazards there (● Fig. 15.33). In 2010, a young girl was rescued after being pinned beneath a 4100 kilogram (9000 lb) boulder that had fallen 15 meters (50 ft) from the nearby steep slope. A larger, fatal rockfall in the park in 1996 moved downslope at an estimated 250 kilometers per hour (160 mph). In that case, the huge mass of moving rock also generated a destructive blast of compressed air that destroyed trees hundreds of meters from the cliff as the rock crashed to the valley floor.

Avalanches An **avalanche** is a type of mass movement in which much of the involved material is *pulverized*—that is, broken into small, powdery fragments—and then flows rapidly as an

airborne density current along Earth's surface. Although the word avalanche may bring to mind billowing torrents of snow and ice roaring down a steep mountainside, *snow avalanches* are not the only kind. Avalanches of pulverized bedrock, called *rock avalanches*, and those of a very poorly sorted mixture of gravel, sands, silts, and clays, called *debris avalanches*, are also common and have caused considerable loss of life and destruction in mountain communities around the world.

Many avalanches begin with rockfalls, debris falls, or snow and ice falls that pulverize upon impact with a lower surface. Snow avalanches, the type of avalanche best known to the public, present serious hazards in areas with steep slopes and deep accumulations of snow (● Fig. 15.34). The 2015 earthquake in Nepal triggered numerous deadly slope failures in that mountainous country,

FIGURE 15.33 A large rockfall at Ahwiyah Point in Yosemite National Park, California, fell about 550 meters (1800 ft) onto a hiking trail and the adjacent forest.

including an avalanche that tore through South Base Camp of Mount Everest, killing at least 19 people. One year earlier, a massive block of ice fell and initiated an avalanche near the same base camp, resulting in 16 deaths. In some more accessible regions, public safety personnel try to limit snow avalanche hazards by purposefully triggering small avalanches and restricting public access to areas where avalanches seem likely. Regardless of the specific type of Earth material involved, avalanches are powerful and dangerous, traveling up to 100 kilometers per hour (60 mph). They easily knock down trees and demolish buildings and have destroyed entire towns.

Slides In **slides**, a cohesive or semicohesive unit of Earth material slips downslope in continuous contact with the land surface. Water plays a somewhat greater role in most slides than it does in falls or avalanches. Slides of all kinds threaten the lives and property of people who live in regions with considerable sloping terrain along with such characteristics as tilted layers of alternating strong and weak rocks.

Slides of large sections of bedrock are called *rockslides*, and they are common in mountainous terrain where originally horizontal sedimentary rock layers have been tilted by tectonism. The importance of water in reducing the resisting forces of rock strength and friction is seen in the fact that rockslides are most common in wet years, or after a rainstorm or snowmelt. As weathering and erosion by water weaken contacts between successive rock layers, the force of gravity can exceed the strength of the bonds between two of the rock layers. When this happens, a segment of rock, often of massive size, detaches and slides along the tilted planar surface of the contact (Fig. 15.35). Rockslides sometimes end as rockfalls if the topography is such that free fall is needed to transport the rock to a stable position on more level ground. Deposits from rockslides tend to consist of larger blocks of broken rock than those that make up rockfall deposits.

Some rockslides are enormous, with volumes measured in cubic kilometers. Anything in their path is obliterated. Rockslide deposits in canyons and river valleys may act as a temporary dam that blocks drainage and causes a lake to form. When the lake becomes deep enough, it might wash out the rockslide dam, producing a sudden and disastrous downstream flood. Thus, immediately after this kind of major rockslide occurs, engineers work to stabilize the resulting rockslide dam and control the overflow outlet of the newly formed lake. This was done successfully after the Hebgen Lake slide in southwestern Montana in 1959 (Fig. 15.36). Triggered by an earthquake, that rockslide, one of the largest in North American history, killed 28 people camped along the Madison River.

(a)

(b)

FIGURE 15.34 (a) A snow avalanche in Utah. Because pulverized snow is denser than air, avalanches stay close to Earth's surface. People are often misled by the billowy cloudlike appearance of many snow avalanches. (b) Avalanches are powerful agents that plow through and incorporate whatever lies in their paths, often leaving heavy, densely packed deposits.

● **FIGURE 15.35** Rock units that dip in the same direction as the topographic slope of the land are especially susceptible to rockslide, as recognized along this stretch of highway in Wyoming.

● **FIGURE 15.36** The 1959 earthquake-induced rockslide on the Madison River in Montana completely blocked the river valley and created a new body of water, Earthquake Lake, seen in the background. The massive slide killed 28 people in a valley campground.

Why can earthquakes trigger landslides?

Huge rockslides can result from instability related to rock structure and from the undercutting of slopes by streams, glaciers, or waves. Today, there are many locations in mountain regions where enormous slabs of rock supported by weak materials are poised on the brink of detachment, waiting only for an unusually wet year or a jarring earthquake to set them in motion.

Rock is not the only Earth material prone to mass wasting by sliding. *Debris slides*, which contain a poorly sorted mixture of gravel and fine-grained sediment, and *mudslides*, which are dominated by wet silts and clays, are also common (● Fig. 15.37).

Slumps are rotational slides where earth—a thick block of fine-grained material—moves along a concave, curved surface. Because of this curved surface of failure, slump blocks undergo a backward rotation as they slide (● Fig. 15.38), causing what was initially the level ground surface at the top of the slump to tilt backward. Slumps are most common in wet years and during wet seasons in many regions with substantial relief, including the Appalachians, New England, and mountainous parts of the western United States. During exceptionally wet winters in Mediterranean climate regions, like California, slumps often damage hillside homes. Slumps can be triggered by earthquakes, which greatly reduce the material's friction and strength. People contribute to the occurrence of slumps and the other types of slide by purposefully or accidentally adding water to hillslope sediments and by increasing slope angles by excavating for construction purposes.

During the last several decades, **landslide** has become a general term popularly used to refer to any form of rapid mass movement. Earth scientists, however, restrict the term to large, rapid mass wasting events that are difficult to classify because they contain elements of more than one category of motion or because multiple types of material—rock, debris, earth, soil, and mud—are involved in a single massive slide. Such large slides are relatively rare but are often newsworthy because of their destructive qualities (● Fig. 15.39).

Flows In slope-failure terminology, **flows** are masses of water-saturated unconsolidated sediments that move downslope because of the pull of gravity. Mass wasting flows carry water in moving

FEMA/Dave Gatley

● **FIGURE 15.37** Mudslide damage in California. Mudslides can be generated by heavy rains or rapid snowmelt on steep hillslopes that have predominantly fine-grained sediments at and near the surface. Like rockslides, mudslides and debris slides travel along an approximately linear, inclined path.

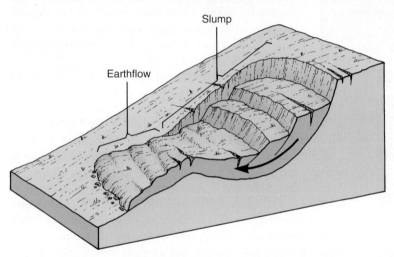

Slump

Earthflow

● **FIGURE 15.38** Slump is the common name for a rotational earthslide. Many slumps transition into the more fluid motion of an earthflow in their lower reaches.

How does the earthflow component differ from the slump component?

Spc. S. Ciaramitara, Washington National Guard

● **FIGURE 15.39** The upper part of a large, deadly landslide that in 2014 devastated a community near Oso, Washington. Once the slope gave way, it took only about 60 seconds for the wet mass of sediment to wipe away or bury 40 homes, resulting in the loss of 43 lives.

The Frank Slide

Many mass wasting events are compound, having elements of more than one type of motion. Avalanches and flows often begin as falls or slides. Slumps commonly grade into earthflows in their lower reaches. The massive and deadly slope failure that occurred more than a century ago at Turtle Mountain in the Canadian Rockies appears to have comprised two of the most catastrophic types of mass movement: rockfall and rock avalanche. This 1903 Turtle Mountain failure is known as the Frank Slide after the town of Frank, Alberta, a portion of which was obliterated by the very rapidly moving 30 million cubic meters (82 million tons) of rock, resulting in the loss of an estimated 70 lives.

Frank was situated at the base of Turtle Mountain along the Canadian Pacific Railroad line. Many of the town's 600 citizens worked as underground coal miners within the steep mountain. Most of the townsfolk didn't know what hit them as they lay sleeping at 4:10 a.m. when the mountain gave way. Others were at work in the mine inside the mountain when the force of gravity overcame the strength of the 1-kilometer (3280-ft) wide, 425-meter (1395-ft) high, and 150-meter (490-ft) thick mass of limestone. The resistance of the rock mass was weakened by underground mining activities, including blasting, and weathering and erosion along fractures near the mountain summit. Severe weather conditions may have also

played a role. In less than 2 minutes, the rockfall that became an avalanche on impact destroyed homes, buildings, roads, and the railroad line in its path and left a huge expanse of broken rock that extends across to the far side of the valley. Amazingly, 17 miners survived and dug their way out of the mountain through the rubble.

Despite having occurred more than a century ago, tremendous evidence of the huge rock failure still exists in the landscape today. The scar on the flank of Turtle Mountain and the rock rubble strewn over more than 3 square kilometers (1.2 sq mi) across the valley floor serve as reminders of the incredible and deadly power that can be unleashed by the force of gravity.

The massive 1903 rockfall–avalanche known as the Frank Slide left a huge scar on Turtle Mountain, Alberta, which remains very obvious in the landscape more than a century later.

Rubble from the rapidly moving rockfall–avalanche was strewn across the valley floor, well beyond the partly buried town of Frank.

sediments, whereas rivers carry sediments in moving water. Compared with slides, which tend to move as cohesive units, flows involve considerable churning and mixing of the materials as they move.

When a relatively thick unit of predominantly fine-grained, unconsolidated hillside sediment or shale becomes saturated and mixes and tumbles as it moves, the mass movement is an *earthflow*. Earthflows occur as independent gravity-induced events or in association with slumps in a compound feature called a *slump-earthflow* (see Fig. 15.38). A slump-earthflow moves as a cohesive

unit along a concave surface in the middle and upper reaches of the failure, the slump part. Downslope of the failure plane, the mass continues to move but in the more fluid-like, less cohesive manner of an earthflow.

Debris flows and **mudflows** differ from each other primarily in grain size and sediment attributes. Both flow faster than earthflows, often move down gullies or canyon stream channels for at least part of their travel, create raised channel rims called *debris-flow* or *mudflow levees*, and leave lobate (tongue-shaped) deposits where they spill out of the channel. They result from torrential rainfall or rapid snowmelt on steep, poorly vegetated slopes, and they are the most fluid of all mass movements. Debris flows transport more coarse-grained sediment than mudflows do.

Debris flows often originate on steep slopes, especially in arid or seasonally dry regions. In humid regions they can occur on steep slopes that have been deforested by human activity or wildfire. In arid and in humid settings, rain or meltwater flush weathered rock material from the steep slopes into canyons, where it acquires additional water from surface runoff. The result is a chaotic, saturated mixture of fine and coarse sediment, ranging in size from tiny clays to large boulders. As it flows down stream channels, some of the debris is piled along the sides as levees. Where a flow spills out of the channel, the unconfined mass spreads out and velocity decreases, resulting in deposition of a lobe of sediment (● Fig. 15.40). Debris flows are powerful mass wasting events that can destroy bridges, buildings, and roads (● Fig. 15.41). With dry summers and wet winters, Mediterranean climate regions are particularly susceptible to mudflows and debris flows, especially when rainy winters follow dry seasons in which considerable vegetation was destroyed by fires.

Serious mudflow hazards also exist in many active volcanic regions. Here, steep slopes may be covered with hundreds of meters of volcanic ash that can become saturated when erupted steam cools and falls as rain. Mudflows composed of volcanic ash are known as **lahars**. Of particular concern are high volcanic peaks capped with glaciers and snowfields. Should an eruption melt the ice and snow, rapid and catastrophic lahars could rush down the mountains with little warning and bury entire valleys and towns. In the United States there is concern about the risk of volcanic eruptions and associated lahars from some of the high volcanoes in the Cascade Range of the Pacific Northwest. Lahars accompanied the 1980 eruption of Mount St. Helens, and Mounts Rainier, Baker, Hood, and Shasta all have the conditions in place, including nearby populated areas, for potentially disastrous lahars to occur (● Fig. 15.42).

● **FIGURE 15.41** A 1995 debris flow in La Conchita, California, destroyed several homes. Steep slopes consisting of weak, unstable sediments of a variety of grain sizes failed during a period of heavy rainfall. A similar precipitation event triggered massive movement there again 10 years later, damaging 36 homes and killing 10 people.

Why might a specific site experience repeated slope failures over time?

● **FIGURE 15.40** Although small, this debris flow in western Utah left well-developed levees on either side of the fresh channel and deposited a tongue-shaped mass (lobe) of sediment where the flow spread out at the base of the slope.

What evidence is there to indicate this is a site of repeated debris flows?

● **FIGURE 15.42** The violent 1980 eruption of Mount St. Helens in Washington generated lahars—mudflows consisting of volcanic ash. This house was half buried in lahar deposits associated with that volcanic eruption.

THINKING GEOGRAPHICALLY

What type of mass wasting (material and motion) is most prominent in this photograph? What circumstances discernible from the photograph might have contributed to the occurrence of this mass wasting event?

Weathering, Mass Wasting, and the Landscape

In this chapter we have concentrated on the exogenic processes of weathering and mass movement. Although neither weathering nor the slower forms of mass movement usually attract much attention from the general public, they are critical to soil formation and, like faster forms of gravity-induced motion, they are significant factors in shaping the landscape. Weathering and mass wasting processes also significantly affect people as well as buildings and infrastructure. The relationship, moreover, is reciprocal; human actions accelerate some weathering processes and help induce mass movements.

Every slope reflects the local weathering and mass wasting processes that have acted on it. These, in turn, are largely determined by the properties of the rocks and the local climate factors. Slow weathering of resistant rocks leaves steep hillslopes, whereas rapid weathering of weak rocks produces gentle hillslopes that are typically blanketed by a thick mantle of soil or regolith. Differential weathering and erosion in areas of multiple rock types or variations in structural weakness produce complex landscapes of variable slopes.

Weathering proceeds rapidly in warm, humid climates where heat and moisture accelerate the reactions that cause minerals and rocks to decompose chemically. Deep mantles of soil and regolith, and rounded forms subject to creep, dominate these regions. In contrast, rocks in arid and cold climates weather more slowly, mainly by physical processes. Arid region slopes typically have thin, discontinuous, sparsely vegetated accumulations of predominantly coarse-grained regolith that are easily mobilized during intense precipitation events. A tendency toward fast mass wasting is reflected in the angular slopes that are common in many arid regions of mountainous terrain.

In reading the following chapters it will be important to remember the key role that weathering plays in preparing Earth materials for erosion, transportation, and deposition by the geomorphic agents. Because weathered rock fragments are often delivered to a geomorphic agent by gravity-induced movement from adjacent slopes, mass wasting is also important in preparing sediment for redistribution by streams, wind, ice, and waves.

CHAPTER 15 ACTIVITIES

■ TERMS FOR REVIEW

weathering	solution	permafrost
mass wasting	carbonation	active layer
physical (mechanical) weathering	hydrolysis	fall
chemical weathering	joint set	rockfall
biological weathering	spheroidal weathering	talus (talus slope, talus cone)
unloading	differential weathering and erosion	avalanche
exfoliation	soil (as a mass wasting material)	slide
exfoliation dome	earth (as a mass wasting material)	slump
thermal expansion and contraction	debris	landslide
granular disintegration	mud	flow
freeze–thaw weathering	slow mass wasting	debris flow
salt crystal growth	fast mass wasting	mudflow
hydration	creep	lahar
clay mineral	heaving	
oxidation	solifluction	

■ QUESTIONS FOR REVIEW

1. How is weathering important to the processes of erosion, transportation, and deposition?
2. In what ways is mass wasting similar to, yet different from, the action of the geomorphic agents?
3. What are four processes involving expansion and contraction that contribute to rock weathering?
4. Distinguish between hydration and hydrolysis.
5. Why is chemical weathering more prevalent in humid climates than in arid climates?
6. How are joints, fractures, and other voids in a rock related to the rate at which weathering takes place?
7. What are the impacts of differential weathering and erosion on shaping landforms?
8. What are the distinguishing characteristics of soil, earth, debris, and mud as materials moved in mass wasting?
9. What is soil creep and what factors facilitate it?
10. Describe the principal differences between (a) a rockslide and a rockfall, and (b) a slump and a debris flow.

■ CONSIDER AND RESPOND

1. What would you recommend as a solution to prevent the loss of valuable historical monuments to weathering processes?
2. What are some ways in which organisms influence weathering and mass wasting?
3. If you were an urban planner in a city with numerous steep slopes, what major hazards would you have to plan for? What recommendations would you make to reduce these dangers to the community?

PRACTICAL APPLICATIONS

1. A mass of wet sediment, including numerous large boulders, traveled rapidly down a desert canyon. The mass stopped moving when it spread out downslope of the canyon mouth, leaving a deposit in the form of a lobe. Laboratory analyses revealed that the sediment consisted of 18% clay, 29% silt, 27% sand, and 26% gravel. Based on this information, what specific type of mass wasting was it?

2. Find the mean annual temperature and annual rainfall data for the city or region where you live. Using that data, determine which theoretical weathering region you live in from the graph in Figure 15.16. Based on your observational evidence of weathering in the local environment (perhaps of natural rocks, tombstones, or building materials like stone, asphalt, and so on), write a short explanation and cite examples of why the rock weathering in your area fits or does not fit the theoretical weathering region from Figure 15.16.

LOCATE AND EXPLORE

1. Using Google Earth, identify the landforms at the following locations (latitude, longitude) and provide a brief discussion of how the landform developed and why the landform is found in that general area.

   a. 33.805°N, 84.145°W
   b. 51.56°N, 116.36°W
   c. 39.134°N, 113.441°W
   d. 37.740°N, 119.609°W
   e. 30.506°N, 98.818°W
   f. 49.305°N, 121.241°W

 MindTap—Make the most of your study time by accessing everything you need to succeed in one place. Read your textbook, take notes, review flashcards, watch videos, complete activities, take practice quizzes, and more online with MindTap. Log in at **www.cengagebrain.com**.

SUBSURFACE WATER AND KARST

OBJECTIVES

WHEN YOU COMPLETE THIS CHAPTER YOU SHOULD BE ABLE TO:

- 16.1 Recall the general distribution of Earth's freshwater resources.
- 16.2 Outline the principal differences among the subsurface water zone of aeration, the intermediate zone, and the zone of saturation.
- 16.3 Explain how water enters the subsurface and what facilitates its ability to move through the rock matter found there.
- 16.4 Describe the principal factors responsible for variations in groundwater supplies over space and time.
- 16.5 Draw a cross-sectional sketch showing a combined topographic and subsurface setting that would lead to an artesian spring.
- 16.6 Recount why underground water is susceptible to pollution.
- 16.7 Discuss why karst landforms are more common in warm and humid climates than in cold or dry climates.
- 16.8 Distinguish among the major landforms produced by solution of rock.
- 16.9 Understand what caverns are and how they are formed.
- 16.10 Explain how water dripping in caverns leads to the development of stalactites and stalagmites.

AS PART OF OUR UNDERSTANDING of the distribution and effects of flowing water on Earth, we must consider the portion of the hydrologic cycle that operates beneath the surface in addition to considering surface water. Like water flowing at the surface, water beneath Earth's surface moves, carries other substances, influences the form and appearance of the landscape, and represents an important source of freshwater for human use.

Freshwater is a precious and limited natural resource. There is much concern today throughout the world about the quantity and quality of our freshwater resources. Many populated regions have limited supplies of freshwater, whereas, ironically, in some sparsely populated or uninhabited areas, such as tundra and tropical rainforest regions, potable water is plentiful. Ice and snow in the polar regions represent 70% of the freshwater on Earth, but it is generally unavailable for human use. When most people think of freshwater, they typically envision rivers and lakes, both of which are important sources. Together, however, rivers and lakes represent less than 1% of this crucial resource. The remaining freshwater, nearly 30%,

Underground water first dissolves limestone to form caverns. Later, slow deposition of calcium carbonate decorates them—with particular beauty at Carlsbad Caverns, New Mexico. ©Natalia Bratslavsky/Shutterstock.com

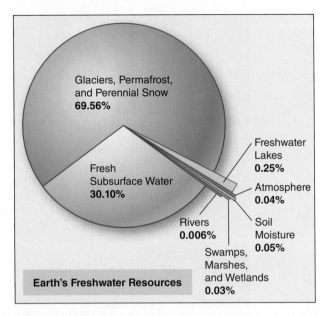

Glaciers, Permafrost, and Perennial Snow
69.56%

Fresh Subsurface Water
30.10%

Freshwater Lakes
0.25%

Atmosphere
0.04%

Rivers
0.006%

Soil Moisture
0.05%

Swamps, Marshes, and Wetlands
0.03%

Earth's Freshwater Resources

● **FIGURE 16.1** With 97% of Earth's water existing as saltwater in the oceans, Earth's freshwater resources are very limited. Almost 70% of Earth's freshwater is stored as glacial ice; the largest supply in liquid form lies underground.

What percentage of liquid freshwater flows in rivers?

lies close to, but beneath, Earth's surface. These underground resources make up an impressive 90% of the freshwater that is readily available for human use (● Fig. 16.1).

During the last 100 years, humanity's use of freshwater has grown twice as fast as the human population. At the same time, the quality of our freshwater resources has been declining as a result of pollution of both surface and subsurface sources of water. Because of the severity of the global freshwater problem and the crucial role of water for life on Earth, the United Nations has followed up its special 2005–2015 Decade for Action: Water for Life with a continued commitment to promote development, conservation, and wise use of water resources. Understanding the nature and distribution of one of our largest freshwater resources— underground water—is critical to maintaining enough water of suitable quality for domestic, agricultural, and industrial purposes and to maintaining environmental quality in general. Thus, in this chapter we investigate the nature and distribution of underground water and its impact on the landscape.

Nature of Underground Water

Subsurface water is a general term encompassing all water that lies beneath Earth's surface. It includes water found in association with soil, sediments, and rock. Most of this water arrives in the subsurface by means of precipitation from the atmosphere. Water from precipitation, or meltwater from frozen precipitation, that soaks into the ground does so by the process of **infiltration**. During infiltration, water moves from the ground surface into void spaces in soil and loose sediments, as well as into cracks, joints, and other fractures in rock. Infiltration **recharges**, that is, it replenishes or adds to, the amount of water in subsurface locations. Water from underground sources, in turn, reaches the surface in seeps, springs, and wells, and it contributes substantially to water in streams and in standing water bodies, such as lakes and ponds.

Some underground water resources tapped and used by people today are irreplaceable because they accumulated during previous wetter times in geologic history. A small portion of subsurface water is so deep beneath Earth's surface that it may never have been part of the hydrologic cycle. Elsewhere, because of changes in Earth's surface, subsurface water has been cut off from the hydrologic cycle for a long time. This water is contained deep within sediment layers that were deposited by ancient rivers or seas. Future changes in the lithosphere could release these trapped waters and return them to the hydrologic cycle. For example, some of this water can be released in the form of steam during volcanic eruptions and as steam or hot water in geysers and hot springs.

Subsurface Water Zones and the Water Table

Organized by their depth and water content, three distinct subsurface water zones exist in humid regions (● Fig. 16.2). Under conditions of moderate precipitation and good drainage, water infiltrating into the ground first passes through a layer called the **zone of aeration**, in which pore spaces in the soil and rocks almost always contain some air and some water. This uppermost zone only rarely becomes saturated. If all the pore spaces do become filled with water because of a large rainfall or snowmelt event, it is temporary. Water will soon drain downward by gravity beyond the zone of aeration to lower levels by the process of **percolation**. Water in the zone of aeration is known as **soil water**.

In the lowest of the three layers of underground water, the spaces between sediment particles and cracks and openings within rocks are completely filled with water. The water in this **zone of saturation** is called **groundwater**. The top of the zone of saturation is the **water table**. The water table does not remain at a fixed depth below the land surface, but in any given area it fluctuates with the quantity of recent precipitation, loss by outflow to the surface, and the amount of removal by pumping. After an unusually wet period, the water table rises. Because the depth to the water table generally reflects the precipitation amount for a given location (minus evaporation and other losses), it typically lies closer to the surface in humid regions and deeper underground in arid regions.

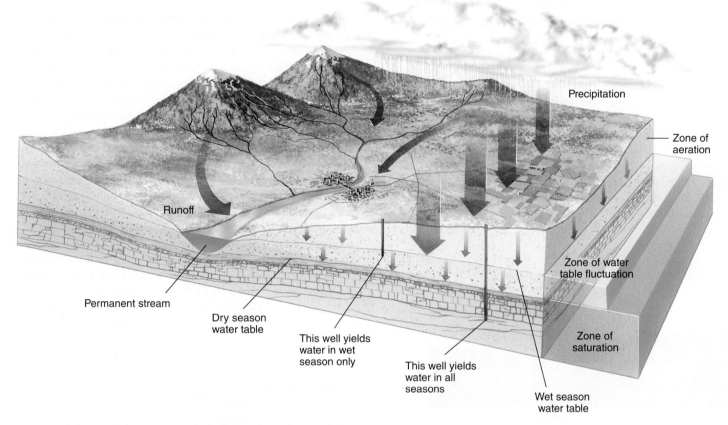

● FIGURE 16.2 The underground subsystem of the hydrological cycle. Water enters the subsurface primarily through infiltration of precipitation and snowmelt. Air is almost always present in void spaces in the zone of aeration; water occupies all void spaces in the zone of saturation (groundwater zone), the top of which is the water table. Infiltrated water percolates down beyond the zone of aeration to the transitional intermediate zone and eventually into the zone of saturation. Water exits the subsurface by direct evaporation from near-surface locations, through transpiration by plants, by seeping into streambeds and lakes, as natural springs, or by flowing into wells. Water table depth responds to changes in infiltration and outflow, falling during dry seasons or years and rising during wet seasons or years.

Between the zones of aeration and saturation lies the **intermediate zone**, which is saturated during periods of ample precipitation but not saturated during intervals of low precipitation. The water table fluctuates through this middle layer, which alternates between unsaturated and saturated conditions. A well or spring originating within the zone of saturation will always bear water, but one originating in the intermediate zone of fluctuation will run dry when the water table falls below it (see Fig. 16.2).

In some desert regions there is no saturated zone at all because water at or just below the surface evaporates soon after rainstorms. In many arid and semiarid landscapes, if considerable groundwater is present, it may be very old, having accumulated during a past period of greater precipitation. Groundwater that people extract from wells in these regions is not replaced by precipitation under current conditions of aridity, and if this extraction continues, the water table will fall lower and lower. Withdrawing this ancient water from the subsurface faster than it is replenished through recharge is called **water mining**, to emphasize that these groundwater resources are of limited supply and will not last indefinitely.

Despite its name, the water table is not typically level but tends to vary in elevation laterally. This lateral variation in the position of the water table generally follows the trend of highs and lows of the surface terrain but in a muted, more gradual fashion. Under elevated topography such as hills, the water table lies at a higher elevation than it does under nearby depressions, such as valleys. Affected by gravitational force, groundwater at higher elevations flows within the zone of saturation toward areas where the water table occupies a lower elevation. As a result, in a region with hilly terrain, the water table is usually closer to the ground surface under low places than it is under high places.

In humid regions of low relief, the water table may be so high that it intersects the ground surface, producing lakes, ponds, or marshes, such as those common in New England and along the Gulf Coast from Louisiana to Florida. Where the landscape is one of hills and narrow valleys, the lowest positions of the water table are often indicated by the location of stream channels on the valley floors. The base of most humid region stream channels lies below the elevation of the water table. In these cases, water from the zone of saturation seeps directly into the channel beneath the stream's water surface, thereby supplying the stream with *base flow*. This **effluent** condition, where groundwater is seeping into a stream, keeps the stream flowing between rains or during dry seasons (● Fig. 16.3a).

Many streams in semiarid and arid regions flow only seasonally or immediately following a significant rainfall. In semiarid regions, the water table typically lies some distance below the streambed during dry periods and rises to intersect the streambed during wet periods. These streams receive groundwater only during wet seasons; during dry, **influent** periods the streams lose water by seepage into the ground below (Fig. 16.3b). In most truly arid regions, surface streams flow only during and immediately after rains, which, although not common, are often torrential when they do occur. With the water table deep beneath the channel throughout the year, only influent flow occurs; these streams lose water by infiltration into the channel bed whenever they flow (Fig. 16.3c).

Groundwater Storage and Movement

The amount and availability of groundwater in an area depend on a variety of factors. Most fundamental is the amount of precipitation that falls in the area and the amount that falls in adjacent regions that drain into it. A second important factor is the rate of evaporation. Third is the capacity of the ground to accept and transmit water to the subsurface through infiltration. A fourth factor involves the amount and type of vegetation cover. Although dense vegetation transpires great amounts of moisture back to the atmosphere, it also inhibits rapid surface runoff of rainfall, encourages infiltration of water into the ground, and lowers evaporation rates by providing shade. Thus, the overall effect of vegetation in humid regions is to increase the supply of groundwater.

Two additional factors that affect the amount and availability of groundwater are the porosity and permeability of the sediments and rocks (● Fig. 16.4). **Porosity** refers to the amount of space between the particles that make up the sediments or rocks. Porosity is expressed as the volume of voids compared to the total volume of the material (including voids). Sediments and rocks consisting of clay-sized clasts, perhaps surprisingly, have a relatively high porosity and therefore can contain considerable amounts of water in the

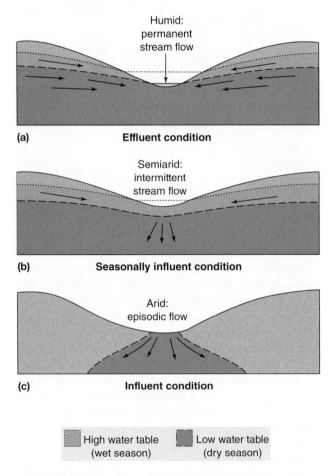

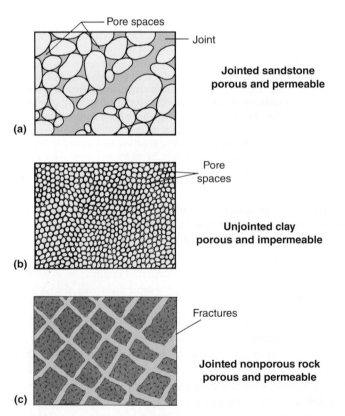

● **FIGURE 16.3** Water can enter or exit the subsurface system through stream channels. (a) In humid regions, groundwater seeps out (effluent condition) into major stream channels all year, providing them with a source of continuous flow. (b) In semiarid or seasonally dry regions, the water table may fall below the stream bed, causing the stream to dry up until the next wet period (seasonally influent condition). (c) In arid regions, significant depth to the water table means that streams flow only immediately after a rainfall event, and water seeps from the streambed into the subsurface (influent condition).

● **FIGURE 16.4** The relationship between porosity and permeability. (a) Sandy sediment and unjointed sandstones tend to have low porosity and high permeability, but fractures (joints) make this sandstone porous and highly permeable. (b) Unjointed shale (clay-sized clasts) has a high porosity, but the pores are small and poorly connected, making for a low permeability. (c) Jointed rock that is nonporous gains porosity and permeability through its fractures.

Which of these three rock types would be best for obtaining water?

large number of very tiny pores. Sand and gravel have comparatively low porosity, but the actual value depends on the packing of grains and uniformity of the grain sizes present. **Permeability** expresses the relative ease with which water flows through void spaces in Earth material. Despite the high porosity of clay sediments and shale, the pore spaces are poorly interconnected, giving them a low permeability. It is therefore very difficult to obtain the water held within unjointed clays and shales. Permeability increases significantly if these materials have joints or fractures that provide interconnections to facilitate flow. The inherently high permeability of sands and gravels often offsets their low porosity to make them good sources from which to obtain water.

Rocks that are composed of interlocking crystals, such as granite, have virtually no pore space and can hold little water within the rock itself. These crystalline rocks, however, may contain water within joints, which allow the passage of groundwater rather freely. Thus, jointed granite can be described as permeable, even though the rock itself is not porous. In contrast, basaltic lavas often contain considerable pore space formed by gas bubbles frozen into the rock, but typically these holes are not interconnected, so the rock itself is porous but not permeable. The presence of fractures is critical for providing permeability in many areas of igneous rocks.

It is important to understand the distinction between porosity and permeability. *Porosity* controls the potential amount of groundwater storage by providing available spaces for the water to be held. *Permeability* affects the rates and volumes of groundwater movement and is facilitated by the presence of large pore spaces, bedding planes, joints, faults, and even caverns. Porosity and permeability both influence the availability of groundwater resources.

An **aquifer** (from Latin: *aqua*, water; *ferre*, to carry) is a sequence of porous and permeable layers of sediment or rock that acts as a storage medium and transmitter of water (● Fig. 16.5). Although any rock material that is sufficiently porous and permeable can serve as an aquifer, most aquifers that supply water for human use are sandstones, limestones, or deposits of loose, coarse sediment (sand and gravel). A rock layer that is relatively impermeable, such as shale or slate, restricts the passage of water and therefore is called an **aquiclude** (from Latin: *claudere*, to close off).

Sometimes an aquifer exists between two aquicludes. In this case, water flows in the aquifer much as it would in a water pipe or hose; water moves within the aquifer but does not escape into the enclosing aquicludes. An aquiclude can also prevent downward-percolating soil water from reaching the zone of saturation. An accumulation of groundwater above an aquiclude is called a **perched water table** (see Fig. 16.5). If people drill down through a perched water table and through the aquiclude supporting it, the water will drain farther down into the subsurface, where it will require much greater effort to obtain.

Springs

Springs are natural outflows of groundwater to the surface (● Fig. 16.6). They mark sites where an aquifer intersects the ground surface and releases water that entered the water-bearing strata elsewhere at a higher elevation. Spring water can seep, trickle, pour, or gush from a single localized spot, along a section of a lineation (fissure, fault, contact), or over a diffuse area. Springs form damp ground, pools, ponds, or irregular wetlands, and they

● **FIGURE 16.5** An aquifer is a natural underground storage medium for groundwater. A perched water table can develop where impermeable rock lies between a permeable layer and the regional water table. Beneath the water table, water in the regional zone of saturation flows toward the nearby river.

Is a perched water table a reliable source of groundwater?

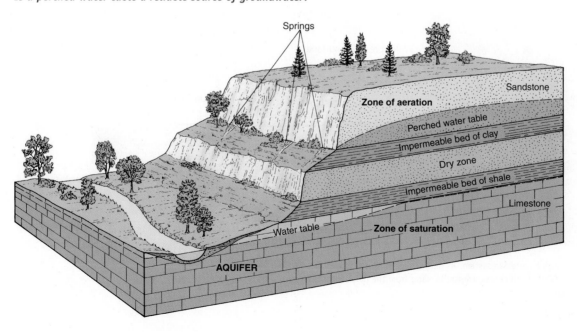

are commonly the point of origin for small streams. A spring flows continuously if the water table always remains at an elevation above the spring's outlet; otherwise the spring flow is intermittent, flowing only when the water table is at a level high enough to feed water to the outlet.

The locations of springs are determined by landform configuration, bedrock structure, elevation of the water table, and the relative position of aquifers and aquicludes. For example, with a perched water table, a spring will occur on a valley side wall where a stream has cut down below the water-bearing layer. The impermeable aquiclude below the perched water table prevents further downward percolation of the water, forcing the water to flow horizontally until it reaches its outlet on the exposed valley wall (see Fig. 16.5). Springs commonly lie along fault zones or in other structural settings where a water-bearing permeable rock layer tilts downslope into a laterally adjacent impermeable rock layer, causing the water level to rise in the aquifer.

Springs are commonly classified as nonthermal or thermal springs based on the temperature of the emerging groundwater. Water temperature of a *nonthermal spring*, also known as a *cold spring*, is at or below the mean annual temperature of the atmosphere in the area of the spring. **Thermal springs** have water temperatures that are higher than the surrounding mean annual atmospheric temperature and are further separated into *warm springs* and *hot springs* by the temperature of 36.7°C (98°F) (● Fig. 16.7).

Artesian Springs

In some cases, groundwater exists in *artesian* conditions, meaning the water is under so much pressure that if it finds an outlet it will flow upward to a level above the local water table. Where water under these conditions achieves natural outflow to the surface, it forms an **artesian spring**. The word *artesian* is derived from the Artois region of France, where people were tapping into pressurized groundwater resources as early as the Middle Ages.

Certain conditions are required for artesian water flow (● Fig. 16.8). First, a permeable aquifer, often sandstone or limestone,

● FIGURE 16.6 Groundwater emerging at the surface as a natural spring.

What evidence indicates that this is a spring rather than a puddle or small pond that accumulated from rainwater?

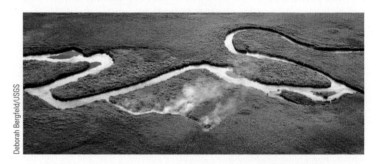

● FIGURE 16.7 Steam indicates the location of a thermal spring that is contributing surface flow to Hot Springs Creek on Akutan Island, Alaska.

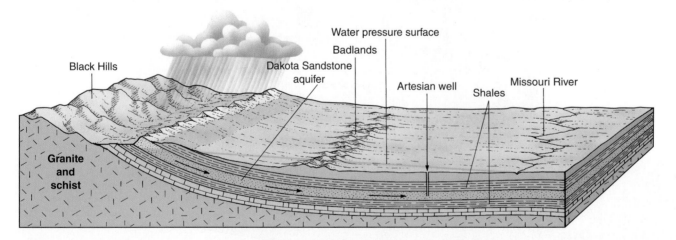

● FIGURE 16.8 Special conditions produce an artesian system. As part of the Northern High Plains Aquifer, the Dakota Sandstone, which averages 30 meters (100 ft) in thickness, transmits water from the Black Hills to locations more than 320 kilometers (200 mi) eastward beneath South Dakota.

Why is the water in this aquifer under pressure?

must be exposed at the surface in an upland area that receives high recharge by precipitation. The aquifer must receive water from the ground surface by infiltration, incline downward often hundreds of meters below the surface, and be confined between impermeable layers that prevent escape of the water except to the artesian springs. These conditions cause the aquifer to act as a pipe that conducts water through the subsurface. Water at a given point in the "pipe" is under pressure from the water that lies in the same aquifer but upslope, closer to the recharge area. As a consequence of this pressure, water flows toward any available natural outlet. The pressure increases with greater amounts of water in the aquifer, more steeply dipping strata forming the aquifer, and fewer available outlets to the surface. In addition to the Artois region of France, well-known artesian systems are found in the American Great Plains and Pacific Northwest, the western Sahara, and eastern Australia's Great Artesian Basin, which is the largest artesian system in the world.

Using Groundwater Resources

Groundwater is a critical natural resource for large numbers of people and wildlife. Half the population of the United States derives its drinking water from groundwater, and some states, such as Florida, draw almost all of their drinking water from this source. In many regions, productive agriculture, and thus food supplies and economic benefits, depends heavily on the extensive use of groundwater for irrigation.

Groundwater also plays a major role in supporting many wetlands and in forming shallow lakes and ponds, all of which provide crucial habitat for a rich array of organisms, including resident and migratory birds. Survival of the Everglades in southern Florida, for example, relies on adequate groundwater flow. This "river of grass," its great variety of birds, and its many other animals are totally dependent on the continued southward flow of groundwater through the region (● Fig. 16.9). As with other invaluable

natural resources, we need to use groundwater wisely and protect it from pollution to ensure continued availability of high-quality supplies. We also need to protect current supplies and recharge areas from pollution.

Wells

Early in human history, people learned to augment natural outflows of groundwater by digging wells. **Wells** are artificial openings dug or drilled down to a point below the water table to extract groundwater (● Fig 16.10). Water is drawn from wells by lifting devices ranging from simple rope-drawn water buckets to pumps powered by gasoline, electricity, or wind. In many shallow wells, the supply of water varies with fluctuations of the water table. Deeper wells that penetrate into aquifers beneath the zone of water table fluctuation provide sources of water that are more reliable because they are less affected by seasonal periods of drought.

Wells can be used to access the local groundwater system whether that system is artesian or nonartesian. As with artesian springs, groundwater in *artesian wells* rises to a level above the local water table. The height to which the water rises in an artesian well depends on the amount of natural pressure exerted on the water. As with artesian springs, pressure on the water in an artesian well is greater for greater amounts of water in the aquifer up-flow from the well, a steeper angle of incline of the aquifer strata, and fewer available outlets for the water—in this case usually other wells. In **flowing artesian wells**, water tapped by the well rises to the ground surface and flows out onto the surface under its own pressure, without pumping. If water in a well rises above the local water table but not to the point of flowing out of the well and onto the ground surface, it is a **nonflowing artesian well**. Pumps are commonly used to bring water in nonflowing artesian wells to the ground surface. An excellent example of wells being supported by a large artesian groundwater system exists in South

● **FIGURE 16.9** Groundwater withdrawals have lowered the regional water table and threaten the survival of the Everglades, an ecologically important wetland area in southern Florida.

● **FIGURE 16.10** This historic hand-dug well supported a small desert community in Turkmenistan for many years.

Dakota, where sandstone exposed at the surface along the flanks of the Black Hills in the western part of the state transmits water eastward to artesian wells as far as 320 kilometers (200 mi) away (see Fig. 16.8).

Depletion of Groundwater Reserves

In areas where there are many wells or a limited groundwater supply, the rate of groundwater removal can exceed the rate of its natural replenishment through groundwater recharge. In many areas that are irrigated from wells, the water table has fallen below the depth of the original wells (● Fig. 16.11). Progressively deeper wells must be dug, or the old ones must be extended, to reach the supply of water. In the Ganges River Valley of northern India, the excavation of deep modern wells to replace shallow hand-dug wells has increased the amount of groundwater being brought to the surface, but this greater usage has lowered the water table significantly.

Irrigation consumes the bulk of the groundwater—more than two-thirds—used in the United States today. One of the largest aquifers supplying groundwater for irrigation is the Ogallala Aquifer, an artesian system also known as the High Plains aquifer, which underlies the Great Plains from west Texas northward to South Dakota (● Fig. 16.12). The Ogallala Aquifer alone supplies more than 30% of the groundwater used for irrigation in the United States. The aquifer, however, is experiencing alarming declines in water level, a condition primarily attributed to large withdrawals for agricultural irrigation (● Fig. 16.13). Serious concern exists about the future of the Ogallala Aquifer, particularly because much of the water withdrawn from it accumulated thousands of years ago. Under the present semiarid climate of the region, recharge of the aquifer is limited and has not kept pace with withdrawals.

In certain environments, particularly where high groundwater demand has led to extensive pumping, a sinking of the

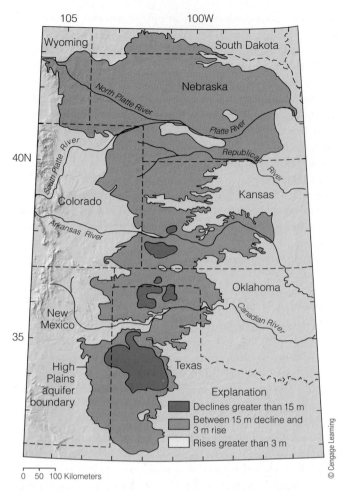

● **FIGURE 16.12** The Ogallala Aquifer supplies water to a large, semiarid area of the High Plains. Much of the water in the Ogallala Aquifer, the largest freshwater aquifer in the United States, accumulated during wetter times thousands of years ago, and the water table is now falling.

Why do you think the drop in water supply has been greatest in the southern part of the aquifer?

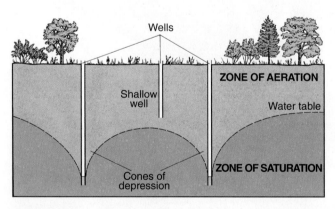

● **FIGURE 16.11** Cones of depression can develop in the water table as a result of pumping water from wells. In areas with many wells, adjacent cones of depression intersect, lowering the regional water table and causing shallow wells to go dry.

What impact might this scenario have on some of the local natural vegetation?

● **FIGURE 16.13** A high-capacity well irrigating crops with water pumped from the Ogallala Aquifer.

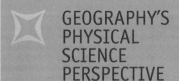
Acid Mine Drainage

Acid mine drainage typically occurs when subsurface water flowing through mines or mine tailings undergoes chemical reactions that leave the water highly acidic (low pH) and metal rich. Acid mine drainage is a serious environmental concern in some coal- and metal-mining regions, including much of the coal-mining area of the eastern United States and parts of Australia, South America, and South Africa. When sufficient quantities of the affected water flow onto the surface as springs or seep into streams, lakes, or ponds, the mineral content and low pH endanger aquatic organisms and make the water unsuitable for human consumption. Fish have completely disappeared from some streams with very low pH and high metal content because of acid mine drainage.

The chemical reactions involved in the formation of acid mine drainage occasionally happen under natural conditions but at much slower rates and in much smaller amounts than on disturbed lands. Mining greatly increases the permeability of susceptible rocks, allowing much more water to undergo the chemical reactions. Susceptible rocks contain pyrite (FeS_2), a common substance in the extensive coals of Pennsylvanian age in the eastern United States. When water containing oxygen comes into contact with pyrite in an abandoned underground coal mine, the pyrite oxidizes readily. Iron, sulfate (SO_4), and hydrogen ions are released into the water as a result of the chemical weathering; it is the presence of hydrogen ions that increases the water's acidity. Additional chemical reactions lead to hydrolysis of the iron ions in the water, which releases more hydrogen ions, further increasing acidity. The rate of hydrogen production is greatly accelerated if certain microorganisms that thrive in conditions of low pH are present.

Mines that lie in the zone of the fluctuating water table between the soil water zone and the zone of saturation, like many mines in the Appalachian coal fields, are particularly susceptible to acid mine drainage because of the frequent introduction of new, moving, oxygenated water. Pumping removes water from these mines while they are operational. When they are abandoned, however, flowing underground water returns to the now highly permeable and chemically reactive environment and produces acid mine drainage.

Controlling the flow of underground water represents a principal way to limit production of acid mine drainage. Approaches include locking susceptible mines out of subsurface water circulation by sealing water out or flooding them with very slow moving, stagnant groundwater that has little opportunity to obtain new oxygen. Prevention and reduction of acid mine drainage remains an active field of research, and understanding the chemistry and circulation of underground drainage is critical to the ongoing investigation.

D. Sack

Orange-colored stream water and the orange coating on channel rocks result from acid mine drainage in this Appalachian coal-mining region. Acid mine drainage lowers the pH of local streams, which encourages precipitation of these orange, oxidized iron compounds.

land, called **subsidence**, can occur as a result of compaction related to the water withdrawal. Mexico City, Venice (Italy), and the Central Valley of California, among many other places, have subsidence problems related to groundwater withdrawal. In parts of Southern California, withdrawn groundwater has been replaced artificially by diverting streams so that they flow over permeable deposits. This process is known as **artificial recharge**.

Groundwater Quality

Because most subsurface water percolates down through a considerable amount of soil and rock, by the time it reaches the zone of saturation it is mostly free of clastic sediment. However, it often carries a large amount of minerals and ions dissolved from the materials through which it passed. As a result of this large mineral content, groundwater is often described as "hard water," in comparison with "softer" (less mineralized) rainwater. Moreover, just as increases in population, urbanization, and industrialization have resulted in the pollution of some of our surface waters, they have also resulted in the pollution of some of our groundwater supplies. For example, subsurface water becomes highly acidic as it moves through underground mines in certain types of coal deposits that are widespread in the eastern United States. If this **acid mine drainage** reaches the surface, it can have serious detrimental effects on the local aquatic organisms.

Other dangers to groundwater quality stem from the introduction of toxic substances or salt water into the zone of saturation. Excessive applications of pesticides and incompletely sealed surface or subsurface storage facilities for toxic substances, including gasoline and oil, are situations that can lead to groundwater pollution through downward percolation. In coastal regions with excessive groundwater pumping, denser salt water from the ocean seeps inland to replace freshwater withdrawn from the zone of saturation. This problem of saltwater replacing freshwater in the groundwater zone has occurred in many coastal localities, notably in southern Florida, New York's Long Island, and Israel, and the problem is expected to expand into new groundwater areas as sea level rises in response to global warming.

Geothermal Water

Water heated by contact with hot rocks in the subsurface is referred to as **geothermal water**. Hot springs form where geothermal waters flow out onto Earth's surface fairly continuously (● Fig. 16.14). Where geothermal water flow is intermittent and somewhat eruptive, it produces a **geyser**, an impressive phenomenon with sporadic bursts of steam and hot water expelled from a fissure or vent. The Old Faithful geyser in Yellowstone National Park, Wyoming, is a well-known example (● Fig. 16.15). Geysers erupt when temperature and pressure of the water at depth reach critical levels, forcing a column of superheated water and steam

Bob Lindstrom, NPS

● **FIGURE 16.14** In addition to high temperature and high mud content, the water in this hot spring in Yellowstone National Park, Wyoming, is highly acidic, with a pH of 2.

What might be some reasons for the lack of vegetation in the tan zone surrounding this spring?

out of the fissure in an explosive manner. The word geyser is an Icelandic term for these intermittent eruptions that are so common on that volcanic island.

Hot springs and geysers are intriguing groundwater-related phenomena. Most hot springs and geysers contain significant amounts of minerals in solution. These minerals, typically rich in calcium carbonate or silica, precipitate out of the water, often forming colorful terraces or mounds around the spring or vent (● Fig. 16.16).

George Marler/National Park Service

● **FIGURE 16.15** One of the world's most famous geysers is Old Faithful in Yellowstone National Park, Wyoming.

How do geysers differ from hot springs?

● **FIGURE 16.16** Because of its high mineral content, this hot spring has left extensive calcareous deposits, known as travertine.

R. Gabler

Geothermal water is usually associated with areas of tectonic and volcanic activity, especially along lithospheric plate boundaries and over hot spots. In several of these locations geothermal water has been used to produce electricity. The best geothermal water for energy uses is not only very hot, but also clean, that is, relatively free of dissolved minerals that can clog pipes and generating equipment.

Landform Development by Solution

In areas where the bedrock is soluble in water, subsurface water is an important agent in shaping landform features at the surface and underground. Underground water is a vital ingredient in subsurface chemical weathering processes, and, like surface water, subsurface water dissolves, removes, transports, and deposits rock-forming materials.

The principal mechanical role of subsurface water in landform development is to encourage mass movement by adding weight and reducing the strength of soil and sediments, thereby contributing to slumps, mud or debris slides, mud or debris flows, and landslides. Chemically, subsurface water also contributes to processes that shape landforms. Through the chemical breakdown of rock materials by carbonation and other forms of solution, and the deposition of those dissolved substances elsewhere, underground water is an effective land-shaping agent, especially in areas where limestone is present. Like surface water, subsurface water can dissolve limestone through carbonation or simple solution in acidic water. In many of these areas, surface outcrops of limestone are pitted and pockmarked by chemical solution, especially along joints, sometimes forming large, flat, furrowed limestone platforms (● Fig. 16.17). Wherever water acts on any rock type that is significantly soluble in water, a distinctive landscape develops.

J. Petersen

● **FIGURE 16.17** Vertical joints cutting through this limestone platform are subject to intense solution as water from the surface infiltrates down them into the subsurface.

Karst Landforms

The most common soluble rock is limestone, the chemical precipitate sedimentary rock composed of calcium carbonate ($CaCO_3$). Landform features created by the solution and reprecipitation (redeposition) of calcium carbonate by surface or subsurface water are found in many parts of the world. The north-central Mediterranean region, in particular, exhibits large-scale limestone solution features. These are most clearly developed on the Karst Plateau along Croatia's scenic Dalmatian Coast. Landforms developed by solution are called **karst** landforms after this impressive locality. Other extensive karst regions are located in Mexico's Yucatán Peninsula, the larger Caribbean islands, central France, southern China, Laos, Vietnam, and many areas of the United States (● Fig. 16.18).

The development of a classic karst landscape, one in which solution has been the dominant process in creating and modifying landforms, requires several special circumstances. A warm, humid climate with ample precipitation is most conducive to karst development. In arid climates, karst features are typically absent or are not well developed, but some arid regions have karst features that originated during previous periods of geologic time when the climate was much wetter than it is today. Compared with colder humid climates, warmer humid climates have greater amounts of vegetation, which supplies carbon dioxide to subsurface water. Carbon dioxide is necessary for carbonation of limestone and it increases the acidity of water, which encourages solution in general.

Another important factor in the development of karst landforms is movement of subsurface water. This movement allows water that has become saturated with dissolved calcium carbonate to flow away and be replaced by water unsaturated with calcium carbonate and therefore capable of dissolving additional limestone. Groundwater moves vigorously where it flows toward a low outlet, as might be provided along a deeply cut stream valley or a tectonic depression. In addition, everything else being equal, faster groundwater flow occurs in Earth material that has a greater permeability.

Because infiltration of water into the subsurface tends to be concentrated where cross-cutting sets of joints intersect, in limestone these intersections are subject to intense solution. Such concentrated solution may produce roughly circular surface depressions, called **sinkholes** or *dolines*, which are prominent features of many karst landscapes (● Figs. 16.19a and b). Two basic types of sinkholes are distinguished by differences in

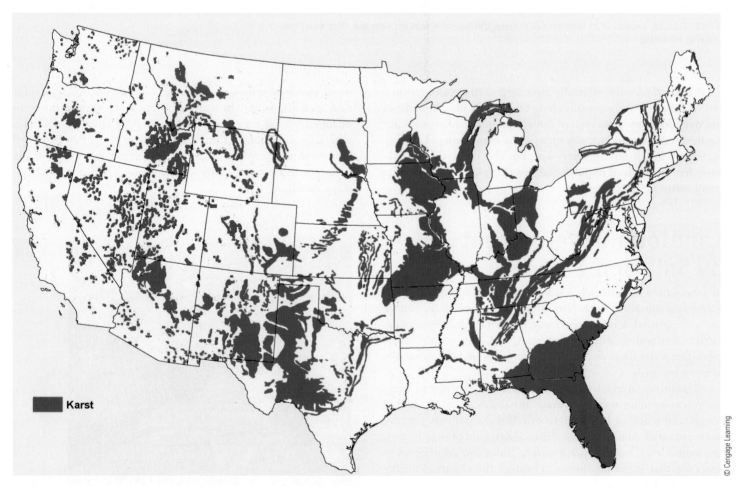

Karst

● **FIGURE 16.18** The distribution of limestone in the conterminous United States indicates where varying degrees of karst landform development exists, depending on climate and local bedrock conditions.

Where is the karst area closest to where you live?

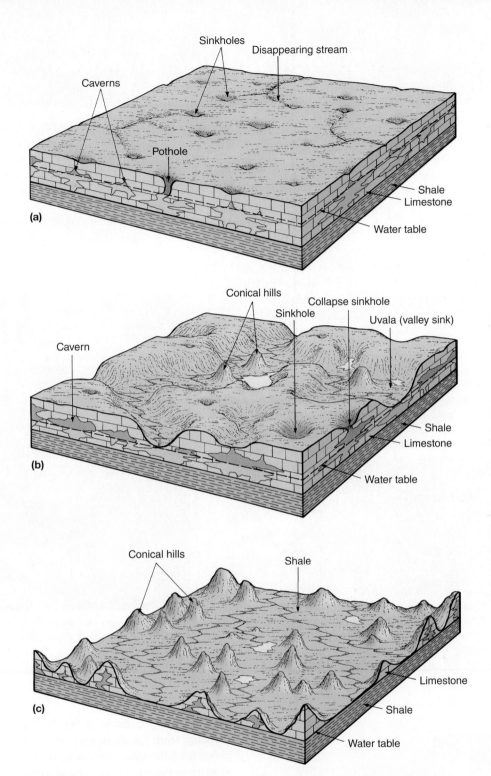

● **FIGURE 16.19** Examples of karst landscapes. (a) Past groundwater solution along limestone fractures and bedding planes formed caverns. Solution at joint intersections and collapse of ceilings in caverns below encourage development of sinkholes (dolines) at the surface. Surface streams can disappear into sinkholes to join the groundwater flow. (b) Some limestone landscapes have more relief, with merged sinkholes that create uvalas (karst valleys), and some conical (haystack) hills. (c) Areas with especially intense solution may be dominated by conical hills that are isolated above an exposed surface of insoluble rock, such as shale.

Why are there no major depositional landforms created at the surface in areas of karst terrain?

Although many sinkholes develop slowly through solution and infiltration of water along joint intersections in soluble rocks, others appear almost instantaneously as the ground unexpectedly collapses into subsurface voids, including caverns. The sudden collapse of the ground caused by the creation of a sinkhole is a problem in many areas of the United States that have soluble rocks at or near the surface. Two of the accompanying photographs show a road collapse, and the third depicts the sinkhole collapse at the National Corvette Museum, all in the karst-rich landscape near Bowling Green, Kentucky.

Earth scientists recognize two types of processes that form collapse sinkholes: regolith collapse (or cover collapse) and bedrock collapse. In both cases, subsurface solution begins long before the surface sinkhole appears.

Sinkhole formation by sudden regolith collapse occurs where vertical gaps in the subsurface rock exist as a result of solution along joints or along other forms of weakness in the rock. The gaps often open to larger passageways at depth. Regolith and soil lie on top of the bedrock and also cross over the gaps as regolith bridges or arches. Initially, the water table lies within the regolith layer that overlies the rock. Sudden regolith collapse is most common

Sinkhole collapse at Dishman Lane near Bowling Green, Kentucky, caused severe road damage in an area larger than a football field. Fixing the problem required completely filling in the sinkhole with rock and repaving the road at a cost of more than $1 million.

Several vehicles were damaged by the sudden collapse of the Dishman Lane sinkhole, which occurred during rush hour. Luckily no one was injured.

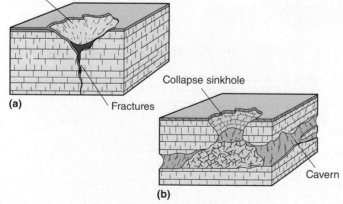

their mode of formation (● Fig. 16.20). If the depressions result primarily from the surface or near-surface solution of rock and removal of the dissolved materials by water infiltrating downward into the subsurface, the depressions are *solution sinkholes* (● Fig. 16.21). *Collapse sinkholes* form when the land surface caves into large subsurface voids created by solution of bedrock some distance beneath the surface (● Fig. 16.22).

● **FIGURE 16.20** The two major types of sinkholes (dolines) based on principal mode of formation. (a) Solution sinkholes develop gradually where surface water funneling into the subsurface dissolves bedrock to create a closed surface depression. (b) Collapse sinkholes form when either the bedrock or the regolith above a large subsurface void fails, falling into the void.

after the water table subsequently falls below the subsurface gaps, such as during drought periods or by excessive pumping from wells. The drop in the water table leaves the regolith in a zone in which some pore spaces fill with air and through which water percolates downward toward the air-filled gap or cavern below. If the force of gravity pulling downward on the soil and regolith exceeds the frictional resistance and cohesive strength holding up the regolith bridge, the regolith collapses into the gap, creating a sinkhole at the ground surface. Adding additional weight to the regolith and soil layer by constructing roads and buildings or by impounding water at the surface increases the force of gravity and therefore the likelihood that the regolith bridges will suddenly collapse.

The formation of a sinkhole by bedrock collapse is associated with a growing underground cavern relatively near the ground surface. Caverns can grow vertically as well as horizontally. If continued solution makes the rock roof of the cavern thinner and thinner, eventually the rock can become so thin that it fails, crashing into the cavern void below. A sinkhole is formed when collapse of the bedrock cavern roof causes overlying Earth material up to the ground surface to lose its support and also slump into the hole.

Physical geographers who have examined the Dishman Lane sinkhole in Bowling Green, Kentucky, report evidence of cave roof collapse in this case. It is likely that much of the roof collapsed thousands of years ago, but the fact that limestone blocks were found in the rubble indicated that part of the cave roof collapsed in this event. The resulting sinkhole affected over 0.4 hectares (1 acre) of surface area and caused considerable property damage.

More costly than the Dishman collapse was the 2014 sinkhole collapse at the National Corvette Museum in Bowling Green, Kentucky. Eight of the museum's prized vehicles fell into the 9-meter (30-ft) deep hole that was roughly 18 meters (60 ft) long and 14 meters (45 ft) wide.

The two processes, solution and collapse, actually cooperate to create most sinkholes in soluble rocks. Whether the depressions are termed solution or collapse sinkholes depends on which of these two processes was dominant in their formation. Solution and collapse sinkholes often occur together in a region. Sinkhole form varies greatly, but solution sinkholes tend to be funnel-shaped whereas collapse sinkholes tend to have steep walls.

Sudden collapse of sinkholes is a significant natural hazard that every year causes extensive property damage, human injury, and even death. The rapid formation of sinkholes may result from excessive groundwater withdrawal for human use or from significant drought periods. Either of these conditions lowers the water table, causing a loss of buoyant support for the ground above, followed by collapse. Rapid sinkhole collapse has damaged roads and railroads and has even swallowed vehicles and buildings, including their occupants.

Sinkholes can grow in area and merge over time to form larger karst depressions called **uvalas** (from Croatian: *uvala*, hollow or depression) or *valley sinks* (see Fig. 16.19b). In many areas, uvalas are linearly arranged along former routes of subsurface water flow.

Despite their humid climates, many karst regions have few continually flowing surface streams. Surface water seeping into a fracture in limestone widens the fracture by solution (see Fig. 16.17). This widened avenue for water flow increases the downward permeability, accelerates infiltration, and directs water rapidly toward the zone of saturation. In some cases, surface streams flowing on

D. Sack

● **FIGURE 16.21** A solution sinkhole in southern Indiana.

What limitations might solution sinkholes place on the type of agricultural activity undertaken on this farm?

© St. Petersburg Times

● **FIGURE 16.22** This large collapse sinkhole appeared in Winter Park, Florida, when a drought-induced lowering of the water table caused the surface to collapse into an underground cavern system.

What human activities might contribute to the occurrence of such hazards?

P. Metz/USGS

(a)

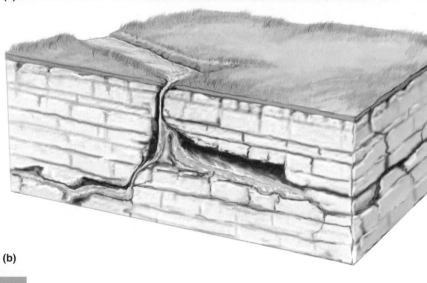

(b)

bedrock of low permeability upstream will encounter a large fracture, joint intersection, or highly permeable rock downstream, where they rapidly lose all their surface flow by infiltration. These are known as **disappearing streams** because they "vanish" from the surface as the water flows into the subsurface (● Fig. 16.23).

Water moving along joints and bedding planes below the surface can also dissolve limestone, sometimes creating a system of connected passageways within the soluble bedrock (see Figs. 16.19a and b). If the water table falls, leaving these passageways above the zone of saturation, they are called **caverns** or *caves*.

In many well-developed karst landscapes with caverns, a complex underground drainage system all but replaces surface water flow. In these cases, the surface terrain includes many large valleys that contain no streams. Surface streams originally existed and excavated the valleys, but the water flow was eventually diverted to underground conduits within the cavern system (● Fig. 16.24). The site where a surface stream disappears into the cavern system is referred to as a **swallow hole** (● Fig. 16.25). In some cases these underground-flowing "lost rivers" reemerge at the surface as springs where they encounter impervious beds below the limestone.

● **FIGURE 16.23** Many karst areas, such as along the Peace River, Florida, have disappearing streams that (a) seem on the surface to stop abruptly, but (b) the water flow is actually funneled down to the subsurface.

● **FIGURE 16.24** In many cavern systems, the underground drainage takes the form of small subterranean lakes and streams. This photo shows an underground stream in Wind Cave, South Dakota.

(a)

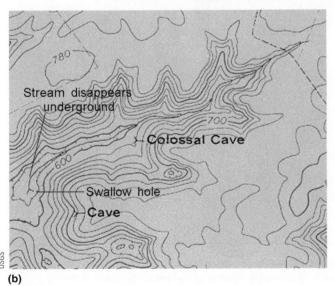

Stream disappears underground

Colossal Cave

Swallow hole

Cave

(b)

● **FIGURE 16.25** Disappearing streams vanish at swallow holes, which divert the stream's flow to the subsurface. (a) A solution-widened swallow hole (at base of boulder) exposed during low flow along a Florida streambed. (b) A disappearing stream and swallow hole as represented on a topographic map. The hachured contours indicate the closed depression into which the stream is flowing.

After intense and long-term karst development, especially in wet tropical conditions, only limestone remnants are left standing above insoluble rock below. These remnants usually take the form of small, steep-sided, and cave-riddled karst hills called **conical hills**, *haystack hills*, or *hums* (see Figs. 16.19b and c). Multiple terms exist for many karst features because similar features of slightly different form were independently named in different parts of the world. Terrain dominated by numerous conical hills have been described as "egg box" landscapes because an aerial view of the numerous haystack hills and sinkholes resembles the shape of an egg carton (● Fig. 16.26). Excellent examples of these steep, karstic hills are found in Puerto Rico, Cuba, and Jamaica. If the limestone hills are particularly high and steep-sided, the landscape is called **tower karst**. Spectacular examples of tower karst landscapes are found in southern China and Southeast Asia (● Fig. 16.27).

Limestone Caverns and Cave Features

Of the various types of landforms created by limestone solution, caverns are the best known and most spectacular. Globally, several cavern systems have been designated as, or are part of, UNESCO World Heritage sites because of their outstanding natural features. Many other beautiful caverns exist in a variety of countries, attracting large numbers of visitors each year. Examples of famous limestone caverns in the United States are Carlsbad Caverns in New Mexico, Mammoth and Colossal Caves in Kentucky, and Luray and Shenandoah Caverns in Virginia. In fact, 34 states have caverns that are open to the public. Some are quite extensive with rooms more than 30 meters (100 ft) high and with kilometers of connecting passageways.

Limestone caverns originate when subsurface water, sometimes flowing much like an underground stream, dissolves rock, leaving networks of passageways. If the water table drops to the floor of the cavern or lower, typically as a result of climate change or tectonic uplift, the cave will become filled with air. Interaction between the cave air and mineral-saturated subsurface water percolating down to the cavern from above will then begin to precipitate minerals, especially calcium carbonate, on the cave ceiling, walls, and floor, decorating them with often-intricate depositional forms.

The nature of any fracturing that exists in soluble bedrock exerts a strong influence on cavern development in karst regions. Groundwater solution widens joints, faults, and bedding planes to produce passageways. The relationship between caverns and fracture distributions is evident on cave maps that show linear and parallel patterns of cave passageways (● Fig. 16.28).

Limestone caverns vary greatly in size, shape, and interior character. All caves formed by solution, however, show some evidence of previous water flow, such as deposits of clay and silt on the cavern floor; in some caves water still actively flows. Continued subterranean flow of water further deepens some caverns, and ceiling collapse enlarges them upward. Many cavern systems have several levels and are almost spongelike in the pattern of their passageways. Some caves contain elaborate deposits of chemical precipitates. The

Courtesy Parris Lyew-Ayee

● **FIGURE 16.26** Intense solution in wet tropical environments can produce a karst landscape consisting of a maze of conical hills and intergrown sinkholes.

variations in cavern size and form can indicate differences in mode of origin. Some small caves might have formed above the water table by water percolating downward through the zone of aeration. The majority of caverns, however, develop just below the water table, where the rate of solution is most rapid. A subsequent lowering of the water table level, caused by the incision of surface streams, climate change, or tectonic uplift, fills the cavern with air, allowing calcium carbonate deposition to begin.

Speleothem is the generic term for any chemical precipitate feature deposited in caves. Speleothems, also referred to as *cave decoration*, develop in a great variety of textures and shapes, and many are delicate and ornate. Speleothems originate when previously dissolved substances, particularly calcium carbonate, precipitate out of subsurface water to produce some of the most beautiful and intricate forms found in nature. The dripping water leaves behind a deposit of calcium carbonate termed *travertine*, or *dripstone*. As these travertine deposits grow

Jeffrey Alford/Asia Access

● **FIGURE 16.27** Guilin, a limestone region in southern China, is famous for its beautiful tower karst.

How does tower karst differ from conical hills?

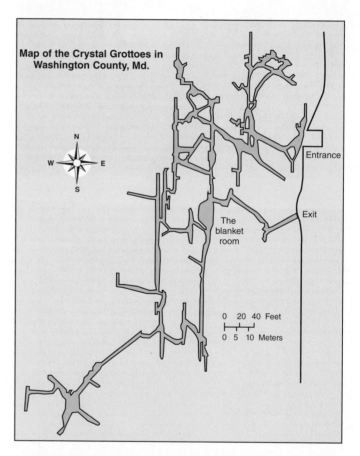

● **FIGURE 16.28** This map of the Crystal Grottoes in Washington County, Maryland, illustrates the influence of fractures on the growth of the cavern system, as evidenced by the geometric arrangement and spacing of passageways. Groundwater flow widens fractures and other zones of weakness to develop a cavern system.

How difficult do you think it would be to create an accurate map of a cavern system?

● **FIGURE 16.29** Stalactites form where mineralized water percolates down to the cave and deposits some of its dissolved minerals, typically calcium carbonate, as hanging features.

● **FIGURE 16.30** Speleothems in Oregon Caves illustrate how stalactites and stalagmites join to create a column.

downward, they form icicle-like spikes called **stalactites** that hang from the ceiling (● Fig. 16.29). Water saturated with calcium carbonate dripping onto the floor of a cavern builds up similar but more massive structures known as **stalagmites**. Stalactites and stalagmites often meet and continue growing to form pillar-like **columns** (● Fig. 16.30).

People who have not studied caves often think that speleothems are made when water evaporates into the cavern air, leaving behind deposits of calcium carbonate, but that is not the case. Caves that foster active development of speleothems typically have air that is fully saturated with water, having a relative humidity near 100%, so evaporation is minimal. Instead, much of the water that percolates into the cave from above first acquires carbon dioxide from the soil and then uses it to dissolve calcium carbonate (by carbonation) from the rocks and sediment it encounters on its way to the cave. Cave air, in contrast, contains a comparatively low amount of carbon dioxide. When the dripping water contacts the cave air, it therefore releases carbon dioxide gas to the air. This degassing of carbon dioxide from the water essentially reverses the carbonation process, causing the water to precipitate calcium carbonate. Eventually, enough calcium carbonate is deposited in this way to form a stalactite, stalagmite, or other depositional cave feature.

Cavern development is a complex process, involving such variables as rock structure, groundwater chemistry, hydrology, and the regional tectonic and erosional history. As a result, the scientific study of caverns, **speleology**, is particularly challenging. Adding to the challenge are the field conditions. Much of our knowledge of cavern systems has come from explorations as deep as hundreds of meters underground by individuals making scientific observations while crawling through mud, water, and even bat droppings in dark, narrow passages (● Fig. 16.31). Rock-climbing expertise is needed to navigate through some cavern systems, whereas exploration and mapping of

THINKING GEOGRAPHICALLY

A cement pad directs rainwater into a storage cistern in a region of intense karst in Jamaica. What physical characteristic of the cement makes it useful for collecting rainwater? Why might this water collection system be needed by this small community which has high annual rainfall?

● **FIGURE 16.31** Cave exploration, called caving or spelunking, can be exciting but is also hazardous.

● **FIGURE 16.32** These spelunkers (also known as cavers) are using their rock-climbing skills to navigate a deep vertical drop. The view is looking down into the cave.

water-filled caves requires scuba diving in conditions that are often extremely dangerous (● Fig. 16.32). Like most landforms, caverns and the surface terrain features formed by solution processes can be hazardous as well as beautiful.

The widespread occurrence on Earth of surface and subsurface karst features highlights the power of water as an agent of chemical erosion as well as the abundance of soluble rocks, particularly limestone. Water that infiltrates into the subsurface, however, is important not only because it can dissolve and precipitate rock-forming minerals. Soil water is used by growing plants, and water that percolates from the soil water zone through the zone of aeration to the zone of saturation recharges groundwater supplies. Groundwater is a critical resource of freshwater that moves slowly from areas where the water table is higher to areas where it is lower. Groundwater returns to be available to people and animals at the surface through natural means at springs and by seeping into most humid-region streams and lakes below their water level. People tap into groundwater reserves artificially by digging or drilling wells into it, but sometimes withdraw more groundwater than is being replaced through natural recharge. People's actions also lead to pollution of subsurface water. Understanding the basic nature, distribution, and availability of subsurface water, including its importance to humans as a source of freshwater, is crucial to whether it will continue to be available in adequate quality and quantity.

CHAPTER 16 ACTIVITIES

■ TERMS FOR REVIEW

subsurface water	permeability	geyser
infiltration	aquifer	karst
recharge	aquiclude	sinkhole (doline)
zone of aeration	perched water table	uvala (valley sink)
percolation	spring	disappearing stream
soil water	thermal spring	cavern (cave)
zone of saturation	artesian spring	swallow hole
groundwater	well	conical hill (haystack hill or hum)
water table	flowing artesian well	tower karst
intermediate zone	nonflowing artesian well	speleothem
water mining	subsidence	stalactite
effluent	artificial recharge	stalagmite
influent	acid mine drainage	column
porosity	geothermal water	speleology

QUESTIONS FOR REVIEW

1. Why is it important to understand the basic nature and properties of underground water?
2. What is the difference between infiltration and percolation?
3. When water soaks into the subsurface, what two zones does it pass through on its way to the zone of saturation? How is the zone of saturation related to groundwater and the water table?
4. How does porosity differ from permeability? How are both of these properties related to groundwater supply?
5. What is the difference between a spring and a well? Under what conditions is a spring or a well correctly described as artesian?
6. What is an aquifer? Describe the conditions necessary for an aquifer.
7. What is karst? What conditions encourage the development of karst features?
8. Describe a sinkhole. How are sinkholes formed?
9. How do caverns form?
10. How is calcium carbonate deposited by subsurface water to form a stalactite in a humid cave?

CONSIDER AND RESPOND

1. If you were a water resources geographer and had to plan for the development of groundwater resources for a community, what would be your major considerations?
2. What are some human impacts and environmental problems related to groundwater resources?
3. Describe the major landform features in a region of karst topography. Over time, what changes might be expected in the landscape?

PRACTICAL APPLICATIONS

1. Drilling at seven sites spaced every 5 kilometers along a straight-line transect heading from west to east has provided the following data on depth to the water table and the rock type present just below the water table: (a) 15 meters in sandstone, (b) 14.5 meters in sandstone, (c) 14 meters in sandstone, (d) 5 meters in limestone, (e) 13 meters in sandstone, (f) 12.5 meters in sandstone, and (g) 12 meters in sandstone.

What is the regional slope of the water table? Provide a reasonable explanation for the data obtained from site d.

2. Scientists studying an air-filled cavern located in an arid region determined that the speleothems present began forming 20,000 years ago and stopped growing 12,000 years ago. Suggest a reasonable history of climatic and hydrologic events related to the formation of the cavern evidence.

LOCATE AND EXPLORE

1. Using Google Earth and Figure 16.18 from this chapter, examine the spatial distribution of karst regions in the United States. The development of karst landscapes depends on geology, precipitation, topography, tectonic activity, vegetation cover, and climate change history. Provide an explanation for karst development in Kentucky, Texas, and Florida.
2. Using Google Earth, identify the landforms at the following locations (latitude, longitude) and provide a brief discussion of how the landform developed. Include a brief discussion of why the landform is found in that general area of the United States or the world.
   a. 18.40°N, 66.53°W
   b. 25.05°N, 110.37°E
   c. 29.658°N, 81.861°W
   d. 38.207°N, 90.175°W
   e. 20.87°N, 107.19°E
   f. 39.4032°N, 113.8625°W

 MindTap—Make the most of your study time by accessing everything you need to succeed in one place. Read your textbook, take notes, review flashcards, watch videos, complete activities, take practice quizzes, and more online with MindTap. Log in at **www.cengagebrain.com**.

Map Interpretation

KARST TOPOGRAPHY

The Map

The Interlachen area is in northeastern Florida. Florida's peninsula is the emerged portion of a gentle anticline called the Peninsular Arch. The region is underlain by thousands of meters of marine limestones and shales. This great thickness of marine sediments originated in the Mesozoic Era when Florida was a marine basin. As the arch rose, Florida became a shallow shelf and was eventually elevated above sea level.

Although nonresidents think of Florida mainly as a state with magnificent beaches and warm winter weather because of its humid subtropical climate, it is also a state with hundreds of lakes dotting its center. The lake region is formed on the Ocala Uplift, a gentle arch of limestone that reaches to 46 meters (150 ft) above sea

level. Lake Okeechobee is the largest of these lakes and has an average depth of less than 4.5 meters (15 ft). Most of the lakes, such as those in the Interlachen map area, are much smaller.

Florida's central lake region is an ideal area for studying karst topography. Both the surface and the subsurface features express the geomorphic effects of subsurface water. Extensive cavern systems exist beneath the surface. Much of the state's runoff is channeled through huge aquifers, and springs are quite common.

You might find it interesting to compare this topographic map to the Google Earth presentation of the area. Find the map area by zooming in on these latitude and longitude coordinates: 29.640425°N, 81.935683°W.

Interpreting the Map

1. As a whole, this area would be classified as what major type of landform (for example, mountain)? What is its average elevation?
2. What contour interval is used on this map? Why do you think the cartographers chose this interval?
3. On what type of bedrock is the map area situated? Do you think the climate has any influence on the landforms in the area? Explain.
4. What landform features on the map indicate that this is a karst region?
5. What type of landform are the round, steep depressions? Why do lakes occupy some of the depressions?
6. Locate Clubhouse Lake on the full-page map (scale 1:62,500) and the smaller map (1:24,000). What is the elevation of Clubhouse Lake? What is its maximum width?

7. What type of feature is the area north of Lake Grandin?
8. What is the approximate elevation of the water table? (Note: You can determine this from the elevation of the lakes' water surface.)
9. Underground, the water flows through an aquifer. Define an aquifer and list the characteristics an aquifer must have. What is the general direction of groundwater flow in the aquifer underlying the Interlachen area?
10. Because much of central Florida is rapidly urbanizing, what problems and hazards do you anticipate in this karst area?

Putnam Hall, Florida
Scale 1:24,000
Contour interval 10 ft

Interlachen, Florida
Scale 1:62,500
Contour interval = 10 ft

USGS

FLUVIAL PROCESSES AND LANDFORMS

◼ OBJECTIVES

WHEN YOU COMPLETE THIS CHAPTER YOU SHOULD BE ABLE TO:

- ◼ 17.1 Describe how surface runoff is generated and how stream channels are initiated.
- ◼ 17.2 Understand how stream systems are organized within drainage basins and how streams and drainage basins are ordered.
- ◼ 17.3 Relate how the amount of water flowing in a stream channel is determined.
- ◼ 17.4 List the variables that affect stream energy.
- ◼ 17.5 Recall the geomorphic work that can be accomplished by stream energy.
- ◼ 17.6 Recognize the major landforms associated with meandering stream systems.
- ◼ 17.7 Recount how an ideal meandering stream system varies from source to mouth.
- ◼ 17.8 Explain how tectonism and climate change can affect stream systems.
- ◼ 17.9 Summarize the changes that a stream undergoes during a flood and the notion of flood frequency.
- ◼ 17.10 Discuss the value of quantitative methods for analyzing stream systems.

FLOWING WATER IS MORE INFLUENTIAL in shaping the surface form of our planet than any other exogenic geomorphic process, primarily because of the sheer number of streams on Earth. Through erosion and deposition, water flowing downslope over the land surface, particularly when concentrated in channels, modifies existing landforms and creates others. Nearly every region of Earth's land surface in arid as well as humid climates exhibits at least some topography that has been shaped by the power of flowing water, and many regions exhibit considerable evidence of stream action. Polar landscapes buried under thick accumulations of perennial ice are the major exception to Earth's extensive areas of stream-dominated topography.

The study of flowing water as a land-shaping process, together with the study of the resulting landforms, is **fluvial geomorphology** (from Latin: *fluvius*, river). Fluvial geomorphology includes the action of channelized and unchannelized flow moving downslope because of the pull of gravity.

Stream is the general term for natural, channelized flow. In the Earth sciences, the word stream pertains to water flowing in a channel of any size, even though in

Streams are the dominant geomorphic agent on Earth's surface, and many are also beautiful landscape features. National Park Service

general usage we refer to large streams as rivers and use local terms, such as *creek*, *brook*, *run*, *draw*, and *bayou*, for smaller streams. The land between adjacent channels in a stream-dominated landscape is the **interfluve** (from Latin: *inter*, between; *fluvius*, river).

Because of the common and widespread occurrence of stream systems and their key role in providing freshwater for people and our agricultural, industrial, and commercial activities, a substantial portion of the world's population lives in close proximity to streams. This makes understanding stream processes, landforms, and hazards fundamental for maintaining human safety and quality of life.

Most streams occasionally expand out of the confines of their channel. Although these floods typically last only a few days at most, they reveal the tremendous—and often very dangerous—geomorphic power potential of flowing water.

This chapter begins by considering how water accumulates into organized channel systems at Earth's surface. We then investigate factors that influence the amount and variability of water flow at any given channel reach and how those flow conditions affect a stream's ability to erode, transport, and deposit rock material. As streams perform that erosion, transportation, and deposition, they create distinctive landforms and landscape features that over long periods may also be influenced by tectonic and climate-change factors. Like short-term floods, the long-term effects of stream flow, whether dominated by erosion or deposition, may also be dramatic. Two impressive examples in the United States that illustrate the effectiveness of flowing water in creating landforms are the Black Canyon of the Gunnison River, carved by long-term river erosion into the Rocky Mountains, and the Mississippi River delta, where fluvial deposition is building new land into the Gulf of Mexico (● Fig. 17.1).

● **FIGURE 17.1** (a) Long-term fluvial erosion by the Gunnison River, with the aid of weathering and mass wasting, has carved the Black Canyon of the Gunnison in the Rocky Mountains of Colorado. The narrow canyon is 829 meters (2722 ft) deep, with resistant rocks forming the steep walls. (b) Part of the extensive fluvial deposits of the low-gradient Mississippi River delta in the Gulf of Mexico. Muddy water on this false-color image appears light blue; clearer, deeper water is dark blue.

(a)

(b)

Surface Runoff

Liquid water flowing over the surface of Earth, which is known **surface runoff**, or *overland flow*, can originate as ice and snow melt or as outflow of subsurface water, such as at springs, but most runoff originates in direct response to precipitation arriving as rain. When rain strikes the ground, several factors interact to determine whether surface runoff will occur. Runoff is generated when the amount of precipitation exceeds the ability of the ground to soak up the moisture. Because the process of water soaking into the ground is *infiltration*, the amount of water that can enter the voids and pore spaces in the soil and surface sediments at a given location is described as its *infiltration capacity*. A portion of infiltrated water percolates (seeps down) to lower subsurface positions and reaches the zone of saturation beneath the water table; much of the rest eventually returns to the atmosphere by evaporation from the soil or by transpiration from plants. When the amount of precipitation is greater than can be infiltrated into the ground, the excess water flows downslope because of gravity as surface runoff.

Various factors act individually or together to either enhance or inhibit the generation of surface runoff. Greater infiltration to the subsurface, and therefore less runoff, occurs under conditions of permeable surface materials, deeply weathered sediments and soils, gentle slopes, dry initial soil conditions, and a dense cover of vegetation. **Interception** of precipitation by vegetation allows greater infiltration by slowing down the rate of delivery of precipitation to the ground. Vegetation makes space in soil pores for additional infiltration when it takes up soil water and returns it to the atmosphere through transpiration. Given the same precipitation event, surface materials of low permeability and limited weathering, thin soils, steep slopes, preexisting soil moisture, and sparse vegetation, each contribute to increased runoff by decreasing infiltration (● Fig. 17.2). Human activities can affect many of these variables, and in some places the generation of runoff has been greatly modified by urbanization, mining, logging, or agriculture.

When surface runoff begins, it initially flows downslope as a thin sheet of unchannelized water, known as **sheet wash**, or *unconcentrated flow*. Because of gravity, after a short distance the sheet wash begins to move preferentially into any preexisting swales or depressions in the terrain. This concentration of flow leads to the formation of tiny channels, called **rills**, or somewhat larger channels, called **gullies**. Rills are on the order of a couple of centimeters deep and a couple of centimeters wide (● Fig. 17.3), whereas gully depth and width may approach as much as a couple of meters.

Water does not flow in rills and gullies all the time but only during and shortly after a precipitation (or snowmelt) event. Channels that are empty of water much of the time like this are described as having **ephemeral flow**. In the downslope direction, the smallest rills typically join to form slightly larger rills, which can join to make gullies. In humid climates, following these successively larger channels of ephemeral flow downslope will eventually lead us to a point where we

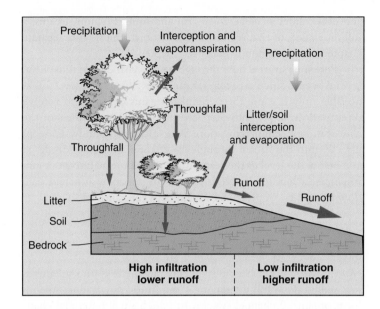

● **FIGURE 17.2** The amount of surface runoff depends on the intensity and duration of a rainstorm, surface features that affect infiltration, and the amount of evapotranspiration. Deep soil, dense vegetation, fractured bedrock, and gentle slopes encourage infiltration, thereby reducing runoff. Thin or absent soils, sparse vegetation, and steep slopes limit infiltration, which enhances surface runoff.

How does vegetation encourage infiltration?

first encounter **perennial flow**. Streams with perennial flow have water present all year, but not always in the same volume or at the same velocity. Most arid region streams flow on an ephemeral basis, although some may instead have **intermittent flow**, which lasts for a couple of months in response to an annual rainy season or spring snowmelt. Because of this contrast in flow duration,

● **FIGURE 17.3** Rills are the smallest type of channel. They are commonly visible on unvegetated surfaces, like these mine tailings.

Is there a gully in this photo?

D. Sack

and other differences between arid and humid region streams, a full discussion of arid region stream systems appears in a separate chapter on arid region landforms, Chapter 18.

Perennial streams flow throughout the year even if it has been several weeks since a precipitation event. In most cases, this is possible only because the perennial streams continue to receive direct inflow of groundwater (see Chapter 16) regardless of the date of most recent precipitation. Slow-moving groundwater seeps directly into the stream through the channel bottom and sides at and below the level of the water surface as *base flow*. Except in rare instances, a humid climate is needed to generate sufficient base flow to maintain a perennial stream between rainstorms.

The Stream System

Most flowing water becomes quickly channelized into streams as it is pulled downhill by gravity. Continuing downslope, streams form organized channel systems in which small perennial channels join to make larger perennial channels, and larger perennial channels join to create even bigger streams. Smaller streams that contribute their water and sediment to a larger one in this way are **tributaries** of the larger channel, which is called the *trunk stream* (● Fig. 17.4).

Drainage Basins

Each individual stream occupies its own **drainage basin** (also known as *watershed* or *catchment*), the expanse of land from which it receives runoff. *Drainage area* refers to the measured extent of a drainage basin and is typically expressed in square kilometers or square miles. Because the runoff from a tributary's drainage basin is delivered by the tributary to the trunk stream, the tributary's drainage basin also constitutes part of the drainage basin of the trunk stream. In this way, small tributary basins are nested within, or are *subbasins* of, a succession of larger and larger trunk stream drainage basins. For example, the Ohio River is a tributary of the Mississippi River, thus the entire drainage basin of the Ohio River forms a portion of the Mississippi River watershed. Large river systems drain spatially extensive watersheds that consist of numerous inset subbasins.

Drainage basins are open systems that involve inputs and outputs of water, sediment, and energy. Knowing the boundaries of a drainage basin and its component subbasins is critical to properly managing the water resources of a watershed. For example, pollution discovered in a river almost always comes from a source within the river's drainage basin, entering the stream system either at the point where the pollutant was first detected or at a location upstream from that site. This knowledge helps us track, detect, and correct sources of pollution.

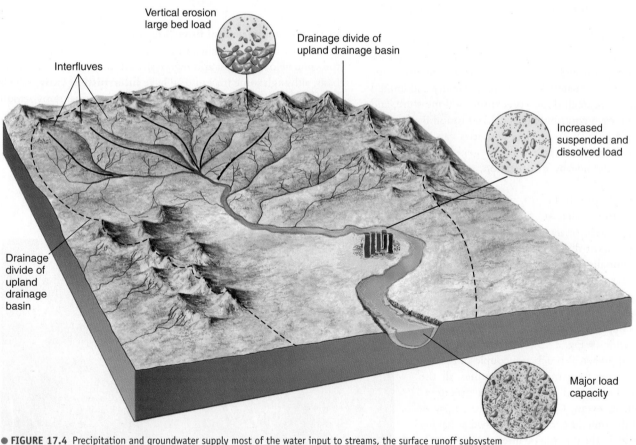

● **FIGURE 17.4** Precipitation and groundwater supply most of the water input to streams, the surface runoff subsystem of the hydrologic cycle. Water output from streams consists of evaporation to the atmosphere, infiltration to the subsurface, and delivery to the ocean at the river mouth. Materials transported by streams, known as load, enter the stream system by erosion and mass movement, particularly in upstream locations. The amounts of water and load increase dramatically downstream as tributaries deliver their inputs to the main stream. Load exits the stream system at the river mouth and, during floods, in terrain adjacent to the channel.

The **drainage divide** represents the outside perimeter of a drainage basin and thus also the boundary between it and adjacent basins (● Fig. 17.5). The drainage divide follows the crest of the interfluve between two adjacent drainage basins. In some places, this crest is a definite ridge, but the higher land that constitutes the divide is not always ridge-shaped, nor is it necessarily much higher than the rest of the interfluve. Surface runoff generated on one side of the divide flows toward the channel in one drainage basin, whereas runoff on the other side travels in a very different direction toward the channel in the adjacent drainage basin. The *Continental Divide* separates North America into a western region where most runoff flows to the Pacific Ocean and an eastern region where runoff flows to the Atlantic Ocean. The Continental Divide generally follows the crest line of high ridges in the Rocky Mountains, but in some locations the highest point between the two huge basins lies along the subtle crest of gently sloping high plains.

A stream, such as the Mississippi River, that has a very large number of tributaries and encompasses several levels of nested subbasins, will differ in some major ways from a small creek that has no perennial tributaries and lies high up in the drainage basin near the divide. Knowing where each stream lies in the hierarchical order of tributaries helps Earth scientists make more meaningful comparisons among different streams. **Stream ordering** is the technique used to describe quantitatively the position of a stream and its drainage basin in the nested hierarchy of tributaries.

First-order streams have no perennial tributaries. Even though they are the smallest perennial channels in the drainage basin, first-order channels are evident on large-scale topographic maps. Most first-order streams lie high up in the drainage basin near the drainage divide, the *source* area of the stream system. Two first-order streams must meet to form a *second-order stream*, which is larger than each of the first-order streams. It takes the intersection of two second-order channels to make a *third-order stream*, regardless of how many first-order streams might independently join the second-order channels. The ordering system continues in this way, requiring two streams of a given order to combine to create a stream of the next higher order. The order of a drainage basin derives from the largest stream order found within it (● Fig. 17.6). For example, the Mississippi is a tenth-order

(a)

(b)

Texas Natural Resources Information System

● **FIGURE 17.5** (a) A drainage divide marks the boundary between adjacent drainage basins and lies along the high ground between the two stream systems. (b) Aerial view of the upper portions of two major drainage basins and the major drainage divide (with roads) that separates them. Headwaters of many small tributaries lie on either side of the large divide. Each tributary occupies its own small drainage basin (bounded by its own drainage divide), and these small basins combine to form the larger drainage basins.

Can you mentally trace the outline of some of the numerous small drainage basins appearing in the photo?

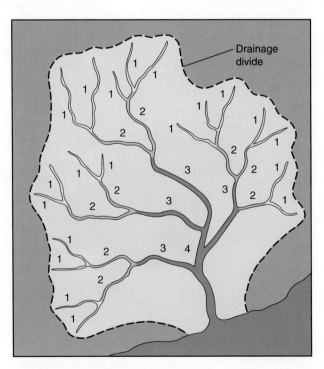

● **FIGURE 17.6** The concept of stream order illustrated by the channels of a fourth-order drainage basin. Stream order changes (increases) only when two streams of equal order join.

What is the highest-order stream in a selected drainage basin called?

Watersheds as Critical Natural Regions

Perhaps the most environmentally logical way to divide Earth's land surface into regions is by the drainage basins for stream systems. Virtually all of Earth's land surface constitutes part of a drainage basin. Like the channel systems that they contain, drainage basins are hierarchical and are conveniently subdivided into smaller subbasins for local studies and management. At the same time, higher-order watersheds of large river systems are subject to broad, integrated, regional-scale analyses.

The various components of a drainage basin are strongly interrelated. Problems in one part of a drainage basin system are likely to cause problems in other parts of the system. Along with the channel network that occupies each watershed, drainage basins consist of water, soil, rock, terrain, vegetation, people, wildlife, and domesticated animals that together form complex natural biotic habitats. Monitoring and managing these complex watershed systems requires an interdisciplinary effort that considers aspects of all four of the world's major spheres.

Surface water from watershed sources provides much of the potable water resources for the world's population. Increasing human populations and land-use intensities in previously natural drainage basins place pressures on these habitats and their water quality. Maintaining water quality within local subbasins contributes to maintaining water quality of larger river systems at the regional scale. Because a stream system is a network of water flowing in one direction (downstream), identifying sources of pollution involves upstream tracing of pollutants to the source and downstream spreading of pollution problems. Humans have a vital responsibility to monitor, maintain, and protect the quality of freshwater resources, and doing so at the local watershed scale has far-reaching and local benefits.

In recent years many governmental agencies have designated watersheds as critical regions for environmental management. Some of these efforts involve the establishment of locally administered river conservation districts representing relatively small-order drainage basins in which community involvement plays a vital

role. There are several sound reasons, including the connectivity and hierarchical nature of watersheds, for this spatial strategy.

Watersheds are clearly defined, well-integrated, natural regions of critical importance to life on Earth, and they make a logical spatial division for environmental management. However, a stream system may flow through many cities, counties, states, and countries, and management problems can arise where regional, political, and administrative boundaries do not coincide with the divide that defines the edge of a watershed. Each of these political jurisdictions may have different needs, goals, and strategies for using and managing their part of a watershed, and often a variety of strategies results in conflict. Still, most administrative units recognize that cooperative management of the watershed system as a whole is the best approach. Many cooperative river basin authorities have been established to encourage a united effort to protect a shared watershed. This watershed-oriented management strategy, based on a fundamental natural region, is an important step in protecting our freshwater resources.

The Mississippi River drainage basin covers an extensive area of the United States.

D. Sack

As a tributary of the Ohio River, the fourth-order Hocking River watershed in southeastern Ohio is a subbasin within the Mississippi River system. Community involvement in improving and maintaining the health of the Hocking River watershed, and other drainage basins, has regional as well as local benefits.

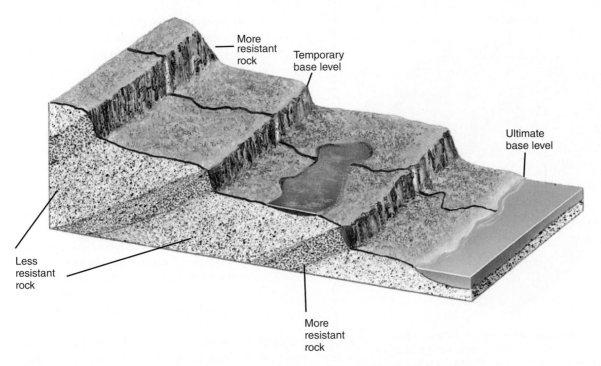

More resistant rock

Temporary base level

Ultimate base level

Less resistant rock

More resistant rock

● **FIGURE 17.7** The lowest point to which a stream can flow is its base level. Most humid region streams have sufficient flow to reach the ocean; sea level is the ultimate base level for all streams. Many arid region streams only reach a regional base level, a topographic basin on the continent. A temporary base level occurs where a rock unit lying in the pathway of a stream is significantly more resistant than the rock upstream from it. The stream will not be able to cut into the less resistant rock any faster than it can cut into the resistant rock of the temporary base level.

drainage basin because the Mississippi River is a tenth-order stream. Stream ordering allows us to compare streams and their various attributes quantitatively by relative size, which helps us better understand how stream systems work. Among other things, comparing streams on the basis of order has shown that as stream order increases, basin area, channel length, channel size, and amount of flow also increase.

Water in a stream system proceeds downslope through a succession of channels of ever-increasing order toward the downstream end, or *mouth*, of the stream. Most humid region, perennial stream systems end at, and thus deliver their water to, the ocean. Drainage basins with channel systems that convey their water to the ocean are described as having **exterior drainage**. Many arid region streams have **interior drainage** because they do not have enough flow to reach the ocean, but terminate instead in local or regional areas of low elevation.

Every stream has a **base level**, the elevation below which it cannot flow. Sea level is the *ultimate base level* for virtually all stream action. Streams with exterior drainage reach ultimate base level; the low point of flow for a stream with interior drainage is referred to as a *regional base level*. In some drainage basins, a resistant rock layer located somewhere upstream from the river mouth can act as *temporary base level*, temporarily controlling the lowest elevation of the flow upstream from that point until the stream is finally able to cut down through the resistant layer (● Fig. 17.7).

Drainage Density

Each system of stream tributaries exhibits spatial characteristics that provide important information about the nature of the drainage basin. The extent of channelization can be represented by measuring **drainage density** (D_d), where $D_d = L/A_d$, the total length (L) of all channels in the drainage basin, divided by the area of the drainage basin (A_d). Because drainage density indicates how dissected the landscape is by channels, it reflects both the tendency of the drainage basin to generate surface runoff and the erodibility of the surface materials (● Fig. 17.8). Regions with high drainage densities have limited infiltration, considerable runoff, and at least moderately erodible surface materials.

In addition to the factors noted previously that reduce infiltration and promote runoff—impermeable sediments, thin soils, steep slopes, sparse vegetation—the ideal climate for high drainage densities is one that is semiarid. Humid climates encourage extensive vegetation cover, which promotes infiltration because of interception and inhibits channel formation by holding soils and surface sediment in place. In arid climates, although the vegetation cover is sparse, low amounts of precipitation do not supply enough runoff to carve many channels. Semiarid climates have sufficient precipitation input to produce overland flow but not enough to support an extensive vegetative cover. For example, the easily eroded Dakota Badlands, which are located in a steppe climate, have an extremely high drainage density of more than

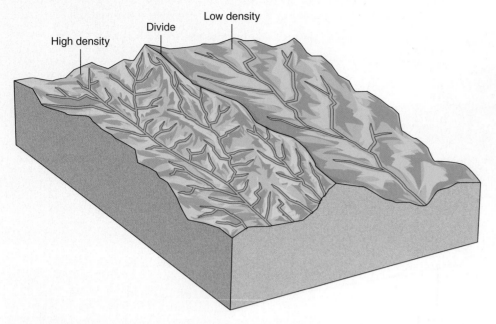

High density   Divide   Low density

● **FIGURE 17.8** Drainage density—the length of channels per unit area—varies with such factors as the ground's erodibility, permeability, slope, and vegetation cover. Ground areas with greater surface runoff and less infiltration tend to have higher drainage densities than areas with less runoff and greater infiltration.

What kind of drainage density would you expect in an area of steep slopes and sparse vegetation cover?

125 kilometers per square kilometer (75 mi/mi²), which means that, on average, 1 square kilometer of the drainage basin contains 125 kilometers of channel. On the other hand, a region with weathering-resistant granite as bedrock and a humid climate may have a drainage density of only 5 kilometers per square kilometer (3 mi/mi²).

Another way to understand the concept of drainage density is to think about what would happen on a hillside in a humid climate if the natural vegetation were burned off in a fire. Erosion would rapidly cut gullies, creating more channels than previously existed there. In other words, the drainage density would increase. We could use the quantitative measure of $D_d = L/A_d$ to determine the change in channelization precisely and to monitor it over time.

Drainage Patterns

When viewed on maps or by looking down at Earth's surface from the air, networks of stream tributaries display distinct spatial arrangements called **drainage patterns**. Two primary factors that influence drainage patterns are bedrock structure and surface topography. A *dendritic* (from Greek: *dendros*, tree) drainage pattern (● Fig. 17.9a) resembles a branching tree with tributaries joining larger streams at acute angles (less than 90°). Dendritic patterns are common and develop where the rocks have a roughly equal resistance to weathering and erosion and are not intensely jointed. A *trellis* pattern consists of long, parallel streams with short tributaries entering at right angles (Fig. 17.9b). Trellis drainage indicates folded terrain where short streams flow down the sides of resistant rock ridges into a larger stream that

occupies the adjacent valley that consists of erodible rock, as in the Ridge and Valley region of the Appalachians. A *radial* pattern develops where streams flow away from a common high point on cone- or dome-shaped geologic structures, such as volcanoes (Fig. 17.9c). The opposite pattern is *centripetal*, with the streams converging on a central lowland, as in an arid region basin of interior drainage (Fig. 17.9d). *Rectangular* patterns occur where streams follow sets of intersecting fractures to produce a blocky network of straight channels with right-angle bends (Fig. 17.9e). In some regions that were recently covered by extensive glacial ice, streams flow on low-gradient terrain left by the receding glaciers, wandering between marshes and small lakes in a chaotic pattern called *deranged* drainage (Fig. 17.9f).

Many streams follow the "grain" of the topography or the bedrock structure, eroding valleys in weaker rocks and flowing away from divides formed on resistant rocks. A number of examples also exist of streams that flow across—that is, transverse to—the structure, cutting a gorge or canyon through mountains or ridges. These **transverse streams** can be puzzling, giving rise to questions as to how a stream can cut a gorge through a mountain range or otherwise cross from one side of a mountain range to the other. Such streams are either antecedent or superimposed.

Antecedent streams exist before the establishment of the mountains that they flow across. As gradual mountain building takes place across their paths, antecedent streams are able to maintain their original course by cutting ever-deepening canyons. The Columbia River Gorge through the Cascade Range in Washington and Oregon and many of the great canyons in the Rocky Mountain region, such as Royal Gorge and the Black Canyon of the Gunnison, both in Colorado, probably originated as antecedent streams (● Fig. 17.10a). Antecedent streams (from Latin: *ante*, before; *cedere*, to go) establish their flow paths before existence of the structure that they now cut across.

In other cases of transverse streams, the mountain or ridge predates the *superimposed stream* course that now cuts across it. In the geologic past, the mountain or ridge, typically partially eroded, became buried beneath more recent rocks. The stream developed its course on the younger strata independent of the characteristics of the buried structure. As the younger rocks were eventually stripped away by erosion, the stream, still following its original course, was set down (superimposed) onto the rocks of the older, lower structure (Fig. 17.10b). This sequence explains why, in many instances, such as the central Appalachians, rivers flow across folded rock structure, creating water gaps through mountain ridges. Examples include the Cumberland Gap and the gap formed by the Susquehanna River in Pennsylvania, both of which were important travel routes for the first European settlers crossing the Appalachians.

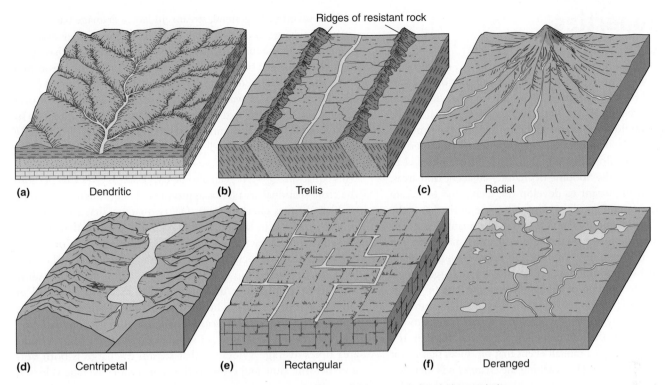

● FIGURE 17.9 Drainage patterns often reflect bedrock structure. (a) The dendritic pattern is found where rocks have uniform resistance to weathering and erosion. (b) A trellis pattern indicates parallel valleys of weak rock between ridges of resistant rock. (c) Multiple channels trending away from the top of a domed upland or volcano form the radial pattern. (d) The centripetal pattern shows multiple channels flowing inward toward the center of a structural lowland. (e) Rectangular patterns indicate sets of linear joints in the bedrock structure. (f) A deranged pattern typically forms following the retreat of continental ice sheets; it is characterized by a chaotic arrangement of channels connecting small lakes and marshes.

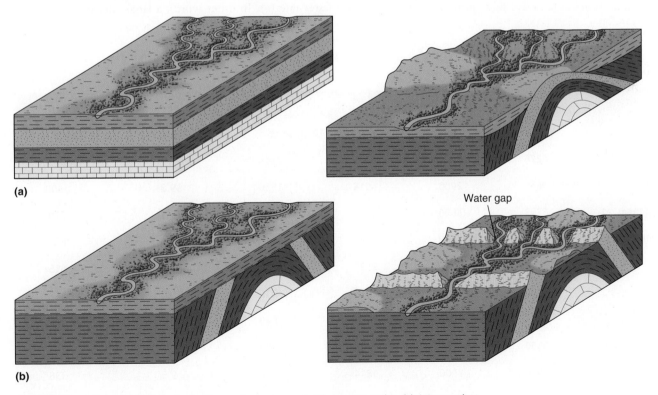

● FIGURE 17.10 Transverse streams form valleys and canyons across ridges or mountains. (a) An antecedent stream maintains its course by cutting a valley through a mountain that is being uplifted gradually over much time. (b) A superimposed stream has uncovered and excavated ancient structures that were buried beneath the surface. As the stream erodes the landscape downward, it cuts across and through the ancient structures.

Flow Properties

Stream Discharge

The amount of water flowing in a stream depends not only on the impact of recent weather patterns but also on such drainage basin factors as its size, relief, climate, vegetation, rock types, and land-use history. Stream flow varies considerably from time to time and place to place. Most streams experience occasional brief periods when the amount of flow exceeds the ability of the channel to contain it, resulting in the flooding of channel-adjacent land areas.

Just as it was important to develop the technique of stream ordering to indicate numerically a channel's place in the hierarchy of tributaries, it is also crucial to be able to describe quantitatively the amount of flow being conveyed through a stream channel. **Stream discharge** (Q) is the volume of water (V) flowing past a given channel cross section per unit time (t): $Q = V/t$. Discharge is most commonly expressed in units of cubic meters per second (m^3/s) or cubic feet per second (ft^3/s). A cross section is essentially a thin slice extending from one stream bank straight across the channel to the other stream bank and oriented perpendicular to the channel. If a drainage basin experiences a rainfall event that produces significant runoff, the volume of water (V) reaching the channel will increase. Notice from the discharge equation that this increase in volume (V) will cause an increase in stream discharge (Q).

It is important to collect and analyze discharge data for several reasons. These data can be used to compare the amount of flow carried in different streams, at different sites along one stream, or at different times at a single cross section. Discharge data indicate the size of a stream and, in times of excessive flow, provide an index of flood severity. In general, streams in larger drainage basins have greater discharge and are longer than streams in smaller drainage basins. Among the major rivers of the world, the Amazon has by far the largest drainage basin and the greatest discharge; the Mississippi River system is ranked fourth in terms of discharge (Table 17.1).

The volume of water rushing through a cross section of a stream per second is extremely difficult to measure directly. In reality, discharge is determined not by measuring $Q = V/t$ directly but by using the fact that discharge (Q) is also equal to the area of the cross section (A) times the average stream velocity (v). This equation, $Q = Av$, can also be expressed as $Q = wdv$ because the cross-sectional area (A) is approximately equivalent to channel width (w) times channel depth (d), two factors that are relatively easy to measure in the field (● Fig. 17.11). Notice that because cross-sectional area (A) is measured in square meters or square feet and average velocity is measured in meters per second or feet per second, solving the equation $Q = Av$ yields discharge values in units of volume per unit time (cubic meters per second or cubic feet per second). This analysis of the measurement units, a procedure known as *dimensional analysis*, should convince you that volume per unit time is indeed equivalent to cross-sectional area times average velocity—that stream discharge is both volume per unit time and cross-sectional area times average velocity, $Q = V/t = Av$.

As is true for any equation, a change on one side of the discharge equation must be accompanied by a change on the other. If discharge increases because a rainstorm delivers a large volume of water to the stream, that increase in volume (V) will occupy a larger cross-sectional area (A) and flow through the cross section at a faster rate (v). In other words, a large rainstorm causes the

TABLE 17.1
Ten Largest Rivers of the World

	Length		Area of Drainage Basin		Discharge	
	km	mi	sq km (×1000)	sq mi	1000 m³/s	1000 cfs
Amazon	6276	3900	6133	2368	112–140	4000–5000
Congo (Zaire)	4666	2900	4014	1550	39.2	1400
Chang Jiang (Yangtze)	5793	3600	1942	750	21.5	770
Mississippi–Missouri	6260	3890	3222	1244	17.4	620
Yenisei	4506	2800	2590	1000	17.2	615
Lena	4280	2660	2424	936	15.3	547
Paraná	2414	1500	2305	890	14.7	526
Ob	5150	3200	2484	959	12.3	441
Amur	4666	2900	1844	712	9.5	338
Nile	6695	4160	2978	1150	2.8	100

Source: Adapted from Morisawa, *Streams: Their Dynamics and Morphology.* New York: McGraw-Hill Book Company, 1968.

● **FIGURE 17.11** These university students are measuring a stream's velocity, depth, and width to find discharge, the volume of water flowing through a given cross section in a stream per unit time.

level of a stream to rise and the water to flow faster. As the water level (flow depth, d) rises, streams also experience an increase in width (w) because most channels flare out a bit as they rise higher up on their banks. Stream channels continually adjust their cross-sectional area and flow velocity in response to changes in flow volume. Understanding the relationships among the factors involved in discharge is very important for understanding how streams work.

Stream Energy

When people have more energy, they can accomplish more, or do more work, than when they have less energy. The same is true for streams and the other geomorphic agents. The ability of a stream to erode and transport sediment—that is, to perform geomorphic work—depends on its available energy. Picking up pieces of rock and moving them require the stream to have *kinetic energy*, which is the energy of motion. When a stream has more kinetic energy, it can pick up and move more clasts (rock particles) and heavier clasts than when it has less energy. The kinetic energy (E_k) equation, $E_k = \frac{1}{2}mv^2$, shows that the amount of energy a stream has depends on its mass (m), but especially on its velocity of flow (v), because kinetic energy varies with velocity squared (v^2). Thus, stream velocity is a critical factor in determining the amount of geomorphic work that can be accomplished by a stream. As we saw in the previous section, when stream discharge (Q) varies because of changes in runoff (V), so does stream velocity (v) because $Q = V/t = wdv$.

Besides discharge variations, another major way to alter stream velocity, and thereby stream energy, is through a change in the slope, or gradient, of the stream. *Stream gradient* is the drop in stream elevation over a given horizontal distance along the stream course, and is usually expressed in meters per kilometer (m/km) or feet per mile (ft/mi). Everything else being equal, water flows faster down channels with steeper slopes and slows down over gentler slopes. Steeper channels typically occupy locations that are

farther upstream and higher in the drainage basin as well as places where the stream flows over rock types that are more resistant to erosion. Gentler stream gradients tend to occur closer to the mouth of a stream and where the stream crosses easily eroded rock types. As water moves through a channel system, its ability to erode and transport sediment—that is, its energy—is continuously changing as variations in stream gradient and discharge cause changes in flow velocity.

Different factors work to decrease the energy of a stream. Friction along the bottom and sides of a channel and even between the stream surface and the atmosphere slows down the stream velocity and therefore contributes to a decrease in stream energy. A channel with numerous substantial irregularities, such as those from large rocks and vegetation sticking up into the flow, has a great amount of **channel roughness**, which causes a considerable decrease in stream energy because of the frictional effects. Smooth, bedrock channels without these irregularities have lower channel roughness and therefore a smaller loss in stream energy resulting from friction. In all cases, however, friction along the bottom and sides of a channel slows down the flow and slows it down the most at the channel boundaries so that the maximum velocity occurs somewhat below the stream surface at the deepest part of the cross section. In addition to the external friction at the channel boundary, streams lose energy because of internal friction in the flow related to eddies, currents, and the interaction among water molecules. About 95% of a stream's energy is consumed in overcoming all types of frictional effects. Only the remaining 5% is available for doing the work of eroding and transporting sediment.

Stream gradients are usually steepest at the headwaters and in new tributaries and diminish in the downstream direction (● Fig. 17.12). Discharge, however, increases downstream in humid regions as the size of the contributing drainage basin increases. As the water originally conveyed in numerous small

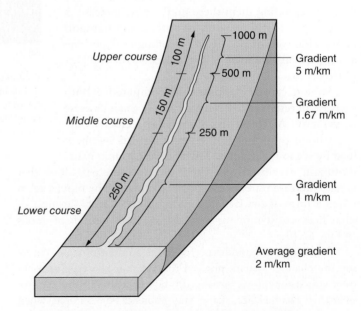

● **FIGURE 17.12** Stream gradient decreases from source to mouth for the ideal stream.

channels high in the drainage basin is collected in the downstream direction into a shrinking number of larger and larger channels, flow efficiency improves and frictional resistance decreases. As a result, flow velocity tends to be higher in downstream parts of a channel system than over the steep gradients in the headwaters. The ability of the stream to carry sediment, therefore, may be equivalent or even greater downstream than upstream.

Sediment being transported by a stream is called the **stream load**. Carrying sediment is a major part of the geomorphic work accomplished by a stream. In order from smallest to largest, the size of sediment that a stream might transport includes clay, silt, sand, granules, pebbles, cobbles, and boulders. Sand marks the boundary between fine-grained (small) clasts and coarse-grained (large) clasts. Gravel is a general term for any sediment larger in size than sand. The maximum size of rock particles that a given stream is able to transport, referred to as the **stream competence** (measured as particle diameter in centimeters or inches), and the total amount of load being moved by the stream, termed **stream capacity** (measured as weight per unit time), depend on available stream energy and thus on the flow velocity.

As a stream flows along, its velocity is always changing, reflecting the constant variations in stream discharge, stream gradient, and frictional resistance factors. As a result, the size and amount of load that the stream can carry is also changing constantly. If the material is available when flow velocity increases, the stream will pick up from its own channel bed larger clasts and more load. When its velocity decreases and the stream can no longer transport such large clasts and so much material, it drops the larger clasts onto its bed, depositing them there until a new increase in velocity provides the energy to entrain and transport them. Thus, whether a stream is eroding or depositing at any particular moment is determined by the complex of factors that control its energy.

Stream capacity and stream competence both increase in response to even a relatively small increase in velocity. A stream that doubles its velocity during a flood can increase the amount of its sediment load by six to eight times. The boulders seen in many mountain streams arrived there during some past flood that greatly increased stream competence; they will be moved again when a flow of similar magnitude occurs. Rivers do most of their heavy clast-moving work during short periods of floods (● Fig. 17.13).

Streams have no influence over the amount of water entering the channel system, nor can they change the type of rocks over which they flow. Streams do, however, have some control over their channel size, shape, and gradient. For example, when a stream undergoes a decrease in energy so that it deposits some

(a)

(b)

● **FIGURE 17.13** (a) Fast-moving floodwaters can break up and move large, heavy objects that lie in the path of the torrent. This bridge in Wimberley, Texas, was destroyed by catastrophic flooding in 2015. (b) A flooding stream can also ruin buildings and people's belongings by pushing objects around, carrying away floating debris, depositing large amounts of sediment, and causing mold and mildew.

of its sediment, the deposit raises the channel bottom and locally creates a steeper gradient downchannel from the deposit. With continued deposition, the location will eventually attain a slope steep enough to cause a sufficient increase in velocity so that the flow again entrains and moves that deposited sediment. Streams tend to adjust their channel properties so they can move the sediment supplied to them. *Dynamic equilibrium* is maintained by adjustments among channel slope, shape, and roughness; amount of load eroded, carried, or deposited; and the velocity and discharge of stream flow.

Fluvial Processes

Erosion

Fluvial erosion is the removal of rock material by flowing water. The process consists of the chemical removal of ions from rocks and the physical removal of rock fragments (clasts). Physical removal of rock fragments includes breaking off new pieces of bedrock from the channel bed or sides and moving them as well as picking up and removing preexisting clasts that were temporarily resting on the channel bottom. The breaking away of new pieces of bedrock proceeds very slowly on highly resistant rock types.

The removal of rock material by erosion does not necessarily mean that the landscape is undergoing long-term lowering. If sediments eroded from the bed of a stream channel are replaced by the deposition of other fragments transported in from upstream, there will be no net drop in the position of the channel bottom. Such lowering, known as channel incision, occurs only when there is net erosion compared with deposition. Net erosion results in the lowering of the affected part of the landscape and is termed **degradation**. Net deposition of sediments results in a building up, or **aggradation**, of the landscape.

One way that streams erode occurs when stream water chemically dissolves rock material and then transports the ions away in the flow. This fluvial erosion process, called **corrosion**, has a limited effect on many rocks but can be significant in certain rock types, such as limestone.

Hydraulic action refers to the physical, as opposed to chemical, process of stream water alone removing pieces of rock. As stream water flows downslope, it exerts stress on the streambed. Whether this stress results in entrainment and removal of a preexisting clast currently resting on the channel bottom, or even the breaking off of a new piece of bedrock from the channel, depends on several factors. These include the volume of water moving in the channel, flow velocity, flow depth, stream gradient, friction with the streambed, the strength and size of the rocks over which the stream flows, and the degree of stream turbulence.

Turbulence is chaotic flow that mixes and churns the water, often with a significant upward component. It greatly increases the rate of erosion as well as the load-carrying capacity of the stream. Turbulence is influenced by channel roughness and the gradient over which the stream is flowing. A rough channel bottom increases the intensity of turbulent flow. Likewise, even a small increase in velocity caused by a steeper gradient can result in a significant increase in turbulence. Turbulent currents contribute to erosion by hydraulic action when they wedge under or pound away at rock slabs and loose fragments on the channel bed and sides, dislodging clasts that are then carried away in the current. *Plunge pools* at the base of waterfalls and in rapids reveal the power of turbulence-enhanced hydraulic action where it is directed toward a localized point (● Fig. 17.14).

As soon as a stream begins carrying rock fragments as load, it can start to erode by **abrasion**, a process even more powerful than hydraulic action. As rock particles bounce, scrape, and drag along the bottom and sides of a stream channel, they break off additional rock fragments. Because solid rock particles are denser

● **FIGURE 17.14** A plunge pool has been eroded at the base of Blue Hen Falls in Cuyahoga Valley National Park, Ohio.

Why do deep plunge pools form at the base of most waterfalls?

than water, the impact of clastic load being thrown against the channel bottom and sides by the current is much more effective than the impact of water alone.

Under some conditions, stream abrasion makes distinctive round depressions called *potholes* in the rock of a bedrock streambed (● Fig. 17.15). Potholes generally originate in specific ideal locations, such as below swirling rapids, at the base of waterfalls, or at points of structural weakness, which include joint intersections in the bedrock on the channel bed. Potholes range in diameter and depth from a few centimeters to many meters. If you peer into a pothole, you can often see one or more round stones at the bottom. These are the abraders, or *grinders*. Swirling whirlpool movements of the stream water cause such stones to grind the bedrock and enlarge the pothole by abrasion, whereas finer sediments are carried away in the current.

● **FIGURE 17.15** These potholes were cut into a streambed of solid rock during times of high flow. Potholes result from stream abrasion and the swirling currents in a fast-flowing river.

In a process related to abrasion, rock fragments traveling downstream as load are gradually reduced in size and their shape changes from angular to rounded. This wear and tear experienced by sediments as they tumble and bounce against one another and against the stream channel is called **attrition**. Attrition explains why gravels found in streambeds are smooth and rounded and why the load carried in the lower reaches of most large rivers is composed primarily of fine-grained sediments and dissolved minerals.

Stream erosion widens and lengthens stream channels and the valleys they occupy. Lengthening occurs primarily at the source through *headward erosion*, accomplished partly by surface runoff into the stream and partly by springs undermining the slope. The lengthening of a river's course in an upstream direction is particularly important where erosional gullies are rapidly dissecting agricultural land. Such gullying may be counteracted by soil conservation practices to reduce erosional soil loss. Channel lengthening also occurs if the path of the stream channel becomes more winding, or *sinuous*, and this will cause a decrease in stream gradient.

Transportation

Streams directly erode from their bed and banks some of the clastic sediment that they contain, and most chemical sediments are delivered to channels from the groundwater system as base flow. Far greater than either of these two sources is the proportion of fluvial load delivered to stream channels by surface runoff and mass movement. Another aspect of the geomorphic work of streams is moving these materials.

Streams transport their load in several ways (● Fig. 17.16). Some minerals are dissolved in the water and are thus carried in the transportation process of **solution**. The finest solid particles are carried in **suspension**, buoyed by vertical turbulence. Such small grains can remain suspended in the water column for long periods, as long as the force of upward turbulence is stronger than the downward settling tendency of the particles. Some grains that are too large and heavy to be carried in suspension bounce along the channel bottom in a process known as **saltation** (from French: *sauter*, to jump).

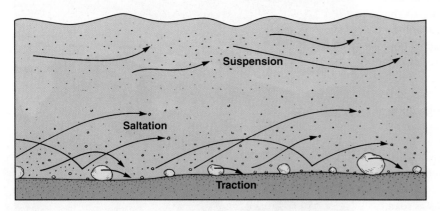

● **FIGURE 17.16** Transportation of solid load in a stream. Clay and silt particles are carried in suspension. Sand typically travels by suspension and saltation. The largest (heaviest) particles move by traction.

What is the difference between traction and saltation?

Finally, particles that are too large and heavy to move by saltation, but still small enough to move, slide and roll along the channel bottom in the transportation process of **traction**.

Fluvial geomorphologists recognize three main types of stream load. Ions of rock material held in solution constitute the **dissolved load**. **Suspended load** consists of the small clastic particles being moved in suspension. Larger particles that saltate or move in traction along the streambed represent the **bed load**. The total amount of load that a stream carries is expressed in terms of the weight of the transported material per unit time.

The relative proportion of each load type present in a given stream varies with such drainage basin characteristics as climate, vegetative cover, slope, rock type, and the infiltration capacities and permeabilities of the rock and soil types. Dissolved loads are larger than average in basins with high amounts of infiltration and base flow because slow-moving groundwater that feeds base flow acquires ions from the rocks through which it moves. The intense weathering in humid regions produces much fine-grained sediment that is incorporated into streams as suspended load. Large amounts of suspended load give streams a characteristically muddy appearance (● Fig. 17.17). The Huang He in northern China, known as the "Yellow River" because of the color of its silty suspended load, carries a huge amount of sediment in suspension, with more than 1 million tons of suspended load per year. Compared with the "muddy" Mississippi River, the Huang He transports five times the suspended sediment load with only 20% of the discharge. Streams dominated by bed load tend to occur in arid regions because of the limited weathering rate in arid climates. Limited weathering leaves considerable coarse-grained sediment in the landscape available for transportation by the stream system.

Deposition

Ions and sediment picked up by and delivered to the stream system are transported downstream and deposited in temporary or permanent storage in the channel, adjacent to the channel, or at the mouth of the stream. Because the capacity and competence of a stream to carry material depend on flow velocity, a decrease in velocity will cause a stream to reduce its load through deposition. Velocity decreases over time when flow subsides—for example, after the impact of a storm—but it also varies from place to place along the stream. A distinctly shallower part of a channel cross section that lies far from the deepest and fastest flow typically experiences low flow velocity and becomes the site of recurring deposition. The resulting accumulation of sediment, like what forms on the inside of a channel bend, is referred to as a **bar**. Sediment also collects in locations where velocity falls because of a reduction in stream gradient, where the river current meets the standing body of water at its mouth, and during floods on the land adjacent to the stream channel.

Alluvium is the general term for fluvial deposits, regardless of the type or size of material. Alluvium is recognized by the characteristic sorting and rounding of sediments that streams perform. A stream sorts particles by size, transporting the sizes

● **FIGURE 17.17** The Mississippi River carries a tremendous load of sediment in suspension, which is evident from its muddy appearance. It obtains much of this load from its major tributaries.

What are some of the major tributaries that enter the Mississippi River?

that it can and depositing larger ones. As velocity fluctuates because of changes in discharge, channel gradient, or roughness, particle sizes that can be picked up, transported, and deposited vary accordingly (● Fig. 17.18). The alluvium deposited by a stream that has fluctuating velocity exhibits alternating layers of coarser and finer sediment.

When streams leave the confines of their channels during floods, the channel cross-sectional width suddenly increases so much that flow velocity must slow down to counterbalance it ($Q = wdv$). The velocity decrease reduces stream competence and capacity, causing deposition of sediment on the flooded land adjacent to the channel. This sedimentation is greatest right next to the channel where aggradation constructs channel-bounding ridges known as **natural levees**, but some alluvium will be left behind wherever load settled out of the receding flood waters.

Floodplains constitute the often extensive, low-gradient land areas composed of alluvium that lie adjacent to many stream channels (● Fig. 17.19). Floodplains are aptly named because they are inundated during floods and because they are at least partially

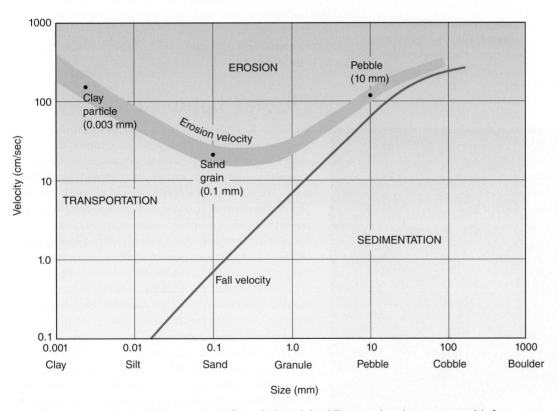

● **FIGURE 17.18** The relationship between stream flow velocity and the ability to erode and transport material of varying sizes. When a stream loses the ability to transport particles of a particular size, those particles are deposited. Small pebbles (particles with a diameter of 10 mm, for example) need a high stream flow velocity to be moved because of their size and weight. Silts and clays (smaller than 0.05 mm) also need high velocities for erosion because they stick together cohesively. Sand-sized particles (between 0.05 and 2.0 mm) are easier to erode (are picked up at a lower flow velocity) than clasts from smaller or larger size categories.

● **FIGURE 17.19** During floods, sediment-laden water that flows over the stream's banks inundates the adjacent, low floodplain areas and deposits alluvium, mainly silts and clays. This is a flooded farm in Wisconsin.

What would the floodwaters leave behind on the ground and in flooded homes and businesses after the water recedes?

composed of *vertical accretion deposits*, the sediment that settles out of slowing and standing floodwater. Most floodplains also contain *lateral accretion deposits*. These are generally channel bar deposits that get left behind as a stream gradually shifts its position in a sideways fashion (laterally) across the floodplain (● Fig. 17.20).

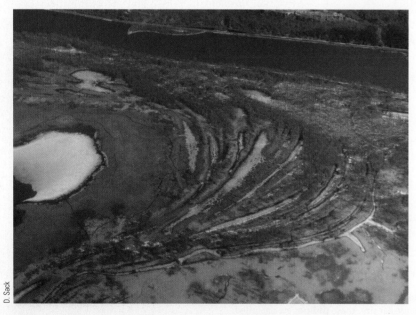

● **FIGURE 17.20** Sediment deposited in a bar on the inside of a channel bend becomes part of the floodplain alluvium if the stream migrates away laterally, leaving the bar deposits behind. Here, winter ice fills swales between ridges that mark the crests of successive bar deposits.

Channel Patterns

Three principal types of stream channels have traditionally been recognized when considering the form of a given channel segment in map view. Although *straight channels* may exist for short distances under natural circumstances, especially along fault zones, joints, or steep gradients, most channels with parallel, linear banks are artificial features that were totally or partially constructed by people.

If a stream has a high proportion of bed load in relation to its discharge, it deposits much of its load as sand and gravel bars in the streambed. The stream flows in multiple interweaving strands that split and rejoin around the bars to give a braided appearance, and indeed these are called **braided channels** (● Fig. 17.21). This channel pattern develops in settings that provide considerable amounts of loose sand and gravel bed load to the stream system and thus tend to be found in arid regions and in glacial meltwater zones. Braided streams are common on the Great Plains (for example, the Platte River), in the desert Southwest, in Alaska, and in Canada's Yukon.

The most common channel pattern in humid climates displays broad, sweeping bends in map view. Over time, these sinuous, **meandering channels** also wander from side to side across their low-gradient floodplains, widening the valley by lateral erosion on the outside of meander bends and leaving behind lateral accretion (bar) deposits on the inside of meander bends (● Fig. 17.22). These streams and their floodplains have a higher proportion of fine-grained sediment and thus greater bank cohesion than the typical braided stream.

Fluvial Landscapes

One way to understand the variety of landform features resulting from fluvial processes is to examine the course of an idealized river as it flows from its headwaters in the mountains to its mouth at the ocean, where it reaches base level. The gradient of this river would diminish continually downstream as it flows from its source toward base level. In nature, exceptions exist to this idealized profile because some streams flow entirely over a low gradient (● Fig. 17.23), whereas other streams, particularly small ones on mountainous coasts, flow down a steep gradient all the way to the standing body of water at their mouth. Rather than having a smoothly decreasing slope from headwaters to mouth, we would expect real streams to have some irregularities in this *longitudinal profile*, the stream gradient from source to mouth (see Fig. 17.12).

The following discussion subdivides the ideal perennial stream course into upper, middle, and lower river sections flowing over steep, moderate, and low gradients, respectively, and ending where it reaches base level at the ocean. Fluvial erosion processes dominate in the steep upper course, whereas deposition predominates in the lower course. The middle course displays important elements of both fluvial erosion and deposition.

● **FIGURE 17.21** Braided channels result from an abundant bed load of coarse sediment. When the stream sets down part of its coarse load, the water splits and flows around the deposit, eventually separating the main stream into numerous strands.

What does the common occurrence of braided channels just downstream from glaciers tell you about the sediment transported and deposited by moving ice?

Upper Stream Course

At the headwaters in the ideal upper course of a river, the stream primarily flows in contact with bedrock. Over the steep gradient high above its base level, the stream works to erode vertically downward by hydraulic action and abrasion. Erosion in the upper course creates a steep-sided valley, gorge, or ravine as the stream channel in the bottom of the valley cuts deeply into the land. Little if any floodplain is present, and the valley walls typically slope directly to the edge of the stream channel. Steep valley sides encourage mass movement of rock material directly into the flowing stream. Valleys of this type, dominated by the downcutting activity of the stream, are often called *V-shaped valleys* because with their steep slopes they attain the form of the letter V (● Fig. 17.24).

The effects of *differential erosion* can be significant in the upper course if the stream cuts through rock layers of varying resistance. The stream will have a steeper gradient where it flows over more resistant rock than where it encounters weaker rock. A steep gradient gives the stream flow more energy, which the stream needs so that it can erode the resistant rock. Rapids and waterfalls often mark the location of resistant materials in a stream's upper course. Where rocks are particularly resistant to weathering and erosion, valleys are narrow, steep-sided gorges or canyons; where rocks are less resistant, valleys tend to be more spacious.

Flowing over terrain composed of alternating strong and weak rocks can place a series of temporary base levels in the path of the stream. Water tends to pond into lakes on the less resistant rocks lying upstream from the resistant rock that forms the temporary base level. Many streams in this type of setting spill from lake to lake in their upper courses, either over open land (like the Niagara River at Niagara Falls, between Lake Erie and Lake Ontario) or through narrow gorges. In either case, the lake will eventually be eliminated—it will become a regular segment of the stream channel—if the outflowing stream lowers the lake basin outlet enough by erosion or if fluvial deposition at the inflow point fills in the lake basin with sediment.

Middle Stream Course

In the middle section of the ideal longitudinal profile, the stream flows over a moderate gradient and on a moderately smooth channel bed. Here, the stream valley includes a floodplain, but remaining ridges beyond the floodplain still form definite valley walls. The stream lies closer to its base level, flows over a gentler gradient, and thus directs less energy toward vertical erosion than in its upper course. The flow still has considerable energy, however, because of the downstream increase in the related variables of water volume, discharge, and velocity. The river now uses

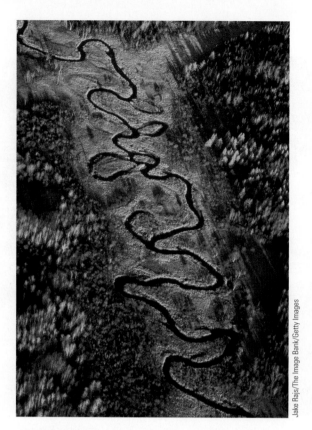

● **FIGURE 17.23** Not all streams fit the generalized pattern of characteristics for upper, middle, and lower stream segments. The Mississippi River, here a relatively small stream, meanders on a low gradient near its headwaters in Minnesota, far upstream from its mouth.

What evidence do you see that indicates lateral changes in the stream's position on its floodplain?

● **FIGURE 17.22** A colorized radar image of meanders of the Missouri River. Where the outside of a meander bend on the floodplain *(purple and blue tones)* impinges on the edge of higher terrain *(orange areas)*, stream erosion undercuts the valley side wall and, with the assistance of mass wasting, contributes to floodplain widening.

much of its available energy for transporting the considerable load that it has accumulated and for lateral erosion of the channel sides.

Because of lateral erosion, the middle course of the stream develops a definite meandering channel pattern with its sinuous bends that wander over time across the valley floor. The stream erodes a **cut bank** on the outside of the meander loops, where the channel is deep and centrifugal force accelerates stream velocity. The cut bank is a steep slope, and slumping may occur there,

particularly when there is a rapid fall in water level. Slumping on the outside of these meander bends contributes to the effect of lateral erosion by the stream and adds load to the stream. In the low velocity and shallow flow on the inside of the meander bends, the stream deposits a **point bar** (● Fig. 17.25). Erosion on the outside and deposition on the inside of stream meander bends result in the sideways displacement, or **lateral migration**, of meanders. When the laterally shifting channel reaches and undercuts the confining valley walls, it widens the gently sloping floodplain. Tributaries flowing into a larger stream also aid in widening the valley through which the trunk stream flows. Although flooding of the valley floor is always a potential hazard, flood sediments bring in nutrients and contribute to the agricultural richness of floodplain soils.

Lower Stream Course

The minimal gradient and close proximity to base level along the ideal lower river course make downcutting virtually impossible. Stream energy, now derived almost exclusively from the higher discharge rather than the downslope pull of gravity, leads to considerable lateral shifting of the stream channel and creation of a large depositional plain. The lower floodplain of a major river is much wider than the width of its meander belt and shows evidence of many changes in course (● Fig. 17.26). The stream

● **FIGURE 17.24** Where the upper course of a stream lies in a mountainous region, it might have rapids and waterfalls, but its valley typically has a characteristic V shape, dramatically represented in Yellowstone Canyon, Wyoming.

How does the gradient of the Yellowstone River compare with that of the stretch of the Mississippi River shown in Figure 17.23?

migrates laterally through its own previously deposited sediment in a channel composed exclusively of alluvium. During floods, these extensive floodplains, called **alluvial plains**, become inundated with sediment-laden water that contributes vertical accretion deposits to the large natural levees and to the already thick alluvial valley fill of the floodplain in general. Natural levees along the Mississippi River rise up to 5 meters (16 ft) above the rest of the floodplain.

A common landform in this deposition-dominated environment provides evidence of the meandering of a river over time. Especially during floods, **meander cut-offs** occur when a stream seeks a shorter, steeper, and straighter path; breaches through the levees; and leaves a former meander loop isolated from the new channel position (● Fig. 17.27). If the cut-off meander remains filled with water, which is common, it forms an **oxbow lake** (● Fig. 17.28).

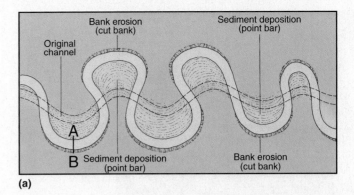

(a)

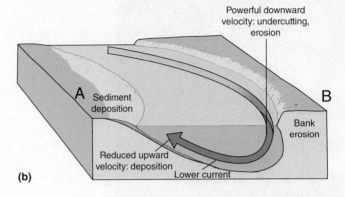

(b)

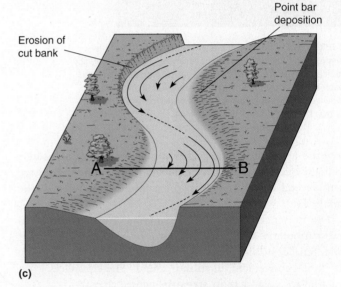

(c)

● **FIGURE 17.25** Characteristics of a meandering river channel. Note that water flowing in a channel has a tendency to flow downstream in a helical, or corkscrew, fashion, which moves water against one side of the channel and then to the opposite side. The up-and-down motion of the water contributes to the processes of erosion, transportation, and deposition.

Sometimes people attempt to control streams by building up levees artificially to keep the river in its channel. During times of reduced discharge, however, when a river has less energy, deposition occurs in the channel. Thus, in an artificially constrained channel, a river may raise the level of its channel bed. In some instances, as in China's Huang He and the Yuba River in northern California, deposition has raised the streambed above the surrounding floodplains. Flooding presents an especially serious danger in this situation with much of the floodplain lying

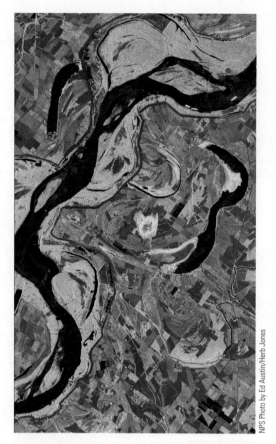

NPS Photo by Ed Austin/Herb Jones

● **FIGURE 17.26** This colorized radar image shows part of the Mississippi River floodplain along the Arkansas-Louisiana-Mississippi state lines. The colors are used to enhance landscape features such as water bodies *(dark)*, field patterns, and forested areas *(green)*. Abandoned portions of former meanders visible on the floodplain show that the river has changed its channel position many times.

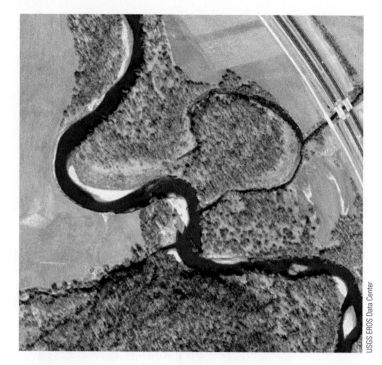

USGS EROS Data Center

● **FIGURE 17.27** A color infrared aerial photograph showing a recently cut-off meander loop. Vegetation appears on this image in shades of red.

Do you see any point bars on this photo?

below the level of the river. Unfortunately, when floodwaters eventually overtop or breach the levees, they can be even more extensive and destructive than they would have been in the natural case.

The presence of levees—both natural and artificial—can prevent tributaries in the lower course from joining the main stream. Smaller streams are forced to flow parallel to the main river until a convenient junction is found. These parallel tributaries are called **yazoo streams**, named after the Yazoo River, which parallels the Mississippi River for more than 160 kilometers (100 mi) until it finally joins the larger river near Vicksburg, Mississippi.

Deltas

Where a stream flows into a standing body of water, such as a lake or the ocean, the flow is no longer confined in a channel. The current expands in width, causing a reduction in flow velocity and thus a decrease in load competence and capacity. If the stream is carrying much load, the sediment will begin to settle out, with larger particles deposited first, closer to the river mouth, and smaller particles deposited farther out in the water body. With continued aggradation, a distinctive landform, called a **delta** because the map view shape of some resembles the Greek letter delta (Δ), will be constructed.

Deltas form at the interface between fluvial systems and coastal environments of lakes or the ocean, and therefore originate in part

● **FIGURE 17.28** Features of a large floodplain common in the lower courses of major rivers. Backswamps are low marshy or swampy parts of the floodplain, generally at the water table.

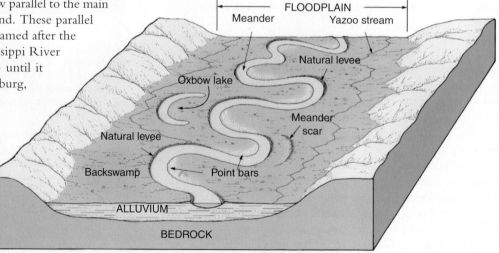

from fluvial and in part from coastal processes. Deltas have a subaqueous (underwater) coastal component, called the *prodelta*, and a fluvial part, the *delta plain*, that exists at, to slightly above, the lake level or sea level. Deltas form only at those river mouths where the fluvial sediment supply is high, the underwater topography does not drop too sharply, and waves, currents, and tides cannot transport away all the sediments delivered by the river. These circumstances exist at the mouths of many, but not all, rivers.

Delta construction is a slow, ongoing process. A river channel that approaches its base level at a large standing body of water typically has a very low gradient. Lacking the ability to incise its channel below base level, the stream may divide into two channels, and can do so multiple times, to convey its water and load to the lake or ocean. These multiple channels flowing out from the main stream on the delta plain are **distributaries**, and they help direct flow and sediment toward the lake or ocean. Natural levees accumulate along the banks of these distributary channels.

Continued deposition and delta formation extend the delta plain and create new land beyond the original shoreline. Rich alluvial deposits and the abundance of moisture allow vegetation to become established quickly on these fertile deposits and further secure the delta's position. Delta plains, such as those of the Mekong, Indus, and Ganges Rivers, form important agricultural areas that feed the dense populations of many parts of Asia. Because of their very low elevation, extremely gentle gradient, and minimal relief, however, delta plains are especially susceptible to extensive flooding, and great loss of life has occurred on densely populated delta plains struck by tropical cyclones and storm surges. The same three factors of low elevation, gradient, and relief make delta plains vulnerable to effects of rising sea level.

Different types of deltas form in different settings. Where the Mississippi River flows into the Gulf of Mexico, the river has constructed the type of delta called a *bird's-foot delta* (● Fig. 17.29a). Bird's-foot deltas are constructed in settings where the influence of the fluvial system far exceeds the ability of waves, currents, and tides of the standing water body to rework the deltaic sediment into coastal landforms or to transport it away. Numerous distributaries slightly above sea level extend far out into the receiving water body. Occasional changes in the distributary channel system occur when a major new distributary is cut that siphons flow away from a previous one, causing the center of deposition to switch to a new location far from its previous position. The appearance in map view of the natural levees extending toward the present and former depositional centers leaves the delta resembling a bird's foot.

Other types of deltas do not project into their receiving water bodies as far or with as much spatial irregularity as bird's foot deltas. *Arcuate deltas*, like that of the Nile River, project to a limited extent into the receiving water body and have a smooth laterally continuous seaward edge (Fig. 17.29b). The shape of arcuate deltas shows moderate reworking of the fluvial deposits by waves and currents—much more so than in the case of the bird's-foot delta. Coastal processes play a strong role in the formation of *cuspate deltas* by pushing the sediments back toward the mainland and reworking them into beach ridges on either side of the river mouth. An example of a cuspate delta is the São Francisco in Brazil.

● **FIGURE 17.29** Satellite views of two types of deltas. (a) The unusual shape of the Mississippi River delta resembles a bird's foot. Waves, currents, and tides in the Gulf of Mexico do little to change the visible shape of this type of delta. (b) The arcuate Horton River delta along the Arctic coast of Canada.

Why are the shapes of some deltas controlled more by fluvial processes, whereas the shapes of others are strongly influenced by coastal processes?

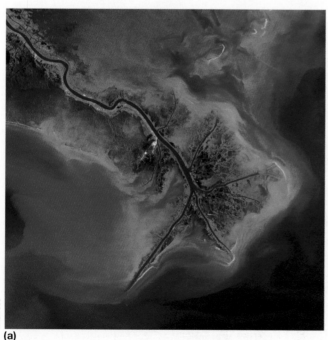

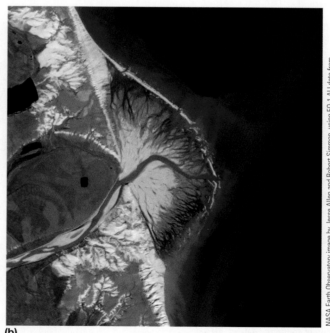

(a)

(b)

Construction of Hoover Dam on the Colorado River was completed in 1936. What effects would the dam have had on the segment of the river just upstream from the dam? How would the dam have affected the river immediately downstream? Do you think the stream system is still adjusting to the presence of the dam?

Lynn Betts/NRCS

Base-Level Changes and Tectonism

A change in elevation along a stream's longitudinal profile will cause an increase or decrease in the stream's gradient, which in turn has an impact on the stream's energy. Elevation changes can occur anywhere along a stream as a result of tectonic uplift or depression; they can also result from the rising or falling of base level at the river mouth. Variations in base level for basins of exterior drainage result principally from climate change. Sea level lowers in response to large-scale growth of glaciers and rises with substantial glacier shrinking. Tectonic uplift along the longitudinal profile, or a fall in base level at the stream's mouth, gives the stream a steeper gradient and increased energy for erosion and transportation. The landscape and its stream are then said to be *rejuvenated* because the stream uses its renewed energy to incise its channel. Waterfalls and rapids might develop as rejuvenated channels are deepened by erosion. For cases of falling base level, the stream will incise to that new elevation. Tectonic depression of the drainage basin or a rise in sea level reduces the stream's gradient and energy, enhancing deposition.

If new uplift occurs gradually in an area where stream meanders have formed, these meanders may become *entrenched* as the stream deepens its valley (● Fig. 17.30). Now, instead of eroding the land laterally, with meanders migrating across an alluvial plain, the rejuvenated stream's primary activity is vertical incision.

Virtually all rivers reaching the sea incised deep valleys near their mouths during the Pleistocene in response to the substantial lowering of their base level (sea level) associated with continental glaciation. The accumulation of water on the land in the form of

Rainer Duttmann

● **FIGURE 17.30** A spectacular example of an entrenched meander along the San Juan River in the Colorado Plateau region of southern Utah.

What type of load is this stretch of the river transporting?

glacial ice during the Pleistocene caused sea level to drop as much as 120 meters (400 ft), lowering the base level of streams of exterior drainage by this amount. Consequently, near their mouths, the streams eroded deep valleys for their channels. Eventual melting of the glaciers returned water to the ocean, which raised sea level. Readjustment to this base-level rise caused the streams to deposit large amounts of their load, filling the Pleistocene valleys near their mouths with sediment. These consecutive changes in base level produced broad, flat, alluvial floodplains above infilled valleys that were cut far below today's sea level.

A drop in base level causes downcutting and a rise causes deposition; interestingly, an upset of stream equilibrium resulting from sizable increases or decreases in discharge or load can have similar effects on the landscape. Variations in base level, tectonic movements, and changes in discharge and load can each cause downcutting in stream valleys, so the valley is slightly deepened, and remnants of the older, higher valley floor are preserved in stairstepped banks along the walls of the valley. These remnants of previous valley floors are **stream terraces**. Multiple terraces are a consequence of successive periods of downcutting and deposition (● Fig. 17.31). Stream terraces provide a great deal of evidence about the geomorphic history of the river and its surrounding region.

Rivers, Lakes, and People
Stream Hazards

There are many benefits to living near streams, but doing so also has its risks, particularly in the form of floods. Variability of stream flow constitutes the greatest problem for life along rivers and is also an impediment to their use. Most stream channels can hold the maximum flows that occur once every year or two. Larger maximum flows that are probable over longer periods of 5, 10, 100, or 1000 years overflow the channel and inundate the surrounding

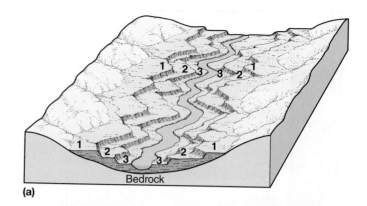

(a)

(b)

Courtesy Shelia Brazier

● **FIGURE 17.31** (a) Diagram of a stream valley displaying two sets of alluvial stream terraces (*labeled 1 and 2*) and the present floodplain (*labeled 3*). The stream terraces are remnants of previous positions of the valley floor, with the higher (1) being the older. (b) River terraces in the Tien Shan Mountains of China.

How many terraces can you identify in this photo?

Sacramento Bee/John Trotter/MCT/Landov

(a)

FEMA

(b)

● **FIGURE 17.32** Living on a floodplain has its risks. (a) The Coast Guard rescues a person stranded on a rooftop after a levee along the Feather River ruptured, sending hectares of water into a community in Sutter County, California. (b) Floodplains are also hazardous places for animals to live, and teams are often deployed to rescue animals stranded by floodwaters.

What can be done to prepare for or to avoid flood problems?

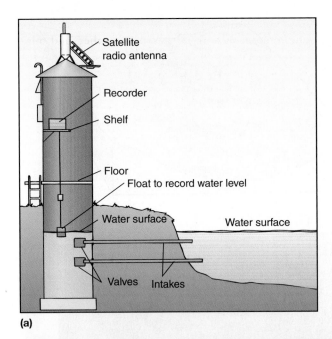

(a)

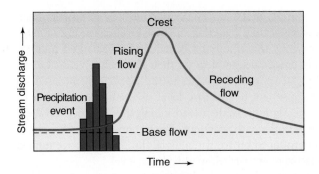

● **FIGURE 17.34** A stream hydrograph depicts changes in discharge, which imply changes in flow depth, for a selected period of time as recorded by a gaging station. This hydrograph shows the rise in a river in response to flood runoff, the flood peak, and the recession of flow waters following the end of the storm.

Why is there a delay between when precipitation starts and when the river rises?

A **stream hydrograph** is the record of discharge changes in a stream over time (● Fig. 17.34). Hydrographs cover a day, a few days, a month, or even a year depending on a scientist's purpose. Because of the relationship of discharge to water depth and velocity ($Q = wdv$), hydrographs are often used to indicate how high and fast the water level rises in response to a precipitation event.

During and just after a runoff-producing rainfall, the stream level rises in response to the increased discharge. The water level peaks, or crests, at the time of maximum discharge and then falls as the river eventually returns to a more average amount of flow. At a gaging station, the rise, peak, and fall in discharge over time are recorded as a climbing, peaking, and then descending curve on the hydrograph.

The shape of the hydrograph curve can be used to understand a great deal about how a watershed and a stream channel respond to an increase in runoff, particularly during and after floods. If conditions in the watershed promote rapid runoff, the stream will rise quickly during a storm. That rapid increase in stream discharge will be represented on the hydrograph curve by a rising limb that climbs steeply to peak discharge. Rising limbs that are represented by very steep curves indicate flash flood conditions. The greater the discharge at the flood crest, the higher the peak of the hydrograph curve. Compared with the rapid rise to peak discharge, the return to a more average flow typically takes longer because water continues to seep into the channel from the precipitation-saturated ground of the watershed. The falling limb of the hydrograph, therefore, usually plots as a comparatively gentle curve. How high a river rises, how fast it reaches peak flow, and how slowly the water recedes in response to a certain amount of precipitation are all important factors involved in preparing for future floods.

Urbanization of a watershed (particularly small watersheds) tends to make the flood peak rise higher and faster than was the case before human development in the drainage basin. Reduction of vegetation and its replacement with impermeable surfaces, like roads, roofs, and parking lots, contribute to higher rates of runoff and increased amounts of runoff, two conditions that directly affect the flow of the stream (● Fig. 17.35). When a drainage basin becomes more highly developed and populated, the potential flood size, the rapidness of flood onset, and the flood hazard all tend to increase.

(b)

● **FIGURE 17.33** U.S. Geological Survey (USGS) stream gaging station. (a) As illustrated here, when the river level rises or falls, so does the water in the lower part of the station, connected to the channel by intake pipes. A gaging float moves up and down with the water level, and the motion is measured, recorded, and transmitted to a satellite allowing real-time monitoring of stream behavior. (b) Gaging stations are commonly located where bridges cross streams.

land, sometimes with disastrous results (● Fig. 17.32). Similarly, exceptionally low flows may produce crises in water supply and bring river transportation to a halt.

The U.S. Geological Survey maintains more than 6000 gaging stations for the measurement of stream discharge in the United States (● Fig. 17.33). Many of these gaging stations operate on solar energy, measure stream discharge in an automated fashion, and send their data to a satellite that relays them to a receiving station. With this system, stream flow changes are monitored at the time they occur, which is critical for issuing flood warnings and for understanding how streams change in response to variations in discharge.

● **FIGURE 17.35** Most aspects of urbanization and suburbanization, as in this area near Boston, increase the extent of impermeable cover within the drainage basin. As a result, the amount and rate of runoff from urbanized areas increase compared to their presettlement state.

What features of the urbanized landscape shown here enhance runoff?

To determine how often we can expect flows of a given magnitude to occur in a particular river, especially for those discharges that cause flooding, Earth scientists assess the river's previous flow history using maximum annual discharge data. These data are used to calculate different **recurrence intervals**, the average number of years between past flow events that equaled or exceeded various discharges. For example, if flooding along a river resulted from a discharge of 7000 cubic meters per second (250,000 ft³/s) and larger in 2 out of 100 years of data, that discharge has a recurrence interval of 50 years—it is the 50-year flood—even if those flows occurred in two consecutive years. Because that discharge (and possibly greater) happened twice in the past 100 years, there is an estimated 2 out of 100, or 2%, probability that it will happen in any given year. Because recurrence intervals are averages determined from historic data of limited duration, the size of a 10-, 50-, 100-year, or other flood is not absolute but rather an estimate subject to change, and the length of time between 50-year floods, for example, will not necessarily be 50 years.

Importance of Rivers and Lakes

Historically, people have used rivers and smaller streams for a variety of purposes. The settlement and growth of the United States would have been very different without the Mississippi River and its far-reaching tributary system that drains most of the country between the Appalachians and the Rockies. The Mississippi, like many other rivers, has been used for exploration, migration, and settlement, and the number of major cities along it illustrates people's tendency to settle along rivers. Navigable rivers

still compete successfully with railroads and trucks for transportation of bulk cargo, grain, lumber, and mineral fuels, and streams have long been used to generate power. They supply irrigation water and agriculturally productive floodplain soils. We also use streams as direct sources of food and water, for recreation, and as a depository for waste.

Lakes are natural standing bodies of inland water. Lakes form wherever the water supply is adequate to accumulate in topographic depressions on the land surface. Many of the world's lake basins, such as those of the North American Great Lakes, are products of glaciation, but rivers, tectonic activity, and volcanism also produce lake basins. With a depth of more than 1600 meters (5250 ft), Lake Baikal in Russia, occupying a tectonic depression, is the world's deepest lake. Crater Lake in Oregon is the deepest lake in the United States with a depth of 592 meters (1943 ft). It lies in the caldera that formed about 7700 years ago by the collapse of a composite volcano in the Cascade Range.

Streams flow both into and out of most of the world's lakes so that the lakes provide temporary storage of freshwater along stream systems. Some lakes, however, such as the Caspian Sea, Dead Sea, and Great Salt Lake in Utah, contain saltwater because they exist in closed basins of interior drainage with inflowing streams but no outflowing streams. Evaporation of water to the atmosphere from those lakes leaves behind the salts, which constituted the dissolved load of the inflowing streams.

Most lakes are temporary features on Earth's surface. Few have been in existence for more than 10,000 years, and the majority are very recent in terms of Earth history (● Fig. 17.36). Sedimentation, biological activity, or downcutting of an outlet by a stream eventually lead to the destruction of most lakes.

● **FIGURE 17.36** Grizzly Lake in Yellowstone National Park, Wyoming, receives inflow from, and outflows to, Straight Creek. Eventually, outflow will erode the temporary base level to the point where there is no longer containment for the lake, and the site will become just another segment of the stream.

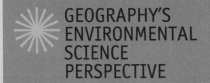
Dams and Dam Removal

Dams of many types and sizes have been built across America's streams for various reasons, including controlling floods, maintaining water supplies through drought periods, generating hydroelectric power, improving river navigation, providing water for irrigation, making drinking ponds for stock, and supplying recreational facilities. Some dams in the United States are very old, but most in existence today date from a period of extensive dam building in the mid-twentieth century. Researchers estimate that by the end of that century, U.S. streams were crossed by more than 80,000 dams. Although the impact of individual large dams in particular was occasionally questioned by individuals at various times throughout history, the existence of many dams is now being questioned along several broad scientific fronts, including physical geography. Reevaluation of the wisdom of having so many dams is encouraged in part by the fact that dams from the mid-twentieth century are getting old, but also by the current trend toward restoring streams to more natural conditions. Because of our knowledge of the many and varied aspects of Earth surface systems, physical geographers are well placed to contribute to this discussion.

There are many different types and sizes of dams, and some are privately owned, but constructing an artificial flow-retarding barrier across any stream changes the fluvial dynamics. Reservoirs behind dams eventually fill with sediment that the water naturally would have transported downstream and reduced in size by attrition. Underloaded flows released from reservoirs have increased erosive power that can scour the downstream channel segment of biologically important sediment sizes. Dams block the migration path of anadromous fish that must swim upstream to spawn, and fish ladders designed to provide an artificial migration route around the dam are not always successful. Acting as a temporary base level, dams interrupt the natural behavior of streams for some distance upstream as well as downstream of the dam site. Assessing the extent of the environmental disruption versus the benefits derived by a dam is typically a complex issue involving economic, social, and cultural factors and knowledge of the physical, chemical, and biological systems. Furthermore, different stakeholders will have different goals and perspectives.

The Elwha River, Washington.

National Park Service

With many aging dams in the United States needing expensive repair, the notion of dam removal is now being considered—and acted on—more frequently than ever before. Natural scientists are working to gain a better understanding of the impact of dam removal on the living and nonliving aspects of the stream system. Sudden release of large amounts of reservoir-stored sediment into the stream channel influences the behavior of the stream flow and can have its own ecological impacts. For some streams, impounded sediment must be dredged and hauled away rather than unleashed into the channel upon dam removal. At other locations, the newly undammed stream is capable of flushing the sediment through the system relatively quickly and without harm to the stream's organisms. Some of the factors to consider when designing how a particular dam will be removed include the amount and grain size characteristics of the impounded sediment, presence of toxic materials, ecological and environmental requirements of the stream's biota, and estimated time for the stream to flush out the sediment naturally, which depends on the stream's gradient, discharge, and nature of the sediment. Each individual dam and its stream must be evaluated carefully when planning a dam removal.

In 2014, the largest dam removal in American history to date was completed on the Elwha River in the state of Washington under the direction of the National Park Service. The massive project removed two dams, the 33-meter (108-ft) high Elwha and 64-meter (210-ft) high Glines Canyon dams, construction of which was completed in 1913 and 1927, respectively. After careful study, flushing of the sediment that had accumulated in the reservoirs was allowed to proceed naturally, accompanied by extensive revegetation of the dam and reservoir areas with native plant species. Reestablishing throughflow of the river's sediment from its source in Olympic National Park to its mouth on the Strait of Juan de Fuca is reinvigorating the coastal environment and valuable wetlands in the vicinity of the river's mouth. Salmon are returning to the river. Scientists continue to carefully monitor and study how the river and its subsystems are responding to and recovering from a century of artificial control. The success of the ambitious Elwha River dam removal project and the knowledge gained from the accompanying research will undoubtedly lead to serious consideration for the decommissioning of other large dams. The environmental costs and benefits of dam removal must be evaluated for each case. Understanding how streams erode, transport, and deposit sediment, and the role of stream energy, velocity, discharge, and gradient on those processes, are critical components for making those dam removal assessments.

Deconstruction of Glines Canyon Dam on the Elwha River, Olympic National Park, Washington. In this dam removal, flushing out the sediment that had accumulated behind the dam was left to the river.

● **FIGURE 17.37** The multipurpose Lookout Point Dam on the Middle Fork of the Willamette River, Oregon.

What are some of the multiple purposes that dams serve?

Lakes are important to humans for more than their scenic appeal and their value for fishing or recreational activities. Like oceans, lakes affect the nearby climates, particularly by reducing daily and seasonal temperature ranges and by increasing humidity. The moderating effect on temperature allows land adjacent to many large lakes to support agricultural activities that are not otherwise possible in the regional climate. Lakes can also cause a downwind increase in precipitation—snow or rain generated by the *lake effect*, in which storms either pick up more moisture from the lake or undergo uplift as they move over a relatively warm body of water.

Because stream flow can be so variable and in some cases unreliable, people have tried to regulate most rivers in some way. Many river systems now consist of a series of **reservoirs**—artificial lakes impounded behind dams. The dams hold back potentially devastating floodwaters and store the discharge of wet periods to make the water available during dry seasons or drought years (● Fig. 17.37). Some dams are also used to generate hydroelectric power. Many rivers have been transformed by dam construction into a series of reservoirs. Unfortunately, the life span of these reservoirs, like that of almost any lake, is only a few centuries at most because they will gradually fill with the sediment delivered to them by their inflowing streams. Structural problems can also plague older dams. To help protect the natural flow and environments of the few remaining undeveloped streams and rivers in the United States, Congress enacted the Wild and Scenic Rivers Act in 1968. More recently, as part of a trend in the natural sciences aimed at mitigating effects of human disturbance on streams and other natural systems, increasing efforts are being directed toward assessing the impacts of dams and dam removal on stream geomorphology and ecology.

Quantitative Fluvial Geomorphology

Quantitative methods are important in studying virtually all aspects of the Earth system, and they are used by climatologists, meteorologists, biogeographers, soil scientists, and hydrologists, as well as geomorphologists. The value of quantitative methods to the objective analysis of fluvial systems in particular can be inferred from the material presented in this chapter. Streams are complicated, dynamic systems with input, throughput, and output of energy and matter that depend on numerous, often-interrelated variables.

Geomorphologists routinely measure stream channel, drainage basin, and flow properties and analyze them using statistical methods and the principles of fluid mechanics so that they can describe, compare, monitor, predict, and learn more about streams and the geomorphic work that they perform. Drainage area, stream order, drainage density, stream discharge, stream velocity, channel width, and channel depth are just some of the numeric data collected in the field, on topographic maps, from digital elevation models, or from remotely sensed imagery to facilitate the study of stream systems. Extensive efforts are made to gather and analyze numeric stream data because of the widespread occurrence of streams and their great importance to human existence. Quantitative studies not only help us better understand the origins and formational processes of landforms and landscapes, but also help us better predict water supplies and flood hazards, estimate soil erosion, and trace sources of pollution.

CHAPTER 17 ACTIVITIES

■ TERMS FOR REVIEW

fluvial geomorphology
interfluve
surface runoff
interception
sheet wash
rill
gully
ephemeral flow
perennial flow
intermittent flow
tributary
drainage basin (watershed, catchment)
drainage divide
stream order
exterior drainage
interior drainage
base level
drainage density
drainage pattern
transverse stream

stream discharge
channel roughness
stream load
stream competence
stream capacity
degradation
aggradation
corrosion
hydraulic action
turbulence
abrasion
attrition
solution
suspension
saltation
traction
dissolved load
suspended load
bed load

bar
alluvium
natural levee
floodplain
braided channel
meandering channel
cut bank
point bar
lateral migration
alluvial plain
meander cut-off
oxbow lake
yazoo stream
delta
distributary
stream terrace
stream hydrograph
recurrence interval
reservoir

QUESTIONS FOR REVIEW

1. Why are streams considered such an important geomorphic agent?

2. What is the relationship between infiltration and surface runoff? What are some of the factors that enhance infiltration? What are some of the factors that enhance surface runoff?

3. Define the concept of stream and drainage basin order. How would a typical first-order channel differ from a typical fifth-order channel in the same drainage basin in terms of tributaries, length, gradient, cross-sectional area, discharge, and velocity?

4. What factors affect the discharge of a stream? What are the two very different equations that can be used to represent stream discharge?

5. Why do drainage basins in semiarid climates tend to have greater drainage density than those in humid climates?

6. Which is the most effective fluvial erosion process? Why is it so effective?

7. What are the differences among the fluvial transportation processes? Which moves the largest particles?

8. How do braided channels differ from meandering channels?

9. Describe the principal landforms of a meandering stream system and the development of natural levees.

10. Explain how tectonism and climate change can each contribute to changes in a stream's base level.

11. Why is it important to collect and study stream flow data from gaging stations?

12. What is the relationship between rivers and lakes, and how do the two types of features interact?

CONSIDER AND RESPOND

1. How do streams represent the concept of dynamic equilibrium? What are some examples of negative feedback in a stream system?

2. In the study of a drainage basin, what types of geographic observations and quantitative data might prove useful in planning for flood control and water supply?

3. How does urbanization affect runoff, discharge, and flood potential in a drainage basin?

■ PRACTICAL APPLICATIONS

1. A measured cross section of a stream has a width of 6.5 meters, a depth of 2.5 meters, and a flow velocity of 2.0 meters per second. What is the area of the cross section? What is the stream discharge at the cross section?

2. A stream channel lies at an elevation of 1200 meters above sea level at one location and at 1075 meters above sea level at a site 11 kilometers downstream. What is the channel gradient between those two locations?

3. One hundred years of maximum annual flow data show that the stream under study had a maximum annual discharge of 3700 cubic meters per second or greater five times. What is the recurrence interval of this flow? What is the probability that a discharge of 3700 cubic meters per second will occur this year?

■ LOCATE AND EXPLORE

1. Using Google Earth, identify the landforms at the following locations (latitude, longitude) and provide a brief discussion of how the landform developed. Include a brief discussion of why the landform is found in that general area of the United States or the world.
   a. 43.73°S, 172.02°E
   b. 52.110°N, 28.571°E
   c. 30.410°N, 85.015°W
   d. 45.120°N, 111.672°W
   e. 37.20°N, 110.00°W
   f. 45.36°N, 111.71°W
   g. 39.27°N, 119.55°W
   h. 30.57°N, 87.19°W

2. Using Google Earth identify the following features. Rank the features from youngest to oldest with an explanation for your order.
   a. 34.543999°N, 90.568936°W
   b. 34.659545°N, 90.480867°W
   c. 34.930455°N, 90.333536°W
   d. 34.642015°N, 90.411898°W
   e. 34.520741°N, 90.254868°W

 MindTap—Make the most of your study time by accessing everything you need to succeed in one place. Read your textbook, take notes, review flashcards, watch videos, complete activities, take practice quizzes, and more online with MindTap. Log in at **www.cengagebrain.com**.

Map Interpretation

FLUVIAL LANDFORMS

The Map

Campti is in northwestern Louisiana on the Gulf Coastal Plain. The coastal plain region stretches westward from northern Florida to the Texas–Mexico border and extends inland in some places for more than 320 kilometers (200 mi). Elevations on the Gulf Coastal Plain gradually increase from sea level at the shoreline to a few hundred meters far inland. The region is underlain by gently dipping sedimentary rock layers. Surface material includes marine sediments and alluvial deposits from rivers that cross the coastal plain, especially those of the Mississippi drainage system. This is a landscape of meanders, natural levees, and bayous.

The headwaters of the Red River are located in the semiarid plains of the Texas Panhandle, but the river flows eastward toward an increasingly more humid climate. About midcourse, the Red River enters a humid subtropical climate region, which supports rich farmland and dense forests. The river flows into the Mississippi in southern Louisiana, about 160 kilometers (100 mi) downstream of the map area. The Red River is the southernmost major tributary of the Mississippi.

Louisiana has mild winters with hot, humid summers. Annual rainfall totals for Campti average about 127 centimeters (50 in.). The warm waters of the Gulf of Mexico supply vast amounts of atmospheric energy and moisture that contribute to a high incidence of thunderstorms as well as occasional tornadoes and, at times, hurricanes that strike the Gulf Coast.

Interpreting the Map

1. How would you describe the general topography of the Campti map area? What is the local relief? What is the elevation of the banks of the Red River at the town of Campti?
2. What kind of landform is the low-relief surface on which the river is flowing? Is this mainly an erosional or depositional landform?
3. In what general direction does the Red River flow? Is the direction of flow easy or hard to determine just from the map area? Why?
4. Does the Red River have a gentle or steep gradient? Why is it difficult to determine the river gradient from the map area?
5. What is the origin of Smith Island? Adjacent to Smith Island is Old River; what is this type of feature called?
6. Explain the stippled brown areas in the meanders south of Campti. Are these areas on the inside or the outside of the meander bend?
7. How would you describe the features labeled "bayou"?
8. Is this map area more typical of the upper, middle, or lower course of a river?
9. Although this is not a tectonically active region, how would the river change if it experienced tectonic uplift?

Healthy green vegetation appears red on color infrared images, here showing the Red River and Smith Island, Louisiana.

U.S. Department of Agriculture

Opposite:
Campti, Louisiana
Scale 1:62,500
Contour interval = 20 ft

ARID REGION AND EOLIAN LANDFORMS

18

OBJECTIVES

WHEN YOU COMPLETE THIS CHAPTER YOU SHOULD BE ABLE TO:

- 18.1 Appreciate the relative role of water and wind in creating landforms within the world's arid regions.
- 18.2 Apply drainage basin, base level, and stream channel concepts appropriately to arid region landscapes.
- 18.3 Draw sketches that illustrate distinguishing features among principal arid region landforms of fluvial erosion.
- 18.4 Compare and contrast pediments and alluvial fans and the specific settings in which each occurs.
- 18.5 Describe the general nature of a playa, and distinguish between the two major types of playa.
- 18.6 Discuss the ways by which wind erodes and transports sediment.
- 18.7 Provide examples of landforms made by eolian erosion.
- 18.8 Explain why and how some sand dunes are stabilized and what their stabilization can mean.
- 18.9 Discuss the major types of sand dunes and the situations under which each forms.
- 18.10 Understand the origin and importance of the world's large loess deposits.

BECAUSE OF THEIR LOW AMOUNTS of precipitation, arid region landscapes look quite different from those of other climatic environments in many ways. The limited water supply of arid climates restricts rock weathering and the amount of vegetation present. Without extensive vegetation to hold weathered rock matter (regolith) in place, the weathered rock particles that are produced are often stripped away when storms do occur. As a result, whereas hillslopes in humid regions tend to be rounded and mantled by soil, mountains and hillslopes in arid regions are generally angular, with extensive, barren exposures of bedrock. Desert lowlands may be filled in with sediments eroded from uplands, or they can consist of just a thin cover of sediments overlying rock strata.

Many desert landscapes have a majestic beauty that derives from the stark display of the colors, characteristics, and structure of the rocks that make up the area. The desert's barrenness reveals evidence about landforms and geomorphic processes that is much more difficult to observe in humid environments, with their cover of soil and vegetation. Much of our understanding of how landforms and landscapes develop has come from scientific studies conducted in desert regions.

Petrified Forest National Park, Arizona, illustrates the dramatic beauty of desert landscapes. T.Scott Williams/NPS

505

The wind plays an important role in arid region geomorphology, but running water does more geomorphic work than the wind does in arid regions. Wind erosion is mainly confined to picking up fine, dust-sized (clay and silt) particles from desert regions and to dislodging loose rock fragments of sand-sized materials. Still, we tend to associate arid environments with wind-derived (eolian) geomorphic processes because of the notable accumulations of wind-deposited sediment found in some desert areas, usually in the form of sand dunes. With the sparse vegetation and other environmental characteristics, the geomorphic work of the wind reaches its optimum in arid environments. However, because air has a much lower density than water, even in deserts the wind is outmatched by fluvial geomorphic processes. We should understand, too, that landforms fashioned by the wind are not confined to arid regions; they are also conspicuous in many coastal areas and in any area where loose sediments—including snow—are frequently exposed to winds strong enough to move them.

J. Petersen

● **FIGURE 18.1** In almost all deserts the effects of running water are prominent in the landscape.

Why do you think this terrain is so intensely channelized?

Surface Runoff in the Desert

Landforms, rather than vegetation, typically dominate desert scenery. The precipitation and evaporation regimes of an arid climate result in a sparse cover of vegetation and, because many weathering processes require water, relatively low rates of weathering. Due to low weathering rates, insufficient vegetation to break the impact of raindrops, and a lack of extensive plant root networks to help hold rock fragments in place, a blanket of moisture-retentive soil cannot accumulate on slopes. Soils tend to be thin, rocky, and discontinuous. This absence of a continuous vegetative and soil cover gives desert landforms their unique character. Under these surface conditions of very limited interception and low permeability, much of the rain that falls in the desert quickly becomes surface runoff available to perform fluvial geomorphic work. With little to hold them in place, any grains of rock that have been loosened by weathering could be swept away in surface runoff produced by the next storm. Ironically, although desert landscapes strongly reflect a deficiency of water, the effects of running water are widely evident on slopes as well as in valley bottoms (● Fig. 18.1). Where vegetation is sparse, running water, when it is available, is extremely effective in shaping the land.

Desert climates characteristically receive small amounts of precipitation and they experience high rates of potential evapotranspiration. In exceptional circumstances in some desert areas, years may pass without any rain. Most desert locations, however, receive some precipitation each year, but the frequency and amount are highly unpredictable, and rains that do fall are often quite intense. The most important impact of rain on landform development in deserts is that when rainfall does occur, much of it falls on impermeable surfaces, producing intense runoff, often generating flash floods, and operating as a powerful agent of erosion.

The visible evidence of water as a geomorphic agent in arid regions stems not only from the climate of those areas today but also from past climates. Paleogeographic studies reveal that many deserts have not always had the arid climate that exists there today. Geomorphologists studying arid regions have found landforms in some locations that are incompatible with the present arid climate and that can be attributed to the work of water under earlier, wetter climates. For midlatitude and subtropical deserts, the most recent major wet period was during the Pleistocene Epoch. While glaciers were advancing at high latitudes and in high-elevation mountain regions during the Pleistocene, precipitation was also greater than it is today in the now-arid basins, valleys, and plains found in parts of the middle and subtropical latitudes. At the same time, cooler temperatures for these regions meant that they also experienced lower evaporation rates. In many deserts, evidence of past wet periods includes deposits and wave-cut shorelines of now-extinct lakes (● Fig. 18.2) and immense canyons occupied by streams that today are too small to have eroded such large valleys.

Running water is a highly effective geomorphic agent in deserts even though it operates only occasionally. In most desert areas, running water flows only during and shortly after rainstorms.

● **FIGURE 18.2** Many desert basins display remnant shorelines of ancient lakes they contained during wetter climates of the Pleistocene. The linear feature extending across this desert mountain front, below which is a smoother slope with smaller channels, was the highest shoreline of one of these ancient lakes.

Why does the terrain look so much smoother below than above the highest shoreline?

● **FIGURE 18.3** A braided stream in Canyon de Chelly National Monument near Chinle, Arizona, splits and rejoins multiple times as it works to carry its extensive bed load of coarse sand.

Why do the number and position of the multiple channels sometimes change rapidly?

Desert streams, therefore, typically have *ephemeral flow*, containing water only for brief intervals and remaining dry the rest of the time. In contrast to streams with *perennial flow*, those that flow all year and are typical of humid environments, streams with ephemeral flow do not receive seepage from groundwater to sustain them between episodes of surface runoff. Ephemeral streams instead lose water to the groundwater system through infiltration into the channel bed. Because of the low weathering rates in desert environments, most arid region streams receive an abundance of coarse sediment that they must transport as bed load. As a result, *braided channels*, in which multiple threads of flow split and rejoin around temporary deposits of coarse-grained sediments, are common in deserts (● Fig. 18.3).

Unlike the typical situation for humid region streams, many desert streams undergo a downstream decrease, rather than increase, in *discharge*, the volume of water flowing past a given point per unit time. A discharge decrease occurs for two major reasons: infiltration into the gravelly streambed continues downstream, removing more and more water from the channel, and evaporation losses increase downstream because of warmer temperatures at lower elevations. As a result of the diminishing discharge, many desert streams terminate before reaching the ocean. The

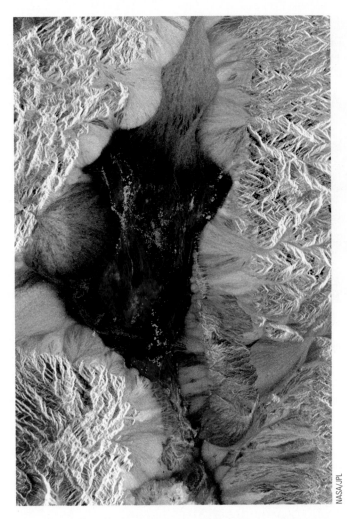

● **FIGURE 18.4** Death Valley, California, here depicted on a radar image, is a basin of interior drainage. Water accumulates on the basin floor from surface inflow, groundwater sources, and rarely by direct precipitation. No streams flow out of the basin.

● **FIGURE 18.5** This stream flows out of a deep gorge in the Atlas Mountains of Morocco, on the arid, rain-shadow side of the mountains, facing the Sahara. The stream loses water by infiltration and evaporation and disappears into the Sahara.

same mountains that contribute to aridity through the rain-shadow effect can block desert streams from flowing to the sea. Without sufficient discharge to reach that ultimate base level, desert streams end in depressions in the continental interior, where the episodically delivered stream water tends to form shallow, ephemeral lakes. Ephemeral lakes evaporate and disappear only to reappear when rain provides another episode of inflow. During the cooler and wetter times of the Pleistocene Epoch, so much water accumulated in some of the topographically blocked, closed basins of now-arid regions that large perennial freshwater lakes existed instead of the shallow ephemeral lakes that they contain today.

Where surface runoff drains into closed desert basins, sea level does not govern erosional base level as it does for streams that flow into the ocean and thereby attain *exterior drainage*. Desert drainage basins characterized by streams that terminate in closed interior depressions are known as basins of *interior drainage* (● Fig. 18.4); such streams are controlled by a **regional base level** instead of ultimate base level. When sedimentation raises the elevation of the desert basin floor located at the stream's terminus, the stream's base level rises, causing a decrease in the stream's slope, velocity, and energy. If tectonic activity lowers the basin floor, the regional base level falls, which could rejuvenate the desert stream. Tectonism has even created some desert basins of interior drainage with floors below sea level, as in Death Valley, California; the Dead Sea Basin in the Middle East; the Turfan Basin in western China; and Australia's Lake Eyre.

Many streams found in deserts originate in nearby humid regions or in cooler, wetter mountain areas adjacent to the desert (● Fig. 18.5). Even these, however, rarely have sufficient discharge to sustain flow across a large desert. With few tributaries and virtually no inflow from groundwater, arid region stream water lost to evaporation and underground seepage is not replenished. In most cases, the flow dwindles and finally disappears. The Humboldt River in Nevada is an outstanding example; after rising in the mountains of central Nevada and flowing 465 kilometers (290 mi), the river disappears into the Humboldt Basin, a closed depression.

Only a few large rivers that originate in humid uplands have enough discharge to survive a long journey across hundreds of kilometers of desert to reach the sea (● Fig. 18.6). These **exotic streams**, like the Nile River in Sudan and Egypt, provide some desert areas with exterior drainage. Under natural conditions, the Colorado River of the United States and Mexico would reach ultimate base level in the Gulf of California, but it no longer does so because of the huge volume of water that people withdraw from the river for agricultural and household use.

Water as a Geomorphic Agent in Arid Lands

When rain falls in the desert, sheets of water run down unprotected slopes, picking up and moving sediment. Dry channels quickly change to flooding streams. The material removed by sheet wash and surface streams is transported,

● FIGURE 18.6 The exotic Nile River meandering across the Sahara in Egypt on its way to the Mediterranean Sea. The river survives across the desert because of the amount of water supplied by the wetter climates of its headwaters in the Ethiopian Highlands and lakes of the East African Rift Zone. In this false-color satellite image, irrigated croplands adjacent to the river contrast with the lighter tones of the surrounding desert terrain.

● FIGURE 18.7 This dry streambed, or wash, has a channel of coarse alluvium and conveys water only during and just after a rainstorm.

Why would this desert stream channel have a high risk for flash floods?

just as in humid lands, until flow velocity decreases sufficiently for deposition to occur. Eventually, flow in these arid region streams disappears when seepage and evaporation losses deplete their discharge. Huge amounts of sediment can be deposited along the way as a stream loses volume and velocity. The processes of erosion, transportation, and deposition by running water are essentially the same in both arid and humid lands. However, the resulting landforms differ because of the sporadic nature of desert runoff, the scarcity of vegetation to protect surface materials against rapid erosion, and the common occurrence of streams that do not reach the sea.

Arid Region Landforms of Fluvial Erosion

Among the most common desert landforms created by surface runoff and erosion are the channels of ephemeral streams. Commonly known as **washes** or *arroyos* in the southwestern United States and *wadis* in North Africa and southwestern Asia, these channels usually form where rushing surface waters cut into unconsolidated sediment or easily eroded rock (● Fig. 18.7). These typically gravelly and braided channels are prone to flash floods, which makes them potentially dangerous sites. Although it may sound strange, many people have drowned in the desert—during flash floods.

In areas of weak, easily eroded clays or shales, rapid erosion from surface runoff can produce barren slopes and ridges dissected by a dense maze of steep gullies and ravines. Such areas of rugged, barren, and highly dissected terrain are known as **badlands** (● Fig. 18.8). Badlands have an extremely high *drainage density*, defined as the length of stream channels per unit area of the drainage basin. Extensive badlands exist in North and South Dakota; Death Valley National Park in California; Big Bend National Park in Texas; and southern Alberta, Canada. Badlands do not generally form under natural conditions in humid climates because the vegetation in those regions inhibits runoff and erosion, leading to lower drainage densities. Removing vegetation from clay or shale areas by overgrazing, mining, or logging, however, can cause badlands to develop even in humid environments.

A **plateau** is an extensive, elevated region with a fairly flat upper surface. Most plateaus consist of tectonically uplifted horizontal rock layers. Many striking plateaus exist in the desert and semiarid regions of the world, including the extensive tectonically uplifted Colorado Plateau, centered on the Four Corners area of Arizona, Colorado, New Mexico, and Utah. In such desert plateau regions, streams responded to the uplift by cutting narrow, steep-sided canyons. Where the canyon walls consist of horizontal layers of alternating resistant and erodible rocks, differential weathering and erosion create stair-stepped walls, with near-vertical cliffs made from resistant rock layers (ordinarily sandstone, limestone, or basalt) and weaker strata (often

Andrew V. Kearns/NPS

● **FIGURE 18.8** In badlands, such as these in Arizona, impermeable clays that lack a soil cover generate rapid runoff, leading to intensive gully erosion and a high drainage density.

Why is it hard for vegetation to become established on badland slopes?

shales) forming gentler slopes (● Fig. 18.9). The cliff-forming, resistant, horizontal rock layer that forms the top of, or caps, the sequence is the **caprock**. Such sequences of alternating strong and weak horizontal rock layers provide the distinctive appearance of much of the landscape of the Colorado Plateau, including the Grand Canyon.

Caprocks top plateaus and constitute canyon rims, but they also form the summits of other, smaller kinds of flat-topped landforms that, although they are found in many climate regions, are most characteristic of deserts. Weathering and erosion will eventually reduce the extent of a caprock until only flat-topped, steep-sided **mesas** remain (from Spanish: *mesa*, table). A mesa has a smaller surface area than a plateau and is roughly as broad across as it is tall. Mesas are relatively common landscape features in the Colorado Plateau region. Through additional erosion of the caprock from all sides, a mesa is reduced to a **butte**, which is a

● **FIGURE 18.9** On the Colorado Plateau, differential weathering and erosion of rock strata of varying resistance, in this case in southeastern Utah, have created a stair-stepped topography that is easily seen in the sparsely vegetated, arid landscape.

National Park Service

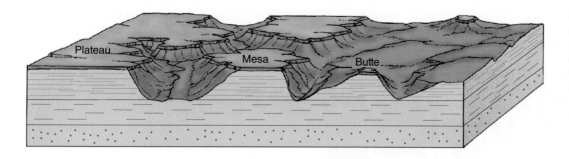

● **FIGURE 18.10** Plateaus, mesas, and buttes develop by weathering and erosion in areas, such as the Colorado Plateau, of horizontal rock layers with a resistant caprock.

similar, flat-topped erosional remnant but with a smaller surface area (● Fig. 18.10). Mesas and buttes in a landscape are generally evidence that uplift occurred in the past and that weathering and erosion have been extensive since that time. Variations in the form of the slope extending down the sides of buttes, mesas, and plateaus are related to the strength and thickness of the caprock compared to the strength and thickness of the exposed rock layers lying beneath the caprock. Monument Valley, in the Navajo Tribal Reservation on the Utah–Arizona state line, is an exquisite example of such a landscape formed with a caprock that is particularly thick, contributing to the distinctive scenery (● Fig. 18.11). Many classic western movies, most notably by mid-twentieth century director John Ford, have been filmed in Monument Valley and in nearby sections of the Colorado Plateau because of the striking, colorful, and photogenic desert landscape.

Over time, sheet wash and gully development can accomplish extensive erosion of mountainsides and hillslopes fringing a desert basin or plain. Particularly in desert regions with exterior drainage or a sizable trunk stream on the basin or plain, this fluvial action, aided by weathering, may lead to the gradual, erosional retreat of the position of the mountain front. As the bedrock slope of the mountain front migrates slowly away from the basin floor or plain, it leaves behind a gently sloping surface of the eroded bedrock, known as a **pediment** (● Fig. 18.12). Characteristically in desert areas, a sharp break in slope marks where the steep mountain front, often rising at angles of 20° to 30° or steeper, meets the gentle slope of the bedrock pediment, which is usually 2° to 7°. Resistant knobs of the bedrock sometimes remain projecting up above the surface of some pediments. These remnant bedrock knobs are termed **inselbergs** (from German: *insel*, island; *berg*, mountain).

● **FIGURE 18.11** The caprock in Monument Valley, Arizona, is particularly thick and represents a rock layer that once covered the entire region. The prominent buttes are erosional remnants of that layer.

Library of Congress Prints and Photographs Division [LC-DIG-highsm-12770]

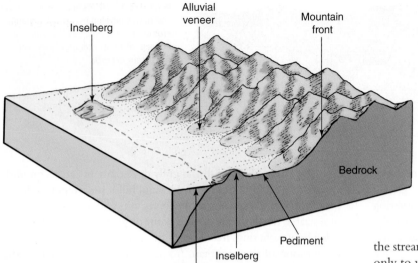

Inselberg

Alluvial veneer

Mountain front

Bedrock

Pediment

Inselberg

Alluvial plain

● **FIGURE 18.12** Pediments are gently sloping erosion surfaces, often thinly covered by alluvium, cut into bedrock downslope from the present mountain front. They are most common in desert areas with exterior drainage that can remove some of the erosion products. Any remaining resistant knobs of bedrock sticking up above the pediment surface are inselbergs.

Multiple explanations of pediment formation have been suggested, and different processes may be responsible for their formation in different regions. There is general agreement that most pediments are erosional surfaces created or partially created by the action of running water. In some areas, weathering, perhaps when the climate was wetter in the past, also appears to have played a strong role in the development of pediments.

Arid Region Landforms of Fluvial Deposition

Deposition is as important as erosion in creating landform features in arid regions, and in many areas sedimentation by water does as much to level the land as erosion does. Many desert areas have wide expanses of alluvium (stream sediments) deposited either in closed basins or at the base of mountains by streams as they lose water in the arid environment. As the flow of a stream diminishes, so does its *capacity*—the amount of load it can transport. Most landforms made by fluvial deposition in arid lands are not exclusive to desert regions but are particularly common and visible in drylands because of the thin soil and sparse vegetation.

Alluvial Fans Whether arid land streams are ephemeral (sporadic), *intermittent* (seasonal), or even perennial, where they flow out of narrow, constricting, upland canyons onto the less confining, open lowland plains, the channels become wider and shallower. You may recall from Chapter 17 that stream discharge *(Q)* equals channel cross-sectional

width *(w)* and depth *(d)* multiplied by flow velocity *(v)*: $Q = wdv$. Because of this relationship, an increase in stream width beyond the canyon mouth will have to be balanced by a decrease in the other two variables, depth and velocity. The decrease in velocity reduces the stream's competence and its capacity—that is, the size and amount, respectively, of load it can transport—thus causing deposition. Furthermore, discharge decreases downstream as water seeps into the coarse channel alluvium. As a result, most of the sediment load carried by streams emerging from mountain-front canyons is deposited along the base of the mountain front.

Upstream of the mouth of the canyon, the channel is constrained by bedrock valley walls, but where the stream flows beyond the canyon mouth, the channel is free not only to widen but also to shift its position laterally. Sediments are initially deposited just beyond the canyon mouth, building up the lowland area near the mountain front. Eventually this aggradation causes the channel to shift laterally, where it begins to deposit and build up another zone close to the mountain front and adjacent to the first aggraded area. The canyon mouth serves as a pivot point anchoring the channel as it swings back and forth over the lowlands near the mountain front, leaving alluvium behind. This creates a fan-shaped depositional landform, called an **alluvial fan**, in which deposition takes place radially away from that pivot point, or **fan apex** (● Fig. 18.13).

An important characteristic of an alluvial fan is the sorting of sediment that occurs on its surface. Coarse sediments, like boulders and cobbles, are deposited near the fan apex where the stream first undergoes a decrease in competence and capacity as it emerges from the confinement of the canyon. In part because of the large size of the clasts deposited there, the slope of an alluvial fan is steepest at its apex and gradually diminishes, along with

● **FIGURE 18.13** Where streams come out of confined mountain canyons and widen, they deposit alluvium that, over time, takes on a fan shape with the fan apex at the canyon mouth.

How do alluvial fans differ from pediments?

grain size, with increasing distance downstream from the canyon mouth. In areas where the uplands generate debris flows rather than stream flows, **debris flow fans** or mixed debris flow and alluvial fans are constructed instead of purely alluvial fans. Debris flow fans tend to be steeper than alluvial fans and do not display the same degree of downslope sorting of grain sizes shown by the fluvial counterpart.

Although they can be found in mountainous areas of almost any climate, alluvial fans are particularly common where ephemeral or intermittent streams laden with coarse sediment flow out of a mountainous region onto desert plains or into arid interior basins. In the western United States, alluvial fans are a major landform in landscapes consisting of fault-block mountains and basins, as in the Great Basin of Nevada, Utah, and California (● Fig. 18.14). Here,

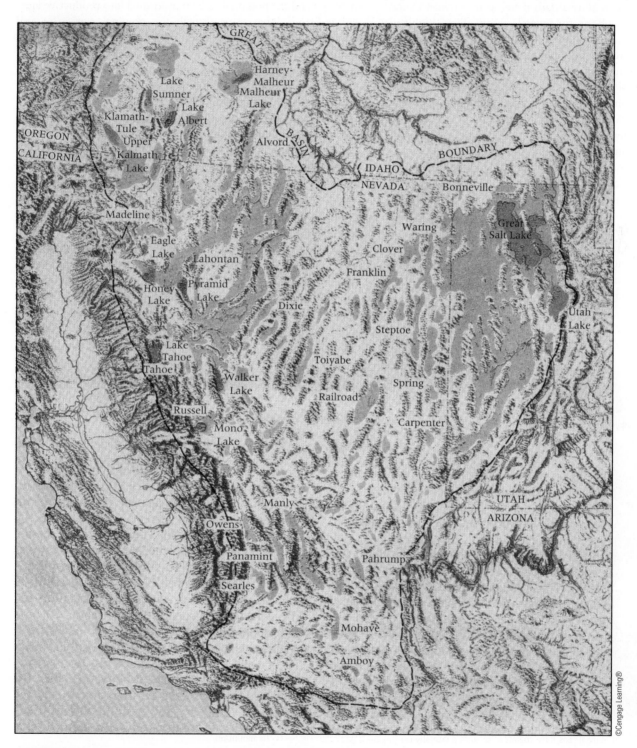

● **FIGURE 18.14** Map of the Great Basin of the western United States showing major lakes that existed during glacial times of the Pleistocene Epoch. Most of these fault-block basins have alluvial fans.

Why are there only a few lakes in this region today?

streams laden with sediment periodically rush from canyons cut into uplifted fault-block mountains and deposit their load in the adjacent desert basins but close to the mountain front. Everything else being equal, fans associated with the larger drainage basins within the uplifted fault-block mountains tend to have greater area and are less steep than fans developed from streams draining smaller upland drainage basins.

Large, conspicuous alluvial fans develop in environmental settings like the Great Basin for several reasons. First, highland areas in desert regions are subject to intense erosion, primarily because of the low density of vegetative cover, steep slopes, and orographically intensified downpours that can occur over mountains. In addition, streams in arid regions tend to carry a greater concentration of sediment load than comparable streams in more humid climates. As the dryland streams flow from confined mountain canyons into desert basins, they deposit most of their coarse sediment near the canyon mouth at the mountain front. Flowing into the desert basin, stream width increases, depth and velocity decrease, and flow volume is significantly reduced through infiltration into the alluvial channel bed. Not far from the canyon, the stream itself may disappear, or it may occasionally reach the desert basin floor where it deposits its remaining load, the silts and clays. Extensive alluvial fans are not as common in humid as in arid regions because most highland streams in moist climates are perennial and have sufficient flow to continue across adjacent lowlands.

Along the base of mountain fronts in arid regions, alluvial fans can become so large that adjacent fans join along their sides to form a continuous ramplike slope of alluvium called a **bajada** (● Fig. 18.15). A bajada consists of adjoining alluvial fans that have coalesced to form an "apron" of alluvium along the mountain base.

Where extensive fans coalesce over very wide areas, they form a **piedmont alluvial plain**, like the area surrounding Phoenix, Arizona (● Fig. 18.16). Piedmont alluvial plains generally have rich soils and the potential to be transformed into productive agricultural lands. The major obstacle for growing crops in an arid environment is inadequate water supply. In many world regions, arid alluvial plains are irrigated with water diverted from mountain areas or obtained from reservoirs on exotic streams.

Where a veneer of alluvium has been deposited on a pediment, the land surface may closely resemble a water-deposited alluvial fan. In some situations, it may not be possible to determine the existence of an underlying pediment without either excavating through the surface alluvium or finding the erosional pediment surface exposed in the walls of washes or gullies. In locations where no extensive pediment exists, alluvial deposits beneath the fans can be tens or even hundreds of meters thick. In contrast, the layer of alluvium overlying a pediment is relatively thin, no more than a few meters deep and sometimes much less.

Playas Desert basins of interior drainage surrounded by mountains are topographically closed basins, or **bolsons**. Most bolsons were formed by faulting that created basins between uplifted mountains. The lowest part of most bolsons is occupied by a

● **FIGURE 18.15** Adjacent alluvial fans have coalesced to form a laterally continuous bajada that connects the mountain upland with the basin lowland.

Why would a series of alluvial fans have a tendency to eventually join to form a bajada?

● **FIGURE 18.16** A large piedmont alluvial plain in Arizona that is extensively urbanized.

landform called a **playa** (from Spanish: *playa*, beach or shore), which is the fine-grained bed of an ephemeral lake. Occasionally, large rainfall (or snowmelt) events or wet seasons cover the playa with a very shallow body of water, called a **playa lake**. Direct precipitation onto the playa, inflow from surface runoff, and discharge from the groundwater zone can contribute water to the playa lake. The playa lake can persist for a day or for a few months (● Fig. 18.17). Wind blowing over the playa lake moves the shallow water, along with its suspended and dissolved load, around on the playa surface. This helps to fill in any low spots on the playa and contributes to making playas one of the flattest of all landforms on Earth. Playa lakes lose most of their water by evaporation to the desert air.

Despite being so flat, considerable variation exists in the nature of playa surfaces. Playas that receive most of their water from surface runoff typically have a smooth clay surface, sometimes called a **clay pan**, that is hard when dried out by the desert sun but extremely gooey and slippery

● **FIGURE 18.17** The floor of most bolsons contains a large playa, which only occasionally holds water.

What evidence suggests that the playa in this photo is partially wet?

Alan Van Valkenburg, NPS

● **FIGURE 18.18** Salt deposits accumulate on playas that evaporate considerable groundwater.

when wet. In contrast, **salt-crust playas**, also known as **salt flats**, receive much of their water from groundwater, are damp most of the time, and are encrusted with salt mineral deposits crystallizing out of the evaporating groundwater (● Fig. 18.18). Some salt-crust playas are the floors of the desiccated ancient lakes that existed in now-desert basins when they experienced more moderate climates in the recent geologic past.

Playas are useful in several ways. Mineral companies extract the rich deposits of evaporite minerals—including such important industrial chemicals as potassium chloride, sodium chloride, sodium nitrate, and borates—that have been deposited in some playa beds. Also, the extensive, flat playa surface makes some of them suitable as racetracks and airstrips. The Bonneville Salt Flats are the bed of an extinct Pleistocene lake, called Lake Bonneville, of which Great Salt Lake is a small remnant. The western portion of that salt-crust playa, where world land-speed records are set, often floods to a depth of 30 to 60 centimeters (1–2 ft) in the winter. Hard, flat clay-pan playas at Edwards Air Force Base, in California's Mojave Desert, have been used for many decades as landing fields for military aircraft, and 54 space shuttle missions landed there between 1981 and 2011. Aircraft landings are occasionally disrupted there because of playa flooding.

Wind as a Geomorphic Agent

Overall, wind is less effective than mass movement, running water, moving ice, or waves in accomplishing geomorphic work. Under certain circumstances, however, wind can be a significant agent in the modification of topography. Landforms—whether in the desert or elsewhere—that are created by wind are called **eolian** (or aeolian) landforms, after Aeolus, the god of winds in classical Greek mythology (● Fig. 18.19). The three principal conditions necessary for wind to be effective as a geomorphic agent are a sparse vegetative cover; the presence of dry, loose materials at the surface; and a wind velocity that is high enough to pick up and move those surface materials. These three conditions occur most commonly in arid regions and in coastal locations, but areas of recent fluvial or glacial deposits, snow, newly plowed fields, and overgrazed lands are also subject to significant wind action (● Fig. 18.20).

A dense cover of vegetation reduces wind velocity near the surface by providing frictional resistance. It also prevents wind from being directed against the land surface and holds materials in place with its root network. Without such a protective cover, fine-grained and sufficiently dry surface materials are subject to removal by strong gusts of wind. If surface particles are damp, they tend to adhere together in wind-resistant aggregates as a result of increased cohesion provided by the water. The arid conditions of deserts, therefore, make these regions very susceptible to wind erosion.

Eolian processes have many things in common with fluvial processes because air and water are both fluids. Some important contrasts also exist, however, as a result of fundamental differences between gases and liquids. For example, rock-forming materials cannot dissolve in air, as some can in water; thus air does not erode by corrosion or move load in solution. Otherwise, the wind detaches and transports rock fragments in ways comparable to flowing water, but it does so with less overall effectiveness

National Park Service

● **FIGURE 18.19** Colorado's Great Sand Dunes illustrate the stunningly beautiful landscapes that can be created by eolian processes.

In what other environments besides arid lands might sand dunes be found?

A principal similarity between the geomorphic properties of wind and running water is that flow velocity controls their competence—that is, the size of particles each can pick up and carry. However, because of its low density, the competence of moving air is generally limited to rock fragments that are sand sized or smaller. Wind erosion selectively entrains small particles, leaving behind the coarser and heavier particles that the wind is not able to lift. Like water-laid sediments, wind deposits are stratified as a result of changes in the fluid's velocity although within a much narrower range of grain sizes than occurs with most alluvium.

Wind Erosion and Transportation

Strong winds blow frequently in arid regions, whipping up loose surface materials and transporting them within turbulent air currents. The finest particles transported by winds, the clays and silts, are moved in *suspension*, buoyed up by vertical currents (● Fig. 18.21). Such particles essentially form a fine dust that will remain in suspension as long as the strength of upward air currents exceeds the tendency of the particles to settle out onto the ground because of gravity. The sediments carried in suspension by the wind make up its suspended load.

If the wind velocity surpasses 16 kilometers (10 mi) per hour, surface sand grains can be put into motion. As with fluvial transportation, particles that are too large to be carried in eolian suspension are bounced along the ground as part of the bed load in the transportation process of *saltation*. When particles moving in eolian saltation, which are typically sand sized, bounce on the ground, they generally dislodge other particles that are then added to the wind's suspended or saltating load. Even larger sand grains too heavy to be lifted into the air are pushed forward along the ground surface by the impact of the saltating grains in a process called **surface creep**. Grains moving along the ground in surface creep often become organized into small wave forms termed **ripples** (● Fig. 18.22).

● **FIGURE 18.20** Wind-blown snow in Shenandoah National Park, Virginia.

What two pieces of evidence observable in this photo indicate that a strong wind has been blowing?

because air has a much lower density than water. Another difference is that, compared to streams, the wind has fewer lateral or vertical limitations on movement. As a result, the dissemination of material by the wind can be more widespread and unpredictable than that by streams.

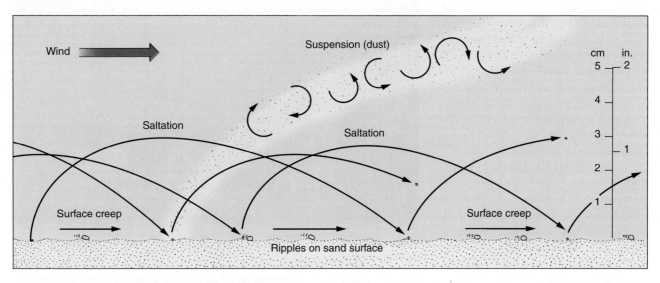

● **FIGURE 18.21** Wind moves sediment in the transportation processes of suspension, saltation, and surface creep.

Why are grains larger than sand not generally moved by the wind?

National Park Service

● FIGURE 18.22 Sandy bed load on dune surfaces is commonly worked into numerous ripples that travel forward during strong winds.

Wind erodes surface materials by two main processes. **Deflation**, which is similar to the hydraulic force of running water, occurs when wind blowing fast enough or with enough turbulence over an area of loose sediment is able on its own to pick up and remove small fragments of rock. As with all of the geomorphic agents, once the wind obtains some load, it also erodes by the process of *abrasion*. In eolian abrasion, particles already being transported by the wind strike rocks or sediment and break off or dislodge additional rock fragments. Wind-driven solid particles are more effective than the wind alone in dislodging and entraining other grains and in breaking off new fragments of rock. Most eolian abrasion is quite literally sandblasting, and quartz sand, which is common in many desert areas, can be a very effective abrasive agent in eolian processes. Yet, sand grains are typically the largest size of clast that the wind can move, and they are rarely lifted higher than 1 meter (3 ft) above the surface. Consequently, the effect of this natural sandblast is limited to a zone close to ground level.

Where loose fine-grained particles exist on the land surface, strong winds will pick them up primarily by deflation and carry them off in suspension. Sometimes the result is a thick, dark, swiftly moving dust cloud that swirls over the land and greatly reduces visibility. These *dust storms* are sometimes so severe that visibility drops to nearly zero and almost all sunlight is blocked

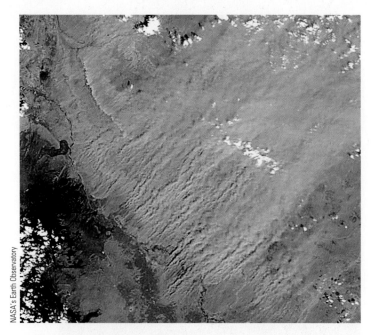

NASA's Earth Observatory

● FIGURE 18.23 Dust storms occur when strong winds entrain and suspend a large amount of silt and clay particles. This dust storm in northeastern Arizona was so large that it is seen on this satellite image.

If north is at the top of the image, what direction were the winds blowing that entrained so much dust?

(● Fig. 18.23). Dust storms are often highly destructive, removing layers of surface materials in one area and depositing them elsewhere over the course of a few hours. The infamous Dust Bowl era of the 1930s particularly affected the southern Great Plains of the United States in this way when devastating dust storms were brought about by years of drought without adequate modification of agricultural practices (● Fig. 18.24).

Library of Congress Prints and Photographs Division [LC-DIG-fsa-8b31257]

● FIGURE 18.24 A dust storm arriving at Elkhart, Kansas, during the height of the Dust Bowl era of the 1930s.

Can you suggest a continent that might be a source of major dust storms in the world today?

Sandstorms can occur in areas where sand-sized sediment is abundant at the surface. Because sand grains are larger and heavier than silts and clays, most sandstorms are restricted to a zone near the ground, where objects are severely abraded by the abundant saltating sand. Evidence of the restricted height of desert sandstorms is visible on vehicles that have traveled through the desert and on fence posts, utility poles, and other structures, all of which are more intensely pitted and abraded in their lower than in their higher sections.

Erosion by deflation can produce shallow depressions in a barren surface of unconsolidated materials. These depressions, which vary in diameter from a few centimeters to a few kilometers, are called **deflation hollows**. Deflation hollows are particularly common in nonmountainous arid regions. They tend to collect rainwater and may hold water for a time, depending on permeability and evaporation rates. Thus, deflation hollows, like bolsons, frequently contain playas. Many thousands of deflation hollows that contain playas occur in the semiarid southern High Plains of western Texas and eastern New Mexico in the United States. Often deflation hollows form at sites that were already exhibiting a slight depression or where vegetation cover has been breached by overgrazing, fire, or other means.

Deflation has traditionally been considered one of several possible factors that help to produce **desert pavement** (or *reg*), a close-fitting mosaic of gravel-sized stones that overlies a deposit of mostly finer-grained sediments. Desert pavement is common in many arid regions, particularly in parts of the Sahara, interior Australia, the Gobi in central Asia, and the American Southwest. If deflation selectively removes the smaller particles from a desert surface of mixed particle sizes, the gravel-sized clasts left behind form a concentration of stones at the surface overlying the mixed grain sizes below (● Fig. 18.25). Sheet wash (unchannelized running water) also contributes to the formation of some desert pavements

by selectively eroding only the fine-grained clasts from an area of mixed grain sizes. Research has shown that other occurrences of pavement form by eolian deposition, that is, by accumulation of wind-deposited fine-grained sediments that subsequently fall, wash, and sift down beneath the stony surface layer. Regardless of its origin, once formed, desert pavement helps to stabilize desert surfaces by protecting the fine-grained material below the pavement from erosion. Desert pavement is fragile, however, and easily disturbed by vehicles. Use of off-road vehicles in the desert severely disrupts the desert pavement and contributes greatly to accelerated erosion, the amount of wind-blown dust in the atmosphere, the occurrence of dust storms, and damage to desert ecological systems.

Like deflation at natural rates not accelerated by humans, eolian abrasion is also responsible for creating interesting desert landform features. Where the land surface is exposed bedrock, wind abrasion can polish, groove, or pit the rock surface and in some cases produces **ventifacts**, which are individual wind-fashioned stones. A ventifact is a rock that has been trimmed back to a smooth slope on one or more sides by sandblast. Because of frictional effects at the surface, the ability of the wind to erode by abrasion increases with increasing distance from the ground surface, at least up to a certain height. Thus, abrasion carves the windward side of a rock into a smooth, sloping surface, or face. Ventifacts subjected to multiple sand-transporting wind directions have multiple faces, called *facets*, which meet along sharp edges (● Fig. 18.26).

A distinctive feature sometimes attributed incorrectly to wind abrasion is the pedestaled, or balanced, rock, in which a larger section of rock sits on a narrower lower part, the pedestal, below. It is tempting to think that eolian abrasion wore away the lower part faster than the upper part. Actually, such forms result from various physical and chemical weathering processes in the damper

● **FIGURE 18.25** The surface of desert pavement is a mosaic of gravel. Immediately underneath the gravel lies a zone of fine-grained sediment with little gravel.

What problem might result if the surface stones are removed from a large area?

● **FIGURE 18.26** A ventifact on Mars. With high wind speeds, plenty of loose surface materials, and no vegetation, rocks on the Martian surface display facets, pits, and other strong evidence of eolian abrasion.

● **FIGURE 18.27** Rocks with pedestals are also known as rock mushrooms. Eolian deflation and abrasion may contribute to the narrowing of some pedestals, but weathering processes and sheet wash are considered to be more important for their formation.

How might the rock on top of the pedestal differ from the rock forming the pedestal?

environment at the base of an exposed rock and are not usually related to eolian abrasion (● Fig. 18.27).

Rates of eolian erosion in arid regions often reflect the strength and kinds of the exposed surface materials. Where eolian abrasion affects rocks of varying resistance, differential erosion etches away softer rocks faster than the more resistant rocks. Even in desert locations of extensive soft rock, such as shale, or semiconsolidated sediments, like ancient lake deposits, abrasion may not act in a uniform fashion over the entire exposure. A **yardang** is a wind-sculpted remnant ridge, often of easily eroded rock or semilithified sediments, left behind after the surrounding material has been eroded by abrasion (● Fig. 18.28), perhaps with deflation assisting in removal of fine-grained fragments. Everything else being equal, abrasion and deflation are most effective where rocks are soft or weak.

Wind Deposition

All materials transported by the wind are eventually deposited whenever and wherever the wind energy falls below the minimum amount needed to keep the grains moving. The nature of the resulting eolian depositional landform depends primarily on the characteristics of the wind, which is commonly affected by topography and vegetation, and the grain size of the sediments. Fine-grained particles (silts and clays) are often transported in suspension long distances from their source area before being deposited, sometimes quite thickly, as a blanket covering the preexisting topography. In comparison, the coarser, sand-sized particles, transported in saltation and surface creep, accumulate much closer to their source area.

Sand Dunes

In some circumstances, eolian sand deposits form a **sand sheet**, a rather uniform cover of sand-sized particles displaying little surface relief. More often, eolian sand deposits accumulate in the shape of hills, mounds, or ridges, called **sand dunes**. In fact, to many people, the word desert evokes an image of endless sand dunes, perhaps accompanied by blinding sandstorms, a blazing sun, mirages, and an occasional palm oasis. These features do exist, particularly in parts of North Africa and the Arabian Peninsula, but most areas of the world's deserts have rocky or gravelly surfaces, scrubby vegetation, and few or no sand dunes. Nevertheless, sand dunes are the most spectacular landforms of wind deposition, whether they occur as seemingly endless dune regions called **sand seas** (or *ergs*), as small dune fields, or as sandy ridges inland from beaches (● Fig. 18.29).

Dune topography is highly variable. Dunes in the great sand seas of the Sahara and Arabian Deserts look like rolling ocean waves. Others have aerodynamic crescent shapes. The specific type of dune that forms depends on the amount of sand available, the strength and direction of sand-transporting winds, and the amount of vegetative cover. As wind that is carrying sand encounters surface obstacles or topographic obstructions that decrease its

(a)

(b)

● **FIGURE 18.28** Eolian erosion carves yardangs, aerodynamically shaped ridges. (a) A single, large yardang in the Kharga Depression of Egypt. (b) Multiple yardangs in the Ydang Qaidam Basin of China.

National Park Service

● **FIGURE 18.29** These coastal dunes lie just inland from the beach on Padre Island, along the Gulf Coast of Texas. In coastal dune regions landscape change is common as dunes move inland and are subsequently invaded by plants.

Why are many coastlines such good locations for sand dune formation?

velocity, the sand is deposited and piles up in drifts. Once initiated, these sand piles themselves interfere with wind velocity and the sand-transporting capabilities of the wind. Dunes grow larger until equilibrium is reached between dune size and the ability of the wind to deliver sand grains to the dune.

Sand dunes are either *active* or *stabilized* (● Fig. 18.30). Active dunes change their shape and advance downwind as a result of wind action. Dunes change shape with variations in wind direction or wind strength. Many active dunes travel, or *migrate*, when wind-eroded sand from their upwind (windward) slope moves by saltation and surface creep over the dune crest and onto the steep leeward (downwind) slope, which is the **slip face**. The slip face of a dune lies at the *angle of repose* for dry, loose sand. The angle of

repose—the steepest slope that a pile of dry, loose material can maintain without experiencing slipping, sliding, or flow down the slope—is about 34° for sand. When wind direction and velocity are relatively constant, a dune can move forward while preserving its general form by this downwind transfer of sediment from the windward slope to the slip face (● Fig. 18.31).

The speed at which active dunes move downwind varies greatly, but as with many processes, the movement is episodic; dunes advance only when the wind is strong enough to pick up and transport sand from the upwind to the downwind side. Because of their greater height and especially their greater volume of sand, large dunes travel more slowly than smaller dunes, which can migrate up to 40 meters (130 ft) a year. During sandstorms, a dune can migrate more than 1 meter (3 ft) in a single day. In places where plants, including trees, lie in the path of an advancing dune, the sand cover can move onto and smother the vegetation. Some active dunes are affected by seasonal wind reversals so that they do not advance, but the crest at the top moves back and forth annually under the influence of seasonally opposing winds.

Stabilized dunes maintain their shape and position over time. Vegetation normally stabilizes dunes. Vegetation can stabilize an active sand dune if plants gain a foothold and send roots down to moisture deep within the dune. This task is difficult for most plants because sand offers limited nutrients and has high permeability. If the vegetation cover on a stabilized dune is breached, perhaps from damage by range animals or off-road vehicles, the wind can then remove some of the sand, creating a local **blowout**. Where invading dunes and blowing sands are a problem, attempts are frequently made to plant grasses or other vegetation to stabilize the dunes, halting their advance (● Fig. 18.32).

D. Sack

(a)

D. Sack

(b)

● **FIGURE 18.30** Active and stabilized dunes. (a) Sand moves freely in active dunes, which typically have little or no vegetation. (b) Vegetation reduces wind speed over the dune and helps increase material cohesion, leading to dune stabilization.

What effect might the trails that are visible on the stabilized dunes have on the dune system?
What factors play a role in which type of dune will be found in a region?

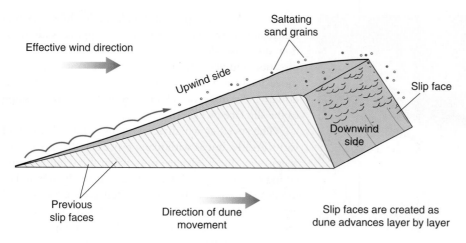

Why does the inside of the migrating dune consist of former slip faces?

Slip faces are created as dune advances layer by layer

One extensive area of stabilized dunes in North America is the Sand Hills of Nebraska. This region features large dunes that formed during a drier period between glacial advances in the Pleistocene Epoch. Subsequent climate change affected sand supply, wind patterns, and moisture availability so that these impressive dunes are now stabilized naturally by grasses (● Fig. 18.33). Similar stabilized dunes are found along the southern edge of the Sahara, where sand dunes that were active during drier conditions in the recent geologic past are now anchored by vegetation.

Types of Sand Dunes

Different types of sand dunes are distinguished on the basis of their shape and on their orientation relative to the wind direction. The different types are also related to the amount of available sand, which affects not only the size but also the shape of sand dunes.

Barchans are a kind of crescent-shaped dune (● Fig. 18.34). The two *arms* of the crescent, also called the dune's *horns*, point downwind (● Fig. 18.35a). The main body of the crescent lies on the upwind side of the dune. From the desert floor at its upwind edge, the dune rises as a gentle slope up which the sand moves until it reaches the highest point, or crest, of the dune and, just beyond that, the slip face at the angle of repose. The slip face is oriented perpendicular to the barchan's arms. The arms extend downwind beyond the location of the slip face. Barchans form in areas of minimal sand supply where winds are strong enough to move sand downwind in a single prevailing direction. They may be most common in smaller desert basins surrounded by highlands where they tend to form near the downslope, sandy edge (toe) of alluvial fans, and adjacent to small playas. Although they may form as isolated dunes, barchans often appear in small groups, called *barchan fields*.

Another type of crescent-shaped dune is the **parabolic dune**, but the orientation of the crescent in this case is reversed from that of a barchan (Fig. 18.35b). With parabolic dunes, the arms of the crescent are stabilized by vegetation, are long, and point

● FIGURE 18.32 Grass has been planted on this dune ridge along the Maryland coast to help stabilize the sand.

Explain how plants can stabilize dunes.

● FIGURE 18.33 The rolling grazing lands of the Sand Hills in central Nebraska were once a major region of active sand dunes.

Why are these dunes no longer active?

● **FIGURE 18.34** A small barchan in southern Utah.

Why do smaller barchans migrate faster than larger barchans?

D. Sack

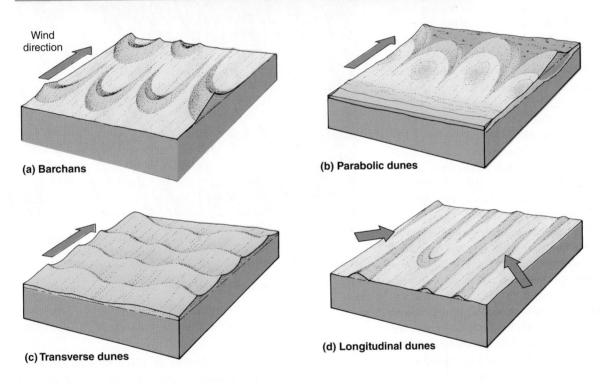

Wind direction

(a) Barchans

(b) Parabolic dunes

(c) Transverse dunes

(d) Longitudinal dunes

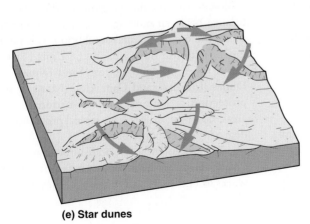

(e) Star dunes

● **FIGURE 18.35** Five principal types of sand dunes. Arrows indicate the dominant wind direction for each.

What factors influence which type of dune will be found in a region?

upwind, trailing behind the unvegetated main body and crest of the dune rather than extending downwind from it. The main body of a parabolic dune points downwind, and the slip face along its downwind edge has a convex shape when viewed from above. Parabolic dunes commonly occur just inland of beaches and along the margin of active dune areas in deserts.

Transverse dunes are created where sand-transporting winds blow from a constant direction and the supply of sand is abundant (Fig. 18.35c). As with barchans, the upwind slope of a transverse dune ridge is gentle, and the steeper downwind slip face lies at the angle of repose. In the downwind direction, transverse dunes form ridge after ridge separated from each other by low swales in a repeating wavelike fashion. Each dune ridge is laterally extensive perpendicular to the sand-transporting wind direction. The dune ridges, slip faces, and interdune swales trend perpendicular to the direction of prevailing winds, hence the name *transverse*. Abundant sand supply derives from such sources as easily eroded sandstone bedrock, sandy alluvium deposited by exotic streams or during wetter climates in the Pleistocene Epoch, or sandy deltas and beaches left in the landscape after the desiccation of ancient lakes.

Long dunes aligned parallel to the average wind direction are **longitudinal dunes** (Fig. 18.35d). There is no consistent distinction between the back slopes and slip faces of these dunes, and their summits may be either rounded or sharp. Strong winds are important to the formation of most longitudinal dunes, which do not migrate but instead elongate in the downwind direction. Small longitudinal dunes, such as those found in North America, can simply represent the long trailing ridges of breached parabolic dunes or a sand streak extending from a somewhat isolated source of sand. Much higher and considerably longer than these are the impressive longitudinal dunes that cross vast areas of the flatter, more open desert topography of North Africa, the Arabian peninsula, and interior Australia (● Fig. 18.36). These dunes develop under bidirectional wind regimes, where the two major sand-transporting wind directions come from adjacent quadrants, such as a northwesterly and southwesterly wind. A type of large longitudinal dune called a **seif** (pronounced *safe*; from the Arabic word for sword) is found in the deserts of Arabia and North Africa. Seif dunes are huge, sinuous (rather than linear), sharp-crested dunes, sometimes hundreds of kilometers long, whose troughs are almost free of sand. They attain heights up to 180 meters (600 ft).

The large, widely spaced **star dunes** have a pyramidal shape in which ridges of sand radiate out from a high center point to resemble a star in map view (Fig. 18.35e). These dunes are most common in areas where there is a great quantity of sand, changing wind directions, and an extremely hot and dry climate. Star dunes are stationary, but ridges and slip faces shift orientation with wind variations.

Dune Protection

Many people consider sand dunes one of nature's most beautiful landforms, and they are attractive sites for recreation. They seem particularly inviting to off-road vehicle enthusiasts. However, although dunes appear to be indestructible and rapidly changing

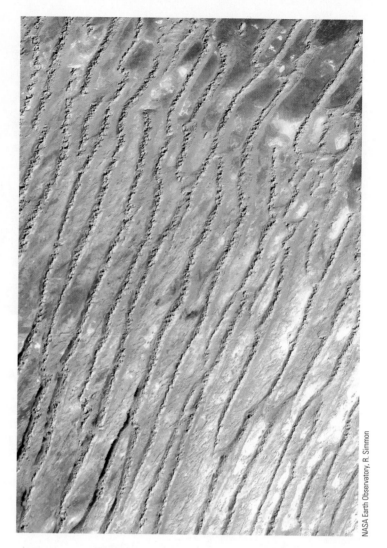

● **FIGURE 18.36** A satellite view of a small part of an extensive longitudinal dune area in the Sahara. The width of this image represents approximately 35 kilometers (21 mi) of terrain.

What is the approximate ground length of the dunes seen on this image?

environments that do not damage easily, this is far from the truth. Dunes are fragile environments with delicate, easily affected ecologies. Because dune regions are the result of an environmental balance between moving dunes and the plants trying to stabilize them, the environmental equilibrium is easily upset. Some of the most spectacular dune areas in the United States, including White Sands (New Mexico), Great Sand Dunes (Colorado), Indiana Dunes (Indiana), and Cape Cod (Massachusetts), lie within the protection of national parks or national seashores. Many dune areas, however, do not have special protection, and environmental degradation is a constant threat.

Dunes should be left undisturbed for many reasons. Near the shoreline, dunes help protect adjacent inland areas from storm waves (● Fig. 18.37). In this regard, they are particularly important

Off-Road Vehicle Impacts on Desert Landscapes

Whether you prefer to call them off-road vehicles (ORVs) or all-terrain vehicles (ATVs), driving motorized vehicles of any kind over the desert landscape off established roadways has tremendous negative impacts on desert flora and fauna, the habitat of those organisms, the stability of landforms and landforming processes, and the aesthetic beauty of the natural desert environment. Deserts are particularly fragile environments, primarily because of the low amount and high variability of precipitation. The climatic regime leads to slow rates of weathering, slow rates of soil formation, low density of vegetation, unusual species of plants and animals specially adapted to the hostile conditions of severe moisture stress, and rapid generation of surface runoff when precipitation does occur.

Off-road vehicles damage desert biota in several ways. They kill and injure plants, animals (for example, birds, badgers, foxes, snakes, lizards, and tortoises, to name just a few), and insects when they hit or ride over the top of them. Even careful ORV riders who drive slowly cannot avoid crushing small animals and insects that lie hidden under a loose cover of sand or shallow roots that extend out far from a plant. Driving ORVs at night is particularly deadly for desert animals, which tend to be nocturnal. Off-road vehicles do not have to hit or ride over organisms to cause populations to decrease. Hearing loss experienced by animals in areas frequented by ORVs puts them at a major disadvantage for feeding, defending themselves, and mating. These vehicles crush and destroy burrows and other animal and insect homes and nesting sites, including plants. Oil and gas leaking from poorly maintained vehicles are another danger for desert dwellers, as are grass and range fires inadvertently started from such vehicles. No desert subenvironment, not even sand dunes, is immune from these negative impacts.

In addition to the direct impacts on organisms, physical properties of the landscape are also negatively affected by ORVs. One major problem is that ORVs compact the desert surface sediments and soil. Compaction causes a decrease in permeability, which greatly restricts the ability of water to infiltrate into the subsurface. More surface runoff translates to greater erosion of the already thin desert soils. In addition, even long after ORV use ceases in a disturbed area, soil compaction makes it very difficult for desert plants, animals, and insects to become reestablished there. Denuded, eroded swaths made by ORVs often remain visible as ugly scars in the landscape for decades and act as sources of sediment, adding to the severity of dust storms. Where ORV trails traverse steep slopes, they have contributed to the occurrence of debris flows, mudflows, and other forms of mass wasting.

The attributes of desert environments that give them such stark beauty also make them very vulnerable to disturbance. With limited but often torrential precipitation, sparse and slow-growing vegetation, and slow rates of weathering and soil formation, deserts require long periods of time to recover from environmental disturbance. Recovery, moreover, will likely not return the landscape to the condition it would have been in if the disturbance had not occurred.

Motorcycles ridden on a desert hillslope in Utah have stripped the vegetation, compacted the soil, instigated accelerated erosion, and seriously marred the aesthetic beauty of the landscape.

Riding off-road vehicles on active sand dunes harms the sensitive organisms that live there and interferes with the natural dune processes by compacting the sand.

Copyright and photograph by Dr. Parvinder S. Sethi

● **FIGURE 18.37** A sign directs visitors on how to help protect this dune area along the South Carolina coast.

Why should some dune areas be protected from human activities such as driving dune buggies and other recreational vehicles?

J. Petersen

● **FIGURE 18.38** Loess, seen here near Ogden, Utah, is wind-deposited silt.

to communities along the low-lying Gulf and Atlantic coasts of the United States, where hurricanes and midlatitude cyclones batter the coastlines. In the Netherlands, coastal dunes are extremely important because the land behind them is below sea level; thus a breach through the dunes could mean disastrous inundation. Desert and coastal dune regions both serve as critical wildlife habitats.

Loess Deposits

The wind carries in suspension dust-sized particles of silt and clay, entrained by deflation, for hundreds or thousands of kilometers before depositing them. Eventually these particles settle out to form a tan, yellowish, or grayish blanket of **loess** that may cover or bury the existing topography over widespread areas (● Fig. 18.38). These deposits vary in thickness from less than a centimeter to more than 100 meters (330 ft). In northern China, downwind from the Gobi Desert, the loess is 30 to 90 meters (100–300 ft) thick (● Fig. 18.39).

Deserts are major sources of dust deposited in downwind locations as loess, but loess sediments also originate in other sparsely vegetated areas. The widespread loess deposits of the American Midwest and Europe were derived from extensive glacier and glacial meltwater deposits during ice ages of the recent geologic

past. As winds blew across the barren glaciated regions or glacial meltwater areas, they picked up a large load of fine sediment and deposited it as loess in downwind regions.

Where loess deposits lie at Earth's surface, they affect the appearance of the landscape. Grassy rolling hills composed of loess often look soft and inviting from the air. Although fine and dusty to the touch, because of its high cohesion, particularly when damp, many extensive loess deposits maintain vertical walls when cut through naturally by a stream or artificially as for a road. Historically, this quality led people to carve caves and tunnels in especially cohesive loess to make permanent dwellings or for temporary living quarters during wars. Sometimes slumping occurs on steep loess slopes, giving a steplike profile to loess bluffs (● Fig. 18.40). Overland flow of water often gullies loess slopes that are sparsely vegetated.

Because of its high calcium carbonate content, loess in regions of moderate rainfall forms the parent material for many of Earth's most fertile agricultural soils. Extensive loess deposits are found in northern China, the Pampas of Argentina, the North European Plain, Ukraine, and Kazakhstan. In the United States, Midwest plains, the Mississippi Valley, and the Palouse region of eastern Washington are underlain by rich loess soils (● Fig. 18.41). All these areas are extremely productive grain-farming regions.

● **FIGURE 18.39** A steep gully has eroded into the thick loess deposits of northern China.

Where did the sediments found in this loess likely originate?

● **FIGURE 18.40** The steep and unstable loess bluffs of Vicksburg, Mississippi.

Why might instability of loess cliffs be a problem?

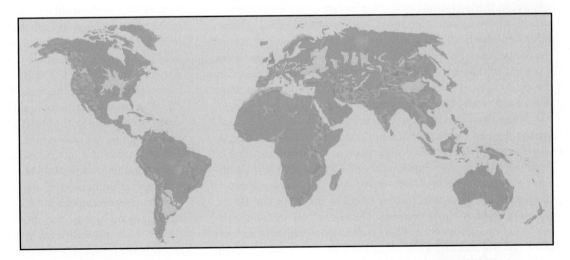

● **FIGURE 18.41** The major loess regions of the world, shown in yellow. Most loess deposits are peripheral to deserts or recently glaciated regions.

Aerial view of a desert plain in the central Asian country of Turkmenistan. An east–west flowing stream extends across the photo center. What is the dominant visible difference between the sand dune area south of the stream zone and the dune area north of it? What might be causing the difference? Describe the stream and how it appears to interact with the dunes.

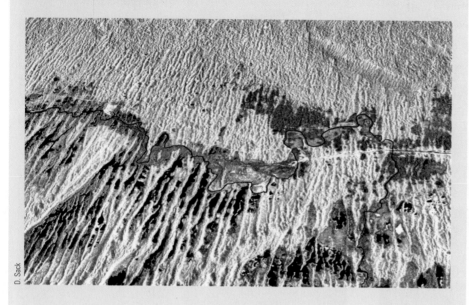

D. Sack

Landscape Development in Deserts

Geomorphic landscape development in arid climates is comparable in many ways to that in humid climates, but not in all ways. Weathering and mass movement processes operate, and fluvial processes predominate, in both environments but at different rates and with the tendency to produce some different landform and landscape features in the two contrasting environments. Arid environments experience the added regionally or locally important effect of eolian processes, which are not very common in humid settings beyond the coastal zone. The major differences in the results of geomorphic work in arid climates, as compared to humid climates, are caused by the great expanses of exposed bedrock, a lack of continuous water flow, and a more active role of the wind in arid regions.

Some desert landscapes, such as most of those in the southwestern United States, are found in regions of considerable topographic relief, whereas others, like much of the Sahara and interior Australia, occupy huge expanses of open terrain with few mountains. An excellent example of a typical desert landscape in a region of considerable structural relief is found in the Basin and Range region of western North America. The region extends from western Texas and northwestern Mexico to eastern Oregon. It includes all of Nevada and large portions of New Mexico, Arizona, Utah, and eastern California. Here, more than 200 mountain ranges, with basins between them, dominate the topography. The Great Basin—a large subregion centered on Nevada and characterized by interior drainage, numerous alternating mountain ranges and basins, and active tectonism—occupies much of the central and northern part of the Basin and Range.

Fault-block mountains in the Basin and Range region rise thousands of meters above the desert basins, and many form continuous ranges (● Fig. 18.42). These high ranges, such as the Guadalupe Mountains, Sandia Mountains, Warner Mountains, and the Panamint Range, to name just a few, encourage orographic rainfall. Fluvial erosion dissects the mountain blocks to carve canyons between peaks and cut washes between interfluves. Where active tectonism continues so that the uplift of mountain ranges matches or exceeds the rate of their erosion, as in the Great Basin, fluvial deposition constructs alluvial fans extending from canyon mouths outward toward the basin floor. In many basins, the alluvial fans have coalesced to create a bajada. In these tectonically active basins with interior drainage, playas often occupy the lowest part of the basin, beyond the toe of the alluvial fans (● Fig. 18.43a). The mountain ranges create roughness, but also funnel the wind, which picks up and transports dust from basin floors; sandy alluvium from the toe areas of alluvial fans; sand-sized aggregates of playa sediments; and, in some cases, sandy sediment from beaches left behind by ancient perennial lakes. The sand-sized sediments are transported relatively short distances before being deposited in local dune fields.

Pediments lie along the base of some mountains, particularly in the tectonically less active areas of exterior drainage beyond the southern boundary of the Great Basin. In these areas, mountains are being lowered and the pediments are extending. Resistant inselbergs remain on some of the pediments. Sandy alluvium along streams and sandy deposits left by ancient rivers and deltas provide sediment to be reworked by the wind into occasional dune fields. The landscapes of the Mojave Desert in California and parts of the Sonoran Desert in Arizona have localities where uplift along faults has been inactive long enough for the landscape to be dominated by extensive desert plains interrupted by a few isolated inselbergs as reminders of earlier, tectonically active, mountainous landscapes (Fig. 18.43b).

The geologic structures and geomorphic processes found in desert areas are, for the most part, the same as those found in humid regions. Variations in the effects and rates of these processes are what make the desert landscape distinctive. Although fault-block mountains and fault-block basins dominate the geologic structure of immense regions of the American West and other arid locations around the world, it is important to know that desert landscapes are as varied as those of other climatic environments. Deserts exist at localities where the landscape has developed in nearly every imaginable geologic setting, including volcanoes, ash deposits and lava flows, folded rocks forming ridges and valleys, horizontal strata, and exposures of massive intrusive rocks. Where

● **FIGURE 18.42** Satellite image of fault-block mountains and basins in northeastern Nevada.

What landforms appear as white areas along the right edge of the image?

arid climates occur in expansive regions of largely open, low-relief terrain, as in the ancient and geologically stable deserts of Australia and the Sahara, common landforms include inselbergs surrounded by extensive desert plains (● Fig. 18.44); deflation hollows; playas; washes; the channels of ancient streams; and the beds of shallow,

ancient lakes. With limited terrain roughness, areas of large longitudinal dunes can develop in these settings.

Arid landscapes and eolian landscapes in any climatic environment can be beautiful and stark, but those who are unfamiliar with these environments too often misunderstand them. Deserts

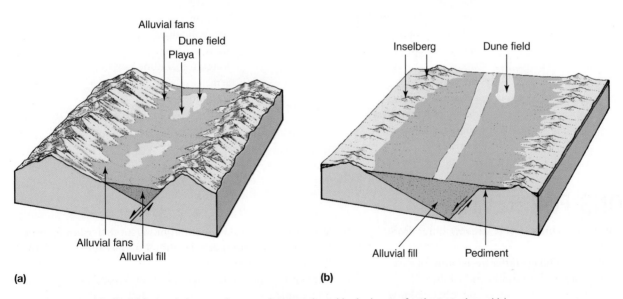

(a)   **(b)**

● **FIGURE 18.43** (a) Alluvial fans and playas tend to occur in mountain and basin deserts of active tectonism, which typically have interior drainage. (b) Pediments and inselbergs are common in desert areas that have been tectonically stable since a distant period of mountain formation. Development of exterior drainage helps transport sediment eroded from the mountains out of the basin.

● **FIGURE 18.44** Uluru (Ayers Rock) is a striking sandstone inselberg rising above the arid, flat interior of the Australian Outback.

are not wastelands, and dunes are not merely piles of sand. Most desert and eolian areas have unique characteristics and scenic beauty. The austere, angular character of their landforms, fragility of their environments, special nature of their biota, and opportunities they provide for learning about how certain Earth systems operate fully qualify them for preservation and protection. One only needs to count the number of national parks, national monuments, and other scenic attractions in the arid southwestern United States and in regions of sand dunes to find ample support for their survival. Deserts and dune localities are places with a beauty all their own and are areas worthy of appreciation and appropriate environmental protection.

CHAPTER **18** ACTIVITIES

■ TERMS FOR REVIEW

regional base level	piedmont alluvial plain	sand sheet
exotic stream	bolson	sand dune
wash (arroyo, wadi)	playa	sand sea (erg)
badlands	playa lake	slip face
plateau	clay pan	blowout
caprock	salt–crust playa (salt flat)	barchan
mesa	eolian (aeolian)	parabolic dune
butte	surface creep	transverse dune
pediment	ripple	longitudinal dune
inselberg	deflation	seif
alluvial fan	deflation hollow	star dune
fan apex	desert pavement (reg)	loess
debris flow fan	ventifact	
bajada	yardang	

■ QUESTIONS FOR REVIEW

1. What are some ways in which desert streams differ from humid region streams?
2. How do basins of interior drainage differ from basins of exterior drainage, and why are both found in arid lands?
3. What distinguishes a plateau from a mesa and a butte? How are they related?
4. How does an alluvial fan differ from a pediment?
5. What are playas, and why are they commonly found in desert bolsons?

6. Why is wind a weaker geomorphic agent than running water?
7. How does formation of a deflation hollow differ from formation of a ventifact?
8. Explain the main differences between a barchan and a parabolic dune.
9. What is loess? What is an important economic activity related to loess in many regions?
10. What are some ways in which climate and vegetation affect landforms in the desert?

CONSIDER AND RESPOND

1. How are eolian erosion and transportation processes similar to fluvial erosion and transportation processes, and how do they differ? What are the main reasons for these similarities and differences? Is eolian deposition similar to fluvial deposition? If so, what are some of the similarities?

2. What physical geographic factors contribute to create so many basins of interior drainage in arid regions? What are some ways in which base-level changes can occur in basins of interior drainage?

3. Name several types of arid region landforms commonly found in tectonically active desert regions, such as the American Great Basin. Name several types of arid region landforms commonly found in deserts, such as interior Australia, that have been tectonically stable for a very long time.

4. Deserts are considered fragile environments. What are some ways in which desert landforms fit or contribute to this designation?

5. Why are there so many national parks and monuments in the arid parts of the American Southwest? What landform characteristics and other factors of desert landscapes draw people to these locations?

PRACTICAL APPLICATIONS

1. A stream flowing out of a canyon mouth onto the apex of an alluvial fan has a discharge of 12 cubic meters per second, flow depth of 2 meters, and flow width of 3 meters. What is the flow velocity at the fan apex? If the velocity decreases downfan by 0.1 meter per second, at what distance from the apex will it reach a velocity of zero?

2. A barchan in Arizona migrated the following total annual amounts (in meters) in each of the last 13 years, listed in order from the oldest to the most recent year: 9, 12, 14, 15, 15, 19, 22, 23, 26, 28, 29, 29, and 30. What was the total distance that the dune migrated during the 13 years, and what was its average annual migration rate for the study period? How would you describe the overall trend in dune migration for this period? What factor or factors might be responsible for the observed trend?

LOCATE AND EXPLORE

1. Using Google Earth, identify the landforms at the following locations (latitude, longitude) and provide a brief discussion of how the landform developed. Include a brief discussion of why the landform is found in that general area of the United States or the world.
   a. 36.415°N, 116.810°W
   b. 37.66°N, 117.625°W
   c. 36.25°N, 116.82°W
   d. 23.42°S, 14.77°E
   e. 36.96°N, 110.11°W
   f. 25.35°S, 131.03°E
   g. 28.07°N, 114.06°W
   h. 36.25°N, 116.82°W
   i. 36.32°N, 116.94°W
   j. 32.77°N, 106.21°W

2. The following are latitude and longitude points for the center of major world deserts. Using Google Earth to view their global position and setting, provide an explanation of why deserts are located in these areas. Consider such characteristics as mountain barriers, continentality, cold ocean currents, and global-scale pressure and wind belts.
   a. Sahara Desert (22.4°N, 12.6°E)
   b. Atacama Desert (23.7°S, 68.8°W)
   c. Simpson Desert (25.3°S, 133.8°E)
   d. Mojave Desert (35.7°N, 115.1°W)
   e. Gobi Desert (38.9°N, 82.6°E)
   f. Namib Desert (24.3°S, 15.1°E)

 MindTap—Make the most of your study time by accessing everything you need to succeed in one place. Read your textbook, take notes, review flashcards, watch videos, complete activities, take practice quizzes, and more online with MindTap. Log in at **www.cengagebrain.com**.

Map Interpretation

DESERT BASIN LANDFORMS

The Map

The map shows a section of Death Valley National Park, California. This is part of the Basin and Range region, characterized by fault-block mountains (ranges) separated by down-faulted valleys (basins). As the rugged ranges erode, the basins fill with sediment carried by infrequent flash floods.

The Panamint Range forms the mountain block along the western slope of Death Valley in the map area. The highest summit in that range, Telescope Peak, reaches an elevation of 3368 meters (11,049 ft). The Amargosa Range forms the eastern boundary. Death Valley's lowest elevation is 86 meters (282 ft) below mean sea level.

The present climate of the Basin and Range is arid, except for high mountains that receive more precipitation. With an average annual rainfall of 3.5 centimeters (1.7 in.) and a potential evapotranspiration that may exceed 380 centimeters (150 in.), Death Valley has the most extreme aridity in the region. Summer temperatures commonly exceed 40°C (104°F), and it holds the world record for maximum measured temperature at 57°C (134°F). In winter, much of the Basin and Range region has freezing nighttime temperatures.

During the Pleistocene Epoch, deep lakes filled numerous basins in the region. Death Valley was occupied by Lake Manly, and traces of this 180-meter (600-ft) deep lake's shoreline can be seen from the valley floor. This is a classic site for studying desert landforms.

Interpreting the Map

1. Describe the general topography of the map area.
2. What is the lowest elevation on the map? How is it possible for the floor of Death Valley to lie below sea level yet not be submerged under ocean water?
3. What specific type of arid region landform is represented by the blue-striped area located in the zone labeled depression? What evidence did you use to come to that conclusion?
4. What surface materials likely compose the feature in Question 3?
5. What is a likely cause of the salinity of Salt Creek?
6. Consider the stream system that flows onto the map area from the northeastern (upper right) corner of the map. What factors cause that stream to end? Do you see other streams on the map that just stop like that?
7. Do you think the streams portrayed on this map have perennial, intermittent, or ephemeral flow? Why?
8. Consider the large zone of almost parallel, curved contours in the south-central part of the map. Does that feature appear to be of fluvial or eolian origin? What specific landform is it? How many other examples of this landform appear on the map?
9. The map region lies at about 36°N. Considering its latitude and the region's topography, why does Death Valley have such a hot and dry climate?
10. Sketch a general east–west profile from the benchmark (elevation −270 ft) at Devil's Speedway to the 2389-foot benchmark on the mountain straight to the west. Label the following landforms: mountain front, mountain peak, alluvial fan, and basin.

A digital terrain model showing the topography of Death Valley.

Opposite:
Furnace Creek, California
Scale 1:62,500
Contour interval = 80 ft

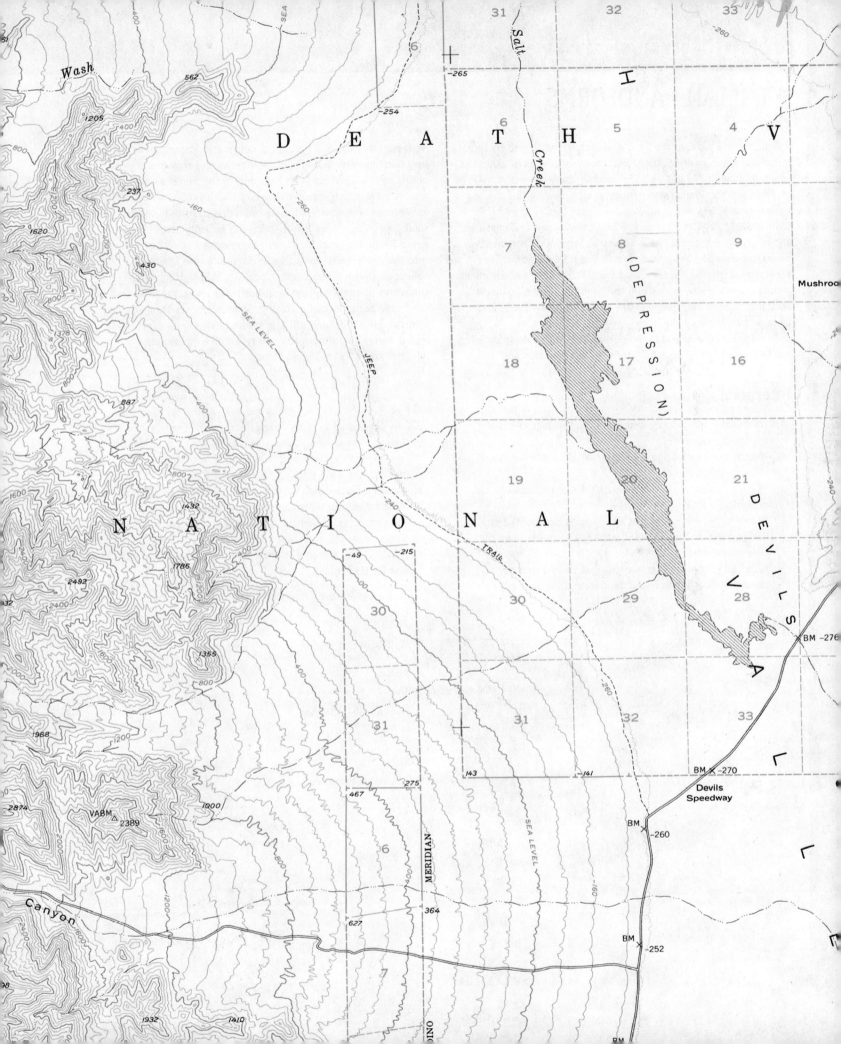

Map Interpretation

■ EOLIAN LANDFORMS

The Map

Eolian processes formed the Sand Hills region, the largest expanse of sand in North America. The region covers more than 52,000 square kilometers (20,000 sq mi) of central and western Nebraska.

The Sand Hills region was part of an extensive North American desert some 5000 years ago. The sand dunes here reached more than 120 meters (400 ft) high and inundated postglacial peat bogs and rivers. As the climate became wetter, vegetation invaded the dunes, greatly reducing eolian erosion and transportation. The vegetation anchored the sand, and the stabilized dunes developed a more rounded form. Underlying the Sand Hills is the Ogallala aquifer. The high water table of the aquifer supports the many lakes nestled between the dunes.

The Sand Hills area has a midlatitude steppe climate (BSk) and receives about 50 centimeters (20 in.) of precipitation annually.

Temperatures have a great annual range, from freezing winters to very hot summers. During summer months, the region is often pelted by thunderstorms and hail; during the winter months, the area is subject to blinding blizzards.

Vegetation is mainly bunch grasses that can survive on the dry, sandy, and hilly slopes. Some species of bunch grasses have extensive root systems that may extend more than 1 meter (3 ft) into the sandy soil. The lakes and marshes in the interdunal valleys support a marsh plant community that in turn supports thousands of migratory and local birds. Currently, the main land use in the region is cattle grazing. Some scientists are predicting that the Sand Hills area will lose its protective grass cover if global warming continues and will revert to an active area of migrating desert sand dunes.

Interpreting the Map

1. What is the approximate relief between the dune crests and the interdunal valleys?
2. What is the general linear direction in which the dunes and valleys trend?
3. To which direction do the steepest sides of the dunes generally face?
4. Knowing that the slip face is the steepest slope of the dunes, determine what the prevailing wind direction was when the dunes were active.
5. Based on your answers to the previous three questions, determine what type of sand dune formed the Sand Hills.

6. Use the elevation of the lakes to determine the water table elevation at several points. Use that information to identify the general direction of groundwater flow in the aquifer beneath the Sand Hills.
7. Sketch a north–south profile across the middle of the map from School No. 94 to the eastern end of School Section Lake (in section 16). Label the following landform features: dune crests, dune slip faces, interdunal valleys, and lakes.
8. What cultural features on the map indicate the dominant land use for the region?

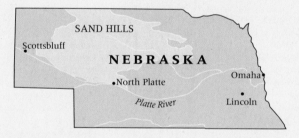

Location map of the Nebraska Sand Hills, which cover almost one third of the state.

Opposite:
Steverson Lake, Nebraska
Scale 1:62,500
Contour interval = 20 ft
USGS

GLACIAL SYSTEMS AND LANDFORMS

19

OBJECTIVES

WHEN YOU COMPLETE THIS CHAPTER YOU SHOULD BE ABLE TO:

- 19.1 Appreciate the role of glaciers in Earth's hydrologic budget.
- 19.2 Explain what a glacier is and how glacial ice forms.
- 19.3 Differentiate between the different types of alpine and continental glaciers.
- 19.4 Discuss the ways in which glaciers flow.
- 19.5 Draw a cross section of an alpine glacier and identify its major surface and subsurface zones.
- 19.6 Understand the processes of glacial erosion.
- 19.7 Define each of the major glacial erosional landforms.
- 19.8 Relate the principal characteristics of glacial sediments.
- 19.9 Distinguish among the various glacial depositional landforms.
- 19.10 Discuss what Earth's glacial environment was like during the Pleistocene Epoch and how that environment influences people today.

LARGE MASSES OF FLOWING ICE, which are termed **glaciers**, play several important roles in the Earth system. They are excellent climate indicators because certain environmental conditions are required for glaciers to exist and they respond visibly to climate variation. Glaciers become established, expand, contract, and disappear in response to changes in climate. Their long-term storage of freshwater as ice has a tremendous impact on the hydrologic cycle and the oceans, and the accumulation of ice by glaciers provides a record of past climates that can be studied from ice cores. Where glaciers once existed or where they were once considerably larger than they are today, much evidence can be found concerning past climate conditions.

The processes of erosion, transportation, and deposition by glaciers, whether ongoing or in the past, leave a distinctive stamp on a landscape. The term **glaciation** refers to the existence and actions of moving ice as well as its effects on the landscape. Some of the most beautiful and rugged terrain in the world exists in mountainous and other highland regions that have been sculpted by glaciers. Virtually every high-mountain region in the world displays glacial landscapes,

Ruth Glacier, Denali National Park, Alaska.   National Park Service

including the Alps, the Rocky Mountains, the Himalayas, and the Andes. Glaciers have also carved impressive steep-sided coastal valleys in Norway, Chile, New Zealand, and Alaska. Rugged mountain peaks rising high above lake-filled valleys or narrow and deep oceanic embayments create the ultimate in scenic appeal for many people.

Masses of moving ice have transformed the appearance of high mountains and large portions of continental plains into distinctive glacial landscapes. The flowing ice of glaciers is an effective and spectacular geomorphic agent on major portions of Earth's surface.

Glacier Formation and the Hydrologic Cycle

Glaciers are masses of flowing ice that have accumulated on land in areas where the annual input of frozen precipitation, especially snowfall, has exceeded its yearly loss by melting and other processes. Snow falls as hexagonal ice crystals that form flakes of intricate beauty and variety. Snowflakes have a low density (mass per unit volume) of about 0.1 gram per cubic centimeter (0.06 oz/in.³). Once snow accumulates on the land it becomes transformed by compaction, along with melting and refreezing at pressure points, into a mass of smaller, rounded grains (● Fig. 19.1).

● FIGURE 19.1 The transformation of frozen water from snow to glacial ice changes the size and shape of the crystals and greatly increases the material density.

How does firn differ from snow?

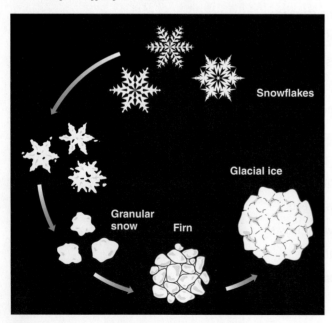

Density increases as the air space around this more granular snow continues to decrease by compaction and melting and refreezing. Through melting, refreezing, and pressure caused by the increasing weight from burial under newer snowfalls, the granular snow compacts further into a denser, crystalline granular stage, known as **firn**, with a density of about 0.5 gram per cubic centimeter (0.29 oz/in.³). Over time, the small firn granules grow together into larger interlocked ice crystals through pressure, partial melting, and refreezing. When the ice is deep enough and has a density up to about 0.9 gram per cubic centimeter (0.52 oz/in.³), it becomes glacial ice. Pressure from burial under many layers of snow, firn, and ice causes the glacial ice below to become plastic and flow outward or downward away from the area of greatest snow and ice accumulation.

Glaciers are open systems with input, storage, and output of material. Any addition of frozen water to a glacier is termed **accumulation**. Most accumulation occurs in winter and consists of snowfall onto a glacier, but there are many ways that frozen water can accumulate, including other forms of precipitation onto the ice, atmospheric water vapor freezing directly onto the ice (gas to solid), and transportation of snow and ice to a glacier surface from surrounding terrain by the wind as *snowdrift* and by avalanches. Snowdrift has been an important source of frozen water to glaciers in such highlands as the Colorado Rockies and the Ural Mountains in Russia.

Ablation is the opposite of accumulation, representing any removal of frozen water mass from a glacier. Although most ablation is accomplished in summer through melting, glaciers also lose mass by direct change from ice to water vapor (solid to gas) and by other processes that may be at work. **Calving**, for example, refers to the loss of large chunks of ice from a glacier to an adjacent lake or ocean, where the chucks float away as icebergs (● Fig. 19.2).

A glacial system is controlled by two basic climatic conditions: frozen precipitation and freezing temperatures. First, to establish a glacier, there must be sufficient mass input (accumulation) to exceed the annual loss through ablation. Mountains along midlatitude coastlines and high mountains near the equator can support glaciers because of heavy orographic snowfalls, despite intense sunshine and warm climates in the surrounding lowlands. Some very cold polar regions in subarctic Alaska and Siberia and a few valleys in Antarctica have no glaciers because the climate is too dry.

The second climatic condition that affects a glacier is temperature. Summer temperatures must not be high for too long, or all of the accumulation (primarily snowfall) from the previous winter will be lost through ablation (primarily melting). Surplus snowfall is essential for glacial formation because it allows the pressure from years of accumulated snow layers to transform older buried snow first into firn and then into glacial ice. When the ice reaches a depth of about 30 meters (100 ft), a pressure threshold is reached that enables the solid, glacial ice to flow.

Glaciers are an important part of Earth's hydrologic cycle and are second only to the ocean in the amount of water they contain. Approximately 2.25% of Earth's total water is currently frozen in glaciers. This frozen water, however, makes up about 70% of the world's freshwater, with the vast majority stored on Greenland and

NASA/GSFC/MITI/ERSDAC/JAROS, and U.S./Japan ASTER Science Team

(a)

Ted Scambos and Rob Bauer, NSIDC

(b)

● **FIGURE 19.2** Icebergs originate as large chunks of ice that calve (break off) from the leading edge of a glacier into the ocean or a large lake. (a) Ellesmere Island, Canada. (b) Melville and Mapple Glaciers, Antarctica.

Antarctica. The total amount of ice is even more impressive if we estimate the water that would be released if all of the world's glaciers were to melt. Sea level would rise about 65 meters (215 ft). This would change the geography of the planet considerably. In contrast, if another ice age were to occur, sea level would drop drastically. During the last major glacial expansion, which occurred late in the Pleistocene Epoch, sea level fell about 120 meters (400 ft).

Unlike the water in a stream system, much of which returns rapidly to the sea or atmosphere, water that becomes glacial ice is stored for a long time in a system that flows much more slowly. Glaciers may store water as ice for hundreds or even hundreds of thousands of years before it is released as meltwater. Yet, glacial ice is not stagnant. It moves slowly, but with tremendous energy, across the land. Glaciers reshape the landscape by engulfing,

● **FIGURE 19.3** Landform evidence of past glaciations includes broad, trough-shaped valleys that no longer contain the glaciers that carved them. Glaciers that eroded this trough originated at higher elevations in the Beartooth Mountains of Montana and Wyoming.

eroding, pushing, dragging, carrying, and finally depositing rock debris, often in places far from its original location. Long after glaciers recede from a landscape, glacial landforms remain as a reminder of the energy of the glacial system and as evidence of past climates (● Fig. 19.3).

Glaciers have not existed on the planet during most of Earth history. A period during which significant areas of the middle latitudes are covered by glaciers is an *ice age*. Today glaciers cover about 10% of Earth's land surface at high latitudes and high elevations on all continents except Australia. During recent Earth history, from about 2.6 million years ago to about 10,000 years before the present, which is the Pleistocene Epoch, glaciers periodically covered nearly a third of Earth's land area. Other ice ages occurred in the much more distant geologic past.

Types of Glaciers

The two major categories of glaciers are alpine and continental. **Alpine glaciers** exist where the precipitation and temperature conditions required for glacier formation result from high elevation. Alpine glaciers are fed by ice and snow in mountain areas and usually occupy valleys that were previously created by stream erosion. The ice masses flow downslope because of their own weight, that is, by the force of

gravity. Alpine glaciers that occupy former stream valleys are **valley glaciers** (● Fig. 19.4). They are known as **piedmont glaciers** when the ice extends to lower elevations beyond the mouth of a canyon, where the ice spreads out over more open terrain. The smallest type of glacier are the alpine **cirque glaciers**. These are restricted to distinctive steep-sided amphitheater-like depressions, called **cirques**, which are eroded by the ice near high peaks at valley heads (● Fig. 19.5). Alpine glaciers begin as cirque glaciers at the start of an ice age, become valley glaciers when they expand into the valley below the cirque, and may eventually become piedmont glaciers as the ice age intensifies. Most cirque glaciers today represent small remnants of previously larger alpine glaciers.

Alpine glaciers created the characteristic rugged scenery of much of the world's high-mountain regions. Today alpine glaciers are found in the Rockies, the Sierra Nevada, the Cascades, the Olympic Mountains, the Coast Ranges, and numerous Alaskan ranges of North America. They also exist in the Andes, the Alps, the Southern Alps of New Zealand, the Himalayas, the Pamirs, and other high Asian mountain ranges. Small alpine glaciers are even found at high elevations on tropical mountains in New Guinea and in East Africa on Mounts Kenya and Kilimanjaro, although most are shrinking rapidly. The largest alpine glaciers currently in existence are located in Alaska and the Himalayas, where some reach lengths of more than 100 kilometers (62 mi).

● **FIGURE 19.4** These valley glaciers in Alaska occupy preglacial stream valleys, with tributary glaciers merging together downslope like the preglacial streams did.

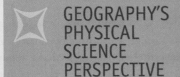
Glacial Ice Is Blue!

When we make ice in our freezers, clear colorless water turns to relatively clear ice cubes. The ice cubes may contain some white crystalline forms and air bubbles, but in general the ice is clear. In nature, the process of making ice is very different from that of an icemaker in a refrigerator–freezer. As snow falls at colder, higher latitudes and elevations, it forms a layer of snow on the surface. Each successive snowfall makes another layer as it piles onto the previous snowpack. The weight of the successive layers of snow creates pressure that compresses the older layers beneath. Through time, the layers of low-density snow become layers of much denser solid ice. Some of this change is the result of compaction, but pressure also causes some melting and refreezing of the ice. The temperature at which ice melts is 0°C (32°F) at atmospheric pressure, but ice can melt at lower temperatures if it is under enough pressure. Both compaction and pressure melting and refreezing work to reduce the amount of air in the frozen mass and thereby increase the material's density.

Objects that appear white to the human eye reflect all wavelengths of light with equal intensity, and this is what the hexagonal crystalline structure of a snowflake does. As the snow strata under great pressure in a glacier become compacted over the years (sometimes hundreds or thousands of years) the ice becomes denser. Basically, under this pressure, more ice crystals are squeezed into the same volume. As the density of ice increases, it reflects increasing amounts of shorter wavelengths of light, which is the blue part of the spectrum. The denser the ice, the bluer it appears. Ice density can be influenced by factors other than time, though, so we must be careful not to assume that bluer layers in a glacier are necessarily the older layers. For example, the packing of higher-density wetter snow as opposed to lower-density drier snow can affect the density of specific layers. Nevertheless, what is certain when looking at massive glacial ice accumulations in nature, such as in ice caps and ice sheets, is that the ice will appear as shades of blue.

NASA/Jim Rossi

An iceberg revealing very old layers of glacial ice.

The second category of glacier, **continental glaciers**, are much larger and thicker than the alpine types and exist where the appropriate conditions for ice formation occur because of high latitude (● Fig. 19.6). Like alpine glaciers, continental glaciers were much more extensive in the Pleistocene Epoch than they are today, at one time covering as much as 30% of Earth's land area. Continental glaciers are subdivided by size. Earth's two massive polar **ice sheets**, the largest type of glacier, bury Greenland and Antarctica to a maximum depth of at least 3 kilometers (2 mi). Ice masses similar to ice sheets but smaller than 50,000 square kilometers (19,000 sq mi) in extent are **ice caps**, which are present on Iceland and some Arctic islands. In contrast to alpine glaciers, ice sheets and ice caps more or less drown the underlying topography in ice, rather than being confined or directed by the topography. Direction of movement within ice sheets and ice caps is from thicker to thinner ice, which is radially outward in all directions from a central source area of maximum ice thickness.

● **FIGURE 19.5** A cirque glacier in Alaska.

Why is only part of the ice a bright white color?

● **FIGURE 19.6** Only the highest mountain peaks and ridges, such as these in Antarctica, project above the ice of continental glaciers.

How Do Glaciers Flow?

Like the slow forms of mass wasting, we normally cannot view glacier movement directly. Nevertheless, flowing ice has a tremendous geomorphic impact on the landscape. Most glaciers move through a combination of processes, but *internal plastic deformation* is the dominant process; virtually all moving glaciers experience this type of flow. Glaciers move in this plastic way when the weight of overlying ice, firn, and snow causes ice crystals at depth to arrange themselves in parallel layers that glide over each other, much like spreading a deck of cards (● Fig. 19.7). This internal plastic deformation happens when a threshold pressure (weight per unit area) from the overlying mass is exceeded. The threshold pressure is achieved at an ice thickness of about 30 meters (100 ft), and the zone experiencing plastic flow extends from that ice depth to the base of the glacier. The speed at which the ice flows increases as pressure from overlying material grows and with steeper slopes. Pressure is greater under thicker accumulations of ice and on the upflow side of obstacles at the base of a glacier. Internal plastic deformation causes continental glaciers to flow radially outward from central areas of thicker ice (higher pressure) to marginal areas of thinner ice (lower pressure).

In addition to internal ice movement by plastic deformation, many glaciers also move downslope by processes concentrated at the base of the ice mass. Many temperate glaciers—those with temperatures at and near the melting point—undergo *basal sliding*. In this process, meltwater at the base of the glacier reduces friction between the ice and ground by means of lubrication and hydrostatic pressure (see Fig. 19.7). When frictional resistance becomes too small to resist the downslope pull of gravity on the ice mass, the affected part of the glacier jerks forward. This type of motion is most common in midlatitude glaciers on steep slopes, particularly during summer when much of the glacier is near the melting point and meltwater is available. Little if any basal sliding occurs during winter and in the colder, polar glaciers, which have little available meltwater. A different type of basal ice movement involves local melting at the base of the glacier, downslope flow of the meltwater, and then its refreezing onto the glacier base.

Lacking sufficient weight of overlying ice, ice at and near the surface of a glacier cannot deform plastically and behaves instead in a brittle manner. This upper, brittle layer of a glacier moves by being carried along with the plastically flowing ice beneath it. Ice in the brittle zone fractures and cracks as it is carried along. These ice cracks, called **crevasses**, are common wherever a glacier becomes stretched, experiencing tensional stress, particularly where it moves over a break in slope (● Fig. 19.8). Where a glacier locally flows over a steep descent, such as over a subglacial cliff, an *icefall* develops in the brittle upper zone (● Fig. 19.9). Here, intersecting crevasses break the brittle ice into a morass of unstable blocks that ride on rapid flowing ice below. Icefalls and crevasses are extremely dangerous areas for anyone venturing onto the ice.

Glacier flow rates vary from imperceptible fractions of a centimeter per day to as much as 30 meters (100 ft) per day. Glaciers flow more rapidly where the slope is steeper, ice is thicker, and temperatures are warmer. The Nisqually Glacier, on the steep slopes of Washington's Mount Rainier, flows 38 centimeters

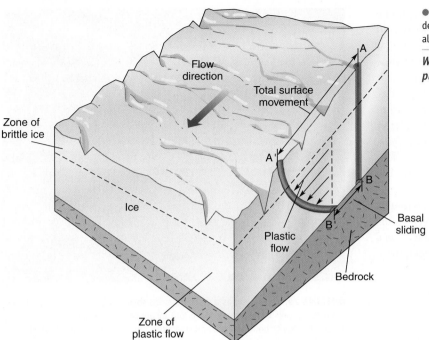

● **FIGURE 19.7** Most glacier movement is by internal plastic deformation of ice at depth, but glaciers with meltwater at their base also move by basal slippage.

Why does surface ice move farthest even though internal plastic deformation is only occurring at depth in the ice?

● **FIGURE 19.9** Icefalls are the glacial equivalent of rapids or waterfalls in a river and are riddled with crevasses that break the ice into huge, unstable blocks.

● **FIGURE 19.8** A large crevasse on the Yanert glacier in Denali National Park, Alaska.

What type of force causes crevasses, compressional or tensional?

M. Trapasso

Copyright and photograph by Dr. Parvinder S. Sethi

(15 in.) per day in summer. As a general rule, temperate alpine glaciers, like the Nisqually Glacier, flow much faster than cold polar continental glaciers.

The flow of any individual glacier varies from time to time with changes in its dynamic equilibrium and from place to place because of variations in the gradient over which it flows or differences in the friction encountered with adjacent rock. Within an alpine glacier, the rate of movement is greatest on the glacier surface toward the middle of the ice because this location experiences accumulated movement from layers of plastic flow below and is farthest from frictional resistance encountered along the valley sides.

Sometimes a glacier's velocity will increase by many times its normal rate, causing the glacier to flow hundreds or even thousands of meters per year. The reasons for such enormous *glacial surges* are not completely clear, although meltwater reducing friction at the base of the ice is probably involved.

Glaciers as Geomorphic Agents

Because a deep, and therefore heavy, accumulation of ice is required for glaciers to flow, even the smaller alpine glaciers are particularly powerful geomorphic agents able to perform great amounts of geomorphic work. Whether it is an alpine glacier carving out a trough-shaped valley or a continental glacier gouging out the basins of the North American Great Lakes, the work done by glaciers is impressive.

Glaciers remove and entrain rock particles by two erosional processes. Glacial **plucking** is the process by which moving ice freezes onto loosened rocks and sediments, incorporating them into the flow. Weathering, particularly the freezing of water in bedrock joints and fractures, breaks rock fragments loose, encouraging plucking. Once load is entrained at the base and sides of the ice, moving glaciers are armed with clastic particles that are very effective tools for scraping and gouging out more rock material by the erosional process of *abrasion* (● Fig. 19.10). Bedrock obstructions subjected to intense glacial abrasion are typically smoother and more rounded than those produced by plucking.

Unlike the situation with liquid water in streams, volume and velocity of flow do not directly determine the particle sizes that plastically flowing, solid ice can erode and transport. Plucking and abrasion provide the bottom and sides of glaciers with a chaotic load of rock fragments of all sizes, from clay-sized crushed rock, called *rock flour*, to giant boulders. Mass wasting along steep mountain slopes, especially above alpine glaciers, contributes sediment, also of a variety of grain sizes, to the ice surface and sides. Also in contrast to streams, little sorting of sediment by size is accomplished by glaciers during transportation and deposition. This lack of sorting makes glacial deposits look very different from accumulations of stream deposits. Because of this contrast in the two types of sediment, it is logical that they are referred to by two different terms. Whereas stream-deposited sediment is called *alluvium*, sediment deposited directly by moving ice is **till**.

● **FIGURE 19.10** As soon as a glacier obtains some load, those clastic particles are used as tools to help erode more rock by abrasion. Here, a cobble from a glacial deposit shows scrapes and scratches obtained by grinding against bedrock and other particles as it was carried in the ice.

Alpine Glaciers

From a mass balance perspective, alpine glaciers consist of two functional parts, or zones (● Fig. 19.11). The colder, snowier upslope portion of a glacier, where annual accumulation (input) exceeds annual ablation (output), is the **zone of accumulation**. In contrast, annual ablation exceeds annual accumulation in the warmer downslope portion of an alpine glacier, the **zone of ablation**. Winter is the dominant accumulation season and summer is the dominant ablation season. Alpine glaciers change size, sometimes dramatically, over the course of a year. The toe of a glacier occupies its farthest downvalley position near the end of winter and its farthest upvalley position at the end of the ablation season, near the end of summer.

The **equilibrium line** marks the boundary between the zones of accumulation and ablation on an alpine glacier. It indicates the position across the width of the glacier surface, at which the annual accumulation equals the annual ablation for the ice mass (● Fig. 19.12). At elevations above the equilibrium line, fresh snow and ice added to the surface of the glacier during the cooler accumulation season does not completely disappear during the warmer ablation season. Below the equilibrium line, the winter's input of fresh ice and snow ablates away in the summer. The equilibrium line differs from the constantly changing *snow line*, which is the minimum elevation at which snow cover lies on a landscape. The snow line changes position multiple times throughout the year and even daily, such as in response to individual snowstorms. The equilibrium line is the highest elevation achieved by the snow line on the glacier during the year.

Several factors influence the location of the equilibrium line. The interaction between latitude and elevation, both of which affect temperature, is an important factor. On mountains near the

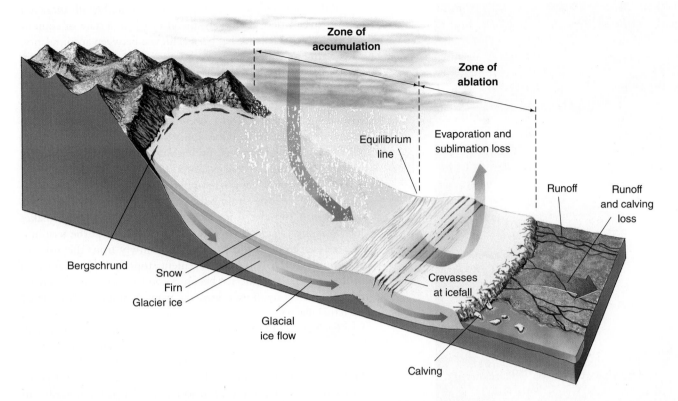

● FIGURE 19.11 Glaciers are considered a subsystem of the hydrologic system. The glacial zone of accumulation experiences greater annual input of frozen water (accumulation) than it loses (ablation) during the year. At lower elevations, in the zone of ablation, annual loss exceeds accumulation. The equilibrium line marks the elevation on the glacier where annual accumulation equals annual ablation. If the glacier as a whole experiences more accumulation than ablation in a year, ice thickness increases and the toe of the glacier advances farther downvalley. If ablation exceeds accumulation for the year, the glacier loses mass and retreats. Equilibrium exists if annual accumulation equals annual ablation for the glacier as a whole. Whether the glacier is advancing, retreating, or in equilibrium, internal plastic flow transports ice from the accumulation zone to the ablation zone.

equator, the equilibrium line lies at very high elevations. Elevation of the equilibrium line decreases with increasing latitude until it coincides with sea level in the polar regions. Equally important to temperature in determining the position of the equilibrium line is the amount of snowfall received during winter. With colder temperatures and greater snowfall, the equilibrium line will decrease in elevation; it retreats to higher elevations if the climate warms. Other attributes causing variations in the equilibrium-line elevation include the amount of insolation. A shady mountain slope will have a lower equilibrium line than one that receives more insolation. Wind is another factor because it produces snowdrifts on the leeward side of mountain ranges. In the midlatitudes of the Northern Hemisphere, the equilibrium line is lower on the north (shaded) and east (leeward) slopes of mountains. Consequently, the most significant glacier development in this region is on north-facing and east-facing slopes.

At the farthest upslope edge of an alpine glacier, the upslope end of the zone of accumulation, lies the *glacier's head*. It abuts the steep bedrock cliff that comprises the *cirque headwall*. Ice within an alpine glacier flows downslope from the zone of accumulation to the zone of ablation. This downslope movement is sometimes evidenced by a large crevasse, known as a *bergschrund*,

that may develop between the head of the glacier and the cirque headwall (see Fig. 19.11). Presence of a bergschrund shows that the ice mass is pulling away from the confining rock walls of the cirque. The downslope end of a glacier is called its *terminus*, or the *glacier toe*.

Equilibrium and the Glacial Budget

Because the location of the toe of a glacier shifts throughout the year with accumulation and ablation, to determine if an alpine glacier is growing or shrinking over a period of years requires annually noting the location of its terminus at the same time of year (● Fig. 19.13). Typically this is done at the end of the ablation season when the glacier is at its minimum size for the year. If a glacier received more input of frozen water (accumulation) during a year than was removed from it (ablation) that year, it experienced net accumulation. The result of net accumulation is a larger glacier, and as alpine glaciers grow they advance; that is, their toes extend farther downvalley. A glacier that undergoes net ablation, more removal than addition of frozen water for the year, shrinks in size causing the toe to retreat upvalley. These are referred to as **glacier advance** and **glacier retreat**, respectively.

If the annually measured toe of a glacier neither advances nor retreats over a period of years, the glacier is in a state of equilibrium in which a balance has been achieved between accumulation and ablation of ice and snow. As long as equilibrium is maintained, the location of the glacier's toe at the end of the ablation season will remain constant.

It is crucial to understand that whether an alpine glacier is advancing, retreating, or in a state of equilibrium, the ice comprising it continues to flow downslope. Even a glacier that is retreating over the long term will receive winter snow, with more at its higher than at its lower elevations. The weight per unit area of this frozen water drives glacier movement through internal plastic deformation. Winter snow buried by additional snow eventually turns to firn and glacial ice and over a period of decades makes its way along the glacier to its terminus. Most ice at the terminus of a glacier today made its way there slowly from the zone of accumulation. Downslope movement stops only if net ablation proceeds so far that the ice becomes too thin to maintain plastic flow. In fact, if ice flow ceases, the mass is no longer considered an active glacier.

Each glacier behaves somewhat differently according to details of its geographic position, climatic environment, topographic setting, history, size, and mass balance. For example, Alaska's Hubbard Glacier has been advancing during the last 100 years. In 1986 and 2002, it advanced so rapidly that it temporarily cut off Russell Fjord from the ocean. Just a few hundred kilometers away, the giant Columbia Glacier has been rapidly receding. Most of the world's glaciers have retreated, and some have disappeared entirely, since the early twentieth century. Rates of glacier recession have been especially rapid since the 1980s, and programs to monitor the size of glaciers have intensified. Considerable attention and scientific effort are currently focused on the nature of the relationship between glacier behavior and human-induced global warming.

Erosional Landforms of Alpine Glaciation

Glacial abrasion leaves **striations**—linear scratches, grooves, and gouges—where sharp-edged rocks scrape across bedrock (● Fig. 19.14). Striations indicate direction of ice flow long after the ice has disappeared from the landscape. Abrasion and plucking at the base of a glacier work together to form **roches moutonnées** (from French: *roche*, rock; *mouton*, sheep), asymmetric bedrock hills or knobs that are smoothly rounded on the up-ice side by abrasion, with plucking evident on the abrupt down-ice side (● Fig. 19.15).

When an alpine glacier first develops in a hollow high in the mountains, its small size results in almost rotational flow lines for the moving ice. In the zone of accumulation at the glacier's head, ice movement has a large downward component. For ice to reach the toe in the zone of ablation a short distance away, an upward component to the flow may be required there. Ice movement, accompanied by weathering and mass wasting, steepens the bedrock wall at the head of the small glacier while it deepens the

USGS/Austin Post

● **FIGURE 19.12** A valley glacier on Alaska's Kenai Peninsula displays a blue zone undergoing ablation at lower elevations contrasted with a white zone where ablation has not occurred at the glacier's higher elevations. The visible contact between the two zones on the photograph is not necessarily the glacier's equilibrium line, which represents the elevation at which accumulation equaled ablation for the year.

What additional information would be needed to assess if the boundary between the white and blue zones on this photo is the glacier's annual equilibrium line?

● **FIGURE 19.13** The Jakobshavn Glacier in Greenland has been retreating since scientists began monitoring it in 1850. The colored lines mark former positions of the glacier's terminus.

What was the average annual rate of retreat from 1851 to 2010?

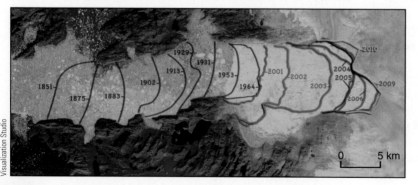

NASA/Goddard Space Flight Center Scientific Visualization Studio

● **FIGURE 19.14** Glacial abrasion produces smooth rock surfaces that are cut by striations (scratches and grooves) oriented parallel to the direction of ice flow.

Can the direction of ice flow be determined with certainty from the evidence in this photograph?

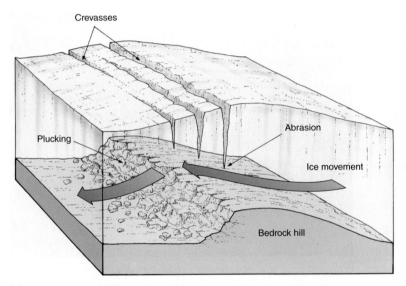

Crevasses

Plucking

Abrasion

Ice movement

Bedrock hill

(a)

(b)

● **FIGURE 19.15** (a) Formation of a roche moutonnée (sheepback rock), by glacial abrasion on the up-ice side of a bedrock hill and plucking on the down-ice side. (b) An example of a roche moutonnée in Yosemite National Park, California.

Why would crevasses form in ice flowing over this feature?

hollow to form the amphitheater-shaped depression that is the cirque. In fact, in many mountain areas cirques that no longer contain ice are formally or informally referred to as "bowls" because of their distinctive shape. Cirque glaciers often deposit a ridge of till at the down-ice edge of the cirque, and this accentuates the bowl-shaped appearance. When the ice disappears as a result of climate change, the erosional cirque is left behind, often forming a natural basin in which water accumulates (● Fig. 19.16a). Lakes that form in cirques are known as **tarns**.

Separate alpine glaciers often develop in adjacent mountainside hollows or valleys. As the cirques or valleys of two adjacent glaciers enlarge, the bedrock ridge between them will be eroded into a jagged, sawtoothed spine of rock, termed an **arête** (from French: *arête*, bone) (Fig. 19.16b). Where three or more cirques surround a mountain summit, headward erosion eventually carves the high ground between them into a pyramid-shaped peak, called a **horn** (Fig. 19.16c). The Matterhorn in the Swiss Alps is a prime example. A **col** (from Latin: *collum*, neck) is a pass formed when headward erosion causes two cirques to intersect, producing a low saddle in a high mountain ridge or arête.

As they expand downslope out of their cirques, alpine glaciers take over the downslope pathways established by streams before the ice accumulated. Steep mountain streams carve valleys that in cross section resemble the letter V. Because glaciers are much thicker than streams are deep, they erode the sides in addition to the bottom of these valleys, bowing the cross-sectional shape out to a U-shaped **glacial trough** (Fig. 19.16d). In addition, a glacier's tendency to flow straight ahead rather than to meander causes it to straighten out the preexisting valley that it occupies.

Landscapes eroded by alpine glaciers show a sharp contrast between the deep, U-shaped glacial troughs scoured smooth by ice flow and the jagged peaks above the former ice levels. The rugged quality of the upper peaks and ridges is caused primarily by mechanical weathering above the ice surface and by powerful glacial erosion below, which steeply undercuts the bedrock to fashion horns and arêtes (● Fig. 19.17).

● FIGURE 19.16 Alpine glaciation produced each of these erosional landforms. (a) A cirque is a bowl-shaped depression carved at the head of the glacier. (b) Jagged, narrow ridges of rock between cirques or glacial valleys are arêtes. (c) A horn, in this case the Matterhorn in the Swiss Alps, forms where several glaciers cut headward into a mountain peak. (d) Little Cottonwood Canyon, in the mountains east of Salt Lake City, is an excellent example of a glacial trough.

Can you identify on the photo in (d) the height to which ice filled the valley?

By preferentially eroding weaker rocks on the valley floor, some alpine glaciers carve a sequence of rock steps and excavated basins. While the ice is still present, crevasses or even icefalls develop on the glacier's surface where it flows over the steps. When the ice recedes, rockbound lakes sometimes fill the basins, often looking like beads connected by a stream flowing down the glacial trough. Because these chains of lakes reminded early glacial scholars of a type of prayer beads, called paternoster beads, they named them **paternoster lakes**.

Most large valley glaciers have smaller *tributary glaciers*. Like the main ice stream, tributary glaciers also carve U-shaped valleys, or troughs (● Fig. 19.18). During peak glacial phases, the surface of a tributary glacier merges smoothly with that of the trunk glacier.

However, because many tributary glaciers have much smaller ice volumes than the main glacier, they are not always able to erode their troughs as deeply as the trunk glacier does. Tributary glacial troughs tend to be smaller and not as deep as the trough of the main glacier. The difference in elevation between the tributary and trunk valley floors is not apparent when the ice is present, but once the ice is gone floors of tributary troughs can be seen perched as **hanging valleys** along the walls of the trunk glacial valley (● Fig. 19.19). A stream that flows down a hanging valley will cascade to the lower glacial valley by a high waterfall or a series of steep rapids. Yosemite Falls and Bridalveil Falls in Yosemite National Park are excellent examples of waterfalls cascading out of hanging valleys into the main glacial trough, which is Yosemite

● **FIGURE 19.17** The alpine glacial topography of the north-central Chugach Mountains, Alaska, includes numerous cirques, arêtes, horns, and deposits of alpine glaciation.

Valley. ● Figure 19.20 shows an example of how alpine glaciation exploits and alters preexisting fluvial topography, leaving behind glacial erosional landforms.

In coastal locations at higher latitudes, many glacial troughs today extend down below sea level. These troughs were carved during the Pleistocene glaciation when sea level was lower than it is today. As sea level rose and the glaciers retreated landward through net ablation because of climate change, the ocean invaded the abandoned glacial troughs, making deep, narrow ocean inlets known as **fjords** (● Fig. 19.21). The fjords of Scotland, Norway, Iceland, and New Zealand show that glaciers in those regions reached the sea during the Pleistocene.

● **FIGURE 19.18** A system of tributary glaciers carved the tributary glacial troughs that exist today in the Ruby Mountains of northeastern Nevada.

● **FIGURE 19.19** At the top of the cliff, Bridalveil Falls in Yosemite National Park cascades off the floor of a U-shaped hanging valley toward Yosemite Valley below.

Depositional Landforms of Alpine Glaciation

Like glacial load, from which it is derived, glacial deposits include clastic sediments of a wide range of sizes, frequently mixed with layers of plant matter and soil. In addition to the poorly sorted till deposited directly by glacial ice, meltwater streams, lakes, and wind occurring in association with glaciers contribute to the deposition of sediments and creation of landforms in glacial terrain. *Glaciofluvial* is used to specify the better sorted and stratified fluvial deposits related to glacial meltwater. All deposits of glacial ice or its meltwater, and therefore including till and glaciofluvial deposits, are included within the more general term **drift** (● Fig. 19.22).

Active alpine glaciers deposit load primarily along the sides and toe of the ice mass. Landforms constructed from glacial deposits, typically ridges of till along these margins of glaciers, are *moraines*. Till deposited as ridges paralleling the side margins of a glacier are **lateral moraines** (● Fig. 19.23a). Where two tributary valley glaciers join, their lateral moraines merge downflow

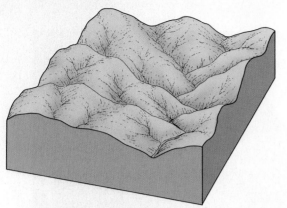

(a) Preglacial fluvial topography

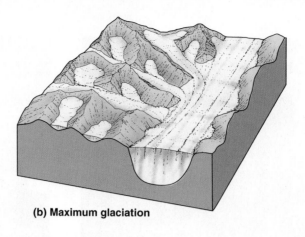

(b) Maximum glaciation

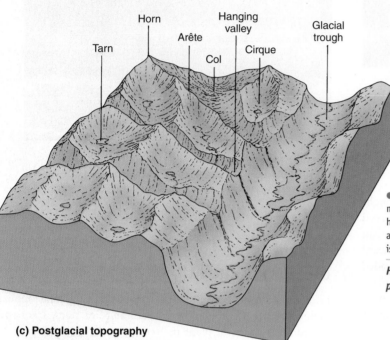

Tarn — Horn — Arête — Col — Hanging valley — Cirque — Glacial trough

(c) Postglacial topography

● **FIGURE 19.20** (a) Preglacial landscape of upland hollows and mountain stream valleys. (b) Alpine glaciers originate in upland hollows and advance down preexisting stream valleys as the ice age intensifies. (c) Geomorphic work accomplished by the glaciers is evident in the postglacial landscape.

How do the stream valley cross sections change from preglacial to postglacial time?

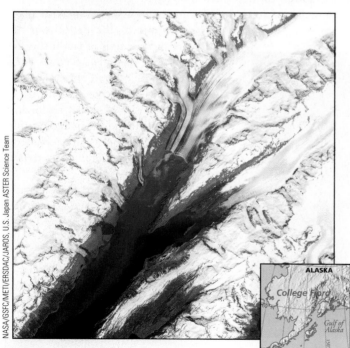

ALASKA
College Fjord
Gulf of Alaska

● **FIGURE 19.21** Glaciers carved the deep College Fjord, seen here in a satellite image, in northwestern Alaska.

How many glaciers do you see on this image?

● **FIGURE 19.22** Glacial till, here deposited by an alpine glacier in the Sierra Nevada, consists of an unsorted, unstratified, rather jumbled mass of gravel, sand, silt, and clay.

Why does till have these disorganized characteristics?

creating a **medial moraine** in the center of the trunk glacier. Medial moraines cause the characteristic dark stripes seen on the surface of many alpine glaciers (Fig. 19.23b). At the toe of a glacier, sediment carried forward by the "conveyor belt" of ice or pushed ahead of the glacier is deposited in a jumbled heap of material of all grain sizes, forming a curved depositional ridge called an **end moraine** (Fig. 19.23c). End moraines that mark the farthest advance of a glacier are *terminal moraines*. A *recessional moraine* is an end moraine deposited as a consequence of a temporary pause by a retreating glacier. A retreating glacier also deposits a great deal of till on the floor of the glacial trough as the ice melts away and leaves its load behind. The hummocky landscape created by these glacial deposits is **ground moraine**.

Braided meltwater streams laden with sediment commonly issue from the glacier terminus (● Fig. 19.24). This sediment is deposited as **glacial outwash** beyond the terminal moraine of the alpine glacier, with larger rocks deposited first, followed downstream by progressively finer particles. Often resembling an alluvial fan confined by valley walls, the depositional landform composed of glacial outwash left by braided streams is known as a **valley train**. Valleys in glaciated regions may be filled with drift deposited in valley trains, moraines, or even moraine-dammed lakes to depths of a few hundred meters.

(a)

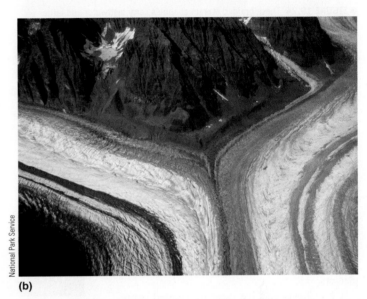

(b)

(c)

● **FIGURE 19.23** (a) Prominent lateral moraines on the Kenai Peninsula of Alaska. (b) Where two valley glaciers flow together, their adjoining lateral moraines merge into a medial moraine, creating a dark stripe of till in the larger glacier that they create. (c) An end moraine of a glacier in Pakistan.

What can we learn from studying moraines?

● **FIGURE 19.24** These braided stream channels carry meltwater and large amounts of outwash derived from glaciers in Denali National Park, Alaska.

Why are braided channels often associated with the deposition of glacial outwash?

Continental Glaciers

In terms of their size and shape, continental glaciers, which consist of ice sheets or the somewhat smaller ice caps, are very different from alpine glaciers. However, all glaciers share certain characteristics and processes, and much of what we have discussed about alpine glaciers also applies to continental glaciers. The geomorphic work of the two categories of glaciers differs primarily in scale, attributable to the enormous disparity in size between continental and alpine glaciers.

Ice sheets and ice caps are shaped somewhat like a convex lens in cross section, thicker in the center and thinning toward the edges. They flow radially outward in all directions from where the pressure is greatest, in the thick, central zone of accumulation, to the surrounding zone of ablation (● Fig. 19.25). Like all glaciers, ice sheets and ice caps advance and retreat by responding to changes in temperature and snowfall. As with alpine glaciers flowing down preexisting stream valleys, movement of advancing continental glaciers takes advantage of paths of least resistance found in preexisting valleys and belts of softer rock.

Existing Continental Glaciers

Glaciers of all categories currently cover about 10% of Earth's land area. In area and ice mass, alpine glaciers are almost insignificant compared to continental glaciers. The huge ice sheets of Greenland and Antarctica alone account for

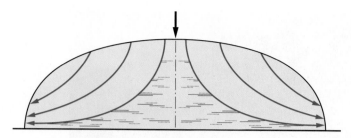

● **FIGURE 19.25** Ice in ice sheets and ice caps flows outward and downward from the center where the glacier is thickest.

How is this manner of ice flow different from and similar to that of an alpine glacier?

96% of the area occupied by glaciers today. Ice caps currently exist in Iceland, on the Arctic islands of Canada and Russia, in Alaska, and in the Canadian Rockies.

The Greenland ice sheet covers the world's largest island with a glacier that is more than 3 kilometers (2 mi) thick at its center. The only land exposed in Greenland is a narrow, mountainous coastal zone (● Fig. 19.26). Where the ice reaches the ocean, it usually does so through fjords. These flows of ice to the sea resemble alpine glaciers and are called **outlet glaciers**. The action of waves and tides on the outlet glaciers breaks off huge ice masses that float away as icebergs. These large, irregular chunks of ice are a hazard to vessels in the North Atlantic shipping lanes south of Greenland. Tragic maritime disasters, such as the sinking of the *Titanic*, have been caused by collisions with icebergs, which are nine-tenths submerged and thus mostly invisible to

THINKING GEOGRAPHICALLY

What various types of landform evidence for glaciation do you see in this photo? Provide a reasonable account of the sequence of general climatic and geomorphic changes that were responsible for creating this landscape as it appears today.

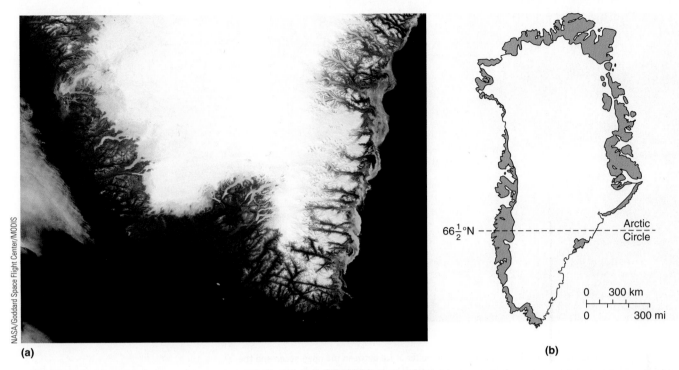

(a)

(b)

● **FIGURE 19.26** (a) Only a narrow coastal zone of Greenland is not covered by the ice sheet, which reaches a maximum thickness of more than 3 kilometers (2 mi). (b) Extent of the Greenland ice sheet.

ships. Today, with iceberg tracking by radar, satellites, and the ships and aircraft of the International Ice Patrol, deadly encounters with icebergs are much less likely.

The Antarctic ice sheet covers some 13 million square kilometers (5 million sq mi), almost 7.5 times the area of the Greenland ice

● **FIGURE 19.27** The Antarctic ice sheet covers an area larger than the United States and Mexico combined. White and blue shades on this satellite image represent ice. The only rocky areas (darker tones on the image) are the Antarctic Peninsula, the Ellsworth Mountains, and the Transantarctic Mountains.

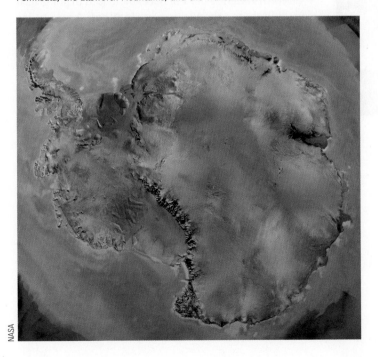

sheet (● Fig. 19.27). As in Greenland, only a small amount of land is exposed in Antarctica, and the weight of the 4.5-kilometer (nearly 3-mi) thick ice in some interior areas has depressed the land well below sea level. Where the ice reaches the sea, it contributes to the **ice shelves**, enormous flat-topped plates of ice attached to land along at least one side (● Fig. 19.28). The flat-topped ice shelves provide the region's icebergs, thus the icebergs in Antarctic waters do not have the irregular surface form of Greenland's icebergs. Because they do not float into heavily used shipping lanes, these extensive tabular-shaped Antarctic icebergs are not as much of a hazard to navigation as the Greenland icebergs (● Fig. 19.29). They do, however, add to Antarctica's accessibility problem. The huge wall of the ice shelf, massive pieces of sea ice (frozen sea water), and the extreme climate combine to make Antarctica inaccessible to all but the hardiest people and equipment. Scientists from many countries, however, travel to Antarctica to study its glaciology, climatology, geology, and ecology.

Pleistocene Glaciation

The Pleistocene Epoch of geologic time was an interval of great climate change that began about 2.6 million years ago and ended around 10,000 years before the present. A great number of glacial fluctuations occurred during the Pleistocene, marked by numerous major advances and retreats of ice over large portions of the world's landmasses. It is a common misconception that the continental ice sheets originated, and remain thickest, at the poles. When the Pleistocene glaciers advanced, ice spread outward from centers in Canada, Scandinavia, eastern Siberia, Greenland, and Antarctica, while alpine glaciers spread to lower elevations. At their maximum extent in the Pleistocene, glaciers covered nearly

© Josh Landis, National Science Foundation

(a)

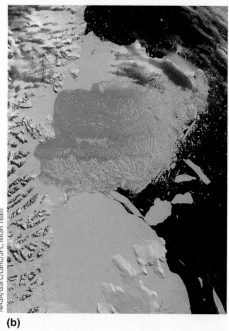

NASA/GSFC/LaRC/JPL, MISR Team

(b)

● **FIGURE 19.28** (a) Antarctica's flat-topped ice shelves form a coastal buffer of ice between the open ocean and the ice-covered continent. (b) The ice shelves have experienced significant destruction in recent years as in this collapse of Antarctica's Larsen B Ice Shelf in 2002.

Hajo Eicke/Alfred Wegener Institute for Polar and Marine Research

● **FIGURE 19.29** Antarctica's flat-topped (tabular) icebergs, here with penguin passengers, look quite different from the irregularly shaped icebergs of the Northern Hemisphere.

What portion of an iceberg is hidden below the ocean surface?

a third of Earth's land surface (● Fig. 19.30). At the same time, the extent of sea ice expanded equatorward. In the Northern Hemisphere, sea ice was present along coasts as far south as Delaware in North America and Spain in Europe. Between each period of maximum ice advance, known as a **glacial** (or *glacial period*), a warmer time called an **interglacial** occurred, during which the enormous continental ice sheets, ice caps, and sea ice

retreated and almost completely disappeared. Studies of glacial deposits and landforms have revealed that within each major glacial advance, many minor retreats and advances occurred, reflecting smaller changes in global temperature and precipitation. During each advance of the ice sheets, alpine glaciers were much more numerous, extensive, and massive in highland areas than they are today, but their total extent was still dwarfed by that of the continental glaciers.

North America and Eurasia experienced major glacial expansion during the Pleistocene. In North America, ice sheets extended as far south as the Missouri and Ohio Rivers and covered nearly all of Canada and much of the northern Great Plains, the Midwest, and the northeastern United States. In New England, the ice was thick enough to overrun the highest mountains, including Mount Washington, which has an elevation of 1917 meters (6288 ft). The ice was more than 2000 meters (6500 ft) thick in the Great Lakes region. In Europe, glaciers spread over most of what is now Great Britain, Ireland, Scandinavia, northern Germany, Poland, and western Russia. The great weight of this ice depressed the land surface several hundred meters. As the ice receded and its weight on the land was removed, the land rose by the process of *isostatic rebound*. Measurable amounts of glacial isostatic rebound are still raising elevations in parts of Sweden, Canada, and eastern Siberia by up to 2 centimeters (1 in.) per year, and it may cause Hudson Bay and the Baltic Sea to emerge someday above sea level. Should Greenland and Antarctica lose their ice sheets, their depressed central land areas would also rise to reach isostatic balance.

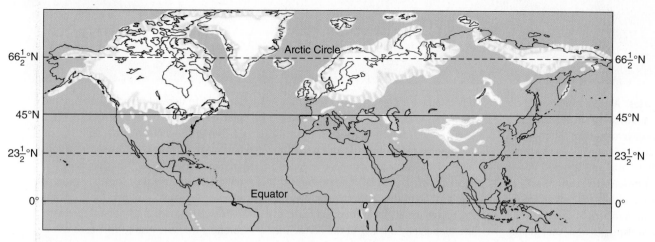

FIGURE 19.30 Glacial ice coverage in the Northern Hemisphere was extensive during the Pleistocene. Glaciers up to several thousand meters thick covered much of North America and Eurasia.

What might be a reason for some areas that were very cold during this time, such as portions of interior Alaska and Siberia, being ice-free?

The geomorphic effects of the last major glacial advance, known in North America as the Wisconsinan stage, are the most visible in the landscape today. The glacial landforms created during the Wisconsinan stage, which ended about 10,000 years ago, are relatively recent and have not been destroyed to any great extent by subsequent geomorphic processes. Consequently, we can derive a fairly clear picture of the extent and actions of the ice sheets and ice caps, as well as alpine glaciers, at that time.

Where did the water locked up in all the ice and snow come from? Its original source was the ocean. During the periods of glacial advance, sea level fell, exposing large portions of the **continental shelves**, which are the edges of the continents and which currently lie below sea level. That lowering of sea level likewise exposed land bridges across the present-day North, Bering, and Java Seas. At the end of the Pleistocene, melting of the glaciers raised sea level by a similar amount—about 120 meters (400 ft). Evidence for this rise in sea level can be seen along many coastlines around the world.

Erosional Landforms of Continental Glaciation

Ice sheets and ice caps erode the land by plucking and abrasion, carving landscape features that have many similarities to those of alpine glaciers but on a much larger scale. Erosional landforms created by ice sheets stretch over millions of square kilometers of North America, Scandinavia, and Russia, and are far more extensive than those formed by alpine glaciation. As ice sheets flowed over the land, they gouged Earth's surface with striations, enlarged valleys that already existed, scoured out rock basins, and smoothed off existing hills. The ice sheets removed most of the soil and then eroded the bedrock below. Today these *ice-scoured plains* are areas of low, rounded hills, lake-filled depressions, and wide exposures of bedrock (● Fig. 19.31).

When ice sheets expand, they cover and totally disrupt the former stream patterns. Because the last glaciation was so recent in terms of landscape development, some of the new drainage systems have not had time to form well-integrated systems of stream channels. In addition to large expanses of exposed gouged bedrock, ice-scoured plains are characterized by extensive areas of standing water, including lakes, marshes, and muskeg (poorly drained areas grown over with vegetation that form in cold climates).

● FIGURE 19.31 Seen from the space shuttle, Lake Manicouagan on the Canadian Shield in Quebec occupies a circular depression that continental glaciers erosionally enhanced in rocks surrounding an asteroid impact structure. The annular (ring-shaped) lake has a diameter of approximately 70 kilometers (43 mi).

The Driftless Area—A Natural Region

The Driftless Area, a region in the upper midwestern United States, mainly covers parts of Wisconsin but also sections of the adjacent states of Iowa, Minnesota, and Illinois. This region was not glaciated by the last Pleistocene ice sheets that extended over the rest of the northern Midwest, so it is free of glacial deposits from that time. *Drift* is a general term for all deposits of glacial ice and its meltwater, thus the regional name is linked to its geomorphic history. There is some debate about whether earlier Pleistocene glaciations covered the region, but the most recent glacial advance did not override this locality.

The landscape of the Driftless Area is very different from the muted terrain and low rounded hills of adjoining glaciated terrain. Narrow stream valleys, steep bluffs, caves, sinkholes, and loess-covered hills produce a scenic landscape with many landforms that would not have survived erosion by a massive glacier.

In some parts of the United States, lobes of the ice sheet extended as much as 485 kilometers (300 mi) farther south than the latitude of the Driftless Area. Why this region remained glacier-free is a result of the topography directly to the north. The Superior Upland, a highland of resistant rock, caused the front of the glacier to split into two masses, called *lobes*, diverting the southward-flowing ice around this topographic obstruction.

The diverging lobes flowed into two valleys that today hold Lake Superior and Lake Michigan. These two lowlands were oriented in directions that channeled the glacial lobes away from the Driftless Area. Although not overridden by these glaciers, the Driftless Area did receive some outwash deposits and a cover of wind-deposited loess, both derived from the surrounding glaciers and their sediments.

The Driftless Area has a unique landscape because of its isolation; it forms an unglaciated "island" or "peninsula" almost completely surrounded by extensive glaciated regions. The rocks, soils, terrain, relief, vegetation, and habitats contrast strongly with the nearby drift-covered terrains and provide a glimpse of what the preglacial landscapes of this midwestern region may have been like. The Driftless Area also provides distinctive habitats for flora and fauna that do not exist in the adjacent glaciated terrain. Unique and unusual aspects of its topography and ecology attract ecotourists and offer countless opportunities for scientific study. For these reasons, there are many protected natural lands in the Driftless Area, an excellent example of a natural region defined by its physical geographic characteristics.

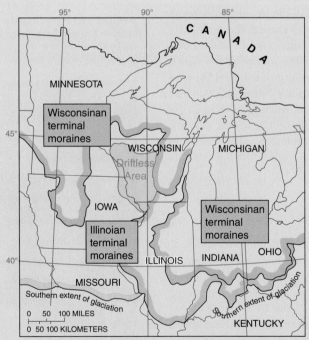

The Driftless Area was virtually surrounded by the advancing Pleistocene ice sheets, yet it was not glaciated.

Depositional Landforms of Continental Glaciation

The sheer disparity in scale causes depositional landforms of ice sheets and ice caps to differ from those of alpine glaciers. Although terminal and recessional moraines, ground moraines, and glaciofluvial deposits are produced by both categories of glaciers, retreating continental glaciers leave significantly more extensive versions of these features than alpine glaciers do (● Fig. 19.32). Because of their size, continental glaciers deposit some ridge and hill forms that alpine glaciers generally cannot.

Moraines and Plains Ranging up to about 60 meters (200 ft) in height, terminal and recessional end moraines deposited by Pleistocene ice sheets form substantial belts of low hills and ridges in areas affected by continental glaciation (● Fig. 19.33). The last major Pleistocene glacial advance through New England left its terminal moraine running the length of New York's Long Island and created the offshore islands of Martha's Vineyard and Nantucket, Massachusetts. Glacial retreat left recessional moraines forming both Cape Cod and the rounded southern end of Lake Michigan. Both terminal and recessional end moraines are usually arc-shaped and convex toward the direction of ice flow. Their pattern and placement indicate that the ice sheets did not maintain an even front but spread out in tongue-shaped lobes channeled by the underlying terrain (● Fig. 19.34). End moraines provide more evidence than just the direction of ice flow. Characteristics of the material deposited in the end moraines also help us detect the sequence of advances and retreats of each successive ice sheet.

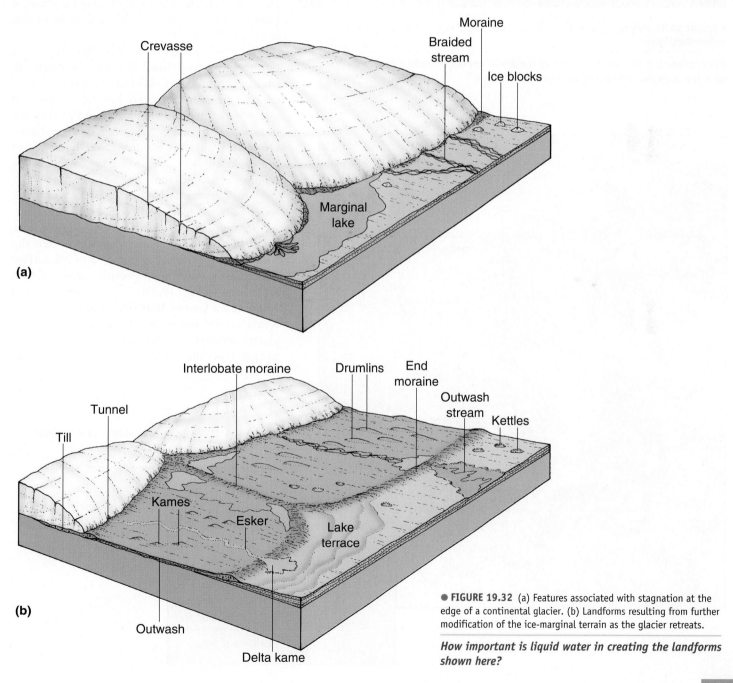

(a)

(b)

● **FIGURE 19.32** (a) Features associated with stagnation at the edge of a continental glacier. (b) Landforms resulting from further modification of the ice-marginal terrain as the glacier retreats.

How important is liquid water in creating the landforms shown here?

● **FIGURE 19.33** Hilly topography of an end moraine deposited by a continental ice sheet in eastern Washington.

What makes the terrain at the left of the photo appear bumpier compared with the smoother surface of the plain at the right?

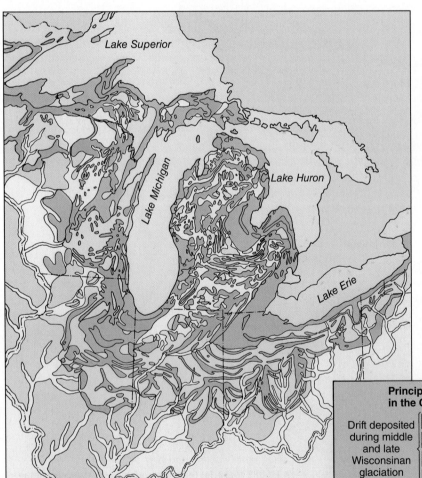

Lake Superior

Lake Michigan

Lake Huron

Lake Erie

Principal glacial deposits in the Great Lakes region

Drift deposited during middle and late Wisconsinan glaciation

- Till plains
- End moraines
- Outwash plains and valley trains
- Glacial lake deposits
- Undifferentiated drift of earlier glaciations
- Driftless regions

Continental glaciers carry massive amounts of poorly sorted sediment, much of it near the base of the ice. As a glacier ablates away at the end of a glacial interval, it abandons its load essentially in place, leaving till strewn over the landscape as a **till plain**. This till often accumulates to depths of 30 meters (100 ft) or more. Because of the uneven nature of deposition from the wasting ice, the topographic configuration of the till-covered land varies from place to place. In some areas, the till is too thin to hide the original contours of the land, whereas in other regions, thick deposits of till make broad, rolling plains of low relief. Small hills and slight depressions, some filled with water, characterize most till plains, reflecting the uneven glacial deposition. The gently rolling till plains of Illinois and Iowa are excellent examples of this glacial landform.

Beyond the belts of hills that represent terminal and recessional moraines lie **outwash plains** composed of meltwater deposits. These extensive areas of relatively low relief consist of glaciofluvial deposits that are sorted as they are transported by meltwater from the ice sheets. Outwash plains, which may cover hundreds of square kilometers, are analogous to the valley trains of alpine glaciers.

Small depressions or pits, called **kettles**, mark some outwash plains, till plains, and moraines. Kettles represent places where blocks of ice were originally buried in glacial deposits. When the blocks of ice eventually melted, they left surface depressions, and many kettles now contain *kettle lakes* (● Fig. 19.35). For example, most of Minnesota's famous 10,000 lakes are kettle lakes. Some kettles occur in association with alpine glacial deposits, but the vast majority are found in landscapes that were occupied by ice sheets or ice caps.

● **FIGURE 19.34** Glacial deposits are widespread in the Great Lakes region.

Why do the many end moraines have such a curved pattern?

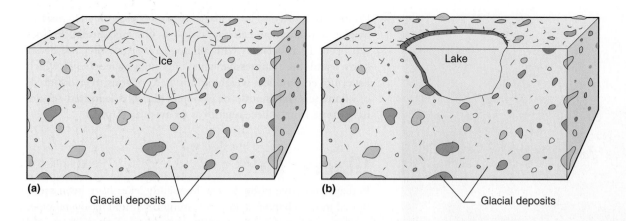

(a) Glacial deposits

(b) Glacial deposits

(c)

● **FIGURE 19.35** Kettle lakes form (a) where large, discrete blocks of ice buried or partly buried in glacial deposits, (b) melt away leaving a depression. (c) This kettle in the Northwest Territories, Canada, now with a kettle lake, formed in an area of ground moraine deposits, as can be seen by the hummocky topography. (d) Numerous large kettle lakes dot parts of Siberia due to Pleistocene continental glaciation.

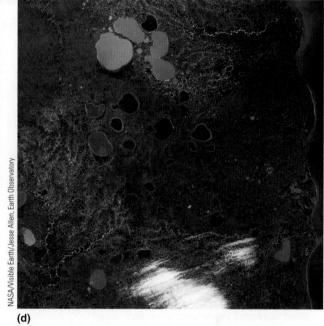

(d)

Ridges and Hills A **drumlin** is a streamlined ridge or hill, typically about 0.5 kilometer (0.3 mi) in length and less than 50 meters (160 ft) high, molded in glacial drift on till plains (● Fig. 19.36). The most conspicuous feature of drumlins is the elongated, streamlined shape that resembles the convex side of a teaspoon. The broad, steep end faces in the up-ice direction, whereas the gently sloping tapered end points in the direction that the ice flowed; thus the geometry of a drumlin is the reverse of that of roches moutonnées. Drumlins are usually found in swarms, called *drumlin fields*, with as many as a hundred or more clustered together. Opinions among scientists differ regarding the origin of drumlins, particularly with respect to the relative importance of ice versus meltwater processes in their formation. Drumlins are well developed in Canada, in Ireland, and in the states of New York and Wisconsin. One of the best-known historical sites in the United States, Bunker Hill in Boston, is a drumlin.

An **esker** is a long and narrow ridge composed of glaciofluvial sands and gravels that in map view winds or meanders along (● Fig. 19.37). Some eskers are as long as 200 kilometers (125 mi), although several kilometers is more typical of esker length. Most eskers probably formed by meltwater streams flowing in ice tunnels at the base of ice sheets. Many eskers are mined for their sand and gravel. Being natural embankments, they are frequently used in marshy, glaciated landscapes as places to build roads. Eskers are especially well developed in Sweden, Finland, and Russia.

Roughly conical hills composed of sorted glaciofluvial deposits are known as **kames**. Kames may develop from sediments that accumulated in glacial ice pits, in crevasses, and among jumbles of detached ice blocks. Like eskers, kames are excellent sources for mining sand and gravel and they are especially common in New England. *Kame terraces* are landforms resulting from accumulations of glaciofluvial sand and gravel along the margins of ice lobes that melted away in valleys of hilly regions. Good examples of kame terraces can be seen in New England and New York.

Large boulders scattered in and on the surface of glacial deposits or on glacially scoured bedrock are called **erratics** if the rock they consist of differs from the local bedrock (● Fig. 19.38). Moving ice is capable of transporting large rocks very far from their source. The source regions of erratics can be identified by rock type, and this provides evidence of the direction of ice flow.

● **FIGURE 19.36** Drumlins, such as this one in Montana, are streamlined hills elongated in the direction of ice flow.

● **FIGURE 19.37** An esker near Albert Lea, Minnesota.

What economic importance do eskers have?

● **FIGURE 19.38** This huge boulder in Yellowstone National Park is a glacial erratic, transported far from its point of origin by a continental glacier.

What does this erratic illustrate about the ability of flowing ice to modify the terrain?

Erratics are known to be of glacial origin because they are marked by glacial striations and are found only in glaciated terrain. Erratics can occur in association with alpine glaciers, but they are better known and more impressive when deposited by ice sheets, which have moved boulders weighing hundreds of tons over hundreds of kilometers. In Illinois, for example, glacially deposited erratics have come from source regions as far away as Canada.

Glacial Lakes

Perhaps you have noticed how frequently lakes have been mentioned in this chapter. Lakes are commonly found in several types of glaciated terrain. Many, many thousands of glacially created lakes exist in depressions within deposits of the continental glaciers that once covered much of North America and Eurasia. The Pleistocene ice sheets created numerous other lake basins by erosion, scooping out deep elongated basins along zones of weak rock or along former stream valleys. New York's Finger Lakes are excellent examples of lakes that have accumulated in elongated, ice-deepened basins (● Fig. 19.39). Lakes are also common in areas that were affected by erosion and deposition from alpine glaciers. Once the glaciers have disappeared, cirques and glacial troughs tend to pond water, especially if those basins are bordered by an end moraine, as in the case of Lakes Maggiore, Como, and Garda in the Italian Alps (● Fig. 19.40). Evidence for many lakes that no longer exist is found in the widespread occurrence of *glaciolacustrine* (*glacial*, ice; *lacustrine*, lake) deposits that allow us to reconstruct the extent and determine the age of those lakes.

Some lakes formed along the front of the Pleistocene ice sheets in areas where glacial deposits disrupted the surface drainage or where a glacier prevented depressions from being drained of meltwater. These lakes usually accumulated where water became trapped between a large end moraine and the ice front or

● **FIGURE 19.39** A satellite image of the Finger Lakes of New York. The lakes occupy linear, glacially eroded basins excavated during the Pleistocene.

What characteristics of the bedrock caused ice to form these narrow lake basins?

From Mortara G., Mercalli L., 2002. Il lago epiglaciale "Effimero" sul ghiacciaio del Belvedere, Macugnaga, Monte Rosa. Nimbus, n. 23–24, p. 10–17

● **FIGURE 19.40** After withdrawal of the ice, end moraines often contribute to the formation of closed depressions in which glacial lakes accumulate.

where the land sloped toward, instead of away from, the ice front (see Fig. 19.32a). In both situations, *ice-marginal lakes* filled with meltwater. They drained and ceased to exist when the retreat of the ice front uncovered an outlet route for the water body.

During their existence, fine-grained sediment accumulated on the floors of these ice-marginal lakes, filling in topographic irregularities. As a result of this sedimentation, extremely flat surfaces characterize glacial plains where they consist of glaciolacustrine deposits. An outstanding example of such a plain is the valley of the Red River in North Dakota, Minnesota, and Manitoba. This plain, one of the flattest landscapes in the world, is of great agricultural significance. The plain was created by deposition in a vast

Pleistocene lake held between the front of the receding continental ice sheet on the north and moraine dams and higher topography to the south. This ancient body of water is named Lake Agassiz for the Swiss scientist who early on championed the theory of an ice age. The Red River flows northward eventually into the last remnant of Lake Agassiz, Lake Winnipeg, which occupies the deepest part of an ice-scoured and sediment-filled lowland.

Another ice-marginal lake in North America produced much more spectacular landscape features but not in the area of the lake itself. In northern Idaho, a glacial lobe moving southward from Canada blocked the valley of a major tributary of the Columbia River, creating an enormous ice-dammed lake known as Lake Missoula. This lake covered almost 7800 square kilometers (3000 sq mi) and was 610 meters (2000 ft) deep at the ice dam. On occasions when the ice dam failed, Lake Missoula emptied in tremendous floods that engulfed much of eastern Washington. The racing floodwaters scoured the basaltic terrain, producing Washington's channeled scablands, which consist of interweaving steep-sided troughs (coulees), dry waterfalls, scoured-out basins, and other features quite unlike those associated with normal stream erosion, particularly because of their gigantic size (● Fig. 19.41).

The Great Lakes of the eastern United States and Canada make up the world's largest lake system. Lakes Superior, Michigan, Huron, Erie, and Ontario occupy former river valleys that were vastly enlarged and deepened by glacial erosion. All the lake basins except that of Lake Erie have been gouged out to depths below sea level and have irregular bedrock floors lying beneath thick blankets of glacial till. The history of the Great Lakes is exceedingly complex, resulting from the back-and-forth movement of the ice front that produced many changes of lake levels and overflow in varying directions at different times (● Fig. 19.42).

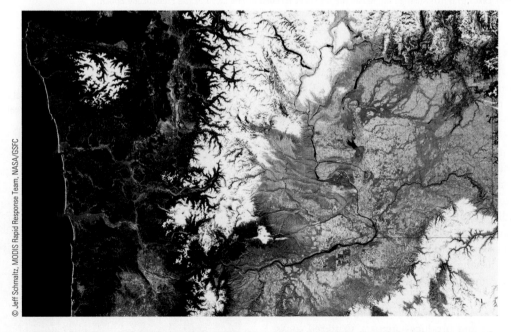

© Jeff Schmaltz, MODIS Rapid Response Team, NASA/GSFC

● **FIGURE 19.41** Much of eastern Washington State *(tan colored)* is called the channeled scablands because of the gigantic, largely abandoned river channels that cross the region *(dark tan)*. These channels represent huge outpourings of water that occurred when glacial ice damming large lakes in the Pleistocene failed, releasing an amount of water perhaps 10 times the flow of all rivers of the world today.

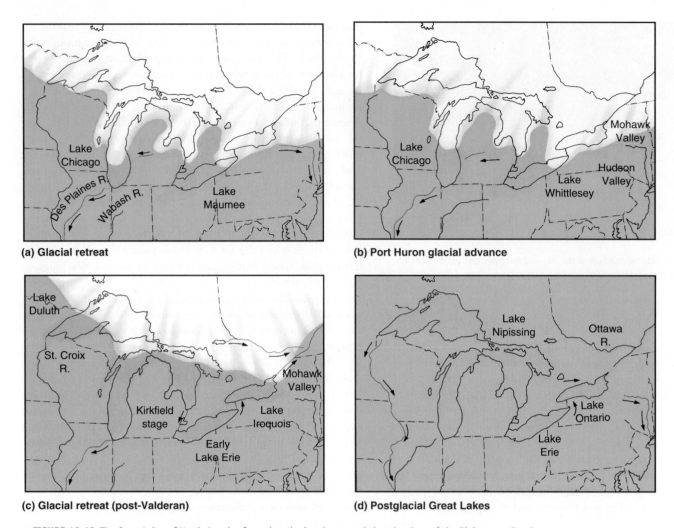

(a) Glacial retreat

(b) Port Huron glacial advance

(c) Glacial retreat (post-Valderan)

(d) Postglacial Great Lakes

● **FIGURE 19.42** The Great Lakes of North America formed as the ice sheet receded at the close of the Pleistocene Epoch.

Periglacial Landscapes

Not all cold regions have sufficient precipitation to lead to permanent accumulation of thick masses of ice. In much of Siberia and interior Alaska, it was too cold and dry during the Pleistocene to generate the massive snow and ice accumulations that occurred elsewhere in North America and northwestern Europe. Instead of glacial processes, these **periglacial** environments (*peri*, near) lacking year-round snow or ice cover undergo intense frost action and are frequently regions of *permafrost* (permanently frozen ground). Large areas of periglacial terrain exist today in Alaska, Canada, Russia, and some areas of high elevation including in mountain and plateau regions of China. During the Pleistocene ice advances, periglacial environments migrated to lower latitudes and lower elevations, leaving relict periglacial features in places, including parts of the Appalachians, where it is no longer actively forming today.

The intense frost action of periglacial landscapes includes freezing of soil moisture and produces angular, shattered rocks. Frost action also causes heaving, thrusting, and size-sorting of stones in the soil that lead to the formation of fascinating repeating designs in *patterned ground* features (● Fig. 19.43).

● **FIGURE 19.43** Intensive frost action in periglacial regions leads to the formation of intricate, repetitive patterned ground features.

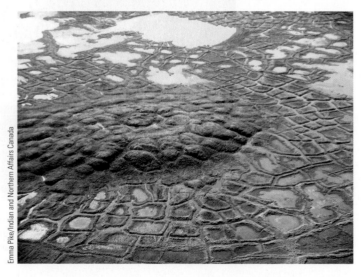

Emma Pike/Indian and Northern Affairs Canada

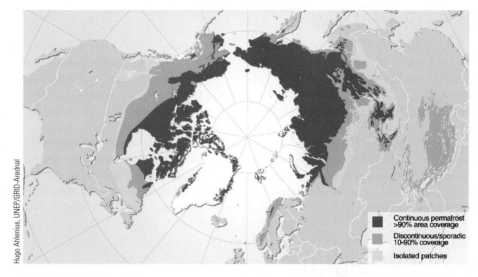

● **FIGURE 19.44** Many periglacial regions contain permafrost. This map shows the distribution of permafrost in the Northern Hemisphere.

Where mean annual temperatures are cold enough, ice-free landscapes develop permafrost (● Fig. 19.44). As discussed in an earlier chapter, extensive areas of permafrost result in considerable solifluction (slowly flowing soil) on slopes when the upper, active layer of the permanently frozen ground thaws in the summer and becomes saturated. Permafrost areas are also prone to the formation of fissures that accumulate ice, leading to the formation of large *ice wedges* in the ground.

Understanding permafrost is important for human activity in periglacial regions. Unless proper construction techniques are used, erecting buildings, roads, pipelines, and other structures on permafrost disrupts the natural thermal environment, often leading to permafrost melting. Saturated ground cannot support the weight of structures resting on top of them. As the ground deforms and slowly flows, the structures are destroyed. To avoid these problems, construction techniques have been developed that keep the permafrost frozen or allow it to experience its natural temperature fluctuations. The latter is generally accomplished by building above the surface so that air can circulate to the permafrost.

CHAPTER 19 ACTIVITIES

■ TERMS FOR REVIEW

glacier	zone of ablation	ground moraine
glaciation	equilibrium line	glacial outwash
firn	glacier advance	valley train
accumulation	glacier retreat	outlet glacier
ablation	striation	ice shelf
calving	roche moutonnée	glacial (glacial period)
alpine glacier	tarn	interglacial
valley glacier	arête	continental shelf
piedmont glacier	horn	till plain
cirque glacier	col	outwash plain
cirque	glacial trough	kettle
continental glacier	paternoster lake	drumlin
ice sheet	hanging valley	esker
ice cap	fjord	kame
crevasse	drift	erratic
plucking	lateral moraine	periglacial
till	medial moraine	
zone of accumulation	end moraine	

QUESTIONS FOR REVIEW

1. How does glacial ice differ from snow?
2. What are the three main types of alpine glacier, and what distinguishes them from one another?
3. Explain how glaciers move.
4. Diagram and label the characteristic parts of an alpine glacier.
5. Define accumulation and ablation, and explain how they are related to glacial advance and retreat.
6. How do glaciers accumulate load? Provide some examples of evidence of glacial erosion and movement.
7. What are some major similarities and some major differences between continental and alpine glaciers?

8. Distinguish till from outwash and drift.
9. Where are the two major existing ice sheets located? How do they compare in area and thickness?
10. How does the present extent of continental ice-sheet coverage compare with the maximum extent of the Pleistocene ice-sheet coverage?
11. How has ice-sheet erosion altered the landscape? What kinds of landscape features are produced by continental glacier deposition and recession?
12. What landscape features would indicate that a region is undergoing periglacial processes?

CONSIDER AND RESPOND

1. How do alpine glaciers differ from streams in terms of flow processes, erosion processes, load characteristics, valley shaping, and nature of tributary valleys?
2. Name four coastal areas of the world where fjords can be found. What is the dominant climate type in coastal areas with fjords? Why are fjords associated with that climate? In what way are fjords related to both alpine and continental glaciers?

3. Describe the ice budget of a glacial system. What two major factors control this budget? How is the ice budget related to the movement of glaciers? Explain how a glacier that is calving can also be undergoing advance.
4. Glaciation is truly an interdisciplinary topic. In addition to physical geographers, what other scientists do you think are involved in the study of glaciers and why?

PRACTICAL APPLICATIONS

1. These tables show the position (elevation) of the toe of a Northern Hemisphere valley glacier as measured yearly on September 30 during the 1970s and then again during the 1990s. Use decadal averages and annual rates of change of position to help you describe how the behavior of the glacier in the 1990s differed from that of the 1970s.

Year	1970	1971	1972	1973	1974	1975	1976	1977	1978	1979
Elevation (meters)	2634	2631	2632	2629	2630	2629	2627	2625	2624	2623

Year	1990	1991	1992	1993	1994	1995	1996	1997	1998	1999
Elevation (meters)	2642	2648	2654	2661	2668	2676	2684	2692	2701	2710

2. Diamonds were recently found at three separate sites in Canada in till deposited by the Pleistocene ice sheet. At Site A, nearby ice-scoured troughs now occupied by lakes are elongated from NNE to SSW. Site B lies 60 kilometers west of Site A, and striations in adjacent bedrock show a NE to SW orientation. Lying 60 kilometers east of Site A, Site C has nearby drumlins with their tail end pointed SSE and their blunt end NNW. Using this information, how might you proceed in narrowing down a search area for the bedrock from which the diamonds originated?

LOCATE AND EXPLORE

1. Using Google Earth, identify the landforms at the following locations (latitude, longitude). Briefly explain how each landform developed and why it is found in that general area.
   a. 59.75°N, 140.59°W
   b. 42.73°N, 76.73°W
   c. 54.06°N, 98.71°W
   d. 61.34°N, 25.47°E
   e. 44.52°S, 168.84°E
   f. 42.35°N, 71.92°W
   g. 58.44°N, 107.58°W

2. Using Google Earth and the Topographic Map Layer for Gilkey Glacier from the USGS (downloadable from the book's CourseMate at www.cengagebrain.com), look at the change in the glacier (58.76°N, 134.57°W) between 1948 and today and answer the following questions:
   a. Use the measuring tool to determine how far the glacier has retreated.
   b. What is the average annual rate of retreat between 1948 and today?
   c. Battle Glacier was formerly joined to Gilkey Glacier, but it has also retreated. What river now occupies the valley formerly filled by Battle and Thiel Glaciers? Why is this type of river commonly associated with glaciers?
   d. What glacial landform is found along the edge of the valley of Battle Glacier (58.743911°N, 134.582156°W)?

 MindTap—Make the most of your study time by accessing everything you need to succeed in one place. Read your textbook, take notes, review flashcards, watch videos, complete activities, take practice quizzes, and more online with MindTap. Log in at **www.cengagebrain.com**.

Map Interpretation

■ ALPINE GLACIATION

The Map

The Chief Mountain, Montana, map area is located in Glacier National Park, which adjoins Canada's Waterton Lakes National Park located across the international border. Alpine glaciation has created much of the spectacular scenery of this region of the northern Rocky Mountains.

Most of the glaciated landscape in this region was produced during the Pleistocene. Today glaciers still exist in Glacier National Park, but because of rapid melting in recent decades, it is estimated that they may be gone by 2050. Prior to glaciation, this region was severely faulted and folded during the formation of the northern Rocky Mountains.

The photograph and map clearly show the rugged nature of the terrain of this map area. Steep slopes, horns, U-shaped valleys, lakes, arêtes, and glaciers are obvious landform characteristics of alpine glaciation. Temperature is a primary control of the highland climate of this region. Elevation, in turn, influences temperature and precipitation amounts.

As you would expect, the rapid decrease in temperature with increasing elevation results in a variety of microclimates within alpine regions. Exposure is also an important highland climate control. West-facing slopes receive the warm afternoon sun, whereas east-facing slopes are sunlit only in the cool of the early morning.

Interpreting the Map

1. What is the approximate amount of local relief depicted on this map? Is the scale of this topographic map larger or smaller than that of the topographic maps in other map exercises in the book?
2. From your examination of the topographic map and photograph of this mountain region, does the landscape seem dominated by glacial erosion or deposition?
3. Locate Grinnell, Swiftcurrent, and Sperry Glaciers. What specific type of glaciers are they?
4. What evidence indicates that the glaciers were once larger and extended farther down the valleys?
5. Note that most of the existing glaciers are located to the northeast of the mountain summits. Explain this orientation. At what elevation are the glaciers found?

6. What types of glacial landform are the following features?
   a. The features occupied by Kennedy, Iceberg, and Ipasha Lakes
   b. The feature occupied by McDonald Lake (in the southwest corner of the map) and Lake Josephine
   c. Mount Gould and Mount Wilbur
7. Along the high ridges runs a dashed line labeled Continental Divide. What is its significance?
8. If you were to hike southeast from Auto Camp on McDonald Creek up Avalanche Creek to the base of Sperry Glacier, how far would you travel and how much elevation change would you encounter?

Glaciated terrain in Glacier National Park.

⊕ In Google Earth, fly to: 48.722°N, 113.737°W.

Opposite:
Chief Mountain, Montana
Scale 1:125,000
Contour interval = 100 ft
USGS

Map Interpretation

CONTINENTAL GLACIATION

The Map

The Jackson, Michigan, map area is located in the Great Lakes section of the Central Lowlands. Massive continental ice sheets covered this region during the Pleistocene. As the glaciers melted and the ice sheets retreated, a totally new terrain, vastly different from that of preglacial times, was exposed. Today evidence of glaciation is found throughout the region. The most obvious glacially produced landforms are the thousands of lakes (including the Great Lakes), the knobby terrain, and moraine ridges. Moraines left by advancing and retreating tongue-shaped ice lobes that extended generally southward from the main continental ice sheet also influenced the shapes of the Great Lakes.

The ice sheet, its deposits of sediment, and its meltwaters also created other, smaller-scale features. The glaciers left a jumbled mosaic of deposits from boulders through smaller gravel to sand, silt, and clay that has produced a hilly and hummocky terrain. The overall relief is low, in part because of glacial erosion. Many landform features of continental glaciation are well illustrated on the Jackson, Michigan, 15' quadrangle map.

This region has a humid continental climate. The summers are mild and pleasant. Excessively warm and humid air seldom reaches this area for more than a few days at a time. Instead, cool but pleasant evening temperatures tend to be the rule in summer. Winters, however, are long and often harsh. Snow can be abundant and on the ground continuously for many weeks or even months at a time. The annual temperature range is quite large; precipitation occurs year-round provided primarily by midlatitude cyclonic storms.

Interpreting the Map

1. Describe the general topography of the map area.
2. What is the local relief? Why is it so difficult to find the exact highest and lowest points on this map?
3. Does the topography of this region indicate glacial erosion or glacial deposition?
4. Does the area appear to be well drained? What are the three main hydrographic features that indicate the drainage conditions?
5. What type of glacial landform is Blue Ridge? What are its dimensions (length and average height)?
6. How is a feature such as Blue Ridge formed? What economic value might it have?
7. What is the majority of the surface material in the map area? What is the term for this type of surface cover?
8. What probably caused the many small depressions and rounded lakes? What are these depressions called?
9. Describe how the topography of the Jackson, Michigan, area appears on the aerial image.
10. What is the advantage of having both aerial image and topographic map coverage of an area of study?

Michigan Center for Geographical Information

Aerial image of the region near Jackson, Michigan.

⊕ In Google Earth, fly to: 42.078°N, 84.422°W.

Opposite:
Jackson, Michigan
Scale 1:62,500
Contour interval = 10 ft
USGS

COASTAL PROCESSES AND LANDFORMS

■ OBJECTIVE

WHEN YOU COMPLETE THIS CHAPTER YOU SHOULD BE ABLE TO:

- 20.1 Sketch the coastal zone and label its principal subdivisions.
- 20.2 Explain how waves are formed and characterized.
- 20.3 Discuss how wind waves transfer and deliver energy to the coastal zone.
- 20.4 Explain why coastlines tend to straighten over time.
- 20.5 Recount how sediment moves along a beach.
- 20.6 Identify coastal erosion processes and the primary landforms found on erosion-dominated coastlines.
- 20.7 Describe the types and sources of sediment found in coastal depositional landforms.
- 20.8 Distinguish among the principal coastal depositional landforms.
- 20.9 Summarize a global-scale and a regional-scale classification of coasts.
- 20.10 Describe the different types of coral reef.

THE WORLD OCEAN COVERS 71% of Earth's surface, and the boundary between the ocean and dry land is of enormous length. A large percentage of the world's population lives near the coast, with cities and towns established there for reasons of transportation, resources, industry, commerce, defense, or tourism and spectacular scenery. Earth's coastlines have tremendous resources and are biologically and geomorphically diverse. They draw more tourists than any other natural environment and continue to attract new residents.

Coastal zones are popular, but they are also subject to an array of natural hazards and human-induced environmental problems. Coastal communities must cope with powerful storms, the influence of tides, waves, currents, and moving sediment. Low-lying coasts are subject to flooding, storm surges, and, in some places, tsunamis. Coasts of high relief are susceptible to rockfalls, landslides, and other forms of mass wasting. Environmental problems stem from rapid urban development, high population densities, and economic and industrial activities ranging from tourism to port operations, offshore oil production, and agricultural runoff. Some of our most polluted waters are found in coastal locations. If global warming significantly reduces the extent of the continental ice sheets, the ensuing rise in sea level will have a profound impact on the human-built infrastructure in coastal regions and on coastal geography and geomorphology. Understanding the natural processes that operate in the coastal zone is fundamental to solving the present and future problems in this dynamic part of Earth's landscape.

The Coastal Zone

Most of the processes and landforms of the marine coastal zone are also found along the coastlines of large lakes. All are considered *standing bodies of water* because the water in each occupies a basin and has an approximately uniform still-water level around the basin. This contrasts with the sloping, channelized flow toward lower elevations that constitutes streams.

The **shoreline** of a standing body of water is the exact and constantly changing contact between the ocean or lake surface and dry land. The position of this boundary fluctuates with incoming waves, with storms, and, in the case of the ocean, with the tides. Over the long term, the position of the shoreline is also affected by tectonic movements and by the amount of water held in the ocean or lake basin. **Sea level** is a complexly determined average position of the ocean shoreline and the vertical position (the reference, or *datum*) above and below which other elevations are measured. The *coastal zone* consists of the general region of interaction between the land and the ocean or lake. It ranges from the inland limit of coastal influence through the average position of the shoreline to the lowest elevation to which the shoreline fluctuates.

As waves approach the mainland from an open body of water, they eventually become unstable and break, sending a rush of water toward land. The *nearshore zone* extends from the seaward or lakeward edge of breakers to the landward limit reached by the broken wave water (● Fig. 20.1). The nearshore zone contains the *breaker zone* where waves break, the *surf zone* through which a bore of broken wave water moves, and, most landward, the *swash zone* over which a thin sheet of water rushes up to the inland limit of water and then back toward the surf zone. This thin sheet of water rushing toward the land is

known as **swash**, and the return flow is **backwash**. The *offshore zone* accounts for the remainder of the standing body of water, that part lying seaward or lakeward of the outer edge of the breaker zone.

Origin and Nature of Waves

Waves are traveling, repeating forms that consist of alternating highs and lows, called *wave crests* and *wave troughs*, respectively (● Fig. 20.2). The vertical distance between a trough and the adjacent crest is *wave height*. *Wavelength* is the horizontal distance between successive wave crests. Other important attributes are **wave steepness**, or the ratio of wave height to wavelength, and **wave period**, the time taken for one wavelength to pass a fixed point.

Waves that have traveled across the surface of a water body are the principal geomorphic agent responsible for coastal landforms. Like streams, glaciers, and the wind, waves erode, transport, and deposit Earth materials, continually reworking the narrow strip of coastal land with which they come in contact. Most of the waves that affect the coastal zone originate in one of three ways. The tides consist of two very long wavelength waves caused by interactions between Earth and the moon and sun. **Tsunamis** result from the sudden displacement of water by movement along faults, landslides, volcanic eruptions, or other impulsive events. Most of the waves that have an impact on the coastal zone, however, are **wind waves**, created when air currents push along the water surface.

Tides

The two long-wavelength waves that comprise the tides always exist on Earth. There are two wave crests (high tides), each followed by a trough (low tide). As a crest and then a trough move slowly through an area of the ocean, they cause a gradual rising and subsequent falling of the ocean surface. Along the marine coastline, the change in water level caused by the tides brings the influence of coastal processes to a range of elevations. Tides are so small on lakes that they have virtually no effect on coastal processes, even in large lakes.

The gravitational pull of the moon, and to a lesser extent the sun, and the force produced by motion of the combined Earth–moon system are the major causes of the tides (● Fig. 20.3). The moon is much smaller than the sun, but because it is significantly closer to Earth, its gravitational influence on Earth exceeds that of the sun. The moon completes one revolution around Earth every 29.5 days, but it does not revolve around the center of Earth. Instead, the moon and Earth are a combined system that as a unit moves around the system's center of gravity. Because Earth's mass is much larger than that of the moon, the combined system's center of gravity occupies a point within Earth on the side that is facing the moon. Being closest to the moon and more easily deformed than land, ocean at Earth's surface above the center of gravity is pulled toward the moon, making the first tidal bulge (high tide). At the same time, ocean water on the opposite side of Earth experiences the outward-flying, or *centrifugal*, force of inertia and forms the other tidal bulge. Troughs (low tides)

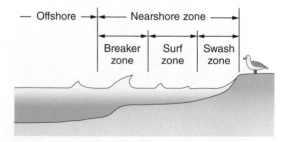

● **FIGURE 20.1** Principal divisions of the coastal zone.
From Komar, P. D., Beach Processes and Sedimentation, © 1979, p. 13. Reprinted by permission of Pearson Education, Inc., Upper Saddle River, NJ.

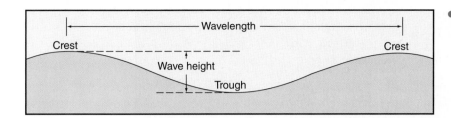

occupy the sides of Earth midway between the two tidal bulges. As Earth rotates on its axis each day, these bulges and troughs sweep across Earth's surface.

The sun has a secondary tidal influence on Earth's ocean waters, but because it is so much farther away, its tidal effect is less than half that of the moon. When the sun, moon, and Earth are aligned, as they are during new and full moons, the added influence of the sun on ocean waters causes higher than average high tides and lower than average low tides. The difference in sea level between high tide and low tide is the *tidal range*. The increased tidal range caused by the alignment of Earth, the moon, and the sun, known as **spring tide**, occurs every 2 weeks. A week after a spring tide, when the moon has revolved a quarter of the way around Earth, its gravitational pull on Earth is exerted at a 90° angle to that of the sun. In this position, the forces of the sun and moon detract from each other. At the time of the first-quarter and last-quarter moon, the counteracting force of the sun's gravitational pull diminishes the moon's attraction. Consequently, the high tides are not as high and the low tides are not as low at those times. This moderated situation, which like spring tides occurs every 2 weeks, is **neap tide** (● Fig. 20.4).

The moon completes its 360° orbit around Earth in a month, traveling about 12° per day in the same direction that Earth rotates daily around its axis. Thus, by the time Earth completes one rotation in 24 hours, the moon has moved on 12° in its orbit around Earth (see Fig. 20.3). To return to a given position with respect to the moon takes the Earth an additional 50 minutes. As a result, the moon rises 50 minutes later every day at any given spot on Earth, the tidal day is 24 hours and 50 minutes, and two successive high tides are ideally 12 hours and 25 minutes apart.

The most common tidal pattern approaches the ideal of two high tides and two lows in a tidal day. This *semidiurnal* tidal regime is characteristic along the Atlantic coast of the United States, for example, but it does not occur everywhere, as seen in ● Figure 20.5. In a few seas that have restricted access to the open ocean, such as the Gulf of Mexico, tidal patterns of only one high and one low tide occur during a tidal day. This type of tide, called *diurnal*, is not very common. A third type of tidal pattern consists of two high tides of unequal height and two low tides, one lower than the other. The waters of the Pacific coast of the United States exhibit this *mixed tide* pattern.

● **FIGURE 20.3** Forces responsible for Earth's two tidal bulges and two troughs. A "tidal day" is 24 hours and 50 minutes long because the moon continues in its orbit around Earth while Earth is rotating.

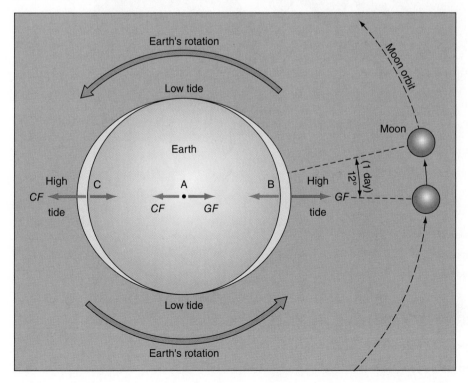

A. Gravitational force (*GF*) and centrifugal force (*CF*) are equal. Thus separation between Earth and the moon remains constant.

B. Gravitational force exceeds centrifugal force, causing ocean water to be pulled toward the moon.

C. Centrifugal force exceeds gravitational force, causing ocean water to be forced outward away from the moon.

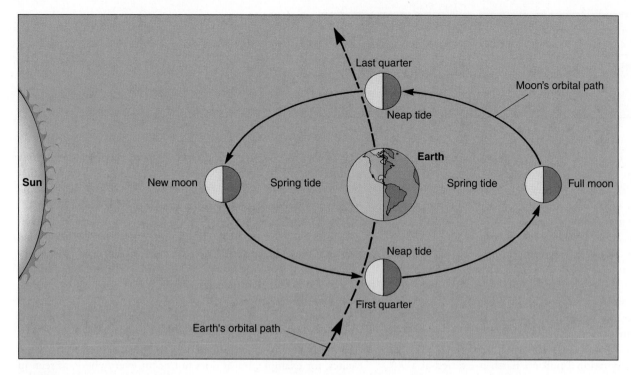

● **FIGURE 20.4** Maximum tidal ranges (spring tides) occur at new and full moon, when the moon and sun are aligned either on the same or opposite sides of Earth, respectively. Minimum tidal ranges (neap tides) occur when the moon and sun act at right angles to each other.

How many spring tides and neap tides occur each month?

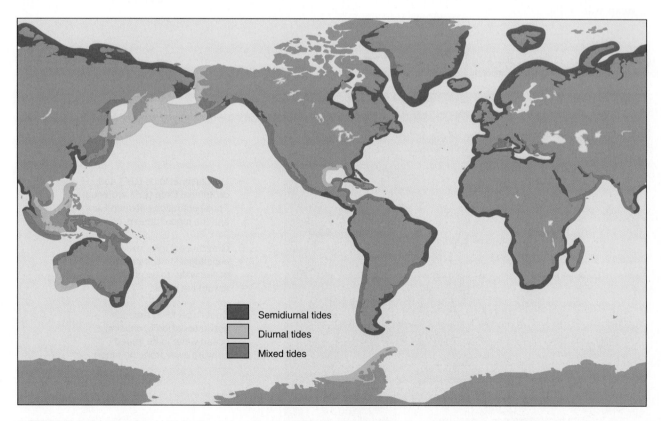

● **FIGURE 20.5** This map of world tidal patterns shows the geographic distribution of diurnal, semidiurnal, and mixed tides.

What is the tidal pattern on the coastal area nearest where you live?

(a)

(b)

● **FIGURE 20.6** The extreme tidal range of the Bay of Fundy in Nova Scotia, Canada, can be seen in the difference between (a) high tide, and (b) low tide at the same point along the coast.

Why does the Bay of Fundy have such a great tidal range?

Tidal range varies from place to place in response to the shape of the coastline, water depth, access to the open ocean, submarine topography, and other factors. The tidal range along open-ocean coastlines, such as the Pacific coast of the United States, averages between 2 and 5 meters (6 and 15 ft). In restricted or partially enclosed seas, such as the Baltic or Mediterranean Sea, the tidal range is usually 0.7 meter (2 ft) or less. Funnel-shaped bays off major oceans, especially the Bay of Fundy on Canada's east coast, produce extremely high tidal ranges. The Bay of Fundy is famous for its enormous tidal range, which averages 15 meters (50 ft) and may have reached a maximum of 21 meters (70 ft) (● Fig. 20.6). Other narrow, elongated coastal inlets that exhibit great tidal ranges are Cook Inlet in Alaska, Washington's Puget Sound, and the Gulf of California in Mexico.

Tsunamis

Tsunamis are long-wavelength waves that form when a large mass of water displaced upward or downward by an earthquake, volcanic eruption, landslide, or other sudden event works to regain its equilibrium condition. Resulting oscillations of the water surface travel outward from the origin as one wave or a series of waves. In deep water, the displacement may cause wave heights of a meter or more that can travel at speeds of up to 725 kilometers (450 mi) per hour yet pass beneath a ship unnoticed. As the long-wavelength waves approach the shallow water of a coastline, their height can grow substantially. As these very large and extremely dangerous waves surge into low-lying land areas, they acquire huge amounts of debris and can cause tremendous damage, injury, and loss of life in addition to erosion, transportation, and deposition of Earth materials.

In 1946, an earthquake in Alaska caused a tsunami that reached Hilo, Hawaii, where it attained a maximum height greater than 10 meters (33 ft) and killed more than 150 people. When the Krakatoa volcano erupted

in 1883, it generated a powerful tsunami 40 meters (130 ft) high that killed more than 37,000 people in the nearby Indonesian islands. In December 2004, the devastating earthquake-generated Indian Ocean tsunami struck the shorelines of Indonesia, Thailand, Myanmar, Sri Lanka, India, Somalia, and other countries, causing an almost unimaginable 230,000 fatalities from the tsunami alone. This tragic event reinforced the importance of tsunami early-warning systems. An early-warning system had been established for the Pacific Ocean, but none was operating in the Indian Ocean in 2004. The United Nations Educational, Scientific, and Cultural Organization (UNESCO) has since helped member nations in the region develop a tsunami warning system. The 2011 tsunami resulting from the 9.0 magnitude Tohoku earthquake near the east coast of the Japanese island of Honshu achieved a height of almost 38 m (124 ft) and extended onto the island by as much as 10 km (6 mi) causing tremendous destruction (● Fig. 20.7).

● **FIGURE 20.7** Example of the horrendous destruction caused by the 2011 tsunami on the east coast of Japan.

ORIGIN AND NATURE OF WAVES

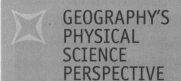
Tsunami Forecasts and Warnings

Tsunamis (from Japanese: *tsu*, harbor; *nami*, wave) are the most dangerous type of wave affecting coastal areas, as evidenced by the devastation caused by the earthquake-induced tsunami in the Indian Ocean in 2004 and in Japan in 2011. The original term *tidal wave* was abandoned decades ago because tsunamis are caused by major, sudden displacements of water and are not related to the tides. The term *seismic sea wave* can also be misleading because not all tsunamis are caused by earthquakes. Submarine landslides, eruption or collapse of underwater volcanoes, and meteor impacts can also cause tsunamis.

The speed at which a tsunami travels across the open ocean is related to ocean depth. In the Pacific Ocean, with an average depth of 4000 meters (13,000 ft), tsunamis can travel up to 800 kilometers (500 mi) per hour. In March 2011, in the city of Sendai, Japan, only 8–10 minutes elapsed between the earthquake and the arrival of the tsunami. Tsunamis also travel great distances. In 1960, a tsunami originating off the coast of Chile traveled more than 17,000 kilometers (10,600 mi) to Japan, where it killed 200 people.

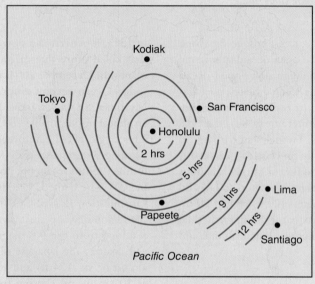

The Pacific Tsunami Warning Center locates earthquake epicenters and estimates arrival times for potential tsunamis in the Pacific region.

Although the dead and missing in Japan from the earthquake and tsunami numbered close to 28,000, a comprehensive system of hazard education, pre-event training, emergency preparedness, and tsunami warnings prevented many more thousands of casualties in that tectonically and volcanically active country.

Wind Waves

Most waves that we see on the surface of standing bodies of water are created by the wind. Where wind blows across the water, frictional drag and pressure differences cause irregularities in the water surface. The wind then pushes on water slopes that face into the wind, transferring energy to the water and building the slopes into larger waves.

If most waves are caused by the wind, why do we see waves at the beach even during calm days? The answer lies in the fact that waves can, and often do, travel very long distances from the storms that created them with limited loss of energy. Waves arriving at a beach on a calm day may have traveled thousands of kilometers to finally expend their energy when they break along the coast.

When a storm develops on the open ocean, gentle breezes first fashion small ripples on the water surface. If the wind increases, it transforms the ripples into larger waves. While under the influence of the storm, waves are steep, choppy, and chaotic and referred to as **sea**. When the waves travel out of the storm area or the wind dies down, the waves become more orderly as they sort themselves into groups of similar speed and length.

Wind waves spawned by a large storm might arrive at the coast with a period of 10 seconds and a wavelength of 150 meters (500 ft), but a tsunami may have a period of an hour and a wavelength of 100 kilometers (60 mi). As a wave moves into shallow coastal water, its speed decreases but its height increases. A tsunami 1 meter (3 ft) high in the open ocean can be 30 meters (100 ft) high at the coast. Because a tsunami typically consists of a series of waves, the danger can last for several hours after the arrival of the first wave.

The tsunamis that struck in the Indian Ocean in December 2004 and Japan in 2011 caused tremendous death, destruction, and human suffering. In part because no sensor-based warning system was in place in the Indian Ocean in 2004, about 230,000 people died and 1.2 million people were left homeless when the ocean surged onshore during that tsunami, in some places with waves as high as 15 meters (50 ft). Detecting a tsunami, determining its speed and direction, and tracking its progress across the ocean are critical for saving lives through tsunami warning systems.

The U.S. National Oceanic and Atmospheric Administration (NOAA) has established an array of instruments (tsunameters) to monitor pressure and temperature on the ocean floor and convert these data to water column height. The array constitutes an important part of a growing international tsunami monitoring network. A 26-nation group, including the Pacific Tsunami Warning Center in Hawaii, share tsunami warnings throughout the Pacific Basin. The Alaska Tsunami Warning Center issues tsunami warnings for the west coast of North America.

Australian Agency for International Development/Robin Davies

Tsunami destruction on the west coast of Aceh province, Indonesia, December 2004.

These gentler, more orderly waves that have traveled beyond the zone of generation are **swell**. It is swell that arrives at coastlines even in the absence of coastal winds.

The energy in a wave is potential energy represented by the wave height. As waves travel they lose a little height, and thus energy, as a result of friction and from the spreading of the wave crest because of the curvature of Earth, but overall they are very efficient means for transporting energy. Three factors that determine the height of wind waves as they form in deep, open bodies of water are (1) wind velocity, (2) duration of the wind, and (3) the area over which the wind blows, known as the **fetch**. Fetch is the expanse of open water across which the wind can blow without interruption. An increase in any of these three factors produces waves of greater height and greater energy.

When swell that, for example, originated in a storm in the South Pacific arrives at the coast of Southern California, it is local water, not water from the South Pacific, that arrives at the shore in the wave. Recall that waves are traveling *forms*. They do not transport water horizontally from one place to another except where they break along a coastline. The movement of waves in the open water body may be considered similar to the movement of stalks of wheat as wind blows across a wheat field, causing wavelike ripples to roll across its surface. The wheat returns to its original position after the passage of each wave. Water particles likewise return to approximately their original position after transmitting a wave.

Deep-water waves are those traveling through water depth (d) greater than or equal to half the wavelength (L), $d \geq L/2$.

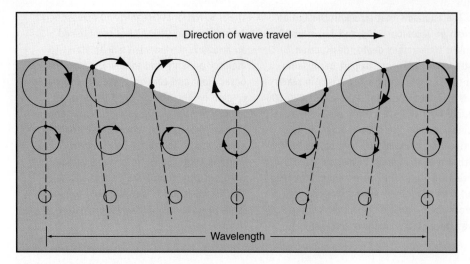

Direction of wave travel →

Wavelength

● **FIGURE 20.8** Orbital paths of water particles cause oscillatory wave motion in deep water. The diameter of the surface orbit equals wave height, which is the vertical distance from trough to crest.

Traveling waves have no impact on what is below that depth. For this reason, the depth $L/2$ is sometimes referred to as **wave base**. ● Figure 20.8 illustrates what happens to surface water during the passage of a wave in deep water. There is little if any net forward motion of water molecules during the passage of the wave. As the crest and trough pass through, the water molecules complete an orbital motion. With increasing depth beneath the water surface, the size of the orbits decreases. By a depth of half the wavelength, the orbits are too small to do any significant work. It is only when the wave enters water of $d < L/2$ that it starts to interact with, or "feel," bottom and become affected by friction with the bed (● Fig. 20.9).

Breaking of Waves

As long as they are in deep water relative to their wavelength, $d \geq L/2$, waves roll along without disturbing the bottom and with little loss of energy. As they approach the coast and enter shallower water, $d < L/2$, friction with the bed causes the waves to undergo a decrease in both velocity and wavelength. The wave bunches up, experiencing an increase in wave height (H). As wave height increases and wavelength decreases, wave steepness ($S = H/L$) increases rapidly to the maximum value of 1/7. At this steepness, the wave will become unstable and break, finally expending the energy it had originally obtained in a storm often hundreds or even thousands of kilometers away. Some breaking waves appear to curl over and crash as though trying to complete one last wave form but with insufficient water available to draw up into that final wave. Once the wave has broken, turbulent surf advances landward, thinning to swash at the water's edge and returning to the surf zone as backwash.

Rip currents (● Fig. 20.10) are relatively narrow zones of strong, offshore-flowing water that occur along some coastal areas. Rip currents are a means for returning broken wave water from the nearshore zone back to deeper water. Rip currents are dangerous. Swimmers who get caught in them often try to swim back to shore against the strong current to keep from being pulled out to deeper water, not always successfully. Rip currents are frequently visible as streaks of foamy, turbid water flowing perpendicular to the shore.

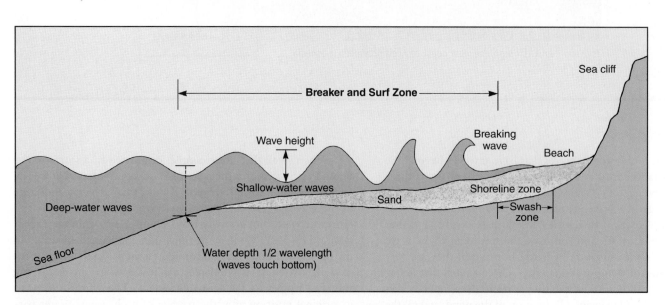

● **FIGURE 20.9** Waves begin to "feel bottom" when the water depth decreases to half the distance between wave crests. As depth gets smaller, the wave velocity and wavelength decrease while the wave height and steepness increase until breaking occurs.

City of Miami Beach, Florida, Public Safety Division

● **FIGURE 20.10** Rip currents move water seaward from a beach. Here, the current can be seen moving offshore, opposite to the wave direction. Rather than fighting the strong current directly, ocean swimmers are often cautioned to swim parallel to shore first to escape the strong rip current zone before turning to swim toward shore.

Why are these currents a hazard to swimmers?

Coastal Geomorphic Processes
Wave Refraction and Littoral Drifting

In map view, or as though looking down from an airplane, we often see parallel, linear wave crests steadily approaching the coastal zone at regular intervals from a uniform direction, probably having originated in the same distant storm. They may approach from directly offshore or at an angle to the trend of the coastline.

Oftentimes successive wave crests each change orientation relative to the coastline as they move through shallower water. **Wave refraction** is this bending of a wave in map view as it approaches a shoreline.

Wave refraction occurs when part of a wave encounters shallow water before other parts. To understand how this happens, imagine an irregular coast of embayments and headlands (● Fig. 20.11). While in deep water, a wave traveling toward the coast from directly offshore has a straight crest in map view. The wave will feel bottom first in the shallower water off the headlands, while off the embayments it is still traveling in the deeper water. This slows the advance of the wave crest toward the headlands while it continues to speed on toward the embayments. This difference in velocity converts the map view trend of the wave crest from a straight line to a curve that increasingly resembles the shape of the shoreline as it gets closer to land. Wave energy is expended perpendicular to the orientation of the crestline. Thus, when the wave breaks, its energy is focused on the headlands and spread out along the embayments. Over time, the headlands are eroded back toward the mainland, while deposition in the low-energy embayments builds those areas toward the water body. Because of wave refraction, coastlines tend to straighten over time (● Fig. 20.12).

Not all waves refract completely before they break (● Fig. 20.13). Crestlines of incompletely refracted waves do not fully conform to the orientation of the shoreline when they break. Incomplete refraction gives a spatial component to sediment transportation within the *littoral* (coastal) zone. This sediment transportation in the coastal zone, called **littoral drifting**, is accomplished in two ways. Both ways are well demonstrated using the example of a sandy beach along a straight coastline that has smooth underwater topography sloping gently into deeper water.

● **FIGURE 20.11** Wave refraction causes wave energy to be concentrated on headlands, eroding them back, whereas in embayments, deposition causes beaches to grow seaward.

How will this coastline change over a long period?

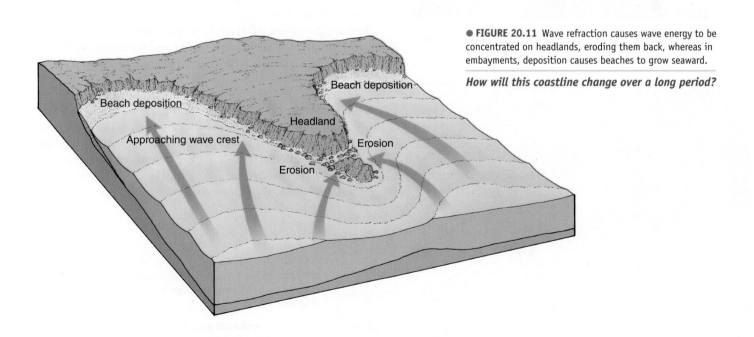

David Messent/Photolibrary/Getty Images

● **FIGURE 20.12** Embayments along this coastline have been building seaward by filling with sediment, whereas wave energy focused on the headlands has been eroding them landward.

What happens to sediment eroded from the headlands?

USGS

● **FIGURE 20.13** Incomplete wave refraction causes waves to break at an angle with respect to the orientation of the shoreline rather than parallel to the shoreline.

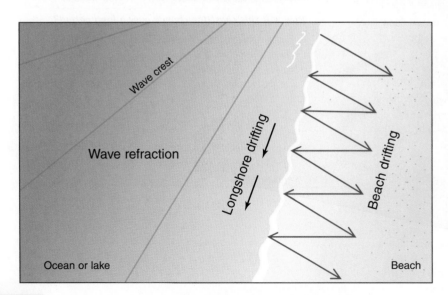

When in map view, a wave crest approaches the straight, gently sloping shoreline at a large angle to the coast (obliquely), it interacts with the bottom and starts to slow down first where it is closest to shore (● Fig. 20.14). This velocity decrease spreads progressively along the crestline as more of the wave enters shallower water. With insufficient time for complete refraction before breaking begins, the crest lies at an angle to the beach, not parallel to it, when it breaks. As a result, the broken wave water, and sediment it has entrained, rushes up the beach face diagonally to the shoreline, rather than directly up its slope. Backwash, however, which also moves sediment, flows straight back down the beach face toward the water body by the force of gravity. In this way, as one incompletely refracted wave after another breaks, sediment zigzags along the beach in the swash zone. **Beach drifting** is this zigzag-like transportation of sediment in the swash zone caused by incomplete wave refraction. Over time, beach drifting transports tons of sediment along the shore.

Another outcome of an incompletely refracted oblique wave is that when the crest arrives at the break point at one location, farther along the beach in the direction the waves are traveling, that same position relative to the shoreline is occupied by a wave trough. This difference in water level initiates a current of water, called the *longshore current*, flowing parallel to the shoreline near the breaker zone. Considerable amounts of sediment suspended when incompletely refracted waves break are transported along the shore in this process of **longshore drifting**.

Coastal Erosion

Because waves and streams both consist of liquid water, similarities exist in how these two geomorphic agents erode rock matter. Like water in streams, water that has accumulated in basins erodes some rock material chemically through *corrosion*. Corrosion is the removal of the ions that have been separated from rock-forming minerals by solution and other chemical weathering processes. Likewise, the power of *hydraulic action* from the sheer physical force of the water alone pounds against and removes coastal rock material, sometimes compressing air or water into cracks to help in the process. The power of storm waves, combined with the buoyancy of water, enables them at times to dislodge and move even large boulders. Once clastic particles are in motion, waves have solid tools to use to perform even more work through the grinding erosive process of *abrasion*. Abrasion is the most effective form of erosion by each of the geomorphic agents, including waves.

● **FIGURE 20.14** Incomplete wave refraction leads to sediment transport along the coast by beach drifting and longshore drifting.

Why is the backwash perpendicular to the trend of the shoreline when the swash is at an angle to it?

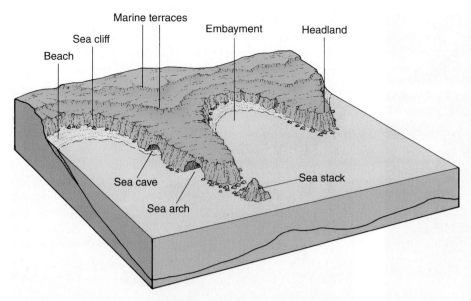

Beach
Sea cliff
Marine terraces
Embayment
Headland
Sea cave
Sea arch
Sea stack

● **FIGURE 20.15** Some of the major landforms found along erosion-dominated coastlines.

Weathering is an important factor in the breakdown of rocks in the coastal zone, as in other environments, preparing pieces for removal by wave erosion. Water is a key element in most weathering processes, and in addition to normal precipitation, rocks near the shoreline are subjected to spray from breaking waves, high relative humidities, and condensation. Salt weathering is particularly significant in preparing rocks for removal through chemical and physical weathering along the marine coast and coasts of salt lakes.

Coasts of high relief are dominated by erosion (● Fig. 20.15). *Sea cliffs* (or *lake cliffs*) are carved where waves pound directly against steep land. If a steep coastal slope continues deep beneath the water, it may reflect much of the incoming wave energy until corrosion and hydraulic action eventually take their toll on the rock. The tides present along marine coasts allow these processes to attack a range of shoreline elevations. Once a recess, or **notch**, has been carved out along the base of a cliff (● Fig. 20.16a), weathering and rockfall within the shaded overhang supply clasts that can collect on the notch floor and be used by the water as tools for more efficient erosion by abrasion. Abrasion extending the notch landward leaves the cliff above subject to rockfall and other forms of mass wasting. Stones used as tools in abrasion quickly become rounded and may accumulate at the base of the cliff as a **cobble beach**. Where the cliffs are well jointed but cohesive, wave erosion can create *sea caves* along the lines of weakness (Fig. 20.16b). *Sea arches* result where two caves meet from each side of a headland (Fig. 20.16c). When the top of an arch collapses or a sea cliff retreats and a resistant pillar is left standing, the remnant is called a **sea stack** (Fig. 20.16d).

Landward recession of a sea cliff leaves behind a wave-cut bench of rock, an **abrasion platform**, that is sometimes visible at lower water levels, such as at low tide (● Fig. 20.17a). Abrasion platforms record the amount of cliff recession. In some cases deposits accumulate as wave-built terraces just seaward of an abrasion platform. If tectonic activity uplifts these wave-cut benches and wave-built terraces above sea level out of the reach of wave action, they become **marine terraces** (Fig. 20.17b). Successive periods of uplift can create a coastal topography of marine terraces that resembles a series of steps. Each step represents a period of time that a terrace was at sea level. The Palos Verdes peninsula just south of Los Angeles has perhaps as many as 10 marine terraces, each representing a period of platform formation separated by episodes of uplift.

Rates of coastal erosion are controlled by the interaction between wave energy and rock type. Coastal erosion is greatly accelerated during high-energy events, such as severe storms and tsunamis. Human actions can also accelerate coastal erosion rates. We commonly do so by interfering with coastal sediment and vegetation systems that would naturally protect some coastal segments from excessive erosion rates. We explore the nature of coastal depositional systems next.

Coastal Deposition

Significant amounts of sediment accumulate along coasts where wave energy is low relative to the amount or size of sediment supplied. Embayments and settings where waves break at a distance from the shoreline, such as areas with gently shelving underwater topography, tend to sap wave energy and encourage deposition. Amount and size of sediment supplied to the coastal zone vary with rock type, weathering rates, and other elements of the climatic, biological, and geomorphic environment.

Sediment within coastal deposits comes from three principal sources. Most of it is delivered to the standing body of water by streams. At its mouth, the load of a stream may be deposited for the long term in a delta or within an **estuary**, a biologically very productive embayment that forms at some river mouths where saltwater and freshwater meet. Elsewhere, stream load may instead be delivered to the ocean or lake for continued transportation. Once in the standing body of water, fine-grained sediments that stay in suspension for long periods may be carried out to deep water where they eventually settle out onto the basin floor. Other clasts are transported by waves and currents in the coastal zone, being deposited when energy decreases and, if accessible, re-entrained when wave energy increases. The same is true of the second major source of coastal sediment, coastal cliff erosion. Of less importance is sediment brought to the coast from offshore sources. Although we may tend to think of sand-sized sediment when we think of coastal deposits, coastal depositional landforms may be composed of silt, sand, or any size classes of gravel from granules and pebbles through cobbles and boulders.

The most common landform of coastal deposition is the **beach**, a wave-deposited feature that is contiguous with the

D.Sack

(a)

NOAA Image Library

(c)

Jim Petersen

(b)

Richard Price/The Image Bank/Getty Images

(d)

● **FIGURE 20.16** Examples of landforms created by wave erosion. (a) A notch exposed near the base of basaltic sea cliffs in Hawaii. (b) A sea cave carved in limestone along Italy's Mediterranean Sea coast. (c) A sea arch in Alaska. (d) Sea stacks along a cliffed coast.

(a)

(b)

Robert Cameron/The Image Bank/Getty Images

D. Sack

● **FIGURE 20.17** (a) An abrasion platform carved on strongly dipping sedimentary rocks in central California. (b) A marine terrace, the wide, beveled plain atop this eroding coastline.

What other coastal erosional landforms do you see in photo (b)?

mainland throughout its length (● Fig. 20.18). Many beaches are sandy, but beaches of other grain sizes are also common—for example, the cobble beach discussed earlier in the section on coastal erosion. In settings with high wave energy, particles tend to be larger and beaches tend to be steeper than where wave energy is low. Beach sediments come in a variety of colors depending on the rock and mineral types represented. Tan quartz, black basalt, white coral, and even green olivine beaches exist on Earth.

Any given stretch of beach may be a permanent feature, but much of the visible sediment deposited in it is not. Individual grains come and go with swash and backwash; wear away through abrasion; are washed offshore in storms; or move into, along, and out of the stretch of beach by littoral drifting. Because waves generated by closer storms tend to be higher than waves generated in distant storms, some beaches undergo seasonal changes in the amount and size of sediment present.

In the midlatitudes, beaches are generally narrower, steeper, and composed of coarser material in winter than they are in summer. The larger winter storm waves are more erosive and destructive, whereas the smaller summer waves, which often travel from the other hemisphere, are depositional and constructive. On the Pacific coast of the United States, summer beaches are generally temporary accumulations of sand deposited over coarser winter

beach materials (● Fig. 20.19). Sand-sized sediment eroded from the beach in winter forms a deposit called a **longshore bar** that lies submerged parallel to shore, with the sediment returning to the beach in summer. On the Atlantic and Gulf coasts of the United States, the late summer to early fall hurricane season is also a time when beach erosion can be severe.

Whereas beaches are attached to the mainland along their entire length, **spits** are coastal depositional landforms connected to the mainland at just one end (● Fig. 20.20). Spits project out into the water like peninsulas of sediment. They form where the mainland curves significantly inland while the trend of the longshore current remains at the original orientation. Sediments accumulate into a spit in the direction of the longshore current (● Fig. 20.21a). Where similar processes form a strip of sediment connecting the mainland to an island, the landform is a **tombolo** (Fig. 20.21b).

Another category of coastal landforms are **barrier beaches**, elongated depositional features constructed parallel to the mainland. Barrier beaches act to protect the mainland from direct wave attack. All barrier beaches have restricted waterways, called **lagoons**, that lie between them and the mainland. Salinity in the lagoon varies from that of the open water body, depending on freshwater inflow and evaporation, and affects organisms living in the lagoon. Like beaches and spits, barrier beaches have a

National Park Service
(a)

NOAA/Captain Albert E. Theberge
(b)

U.S. Coast Guard
(c)

D. Sack
(d)

● **FIGURE 20.18** Beaches are the most common evidence of wave deposition and may be made of any material deposited by waves. (a) A sandy beach on the eastern shore of Lake Michigan. (b) A boulder beach in Arcadia National Park, Maine (note the person for scale). (c) A tropical, white sand beach made of broken coral pieces. (d) A sea turtle enjoying a beach of black basalt on the island of Hawaii.

University of Washington Libraries, Special Collections, John Shelton Collection
(a)

University of Washington Libraries, Special Collections, John Shelton Collection
(b)

● **FIGURE 20.19** Because of seasonal variations in wave energy, the differences in a beach from summer to winter can be striking, particularly in the midlatitudes. (a) Waves in summer are generally mild and deposit sand on the beach. (b) Winter waves from storms nearer to and at the beach remove the sand, leaving boulders and bare bedrock in the beach area.

What attribute of waves represents the amount of energy they have?

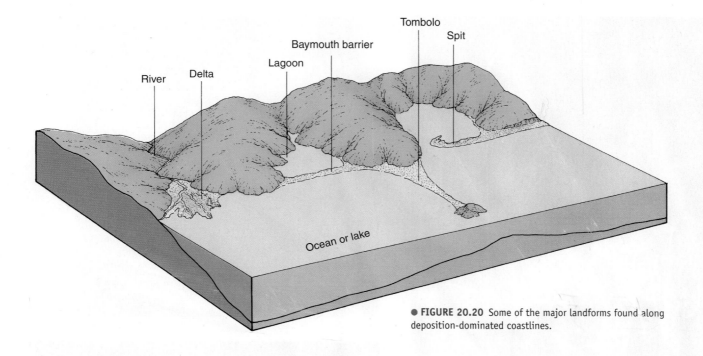

River Delta Lagoon Baymouth barrier Tombolo Spit

Ocean or lake

● **FIGURE 20.20** Some of the major landforms found along deposition-dominated coastlines.

USGS Coastal & Marine Geology Program

(a)

NOAA/Captain Albert E. Theberge

(b)

USGS Coastal & Marine Geology Program

(c)

● **FIGURE 20.21** (a) A spit is attached to the mainland at one end. (b) A tombolo connects a nearby island with the mainland at Point Sur, California. (c) A baymouth barrier crosses the mouth of an embayment connecting to land at both ends.

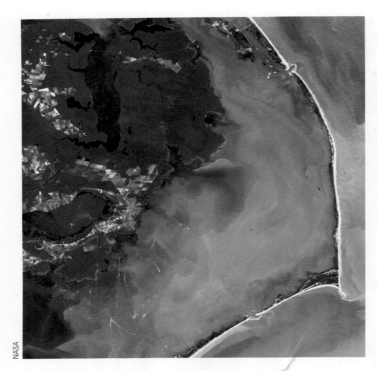

● **FIGURE 20.22** Barrier islands lie parallel to the mainland but are not attached to it. This barrier island is located near Pamlico Sound on the North Carolina coast.

What feature separates a barrier island from the mainland?

submerged part and a portion that is always above water, except in extreme storm conditions or extremely high tide. This contrasts with bars, like the longshore bars discussed previously, which are submerged except in extreme conditions.

There are three kinds of barrier beaches. A **barrier spit** originated as a spit and thus is attached to the mainland at one end but has extended almost completely across the mouth of an embayment to restrict the circulation of water between it and the ocean or lake. If the barrier spit crosses the mouth of the embayment to connect with the mainland at both ends, it becomes a **baymouth barrier** (Fig. 20.21c). Limited connection is maintained between the lagoon and the main water body through a breach or *inlet* cut across the barrier somewhere along its length. The position of inlets can change during storms. **Barrier islands** are likewise elongated parallel to the mainland and separate lagoons and the mainland from the open water body, but they are not attached to the mainland at all (● Fig. 20.22).

Barrier islands are common features of low-relief coastlines. They dominate the Atlantic and Gulf coasts of the United States from New York to Texas. Some excellent examples of long barrier islands are Fire Island (New York), Cape Hatteras (North Carolina), Cape Canaveral and Miami Beach (Florida), and Padre Island (Texas).

Rising sea level since the Pleistocene appears to have played a major role in the formation of barrier islands. They migrate landward over long periods and may change drastically during severe storms, especially hurricanes (● Fig. 20.23).

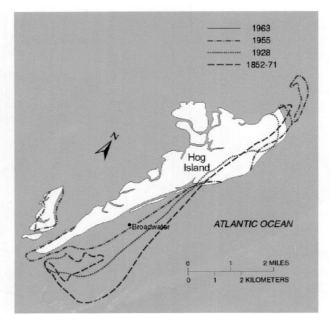

(a)

(b)

(c)

● **FIGURE 20.23** (a) Historic maps show us that barrier islands, like this one off the coast of Virginia, shift their shapes and positions over time but especially during storm events. (b) Homes along the shore of a barrier island before a hurricane. (c) The same homes after a hurricane.

How can this type of damage be prevented in the future?

USGS

● **FIGURE 20.24** A series of groins on the Atlantic shoreline at Norfolk, Virginia, captures sand to maintain the beach along one stretch of the coastline.

How do you think the stretch of coast beyond the last groin would be affected by these structures?

Not only do beaches serve recreational needs but they also help protect coastal settlements from erosion and flooding by storm waves. Beach systems are in equilibrium when input and output of sediment are in balance. People interrupt natural coastal depositional processes for various reasons, such as to keep harbors free of sediment or to encourage growth of recreational beaches. To increase the size of a beach, people build artificial obstructions to the principal direction of littoral drifting. A **groin** is an obstruction, usually a concrete or rock wall, built perpendicular to a beach to inhibit sediment removal by littoral drifting while the input of sediment from littoral drifting remains the same. This obstruction, however, starves the adjacent, downcoast beach area of material input while the rate of sediment removal from that adjacent beach does not change (● Fig. 20.24). Human actions deplete the natural sediment supply to the coast by damming rivers. With decreased sediment supply, beaches become narrower and lose some of their ability to protect the coastal region against storms. In Florida, New Jersey, and California, hundreds of millions of dollars have been spent to artificially replenish sandy beaches that have lost significant amounts of sediment, and thus their recreational and protective abilities, because of human actions.

Types of Coasts

Coasts are spectacular, dynamic, and complex systems that are influenced by tectonics, global sea-level change, storms, and marine and continental geomorphic processes. Because of this complexity, there is no single classification system for coasts. Each major way of classifying coasts focuses on a different characteristic of coastal systems; all aid our understanding of these natural, complex systems.

On a global scale, coastal classification is based on plate tectonic relationships. This system has two major coastal types: passive-margin coasts and active-margin coasts. **Passive-margin coasts** are tectonically quiescent, with little mountain-building or volcanic activity. Like the eastern seaboard of the United States, which is an excellent example, these coasts generally have low relief with broad coastal plains and wide continental shelves (● Fig. 20.25), but passive-margin coasts that are relatively young, such as those of the Red Sea and Gulf of California, may have somewhat greater relief. Passive-margin coasts are typically dominated by depositional landforms and are well represented

● **FIGURE 20.25** The general nature of the boundary between the continents (continental crust) and ocean basins (oceanic crust). The gently sloping continental shelf varies in width along different coasts depending on plate tectonic history and proximity to plate margins.

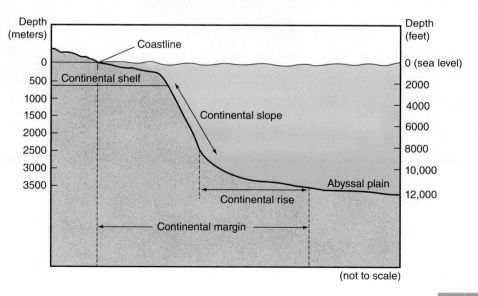

Beach Protection

Beaches act as buffers to help protect the land behind them from wave erosion. Under natural conditions, sediment eroded from the beach in a storm will eventually be replaced through the action of gentle waves and by littoral drifting bringing in new sediment. People interfere with the natural sediment budget when they dam rivers and build on the normally shifting sediment of beach and dune systems in the coastal zone. Intense human development, like that found at many popular tourist beaches, typically interrupts the supply of replacement sediment and leads to more permanent erosion of beaches. Beach erosion leaves the coastal buildings and infrastructure susceptible to damage from storm waves while eliminating the primary attraction, the beach itself.

Beach protection and restoration strategies include the construction of *groins* built perpendicular to the trend of the coastline to slow down the rate of littoral drifting of sediment out of the area. *Jetties* are also built perpendicular to the coastline, but always in pairs, and act to keep sediment from blocking an inlet, such as a river mouth or a channel for boats. *Breakwaters* are walls built parallel to the shoreline in the breaker zone. Large waves expend the bulk of their energy breaking on the structure, thus limiting the amount of beach erosion. Another strategy has been to add sediment to the beach artificially by dredging harbors or reservoirs and trucking or pumping that sediment through pipes to the beach. Unless the factors that are limiting the

natural sediment supply to the beach are addressed, this artificial form of *beach nourishment* will likely have to continue, at least periodically.

Miami Beach, Florida, has long been a popular destination for millions of annual visitors, who are drawn to the magnificent beaches. After decades of development, by 1970 the beaches had mostly disappeared because of erosion. With the beaches gone, shoreline condominiums and hotels were threatened with serious damage from storm waves, and the hotels were half empty as tourists traveled to other destinations.

To help solve the problem, the U.S. Army Corps of Engineers (the federal agency responsible for the development of inland and coastal waterways) chose Miami Beach to experiment with the *beach*

Visitors and residents of Miami Beach benefit from the artificial replacement of sand to the beach.

building method of shoreline protection. The technology involves transferring millions of cubic meters of sand from offshore to replenish existing beaches or build new ones where beaches have eroded away. Huge barge-mounted dredges dig sand from the sea bottom near the shore, and the sand is pumped as a slurry through massive movable tubes to be deposited on a beach.

The initial project cost $72 million, but it was so successful that within 2 years Miami Beach again had a sandy beach 90 meters (300 ft) wide and 16 kilometers (10 mi) long. With the return of the beach, the number of visitors soon grew to three times the number prior to beach building.

Today beach building has become the accepted way to respond to beach erosion. Nearly $1.7 billion was spent on beach restoration during the last decade of the twentieth century. Since 1950, the Corps of Engineers has undertaken beach nourishment projects on more than 565 kilometers (350 mi) of coastline. According to the Corps, these projects are meant primarily to protect buildings on or near the beach, rather than to provide beaches for recreational purposes. In the view of the Corps of Engineers, the true value of beach building can best be measured in savings from storm damage. For example, it has been estimated that beach replenishment along the coast of Miami-Dade County prevented property losses of $24 million during Hurricane Andrew in 1992.

When beaches are rebuilt or widened to protect against storms, continued erosion is inevitable. In deciding how much sand must be pumped in, it is important to determine the beach width necessary to protect shoreline property from storms. This is usually about 30 meters (100 ft). Then an amount of additional "sacrificial" sand is added to produce a beach twice the width of the desired permanent beach. In about 7 years, the sacrificial sand will be lost to wave erosion. If beach rebuilding is repeated each time the sacrificial sand has been removed, the permanent beach will remain in place and the coastal zone will be protected. An estimated $5.5 billion has already been committed to the continuous rebuilding of existing projects over the next 50 years.

Is the battle with the environment at the beachfront worth the price we are paying? Many people are not so sure, but most environmental scientists respond to the question with a resounding no because they are concerned with a price that is not measured solely in appropriated dollars. This price is paid in destroyed natural beach environments, reduced offshore water quality, eliminated or displaced species, and repeated damage to food chains for coastal wildlife each time a beach is rebuilt. Clearly, beach building is a mixed blessing.

Seagulls enjoy the swash on this sand-replenished beach in Alameda, California, on San Francisco Bay.

by the coastal regions of continents along the Atlantic Ocean (● Fig. 20.26). Most major tectonic activity within the Atlantic occurs in the center of the ocean along the Mid-Atlantic Ridge, rather than along its coastlines.

Active-margin coasts are best represented by coastal regions along the Pacific Ocean (● Fig. 20.27). There, most tectonic activity occurs around the ocean margins because of active subduction and transform plate boundaries along the Pacific Ring of Fire. Active-margin coasts are usually characterized by high relief with narrow coastal plains, narrow continental shelves, earthquake activity, and volcanism. These coasts tend to be erosional, having less time within Earth history for the development of marine or continental depositional features. The coast of Japan and the west coast of the United States are excellent examples of active-margin coasts.

On a regional scale, coasts can be classified as coastlines of emergence or coastlines of submergence. **Coastlines of emergence** occur where the water level has fallen or the land has risen in the coastal zone. In either case, land that was once below sea level has emerged above the water. Evidence for emergence includes marine terraces and relict sea cliffs, sea stacks, and beaches found above the reach of present wave action. Many coastlines of emergence would have existed during the glacial phases of the Pleistocene Epoch when sea level was about 120 meters (400 ft) lower than it is today. Features of emergence are common along active-margin coasts such as those of California, Oregon, and Washington, where tectonic uplift has raised coastal landforms as much as 365 meters (1200 ft) above sea level (● Fig. 20.28). Other emergent coastlines, such as around the Baltic Sea and Hudson Bay, are located where isostatic rebound has elevated the land following the retreat of the continental ice sheets.

Stockbyte/Getty Images

● **FIGURE 20.26** A sandy passive-margin coast on the Atlantic Ocean at Marconi Beach, Cape Cod National Seashore, Massachusetts.

Jim Petersen

● **FIGURE 20.27** The rugged coast of Point Lobos, California, exemplifies an active-margin coastline, having experienced much tectonic unrest that includes general uplift in addition to great lateral movement along the San Andreas Fault.

USGS

● **FIGURE 20.28** Cape Blanco, Oregon, represents an emergent coastline. The flat surface on which the lighthouse is built is a marine terrace formed by wave erosion and deposition prior to tectonic uplift. The elevation of the marine terrace is about 61 meters (200 ft) above sea level.

Along **coastlines of submergence** many features of the former shore lie underwater and the present shoreline crosses land areas that are not fully adjusted to coastal processes. Coastlines of submergence were created as global sea level rose in response to the retreat of the Pleistocene ice sheets. Coastlines of submergence also occur where tectonic forces have lowered the level of the land, as in San Francisco Bay. Great thickness of river deposits and compaction of alluvial sediments, as along the Louisiana coast, may also cause coastal submergence. The specific features present along a coastline of submergence are related to the character of coastal lands prior to submergence. Plains, for instance, will produce a far more regular shoreline than will a mountainous region or an area of hills and valleys. When areas of low relief with soft sedimentary rocks are submerged, barrier islands form with shallow bays and lagoons behind them. The classic examples of this type of submerging coastline are the Atlantic and Gulf coasts of the United States.

Two special types of submerged coastlines are ria and fjord coasts. **Rias** are created where river valleys are drowned by a rise in sea level or a sinking of the coastal area (● Fig. 20.29). These irregular coastlines result when valleys become narrow bays and the ridges form peninsulas. The Aegean coast of Greece and Turkey is an outstanding example of a ria coastline. *Fjords*, which are drowned glacial valleys, form spectacular scenic shorelines (● Fig. 20.30). A coastline with fjords is highly irregular, with deep, steep-sided embayments penetrating far inland in U-shaped valleys originally deepened by glaciers. Tributary streams cascade down fjord sidewalls, which may be a few thousand meters high. Fjord coastlines are found in Norway (where the term originated), Chile, New Zealand, Greenland, and Alaska. Canada, however, has more fjords than any other nation. In many fjords, the glaciers have retreated far inland, but in others, especially in Greenland and Alaska, glaciers still reach the fjords calving icebergs into their cold ocean water.

Some coastlines, such as those composed of river deltas, cannot be classified as either submerging or emerging. Actually, most shorelines show evidence of more than one type of development largely because the land elevation and the level of the ocean have changed many times during geologic history. For this reason, features of both submerged and emerged shorelines characterize some coastal areas.

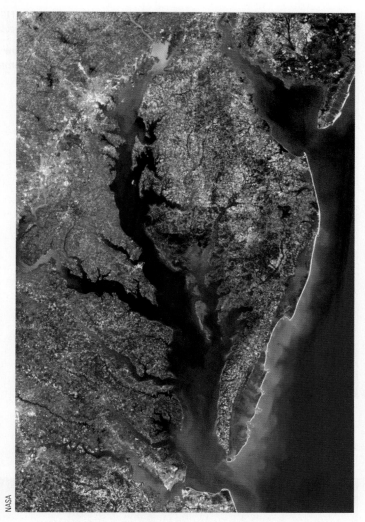

NASA

● **FIGURE 20.29** Submergent coasts like the Chesapeake Bay region are characterized by drowned river valleys, known as rias, that developed as sea level rose at the end of the Pleistocene.

● **FIGURE 20.30** Fjords, like this one in Greenland, are glaciated valleys that were drowned by the sea following the Pleistocene Epoch as the glaciers receded and sea level rose.

© Matt Ebiner

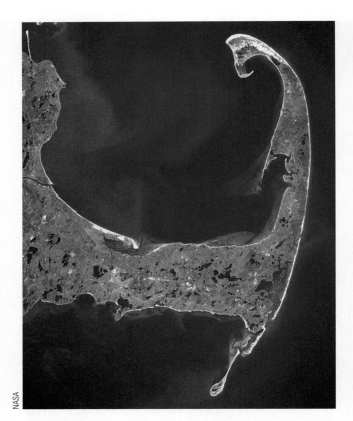

NASA

How has wave action modified the moraine that originally formed Cape Cod?

Because both continental and marine geomorphic processes shape coastlines, another regional classification system recognizes two types of coasts: primary and secondary coastlines. Erosional and depositional processes of the land dominate **primary coastlines**. Primary coastlines result from such rapid changes in the position of the shoreline that coastal processes are not able to have a significant landforming effect. The major types of primary coastlines (with examples) are drowned river valleys (Delaware and Chesapeake Bays), glacial erosion coasts (southeastern Alaska, British Columbia, Puget Sound in Washington, and from Maine to Newfoundland), glacial deposition coasts (Cape Cod, Massachusetts [● Fig. 20.31], and the north shore of New York's Long Island), river deltas (Mississippi Delta, Louisiana), volcanic coasts (Hawaii), and faulted coasts (California).

Secondary coastlines are those formed mainly by coastal geomorphic agents, especially waves, and by aquatic organisms. Sea cliffs, arches, sea stacks, and sea caves generally dominate erosional secondary coasts, like that of Oregon. Depositional secondary coasts typically display barrier beaches and spits, such as the coastline of North Carolina. An example of coasts built by aquatic organisms is the coral reef (● Fig. 20.32). The Florida

● **FIGURE 20.32** Corals building the coastline at Vieques National Wildlife Refuge, Puerto Rico, provide an example of a secondary shoreline.

U.S. Fish and Wildlife Service

Keys were constructed by coral growth. Mangrove trees and salt-marsh grasses also trap sediment to build new land areas in shallow coastal waters.

Islands and Coral Reefs

The perimeters of islands are subject to the same coastal processes as the boundary between a standing body of water and the mainland. Within the ocean there exist three basic types of islands: continental, oceanic, and atolls. **Continental islands** are geologically part of the continent. Continental islands are usually found on the part of a continent submerged by the ocean, the continental shelf. Most large continental islands became separated from the continent because of global sea-level change or regional tectonic activity. A few large continental islands, such as New Zealand and Madagascar, are isolated continental fragments that separated from continents millions of years ago. The world's largest islands—Greenland, New Guinea, Borneo, and Great Britain—are continental. Smaller continental islands include the barrier islands off the Atlantic and Gulf coasts of the United States, New York's Long Island, California's Channel Islands, and Vancouver Island off the west coast of Canada.

Oceanic islands are volcanoes that rise from the deep-ocean floor and are geologically related to oceanic crust, not the continents. Most oceanic islands, such as the Aleutians, Tonga, and the Marianas, occur in island arcs along the edges of the trenches. Others, like Iceland and the Azores, are peaks of oceanic ridges rising above sea level. Many oceanic islands, like the Hawaiian Islands, occur in chains. The oceanic crust sliding over a stationary "hot spot" in the mantle causes these island chains. The Hawaiian Islands are moving northwestward with the Pacific plate and will slowly submerge. A new volcanic Hawaiian Island, named Loihi, is forming to the southeast (● Fig. 20.33). Evidence of the plate motion is indicated by the fact that the youngest islands of the Hawaiian chain, Hawaii and Maui, are to the southeast, whereas the older islands, such as Kauai and Midway, are located to the northwest.

An **atoll** is an island consisting of a ring of coral reefs that have grown up from a subsiding volcanic island and that encircle a central lagoon (● Fig. 20.34a). To understand atolls, we must first consider coral reefs.

Coral reefs are shallow, wave-resistant structures made by the accumulation of remains of tiny sea animals that secrete a skeleton of calcium carbonate. Many other organisms, including algae, sponges, and mollusks, add material to the reef structure. Reef corals need special conditions to grow—clear and well-aerated water, water temperatures above 20°C (68°F), plenty of sunlight, and normal marine salinity. These conditions are found in the shallow waters of tropical regions, including Hawaii, the West Indies, Indonesia, the Red Sea, and the coast of Queensland in Australia. Today coastal water pollution, dredging, souvenir coral collecting, and climate change threaten the survival of many coral reefs.

A **fringing reef** is a coral reef attached to the coast (Fig. 20.34b). Fringing reefs tend to be wider where there is more wave action that brings a continuous supply of well-aerated water and additional nutrients for increased coral growth. They are usually absent near river mouths because the coral cannot grow where the waters are laden with sediment or where river water lowers the salinity of the marine environment.

● **FIGURE 20.33** The oceanic island of Hawaii was formed over a "hot spot" like the other Hawaiian Islands before it. Two large shield volcanoes, Mauna Loa and Mauna Kea, dominate the island. To the south, Loihi, an active submarine volcano, may reach sea level to become the newest Hawaiian Island in 50,000 years.

Tahiti Tourism Board

(a)

David Hiser/Getty Images

(b)

Paul Chelsey/Stone/Getty Images

(c)

● **FIGURE 20.34** The major types of coral reefs are evident in the Society Island chain of French Polynesia. (a) Atolls are islands consisting of a ring of coral with no surface evidence remaining of its former volcanic core. (b) Fringing reefs are attached to mainland or island coasts. (c) Like coastal barriers in general, barrier reefs are separated from dry land by a lagoon.

Sometimes coral forms a **barrier reef**, which lies offshore, separated from the land by a shallow lagoon (Fig. 20.34c). Most barrier reefs occur in association with slowly subsiding oceanic islands, growing at a pace that keeps them above sea level. Other barrier reefs, including Australia's Great Barrier Reef, the Florida Keys, and the Bahamas, were formed on continental shelves and grew upward as sea level rose after the Pleistocene ice age waned. At more than 2000 kilometers (1250 mi) long, the world's largest organic structure is the Great Barrier Reef of Australia.

● Figure 20.35 illustrates the manner in which atolls develop. As a volcanic island subsides, the fringing reef grows upward, keeping pace with the seafloor subsidence, becoming a barrier reef when a lagoon develops and finally an atoll when the only solid material is the outer ring of coral. Charles Darwin proposed this explanation of atoll formation in the 1830s. Evidence from drilling indicates that there have been as much as 1200 meters (4000 ft) of subsidence and an equal amount of reef development in the past 60 million years.

Atolls pose severe challenges as environments for human habitation. First, they have a low elevation above sea level and provide no defense against huge storm waves and tsunamis that can inundate the entire atoll, drowning all its inhabitants. Rising sea level due to global warming also threatens these low islands. Second, there is little freshwater available on the porous coral limestone surface. Third, little vegetation can survive in the lime-rich rock and soil of the atoll islands. The coconut palm is an exception, and coconuts were vital to the survival of early inhabitants of the atoll islands in Polynesia and Micronesia.

Change over Time

Like many other facets of our physical environment, Earth's coastal systems represent beautiful and useful, yet hazardous settings. Knowing the fundamentals of tides, tsunamis, and wind waves; processes and landforms of coastal erosion, transportation, and deposition; and the different types of coastlines helps us appreciate the power of coastal processes as well as the sensitivity of coastal zones to natural and human-induced alteration.

Also like other Earth subsystems, coastal systems are subject to change on multiple scales. For example, although a beach might not appear to change in one day, individual grains come and go in response to such factors as variations in the energy and orientation of waves. Over months and years, the beach can become narrower if the longshore current is interrupted by groins or the clastic load of a nearby stream is trapped behind a dam. Over the past thousands of years, tectonism and climate change have varied the relative position of land and sea along the coast, leaving former shorelines abandoned both above and below the present shoreline position.

Sea level generally rose over the 20th century, and it is expected to rise at a faster rate in the 21st century as climate change takes its toll on the Greenland and Antarctic ice sheets and Earth's ice caps, combined with thermal expansion of the ocean. Studies

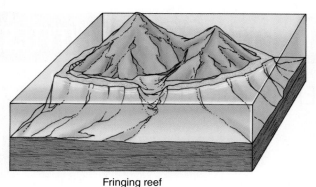

Fringing reef

(a)

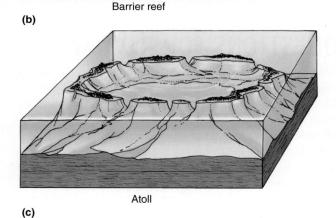

Barrier reef

(b)

Atoll

(c)

● **FIGURE 20.35** Charles Darwin proposed the three-part theory of how coral reefs develop around oceanic islands. (a) First, a fringing reef grows along the shore. (b) Later, as the island erodes and subsides, a barrier reef develops. (c) Further subsidence causes the coral to build upward while the volcanic core of the island is completely submerged below a central lagoon, forming an atoll.

suggest a sea level rise of approximately 32 centimeters (12.6 in.) by 2050. In addition to the huge expense of trying to protect or move structures that people have built close to sea level, this rise in ultimate base level will affect the behavior of streams that reach the ocean, and new spatial configurations of the shoreline will cause new patterns of erosion, transportation, and deposition along many coasts. Change in one part of the Earth system instigates adjustments in other parts. Careful observations of how wave-produced landforms, streams, dunes, terrestrial and marine ecology, storm activity, and many other variables in coastal zones respond to climate change, either directly or indirectly, should reveal much about the nature of the interconnections of these subsystems.

CHAPTER 20 ACTIVITIES

■ TERMS FOR REVIEW

shoreline	littoral drifting	baymouth barrier
sea level	beach drifting	barrier island
swash	longshore drifting	groin
backwash	notch	passive-margin coast
wave steepness	cobble beach	active-margin coast
wave period	sea stack	coastlines of emergence
tsunami	abrasion platform	coastlines of submergence
wind wave	marine terrace	ria
spring tide	estuary	primary coastline
neap tide	beach	secondary coastline
sea	longshore bar	continental island
swell	spit	oceanic island
fetch	tombolo	atoll
wave base	barrier beach	coral reef
rip current	lagoon	fringing reef
wave refraction	barrier spit	barrier reef

■ QUESTIONS FOR REVIEW

1. What is the difference between a shoreline and the coastal zone?
2. Describe the major factors that produce the tides. What are some variations in tidal patterns?
3. How do waves change when they enter shallow coastal waters? What is the main factor causing this change in the waves?
4. What is wave refraction? How is wave refraction related to the shape of the coastline?
5. Describe how sea cliffs form. Name three other coastal erosional landforms typically found in areas with sea cliffs.
6. What are the differences between beach drifting and littoral drifting? What causes both?
7. Describe the similarities and differences between beaches and barrier beaches.
8. How does a longshore bar differ from a barrier?
9. What are the major differences between active-margin coastlines and passive-margin coastlines?
10. What is a coral reef, and how does it form?

■ CONSIDER AND RESPOND

1. What major changes would you expect in the world's coastal zones if sea level rises substantially as a result of global warming? How would the landforms change?
2. Explain some of the various ways that people can influence coastal geomorphology.
3. How are systems of coastal classification useful in studying coastal environments?
4. If you were asked to plan for construction of a major tourist resort on a beautiful tropical atoll, what limitations and concerns would you need to evaluate before construction?

PRACTICAL APPLICATIONS

1. A wind wave with a wavelength of 75 meters and a height of 0.8 meter takes 7 seconds to travel past a given point. What is the wave's steepness? What is its period?

2. If a wave 60 meters long travels through water that is 25 meters deep, is the wave feeling bottom? As the wave comes in toward land, how high can it get before it breaks?

LOCATE AND EXPLORE

1. Using Google Earth and the Tide Layer (download from the book's CourseMate at www.cengagebrain.com), examine the hourly change in tidal elevation over the last week for the coastal sites listed here. Identify the tides as either diurnal, semidiurnal, or mixed and rank the sites by the tidal range. Compare the predicted and observed tidal variations. What, other than tides, causes the water level to vary?
   a. Portland, Maine
   b. Fort Pulaski, Georgia
   c. Key West, Florida
   d. Dauphin Island, Alabama
   e. Rockport, Texas
   f. Los Angeles, California
   g. South Beach, Oregon
   h. Seattle, Washington

2. Using Google Earth, examine the barrier islands along the Maryland coast north and south of 38.325°N, 75.090°W. What is the main direction of sediment transport along shore, and what evidence do you see to support your answer? Why do you think the island to the south is displaced landward of the more developed island to the north?

3. Using Google Earth and Hurricane Ivan LIDAR Layer for Santa Rosa Island in northwest Florida (download from the book's CourseMate at www.cengagebrain.com), view the topographic data from before and after Hurricane Ivan (September 2004). The topography of the island was created using Light Detection and Ranging (LIDAR), an increasingly used mapping technology that provides spatially dense and accurate topographic data. The data are collected with aircraft-mounted lasers capable of recording elevation measurements at a rate of 2000 to 5000 pulses per second and with a vertical precision of 0.15 meter. In the prestorm data you will see that the dunes along the beach are highly variable in height and extent, whereas in the poststorm data you will see that the dunes are gone and replaced by overwash fans of different sizes.
   a. Change the transparency of the pre-Ivan LIDAR image and use the ruler tool to measure the change in the position of the shoreline as a result of Hurricane Ivan. Now make the post-Ivan image transparent and measure the change in shoreline after Hurricane Ivan.
   b. Using both LIDAR images, describe how the overwash fan development is related to prestorm dune height. Can you develop a simple model to predict the impact of a hurricane on a barrier island using the height of the dunes relative to the elevation of the hurricane's storm surge?

 MindTap—Make the most of your study time by accessing everything you need to succeed in one place. Read your textbook, take notes, review flashcards, watch videos, complete activities, take practice quizzes, and more online with MindTap. Log in at **www.cengagebrain.com**.

Map Interpretation

■ ACTIVE-MARGIN COASTLINES

The Map

Point Reyes National Seashore lies north of San Francisco on the rugged California coast. Point Reyes consists of resistant bedrock that is being eroded by coastal processes. Marine life is abundant, including many birds and marine mammals (note Sea Lion Cove below Point Reyes). A hilly area, Punta de Los Reyes, separates Drake's Bay and Point Reyes from Tomales Bay (barely showing in the northeastern corner of the map).

The oblique aerial photo of Point Reyes was taken from a high-altitude NASA aircraft. Healthy vegetation appears red on color infrared images. North is at the top. The San Andreas Fault lies just northeast of the area shown on the image. It separates the Pacific plate, on which Point Reyes is located, from the North American plate, which underlies the area east of the fault. Point Reyes is moving to the northwest along this fault. Emergent coastal features and recent tectonic activity characterize active-margin coastlines.

Point Reyes has a Mediterranean climate, influenced by the cool offshore California Current. Here, the ocean is not hospitable for swimming because of the uncomfortably cool seawater, high surf, and dangerous rip currents. The coastal location and onshore westerly winds create a truly temperate climate. Although very hot and very cold temperatures are rarely experienced, this is one of the windiest and foggiest coastlines in the United States. It is an area of rugged natural beauty with wind-sculpted trees, grasslands, rocky sea cliffs, and long sandy beaches.

Interpreting the Map

1. Which area of the coast is most exposed to wave erosion? What features indicate this type of high-energy activity?
2. Which area of the map is under the influence of strong longshore currents? What is the general direction of flow, and what coastal feature would indicate this flow?
3. Looking into the future, what may happen to Drakes Estero? What would Limantour Spit become?
4. Which area of the map appears to have strong wind activity? What would indicate this? What do you think the prevailing wind direction is?

5. Locate examples of the following coastal erosional landforms:
   a. Headland
   b. Sea stack
   c. Sea cliffs
6. Note that the bays and esteros (Spanish for estuary) have mud bottoms. Because the creeks and streams are very small in the region, what is the probable source and cause for movement of the mud?
7. Note the offshore bathymetric contours (blue isolines). Which has a steeper gradient, the Pacific coast (west side) or Drakes Bay (east side)? Why do you think there is such a great difference?

High-altitude oblique aerial photo of Point Reyes, California.

⊕ In Google Earth fly to: 38.001°N, 122.997°W.

Opposite:
Point Reyes, California
Scale 1:62,500
Contour interval = 80 ft
USGS

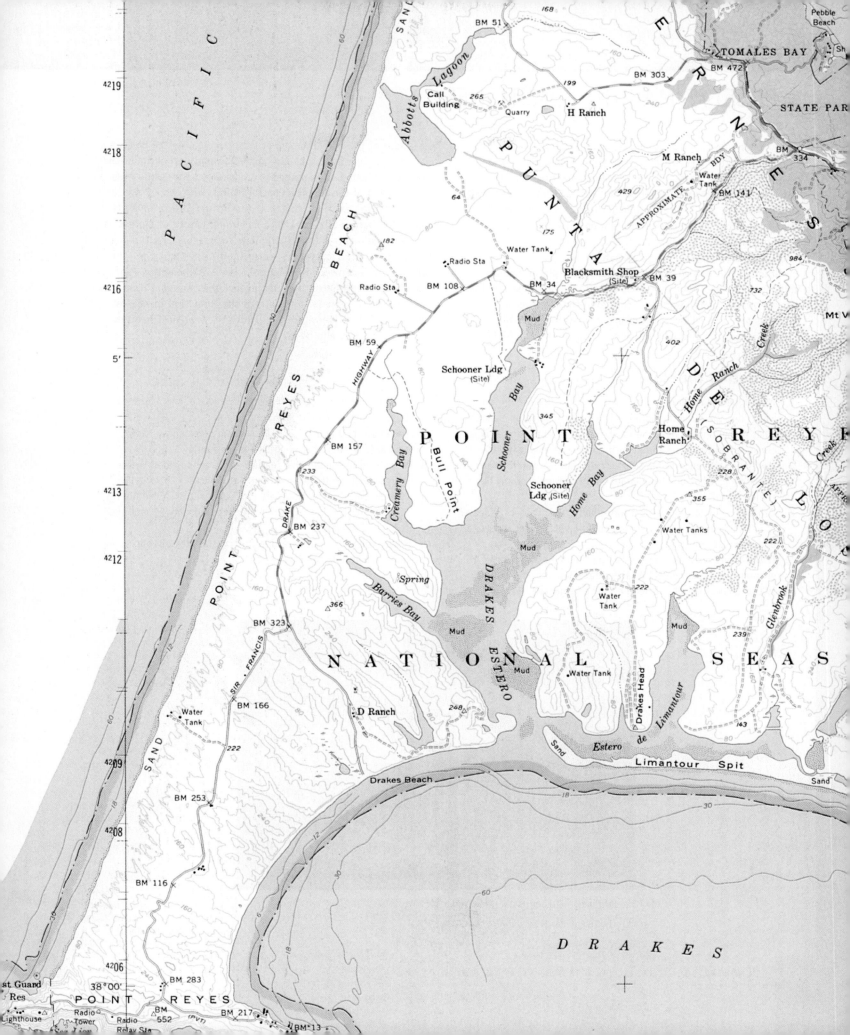

Map Interpretation

PASSIVE-MARGIN COASTLINES

The Map

Eastport, New York, is located on the south shore of Long Island, 115 kilometers (70 mi) east of New York City. Long Island is part of the Atlantic Coastal Plain, which extends from Cape Cod, Massachusetts, to Florida. Water bodies such as Delaware Bay, Chesapeake Bay, and Long Island Sound embay much of the Atlantic Coastal Plain in the eastern United States. The south shore of Long Island has low relief, and its coastal location moderates its humid continental climate.

Although this is a coastal region, its recent glacial history influences the landforms that exist today. Two east-west trending glacial terminal moraines deposited during the Pleistocene glacial advances form Long Island. Between the coast and the south moraine is a sandy glacial outwash plain that forms the higher elevations at the northern part of the map area. As the glaciers melted, sea level rose, submerging the lowland now occupied by Long Island Sound. This water body separates Long Island from the mainland of the Atlantic Coastal Plain.

The Atlantic and Gulf coasts of the United States have nearly 300 barrier islands with a combined length of more than 2500 kilometers (1600 mi). New York, especially the coastal zone of Long Island's south shore, has 15 barrier islands with a total length of more than 240 kilometers (150 mi). Coastal barrier islands protect the mainland from storm waves, and they contribute to the formation of coastal wetlands on their landward side, which are a critical habitat for fish, shellfish, and birds.

Interpreting the Map

1. What type of landform is the entire long, narrow feature on which lies the straight shoreline labeled Westhampton Beach?

2. What is the highest elevation on the linear coastal feature? What do you think comprises the highest portions of this feature?

3. How much evidence of human use exists on the landform in Question 1? What problems might these cultural features be subjected to?

4. Behind the linear feature is a water body. Is it deep or shallow? Is it a high-energy or low-energy environment? What is a water body such as this called?

5. Is the coastline shown on the Eastport map predominantly erosional or depositional? What features support your answer? Is this a coastline of submergence or emergence?

6. Note Beaverdam Creek in the upper middle of the map area. Does it have a steep or gentle gradient?

7. Based on its gradient, does Beaverdam Creek always flow seaward? If not, what might influence its flow?

8. What kind of topographic feature exists as indicated by the map symbols in the area surrounding Oneck?

9. How do you think geomorphic processes will likely modify this map area in the future? Which area will probably be modified the most? Explain.

10. What type of natural hazard is this coastal area of Long Island most susceptible to? Why?

Satellite image of Long Island, New York.

In Google Earth fly to: 40.784°N, 72.680°W.

Opposite:
Eastport, New York
Scale 1:24,000
Contour interval = 10 ft
USGS

Appendix A

International System of Units (SI), Abbreviations, and Conversions

Symbol	Multiply	By	To Find	Symbol
Area				
in.²	square inches	645.2	square millimeters	mm²
ft²	square feet	0.093	square meters	m²
yd²	square yards	0.836	square meters	m²
ac	acres	0.405	hectares	ha
mi²	square miles	2.59	square kilometers	km
ac	acres	43,560	square feet	ft²
mm²	square millimeters	0.0016	square inches	in.²
m²	square meters	10.764	square feet	ft²
m²	square meters	1.196	square yards	yd²
ha	hectares	2.47	acres	ac
km²	square kilometers	0.386	square miles	mi²
Mass				
oz	ounces	28.35	grams	g
lb	pounds	0.454	kilograms	kg
g	grams	0.035	ounces	oz
kg	kilograms	2.202	pounds	lb
Length				
in.	inches	25.4	millimeters	mm
ft	feet	0.305	meters	m
yd	yards	0.914	meters	m
mi	miles	1.61	kilometers	km
ft	feet	5280	mile	mi
mm	millimeters	0.039	inches	in.
m	meters	3.28	feet	ft
m	meters	1.09	yards	yd
km	kilometers	0.62	miles	mi
Volume				
gal	U.S. gallons	3.785	liters	l
ft³	cubic feet	0.028	cubic meters	m³
yd³	cubic yards	0.765	cubic meters	m³
l	liters	0.264	U.S. gallons	gal
m³	cubic meters	35.30	cubic feet	ft³
m³	cubic meters	1.307	cubic yards	yd³
Velocity				
mph	miles/hour	1.61	kilometers/hour	km/h
knot	nautical miles/hour	1.15	miles/hour	mph
knot	nautical miles/hour	1.85	kilometers/hour	km/h
km/h	kilometers/hour	0.62	miles/hour	mph
km/h	kilometers/hour	0.54	nautical miles/hour	knots
Pressure or Stress				
mb	millibars	0.75	millimeters of mercury	mm Hg
mb	millibars	0.02953	inches of mercury	in. Hg
mb	millibars	0.01450	pounds per square inch	(lb/in.² or psi)
lbs/in.²	pounds per square inch	6.89	kilopascals	kPa
in. Hg	inches of mercury	33.865	millibars	mb
kPa	kilopascals	0.145	pounds per square inch	(lb/in.² or psi)

Standard Sea-Level Pressure
29.92 in. Hg
14.7 lb/in.^{2}
1013.2 mb
760 mm Hg

Temperature				
°F	Fahrenheit	(°F − 32)/1.8	Celsius	°C
°C	Celsius	1.8°C + 32	Fahrenheit	°F
K	Kelvin	K = °C + 273	Celsius	°C

Powers of Ten			
nano	one billionth	= 10^{-9}	= 0.000000001
micro	one millionth	= 10^{-6}	= 0.000001
milli	one thousandth	= 10^{-3}	= 0.001
centi	one hundredth	= 10^{-2}	= 0.01
deci	one tenth	= 10^{-1}	= 0.1
hecto	one hundred	= 10^{2}	= 100
kilo	one thousand	= 10^{3}	= 1000
mega	one million	= 10^{6}	= 1,000,000
giga	one billion	= 10^{9}	= 1,000,000,000

Appendix B

Topographic Maps

Mapping has changed considerably in recent years, and with the ever-increasing capabilities of computers to store, retrieve, and display graphics, this trend will continue well into the future. In the United States the U.S. Geological Survey (USGS) produces the vast majority of topographic maps available. These maps have long been the tried-and-true tools of geographers and scientists in many other disciplines who study various aspects of the environment. Today virtually all USGS topographic maps are accessible in digital format, for downloading and printing, or for examining on a computer screen. Computers and the Internet have made a great variety of maps widely available and easily accessible for public and professional use. Today it is easy to print or view maps or map segments at home, school, or work. A logical question then would be, will paper maps become obsolete, given computer displays?

There are several reasons why paper maps will still be popular and useful, whether they are purchased from the source or downloaded and printed. Topographic maps are particularly important in fieldwork. Maps are highly portable, require no batteries or electrical power that could fail, and do not suffer from technology glitches. They are easy to use and they provide a good base for making field notes and marking routes.

Today, the USGS maintains and continues to improve a seamless geospatial database of the United States, with many kinds of maps, images, and data in digital form (The National Map at: http://nationalmap.gov/). Areas that once were split on adjacent maps can be now printed on a single sheet. Topographic quadrangle maps (often called quads), however, will continue to be in use for a long time. These maps are now available free, downloadable as PDF files from the USGS website: (http://viewer. nationalmap.gov/basic/?basemap=b1&category=histtopo,ustopo &title=Map%20View).

There are several standard quadrangles, each with a specific scale, and there are also many other special purpose topographic maps.

7.5-minute quads—these are printed at a scale of 1:24,000 and cover 7.5 minutes of longitude and 7.5 minutes of latitude.

15-minute quads—these are printed at a scale of 1:62,500 and cover 15 minutes of longitude and 15 minutes of latitude.

1° × 2° quads—these are printed at a scale of 1:250,000 and cover 1 degree of latitude and 2 degrees of longitude.

1:100,000 metric quads—these are printed at a scale of 1:100,000, cover 30 minutes of longitude and 60 minutes of latitude, and use metric measurements for distance and elevation.

Brown topographic contours are used to show elevation differences and the terrain. Some rules for interpreting contours are given in Chapter 2.

Determining Distances on a Map, Distances from a Map, or an RF Scale It is important to understand representative fractions (RF), such as 1:24,000, which means that any measurement on the map will represent 24,000 of the same measurements on the ground. This knowledge is particularly significant because reproduced maps may not be printed at the original size. On a map that is reduced or enlarged, the bar scale will still be accurate, but the printed RF scale will not. Enlarging or reducing a map changes the RF scale from the original.

How to Find the RF of a Map (or Air Photo, or Satellite Image) of Unknown Scale

Here is the formula:

$$1/RFD = MD/GD$$

The numerator in an RF scale is always the number 1. The RFD is the denominator of the RF (such as 24,000). MD is map distance measured in any particular units on the map (cm, in.). GD is the true ground distance that the map distance represents (expressed in the same units used to measure the map distance). Never mix units in this calculation, but convert values into desired units afterward.

Example: How long in inches is a mile on a 1:24,000 scale map?

Important information: There are 63,360 inches in a mile.
To find MD, for a known distance (mile) on a map of known scale, use this formula:

$$1/24,000 = MD/63,360 \text{ in.}$$
$$1 \text{ mile} = 2.64 \text{ inches at } 1:24,000$$

This statement has now been converted into a **stated scale** so units may be mixed.

1:24,000 Bar Scale Here is a bar scale that can be used to make distance measurements directly from the 1:24,000 maps printed in the Map Interpretation sections. Note: Check the listed RF, because not all are at a 1:24,000 scale.

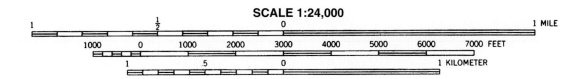

SCALE 1:24,000

BOUNDARIES

National ....

State or territorial ....

County or equivalent ....

Civil township or equivalent ....

Incorporated-city or equivalent ....

Park, reservation, or monument ....

Small park ....

LAND SURVEY SYSTEMS

U.S. Public Land Survey System:

Township or range line ....

Location doubtful ....

Section line ....

Location doubtful ....

Found section corner; found closing corner ...

Witness corner; meander corner ....

Other land surveys:

Township or range line ....

Section line ....

Land grant or mining claim; monument ....

Fence line ....

ROADS AND RELATED FEATURES

Primary highway ....

Secondary highway ....

Light duty road ....

Unimproved road ....

Trail ....

Dual highway ....

Dual highway with median strip ....

Road under construction ....

Underpass; overpass ....

Bridge ....

Drawbridge ....

Tunnel ....

BUILDINGS AND RELATED FEATURES

Dwelling or place of employment: small; large

School; church ....

Barn, warehouse, etc.: small; large ....

House omission tint ....

Racetrack ....

Airport ....

Landing strip ....

Well (other than water); windmill ....

Water tank: small; large ....

Other tank: small; large ....

Covered reservoir ....

Gaging station ....

Landmark object ....

Campground; picnic area ....

Cemetery: small; large ....

RAILROADS AND RELATED FEATURES

Standard gauge single track; station ....

Standard gauge multiple track ....

Abandoned ....

Under construction ....

Narrow gauge single track ....

Narrow gauge multiple track ....

Railroad in street ....

Juxtaposition ....

Roundhouse and turntable ....

TRANSMISSION LINES AND PIPELINES

Power transmission line; pole; tower ....

Telephone or telegraph line ....

Aboveground oil or gas pipeline ....

Underground oil or gas pipeline ....

CONTOURS

Topographic:

Intermediate ....

Index ....

Supplementary ....

Depression ....

Cut; fill ....

Bathymetric:

Intermediate ....

Index ....

Primary ....

Index Primary ....

Supplementary ....

MINES AND CAVES

Quarry or open pit mine ....

Gravel, sand, clay, or borrow pit ....

Mine tunnel or cave entrance ....

Prospect; mine shaft ....

Mine dump ....

Tailings ....

SURFACE FEATURES

Levee ....

Sand or mud area, dunes, or shifting sand ....

Intricate surface area ....

Gravel beach or glacial moraine ....

Tailings pond ....

VEGETATION

Woods ....

Scrub ....

Orchard ....

Vineyard ....

Mangrove ....

COASTAL FEATURES

Foreshore flat ....

Rock or coral reef ....

Rock bare or awash ....

Group of rocks bare or awash ....

Exposed wreck ....

Depth curve; sounding ....

Breakwater, pier, jetty, or wharf ....

Seawall ....

BATHYMETRIC FEATURES

Area exposed at mean low tide; sounding datum ...

Channel ....

Offshore oil or gas: well; platform ....

Sunken rock ....

RIVERS, LAKES, AND CANALS

Intermittent stream ....

Intermittent river ....

Disappearing stream...

Perennial stream ....

Perennial river ....

Small falls; small rapids ....

Large falls; large rapids ....

Masonry dam ....

Dam with lock ....

Dam carrying road ....

Intermittent lake or pond ....

Dry lake ....

Narrow wash ....

Wide wash ....

Canal, flume, or aqueduct with lock ....

Elevated aqueduct, flume, or conduit ....

Aqueduct tunnel ....

Water well; spring or seep ....

GLACIERS AND PERMANENT SNOWFIELDS

Contours and limits ....

Form lines ....

SUBMERGED AREAS AND BOGS

Marsh or swamp ....

Submerged marsh or swamp ....

Wooded marsh or swamp ....

Submerged wooded marsh or swamp ....

Rice field ....

Land subject to inundation ....

Appendix C

Understanding and Recognizing Some Common Rocks

Rocks are aggregates of minerals, and although there are thousands of kinds of rocks on our planet, they can be classified into three fundamental kinds based on their origin: igneous, sedimentary, and metamorphic. The formation of rocks was outlined in Chapter 13. Having a solid knowledge of how the Rock Cycle operates (see Fig. 13.21) as well as its components and its processes is essential to understanding the solid Earth. Specific rock types are mentioned in several chapters of this book. Although making a positive identification of a rock type requires examining several physical properties, having a mental image of what different rocks look like will be an aid to understanding Earth processes and landforms. The following is an illustrated guide to a few common rocks to help in their identification. Intrusive igneous rocks form from a molten state by cooling and crystallizing underground; they generally cool very slowly, which allows coarse crystals to form that are easy to see with the naked eye.

Igneous Rocks

Igneous rocks are subdivided into intrusive and extrusive, depending on whether they cooled within Earth or on its exterior.

Intrusive Igneous

Copyright and photograph by Dr. Parvinder S. Sethi

Granite
Granite forms deep in the crust, and has easily visible intergrown crystals of light and dark minerals, but is dominated by light- colored silicate minerals. Granitic rocks are typically gray or pink, and their mineral composition is similar to that of the continental crust.

Copyright and photograph by Dr. Parvinder S. Sethi

Diorite
Diorite is an intermediate intrusive rock, meaning that it has a roughly equal mix of easily visible light and dark minerals, which gives it a spotted appearance. Generally diorite is dark gray in color.

Copyright and photograph by Dr. Parvinder S. Sethi

Gabbro
Gabbro is a dark, intrusive rock dominated by heavy, iron-rich silicate minerals. The crystals are coarse enough to be easily visible, but because of the overall dark tone, they tend to blend together. Gabbro is black and may contain some very dark green minerals.

Extrusive Igneous

Extrusive igneous rocks cool at or near the surface, and include lavas, as well as rocks made of tephra (pyroclastics), fragments blown out of a volcano. Extrusive rocks sometimes preserve gas bubble holes, and may contain visible crystals, but typically the grains are small. Cooling relatively rapidly at the surface produces fine-grained lavas.

Copyright and photograph by Dr. Parvinder S. Sethi

Rhyolite
Rhyolite is a very thick lava when molten (much like melted glass), light in color and high in silica. Colors of rhyolite vary widely, but light gray, light brown, and pink or reddish are common. Rhyolite is the extrusive equivalent of granite in terms of mineral composition.

Copyright and photograph by Dr. Parvinder S. Sethi

Andesite
Andesite, named for the Andes, is an intermediate lava in terms of both mineral content and color. Associated with composite cone volcanoes, it is relatively thick when molten. Often mineral crystals are visible in a matrix of finer grains, and the color is gray to brown. Andesite is the extrusive equivalent of diorite.

Copyright and photograph by Dr. Parvinder S. Sethi

Basalt
Basalt is dark, typically black, and heavier than other lavas. Associated with fissure flows and shield volcanoes, basalt is relatively low in silica, so it has a lower viscosity than other lavas. Basalt tends to be hotter than other lavas, and is relatively thin when molten, thus it can flow for miles before cooling enough to stop. Basalt is the rock that forms the oceanic crust and is the extrusive equivalent of gabbro.

Copyright and photograph by Dr. Parvinder S. Sethi

Tuff
Tuff is a rock made of fine tephra—volcanic ash—that was blown into the atmosphere by a volcanic eruption, and settled out in layers that blanketed the surface. Tephra (loose fragments) was converted into tuff by burial and compaction, or by being welded together from intense heat. Tuff is gray to tan.

Sedimentary Rocks

Sedimentary rocks can also be divided into two major categories: clastic and nonclastic.

Clastic Sedimentary

Clastic means made up of cemented rock fragments, such as clay, silt, sand, granules, pebbles, cobbles, or even boulders. The sizes and shapes of clasts within a sedimentary rock provide clues about the environments under which the fragments were deposited (fluvial, eolian, glacial, coastal).

Copyright and photograph by Dr. Parvinder S. Sethi

Shale
Shale is a fine-grained clastic rock that contains lithified clays, generally deposited in very thin layers. Shales represent a calm water environment, such as a sea or lake bottom. Shales vary widely in color, but most are gray or black, and they break up into smooth, flat surfaces.

Copyright and photograph by Dr. Parvinder S. Sethi

Sandstone
Sandstone is made of cemented fragments of sand size, typically made of quartz or other relatively hard minerals. Sandstones can be virtually any color, feel gritty, and may be banded or layered. Sandstone may represent ancient beaches, dunes, or fluvial deposits.

Copyright and photograph by Dr. Parvinder S. Sethi

Conglomerate
Conglomerate contains rounded pebbles cemented together by finer sediments. Conglomerate may represent deposits from a river's bed load or a pebble beach.

Copyright and photograph by Dr. Parvinder S. Sethi

Breccia
Breccia is similar to conglomerate, but the cemented fragments are angular. Breccia is associated with mudflows, pyroclastic flows, and fragments deposited by mass wasting.

Nonclastic Sedimentary

Nonclastic sedimentary rocks consist of materials that are not rock fragments. Examples include chemical precipitates, such as limestone, evaporites such as rock salt, and deposits of organic materials, such as coal, or limestones made of shells and coral fragments. Nonclastic rocks also represent the ancient environment under which they were deposited.

Copyright and photograph by Dr. Parvinder S. Sethi

Rock Salt
Rock salt consists of sodium chloride, table salt, often with a mix of other salts. Rock salt represents the deposits left behind by the evaporation of a saline inland lake, or an arm of the sea that was cut off from the ocean by sea-level change or tectonic activity. Common color is white.

Copyright and photograph by Dr. Parvinder S. Sethi

Limestone
Limestone is made of calcium carbonate deposits (lime, $CaCO_3$). Limestone can represent a variety of environments and varies widely in color and appearance. Typical colors are white or gray, and the most common depositional environment was in shallow tropical seas, which were rich in lime and these limestones often contain marine fossils. Many cave and spring deposits are also varieties of limestone.

Copyright and photograph by Dr. Parvinder S. Sethi

Coal
Coal is a rock made of the carbonized remains of ancient plants. Coal deposits typically represent a swampy lowland environment that was invaded by sea-level rise, which killed and buried dense vegetation.

Metamorphic Rocks

Metamorphic rocks can also be divided into two general categories: foliated rocks and nonfoliated rocks.

Foliated

Foliated metamorphic rocks have either wavy, roughly parallel plates, or bands of light and dark minerals that formed under intense heat and pressure. The nature of these foliations indicates the degree of metamorphism or change from the rock's original state.

Copyright and photograph by Dr. Parvinder S. Sethi

Slate
Slate is metamorphosed shale, and looks much like shale, except it is harder and has very thin platy foliation. The most common color is black.

Copyright and photograph by Dr. Parvinder S. Sethi

Schist
Schist has very prominent, wavy, and platy foliation generally covered with mineral crystals that formed during metamorphism. Schist represents a high degree of metamorphism and could originally have been any of a wide variety of rocks, so colors and appearance vary greatly.

Copyright and photograph by Dr. Parvinder S. Sethi

Gneiss
Gneiss (pronounced "nice") is a banded metamorphic rock, with alternating bands of light and dark minerals. Gneiss represents extreme heat and pressure during metamorphism and also may originally have been any of a variety of rocks. Metamorphism of granites commonly produces gneiss.

Nonfoliated

Nonfoliated metamorphic rocks do not display regular patterns of banding or platy foliation. In general, nonfoliated metamorphics represent a rock that has been changed by the fusing and recrystallization of minerals in the original, often identifiable, rock.

Quartzite

Quartzite is metamorphosed quartz sandstone in which the former sand grains have fused together to produce an extremely hard, resistant rock.

Marble

Marble is metamorphosed limestone that has been recrystallized. Many colors and patterns of marble exist, and its relative softness compared to other rocks makes it easy to cut and polish.

Glossary

aa a blocky, angular surface of a lava flow.

abiotic natural, nonliving component of an ecosystem.

ablation any removal of frozen water from the mass of a glacier.

abrasion erosion process in which particles being carried by a geomorphic agent are used as tools to aid in eroding more Earth material.

abrasion platform wave-cut bench of rock just below the water level; indicates the landward extent of coastal cliff erosion.

absolute humidity mass of water vapor present per unit volume of air, in grams per cubic meter, or grains per cubic foot.

absolute location location of an object on the basis of mathematical coordinates on an Earth grid.

absolute zero zero on the Kelvin temperature scale, indicating absence of molecular motion (no temperature).

accretion growth of a continent by adding large pieces of crust along its border by plate tectonic collision.

accumulation any addition of frozen water to the mass of a glacier.

acid mine drainage surface seepage of subsurface water that has become highly acidic by flowing through underground mines.

acid rain rain with a pH of less than 5.6, the pH of natural rain; often linked to pollution associated with the burning of fossil fuels.

active layer the upper soil zone that thaws in summer in regions underlain by permafrost.

active-margin coast coastal region characterized by active volcanism and tectonism.

actual evapotranspiration (actual ET) actual amount of moisture loss through evapotranspiration measured from a surface.

adiabatic heating and cooling changes of temperature within a gas because of compression (resulting in heating) or expansion (resulting in cooling); no heat is added or subtracted from outside.

advection horizontal heat transfer within the atmosphere; air masses moved horizontally, usually by wind.

advection fog fog produced by the movement of warm, moist air across a cold sea or land surface.

aerial photograph photograph of the terrain taken with a camera and film from an aircraft.

aerosols Tiny solid particles or droplets of liquid suspended in the atmosphere.

aggradation building up of an area or landscape that results from more deposition than erosion over time.

air mass huge parcel of air, often subcontinental in size, that can move over Earth's surface as a distinct, relatively homogeneous entity.

albedo percentage of solar radiation received by a surface that is reflected away from that surface.

Aleutian Low center of low atmospheric pressure in the area of the Aleutian Islands, especially persistent in winter.

alfisol soil that has a subsurface clay horizon, is medium to high in bases, and is light colored.

Allen's Rule in warm-blooded species, the relative size of exposed parts of the body decreases with decreasing mean temperatures.

alluvial fan fan-shaped depositional landform, particularly common in arid regions, occurring where a stream emerges from a mountain canyon and deposits sediment on a plain.

alluvial plain extensive floodplain of very low relief.

alluvium general term for clastic particles deposited by a stream.

alpine glacier a mass of flowing ice that exists in a mountain region due to climatic conditions resulting from high elevation.

Altithermal an interval of time about 7000 years ago when the climate was warmer than it is today.

altitude heights of points above Earth's surface, measured from mean sea level.

alto signifies a middle-level cloud (i.e., from 2000 to 6000 m in elevation).

analemma a diagram that shows the declination of the sun throughout the year.

andisol soil that develops on volcanic parent material.

aneroid barometer a device that uses the expansion and contraction of a sealed, accordion-like cell to measure atmospheric pressure.

angle of inclination tilt of Earth's polar axis at an angle of $23\frac{1}{2}°$ from the vertical to the plane of the ecliptic.

angle of repose maximum angle at which a slope of loose sediment can stand without particles tumbling or sliding downslope.

annual march of temperature changes in monthly temperatures throughout the months of the year.

annual temperature lag time lag between the maximum (minimum) of insolation during the solstices and the warmest (coldest) temperatures of the year.

annual temperature range difference between the mean monthly temperatures for the warmest and coolest months of the year.

Antarctic Circle latitude at $66\frac{1}{2}°$S; the northern limit of the zone in the Southern Hemisphere that experiences a 24-hour period of sunlight and a 24-hour period of darkness at least once a year.

anthropogenic change a change in environmental conditions related to or as a direct result of human activities.

anticline the upfolded element of folded rock structure.

anticyclone (high) an area of high atmospheric pressure; also known as a high.

aphelion position of Earth's orbit at farthest distance from the sun during each Earth revolution.

aquiclude rock layer that restricts flow and storage of groundwater; it is impermeable and nonporous.

aquifer rock layer that is a container and transmitter of groundwater; it is both porous and permeable.

Arctic Circle parallel of latitude at $66\frac{1}{2}°$N; the southern limit of the zone in the Northern Hemisphere that experiences a 24-hour period of sunlight and a 24-hour period of darkness at least once a year.

arête jagged, sawtooth spine-like ridge or wall of bedrock separating two glacial valleys or cirques.

arid climate climate region or condition where annual potential evapotranspiration greatly exceeds annual precipitation.

aridisol soil that develops in deserts where precipitation is less than half the potential evaporation.

artesian condition in which groundwater is under sufficient pressure to flow to an outlet if one becomes available.

artesian spring natural flow of groundwater to the surface from below due to pressure.

artesian well *see* flowing artesian well; and nonflowing artesian well.

artificial recharge diverting surface water to permeable terrain for the purpose of replenishing groundwater supplies.

asteroid a solar system body composed of rock and/or metal not exceeding 800 kilometers (500 mi) in diameter.

asthenosphere thick, plastic layer within Earth's mantle that flows in response to convection, instigating plate tectonic motion.

atmosphere blanket of air, composed of various gases, that envelops Earth.

atmospheric disturbance a variation in atmospheric conditions that cannot correctly be classified as a storm—for example, frontal passage, or air mass changes.

atmospheric element refers to the major elements that affect the atmosphere including solar energy, temperature, pressure, winds, and precipitation.

atoll ring of coral reefs and islands encircling a lagoon, with no inner island.

attrition the reduction of size in sediment as it is transported downstream.

aurora colorful interaction of solar radiation with ions in Earth's upper atmosphere; more commonly seen in higher latitudes.

aurora australis the southern equivalent of the aurora borealis (the northern lights).

aurora borealis also called the Northern Lights—illuminations in the night sky, most common in the high latitudes, caused by charged particles interacting with the Earth's magnetic field.

autotroph (producer) organism that, because it is capable of photosynthesis, is at the foundation of a food web and is considered a basic producer.

autumnal equinox the day when the subsolar point is at the equator and that marks the first day of autumn; occurs on the September equinox for the Northern Hemisphere and the March equinox for the Southern Hemisphere.

avalanche density current of pulverized (powdered) Earth material (including ice and snow) traveling rapidly downslope by the pull of gravity.

axis an imaginary line connecting the geographic North Pole and South Pole, around which the planet rotates.

azimuth an angular direction of a line, point, or route measured clockwise from north 0–360°.

azimuthal map projection a projection that preserves the true direction from the map center to any other point on the map.

Azores High *see* Bermuda High.

backwash thin sheet of water from a wave that rushes back down the beach face in the swash zone.

badlands barren region of soft rock material intensely eroded into ridges and ravines by numerous gullies and washes.

bajada an extensive intermediate slope of adjacent, coalescing alluvial fans connecting a steep mountain front with a basin or plain.

bar a shallow, submerged, mounded accumulation of sediment in nearshore locations or in streams.

barchan crescent-shaped sand dune with arms (horns) pointing downwind.

barometer instrument for measuring atmospheric pressure.

barometric pressure (atmospheric pressure) force per unit area that the atmosphere exerts on a surface.

barrier beach a category of elongate depositional coastal landforms that lie parallel to the mainland but separated from it by a lagoon.

barrier island a barrier beach constructed parallel to the mainland, but not attached to it; separated from the mainland by a lagoon.

barrier reef coral reef parallel to the coast and separated from it by a lagoon.

barrier spit a barrier beach attached to the mainland at one end like a spit, but enclosing a lagoon.

basalt a dark-colored, fine-grained extrusive igneous rock generally associated with the oceanic crust and oceanic volcanoes.

basaltic plateau elevated area of low surface relief consisting of horizontal layers of basaltic lava.

base flow groundwater that seeps into stream channels below the water surface; sustains perennial streams between storms.

base level elevation below which a stream cannot flow; most humid region streams flow to sea level (ultimate base level).

base line east-west lines of division in the U.S. Public Lands Survey System.

batholith a large, irregular mass of intrusive igneous rock (pluton) at least 100 km² (40 mi²) in size.

baymouth barrier a barrier beach that extends across the mouth of an embayment, attached to the mainland at both ends, to form a lagoon.

beach coastal landform of wave-deposited sediment attached to dry land along its entire length.

beach drifting littoral drifting in which waves breaking at an angle to the shoreline move sediment along the beach in a zigzag fashion in the swash zone.

bearing an angular direction of a line, point, or route measured from north or from a current location to a desired location (often in 90° compass quadrants).

bed load solid particles moved by wind or water by bouncing, rolling, or sliding along the ground or streambed.

bedding plane boundary between different sedimentary layers.

bedrock solid rock of Earth's crust that underlies soil and other unconsolidated materials.

benthos classification of marine organisms that live on the ocean floor.

Bergeron (ice crystal) process rain-forming process where cloud droplets begin as ice crystals and melt into rain as they fall toward the surface.

Bergmann's Rule Within a warm-blooded species, the body size of the subspecies usually increases with the decreasing mean temperature of its habitat.

bergschrund the large crevasse at the head of a valley glacier, beneath the cirque headwall.

Bermuda High (Azores High) persistent, high atmospheric pressure center located in the subtropics of the north Atlantic Ocean, also called the Azores High.

biogeography The study of the distribution of organisms and their existing and changing relationships.

biological weathering the breakdown of rocks related to vegetational processes or animal activities.

biomass amount of living material or standing crop in an ecosystem or at a particular trophic level within an ecosystem.

biome one of Earth's major terrestrial ecosystems, classified by the vegetation types that dominate the plant communities within the ecosystem.

biosphere the life forms—human, animal, or plant—of Earth that form one of the major Earth subsystems.

blizzard heavy snowstorm with strong winds (35 mph or higher) and snowfall reduces visibility to less than one-quarter mile.

blowout local, wind-eroded surface depression in an area dominated by wind-deposited sand.

bolson desert basin, surrounded by mountains, with no drainage outlet.

boreal forest (taiga) coniferous forest dominated by thin spruce and fir trees growing in subarctic conditions poleward of 50° north latitude.

boulder a rock fragment greater than 256 millimeters (10 in) in diameter.

braided channel stream channel composed of multiple subchannels of simultaneous flow that split and rejoin and frequently shift position.

breaker zone the part of the nearshore area in which waves break.

butte isolated erosional remnant, a hill with a flat summit formed by a caprock and is often bordered by steep-sided escarpments. Buttes are usually found in arid regions of flat-lying sediments and are somewhat taller than they are wide.

buttressed trunk tree trunk with lateral extensions of roots flaring out above the ground from the lower part of a tree.

buttressed trunks Most common in tropical rainforests, they also occur in some midlatitude and subtropical forests with very wet climates.

calcification soil-forming process of subhumid and semiarid climates. Soil types in the mollisol order, the typical end products of the process, are characterized by little leaching or eluviation and by the accumulation of both humus and mineral bases (especially calcium carbonate, $CaCO_3$).

caldera a large depression formed by a volcanic eruption, often a crater formed by the inward collapse of a volcano.

caliche hardened layers of lime ($CaCO_3$) deposited at the surface of a soil by evaporating capillary water.

calorie amount of heat necessary to raise the temperature of 1 gram of water 1°C.

calving breaking off a mass of ice from the toe of a glacier at its junction with the ocean or a lake.

Canadian High high atmospheric pressure area that tends to develop over the central North American continent in winter.

capacity (moisture) the maximum amount of water vapor that can be contained in a given quantity of air at a given temperature.

capillary action the upward movement of water through tiny cracks and pore spaces.

capillary water soil water that clings to soil peds and individual soil particles as a result of surface tension. Capillary water moves in all directions through the soil from areas of surplus water to areas of deficit.

carbonate substance that contains one or more CO_3 ions as part of its chemical formula.

carbonation carbon dioxide in water chemically combining with other substances to create new compounds.

carnivore animal that eats only other animals.

cartography the science of mapmaking.

catastrophism once-popular theory that Earth's landscapes developed in a relatively short time by cataclysmic events.

Celsius (centigrade) scale temperature scale in which 0° is the freezing point of water and 100° its boiling point at standard sea-level pressure.

centrifugal force force that pulls a rotating objects away from the center of rotation.

channel roughness an expression of the frictional resistance to stream flow due to irregularities in a stream channel bed and sides.

chaparral sclerophyllous vegetation growing in Mediterranean climate regions of the western United States; these seasonal, drought-resistant plants are mainly low-growing shrubs, with small, hard-surfaced leaves and deep, water-probing roots.

chemical precipitate sedimentary rock rock created from dissolved minerals that have precipitated out of water.

chemical weathering breakdown of rock material by chemical reactions that change the rock's mineral composition (decomposition).

chinook dry warm wind on the eastern slopes of the Rocky Mountains. *See also* foehn wind.

cinder cone hill composed of fragments of volcanic rock (pyroclastics) erupted from a central vent.

circle of illumination line dividing the sunlit (day) hemisphere from the shaded (night) hemisphere; experienced by individuals on Earth's surface as sunrise and/or sunset.

cirque deep, sometimes steep-sided amphitheater formed at the head of an alpine valley by glacial ice erosion.

cirque glacier a generally small alpine glacier restricted to a high-elevation basin (cirque).

cirro signifies a high-level cloud (i.e., above 6000 m in elevation).

cirrus high, detached clouds consisting of ice particles. Cirrus clouds are white and feathery or fibrous in appearance.

Cl, O, R, P, T Hans Jenny's list of soil formation factors: Climate, Organics, Relief, Parent material, and Time.

clast solid broken piece of rock, bone, or shell.

clastic sedimentary rock rock formed by the compaction and cementation of preexisting rock debris.

clay (clayey) a very fine-grained mineral particle with a size less than 0.004 mm (1/256 in), often the product of weathering.

clay mineral A small (clay-sized) hydrous aluminum silicate material typically created by chemical weathering, especially hydrolysis.

clay pan an ephemeral lake bed composed mostly of fine-grained clastic particles.

climate accumulated and averaged weather patterns of a locality or region; the full description is based on long-term statistics and includes extremes or deviations from the norm.

climate classification system a method for determining categories of climate types based on climatic and weather data and characteristics.

climate region a region defined by the similarity of climatic conditions in the area within its boundaries.

climatology scientific study of climates of Earth and their distribution.

climax community the final step in the succession of plant communities that occupy a specific location.

climograph graphic means of giving information on mean monthly temperature and rainfall for a select location or station.

closed system system in which no substantial amount of materials can cross its boundaries.

cloud forest rainforest that is produced by nearly-constant light rain on the windward slopes of mountains.

coastal zone general region of interaction between the land and a lake or the ocean.

coastline of emergence coast with formerly submerged land that is now above water, either due to uplift of the land or a drop in sea level.

coastline of submergence coastal area that has undergone sinking or subsidence relative to sea level.

cobble rock fragments ranging in diameter from 64 to 256 millimeters (2.5–10 in).

cobble beach beach dominated by cobble-sized rocks deposited by waves along the shoreline, often found along the base of sea cliffs or lake cliffs.

col a glacially eroded pass between two mountain valleys.

cold current a flow of seawater that moves like a river through the ocean and is relatively colder than water in the ocean area that it flows through.

cold front leading edge of a relatively cooler, denser air mass that advances upon a warmer, less dense air mass.

collapse sinkhole topographic depression formed mainly by the cave-in of the land above a cavern.

collision-coalescence process rain-forming process where raindrops form by collision between cloud droplets.

column column-shaped speleothem resulting from the joining of a stalactite and a stalagmite.

columnar joints vertical, polygonal fractures caused by lava shrinking as it cools.

comet a small body of icy and dusty matter that revolves about the sun. When a comet comes near the sun, some of its material vaporizes, forming a large head and often a tail.

composite cone (stratovolcano) volcano formed from alternating layers of lava and pyroclastic materials; generally known for violent eruptions.

compressional tectonic force pushing together from opposite sides (convergence).

compromise projection maps that compromise true shape and true area in order to display both fairly well.

computer-generated model a computer-developed representation of a feature, process, or an environmental system.

conceptual model image in a person's mind of an Earth feature or landscape as derived from personal experiences.

condensation process by which a vapor is converted to a liquid during which energy is released in the form of latent heat.

condensation nuclei minute particles in the atmosphere (e.g., dust, smoke, pollen, and sea salt) on which condensation can take place.

conduction transfer of heat within a body or between adjacent matter by means of internal molecular movement.

conformal map a map using a projection that maintains the true shape of small areas on Earth's surface.

conic projection map projection achieved by transferring a spherical area (typically the geographic grid) onto a cone.

conical hill (haystack hill or hum) erosional remnant hill in a limestone region with a shape similar to cones or haystacks.

continental pertaining to, or having characteristics derived from, large land masses.

continental Antarctic (cAA) very cold, very dry air mass originating from Antarctica.

continental Arctic (cA) very cold, very dry air mass originating from the arctic region.

continental collision the fusing together of landmasses as tectonic plates converge.

continental crust the less dense (average mass of 2.7 g/cm³), thicker portion of Earth's crust; underlies the continents.

Continental Divide line of separation dividing runoff between the Pacific and Atlantic Oceans. In North America it generally follows the crest of the Rocky Mountains.

continental drift theory proposed by Alfred Wegener stating that the continents joined, broke apart, and moved on Earth's surface; it was later replaced by the theory of plate tectonics.

continental glacier a very large and thick mass of flowing ice that exists due to climatic conditions resulting from high latitude. *See also* ice sheet (glacier).

continental island island that is geologically part of a continent and usually located on the continental shelf.

continental Polar (cP) cold, dry air mass originating from landmasses approximately 40° to 60°N or S latitude.

continental shelf the gently sloping margin of a continent overlain by ocean water.

continental shield the ancient part of a continent that consists of crystalline rock.

continental Tropical (cT) warm, dry air mass originating from subtropical landmasses.

continentality the distance a particular place is located away from the ocean; the greater the distance, the greater the continentality.

continuous data numerical representations of phenomena that are present everywhere—such as air pressure, temperature, and elevation.

contour interval vertical distance represented by two adjacent contour lines on a topographic map.

contour map a map that uses contour lines to show differences in continuous data such as elevation, temperature, and air pressure.

controls of temperature factors that control temperatures around the world, such as latitude, proximity to water bodies, ocean currents, altitude, landform barriers, and human activities.

convection process by which a circulation is produced within an air mass or fluid body (heated material rises, cooled material sinks); also, in tectonic plate theory, the method whereby heat is transferred to Earth's surface from deep within the mantle.

convectional precipitation precipitation resulting from condensation of water vapor in an air mass that is rising convectionally as it is heated from below.

convective thunderstorm a thunderstorm produced by the convective uplift mechanism.

convergent wind circulation pressure-and-wind system where the airflow is inward toward the center, where pressure is lowest.

coordinate system a precise system of grid lines used to describe locations.

coral reef shallow, wave-resistant structure made by the accumulation of skeletal remains of tiny sea animals.

core extremely hot and dense innermost portion of Earth's interior; the molten outer core is 2400 kilometers (1500 mi) thick; the solid inner core is 1120 kilometers (700 mi) thick.

Coriolis effect apparent effect of Earth's rotation on horizontally moving bodies, such as wind and ocean currents; such bodies tend to be deflected to the right in the Northern Hemisphere and to the left in the Southern Hemisphere.

corridor in biogeographic terms, any linear feature that crosses the general vegetation cover in an area (for example, rivers, power line clearings, and roads).

corrosion chemical erosion of rock matter by water; the removal of ions from rock-forming minerals in water.

creep slow downslope movement of Earth material involving the lifting and falling action of sediment particles.

crevasse stress crack commonly found along the margins and at the toe of a glacier.

cross bedding thin layers within sedimentary rocks that were deposited at an angle to the dominant rock layering.

crust relatively thin, approximately 8–64 kilometers (5–40 mi) deep, low-density surface layer of Earth.

crustal warping gentle bending and folding of crustal rocks.

cumulus globular clouds, usually with a horizontal base, and undergoing vertical development.

cut bank the steep slope found on the outside of a bend in a meandering stream channel.

cyclone (low) center of low atmospheric pressure.

cyclonic (convergence) precipitation precipitation formed by cyclonic uplift.

cylindrical projection map projection achieved by transferring a spherical area (typically the geographic grid) onto a cylinder.

daily (diurnal) temperature range difference between the highest and lowest temperatures of the day.

daily march of temperature changes in daily temperatures from an overnight low to a daytime high and to the next overnight low.

daily temperature lag the time of offset between solar noon, when solar radiation is most effective in heating, and the warmest time of day (usually 2 to 4 hours after 12 noon).

debris unconsolidated slope material with a wide range of grain sizes including at least 20% gravel (>2 mm).

debris flow rapid, gravity-induced downslope movement of wet, poorly sorted Earth material.

debris flow fan fan-shaped depositional landform, particularly common in arid regions, created where debris flows emerge onto a plain from a mountain canyon.

deciduous a plant, usually a tree or shrub, that drops its leaves (usually seasonally).

decimal degrees angular degrees with fractions expressed as tenths, hundredths, etc., rather than as the 60-based traditional subdivisions of minutes and seconds.

deep-water wave wave traveling in water with a depth greater than or equal to half the wavelength.

deflation entrainment and removal of loose surface sediment by the wind.

deflation hollow a wind-eroded depression in an area not dominated by wind-deposited sand.

degradation landscape lowering that results from more erosion than deposition over time.

delta depositional landform constructed where a stream flows into a standing body of water (a lake or the ocean).

delta plain the portion of a delta that lies above the level of the lake or ocean.

dendritic term used to describe a drainage pattern that is treelike with tributaries joining the main stream at acute angles.

dendrochronology method of determining past climatic conditions using tree rings.

deposition accumulation of Earth materials at a new site after being moved by gravity, water, wind, or glacial ice.

desert climate climate where the amount of precipitation received is less than one-half of the potential ET.

desertification the impact of either climate change toward aridity or from injurious human impacts that produce desert conditions or a desert-like landscape.

desert pavement (reg) desert surface mosaic of close-fitting stones that overlies a deposit of mostly fine-grained sediment.

detritivore (decomposer) organism that promotes decay by feeding on dead plant and animal material and returns mineral nutrients to the soil or water in a form that plants can utilize.

dew tiny droplets of water on ground surfaces, grass blades, or solid objects. Dew is formed by condensation when air at the surface reaches the dew point.

dew point temperature (dew point) the temperature at which an air mass becomes saturated; any further cooling will cause condensation of water vapor in the air.

differential weathering and erosion rock types vary in resistance to weathering and erosion, causing the processes to occur at different rates and often producing distinctive landform features.

digital elevation model (DEM) computer-generated three-dimensional image of topography.

digital image an image made from computer data displayed like a mosaic of tiny squares, called pixels.

digital map layers mapped information and data that are stored in digital form and can be overlaid with other mapped information in a geographic information system.

digital terrain model a computer-generated three-dimensional graphic representation of topography.

dike igneous intrusion with a wall-like shape.

dip inclination of a rock layer from the horizontal; always measured at right angles to the strike.

dip-slip fault a vertical fault where the movement is up and down the dip of the fault surface.

direct (vertical) rays sun's rays that strike Earth's surface at a 90° angle.

disappearing stream stream that has its flow diverted entirely to the subsurface.

discharge see stream discharge.

discrete data numerical or locational representations of phenomena that are present only at certain locations—such as earthquake epicenters, sinkholes, and tornado paths.

dissolved load soluble minerals or other chemical constituents carried in water as a solution.

distributary a smaller stream that conducts flow away from the larger main channel, especially on deltas; the opposite of a tributary.

diurnal synonym for daily, particularly in reference to a day to night pattern of change.

divergent wind circulation pressure-and-wind system where the airflow is outward away from the center, where pressure is highest.

doldrums zone of low pressure and calms along the equator.

doline see sinkhole.

Doppler radar advanced radar system that can detect motion in storms, toward and away from the radar signal.

drainage area the size of a drainage basin, typically expressed in km^2 or mi^2.

drainage basin (watershed, catchment) the region that provides runoff to a stream.

drainage density the summed length of all stream channels per unit area in a drainage basin.

drainage divide the outer boundary of a drainage basin.

drainage pattern the form of the arrangement of channels in a stream system in map view.

drainage wind (katabatic wind) downslope flow of cold, dense air from a high mountain valley, an elevated plateau, or a glacial ice sheet.

draping overlaying a digital image over a digital elevation model of a landscape.

drift sediment deposited in association with glacial ice or its meltwater.

drizzle fine mist or haze of very small water droplets with a barely perceptible falling motion.

drumlin streamlined, elongated hill composed of glacial drift with a tapered end, aligned along the direction of continental ice flow.

dry adiabatic lapse rate rate at which a rising mass of air is cooled by expansion when no condensation is occurring (10°C/1000 m or 5.6°F/1000 ft).

dust storm a moving cloud of wind-blown dust (typically silt).

dynamic equilibrium constantly changing relationship among the variables of a system, which produces a balance between the amounts of energy and/or materials that enter a system and the amounts that leave.

earth (as a mass wasting material) thick unit of unconsolidated, predominantly fine-grained slope material.

Earth system set of interrelated components, variables, processes, and subsystems (e.g., atmosphere, lithosphere, biosphere, and hydrosphere), which interact and function together to make up Earth.

earthquake series of shock waves set in motion by sudden movement along a fault.

earthquake intensity a measure of the impact of an earthquake on humans and their built environment.

earthquake magnitude measurement representing an earthquake's size in terms of energy released.

Earth's energy budget relationship between solar energy input, storage, and output within the Earth system.

easterly wave trough-shaped, weak, low pressure cell that progresses slowly from east to west in the trade wind belt of the tropics; this type of disturbance sometimes develops into a tropical hurricane.

eccentricity cycle the change in Earth's orbit from slightly elliptical to more circular, and back to its earlier shape every 100,000 years.

ecological niche combination of role and habitat as represented by a particular species in an ecosystem.

ecology science that studies the interactions between organisms and their environment.

ecosystem community of organisms functioning together through interdependent relationships with the environment that they occupy.

ecotone transition zone of varied natural vegetation occupying the boundary between two adjacent and differing plant communities.

effective precipitation actual precipitation available to supply plants and soil with usable moisture; does not take into consideration storm runoff or evaporation.

effluent the condition of groundwater seeping into a stream.

effusive eruption streaming flows of molten rock matter pouring onto Earth's surface from the subsurface.

El Niño warm countercurrent that influences the central and eastern Pacific.

El Niño/Southern Oscillation (ENSO) the periodic shift (oscillation) that occurs between El Niño and La Niña conditions.

elastic solid a solid that withstands stress with little deformation until a maximum value is reached, whereupon it breaks.

electromagnetic energy all forms of energy that share the property of moving through space (or any medium) in a wavelike pattern of electric and magnetic fields; also called radiation.

electromagnetic spectrum the full range of wavelengths of electromagnetic energy, portions of which constitute such bands as visible light, ultraviolet radiation, infrared radiation, and others.

elevation vertical distance from mean sea level to a point or object on Earth's surface.

eluviation downward removal of soil components by water.

empirical classification classification process based on statistical, physical, or observable characteristics of phenomena; it ignores the causes or theory behind their occurrence.

end moraine a ridge of till deposited at the toe or terminus of a glacier.

endogenic process landforming process originating within Earth.

energy balance the concept that the energy Earth receives from the sun is equaled (balanced) by return of that energy back to space, thus Earth has not continually increased or decreased in temperature over its history of billions of years.

Enhanced Fujita Scale enhancements made to the tornado intensity scale, which concentrates more specifically on the types of damages that occur.

enhanced greenhouse effect refers to the increased global warming impact by human contributions to the atmosphere, which is intended to differentiate it from the natural greenhouse effect.

entisol soil with little or no development.

entrenched stream a stream that has eroded downward so that it flows in a relatively deep and steep-sided valley or canyon.

environment surroundings, whether of people or of any other living organism; includes physical, social, and cultural conditions that affect the development of that organism.

environmental degradation processes or results, usually related to human activities that are detrimental to an environment.

environmental overshoot a situation where the rate of resource use is unsustainable, given the existing supply or renewability of that resource.

environmental science perspective a viewpoint and a scientific approach that considers the interactions between humans and environmental conditions.

environmental sustainability interacting with the environment in ways that will ". . . provide the best outcomes for the human and natural environments both now and into the indefinite future." (EPA)

eolian (aeolian) pertaining to the land-forming work of the wind.

ephemeral flow describes streams that conduct flow only occasionally, during to shortly after precipitation events or due to ice or snowmelt.

epicenter point on Earth's surface directly above the focus of an earthquake.

epiphyte a plant that grows on another plant, but does not take its nutrients from the host plant.

epipedon surface soil layer that possesses specific characteristics essential to the identification of soils in the National Resources Conservation Service System.

epiphyte a plant that grows on another plant but does not take its nutrients from the host plant.

equal-area map a map created with a projection that preserves correct proportional sizes for any given areas of Earth's surface.

equator great circle of Earth midway between the poles; the zero degree parallel of latitude that divides Earth into the Northern and Southern Hemispheres.

equatorial low (equatorial trough) zone of low atmospheric pressure centered more or less over the equator where heated air is rising. *See also* doldrums.

equidistance a property of some maps that depicts distances equally without scale variation.

equilibrium state of balance between the interconnected components of a system, an organized whole.

equilibrium line balance position on a glacier that separates the zone of accumulation from the zone of ablation.

equinox two times each year (approximately March 21 and September 22) when the sun is directly overhead and its vertical rays strike at the equator; all over Earth, day and night are of equal length.

erosion removal of Earth materials from a site by gravity, water, wind, or glacial ice.

erratic large glacially transported boulder deposited on top of bedrock of different composition.

escarpment (scarp) a steep slope, or cliff, of any height regardless of the geomorphic or geologic process by which it formed.

esker narrow, winding ridge of coarse sediment probably deposited in association with a meltwater tunnel at the base of a continental glacier.

estuary coastal embayment where salt and fresh water mix.

evaporation process by which a liquid is converted to the gaseous state (vapor) by the addition of latent heat.

evaporative cooling the lowered temperature experienced when water evaporates.

evaporite mineral salts that are soluble in water and accumulate when water evaporates.

evapotranspiration combined water loss to the atmosphere from ground and water surfaces by evaporation and, from plants, by transpiration.

evergreen a plant, usually a tree or shrub, does not seasonally drop its leaves.

exfoliation successive removal of outer rock sheets or slabs broken from the main rock mass by weathering.

exfoliation dome large, smooth, convex (dome-shaped) mass of exposed rock undergoing exfoliation due to weathering by unloading.

exfoliation sheet relatively thin outer layer of rock broken from the main rock mass by weathering.

exogenic process land-forming process originating at or very near Earth's surface.

exotic stream stream that originates in a humid region and has sufficient water volume to flow across a desert region.

explosive eruption violent blast of molten and solid rock material into the air.

exposure direction of mountain slopes with respect to prevailing wind direction.

exterior drainage streams and stream systems that flow to the ocean.

extratropical disturbance midlatitude disturbance. *See also* atmospheric disturbance.

extrusive igneous rock rock solidified at Earth's surface from lava; also called *volcanic rock*.

Fahrenheit scale temperature scale in which 32° is the freezing point of water, and 212° its boiling point, at standard sea-level pressure.

fall type of fast mass wasting characterized by Earth material plummeting downward freely through air.

fan apex the most upstream point on an alluvial fan; where the fan-forming stream emerges from the mountain canyon.

fast mass wasting gravity-induced downslope movement of Earth material that people can witness directly.

fault breakage zone along which rock masses have slid past each other.

fault block discrete block-like region of crustal rocks bordered on two opposite sides by normal faults.

fault scarp (escarpment) the steep cliff or exposed face of a fault where one crustal block has been displaced vertically relative to another.

faulting the movement of rock masses past each other along either side of a fault.

feedback sequence of changes in the elements of a system, which ultimately affects the element that was initially altered to begin the sequence.

feedback loop path of change as its effects move through the variables of a system until the effects impact the variable originally experiencing change.

fertilization *see* soil fertilization.

fetch distance over open water that winds blow without interruption.

firn compact granular snow formed by partial melting and refreezing due to overlying layers of snow.

fissure an extensive crack or break in rocks along which lava may be extruded, which is called a *fissure flow*.

fissure flow lava flow that emanates from a crack (fissure) in the surface rather than from the vent or crater of a volcano.

fjord deep, glacial trough along the coast invaded by the sea after the removal of a glacier.

flood stream or river water exceeding the amount that can be contained within its channel.

flood basalt massive outpouring of basaltic lava.

floodplain the low-gradient area adjacent to many stream channels that is subject to flooding and primarily composed of alluvium.

flow rapid downslope movement of wet unconsolidated Earth material that experiences considerable mixing.

flowing artesian well Artificially dug outlet that allows naturally pressurized groundwater to flow to Earth's surface.

fluvial term used to describe landforms and processes associated with the work of streams and rivers.

fluvial geomorphology the study of streams as land-forming agents.

focus point within Earth's crust or upper mantle where an earthquake originated.

foehn warm, dry, downslope wind on lee of mountain range, caused by adiabatic heating of descending air.

fog mass of suspended water droplets within the atmosphere that is in contact with the ground.

folding the bending or wrinkling of Earth's crust due to compressional tectonic forces.

foliation the occurrence of banding or platy structure in metamorphic rocks.

food chain sequence of levels in the feeding pattern of an ecosystem.

food web feeding mosaic formed by the interrelated and overlapping food chains of an ecosystem.

freeze–thaw weathering breaking apart of rock by the expansive force of water freezing in cracks; also called *frost weathering* or *ice wedging*.

freezing rain (glaze) rainfall that freezes into ice upon coming in contact with a surface or object that is colder than 0°C (32°F).

fringing reef coral reef attached to the coast.

front sloping boundary or contact surface between air masses with different properties of temperature, moisture content, density, and atmospheric pressure.

frontal precipitation precipitation resulting from condensation of water vapor by the lifting or rising of warmer, lighter air above cooler, denser air along a frontal boundary.

frontal thunderstorm a thunderstorm produced by the frontal uplift mechanism.

frost frozen condensation that occurs when air at ground level is cooled to a dew point of 0°C (32°F) or below; also any temperature near or below freezing that threatens sensitive plants.

fusion (thermonuclear) reaction the fusing together of two hydrogen atoms to create one helium atom. This process releases tremendous amounts of energy.

galactic movement movement of the solar system related to movements of the Milky Way Galaxy.

galaxy a large assemblage of stars; a typical galaxy contains millions to hundreds of billions of stars.

galleria forest jungle-like vegetation extending along and over streams in tropical forest regions.

gap an area within the territory occupied by a plant community when the climax vegetation has been destroyed or damaged by some natural process, such as a hurricane, forest fire, or landslide.

gelisol soil that experiences frequent freezing and thawing.

generalist species that can survive on a wide range of food supplies.

genetic classification classification process based on the causes, theory, or origins of phenomena; generally ignoring their statistical, physical, or observable characteristics.

geographic grid lines of latitude and longitude form the geographic grid.

geographic information system (GIS) versatile computer software/hardware system that combines the features of cartography and database management to produce maps and other displays of spatial data for solving geographic problems.

geography study of Earth phenomena; includes an analysis of distributional patterns and interrelationships among these phenomena.

geomorphic agent a medium that erodes, transports, and deposits Earth materials; includes water, wind, and glacial ice.

geomorphology the study of the origin and development of landforms.

geostrophic winds upper-level winds in which the Coriolis effect and pressure gradient are balanced, resulting in winds flowing parallel to the isobars.

geothermal water water heated by contact with hot rocks in the subsurface.

geyser natural eruptive outflow of water that alternates between hot water and steam.

giant planet (gas planet) the four largest planets—Jupiter, Saturn, Uranus, and Neptune.

glacial (glacial period) a span of time characterized by the extensive expansion of glaciers in high latitude and high elevation areas, as well as into the middle latitudes. *See also* Ice Age.

glacial outwash the fluvial deposits derived from glacial meltwater streams.

glacial trough a U-shaped valley carved by glacial erosion.

glaciation (glacial period) an interval of glacial activity.

glacier a large mass of ice that flows as a plastic solid.

glacier advance expansion of the toe or terminus of a glacier to a lower elevation or lower latitude due to an increase in size.

glacier head farthest upslope part of an alpine glacier.

glacier retreat withdrawal of the toe or terminus of a glacier to a higher elevation or higher latitude due to a decrease in size.

glacier toe (terminus) farthest downslope part of an alpine glacier.

glaciofluvial deposit sorted glacial drift deposited by meltwater.

glaciolacustrine deposit sorted glacial drift deposited by meltwater in lakes associated with the margins of glaciers.

gleization soil-forming process of poorly drained areas in cold, wet climates. The resulting soils have a heavy surface layer of humus with a water-saturated clay horizon directly beneath.

Global circulation models (GCMs) numerical/computer systems for forecasting climatic impacts and shifts resulting from variations in processes and interactions among the atmosphere, oceans, and other climate-related environmental variables.

Global Navigational Satellite Systems the general term for satellite-based location systems, no matter which country developed them.

Global Positioning System (GPS) United States system that uses satellite data to locate positions and travel routes anywhere on Earth.

global warming climate change that causes Earth's temperatures to rise.

gnomonic projection planar projection with greatly distorted land and water areas; valuable for navigation because all great circles on the projection appear as straight lines.

graben block of crustal rocks between two parallel normal faults that has slid downward relative to adjacent blocks.

graded stream stream where slope and channel size provide a velocity just sufficient to transport the load supplied by the drainage basin; a theoretical balanced state averaged over a period of many years.

gradient a term for slope often used to describe the upstream/downstream angle of a streambed.

granite a coarse-grained intrusive igneous rock generally associated with continental crust.

granular disintegration weathering feature of coarse crystalline rocks in which visible individual mineral grains fall away from the main rock mass.

graphic (bar) scale a ruler-like device placed on maps for making direct measurements in ground distances.

gravel a general term for sediment sizes larger than sand (larger than 2.0 mm)

gravitational water meteoric water that passes through the soil under the influence of gravitation.

gravity the mutual attraction of bodies or particles.

great circle circle in any direction formed by dividing Earth into two equal hemispheres; the plane of a great circle passes through the center of the globe. A route following a great circle that passes through two places marks shortest distance between those locations.

greenhouse effect warming of the atmosphere that occurs because solar radiation is absorbed by the planet's surface and then reradiated as thermal energy, which generates heat, but the loss of thermal radiation is hindered by certain atmospheric gases including those associated with human activity (e.g., CO_2).

greenhouse gases atmospheric gases which hinder the escape of Earth's heat energy.

Greenwich mean time (GMT) time at zero degrees longitude used as the base time for Earth's 24 time zones; also called Universal Time or Zulu Time.

groin artificial structure extending out into the water from a beach built to inhibit loss of beach sediment.

ground-inversion fog *see* radiation fog.

ground moraine irregular, hummocky landscape of till deposited on Earth's surface by a wasting glacier.

groundwater water in the saturated zone below the water table.

gully steep-sided stream channel somewhat larger than rills that even in humid climates flow only in direct response to precipitation events.

gyre broad circular patterns of major surface ocean currents related to the large subtropical high pressure systems.

habitat location within an ecosystem occupied by a particular organism.

hail form of precipitation consisting of pellets or balls of ice with a concentric layered structure usually associated with the strong convection of cumulonimbus clouds.

hanging valley tributary glacial trough that enters a main glaciated valley at a level high above the valley floor.

hardpan dense, compacted, clay-rich layer occasionally found in the subsoil (B horizon) that is an end product of excessive illuviation.

haystack hill (conical hill or hum) remnant hills of soluble rock remaining after adjacent rock has been dissolved away in karst areas.

headward erosion gullying and valley cutting that extends a stream channel in an upstream direction.

heat energy being transferred from one substance or medium to another because of temperature differences.

heat island mass of air overlying urban areas that is warmer than the surrounding suburban and rural countryside.

heaving various means by which particles are lifted perpendicular to a sloping surface, then fall straight down by gravity.

hemisphere half of a sphere; for example, the northern or southern half of Earth divided by the equator or the eastern and western half divided by two meridians, the 0° and 180° meridians.

herbivore an animal that eats only living plant material.

heterotroph (consumer) an organism that consumes organic material from other life forms, including all animals and parasitic plants.

high *see* anticyclone.

highland climate a general climate classification for regions of high, yet varying, elevations.

histosol soil that develops in poorly drained areas.

holistic approach considering and examining all phenomena relevant to a problem.

Holocene the most recent time interval of warm, relatively stable climate that began with the retreat of major glaciers about 10,000 years ago.

horn pyramid-like peak created where three or more expanding cirques meet at a mountain summit.

"horse latitudes" historic term for the latitudinal belt influenced by the subtropical high pressure cells.

horst crustal block between two parallel normal faults that has slid upward relative to adjacent blocks.

hot spot a mass of hot molten rock material at a fixed location beneath a lithospheric plate.

hot spring natural outflow of geothermal groundwater to the surface.

human geography specialization in the systematic study of geography that focuses on the location, distribution, and spatial interaction of human (cultural) phenomena.

humid continental, hot-summer climate climate type characterized by hot, humid summers and mild, moist winters.

humid continental, mild-summer climate climate type characterized by mild, humid summers and cold, and moist winters.

humid subtropical climate climate type characterized by hot humid summers, found in the subtropical regions of the world.

humidity amount of water vapor in an air mass at a given time.

humus organic matter found in the surface soil layers that is in various stages of decomposition as a result of bacterial action.

hurricane severe tropical cyclone of great size with nearly concentric isobars. Its torrential rains and high-velocity winds create unusually high seas and extensive coastal flooding; also called *cyclones (tropical)*, and *typhoons*.

hydration rock weathering due to substances in cracks swelling and shrinking with the addition and removal of water molecules.

hydraulic action erosion resulting from the force of moving water.

hydrologic cycle circulation of water within the Earth system, from evaporation to condensation, precipitation, runoff, storage, and re-evaporation back into the atmosphere.

hydrolysis water molecules chemically recombining with other substances to form new compounds.

hydrosphere major Earth subsystem consisting of the waters of Earth, including oceans, ice, freshwater bodies, groundwater, and water within the atmosphere and biomass.

hygroscopic water water in the soil that adheres to mineral particles.

Ice Age period of Earth history when large areas of Earth's land surface were covered with massive continental ice sheets and other kinds of glaciers. The most recent ice age occurred during the Pleistocene Epoch.

ice cap a continental glacier of regional size, less than 50,000 km².

icefall section of a glacier moving over and down a steep slope, creating a rigid ice cascade, criss-crossed with crevasses.

ice sheet the largest type of glacier; a continental glacier larger than 50,000 km².

ice-sheet climate climate type where the average temperature of every month of the year is below freezing.

ice shelf large, flat-topped plate of ice overlying the ocean but attached to land; a source of icebergs.

Icelandic Low center of low atmospheric pressure located in the north Atlantic, especially persistent in winter.

ice wedge wedge-shaped ice mass that has accumulated in and expanded a ground fissure in a periglacial environment.

iceberg free-floating mass of ice broken off from a glacier where it flows into the ocean or a lake.

ice-marginal lake temporary lake formed by the disruption of meltwater drainage by deposition along a glacial margin, usually in the area of an end moraine.

ice-scoured plain a broad area of low relief and bedrock exposures eroded by a continental glacier.

igneous intrusion (pluton) a mass of igneous rock that cooled and solidified beneath the surface.

igneous processes processes related to the solidification and eruption of molten rock matter.

igneous rock one of the three major categories of rock; formed from the cooling and solidification of molten rock matter.

illuviation deposition of fine soil components in the subsoil (B horizon) by gravitational water.

imaging radar radar systems designed to sense the ground and convert reflections into a map-like image.

inceptisol young soil with weak horizon development.

infiltrate water seeping downward into the soil or other surface materials.

infiltration the process of water seeping downward into the soil or other surface materials.

infiltration capacity the greatest amount of infiltrated water that a surface material can hold.

influent the condition of stream water seeping into the channel bed and adding to groundwater.

inner core the solid, innermost portion of Earth's core, probably of iron and nickel, that forms the center of Earth.

input energy and material entering an Earth system.

inselberg remnant bedrock hill or mountain rising above a stream-eroded plain or pediment in an arid or semiarid region.

insolation incoming solar radiation, that is, energy received from the sun.

instability condition of air when it is warmer than the surrounding atmosphere and is buoyant with a tendency to rise; the lapse rate of the surrounding atmosphere is greater than that of unstable air.

interception the delay in arriving at the ground surface experienced by precipitation that strikes vegetation.

interfluve the land between two stream channels.

interglacial warmer period between glacial advances during which continental ice sheets and many valley glaciers retreat and disappear or are greatly reduced in size.

interior drainage streams and stream systems that flow within a closed basin and thus do not reach the ocean.

intermediate zone subsurface water layer between the zone of aeration above and the zone of saturation below; saturated only during times of ample precipitation.

intermittent flow describes streams that conduct flow seasonally.

International Date Line line roughly along the 180° meridian, where each day begins and ends; it is always a day later west of the line than east of the line.

Intertropical Convergence Zone (ITCZ) zone of low pressure formed along the equator, where air carried by the trade winds from both sides of the equator converges and is forced to rise.

intrusive igneous rock rock that solidified within Earth from magma; also called *plutonic rock*.

inversion *see* temperature inversion.

inversion layer the altitudinal boundary, which can be sharp, between cold air near the surface and warmer air aloft—which are the conditions of a temperature inversion.

ionosphere layer of ionized gasses concentrated between 80 km (50 miles) and outer limits of the atmosphere.

isarithm line on a map that connects all points of the same numerical value, such as isotherms, isobars, and isobaths.

island arc a chain of volcanic islands along a deep oceanic trench; found near tectonic plate boundaries where subduction is occurring.

isobar line drawn on a map to connect all points with the same atmospheric pressure.

isoline a line on a map that represents equal values of some numerical measurement such as lines of equal temperature or elevation contours.

isostasy theory that holds that Earth's crust floats in hydrostatic equilibrium on the denser plastic layer of the mantle.

isotherm line drawn on a map connecting points of equal temperature.

jet stream high-velocity upper-air current with speeds of 120–640 kilometers per hour (75–250 mph).

jetty artificial structure, typically built in pairs, extending into a body of water; built to protect a harbor, inlet, or river mouth from excessive coastal sedimentation.

joint fracture or crack in rock.

joint set system of multiple parallel cracks (joints) in rock.

jungle dense tangle of trees and vines in areas where sunlight reaches the ground surface (not a true rainforest).

kame conical hill composed of sorted glaciofluvial deposits; presumed to have formed in contact with glacial ice when sediment accumulated in ice pits, crevasses, and among jumbles of detached ice blocks.

kame terraces landform resulting from accumulation of glaciofluvial sand and gravel along the margin of a glacier occupying a valley in an area of hilly relief.

Karman Line the Karman line marks the internationally accepted beginning of space at an altitude of 100 km (62 mi)—the realm of astronauts.

karst unique landforms and landscapes derived by the solution of soluble rocks, particularly limestone.

Kelvin scale temperature scale developed by Lord Kelvin, equal to Celsius scale plus 273; no temperature can drop below absolute zero, or 0 degrees Kelvin.

kettle depression formed by the melting of an ice block buried in glacial deposits left by a retreating glacier.

kettle lake a small lake or pond occupying a kettle.

kinetic energy energy of motion; one-half the mass *(m)* times velocity *(v)* squared, $KE = \frac{1}{2}mv^2$.

Koppen system climate classification based on monthly and annual averages of temperature and precipitation; boundaries between climate classes are designed so that climate types coincide with vegetation regions.

krummholz stunted, and crooked trees, growing low to the ground, under very harsh conditions of cold and wind at the elevational limit of tree growth, the boundary between the subalpine and alpine tundra zones.

La Niña cold sea-surface temperature anomaly in the Equatorial Pacific (opposite of El Niño).

laccolith massive igneous rock intrusion that bows overlying rock layers upwards in a domal fashion.

lagoon coastal embayment of water that is blocked from free circulation with the ocean or lake by a landform barrier.

lahar rapid, gravity-driven downslope movement of wet, fine-grained volcanic sediment.

lake-effect snow Enhanced, often heavy, snowfall in an area downwind from a lake, caused when a storm picks up moisture and rises as it passes over the lake.

land breeze–sea breeze air flow at night from the land toward the sea, caused by the movement of air from a zone of higher pressure associated with cooler nighttime temperatures over the land. *See also* sea breeze.

landform a terrain feature, such as a mountain, valley, and plateau.

Landsat a family of U.S. satellites that have been returning digital images since the 1970s.

landslide layperson's term for any fast mass wasting; used by some earth scientists for massive slides that involve a variety of Earth materials.

lapse rate *see* normal lapse rate.

latent heat exchange conversion between sensible heat, which we can feel, and latent heat, which is stored heat energy that we cannot sense, through changes in the state of water—for example evaporation, condensation, and freezing.

latent heat of condensation energy release in the form of heat, as water is converted from the gaseous (vapor) to the liquid state.

latent heat of evaporation amount of heat absorbed by water to evaporate from a surface (i.e., 590 calories/g of water).

latent heat of fusion amount of heat transferred when liquid turns to ice and vice versa; this amounts to 80 calories/gram.

latent heat of sublimation amount of heat that is leased when ice turns to vapor without first going through the liquid phase; this amounts to 670 calories/gram.

lateral migration the sideways shift in the position of a stream channel over time.

lateral moraine a ridge of till deposited along the side margin of a glacier.

laterite iron-, aluminum-, and manganese-rich layer in the subsoil (B horizon) that can be an end product of laterization in the wet-dry tropics (tropical savanna climate).

laterization soil-forming process of hot, wet climates. Oxisols, the typical end product of the process, are characterized by the presence of little or no humus, the removal of soluble and most fine soil components, and the heavy accumulation of iron and aluminum compounds.

latitude angular distance (distance measured in degrees) north or south of the equator.

lava molten (melted) rock matter erupted onto Earth's surface; solidifies into extrusive igneous (volcanic) rocks.

lava flow erupted molten rock matter that oozed over the landscape and solidified.

leaching removal by gravitational water of soluble inorganic soil components from the surface layers of the soil.

leeward located on the side facing away from the wind.

legend key to symbols used on a map.

levee raised bank along the margins of a stream or debris flow channel. *See also* natural levee.

liana woody vine found in tropical forests that roots in the forest floor but uses trees for support as it grows upward toward available sunshine.

lidar an abbreviation for "light detection and ranging," a system that uses lasers to scan and image landscapes—a remote sensing system.

light-year the distance light travels in 1 year—9.5 trillion kilometers (6 trillion mi).

lightning visible electrical discharge produced within a thunderstorm.

lithification the combined processes of compaction and cementation that transform clastic sediments into sedimentary rocks.

lithosphere a general term for the solid (rock) part of the Earth system, in contrast to the atmosphere, hydrosphere, and biosphere.

lithosphere (planetary structure) rigid and brittle outer layer of Earth consisting of the crust and uppermost mantle.

lithospheric plate Earth's exterior is broken into several of these large regions (plates) of rigid and brittle crust and upper mantle (lithosphere).

Little Ice Age a period of generally cooler conditions and growth of glaciers from the mid-16th to the mid-19th centuries.

littoral drifting general term for sediment transport parallel to shore in the nearshore zone due to incomplete wave refraction.

loam a soil with a texture in which none of the three soil grades (sand, silt, or clay) predominate over the others.

loess wind-deposited silt; usually transported in dust storms and derived from arid or glaciated regions.

longitude angular distance (distance measured in degrees) east or west of the prime meridian.

longitudinal dune a linear ridge-like sand dune that is oriented parallel to the prevailing wind direction.

longitudinal profile the changes in stream channel elevation with distance downstream from source to mouth.

longshore bar submerged feature of wave- and current-deposited sediment lying close to and parallel with the shore.

longshore current flow of water parallel to the shoreline just inside the breaker zone; caused by incomplete wave refraction.

longshore drifting transport of sediment parallel to shore by the longshore current.

longwave radiation electromagnetic radiation, generally in the wavelength range beginning with thermal infrared radiation and longer wavelength energy of the electromagnetic spectrum. Often, also a general reference to terrestrial radiation, thermal infrared radiation emitted by Earth that heats the Earth system.

low *see* cyclone.

map scale ratio between a distance as measured on Earth and the same distance as measured on a map, globe, or other representation of Earth.

magma molten (melted) rock matter located beneath Earth's surface from which intrusive igneous rocks are formed.

magnetic declination horizontal angle between geographic north and magnetic north.

mantle moderately dense, relatively thick (2885 km/1800 mi) middle layer of Earth's interior that separates the crust from the outer core.

map projection any presentation of the spherical Earth on a flat surface.

maquis sclerophyllous woodland and plant community, similar to North American chaparral; can be found growing throughout the Mediterranean region.

marine terrace abrasion platform that has been elevated above sea level and thus abandoned from wave action.

marine west coast climate climate type characterized by cool, wet conditions much of the year.

maritime relating to weather, climate, or atmospheric conditions in coastal or oceanic areas.

maritime Equatorial (mE) hot and humid air mass originating in the ocean region straddling the equator.

maritime Polar (mP) cold, moist air mass originating from the oceans around 40 to 60 degrees N or S latitude.

maritime Tropical (mT) warm, moist air mass originating in the tropical ocean regions.

mass a measure of the total amount of matter in a body.

mass wasting gravity-induced downslope movement of Earth material.

mathematical/statistical model quantitative representation of an area or Earth system using statistical data.

matrix the dominant area of a mosaic (ecosystem supporting a particular plant community) where the major plant in the community is concentrated.

meander a broad, sweeping bend in a river or stream.

meander cut-off bend of a meandering stream that has become isolated from the active channel.

meandering channel stream channel with broadly sinuous banks that curve back and forth in sweeping bends.

medial moraine a central moraine in a large valley glacier formed where the interior lateral moraines of two tributary glaciers merge.

Mediterranean climate climate type characterized by warm dry summers and cool moist winters.

mental map conceptual model or mental image of an area or a location in map form that people develop in their minds.

Mercator projection mathematically produced, conformal map projection showing true compass bearings as straight lines.

mercury barometer instrument measuring atmospheric pressure by balancing it against a column of mercury.

meridian a line of longitude that forms half of a great circle on the globe and connects all points of equal longitude; all meridians connect the North and South Poles.

mesa flat-topped, steep-sided erosional remnant reminiscent of a table, broader than tall, and common in arid regions with flat-lying sedimentary rocks.

mesosphere layer of atmosphere above the stratosphere; characterized by temperatures that decrease regularly with altitude.

mesothermal climate climate region or condition with hot, warm, or mild summers that do not have any months that average below freezing.

metamorphic rock one of the three major categories of rock; formed by heat and pressure changing a pre-existing rock.

meteor mass of rock pulled toward Earth by gravity; most meteors burn up before hitting Earth's surface and form a streak of light across the night sky ("shooting star").

meteorite any fragment of a meteor that reaches Earth's surface.

meteoroid stone or iron mass that enters our atmosphere from space, becoming a meteor as it burns up in the atmosphere.

meteorology study of the patterns and causes associated with short-term changes in the elements of the atmosphere.

microclimate climate associated with a small area at or near Earth's surface; the area may range from a few inches to 1 mile in size.

microplate terrane segment of crust of distinct geology added to a continent during tectonic plate collision.

microthermal climates climate regions or conditions with warm or mild summers that have winter months with temperatures averaging below freezing.

midlatitude cyclone (extratropical cyclone) storm system that travels with the westerly winds of the midlatitudes and the polar front, typically consisting of a cold front, a warm front, and three sectors (cold, warm and cool).

midlatitude zone the latitudinal area between $23\frac{1}{2}°$ (N or S), and $66\frac{1}{2}°$ (N or S).

midoceanic ridge linear seismic mountain range that interconnects through the major oceans; it is where new molten crustal material rises through the oceanic crust.

Milankovitch theory the theory that long-term climate change is related to periodic changes in Earth's axis and the planet's motions in space.

millibar unit of measurement for atmospheric pressure; 1 millibar equals a force of 1000 dynes per square centimeter; 1013.2 millibars is standard sea-level pressure.

mineral naturally occurring inorganic substance with a specific chemical composition and crystalline structure.

model simplified representation of a more complex reality that is useful for analysis and prediction.

modified Mercalli scale an earthquake intensity scale with Roman numerals from I to XII used to assess spatial variations in the degree of impact on people and structures that a tremor generates.

Mohorovičić discontinuity (Moho) zone marking the transition between Earth's crust and the denser mantle.

mollisol soil that develops in grassland regions.

moment magnitude the measure of an earthquake's size in terms of energy released.

monsoon seasonal wind that reverses direction during the year in response to a reversal of pressure over a large landmass. The classic monsoons of Southeast Asia blow onshore in response to low pressure over Eurasia in summer and offshore in response to high pressure in winter.

moraine a category of glacial landform, with multiple subtypes, deposited beneath or along the margins of an ice mass.

mosaic a plant community and the ecosystem on which it is based, viewed as a landscape of interlocking parts by ecologists.

mountain breeze–valley breeze air flowing downslope from mountains toward valleys during the night and from the valleys upward to the mountains during the day.

mouth downstream termination of a stream.

mud wet, fine-grained sediment, particularly clay and silt sizes.

mudflow rapid mass wasting of wet, fine-grained sediment; may deposit levees and lobate (tongue-shaped) masses.

multispectral remote sensing combining a number of energy wavelength bands to create images.

muskeg poorly drained vegetation-rich marshes or swamps usually overlying permafrost areas of polar climatic regions.

natural hazards any natural process that has a detrimental impact on humans and the built environment.

natural levee bank of a stream channel (or margin of a mass wasting flow channel) raised by deposition from flood (or flow) deposits; artificial levees are sometimes built along stream banks for flood control.

natural resource any element, material, or organism existing in nature that may be useful to humans.

natural vegetation vegetation that has been allowed to develop naturally without obvious interference from or modification by humans.

navigation the science of location and finding one's way, position, or direction.

neap tide the smaller than average tidal range that occurs during the first- and third-quarter moon.

near Earth objects (NEO) large celestial bodies, such as comets and asteroids, that might come close enough to collide with Earth.

near-infrared (NIR) image photograph or digital picture created by imaging reflected light in wavelengths just longer than red and are not visible to people.

nearshore zone area from the seaward or lake-ward edge of breaking waves to the landward limit of broken wave water.

negative feedback reaction to initial change in a system that counteracts the initial change and leads to dynamic equilibrium in the system.

nekton classification of marine organisms that swim in the oceans.

nimbus (nimbo) term used in cloud description to indicate precipitation; thus cumulonimbus is a cumulus cloud from which rain is falling.

nonflowing artesian well artificially dug outlet into which groundwater flows under its own pressure, but not reaching Earth's surface.

nor'easters a New England term for a storm that stalls over the north Atlantic along the east coast of North America, delivering strong winds from the poleward side of the storm that blow from the northeast; nor'easters typically bring heavy rain or snow and strong waves on the coast as the storm winds blow onshore.

normal fault breakage zone with rocks on one side sliding down relative to rocks on the other side because of tensional forces; footwall up, hanging wall down.

normal lapse rate (environmental lapse rate) decrease in temperature with altitude under normal atmospheric conditions; approximately 6.5°C/1000 meters (3.6°F/1000 ft).

North Atlantic Oscillation (NAO) oscillating (see-saw) pressure tendencies between the Azores High and the Icelandic Low.

North Pole maximum north latitude (90°N), at the point marking the axis of rotation.

North Star (Polaris) the star that is very near to being directly above Earth's north pole of rotation (90°N). Polaris is useful for celestial navigation in the northern hemisphere.

northeast trades *see* trade winds.

notch a recess, relatively small in height, eroded by wave action along the base of a coastal cliff.

nutrient cycle process of moving nutrients from the physical environment to living organisms, and their return of nutrients to the physical environment.

oblate spheroid Earth's shape—a slightly flattened sphere.

obliquity cycle the change in the tilt of the Earth's axis relative to the plane of the ecliptic over a 41,000-year period.

occluded front boundary between a rapidly advancing cold air mass and an uplifted warm air mass cut off from Earth's surface; denotes the last stage of a midlatitude cyclone.

ocean current horizontal flows of ocean water, usually in response to major patterns of atmospheric circulation.

oceanic crust the thinner and denser (averaging 3.0 g/cm³), basaltic portion of Earth's crust; underlies the ocean basins.

oceanic island volcanic island that rises from the deep ocean floor.

oceanic trench (trench) long, narrow depression on the seafloor usually associated with an island arc. Trenches mark the deepest portions of the oceans and are associated with subduction of oceanic crust.

offshore zone the expanse of open water lying seaward or lakeward of the breaker zone.

omnivore animal that can feed on both plants and other animals.

open system system in which energy and materials can freely cross its boundaries.

organic sedimentary rock rock created from deposits of organic material, such as carbon from plants (coal).

orographic precipitation precipitation resulting from condensation of water vapor in an air mass that is forced to rise over a mountain range or other raised landform.

orographic thunderstorm a thunderstorm produced by the orographic uplift mechanism.

outcrop a mass of bedrock exposed at Earth's surface that is not concealed by regolith or soil.

outer core the upper portion of Earth's core; considered to be composed of molten iron liquefied by Earth's internal heat.

outlet glacier a valley glacier that flows outward from the main mass of a continental glacier.

output energy and material leaving an Earth system.

outwash sediments deposited beyond the terminus of a glacier by glacial meltwater.

outwash plain extensive, relatively smooth plain covered with sorted deposits carried forward by the meltwater from a continental glacier.

overthrust low angle fault with rocks on one side pushed a considerable distance over those of the opposite side by compressional forces; the wedge of rocks that have overridden others in this way.

oxbow lake a lake or pond found in a meander cut-off on a floodplain.

oxidation union of oxygen with other elements to form new chemical compounds.

oxisol soil that develops over a long period of time in tropical regions with high temperatures and heavy annual rainfall.

oxide a mineral group composed of oxygen combining with other Earth elements, especially metals.

oxygen-isotope analysis a dating method used to reconstruct climate history; it is based on the varying evaporation rates of water molecules containing different oxygen isotopes and the changing ratio between the isotopes.

ozone gas molecule consisting of three atoms of oxygen (O_3); forms a layer in the upper atmosphere that screens out ultraviolet radiation harmful at Earth's surface.

ozone layer part of the stratosphere that surrounds Earth, with concentrations of ozone that help to shield the surface from harmful shortwave energy from the sun, such as ultraviolet light.

ozonosphere also known as the *ozone layer;* this is a concentration of ozone gas in a layer between 20 and 50 km (13 to 50 miles) above Earth's surface.

Pacific High persistent cell of high atmospheric pressure located in the subtropics of the North Pacific Ocean, also called the *Hawaiian High*.

pahoehoe a smooth, ropy surface of a lava flow.

paleoclimate the climatic conditions of the past; typically refers to prehistoric times.

paleogeography study of the past geographical distribution of environments.

paleomagnetism the historic record of changes in Earth's magnetic field.

palynology method of determining past climatic conditions using pollen analysis.

Pangaea ancient continent that consisted of all of today's continental landmasses.

parabolic dune crescent-shaped sand dune with arms pointing upwind.

parallel circle on the globe connecting all points of equal latitude.

parallelism of the axis tendency of Earth's polar axis to remain parallel to itself at all positions in its orbit around the sun.

parent material residual (derived from bedrock directly beneath) or transported (by water, wind, or ice) mineral matter from which soil is formed.

particulates very small pieces of solid matter that can remain airborne for long time periods.

passive-margin coast coastal region that is far removed from the volcanism and tectonism associated with lithospheric plate boundaries.

patch a gap or area within a matrix (territory occupied by a dominant plant community) where the dominant vegetation is not supported due to natural causes.

paternoster lake chain of lakes connected by a postglacial stream occupying the trough of a glaciated mountain valley.

patterned ground (frost polygons) natural, repeating, often-polygonal designs of sorted sediment seen on the surface of periglacial environments.

pediment gently sloping surface of eroded bedrock, thinly covered with fluvial sediments, found at the base of an arid-region mountain.

perched water table a minor zone of saturation overlying an aquiclude that exists above the regional water table.

percolation subsurface water moving downward to lower zones by the pull of gravity.

perennial flow describes streams that conduct flow continuously all year.

periglacial pertaining to cold-region landscapes that are impacted by intense frost action but not covered by year-round snow or ice.

perihelion position of Earth at closest distance to sun during each Earth revolution.

permafrost permanently frozen subsoil and underlying rock found in climates where summer thaw penetrates only the surface soil layer.

permeability characteristic of soil or bedrock that determines the ease with which water moves through Earth material.

pH scale scale from 0 to 14 that describes the acidity or alkalinity of a substance and that is based on a measurement of hydrogen ions; pH values below 7 indicate acidic conditions; pH values above 7 indicate alkaline conditions.

photosynthesis the process by which carbohydrates (sugars and starches) are manufactured in plant cells; requires carbon dioxide, water, light, and chlorophyll (the green color in plants).

physical geography specialization in the systematic study of geography that focuses on the location, distribution, and spatial interaction of physical (environmental) phenomena.

physical (mechanical) weathering breakdown of rocks into smaller fragments (disintegration) by physical forces without chemical change.

physical model three-dimensional representation of all or a portion of Earth's surface.

physical science perspective a viewpoint that seeks to answer questions according to scientific principles and by applying the scientific method.

phytoplankton tiny plants, algae and bacteria, that float and drift with currents in water bodies.

pictorial/graphic model representation of a portion of Earth's surface by means of maps, photographs, graphs, or diagrams.

piedmont alluvial plain a plain created by stream deposits at the base of an upland, such as a mountain, a hilly region, or a plateau.

piedmont fault scarp steep cliff due to movement along a fault that has offset unconsolidated sediment.

piedmont glacier an alpine glacier that extends beyond a mountain valley spreading out onto lower flatter terrain.

pioneer community a vegetation community that is the first to colonize an unvegetated, barren area.

pixel the smallest area that can be resolved in a digital image. Pixels, short for "picture element," are much like pieces in a mosaic, fitted together in a grid to make an image.

planar projection map projection in which grid lines are transferred onto a plane or flat surface; planar maps are most often used to show the polar regions, with the pole centrally located on the circular map, displaying one hemisphere.

plane of the ecliptic plane of Earth's orbit about the sun and the apparent annual path of the sun along the stars.

planet any of the eight largest bodies revolving about the sun, or any similar bodies that may orbit other stars.

plankton passively drifting or weakly swimming marine organisms, including phytoplankton (plants) and zooplankton (animals).

plant community variety of individual plants living in harmony with each other and the surrounding physical environment.

plastic solid any solid material that changes its shape under stress, and retains that deformed shape after the stress is relieved.

plate convergence movement of lithospheric plates toward each other and colliding.

plate divergence movement of lithospheric plates away from each other.

plate tectonics theory that superseded continental drift and is based on the idea that the lithosphere is composed of a number of segments or plates that move independently of one another, at varying speeds, over Earth's surface.

plateau an extensive, flat-topped landform or region characterized by relatively high elevation, but low relief.

playa dry lake bed in a desert basin; typically fine-grained clastic (clay pan) or saline (salt crust).

playa lake a temporary lake that forms on a playa from runoff after a rainstorm or during a wet season.

Pleistocene Epoch the interval of Earth history immediately before the present (Holocene) epoch that experienced times of glacial advance and lasted from about 2.6 million to 10,000 years ago.

plucking erosion process by which a glacier pulls rocks and sediment from the ground along its bed and into the flowing ice.

plug dome a steep-sided, explosive type of volcano with its central vent or vents plugged by the rapid congealing of its highly viscous, silica-rich lava.

plunge pool depression at the base of a waterfall formed by the impact of cascading water.

pluton *see* igneous intrusion.

plutonic rock *see* intrusive igneous rock.

plutonism the processes associated with the formation of rocks from magma cooling deep beneath Earth's surface.

podzolization soil-forming process of humid climates with long cold winter seasons. Spodosols, the typical end product of the process, are characterized by the surface accumulation of raw humus, strong acidity, and the leaching or eluviation of soluble bases and iron and aluminum compounds.

point bar deposit of alluvium found on the inside of a bend in a meandering stream channel.

polar climate climate region that does not have a warm season and is frozen either for much, or all of the year.

polar easterlies easterly surface winds that move out from the polar highs toward the subpolar lows.

polar front the boundary, an often windy and stormy linear zone of low pressure, between cold air from the high latitudes and warm air from the tropics or subtropics.

polar front jet stream shifting boundary between cold polar air and warm subtropical air, located within the middle latitudes and strongly influenced by the polar jet stream.

polar highs high pressure systems located near the poles where air is settling and diverging.

polar outbreak strong flow of cold polar or arctic air that intrudes into the middle latitudes.

polar zone the latitudinal area poleward of 66½° (N or S).

polarity reversal a time in geologic history when the south magnetic pole became the north magnetic pole and vice versa.

pollution alteration of the physical, chemical, or biological balance of the environment that has adverse effects on the normal functioning of all life forms, including humans.

porosity characteristic of soil or bedrock that relates to the amount of pore space between individual peds or soil and rock particles and that determines the water storage capacity of Earth material.

positive feedback reaction to initial change in a system that reinforces the initial change and leads to imbalance in the system.

potential evapotranspiration (potential ET) hypothetical rate of evapotranspiration if at all times there is a more than adequate amount of soil water for growing plants.

pothole bedrock depression in a streambed drilled by the spinning of abrasive rocks in swirling flow.

prairie grassland regions of the middle latitudes. Tall-grass prairies are native to areas of moderate rainfall; and short-grass prairies are native to semiarid (steppe) environments.

precession cycle changes in the time (date) of the year that perihelion occurs; the date is determined on the basis of a major period 21,000 years in length and a secondary period 19,000 years in length.

precipitation water in liquid or solid form that falls from the atmosphere and reaches Earth's surface.

pressure belts zones of high or low pressure that tend to circle Earth parallel to the equator in a simplified model of world atmospheric pressure.

pressure gradient rate of change of atmospheric pressure horizontally with distance, measured along a line perpendicular to the isobars on a map of pressure distribution.

prevailing wind direction from which the wind for a particular location blows during the greatest proportion of the time.

primary coastline coast that has developed its present form primarily from land-based processes, especially fluvial and glacial processes.

primary productivity rate at which new organic material is created at a particular trophic level. Primary productivity through photosynthesis by autotrophs is at the first trophic level; secondary productivity is by heterotrophs at subsequent trophic levels.

prime meridian half of a great circle that connects the North and South Poles and marks zero degrees longitude. By international agreement the meridian passes through the Royal Observatory at Greenwich, England.

principal meridian one of many lines running N–S (meridians of longitude) designated by the U.S. Public Land Survey System for locating plots of land north or south of these meridians.

prodelta the portion of a delta that lies submerged in the standing body of water.

profile a graph of changes in height over a linear distance, such as a topographic profile.

punctuated equilibrium concept that periods of relative stability in many Earth systems are interrupted by short bursts of intense action causing major change.

pyroclastic flow airborne density current of hot gases and rock fragments unleashed by an explosive volcanic eruption.

pyroclastic material (tephra) pieces of volcanic rock, including cinders and ash, solidified from molten material erupted into the air.

RADAR *RAdio Detection And Ranging.*

radiation emission of electromagnetic wave energy through space. *See also* shortwave radiation and longwave radiation.

radiation fog (ground or temperature-inversion fog) fog produced by cooling of air in contact with a cold ground surface.

rain falling droplets of liquid water.

rain shadow dry, leeward side of a mountain range, resulting from the adiabatic warming of descending air.

range of tolerance the extent of limits on environmental conditions under which an organism can survive (for example: a range of temperature, or a range of annual precipitation).

recessional moraine end moraine deposited behind the terminal moraine marking a pause in the overall retreat of a glacier.

recharge replenishing the amount of stored water, particularly in the subsurface.

recumbent fold a fold in rock pushed over onto one side by asymmetric compressional forces; the axial plane of the fold is horizontal rather than vertical.

recurrence interval average length of time between events, such as floods, equal to or exceeding a given magnitude.

region area identified by certain characteristics it contains that makes it distinctive and separates it from surrounding areas.

regional base level the lowest level to which a stream system in a basin of interior drainage can flow.

regional geography specialization in the systematic study of geography that focuses on the location, distribution, and spatial interaction of phenomena organized within arbitrary areas of Earth space designated as regions.

regolith weathered rock material; usually covers bedrock.

rejuvenated stream a stream that has deepened its channel by erosion because of tectonic uplift in the drainage basin or lowering of its base level.

relative humidity ratio between the amount of water vapor in air of a given temperature and the maximum amount of water vapor that the air could hold at that temperature, if saturated; usually expressed as a percentage.

relative location a description of an object's location, relative to the position of some other object or feature.

relief a measurement or expression of the difference between the highest and lowest location in a specified area.

remote sensing collection of information about the environment from a distance, usually from aircraft or spacecraft—for example, photography, radar, and infrared.

representative fraction (RF) scale a map scale presented as a fraction or ratio between the size of a unit on the map to the size of the same unit on the ground, as in 1/24,000 or 1:100,000.

reservoir an artificial lake impounded behind a dam.

residual parent material rock fragments that form a soil and have accumulated in place through weathering.

resolution (spatial resolution) size of an area that is represented by a single pixel.

reverse fault high angle break with rocks on one side pushed up relative to those on the other side by compressional forces; hanging wall up, footwall down.

revolution motion of Earth along its orbital path around the sun. One complete revolution determines an Earth year.

rhumb line a true compass bearing (heading) displayed as a line on a map.

ria a narrow coastal embayment that occupies a submerged river valley.

rift valley major lowland consisting of one or more crustal blocks down-faulted as a result of tensional tectonic forces.

rill a narrow stream channel that even in a humid climate conducts flow only during precipitation events.

rime ice crystals formed along the windward side of tree branches, airplane wings, and the like, under conditions of supercooling.

rip current strong, narrow surface current flowing away from shore. It is produced by the return flow of water piled up near shore by incoming waves.

ripples small (centimeter-scale) wave forms in water or sediment.

roche moutonnée bedrock hill subjected to intense glacial abrasion on its up-ice side, with some plucking evident on the down-ice side.

rock a solid, natural aggregate of one or more minerals or particles of other rocks.

rock cycle a representation of the processes and pathways by which Earth material becomes different types of rocks.

rock flour rock fragments finely ground between the base of a glacier and the underlying bedrock surface.

rock structure the orientation, inclination, and arrangement of rock layers in Earth's crust.

rockfall nearly vertical drop of individual rocks or a rock mass through air pulled downward by the force of gravity.

rockslide rock unit moving rapidly downslope by gravity in continuous contact with the surface below.

Rossby waves horizontal undulations in the flow of the upper air winds of the middle and upper latitudes.

rotation turning of Earth on its polar axis; one complete rotation requires 24 hours and determines one Earth day.

runoff flow of water from the land surface, generally in the form of streams and rivers.

Saffir–Simpson Hurricane Scale a ranking (1 to 5) of the intensity of a hurricane, based on wind speeds and on the types of damage that occur, or may occur.

salinas *see* salt-crust playa.

salinization soil-forming process of low-lying areas in desert regions; the resulting soils are characterized by a high concentration of soluble salts as a result of the evaporation of surface water.

salt-crust playa (salt flat) an ephemeral lake bed composed mostly of salt minerals.

salt crystal growth weathering by the expansive force of salts growing in cracks in rocks; common in arid and coastal regions.

saltation the transportation by running water or wind of particles too large to be carried in suspension; the particles are bounced along on the surface or streambed by repeated lifting and deposition.

sand (sandy) sediment particles ranging in size from about 0.05 millimeter to 2.0 millimeters.

sand dune mound or hill of sand-sized sediment deposited and shaped by the wind.

sand sea (erg) an extensive area covered by sand dunes.

sand sheet an extensive cover of wind-deposited sand having little or no surface relief.

sandstorm strong winds blowing sand along the ground surface.

Santa Ana wind very dry downslope wind occurring in Southern California. *See also* foehn wind.

satellite any body that orbits a larger primary body, for example, the moon orbiting Earth.

saturation point at which sufficient cooling has occurred so that an air mass contains the maximum amount of water vapor it can hold. Further cooling produces condensation of excess water vapor.

savanna tropical vegetation consisting primarily of coarse grasses, often associated with scattered low-growing trees or patches of bare ground.

scale the ratio between an actual distance measurement and the size of the same distance as represented and measured on a map, globe, or other representation of Earth.

scattering diffusion in all directions of incoming solar radiation by gas molecules, water droplets, or particles in the atmosphere.

scientific method answering questions (hypotheses) by collecting evidence, objectively analyzing the evidence, and proposing answers based on what the evidence supports.

sclerophyllous vegetation type commonly associated with the Mediterranean climate; characterized by tough surfaces, deep roots, and thick, shiny leaves that resist moisture loss.

sea steep, choppy, chaotic waves still forming under the influence of a storm.

sea arch span of rock extending from a coastal cliff under which the ocean or lake water freely moves.

sea breeze air flow by day from the sea toward the land; caused by the movement of air toward a zone of lower pressure associated with higher daytime temperatures over the land.

sea cave large wave-eroded opening formed near the water level in coastal cliffs.

sea cliff (lake cliff) steep slope of land eroded at its base by wave action.

sea level average position of the ocean shoreline.

sea stack resistant pillar of rock projecting above water close to shore along an erosion-dominated coast.

seafloor spreading movement of oceanic crust in opposite directions away from the midoceanic ridges, associated with the formation of new crust at the ridges and subduction of old crust at ocean margins.

secondary coastline coast that has developed its present form primarily through the action of coastal processes (waves, currents, and/or coral reefs).

secondary productivity the formation of new organic matter by heterotrophs, consumers of other life forms.

secondary succession the reestablishment of an ecosystem in an area after having been destroyed or seriously disrupted.

section a square parcel of land with an area of 1 square mile as defined by the U.S. Public Lands Survey System.

sedimentary rock one of three major rock categories; formed by compaction and cementation of rock fragments, organic remains, or chemical precipitates.

seif a large, long, somewhat sinuous sand dune elongated parallel to the prevailing wind direction

seismic wave traveling wave of energy released during an earthquake or other shock.

seismograph instrument used to measure amplitude of passing seismic waves.

selva characteristic tropical rainforest comprising multistoried, broadleaf evergreen trees with relatively little undergrowth.

semiarid climate *see* steppe climate.

sextant navigation instrument used to determine latitude by star and sun positions.

shearing tectonic force a force that works to move two objects past each other in opposite directions.

sheet wash thin sheet of unchannelized water flowing over land.

shield volcano dome-shaped accumulation of multiple successive lava flows extruded from one or more vents or fissures.

shoreline exact contact between the edge of a standing body of water and dry land.

short-grass prairie environment where the dominant vegetation type is short grasses.

shortwave radiation energy radiated in wavelengths of less than about 1 micrometer (one millionth of a meter); includes X-rays, gamma rays, ultraviolet rays, visible light, and near infrared light.

Siberian high intensively developed center of high atmospheric pressure located in northern central Asia in winter.

silicate the largest mineral group, composed of oxygen and silica and forming most of the Earth's crust.

sill a horizontal sheet of igneous rock intruded and solidified between other rock layers.

silt (silty) sediment particles with a grain size between 0.002 millimeter and 0.05 millimeter.

sinkhole (doline) roughly circular surface depression related to the solution of rock in karst areas.

slash-and-burn (shifting cultivation) typical subsistence agriculture of indigenous societies in the tropical rainforest. Trees are cut, the branches are burned, and crops are planted between the larger trees or stumps before rapid deterioration of the soil forces a move to a new area.

sleet form of precipitation produced when raindrops freeze as they fall through a layer of cold air; may also, locally, refer to a mixture of rain and snow.

slide fast mass wasting in which Earth material moves downslope in continuous contact with a discrete surface below.

slip face the steep, downwind side of a sand dune.

slope aspect direction that a slope on a hill or a mountain faces in respect to the sun's rays.

slow mass wasting gravity-induced downslope movement of Earth material occurring so slowly that people cannot observe it directly.

slump thick unit of unconsolidated fine-grained material sliding downslope on a concave, curved slip plane.

small circle any circle that is not a full circumference of the globe. The plane of a small circle does not pass through the center of the globe.

smog combination of chemical pollutants and particulate matter in the atmosphere, typically over urban industrial areas.

snow precipitation in the form of ice crystals.

snow line elevation in mountain regions above which snow covers the land, this elevation changes seasonally, locally, and even daily depending on weather conditions.

snowstorm storm situation where precipitation falls in the form of snow.

soil a dynamic, natural layer on Earth's surface that is a complex mixture of inorganic minerals, organic materials, microorganisms, water, and air.

soil (as a mass wasting material) relatively thin unit of unconsolidated fine-grained slope material.

soil fertilization adding nutrients to the soil to meet the conditions that certain plants require.

soil grade classification of soil texture by particle size: clay (less than 0.002 mm), silt (0.002–0.05 mm), and sand (0.05–2.0 mm) are soil grades.

soil horizon distinct soil layer characteristic of vertical zonation in soils; horizons are distinguished by their general appearances and their specific chemical and physical properties.

soil order largest classification of soils based on development and composition of soil horizons.

soil ped naturally forming soil aggregate or clump with a distinctive shape that characterizes a soil's structure.

soil profile vertical cross-section of a soil that displays the various horizons or soil layers that characterize it; used for classification.

soil survey a publication that describes the soil types in an area and includes maps that show their distribution.

soil taxonomy the classification and naming of soils.

soil texture the distribution of particle sizes in a soil that give it a distinctive "feel."

soil water water in the zone of aeration, the uppermost subsurface water layer.

soil-forming regime processes that create soils.

solar constant rate at which insolation is received just outside Earth's atmosphere on a surface at right angles to the incoming radiation.

solar declination latitude on Earth where the sun is directly overhead at solar noon (will always be within the tropics).

solar energy *see* insolation.

solar noon the time of day when the sun angle is at a maximum above the horizon (zenith).

solar system the system of the sun and the planets, their satellites, comets, meteoroids, and other objects revolving around the sun.

solar wind streams of hot ions (protons and electrons) traveling outward from the sun.

solifluction slow movement of saturated soil downslope by the pull of gravity; especially common in permafrost areas.

solstice (summer and winter) one of two times each year when the position of the noon sun is overhead at its farthest distance from the equator; this occurs when the sun is overhead at the Tropic of Cancer (about June 21) and the Tropic of Capricorn (about December 21).

solution dissolving material in a fluid, such as water, or the liquid containing dissolved material; water transports dissolved load in solution.

solution sinkhole topographic depression formed mainly by the solution and removal of soluble rock at the surface.

sonar a system that uses sound waves for location and mapping underwater.

source location high in a drainage basin, near the drainage divide, where a stream system's flow begins, or the location of the beginning flow from a spring-fed stream.

source region nearly homogeneous surface of land or ocean over which an air mass acquires its temperature and humidity characteristics.

South Pole maximum south latitude (90°S), at the point marking the axis of rotation.

southeast trades *see* trade winds.

Southern Oscillation the systematic variation in atmospheric pressure between the eastern and western Pacific Ocean.

spatial discipline term used when defining geography as the science that examines phenomena as it is located, distributed, and interacts with other phenomena throughout Earth space.

spatial distribution location and extent of an area or areas where a feature exists.

spatial interaction process whereby different phenomena are linked or interconnected, and, as a result, impact one another through Earth space.

spatial pattern arrangement of a feature as it is distributed through Earth space.

spatial resolution size of an area on Earth that is represented by a single pixel.

spatial science perspective the viewpoint of analyzing the distribution, density, and arrangement of phenomena to understand why they exist where they occur, the potential interactions among them, and the significance of these spatial aspects.

spatial science term used when defining geography as the science that examines the locations and distributions of phenomena and their interactions with other phenomena throughout Earth space.

specialist an animal that can survive only on a single or a very limited food type for its nutrition.

specific heat the amount of heat required to raise the temperature of 1 gram of material by 1°C.

specific humidity mass of water vapor present per unit mass of air, expressed as grams per kilogram of moist air.

speleology the scientific study of caverns.

speleothem general term for any cavern feature made by secondary (later) precipitation of minerals from subsurface water.

spheroidal weathering rounded shape of rocks often caused by preferential weathering along joints of cross-jointed rocks.

spit coastal landform of wave- and current-deposited sediment attached to dry land at one end.

spodosol soil that develops in porous substrates such as glacial drift or beach sand.

spring natural outflow of groundwater to the surface.

spring tide the larger than average tidal range that occurs during new and full moon.

squall line narrow line of rapidly advancing storm clouds, strong winds, and heavy precipitation; usually develops in front of a fast-moving cold front.

stability condition of air when it is cooler than the surrounding atmosphere and resists the tendency to rise; the lapse rate of the surrounding atmosphere is less than that of stable air.

stalactite spire-shaped speleothem that hangs from the ceiling of a cavern.

stalagmite spire-shaped speleothem that rises up from a cavern floor.

standard sea level pressure a reference standard for average atmospheric pressure at sea level, 1013.25 millibars, or 29.92 inches of mercury measured by a barometer.

star dune a large pyramid-shaped sand dune with multiple slip faces resulting from convergent wind directions.

stated scale a written or oral reference to a map or image scale by stating how many units of measure in map size represent how much actual distance on the ground. Stated scales often mix units such as; 5.28 inches on the map equals 2 miles on the ground.

stationary front frontal system between air masses of nearly equal strength; produces stagnation over one location for an extended period of time.

steppe climate (semiarid climate) characterized by relatively treeless, midlatitude semi-arid vegetation, and dominated by short bunch grasses.

stock an irregular mass of intrusive igneous rock (pluton) smaller than a batholith.

storm local atmospheric disturbance often associated with rain, hail, snow, sleet, lightning, or strong winds.

storm surge rise in sea level due to wind and reduced air pressure during a hurricane or other severe storm.

storm track path frequently traveled by a cyclonic storm as it moves in a generally eastward direction from its point of origin.

stratification distinct layers within sedimentary rocks, called strata.

strato signifies a low-level cloud (i.e., from the surface to 2000 m in elevation).

stratopause upper limit of stratosphere, separating it from the mesosphere.

stratosphere layer of atmosphere lying above the troposphere and below the mesosphere, characterized by fairly constant temperatures and ozone concentration.

stratovolcano *see* composite cone.

stratus uniform layer of low sheet-like clouds, frequently grayish in appearance.

stream general term for any natural, channelized flow of water regardless of size.

stream capacity the maximum amount of load that a stream can carry; varies with the stream's velocity.

stream competence the largest particle size that a stream can carry; varies with a stream's velocity.

stream discharge volume of water flowing past a point in a stream channel in a given unit of time.

stream gradient *see* gradient.

stream hydrograph plot showing changes in the amount of stream flow over time.

stream load material being transported by a stream at a given instant; includes bed load, suspended load, and dissolved load.

stream order numerical index expressing the position of a stream channel within the hierarchy of a stream system.

stream terrace former floor of a stream valley now abandoned and perched above the present valley floor and stream channel.

stress (pressure) force per unit area.

striation gouge, groove, or scratch carved in bedrock by abrading rock particles imbedded in a glacier.

strike compass direction of the line formed at the intersection of a tilted rock layer and a horizontal plane.

strike-slip fault (lateral fault) a fault with horizontal motion, where movement takes place along the strike of the fault.

subarctic climate high latitude climate type that produces a boreal forest (taiga) landscape. *See also* boreal forest.

subduction process associated with plate tectonic theory whereby an oceanic crustal plate is forced downward into the mantle beneath a lighter continental plate when the two converge.

sublimation direct change of state of a material, such as water, from solid to gas.

subpolar lows east–west trending belts or cells of low atmospheric pressure located in the upper middle latitudes.

subsidence the sinking of an air parcel to a lower altitude, or the sinking or lowering of a land surface.

subsolar point location on Earth, between the lines of the tropics, where the noon sun is directly overhead (90° angle) on a given day. *See also* solar declination.

subsurface horizon a distinct horizontal soil layer that lies beneath the surface, characteristic of vertical zonation in soils.

subsurface water general term for all water that lies beneath Earth's surface, including soil water and groundwater.

subsystem separate system operating within the boundaries of a larger Earth system.

subtropical highs cells of high atmospheric pressure centered over the eastern portions of the oceans in the vicinity of 30°N and 30°S latitude; source of the westerlies poleward and the trades equatorward.

subtropical jet stream high-velocity air current flowing above the sinking air of the subtropical high-pressure cells; most prominent in the winter season.

succession progression of natural vegetation from one plant community to the next until a final stage of equilibrium has been reached with the natural environment.

sunspot visible dark (cooler) spot on the surface of the sun.

supercooled water liquid water that exists with a temperature below the freezing point of 0°C or 32°F.

supervolcano a volcanic center that has experienced an eruptive blast that emitted a volume of volcanic material greater than 1,000 cubic kilometers (240 cubic miles).

surf zone the part of the nearshore area that consists of a turbulent bore of broken wave water.

surface creep wind-generated transportation consisting of pushing and rolling sediment downwind in continuous contact with the surface.

surface inversion (radiation inversion, ground inversion) a temperature inversion that forms close to the ground as a result of radiational cooling of the land surface during clear, calm nights.

surface ocean current an ocean current that flows on the surface, as opposed to a deep-water, undersea current.

surface runoff liquid water flowing over Earth's land surface.

surge (glacial) sudden shift downslope of glacial ice, possibly caused by a reduction of basal friction with underlying bedrock.

suspended load solid particles that are small enough to be transported considerable distances while remaining buoyed up in a moving air or water column.

suspension transportation process that moves small solids, often considerable distances, while buoyed up by turbulence in the moving air or water.

sustainable development development and natural resource use that meets present needs without compromising the ability of future generations to meet their needs.

swallow hole the site where a surface stream is diverted to the subsurface, such as into a cavern system.

swash thin sheet of broken wave water that rushes up the beach face in the swash zone.

swash zone the most landward part of the nearshore zone; where a sheet of water rushes up, then back down, the beach face.

swell orderly lake or ocean waves of rounded form that have traveled beyond the storm zone of wave generation.

symbiotic relationship relationship between two organisms that is mutually beneficial to both organisms.

syncline the downfolded element of folded rock structure.

system group of interacting and interdependent units that together form an organized whole.

systems analysis determining the parts of a system, the processes involved, and studying how changes in interactions among those parts and processes may affect the system and its operation.

systems theory a means of understanding how a system works by examining its component parts, the processes involved, the interactions that take place, how they influence each other and their impact on a system as a whole.

taiga *see* boreal forest.

tall-grass prairie environment where the dominant vegetation type is tall grasses, with a few scattered trees.

talus (talus slope, talus cone) slope (sometimes cone-shaped) of angular, broken rocks at the base of a cliff deposited by rockfall.

tarn mountain lake in a glacial cirque.

tectonic force force originating within Earth that breaks and deforms Earth's crust.

tectonic processes processes that derive their energy from within Earth's interior and serve to create landforms by elevating, disrupting, and roughening Earth's surface.

temperature degree of heat or cold and its measurement.

temperature gradient rate of change of temperature with distance in any direction from a given point; refers to rate of change horizontally; a vertical temperature gradient is referred to as the lapse rate.

temperature inversion reverse of the normal pattern of vertical distribution of air temperature; in the case of inversion, temperature increases rather than decreases with increasing altitude.

temperature lag *see* daily temperature lag and annual temperature lag.

tensional tectonic force force originating within Earth that acts to pull two adjacent areas of rock away from each other (divergence).

terminal moraine end moraine that marks the farthest advance of an alpine or continental glacier.

terra rossa characteristic calcium-rich (developed over limestone bedrock) red-brown soils.

terrestrial planets the four closest planets to the sun—Mercury, Venus, Earth, and Mars.

terrestrial radiation the long wave thermal infrared radiation emitted from Earth as heat.

thematic map a map designed to present information or data about a specific theme, as in a population distribution map, a map of climate or vegetation.

thermal expansion and contraction notion that rocks can weather due to expansion and contraction effects of alternating heating and cooling.

thermal infrared (TIR) images made with scanning equipment that produces an image of heat differences.

thermal spring a groundwater spring flowing with hot or warm water, often with a high dissolved mineral content.

thermosphere highest layer of atmosphere extending from the mesopause to outer space.

Thornthwaite system climate classification based on moisture availability and of greatest use at the local level; climate types are distinguished by examining and comparing potential and actual evapotranspiration.

threshold a limiting condition that if it is exceeded (or not met) will cause a dramatic, often irreversible change in the system.

thrust fault low angle break with rocks on one side pushed over those of the other side by compressional forces.

thunder sound produced by the rapidly expanding, heated air along the channel of a lightning discharge.

thunderstorm intense convectional storm characterized by thunder and lightning, short in duration and often accompanied by heavy rain, hail, and strong winds.

tidal interval the time between successive high tides, or between successive low tides.

tidal range elevation difference between water levels at high tide and low tide.

tide periodic rise and fall of sea level in response to the gravitational interaction of the moon, sun, and Earth.

till sediment deposited directly by glacial ice.

till plain a broad area of low relief covered by continental glacier deposits.

tilted fault block crustal block between two parallel normal faults that has been uplifted along one fault and relatively down-dropped along the other.

time zone Earth is divided into 24 time zones (24 hours) to coordinate time with Earth's rotation.

tolerance the range (and limits) of conditions that can support the ability of a species to survive.

tombolo strip of wave- and current-deposited sediment connecting the mainland to an island.

topographic contour line line on a map connecting points that are the same elevation above mean sea level.

topography the arrangement of high and low elevations in a landscape.

tornado small, intense, funnel-shaped cyclonic storm of very low pressure, violent updrafts, and converging winds of enormous velocity.

tornado outbreak when a thunderstorm(s) produce more than one tornado.

tower karst high, steep-sided hills formed by solution of limestone or other soluble rocks in karst areas.

township a square plot of land 6 miles on a side, further subdivided into 36 sections, laid out within the U.S. Public Lands Survey System.

trace (of rain) less than a measurable amount of rain or snow (i.e., less than 1 mm or 0.01 in.).

traction transportation process in moving water that drags, rolls, or slides heavy particles along in continuous contact with the bed.

trade winds consistent surface winds blowing in low latitudes from the subtropical highs toward the intertropical convergence zone; labeled *northeast trades* in the Northern Hemisphere and *southeast trades* in the Southern Hemisphere.

transform movement horizontal sliding of tectonic plates alongside and past each other.

transpiration transfer of moisture from living plants to the atmosphere by the emission of water vapor, primarily from leaf pores.

transportation movement of Earth materials from one site to another by gravity, water, wind, or glacial ice.

transported parent material rock fragments that form a soil and originated elsewhere and then were transported and deposited in the new location.

transverse dune a linear ridge-like sand dune that is oriented at right angles to the prevailing wind direction.

transverse stream a stream that flows across the general orientation or "grain" of the topography in a valley that cuts through and across mountains or ridges.

travertine calcium carbonate (limestone) deposits forming at springs and in caverns where groundwater is saturated with $CaCO_3$.

treeline elevation in mountain regions above which cold temperatures and wind stress prohibit tree growth.

tributary stream channel that delivers its water to another, larger channel.

trophic level number of feeding steps that a given organism is removed from the autotrophs (e.g., green plant—first level, herbivore—second level, carnivore—third level, etc.).

Tropic of Cancer parallel of latitude at $23\frac{1}{2}°$N; the northern limit to the migration of the sun's vertical rays.

Tropic of Capricorn parallel of latitude at $23\frac{1}{2}°$S; the southern limit to the migration of the sun's vertical rays.

tropical climates climate regions that are warm all year.

tropical easterlies winds that blow from the east in tropical regions.

tropical monsoon climate climate characterized by alternating rainy and dry seasons.

tropical rainforest climate hot, wet climate that promotes the growth of rainforests.

tropical savanna climate warm, semi-dry climate that promotes tall grasslands.

tropical zone region on Earth lying between the Tropic of Cancer ($23\frac{1}{2}°$N latitude) and the Tropic of Capricorn ($23\frac{1}{2}°$S latitude).

tropopause boundary between the troposphere and stratosphere.

troposphere lowest layer of the atmosphere, exhibiting a steady decrease in temperature with increasing altitude and containing virtually all atmospheric dust and water vapor.

trough elongated area or "belt" of low atmospheric pressure; also glacial trough, a U-shaped valley carved by a glacier.

trunk stream the largest channel in a drainage system; receives inflow from tributaries.

tsunami wave caused when an earthquake, volcanic eruption, or other sudden event displaces ocean water; builds to dangerous heights in shallow coastal waters.

tundra high-latitude or high-altitude environments or climate regions that are not able to support tree growth because the growing season is too cold or too short.

tundra climate characterized by treeless vegetation of polar regions and very high mountains, consisting of mosses, lichens, and low-growing shrubs and flowering plants.

turbulence chaotic, mixing flow of fluids, often with an upward component.

typhoon a tropical cyclone found in the western Pacific; the same as a hurricane.

ultisol soil that has a subsurface clay horizon, is low in bases, and is often red or yellow in color.

unconformity an interruption in the accumulation of different rock layers; often represents a period of erosion.

uniformitarianism widely accepted theory that Earth's geological processes operate today as they have in the past.

unloading physical weathering process whereby removal of overlying weight leads to rock expansion and breakage.

uplift mechanisms methods of lifting surface air aloft; methods are orographic, frontal, convergence (cyclonic), and convectional.

upper-level inversion an atmospheric condition where air descending from aloft warms as it compresses during its descent and overrides cooler air below.

upslope fog type of fog where upward flowing air cools to form fog that hugs the slope of mountains.

upwelling upward movement of colder, nutrient-rich, subsurface ocean water, replacing surface water that is pushed away from shore by winds.

urban heat island *see* heat island.

U.S. Public Lands Survey System a method for locating and dividing land, used in much of the Midwest and western United States. This system divides land into 6- by 6-mile-square townships consisting of 36 sections of land (each 1 sq mi) that can be further subdivided into halves, quarter sections, and quarter-quarter-sections.

uvala (valley sink) large surface depression resulting from coalescing of sinkholes in karst areas.

valley breeze *see* mountain breeze–valley breeze.

valley glacier an alpine glacier that extends beyond the zone of high mountain peaks into a confining mountain valley below.

valley train outwash deposit from glacial meltwater, resembling an alluvial fan confined by valley walls.

vent pipe-like conduit through which volcanic rock material is erupted.

ventifact rock displaying distinctive wind-abraded faces, pits, grooves, and polish.

verbal scale stating the scale of a map using words such as "one inch represents one mile."

vernal equinox the day when the subsolar point is at one of the lines of the tropics (23½°N and S) that marks the first days of summer and winter; summer begins on the June solstice for the Northern Hemisphere and winter begins on the December solstice. In the Southern Hemisphere, the June solstice marks the beginning of winter and December marks the beginning of winter.

vertical exaggeration a technique that stretches the height representation of terrain in order to emphasize topographic detail.

vertisol soil that develops in regions with strong seasonality of precipitation.

viscosity a fluid's resistance to flowing.

visual spectrum wavelength range of the electromagnetic spectrum sensed by human vision (0.4 to 0.7 micrometers; violet to red).

visualization models (visualization) a wide array of digital techniques used to vividly illustrate a place or concept, or the illustration produced by one of these techniques.

volcanic ash erupted fragments of volcanic rock of sand size or smaller (< 2.0 mm).

volcanic neck vertical igneous intrusion that solidified in the vent of a volcano.

volcanism the eruption of molten rock matter onto Earth's surface.

volcano mountain or hill created from the accumulation of erupted rock matter.

V-shaped valley the typical shape of a stream valley cross-profile where the stream gradient is steep.

warm current a flow of sea water that moves like a river through the ocean and is relatively warmer than water in the ocean area that it flows through.

warm front leading edge of a relatively warmer, less dense air mass advancing upon a cooler, denser air mass.

warping broad and general uplift or settling of Earth's crust with little or no local distortion.

wash (arroyo, wadi) an ephemeral stream channel in an arid climate.

water budget relationship between evaporation, condensation, and storage of water within the Earth system.

water mining taking more groundwater out of an aquifer through pumping than is being replaced by natural processes in the same period of time.

water table upper limit of the zone of saturation below which all pore spaces are filled with water.

water vapor water in its gaseous form.

wave base water depth equal to half the length of a given wave; at smaller depths the wave interacts with the underwater substrate.

wave crest the highest part of a wave form.

wave height vertical distance between the trough and adjacent crest of a wave.

wave period time it takes for one wavelength to pass a given point.

wave refraction bending of waves in map view as they approach the shore, aligning themselves with the bottom contours in the breaker zone.

wave steepness ratio of wave height to wavelength for a given wave.

wave trough the lowest part of a wave form.

wavelength horizontal distance between two successive crests of a given wave.

weather atmospheric conditions, at a given time, in a specific location.

weather radar radar that is used to track thunderstorms, tornadoes, and hurricanes.

weathering physical (mechanical) fragmentation and chemical decomposition of rocks and minerals at and near Earth's surface.

well artificial opening that reaches the zone of saturation for the purpose of extracting groundwater.

westerlies surface winds flowing from the polar portions of the subtropical highs, carrying fronts, storms, and variable weather conditions from west to east through the middle latitudes.

wet adiabatic lapse rate rate at which a rising mass of air is cooled by expansion when condensation is taking place. The rate varies but averages 5°C/1000 meters (3.2°F/1000 ft).

wind air in motion from areas of higher pressure to areas of lower pressure; movement is generally horizontal, relative to the ground surface.

wind friction force that acts opposite to the direction of movement or flow; for example, turbulent resistance of Earth's surface on the flow of the atmosphere.

wind wave wave on a water body created when air currents push the water surface along.

windward location on a side that faces toward the wind; often refers to mountains, islands, and sand dunes.

xerophytic vegetation (xerophytes) that is able to withstand extended periods of drought.

yardang aerodynamically shaped remnant ridge of wind-eroded bedrock or partly consolidated sediments.

yazoo stream a stream tributary that flows parallel to the main stream for a considerable distance before joining it.

zenith the highest angle above the horizon that the sun reaches at solar noon in a given place and time of year.

zone of ablation the lower elevation part of a glacier; where more frozen water is removed than added during the year.

zone of accumulation the higher elevation part of a glacier; where more frozen water is added than removed during the year.

zone of aeration uppermost layer of subsurface water where pore spaces typically contain both air and water.

zone of depletion top layer, or A horizon, of a soil, characterized by the removal of soluble and insoluble soil components through leaching and eluviation by gravitational water.

zone of saturation subsurface water zone in which all voids in rock and soil are filled with water; the top of this zone is the water table.

zone of transition an area of gradual change from one region to another.

zooplankton tiny animals that float and drift with currents in water bodies.

Index

Lambert, Johann, 70
Lambert's law, 70, 71
Land
　degradation, 304
　distribution, in controls of temperature, 100
　soil formation and, 337
Land breeze–sea breeze, 130–131
Landfall
　defined, 182, 184
　probability maps for hurricanes, 184
　sites of hurricanes, 1950-2013, 186
Landform barriers, 101
Landforms, 385
　sphericity and, 29
Landmass changes, climate and, 218–219
Landslides, 9, 14, 16, 17, 23, 439, 440
La Niña, 136–137
Lapilli, 403
Large-scale maps, 42
Latent heat exchange, 91–92
Latent heat of condensation, 93
Lateral accretion deposits, 486
Lateral fault (strike-slip fault), 393–394
Lateral migration, 488
Lateral moraines, 549, 551
Laterite, 338
Laterization, 338
Latitude, 30–31, 32. *See also* Latitudinal zones
　in controls of temperature, 99
　distribution of precipitation, 163, 166
　lines delimiting solar energy, 75–76
　measuring, 30–31
Latitudinal zones
　air masses in
　　identifying on map, 172
　　North America, 173
　biomes in, 306
　migration with seasons, 124–125
　precipitation in, 163, 166
　variations of insolation in, 76–78
　winds in, 123–126
Laurasia, 368, 380
Lava, 360
Lava flows, 403–404
Laws of thermodynamics, 291, 293
Leaching, 325
Leap year, 69
Leeward, 116
Legend, 40
Lianas, 310
Lidar, 52, 54
Lifting condensation level (LCL), 161
Light
　near-infrared energy as, 52
　shortwave solar radiation as, 90
Light-year, 62
Limestone, 363, 366
Limestone caverns, 463–466
Lithify, 361
Lithosphere, 12
Lithosphere (planetary structure), 357, 379
Lithospheric plates, 371–372, 373
Little Ice Age, 223, 224
Littoral drifting, 579–580
Loam, 328
Local time, 33
Local winds, 130–133
　Chinook, 132
　drainage winds, 132–133
　land breeze–sea breeze, 130–131
　mountain breeze–valley breeze, 131–132
　Santa Ana winds, 132
Location on Earth, 27–54
　coordinate system, 30, 31, 47
　decimal degrees, 31–32
　geographic grid, 32–36

latitude, 30–31, 32
longitude, 31, 32
map projections, 37–40
maps, 28–29, 37, 40–47, 40–49
remote sensing, 50–54
Loess, 526–528
Loma Prieta earthquake, 1989, 397, 399
Longitude, 31, 32
　measuring, 31
　time and, 33–34
Longitudinal dunes, 523, 524
Longitudinal profile, 486
Longshore bar, 583
Longshore current, 580
Longshore drifting, 580
Longwave radiation, 67, 90
Long waves, 127
Lower stream course, 488–490
Luray Caverns, 463
Luster, 358

M

Mafic, 361
Mafic magmas, 403
Magma, 360
Magnetic declination, 43
Magnetic dip, 370
Magnetic north, 42
Mallee, 260
Mantle, 356
Mapmaking, 28, 47–49
Mapping
　of pressure distributions, 114–115
　of pressure systems, 177
Map projections, 37–40
　conic projections, 38
　cylindrical projections, 38
　planar projections, 37, 38
　properties of, 38–40
　　area, 39
　　compromise projections, 40
　　direction, 40
　　distance, 39–40
　　shape, 39
Maps, 28–29, 37, 40–49. *See also* Mapmaking
　advantages of, 37
　direction, 42–43
　legend, 40
　limitations of, 37
　scale, 41–42
　spatial data, 43–44
　thematic, 43–44
　topographic, 44–47
Map scale, 41
Maquis, 260, 312
Marble, 365, 366
Marine ecosystems, 317–319
Marine terraces, 581
Marine west coast climate, 264–268
　climographs of, 265
　clouds and precipitation, 265–266
　defined, 264
　oceanic influences, 264–265
　overview of, 259
　resource potential, 266–268
Maritime, 96
Maritime equatorial *(mE)* air masses, 172
Maritime polar *(mP)* air masses, 172, 173–174
Maritime tropical *(mT)* air masses, 172, 174
Mars, 63, 64
Mass, 62
Mass wasting, 432–443
　defined, 420
　fast mass wasting, 436–443
　landscape and, 443
　materials and motion, 434

overview of, 432–434
　slow mass wasting, 434–436
Mathematical/statistical models, 19
Matorral, 260
Matrix, 298
Mawsynram, 148
Meander cut-offs, 489
Meandering channels, 486
Mechanical weathering. *See* Physical weathering
Medial moraine, 551
Mediterranean climate, 258–262
　agriculture in, 260–261
　characteristics of, 258–260
　climographs of, 260
　defined, 258
　humid subtropical climate compared to, 262–263
　overview of, 259
　special adaptations, 260–262
　vegetation in, 260–261
Mediterranean sclerophyllous woodland, 312
Medium-scale maps, 42
Megapixels, 50
Melting, 92
Mental maps, 19
Mercalli Scale, 396, 397
Mercator, Gerhardus, 38
Mercator projection, 38, 39
Mercury, 63
Mercury barometer, 112
Meridians, 32–33
Mesa, 510
Mesosphere, 89
Mesothermal climates, 206, 207, 210–211, 258–268
　defined, 258
　humid subtropical climate, 262–264
　index map of, 258
　marine west coast climate, 264–268
　Mediterranean climate, 258–262
　overview of, 259
Mesozoic Era, 380
Metamorphic rock, 364–365
Meteorite, 63
Meteoroids, 219, 220–221
Meteorologists, 5
Meteorology, 106
Meteors, 63
Methane as greenhouse gas, 222, 223
Metric system, 94
Mexico City earthquake, 1985, 396, 399–400
Microclimate, 212–213
Microplate terranes, 378
Microthermal climates, 206, 207, 210–211, 268–277
　defined, 268
　humid continental, hot-summer climate, 269–271
　humid continental, mild-summer climate, 271–274
　index map of, 258
　overview of, 269
　subarctic climate, 274–277
Middle stream course, 487–488
Midlatitude cyclone (extratropical cyclone), 178–182
　overview of, 178–180
　surface weather and upper air flow, 181–182
　weather conditions and, 180–181
Midlatitude forests, 312–314
　broadleaf deciduous forest, 312–313
　broadleaf evergreen forest, 313
　coniferous forest, 313–314
　Mediterranean sclerophyllous woodland, 312
　mixed forest, 313
Midlatitude grasslands, 314–316
　short-grass prairies, 315, 316
　tall-grass prairies, 315
Midlatitude zone, 76
Midoceanic ridge, 371
Mie, Gustav, 90
Mie scattering, 90

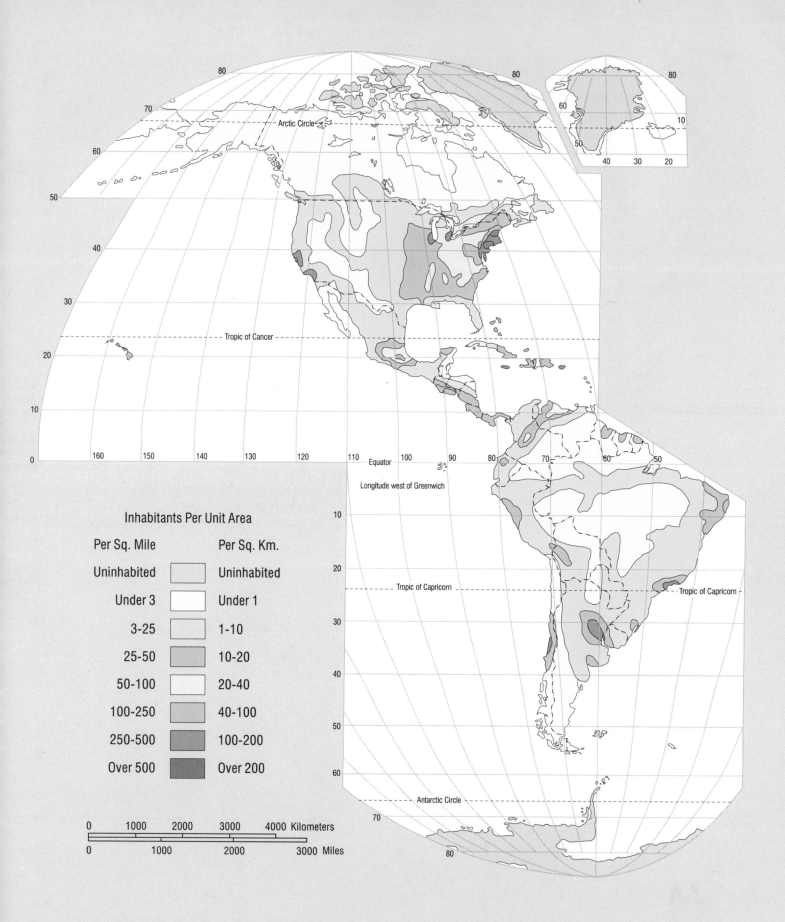

Inhabitants Per Unit Area

Per Sq. Mile		Per Sq. Km.
Uninhabited		Uninhabited
Under 3		Under 1
3-25		1-10
25-50		10-20
50-100		20-40
100-250		40-100
250-500		100-200
Over 500		Over 200

0   1000   2000   3000   4000 Kilometers

0   1000   2000   3000 Miles

World Map of Population Density